ZnO and TiO_2 Based Nanostructures

ZnO and TiO_2 Based Nanostructures

Special Issue Editor

Andrea Lamberti

MDPI • Basel • Beijing • Wuhan • Barcelona • Belgrade

Special Issue Editor
Andrea Lamberti
Politecnico di Torino
Italy

Editorial Office
MDPI
St. Alban-Anlage 66
Basel, Switzerland

This is a reprint of articles from the Special Issue published online in the open access journal *Nanomaterials* (ISSN 2079-4991) from 2017 to 2018 (available at: http://www.mdpi.com/journal/nanomaterials/special_issues/ZnO_TiO2nano)

For citation purposes, cite each article independently as indicated on the article page online and as indicated below:

LastName, A.A.; LastName, B.B.; LastName, C.C. Article Title. *Journal Name* **Year**, *Article Number*, Page Range.

ISBN 978-3-03897-013-2 (Pbk)
ISBN 978-3-03897-014-9 (PDF)

Cover image courtesy of Thomas Kämpfe.

Contents

About the Special Issue Editor

Andrea Lamberti received the PhD degree in Electronic Devices with a thesis entitled "Metal-oxide nanostructures for energy applications" from Politecnico di Torino in 2013. He worked as post-doc researcher at Italian Institute of Technology on the in synthesis and characterization of innovative metal-oxide nanostructured materials and their integration into energy conversion and storage devices. Currently, he is working as Assistant Professor at Department of Applied Science and Technology in Politecnico di Torino on the study of innovative graphene-based nanostructures, metal-oxide nanomaterials and technological processes for the fabrication of energy harvesting and storage devices. He is author and co-author of more than 80 publications on international peer-reviewed journals, three book chapter and two patents. He acts as editor for the Journal of Advances in Materials Science and Engineering (Hindawi), guest editor for Journal of Vacuum Science & Technology B (AVS), guest editor for the journal Nanomaterials (MDPI) and He is member of the executive committee of the Italian Association of Science and Technology (AIV).

nanomaterials

MDPI

Review

ZnO- and TiO_2-Based Nanostructures

Andrea Lamberti [1,2]

1 Department of Applied Science and Technology, Politecnico di Torino, C.so Duca degli Abruzzi 24, 10129 Turin, Italy; andrea.lamberti@polito.it

2 Center for Sustainable Future Technologies, Istituto Italiano di Tecnologia (IIT@Polito), C.so Trento 21, 10129 Turin, Italy

Received: 8 May 2018; Accepted: 10 May 2018; Published: 14 May 2018

Keywords: ZnO; TiO_2; nanostructures; smart materials

Transition-metal oxide (TMO) nanostructures are the focus of current research efforts in nanotechnology since they are the most common minerals on Earth, and also thanks to their special shapes, compositions, and chemical and physical properties. They have now been widely used in the design of energy saving and harvesting devices, such as lithium-ion batteries, fuel cells, solar cells, and even transistors, light emitting devices (LEDs), hydrogen production by water photolysis and its storage, water and air purification by degradation of organic/inorganic pollutants, bio-sensing devices, environmental monitoring by their applications in the fabrication of gas, humidity, and temperature sensors, and photodetectors. TMOs can overcome the limitation imposed by the relatively poor properties of standard electrodes, showing high carrier mobility and significantly low charge recombination rates.

In addition to the great application potentials, TMO-based nanomaterials, such as ZnO and TiO_2, have recently revolutionized nanomaterial research thanks to their outstanding smart properties. They can be produced in different shapes (such as nanowires, nanobelts, nanorods, nanotubes, nanocombs, nanorings, nanohelixes/nanosprings, nanocages and nanosheets, and nanostars) depending on the synthesis routes, resulting in different physicochemical properties.

The present Special Issue covers the most recent advances in ZnO and TiO_2 nanostructures, concerning not only their synthesis and characterization, but also reports of the manner(s) in which their functional and smart properties can be applied in working devices. Applications of nanosized ZnO and TiO_2 can range widely, from biomedical and drug delivery devices to piezoelectric and chemical sensors, and energy harvesting, conversion, and storage devices.

Twenty-seven papers compose this issue, with invited contributions and regular original papers, and both reviews and research articles.

The two first reviews address the application of ZnO nanostructures to chemoresistive sensing (Rackauskas and coworkers [1], invited contribution) and Surface Enhanced Raman Scattering (by Yang et al. [2]).

The last two reviews focus on TiO_2 nanostructures: Wang et al. [3] describe the engineering of the surface/interface structures of titanium dioxide micro- and nano-architectures towards environmental and electrochemical applications while the wet chemical preparation of TiO_2-based composites with different morphologies and their photocatalytic properties are reported by Xiang and Zhao [4].

Several research articles focus on the synthesis and deposition of ZnO and TiO_2 nanostructures. Hsu et al. [5] discuss the effect of process parameters on the sputtering deposition of indium titanium zinc oxide thin film for transistor fabrication. Shih and Wu [6] investigate the growth mechanism of ZnO nanowires, providing experimental observation and providing a short-circuit diffusion analysis. The sol-gel synthesis of ZnO/ZnS heterostructures is discussed by Berbel Manaia et al. [7], focusing on the critical role of thioacetamide concentration. Chen et al. [8] report on the preparation and characterization of ZnO nanoparticles supported on amorphous SiO_2. Rana and coworkers [9] discuss how the growth method and process parameters influence the optical, conductive, and physical properties of solution-grown ZnO nanostructures. Folger and coworkers [10] show how the heat treatments in different environments allow electronic conductivity to be tuned in hydrothermally grown rutile TiO_2 nanowires. Jin et al. [11] present a simple and novel strategy to obtain TiO_2 nanowire networks by titanium substrate corrosion and their application in third generation solar cells. Berthod et al. [12] show the improvement achieved in the fabrication of periodic TiO_2 nanostructures by colloidal lithography approach. Liang and coworkers [13] describe the synthesis and characterization of organozinc precursor-derived crystalline ZnO nanoparticles.

Some contributions discuss the photo-induced smart properties of metal-oxide nanostructures: Hu et al. [14] show how via constructing appropriate heterostructures between mesopore TiO_2 nanospheres and Sn_3O_4 nanoparticles it is possible to enhanced their ultraviolet-visible light photocatalytic activity. The study of the photodynamic activity of N-doped TiO_2 nanoparticles conjugated with aluminum phthalocyanine is reported by Pan et al. [15] Li and coworkers investigate the photoelectrochemical water splitting properties of Ti-Ni-Si-O nanostructures grown on Ti-Ni-Si alloy substrate [16]. The visible light response of mesoporous titania films loaded with silver salts is reported by Crespo-Monteiro et al. [17] for the degradation of methyl bluc.

Other functional properties of ZnO and TiO_2 nanostructures are described: the piezoelectric potential in single-crystalline ZnO nanohelices is studied by finite element analysis by Hao et al. [18] (invited contribution) while the self-cleaning behavior of a nano-TiO_2-coated SiO_2 microsphere composite is reported by Sun et al. [19].

The electrical properties are also analyzed: the memristive response of TiO_2 nanoparticle is investigated by Schmidt et al. [20] while Bruzzi and coworkers [21] report the thermally stimulated current in nanocrystalline titania.

Other contributions addressed biological issues: Yamamoto et al. [22] show novel results on the in vitro sonodynamic therapeutic effect of polyion complex micelles incorporating titanium dioxide nanoparticles. Ancona et al. [23] (invited contribution) report a novel strategy for photodynamic cancer therapy exploiting lipid-coated ZnO nanoparticles. Zhang and coworkers [24] demonstrate the duplex bioactivity of titanium-based materials achieved by oxidation and nitrogen implantation. McCall et al. [25] reported the protective effect of ZnO nanoparticle for RNA, while the interaction of ZnO surface with biomatrices is discussed by Yu et al. [26].

Finally, Casu et al. [27] (invited contribution) showed how heating anodically grown TiO_2 nanotubes in situ in a transmission electron microscope allows for the monitoring of their crystallization from amorphous to polycrystalline with polymorphs dependent on the selected temperature.

I would like to gratefully acknowledge all the authors for their valuable contributions and expertise, as well as the reviewers for their comments and suggestions and the assistant editors for the constant support: all have contributed to the success of this special issue.

Hoping that the special issue contents provide an actual overview of the TiO_2 and ZnO nanostructures synthesis and applications, even if not exhaustive of this huge research field, I wish you a good reading.

References

1. Rackauskas, S.; Barbero, N.; Barolo, C.; Viscardi, G. Zno nanowire application in chemoresistive sensing: A review. *Nanomaterials* **2017**, *7*, 381. [CrossRef] [PubMed]
2. Yang, L.; Yang, Y.; Ma, Y.; Li, S.; Wei, Y.; Huang, Z.; Long, N. Fabrication of semiconductor ZnO nanostructures for versatile SERS application. *Nanomaterials* **2017**, *7*, 398. [CrossRef] [PubMed]
3. Wang, X.; Zhao, Y.; Mølhave, K.; Sun, H. Engineering the surface/interface structures of titanium dioxide micro and nano architectures towards environmental and electrochemical applications. *Nanomaterials* **2017**, *7*, 382. [CrossRef] [PubMed]

4. Xiang, L.; Zhao, X. Wet-chemical preparation of TiO_2-based composites with different morphologies and photocatalytic properties. *Nanomaterials* **2017**, *7*, 310. [CrossRef] [PubMed]
5. Hsu, M.-H.; Chang, S.-P.; Chang, S.-J.; Wu, W.-T.; Li, J.-Y. Oxygen partial pressure impact on characteristics of indium titanium zinc oxide thin film transistor fabricated via RF sputtering. *Nanomaterials* **2017**, *7*, 156. [CrossRef] [PubMed]
6. Shih, P.-H.; Wu, S. Growth mechanism studies of ZnO nanowires: Experimental observations and short-circuit diffusion analysis. *Nanomaterials* **2017**, *7*, 188. [CrossRef] [PubMed]
7. Berbel Manaia, E.; Kiatkoski Kaminski, R.; Caetano, B.; Magnani, M.; Meneau, F.; Rochet, A.; Santilli, C.; Briois, V.; Bourgaux, C.; Chiavacci, L. The critical role of thioacetamide concentration in the formation of zno/zns heterostructures by sol-gel process. *Nanomaterials* **2018**, *8*, 55. [CrossRef] [PubMed]
8. Chen, Y.; Ding, H.; Sun, S. Preparation and characterization of ZnO nanoparticles supported on amorphous SiO_2. *Nanomaterials* **2017**, *7*, 217. [CrossRef] [PubMed]
9. Rana, A.; Lee, J.; Shahid, A.; Kim, H.-S. Growth method-dependent and defect density-oriented structural, optical, conductive, and physical properties of solution-grown ZnO nanostructures. *Nanomaterials* **2017**, *7*, 266. [CrossRef] [PubMed]
10. Folger, A.; Kalb, J.; Schmidt-Mende, L.; Scheu, C. Tuning the electronic conductivity in hydrothermally grown rutile TiO_2 nanowires: Effect of heat treatment in different environments. *Nanomaterials* **2017**, *7*, 289. [CrossRef] [PubMed]
11. Jin, S.; Shin, E.; Hong, J. TiO_2 nanowire networks prepared by titanium corrosion and their application to bendable dye-sensitized solar cells. *Nanomaterials* **2017**, *7*, 315. [CrossRef] [PubMed]
12. Berthod, L.; Shavdina, O.; Verrier, I.; Kämpfe, T.; Dellea, O.; Vocanson, F.; Bichotte, M.; Jamon, D.; Jourlin, Y. Periodic TiO_2 nanostructures with improved aspect and line/space ratio realized by colloidal photolithography technique. *Nanomaterials* **2017**, *7*, 316. [CrossRef] [PubMed]
13. Liang, Y.; Wicker, S.; Wang, X.; Erichsen, E.; Fu, F. Organozinc precursor-derived crystalline ZnO nanoparticles: Synthesis, characterization and their spectroscopic properties. *Nanomaterials* **2018**, *8*, 22. [CrossRef] [PubMed]
14. Hu, J.; Tu, J.; Li, X.; Wang, Z.; Li, Y.; Li, Q.; Wang, F. Enhanced UV-visible light photocatalytic activity by constructing appropriate heterostructures between mesopore TiO_2 nanospheres and Sn_3O_4 nanoparticles. *Nanomaterials* **2017**, *7*, 336. [CrossRef] [PubMed]
15. Pan, X.; Liang, X.; Yao, L.; Wang, X.; Jing, Y.; Ma, J.; Fei, Y.; Chen, L.; Mi, L. Study of the photodynamic activity of N-doped TiO_2 nanoparticles conjugated with aluminum phthalocyanine. *Nanomaterials* **2017**, *7*, 338. [CrossRef] [PubMed]
16. Li, T.; Ding, D.; Dong, Z.; Ning, C. Photoelectrochemical water splitting properties of Ti-Ni-Si-O nanostructures on Ti-Ni-Si alloy. *Nanomaterials* **2017**, *7*, 359. [CrossRef] [PubMed]
17. Crespo-Monteiro, N.; Cazier, A.; Vocanson, F.; Lefkir, Y.; Reynaud, S.; Michalon, J.-Y.; Kämpfe, T.; Destouches, N.; Jourlin, Y. Microstructuring of mesoporous titania films loaded with silver salts to enhance the photocatalytic degradation of methyl blue under visible light. *Nanomaterials* **2017**, *7*, 334. [CrossRef] [PubMed]
18. Hao, H.; Jenkins, K.; Huang, X.; Xu, Y.; Huang, J.; Yang, R. Piezoelectric potential in single-crystalline ZnO nanohelices based on finite element analysis. *Nanomaterials* **2017**, *7*, 430. [CrossRef] [PubMed]
19. Sun, S.; Deng, T.; Ding, H.; Chen, Y.; Chen, W. Preparation of nano-TiO_2-coated SiO_2 microsphere composite material and evaluation of its self-cleaning property. *Nanomaterials* **2017**, *7*, 367. [CrossRef] [PubMed]
20. Schmidt, D.; Raab, N.; Noyong, M.; Santhanam, V.; Dittmann, R.; Simon, U. Resistive switching of sub-10 nm TiO_2 nanoparticle self-assembled monolayers. *Nanomaterials* **2017**, *7*, 370. [CrossRef] [PubMed]
21. Bruzzi, M.; Mori, R.; Baldi, A.; Carnevale, E.; Cavallaro, A.; Scaringella, M. Thermally stimulated currents in Nanocrystalline titania. *Nanomaterials* **2018**, *8*, 13. [CrossRef] [PubMed]
22. Yamamoto, S.; Ono, M.; Yuba, E.; Harada, A. In Vitro sonodynamic therapeutic effect of polyion complex micelles incorporating titanium dioxide nanoparticles. *Nanomaterials* **2017**, *7*, 268. [CrossRef] [PubMed]
23. Ancona, A.; Dumontel, B.; Garino, N.; Demarco, B.; Chatzitheodoridou, D.; Fazzini, W.; Engelke, H.; Cauda, V. Lipid-coated zinc oxide nanoparticles as innovative ROS-generators for photodynamic therapy in cancer cells. *Nanomaterials* **2018**, *8*, 143. [CrossRef] [PubMed]

24. Zhang, P.; Wang, X.; Lin, Z.; Lin, H.; Zhang, Z.; Li, W.; Yang, X.; Cui, J. Ti-based biomedical material modified with TiO_x/TiN_x duplex bioactivity film via Micro-Arc oxidation and nitrogen ion implantation. *Nanomaterials* **2017**, *7*, 343. [CrossRef] [PubMed]
25. McCall, J.; Smith, J.; Marquardt, K.; Knight, K.; Bane, H.; Barber, A.; DeLong, R. ZnO nanoparticles protect RNA from degradation better than DNA. *Nanomaterials* **2017**, *7*, 378. [CrossRef] [PubMed]
26. Yu, J.; Kim, H.-J.; Go, M.-R.; Bae, S.-H.; Choi, S.-J. ZnO interactions with biomatrices: Effect of particle size on ZnO-protein corona. *Nanomaterials* **2017**, *7*, 377. [CrossRef] [PubMed]
27. Casu, A.; Lamberti, A.; Stassi, S.; Falqui, A. Crystallization of TiO_2 nanotubes by in situ heating TEM. *Nanomaterials* **2018**, *8*, 40. [CrossRef] [PubMed]

nanomaterials

MDPI

Review

ZnO Nanowire Application in Chemoresistive Sensing: A Review

Simas Rackauskas *, Nadia Barbero, Claudia Baroloand Guido Viscardi

Department of Chemistry, NIS Interdepartmental Centre and INSTM Reference Centre, University of Turin, Via Pietro Giuria 7, 10125 Turin, Italy; nadia.barbero@unito.it (N.B.); claudia.barolo@unito.it (C.B.); guido.viscardi@unito.it (G.V.)

* Correspondence: simas.rackauskas@gmail.com; Tel.: +39-011-670-5323

Received: 30 September 2017; Accepted: 6 November 2017; Published: 9 November 2017

Abstract: This article provides an overview of the recent development of ZnO nanowires (NWs) for chemoresistive sensing. Working mechanisms of chemoresistive sensors are unified for gas, ultraviolet (UV) and bio sensor types: single nanowire and nanowire junction sensors are described, giving the overview for a simple sensor manufacture by multiple nanowire junctions. ZnO NW surface functionalization is discussed, and how this effects the sensing is explained. Further, novel approaches for sensing, using ZnO NW functionalization with other materials such as metal nanoparticles or heterojunctions, are explained, and limiting factors and possible improvements are discussed. The review concludes with the insights and recommendations for the future improvement of the ZnO NW chemoresistive sensing.

Keywords: ZnO; nanowire; sensors; chemoresistive; biosensing

1. Introduction

ZnO nanowires (NWs) have attracted a great deal of interest due to their exceptional properties. ZnO NWs, because of their biocompatibility, piezoelectricity, and optoelectronic properties, are applicable for various electronic, photonic, biological, and energy related applications [1–3]. The most attractive semiconducting properties of ZnO are wide bandgap (3.37 eV), exciton binding energy (60 meV), high refractive index which is higher than 2 and also many other features of ZnO, including manufacturing techniques for high surface area, feasibility for wet chemical etching, structural stability and resistance to high-energy radiation. Moreover, ZnO has a tendency to form numerous nanostructures, because of the several fast grow directions available, which brings a diversity of methods for ZnO nanostructure synthesis, including by low-cost and low-temperature techniques. The potential is seen for application ultraviolet lasers [4], photodetectors [5], dye-sensitized solar cells [6,7], light-emitting diodes [8], and biomedical applications [9].

The names of ZnO nanomaterials such as nanowires [10] nanorods [11] or nanobelts [12] are used to characterize the shape of the obtained 1D material. ZnO NWs could be also grown in more sophisticated forms such as nanoforests [13], tetrapods [14–17], hierarchical nanowires [18], and many others [19–21] by simply controlling the crystal grow direction. ZnO nanowires are mostly grown by hydrothermal [22,23] or vapor transport methods [24,25]. A direct growth of ZnO NW on Zn metal is also attractive for its simplicity [26,27]. The variety of growth methods and availability of ZnO nanostructures gives a certain freedom to optimize ZnO NW sensors for the best surface reactivity and sensitivity.

ZnO NW application for sensing shows multiple advantages; especially interesting is a high surface ratio which provides an enhancement of the surface effects. Moreover, ZnO NWs are single crystals with a well-defined lattice, guiding to controlled reactions and providing stability. Comparing to polycrystalline structures, ZnO NW response is typically faster due to the reaction on the NW surface and no need for inter-grain diffusion.

For the purpose of higher sensitivity and selectivity, ZnO NW sensor surface can be further modified or alternatively heterojunctions with other materials can be formed. Other forms of modification are NW morphology with the contact types or developing ZnO NW hybrid structures. Moreover, ZnO NW sensors can bring new functionalities with their flexible or stretchable configuration.

2. ZnO Nanowire Sensors

The main principle of sensing is to get a measurable response (mostly electrical) from the added substance. High sensitivity (response to extremely small amounts), selectivity (ability to differentiate between various substances) and linearity (response is linear amount of substance measured, or response can be described by other simple law) are the main parameters for efficient sensors. Additionally, there are practical requirements such as stability, simplicity of manufacture and measurement, low cost, biocompatibility and many others, which depend on the application field.

Recently, sensors which work on the change of electrical parameters (chemoresistive, electrochemical, bio-electrochemical) have drawn much attention, due to their multiple advantages, such as simple measurement and manufacturing, low cost, diverse functionalization possibilities just to name few. This class of sensors, employing ZnO as a base material, will be covered in this review.

In a broad sense, ZnO NW resistive sensors for UV, gas and bio sensing have the same working principle: sensor response is related to the charge accumulated or transferred to NW. The working principle can be explained from the oxygen adsorption and desorption from the surface in gas sensors [28]. The mechanism is also attributed to CuO [29], SnO_2 [16,30] and other metal oxides [31].

Most ZnO NW sensors are n-type semiconductors, since ZnO is an intrinsically stable electron-dominated conductor [32]; it is challenging to incorporate acceptors for stable conduction properties, therefore *p*-type ZnO NWs are rare [33]. Considering the *n*-type case, when ZnO NWs are exposed to atmosphere, oxygen adsorbs on the surface and becomes negatively charged (O^-, O^{2-}, or O_2^-) and extracts electrons from ZnO. The surface region that loses electrons is defined as the surface charge depletion layer (Figure 1). The depletion layer controls the effective conduction channel and increases the energy barrier height of the contact to NW or NW-NW junctions. When exposed to reducing gas, such as H_2S, CO, and NH_3, adsorbed O_2 reacts with these gases and releases electrons back to the ZnO, which increases the electrical conductivity. As the depletion layer thickness is related to the oxygen coverage, the presence of reducing gas can diminish the thickness of the depletion layer, which increases the overall conductivity of the ZnO NWs. In the same manner, the oxidizing gases such as NO_2 and O_3 generally decrease conductance by extracting more electrons from the surface and increasing the depletion layer width [28,34]. Similarly, as in gas sensors, UV sensor response is associated with the depletion layer width. Illuminating ZnO NW with a wavelength, higher than the bandgap, photogenerated holes combine with the negative O_2 ions, inducing desorption and increasing the conductivity. In biosensors, typically the ZnO NW surface is covered with a material (e.g., enzyme) which attaches the targeted substance from the analyte; consequently, charge is transferred to ZnO NW, changing its depletion layer width. However, in the case of biosensors, attention is paid to the immobilization of biomolecules on the surface, as is described in Section 2.2.

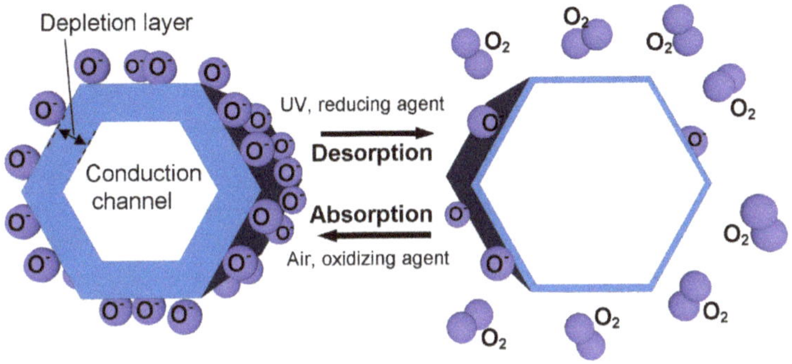

Figure 1. Schematics of unified ZnO nanowire (NW) chemoresistive sensing principle, based on the depletion layer width change with absorption-desorption of oxygen.

According to the device structure and working principle, nanowire sensors can be divided into two types: single nanowire connection and nanowire-nanowire junctions (Figure 2). Single connection devices rely on the response of a single NW, deposited or aligned between two metal contacts. Ohmic contacts are mostly used to contact ZnO NW; however, it is also possible to make Schottky contacts (e.g., contacting with Pt) in order to obtain a rectifying character or conduction [35]. However, most of research is done with NW-NW junction type of contacts, which is used not only because of the ease of sensor manufacture, but also due to the ability to control the barrier height. NW-NW junction sensors are typically more sensitive to small concentration changes, and have high rates of response as the conduction path involves tunneling through the depletion layer (Figure 2) [15]; however, in principle, single NW sensors can cover higher concentration ranges. Moreover, NW-NW junction integration into gas sensing devices is much more simple, as it relies on the randomly built conduction path between multiple NWs [15], whereas for the single NW sensors at least some degree of orientation is needed, wherefore ability to manipulate the NW or contacts is preferred [36].

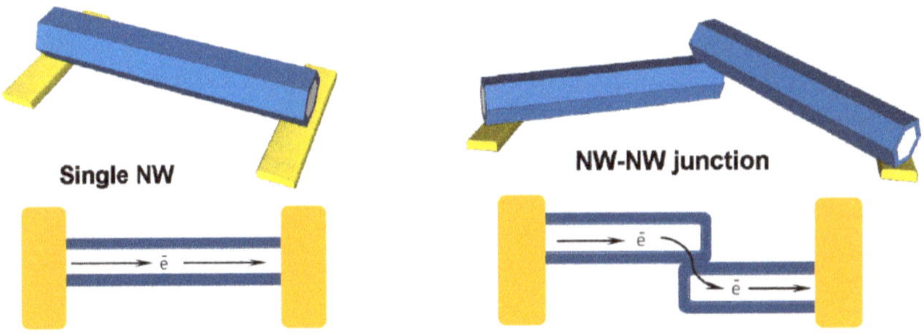

(**a**) (**b**)

Figure 2. Schematics of ZnO NW sensor geometry types: (**a**) single NW; (**b**) NW-NW junction.

Selectivity is another important feature of the ZnO NW sensors, which is connected to the ability of the sensor to discriminate among different types of chemical species. While, in biosensors, different analytes can be targeted with a special biomolecule, it is an especially difficult task to differentiate among the same type of gases (oxidizing or reducing), since depletion layer width change and consequently response will be in the same direction. Selective recognition of the gases can be addressed by variations in chemical adsorption and dissociation of the target gases at the

NW surface; therefore, NW sensor selectivity can be approached by several methods: NW geometry control [36,37], NW functionalization [38,39], selective contact formation [40], heterojunction [41], operating temperature modulation [42], and sensor array formation [43].

2.1. ZnO NW Heterojunction

Heterojunction formation on ZnO NW, or simply a junction with other material, is mainly targeted to improve response and selectivity. A common choice for heterojunction is the use of metal oxides such as SnO_2, because of high sensitivity. It was shown that ZnO NWs uniformly covered with the outer layer of *n*-type SnO_2 nanoparticles [41] considerably improved the sensor response. Comparison of the gas sensor performance between SnO_2, ZnO NW and the formed SnO_2/ZnO NW heterojunction showed that, after functionalization, SnO_2/ZnO NW sensor was responding at a high rate, selectivity and repeatability to *n*-butylamine gas was good, and therefore it can be applied for organic amine sensors. Another study used SnO_2 NWs growth on ZnO NW for volatile organic compound sensing [44]. Advantageous sensor performance comparing to the pure ZnO NWs was explained by potential barriers, which were forming at the SnO_2/ZnO. Moreover, the geometry of SnO_2 nanowires affects the selectivity of triethylamine, toluene, ethanol, acetic acid, acetone, and methanol, allowing ZnO/SnO_2 heterostructures to discriminate acetone from other volatile compounds.

Another option is to use ZnO NW heterojunction with *p*-type materials, forming local p-n junctions. Park et al. studied ZnO NW junction with CuO NW, by growing long crystalline nanowires and forming air-bridge-structured junction [45]. It was shown that formation of nanoscale *p-n* junctions lowers sensor conductance by two orders of magnitude (Figure 3). The current-voltage I-V characteristics of both ZnO NW and CuO NW contacts were symmetric, while ZnO/CuO NW heterocontacts were asymmetric, which indicate the built-in potential established near the *p-n* junction. *P*-type material plays the role of catalyst and expands the adjacent electron depletion layer of ZnO NWs. However, sensitivity toward H_2, CO and NO_2 gases was lower of such *p-n* heterocontacts than *n-n* contact between ZnO NWs.

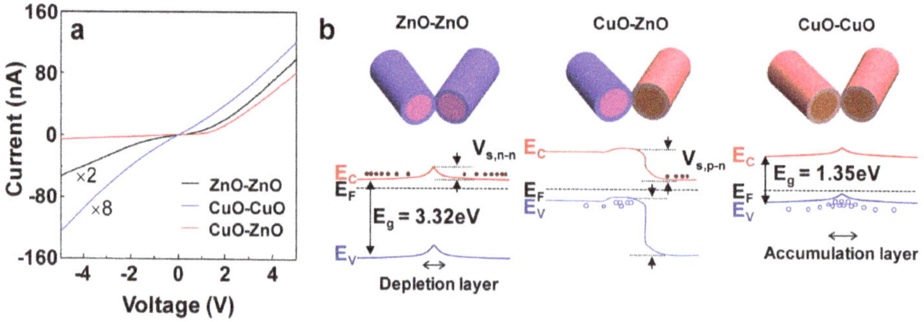

Figure 3. Comparison of contacts between ZnO and CuO NWs: (**a**) current-voltage (I-V) curves obtained at room temperature, demonstrating the characteristic nonlinear Schottky-like transport behavior; (**b**) Energy band diagram for every interconnection type between NWs. Reproduced with permission from [45]. Copyright American Chemical Society, 2013.

Zhang et al. reported an increased sensitivity of ZnO NW with co-precipitated *p*-type CuO flower-type structure for ethanol vapor sensors [46]. The ZnO/uO heterojunction demonstrates 2.5 times higher ethanol response at 300 °C compared to ZnO NWs without CuO. Good selectivity, long-term stability and also response and recovery time of 7 and 9 s, respectively, were demonstrated. The improved ethanol response was also due to a widening of depletion layer on the ZnO/CuO interface.

In order to make a *p-n* heterojunction with ZnO NWs it is also possible to use other *p* type metal oxides, such as NiO [47,48], Cr_2O_3 [49,50], Mn_3O_4 [51] or modify with *p*-typematerials, such as cobalt phthalocyanine (CoPc) [52].

UV sensor performance can also be enhanced applying ZnO NW heterojunctions with *p* type oxides such as CuO [53] or NiO [54]. Ultrafast response to UV in μs range was obtained at ZnO NW contact with *p*-Si [55], which was explained by pyro-phototronic effect [56]. The pyro-phototronic results are from three elements: pyroelectric effect, photonic excitations, and semiconductor properties, the coupling of which paves a way to reach ultrafast photosensing performance with optoelectronic processes. Moreover, the response time of ZnO NW UV sensors decreases with increasing illumination intensity; thus, this could be potentially applied for ultrafast detection of intensive light.

Another interesting option is to decorate ZnO with noble metal particles. Metal nanoparticles can be employed to improve optical absorbance or emission in semiconductors, due to a high plasmon interaction with electromagnetic fields. ZnO NW gas sensing of ethanol can be improved by adding noble metallic nanoparticles, for example Au, Pt [57] or Ag [58]. Alternatively, a selective Pd contact with ZnO NWs [40] can be used for various gases such as H_2, CH_4, H_2S and CO_2 at different operating temperatures, where high efficiency in hydrogen detection was reported. Moreover, Pt contacted ZnO NW networks can be used as self-powered UV sensors, since excellent photoresponse properties to 365 nm UV irradiation was obtained at zero bias [31].

Combination of UV and Gas Sensing in ZnO NW

The high operating temperatures (typically about 350 °C) are essential for gas detection and sensing, which is a major technical limitation in applicability. Moreover, adsorption of water on the ZnO surface leads to a decrease in surface potential, at relative humidity higher than 14% due to adsorbed water molecules increasing the surface electron density [59]. Irradiation of ZnO by photons with an energy greater than the band gap (3.37 eV) changes adsorbed oxygen species on the surface, which is a practical alternative for achieving chemical reactions without the necessary heating. However, the UV activation shows an order of magnitude lower sensitivity compared to the same sensors activated by traditional heating methods [60]. UV activation, in order to increase the response, is used for ZnO films [59,61] and also for other materials [62,63]. A transparent ZnO NW sensor, which detects both UV and ethanol gas, was fabricated and deposited onto a silicon solar cell [64]. In UV sensing, the current rise was obtained in 137 s. In ethanol vapor detection, UV was used for sensitivity improvement. The ZnO NW sensor response increased by 13% with an ethanol gas concentration change of 100 ppm at 53 °C (heat generated by the c-Si solar cell). The sensor response is approximately zero without solar illumination.

Combining the response to chemical substance with a UV illumination, it is possible to get a synergy of these two effects and an increase of the performance of ZnO NW sensors (Figure 4); however, there are still only some studies on the combination of more than two effects on the performance of NW sensors. Synergistic effects of Cr_2O_3 functionalized ZnO NW sensor with UV irradiation were demonstrated for the ethanol gas sensing [65]. The responses at room temperature to ethanol were increased by UV illumination by 3.8 times for the pristine and by 7.7 times for Cr_2O_3 functionalized ZnO NW sensors. Moreover, the Cr_2O_3 modified ZnO NW sensor demonstrated rapid response and recovery; moreover, selectivity was also increased. This shows that combining heterostructures with UV activation has a synergistic effect on sensor performance. The synergistic effects arise from the Cr_2O_3 catalytic oxidation of ethanol and also from conduction channel width change due to Cr_2O_3 nanoparticle effect on ethanol adsorption and desorption under UV illumination in the Cr_2O_3 modified ZnO NW sensor.

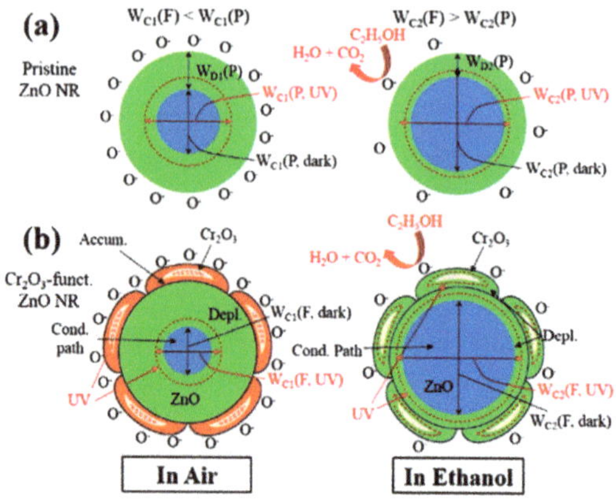

Figure 4. Schematic diagrams: (**a**) pristine ZnO NW; (**b**) Cr_2O_3-modified ZnO NW in the dark and under UV illumination in air and ethanol showing the depletion layer and conduction path. Reproduced with permission from [65]. Copyright American Chemical Society, 2016.

Similarly, gold-decorated ZnO thin films showed improved sensing properties compared to bare ZnO under UV illumination. The sensor showed a selective response to NO_2 gas under green light illumination. Moreover, Au/ZnO sensor can detect SO_2 gas in ppm level in humid conditions [66].

2.2. *ZnO Nanowire Biosensors*

We can identify three generations of biosensors [67]. In the first generation (Figure 5a), an electrical response is generated by the diffusion of the reaction products to the transducer. In the second generation (Figure 5b), instead, an initial redox reaction is performed by a mediator between the enzyme and its substrate. The enzyme and mediator are usually co-immobilized at an electrode surface in the third-generation of biosensors (Figure 5c); in this way, the biorecognition component is an integral part of the electrode transducer [68].

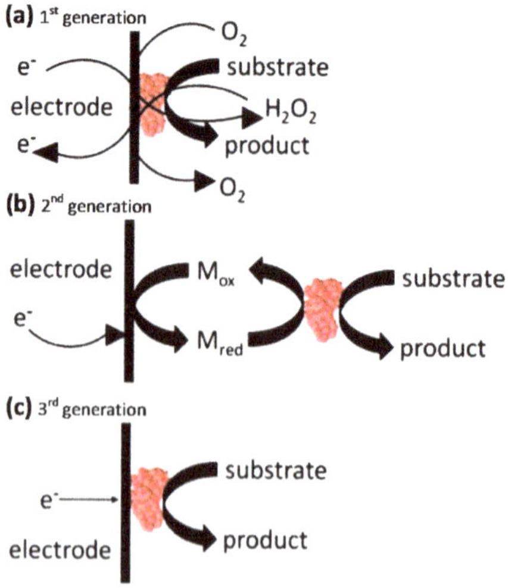

Figure 5. Schematic diagrams of three generations of sensors: (**a**) 1st generation; (**b**) 2nd generation and (**c**) 3rd generation [69].

Electronic devices based on semiconductor NW have emerged as a potential platform for the qualitative and quantitative detection of chemical and biological species due to their ultralow detection limit, fast readout and easy fabrication [70].

In particular, ZnO possesses a series of properties and characteristics (high electron mobility, easiness of fabrication of ZnO, biocompatibility and low toxicity) that makes it a nice candidate for the construction of biosensors [71]. In particular, ZnO NWs, due to their low weight with extraordinary mechanical, electrical, thermal and multifunctional properties along with their high surface area, are suitable for adsorption or immobilization of various biomolecules such as proteins, enzymes or antibodies [9]. ZnO nanowires possess active surfaces that can be modified for the immobilization of a large number of biomolecules. These nanostructures can be perfect transducers for producing signals to interface with macroscopic instruments since they present diameters which are comparable to the sizes of the biological and chemical species being sensed [72]. Based on the specific feature of NW, biosensors may even go down to a single molecule detection [73].

Qualitative and quantitative detection of chemical and biological species is crucial in a huge variety of fields. If the literature on biosensors based on ZnO nanomaterials is quite rich, relatively few examples are focused on ZnO nanowires. Below, we will briefly summarize some recent findings reporting biosensors based on ZnO nanowires.

2.2.1. Urea Biosensors

The importance of urea measurements in blood and serum is important for a certain number of diseases. ZnO nanomaterials comprising nanorods and nanowires have been used for the fabrication of urea biosensors [9]. Well-aligned ZnO NW arrays were fabricated on gold-coated plastic substrates by using a low-temperature aqueous chemical growth (ACG) method and were proved to be sensitive to urea detection at a concentration from 0.1 to 100 mM [74]. Urease was immobilized on the surface of the ZnO NWs using an electrostatic process. More recently, high quality ZnO NWs, synthesized by the chemical vapor deposition (CVD) method, were used for the fabrication of electrical biosensors based on field effect transistor (FET) for the simultaneously low and high concentrations detection of uric acid [73]. The obtained ZnO NW bioFET sensors could easily detect uric acid down to a concentration of 1 pM with 14.7 nS of conductance increase, and the response time turns out to be in the order of milliseconds.

2.2.2. Glucose Biosensors

Glucose is a critical metabolite for living organisms, particularly for patients who are suffering from diabetes, and glucose sensors attracted a huge amount of interest being one of the most important sensing technologies in medical science, clinical diagnostics, and food industry [75].

High-density vertical ZnO NWs were synthesized using the vapor-phase deposition method on patterned Au/glass electrode substrate with and without Au nanoparticle (NP) modification. A huge enhancement of the sensitivity toward glucose was obtained with Au NP modification [76]. A similar approach has been followed for the fabrication of high density, well-aligned ZnO NWs decorated with Pt nanoparticles to fabricate the working electrode for a non-enzymatic glucose biosensor [77]. ZnO NWs were synthesized hydrothermally on a glass substrate. The Pt NPs decoration allowed to enhance the biosensor's glucose sensitivity 10-fold in comparison with the pristine ZnO NWs electrode. The large specific surface area, abundant microspace, small channels, and high isoelectric point (IEP) fracture of ZnO enable effective fluid circulation and good biocompatibility boding well for immobilization of enzymes. These characteristics were exploited for the immobilization of glucose oxidase on a glucose enzymatic biosensor composed of ZnO NWs supported by silicon NWs (ZnO/Si NWs) [78]. A Si NWs/ZnO nanowire nanocomposites enzymatic biosensor exhibited very strong and sensitive amperometric responses to glucose, even in the presence of common interfering species, and showed a high sensitivity of 129 $\mu A \cdot mM^{-1}$, low detection limit of 12 μM, and good stability as well as reproducibility. Very recently, a roll-to-roll flexographic printing technique was used for the fabrication of a three electrode electrochemical enzymatic biosensor consisting of ZnO NWs [79]. This biosensor device showed a typical sensitivity of $1.2 \pm 0.2\ \mu A \cdot mM^{-1} \cdot cm^{-2}$ with a linear response to the addition of glucose over a concentration range of 0.1 to 3.6 mM.

2.2.3. DNA Detection

Nanomaterials can provide good opportunities for sequence-specific target DNA detection as a medium for signal amplification. The possibility of using ZnO NWs to fabricate electrochemical DNA biosensors was explored some years ago. In order to improve the sensitivity, multi-walled carbon nanotubes (MWCNTs) and gold nanoparticles (Au NPs) were employed. The resulting device was able to quantitatively detect target DNA from 1.0×10^{-7} to 1.0×10^{-13} M with a detection limit of 3.5×10^{-14} M [80]. Another ZnO NWs/Au electrode showed to be a suitable platform for the immobilization of DNA for the rapid detection of a sequence specific for the breast cancer 1 (BRCA1) gene [81]. This DNA biosensor has the ability to detect the target sequence in the range of concentration between 10.0 and 100.0 μM with a detection limit of 3.32 μM. A further sensitive and in situ selective label-free DNA sensor, based on a Schottky contacted ZnO NW device, has been fabricated. The performance of this device was significantly enhanced by the presence of piezotronic effect [82].

3. Conclusions

There is a growing interest in ZnO NW application for sensing due to their high temperature stability, biocompatibility, simple synthesis route and sensor manufacture. In this review, we have summarized recent strategies to enhance ZnO NW sensor performance. Several conclusions could be made with the proposal for future trends:

- Many synthetic routes for ZnO NWs are already found, and different ZnO NW structures can be obtained by varying growth conditions. Decoration with metal particles is easily achieved; however, there is still a lack of comparative studies where the sensor performance is related not to the synthesis conditions but to the morphology of the sensor, namely the size of a conductive channel and how it is influenced by absorbed material.
- Heterojunction brings another possibility for the improvement of the sensing which can be achieved with comparingly facile fabrication techniques. Junctions with other p or n materials can be made with controlled surface coverage and thickness; still, there is a need to optimize ZnO NW interface with other materials in order to obtain high efficiency sensing.
- Using UV activation for bio and chemoresistive sensors shows considerable improvement in sensing. However, there is still a lack of understanding for the interplay of several effects (UV, temperature, oxygen adsorption, etc.), especially at the junctions of NW to NW or *p-n* junctions.
- The unique properties of ZnO NWs, simple fabrication and the possibility for suitable surface functionalization of the NWs make them exemplar as biosensor materials for a great variety of applications.
- The devices can be fabricated by roll-to-roll printing, which is suitable for low-cost high-volume production and a spread of large-scale commercialization of the biosensors.

Acknowledgments: The work has received funding from the European Union's Seventh Framework programme for research and innovation under the Marie Sklodowska-Curie grant agreement No 609402–2020 researchers: Train to Move (T2M).

Author Contributions: Simas Rackauskas worked on the ZnO nanowire sensors paragraphs while Nadia Barbero conceived and drafted the biosensor part. Claudia Barolo and Guido Viscardi coordinated and supervised the project. All authors analyzed and approved the final version of the manuscript.

Conflicts of Interest: The authors declare no conflict of interest.

References

1. Zhang, Y. *ZnO Nanostructures*; Royal Society of Chemistry: Cambridge, UK, 2017; ISBN 978-1-78262-741-8.
2. Kołodziejczak-Radzimska, A.; Jesionowski, T. Zinc Oxide—From synthesis to application: A review. *Materials* **2014**, *7*, 2833–2881. [CrossRef] [PubMed]
3. Kaps, S.; Bhowmick, S.; Gröttrup, J.; Hrkac, V.; Stauffer, D.; Guo, H.; Warren, O.L.; Adam, J.; Kienle, L.; Minor, A.M.; et al. Piezoresistive response of quasi-one-dimensional ZnO nanowires using an In Situ electromechanical device. *ACS Omega* **2017**, *2*, 2985–2993. [CrossRef]
4. Chu, S.; Wang, G.; Zhou, W.; Lin, Y.; Chernyak, L.; Zhao, J.; Kong, J.; Li, L.; Ren, J.; Liu, J. Electrically pumped waveguide lasing from ZnO nanowires. *Nat. Nanotechnol.* **2011**, *6*, 506–510. [CrossRef] [PubMed]
5. Soci, C.; Zhang, A.; Xiang, B.; Dayeh, S.A.; Aplin, D.P.R.; Park, J.; Bao, X.Y.; Lo, Y.H.; Wang, D. ZnO nanowire UV photodetectors with high internal gain. *Nano Lett.* **2007**, *7*, 1003–1009. [CrossRef] [PubMed]
6. Rackauskas, S.; Barbero, N.; Barolo, C.; Viscardi, G. ZnO nanowires for dye sensitized solar cells. In *Nanowires—New Insights*; InTech: Rijeka, Croatia, 2017.
7. Znajdek, K.; Sibiński, M.; Lisik, Z.; Apostoluk, A.; Zhu, Y.; Masenelli, B.; Sędzicki, P. Zinc oxide nanoparticles for improvement of thin film photovoltaic structures' efficiency through down shifting conversion. *Opto-Electron. Rev.* **2017**, *25*, 99–102. [CrossRef]
8. Zhang, X.M.; Lu, M.Y.; Zhang, Y.; Chen, L.J.; Wang, Z.L. Fabrication of a high-brightness blue-light-emitting diode using a ZnO-Nanowire array grown on *p*-GaN thin film. *Adv. Mater.* **2009**, *21*, 2767–2770. [CrossRef]
9. Zhang, Y.; Nayak, T.R.; Hong, H.; Cai, W. Biomedical applications of zinc oxide nanomaterials. *Curr. Mol. Med.* **2013**, *13*, 1633–1645. [CrossRef] [PubMed]
10. Rackauskas, S.; Nasibulin, A.G.; Jiang, H.; Tian, Y.; Statkute, G.; Shandakov, S.D.; Lipsanen, H.; Kauppinen, E.I. Mechanistic investigation of ZnO nanowire growth. *Appl. Phys. Lett.* **2009**, *95*, 183114. [CrossRef]
11. Rokhsat, E.; Akhavan, O. Improving the photocatalytic activity of graphene oxide/ZnO nanorod films by UV irradiation. *Appl. Surf. Sci.* **2016**, *371*, 592–595. [CrossRef]
12. Tarat, A.; Majithia, R.; Brown, R.A.; Penny, M.W.; Meissner, K.E.; Maffeis, T.G.G. Synthesis of nanocrystalline ZnO nanobelts via pyrolytic decomposition of zinc acetate nanobelts and their gas sensing behavior. *Surf. Sci.* **2012**, *606*, 715–721. [CrossRef]
13. Ko, S.H.; Lee, D.; Kang, H.W.; Nam, K.H.; Yeo, J.Y.; Hong, S.J.; Grigoropoulos, C.P.; Sung, H.J. Nanoforest of hydrothermally grown hierarchical ZnO nanowires for a high efficiency dye-sensitized solar cell. *Nano Lett.* **2011**, *11*, 666–671. [CrossRef] [PubMed]
14. Rackauskas, S.; Klimova, O.; Jiang, H.; Nikitenko, A.; Chernenko, K.A.; Shandakov, S.D.; Kauppinen, E.I.; Tolochko, O.V.; Nasibulin, A.G. A novel method for continuous synthesis of ZnO tetrapods. *J. Phys. Chem. C* **2015**, *119*, 16366–16373. [CrossRef]
15. Rackauskas, S.; Mustonen, K.; Järvinen, T.; Mattila, M.; Klimova, O.; Jiang, H.; Tolochko, O.; Lipsanen, H.; Kauppinen, E.I.; Nasibulin, A.G. Synthesis of ZnO tetrapods for flexible and transparent UV sensors. *Nanotechnology* **2012**, *23*, 95502. [CrossRef] [PubMed]
16. Mishra, Y.K.; Modi, G.; Cretu, V.; Postica, V.; Lupan, O.; Reimer, T.; Paulowicz, I.; Hrkac, V.; Benecke, W.; Kienle, L.; et al. Direct growth of freestanding ZnO tetrapod networks for multifunctional applications in photocatalysis, UV photodetection, and gas sensing. *ACS Appl. Mater. Interfaces* **2015**, *7*, 14303–14316. [CrossRef] [PubMed]
17. Gedamu, D.; Paulowicz, I.; Kaps, S.; Lupan, O.; Wille, S.; Haidarschin, G.; Mishra, Y.K.; Adelung, R. Rapid fabrication technique for interpenetrated ZnO nanotetrapod networks for fast UV sensors. *Adv. Mater.* **2014**, *26*, 1541–1550. [CrossRef] [PubMed]
18. Bielinski, A.R.; Kazyak, E.; Schlepütz, C.M.; Jung, H.J.; Wood, K.N.; Dasgupta, N.P. Hierarchical ZnO nanowire growth with tunable orientations on versatile substrates using atomic layer deposition seeding. *Chem. Mater.* **2015**, *27*, 4799–4807. [CrossRef]
19. Pugliese, D.; Bella, F.; Cauda, V.; Lamberti, A.; Sacco, A.; Tresso, E.; Bianco, S. A chemometric approach for the sensitization procedure of ZnO flowerlike microstructures for dye-sensitized solar cells. *ACS Appl. Mater. Interfaces* **2013**, *5*, 11288–11295. [CrossRef] [PubMed]

20. Kuo, S.-Y.; Yang, J.-F.; Lai, F.-I. Improved dye-sensitized solar cell with a ZnO nanotree photoanode by hydrothermal method. *Nanoscale Res. Lett.* **2014**, *9*, 206. [CrossRef] [PubMed]
21. Zhang, S.; Chen, H.-S.; Matras-Postolek, K.; Yang, P. ZnO nanoflowers with single crystal structure towards enhanced gas sensing and photocatalysis. *Phys. Chem. Chem. Phys.* **2015**, *17*, 30300–30306. [CrossRef] [PubMed]
22. Lu, C.; Qi, L.; Yang, J.; Tang, L.; Zhang, D.; Ma, J. Hydrothermal growth of large-scale micropatterned arrays of ultralong ZnO nanowires and nanobelts on zinc substrate. *Chem. Commun.* **2006**, *432*, 3551–3553. [CrossRef] [PubMed]
23. Baruah, S.; Dutta, J. Hydrothermal growth of ZnO nanostructures. *Sci. Technol. Adv. Mater.* **2009**, *10*, 013001. [CrossRef] [PubMed]
24. Bae, S.Y.; Seo, H.W.; Park, J. Vertically aligned sulfur-doped ZnO nanowires synthesized via chemical vapor deposition. *J. Phys. Chem. B* **2004**, *108*, 5206–5210. [CrossRef]
25. Kuo, T.J.; Lin, C.N.; Kuo, C.L.; Huang, M.H. Growth of ultralong ZnO nanowires on silicon substrates by vapor transport and their use as recyclable photocatalysts. *Chem. Mater.* **2007**, *19*, 5143–5147. [CrossRef]
26. Rackauskas, S.; Nasibulin, A.G.; Jiang, H.; Tian, Y.; Kleshch, V.I.; Sainio, J.; Obraztsova, E.D.; Bokova, S.N.; Obraztsov, A.N.; Kauppinen, E.I. A novel method for metal oxide nanowire synthesis. *Nanotechnology* **2009**, *20*, 165603. [CrossRef] [PubMed]
27. Kleshch, V.I.; Rackauskas, S.; Nasibulin, A.G.; Kauppinen, E.I.; Obraztsova, E.D.; Obraztsov, A.N. Field emission properties of metal oxide nanowires. *J. Nanoelectron. Optoelectron.* **2012**, *7*, 35–40. [CrossRef]
28. Choi, K.J.; Jang, H.W. One-dimensional oxide nanostructures as gas-sensing materials: Review and issues. *Sensors* **2010**, *10*, 4083–4099. [CrossRef] [PubMed]
29. Lupan, O.; Cretu, V.; Postica, V.; Ababii, N.; Polonskyi, O.; Kaidas, V.; Schütt, F.; Mishra, Y.K.; Monaico, E.; Tiginyanu, I.; et al. Enhanced ethanol vapour sensing performances of copper oxide nanocrystals with mixed phases. *Sens. Actuators B* **2016**, *224*, 434–448. [CrossRef]
30. Paulowicz, I.; Hrkac, V.; Kaps, S.; Cretu, V.; Lupan, O.; Braniste, T.; Duppel, V.; Tiginyanu, I.; Kienle, L.; Adelung, R.; et al. Three-dimensional SnO_2 nanowire networks for multifunctional applications: From high-temperature stretchable ceramics to ultraresponsive sensors. *Adv. Electron. Mater.* **2015**, *1*, 1500081. [CrossRef]
31. Bai, Z.; Yan, X.; Chen, X.; Zhao, K.; Lin, P.; Zhang, Y. High sensitivity, fast speed and self-powered ultraviolet photodetectors based on ZnO micro/nanowire networks. *Prog. Nat. Sci.* **2014**, *24*, 1–5. [CrossRef]
32. Lu, M.-P.; Lu, M.-Y.; Chen, L.-J. *p*-Type ZnO nanowires: From synthesis to nanoenergy. *Nano Energy* **2012**, *1*, 247–258. [CrossRef]
33. Yuan, G.D.; Zhang, W.J.; Jie, J.S.; Fan, X.; Zapien, J.A.; Leung, Y.H.; Luo, L.B.; Wang, P.F.; Lee, C.S.; Lee, S.T. *p*-type ZnO nanowire arrays. *Nano Lett.* **2008**, *8*, 2591–2597. [CrossRef] [PubMed]
34. Chen, J.; Wang, K.; Cao, B.; Zhou, W. Highly sensitive and selective gas detection by 3D metal oxide nanoarchitectures. In *Three-Dimensional Nanoarchitectures*; Springer: New York, NY, USA, 2011; pp. 391–412. ISBN 978-1-4419-9822-4.
35. Bercu, B.; Geng, W.; Simonetti, O.; Kostcheev, S.; Sartel, C.; Sallet, V.; Lérondel, G.; Molinari, M.; Giraudet, L.; Couteau, C. Characterizations of Ohmic and Schottky-behaving contacts of a single ZnO nanowire. *Nanotechnology* **2013**, *24*, 415202. [CrossRef] [PubMed]
36. Schütt, F.; Postica, V.; Adelung, R.; Lupan, O. Single and networked ZnO–CNT hybrid tetrapods for selective room-temperature high-performance ammonia sensors. *ACS Appl. Mater. Interfaces* **2017**, *9*, 23107–23118. [CrossRef] [PubMed]
37. Sysoev, V.V.; Strelcov, E.; Sommer, M.; Bruns, M.; Kiselev, I.; Habicht, W.; Kar, S.; Gregoratti, L.; Kiskinova, M.; Kolmakov, A. Single-nanobelt electronic nose: Engineering and tests of the simplest analytical element. *ACS Nano* **2010**, *4*, 4487–4494. [CrossRef] [PubMed]
38. Lupan, O.; Postica, V.; Gröttrup, J.; Mishra, A.K.; de Leeuw, N.H.; Carreira, J.F.C.; Rodrigues, J.; Ben Sedrine, N.; Correia, M.R.; Monteiro, T.; et al. Hybridization of Zinc oxide tetrapods for selective gas sensing applications. *ACS Appl. Mater. Interfaces* **2017**, *9*, 4084–4099. [CrossRef] [PubMed]
39. Sun, G.-J.; Lee, J.K.; Choi, S.; Lee, W.I.; Kim, H.W.; Lee, C. Selective oxidizing gas sensing and dominant sensing mechanism of *n*-CaO-decorated *n*-ZnO nanorod sensors. *ACS Appl. Mater. Interfaces* **2017**, *9*, 9975–9985. [CrossRef] [PubMed]

40. Kumar, M.; Singh Bhati, V.; Ranwa, S.; Singh, J.; kumar, M. Pd/ZnO nanorods based sensor for highly selective detection of extremely low concentration hydrogen. *Sci. Rep.* **2017**, *7*, 236. [CrossRef] [PubMed]
41. Wang, L.; Li, J.; Wang, Y.; Yu, K.; Tang, X.; Zhang, Y.; Wang, S.; Wei, C. Construction of 1D SnO_2-coated ZnO nanowire heterojunction for their improved *n*-butylamine sensing performances. *Sci. Rep.* **2016**, *6*, 35079. [CrossRef] [PubMed]
42. Li, F.-A.; Jin, H.; Wang, J.; Zou, J.; Jian, J. Selective sensing of gas mixture via a temperature modulation approach: New strategy for potentiometric gas sensor obtaining satisfactory discriminating features. *Sensors* **2017**, *17*, 573. [CrossRef] [PubMed]
43. Fedorov, F.; Vasilkov, M.; Lashkov, A.; Varezhnikov, A.; Fuchs, D.; Kübel, C.; Bruns, M.; Sommer, M.; Sysoev, V. Toward new gas-analytical multisensor chips based on titanium oxide nanotube array. *Sci. Rep.* **2017**, *7*, 9732. [CrossRef] [PubMed]
44. Rakshit, T.; Santra, S.; Manna, I.; Ray, S.K. Enhanced sensitivity and selectivity of brush-like SnO_2 nanowire/ZnO nanorod heterostructure based sensors for volatile organic compounds. *RSC Adv.* **2014**, *4*, 36749–36756. [CrossRef]
45. Park, W.J.; Choi, K.J.; Kim, M.H.; Koo, B.H.; Lee, J.-L.; Baik, J.M. Self-assembled and highly selective sensors based on air-bridge-structured nanowire junction arrays. *ACS Appl. Mater. Interfaces* **2013**, *5*, 6802–6807. [CrossRef] [PubMed]
46. Zhang, Y.-B.; Yin, J.; Li, L.; Zhang, L.-X.; Bie, L.-J. Enhanced ethanol gas-sensing properties of flower-like *p*-CuO/*n*-ZnO heterojunction nanorods. *Sens. Actuators B* **2014**, *202*, 500–507. [CrossRef]
47. Ju, D.; Xu, H.; Qiu, Z.; Guo, J.; Zhang, J.; Cao, B. Highly sensitive and selective triethylamine-sensing properties of nanosheets directly grown on ceramic tube by forming NiO/ZnO PN heterojunction. *Sens. Actuators B* **2014**, *200*, 288–296. [CrossRef]
48. Na, C.W.; Woo, H.-S.; Lee, J.-H.; Yao, P.J.; Nishibori, M.; Wang, D.; Wang, C.; Pinna, N.; Lee, J.H. Design of highly sensitive volatile organic compound sensors by controlling NiO loading on ZnO nanowire networks. *RSC Adv.* **2012**, *2*, 414–417. [CrossRef]
49. Woo, H.-S.; Na, C.W.; Kim, I.-D.; Lee, J.-H. Highly sensitive and selective trimethylamine sensor using one-dimensional ZnO–Cr_2O_3 hetero-nanostructures. *Nanotechnology* **2012**, *23*, 245501. [CrossRef] [PubMed]
50. Park, S.; Kim, S.; Kheel, H.; Lee, C. Oxidizing gas sensing properties of the *n*-ZnO/*p*-Co_3O_4 composite nanoparticle network sensor. *Sens. Actuators B* **2015**, *222*, 1193–1200. [CrossRef]
51. Na, C.W.; Park, S.Y.; Chung, J.H.; Lee, J.H. Transformation of ZnO nanobelts into single-crystalline Mn_3O_4 nanowires. *ACS Appl. Mater. Interfaces* **2012**, *4*, 6565–6572. [CrossRef] [PubMed]
52. Kumar, A.; Samanta, S.; Singh, A.; Roy, M.; Singh, S.; Basu, S.; Chehimi, M.M.; Roy, K.; Ramgir, N.; Navaneethan, M.; et al. Fast response and high sensitivity of ZnO nanowires-cobalt phthalocyanine heterojunction based H_2S sensor. *ACS Appl Mater. Interfaces* **2015**, *7*, 17713–17724. [CrossRef] [PubMed]
53. Wu, J.-K.; Chen, W.-J.; Chang, Y.H.; Chen, Y.F.; Hang, D.-R.; Liang, C.-T.; Lu, J.-Y. Fabrication and photoresponse of ZnO nanowires/CuO coaxial heterojunction. *Nanoscale Res. Lett.* **2013**, *8*, 387. [CrossRef] [PubMed]
54. Li, Y.-R.; Wan, C.-Y.; Chang, C.-T.; Tsai, W.-L.; Huang, Y.-C.; Wang, K.-Y.; Yang, P.-Y.; Cheng, H.-C. Thickness effect of NiO on the performance of ultraviolet sensors with *p*-NiO/*n*-ZnO nanowire heterojunction structure. *Vacuum* **2015**, *118*, 48–54. [CrossRef]
55. Wang, Z.; Yu, R.; Wang, X.; Wu, W.; Wang, Z.L. Ultrafast response *p*-Si/*n*-ZnO heterojunction ultraviolet detector based on pyro-phototronic effect. *Adv. Mater.* **2016**, *28*, 6880–6886. [CrossRef] [PubMed]
56. Wang, Z.; Yu, R.; Pan, C.; Li, Z.; Yang, J.; Yi, F.; Wang, Z.L. Light-induced pyroelectric effect as an effective approach for ultrafast ultraviolet nanosensing. *Nat. Commun.* **2015**, *6*, 8401. [CrossRef] [PubMed]
57. Liu, X.; Zhang, J.; Guo, X.; Wu, S.; Wang, S. Amino acid-assisted one-pot assembly of Au, Pt nanoparticles onto one-dimensional ZnO microrods. *Nanoscale* **2010**, *2*, 1178–1184. [CrossRef] [PubMed]
58. Ding, J.; Zhu, J.; Yao, P.; Li, J.; Bi, H.; Wang, X. Synthesis of ZnO-Ag hybrids and their gas-sensing performance toward ethanol. *Ind. Eng. Chem. Res.* **2015**, *54*, 8947–8953. [CrossRef]
59. Jacobs, C.B.; Maksov, A.B.; Muckley, E.S.; Collins, L.; Mahjouri-Samani, M.; Ievlev, A.; Rouleau, C.M.; Moon, J.-W.; Graham, D.E.; Sumpter, B.G.; et al. UV-activated ZnO films on a flexible substrate for room temperature O_2 and H_2O sensing. *Sci. Rep.* **2017**, *7*, 6053. [CrossRef] [PubMed]

60. Alenezi, M.R.; Alshammari, A.S.; Jayawardena, K.D.G.I.; Beliatis, M.J.; Henley, S.J.; Silva, S.R.P. Role of the exposed polar facets in the performance of thermally and UV activated ZnO nanostructured gas sensors. *J. Phys. Chem. C* **2013**, *117*, 17850–17858. [CrossRef] [PubMed]
61. Fan, S.-W.; Srivastava, A.K.; Dravid, V.P. UV-activated room-temperature gas sensing mechanism of polycrystalline ZnO. *Appl. Phys. Lett.* **2009**, *95*, 142106. [CrossRef]
62. Comini, E.; Cristalli, A.; Faglia, G.; Sberveglieri, G. Light enhanced gas sensing properties of indium oxide and tin dioxide sensors. *Sens. Actuators B* **2000**, *65*, 260–263. [CrossRef]
63. Gromova, Y.; Alaferdov, A.; Rackauskas, S.; Ermakov, V.; Orlova, A.; Maslov, V.; Moshkalev, S.; Baranov, A.; Fedorov, A. Photoinduced electrical response in quantum dots/graphene hybrid structure. *J. Appl. Phys.* **2015**, *118*, 104305. [CrossRef]
64. Lin, C.H.; Chang, S.-J.J.; Chen, W.-S.; Hsueh, T.-J.J. Transparent ZnO-nanowire-based device for UV light detection and ethanol gas sensing on c-Si solar cell. *RSC Adv.* **2016**, *6*, 42–54. [CrossRef]
65. Park, S.; Sun, G.-J.; Jin, C.; Kim, H.W.; Lee, S.; Lee, C. Synergistic effects of a combination of Cr_2O_3—Functionalization and UV-irradiation techniques on the ethanol gas sensing performance of ZnO nanorod gas sensors. *ACS Appl. Mater. Interfaces* **2016**, *8*, 2805–2811. [CrossRef] [PubMed]
66. Gaiardo, A.; Fabbri, B.; Giberti, A.; Guidi, V.; Bellutti, P.; Malagu, C.; Valt, M.; Pepponi, G.; Gherardi, S.; Zonta, G.; et al. ZnO and Au/ZnO thin films: Room-temperature chemoresistive properties for gas sensing applications. *Sens. Actuators B* **2016**, *237*, 1085–1094. [CrossRef]
67. Zhu, P.; Weng, Z.; Li, X.; Liu, X.; Wu, S.; Yeung, K.W.K.; Wang, X.; Cui, Z.; Yang, X.; Chu, P.K. Biomedical applications of functionalized ZnO nanomaterials: From biosensors to bioimaging. *Adv. Mater. Interfaces* **2016**, *3*, 1500494. [CrossRef]
68. Ronkainen, N.J.; Halsall, H.B.; Heineman, W.R. Electrochemical biosensors. *Chem. Soc. Rev.* **2010**, *39*, 1747–1763. [CrossRef] [PubMed]
69. Putzbach, W.; Ronkainen, N.J. Immobilization techniques in the fabrication of nanomaterial-based electrochemical biosensors: A review. *Sensors* **2013**, *13*, 4811–4840. [CrossRef] [PubMed]
70. Zhu, C.; Yang, G.; Li, H.; Du, D.; Lin, Y. Electrochemical sensors and biosensors based on nanomaterials and nanostructures. *Anal. Chem.* **2015**, *87*, 230–249. [CrossRef] [PubMed]
71. Xu, C.X.; Yang, C.; Gu, B.X.; Fang, S.J. Nanostructured ZnO for biosensing applications. *Chin. Sci. Bull.* **2013**, *58*, 2563–2566. [CrossRef]
72. Choi, A.; Kim, K.; Jung, H.I.; Lee, S.Y. ZnO nanowire biosensors for detection of biomolecular interactions in enhancement mode. *Sens. Actuators B* **2010**, *148*, 577–582. [CrossRef]
73. Liu, X.; Lin, P.; Yan, X.; Kang, Z.; Zhao, Y.; Lei, Y.; Li, C.; Du, H.; Zhang, Y. Enzyme-coated single ZnO nanowire FET biosensor for detection of uric acid. *Sens. Actuators B* **2013**, *176*, 22–27. [CrossRef]
74. Ali, S.M.U.; Ibupoto, Z.H.; Salman, S.; Nur, O.; Willander, M.; Danielsson, B. Selective determination of urea using urease immobilized on ZnO nanowires. *Sens. Actuators B* **2011**, *160*, 637–643. [CrossRef]
75. Yu, R.; Pan, C.; Chen, J.; Zhu, G.; Wang, Z.L. Enhanced performance of a ZnO nanowire-based self-powered glucose sensor by piezotronic effect. *Adv. Funct. Mater.* **2013**, *23*, 5868–5874. [CrossRef]
76. Lin, S.Y.; Chang, S.J.; Hsueh, T.J. ZnO nanowires modified with Au nanoparticles for nonenzymatic amperometric sensing of glucose. *Appl. Phys. Lett.* **2014**, *104*, 193704. [CrossRef]
77. Hsu, C.L.; Lin, J.H.; Hsu, D.X.; Wang, S.H.; Lin, S.Y.; Hsueh, T.J. Enhanced non-enzymatic glucose biosensor of ZnO nanowires via decorated Pt nanoparticles and illuminated with UV/green light emitting diodes. *Sens. Actuators B* **2017**, *238*, 150–159. [CrossRef]
78. Miao, F.; Lu, X.; Tao, B.; Li, R.; Chu, P.K. Glucose oxidase immobilization platform based on ZnO nanowires supported by silicon nanowires for glucose biosensing. *Microelectron. Eng.* **2016**, *149*, 153–158. [CrossRef]
79. Fung, C.M.; Lloyd, J.S.; Samavat, S.; Deganello, D.; Teng, K.S. Facile fabrication of electrochemical ZnO nanowire glucose biosensor using roll to roll printing technique. *Sens. Actuators B* **2017**, *247*, 807–813. [CrossRef]
80. Wang, J.; Li, S.; Zhang, Y. A sensitive DNA biosensor fabricated from gold nanoparticles, carbon nanotubes, and zinc oxide nanowires on a glassy carbon electrode. *Electrochim. Acta* **2010**, *55*, 4436–4440. [CrossRef]

81. Mansor, N.A.; Zain, Z.M.; Hamzah, H.H.; Noorden, M.S.A.; Jaapar, S.S.; Beni, V.; Ibupoto, Z.H. Detection of Breast Cancer 1 (BRCA1) gene using an electrochemical DNA biosensor based on immobilized ZnO nanowires. *Open J. Appl. Biosens.* **2014**, *3*, 9–17. [CrossRef]
82. Cao, X.; Cao, X.; Guo, H.; Li, T.; Jie, Y.; Wang, N.; Wang, Z.L. Piezotronic effect enhanced label-free detection of DNA using a schottky-contacted ZnO nanowire biosensor. *ACS Nano* **2016**, *10*, 8038–8044. [CrossRef] [PubMed]

nanomaterials

MDPI

Review

Fabrication of Semiconductor ZnO Nanostructures for Versatile SERS Application

Lili Yang [1,2], Yong Yang [1,*], Yunfeng Ma [1,2], Shuai Li [1,2], Yuquan Wei [1,2], Zhengren Huang [1] and Nguyen Viet Long [3]

1 State Key Laboratory of High Performance Ceramics and Superfine Microstructures, Shanghai Institute of Ceramics, Chinese Academy of Sciences, 1295 Dingxi Road, Shanghai 200050, China; llyang@student.sic.ac.cn (L.Y.); mayunfeng@student.sic.ac.cn (Y.M.); lishuai@student.sic.ac.cn (S.L.); weiyq@shanghaitech.edu.cn (Y.W.); zhrhuang@mail.sic.ac.cn (Z.H.)

2 Graduate University of Chinese Academy of Sciences, No. 19(A) Yuquan Road, Beijing 100049, China

3 Ceramics and Biomaterials Research Group, Ton Duc Thang University, Ho Chi Minh City 800010, Vietnam; nguyenvietlong@tdt.edu.vn

* Correspondence: yangyong@mail.sic.ac.cn; Tel.: +86-21-5241-4321

Received: 1 October 2017; Accepted: 6 November 2017; Published: 19 November 2017

Abstract: Since the initial discovery of surface-enhanced Raman scattering (SERS) in the 1970s, it has exhibited a huge potential application in many fields due to its outstanding advantages. Since the ultra-sensitive noble metallic nanostructures have increasingly exposed themselves as having some problems during application, semiconductors have been gradually exploited as one of the critical SERS substrate materials due to their distinctive advantages when compared with noble metals. ZnO is one of the most representative metallic oxide semiconductors with an abundant reserve, various and cost-effective fabrication techniques, as well as special physical and chemical properties. Thanks to the varied morphologies, size-dependent exciton, good chemical stability, a tunable band gap, carrier concentration, and stoichiometry, ZnO nanostructures have the potential to be exploited as SERS substrates. Moreover, other distinctive properties possessed by ZnO such as biocompatibility, photocatcalysis and self-cleaning, and gas- and chemo-sensitivity can be synergistically integrated and exerted with SERS activity to realize the multifunctional potential of ZnO substrates. In this review, we discuss the inevitable development trend of exploiting the potential semiconductor ZnO as a SERS substrate. After clarifying the root cause of the great disparity between the enhancement factor (EF) of noble metals and that of ZnO nanostructures, two specific methods are put forward to improve the SERS activity of ZnO, namely: elemental doping and combination of ZnO with noble metals. Then, we introduce a distinctive advantage of ZnO as SERS substrate and illustrate the necessity of reporting a meaningful average EF. We also summarize some fabrication methods for ZnO nanostructures with varied dimensions (0–3 dimensions). Finally, we present an overview of ZnO nanostructures for the versatile SERS application.

Keywords: ZnO nanostructures; SERS; versatile substrates; preparation methods; meaningful averaged EFs

1. Introduction

Surface-enhanced Raman scattering (SERS) has attracted great interest as a real-time surface analysis technique with many advantages [1–3] such as ultra-sensitivity, non-destructivity, "fingerprint" ability, and low requirement for samples. It was first observed by Fleishman [4] in 1974 when his research group found that the Raman signal of pyridine on rough Ag electrode was abnormally enhanced by 10^5–10^6. In 1977, the SERS phenomenon was first disclosed by Albrecht [5] and Jeanmaire [6]. Subsequently, scientists from various disciplines began to study the nature and mechanism of SERS and further implement its application. Until now, many achievements have been made and SERS has been successfully applied in many domains, including biomolecule [7,8] and pesticide detection [9–11], molecule imaging [12], identification of cancer cells [13], dynamic study of catalytic reactions [14,15], terrorist threat detection [16,17], food safety [18], etc. A prerequisite for SERS to come into play is the choice, design, and preparation of the substrate materials. The SERS activities of substrates are usually evaluated by an important parameter called the enhancement factor (EF).

The earliest developed and the most widely used SERS substrate materials are noble metallic nanostructures (Au, Ag, Cu) [19,20]. They have a unique superiority as SERS substrates, namely their ultra-high SERS sensitivity with a maximal EF of 10^{14}–10^{15}; thus, they can even be used for single-molecule detection [21–25]. Nevertheless, their disadvantages, such as difficulties in fabricating highly-uniform nanostructures at low cost [26], instability problems including easy aggregation and oxidation during application [27], as well as a limited number of noble metals with excellent SERS activity have hindered the development and wide application of these noble metal substrates.

After being confirmed to have SERS activity themselves [28,29], semiconductors [30–32] are gradually being exploited as promising candidates for SERS substrates owing to their outstanding advantages [33–35] such as abundant active substrates, diverse and mature synthetic techniques at low cost, controllable band structure and photoelectrical properties, and high chemical stability and biocompatibility when compared with noble metals. Additionally, synergistic collaboration between SERS and other properties ranging from photocatalysis [36], magnetism [37], to gas-, bio-, and chemo-sensing [38,39] can be realized in semiconductors.

It is well known that ZnO is a wide band gap (~3.3 eV) metallic oxide semiconductor with distinctive physical, chemical, and photoelectric properties [40–42]. At the same time, ZnO nanomaterials can be grown into numerous morphologies, including nanospheres, nanowires, nanorods, nanoneedles, nanocones, nanobelts, nanocombs, nanorings, nanosprings, and nanocages [42–44]. It is one of the most common and versatile semiconductors, with a wide and critical application in many fields, such as photocatalysis, lithium-ion batteries, dye sensitized solar cells, sensing devices, functional ceramics, and light emitting devices [45–49]. With the development of SERS substrates from noble metals to semiconductor nanomaterials, ZnO nanostructures have tremendous potential to be exploited as active SERS substrates for the following reasons. Firstly, high-refractive-index ZnO has the ability to confine the light to enhance the SERS effect [50]. Secondly, abundant available morphologies of ZnO nanostructures are in favor of the combination with noble metals to improve the SERS activity. Additionally, the material's many advantages, including bio-compatibility, tunable photoelectric properties, high chemical stability, superhydrophobicity, and photocatalytic self-cleaning effect [40,51,52] can be coordinated with the SERS effect on the ZnO nanostructure substrate to achieve versatility and multifunctionality.

However, an inferior EF (10–10^3) is a fatal weakness of the pure ZnO nanomaterials, and has become a bottleneck in the development of semiconductor ZnO as an active SERS substrate. A top priority task is to find the root cause of the great disparity between the EF of noble metals and that of semiconductors. Two types of SERS enhancement mechanisms [53,54]—electromagnetic (EM) and chemical (CM)—have been studied, and although there are difficulties in clearly quantifying the specific contribution of EM and CM mechanisms to the EF [55], it has been revealed that EM can contribute 10–11 orders of magnitude to the EF of noble metals under special circumstances (i.e., "hot spots"), and another 10^3 of EF comes from the CM [56,57]. With regard to semiconductors, the SERS enhancement is dominated by the CM, which usually has a value of 10–10^3 [58,59]. After clarifying the primary cause of the weak enhancement of semiconductor substrates, two specific methods have been put forward to improve the SERS activity of semiconductor ZnO nanostructures, which are heavy elemental doping and combination of ZnO with noble metals.

2. Improved SERS Activity of ZnO Nanostructures: Theoretical Basis and Improved Methods

2.1. Theoretical Basis

EM and CM are two types of important enhancement mechanisms used to explain the SERS phenomenon on noble metals and semiconductors; thus, the SERS activity of semiconductor ZnO can be improved in two different ways: improving the electromagnetic enhancement, and improving the chemical enhancement. In order to make a great breakthrough in elevating the EF of semiconductor ZnO, efforts should be concentrated on improving the electromagnetic enhancement due to the enormous gap between the contributions of EM to noble metals (up to 10^{11}) and semiconductors (little). Further, some progress should also be made in chemical enhancement due to its non-negligible contributions to the Raman enhancement.

2.1.1. Theoretical Basis for Improving the Electromagnetic Enhancement of Semiconductor ZnO

Here a pivotal issue should be raised: why is the electromagnetic enhancement of noble metals much larger than that of semiconductors? The local surface plasmon resonance (LSPR) [60] and "hot spots" effect in metallic nanoparticles may give the answers.

It is well known that the EM is the result of an enhanced local electric field generated by the collective resonance of surface plasmons in metallic nanoparticles under the irradiation of an incident laser. LSPR can occur in noble metal nanoparticles for the reason that the LSPR bands of noble metal nanoparticles are usually located in the visible (VIS) spectral region, which is the prerequisite for strongly absorbing the incident light by metallic nanoparticles. For semiconductors, the LSPR peak of the conduction band is normally centered in the near-infrared (NIR) spectral region due to the low electron density, while the LSPR peak of the valence band is approximately in the ultraviolet (UV) spectral region because of the high electron density (10^{22}–10^{24} cm^{-3}) [61,62], and consequently plasmons in semiconductors have scarcely any contribution to the EM effect, which is dominated by the LSPR under the visible light. Thus, the electromagnetic enhancement of semiconductors is far inferior to that of noble metals.

Additionally, when the gap between the metallic nanoparticles has a close spacing (less than 10 nm), the local electric field in these narrow junctions can get huge enhancement due to the coupling plasma resonance effect under the incident laser. These spots with an enhanced electric field are referred to as "hot spots" [63]. The comparative rarity of these "hot spots" notwithstanding, their remarkable contributions to the Raman enhancement can reach 10–11 orders of magnitude. The "hot spots" effect is usually achieved at the tips and corners on the rugged surface of noble metal nanostructures [64]. In conclusion, both the LSPR and the "hot spots" effect result in a remarkable Raman enhancement on noble metal substrates.

Through analysis of the aforementioned issue, an alternative solution is proposed to introduce electromagnetic enhancement into the semiconductor ZnO by means of tuning the LSPR peak location of semiconductor ZnO nanostructures near to the VIS spectral region and effectively creating "hot spots" in ZnO nanostructures.

Fortunately, the concrete influence of the free carrier density and nanosphere diameter, along with the number of free carriers per quantum dot (QD) on the LSPR frequency has been presented by Joseph M. Luther et al. in Figure 1 [65]. The LSPR frequency of semiconductors is primarily controlled by the free electron density. It can be conceived that the unified LSPR peak of ZnO nanostructures can be shifted to the NIR and even VIS spectral region when the free carrier concentration reaches up to 10^{21}–10^{23} cm^{-3}. It deserves the expectation that the EM can be introduced into semiconductors to improve the SERS activity by regulating the LSPR frequency with the doping level. The LSPR frequency can be calculated according to the equation:

$$\omega_p = \left(\frac{N_h e^2}{\varepsilon_0 m_h}\right)^{1/2} \tag{1}$$

where ω_p is the LSPR frequency and ε_0 is the vacuum permittivity, N_h is the density of free carriers (holes), m_h is the hole effective mass, approximated as 0.8 m_0, where m_0 is the electron mass.

The Drude model [66] has been used by Xiangchao Ma et al. [67] to derive the critical carrier concentration required to make the real part of the permittivity of ZnO negative at 1000 nm according to the following equation:

$$n > \frac{\varepsilon_0 m^*}{e^2} \varepsilon_b \left(\omega_c^2 + \gamma^2 \right) \quad (2)$$

where n is the critical concentration, ε_0 is the permittivity of free space, e is the electron charge, m^* is the carrier effective mass, ε_b is the background permittivity which describes the polarization response of the core electrons, ω_c is chosen to be corresponding to the wavelength of 1000 nm which is relevant to photocatalysis and γ is approximated to be $\gamma = \frac{e}{\mu m^*}$; here, μ is the carrier mobility. For the semiconductor ZnO, the critical carrier concentration is calculated as being 1.62×10^{21} cm^{-3}.

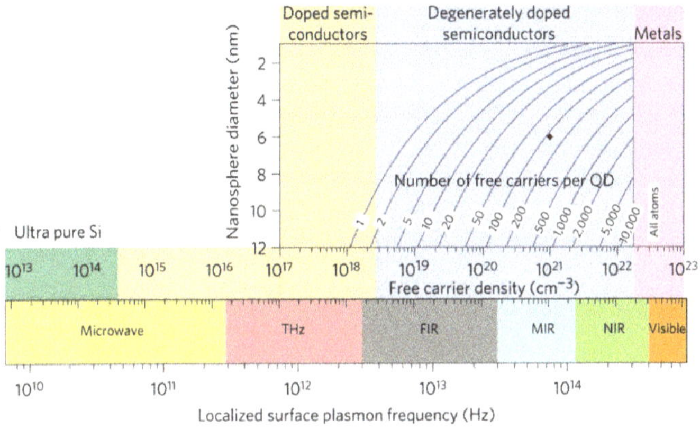

Figure 1. Local surface plasmon resonance (LSPR) frequency dependence on the free electron density for semiconductors, and number of free carriers per quantum dot (QD) required for nanosphere diameter ranging from 2 to 12 nm to achieve a free carrier density between 10^{17} and 10^{23} cm^{-3}. Reproduced with permission from [65]. Copyright Nature Publishing Group, 2011.

The effective creation of "hot spots" has as a means of enhancing Raman scattering, and then concerns should be paid to how to create "hot spots" on ZnO nanostructures. "Hot spots" on noble metals are usually located at the tips and corners, which can provide guidance for the morphology and structural optimization of ZnO nanomaterials. Randomly oriented rather than aligned nanowires (NWs), closely branched nanostructures instead of flat plate structure, and analogous ZnO nanostructures may generate more "hot spots" regions. Muhammad A. Khan et al. [50] have compared the SERS activity of randomly oriented ZnO NWs with partially aligned ZnO NWs to determine the contribution from the cross-junctions between nanowires. They found that the close coupling between these high-refractive-index ZnO NWs indeed improved the SERS activity.

At the same time, the relatively larger resolution limit of the conventional fabrication techniques compared to "hot spots" for metal nanomaterials, the easy aggregation among metallic nanoparticles, and the highly-uniform "hot spots" distribution are key challenges when creating "hot spots" between junctions of noble metals. Particular ZnO nanostructures can solve these problems to a certain extent thanks to their easily available steady nanostructures with crossings and junctions as well as their large plasmon-active surface to deposit and anchor noble metals. The branched ZnO nanostructures could be the backbone for the metallic nanoparticles and "hot spots". For example, the close-coupling between the high-refractive-index ZnO nanowires can not only "trap" the incident light, but can also load metallic nanoparticles to create "hot spots" between these close metallic nanoparticles on the same or adjacent ZnO nanostructures. Muhammad A. Khan's group [50] also deposited Au film on randomly oriented ZnO NWs, and the thickness of the Au film was confirmed to be vital to the Raman enhancement. They found that the moderate thickness—which was conducive to mutual surface plasmon interaction—can produce the strongest Raman enhancement due to the most appropriate gap to generate "hot spots" among the Au islands.

2.1.2. Theoretical Basis for Improving Chemical Enhancement of Semiconductor ZnO

There is only a small difference between the contributions of the CM to noble metals and semiconductors, while its contributions to Raman enhancement cannot be neglected. It is considered that the CM is related to the interaction between molecules and substrates, and is dominated by the photo-induced charge transfer (PICT), which is thermodynamically permitted under the incident light excitation when the highest unoccupied molecular orbital (HOMO) and the lowest unoccupied molecular orbital (LUMO) of the molecules match with the conduction band and valence band of the semiconductors [68–70]. The CM cannot work if the incident laser does not have enough energy or if there is little interaction between the molecules and the substrate. Thus, the CM cannot take effect unless there is a laser with enough energy and a large active surface of ZnO nanostructures to adsorb the analyte.

When the size of the semiconductors becomes comparable to the size of the exciton Bohr radius, the exciton resonance cannot be ignored. The quantum confinement effect [71] can split the exciton levels and make the SERS spectra strongly depend on the size of semiconductor substrates. Compared with the noble metal, the quantum confinement effect of the semiconductor can elevate the EF by 10 [72]. Thus, a suitable size can improve the SERS activity of semiconductors to some extent via the quantum confinement effect. For ZnO nanostructures, the reported exciton Bohr radius are all less than 2.5 nm [73]. Nevertheless, the size effect does not work when the size of the ZnO nanostructures is much larger than the Bohr radius.

2.2. *Improvement Methods*

Based on the theoretical analysis about the source of the huge gap between Raman enhancements of noble metals and semiconductors, two specific methods have been put forward to improve the SERS activity of semiconductor ZnO nanostructures, namely: heavy elemental doping and combination of ZnO nanomaterials with noble metals. Further, morphology, structure, and size optimizations are also important to improve the SERS activity of ZnO nanostructure substrates.

For heavy-doped ZnO nanostructures, the heavy doping degree is difficult to reach, and a giant free carrier density of 10^{23} is scarcely possible to obtain. Mg, Co, H, and some other elements have been doped into ZnO nanostructures, and the SERS activity can be improved to a limited degree. Certainly, it is also helpful to tune the LSPR peak location to the VIS region by decreasing QD sizes and increasing QD numbers according to the Figure 1.

With regard to another improvement method, there are three advantages to the use of noble metal/ZnO composite nanostructures as SERS substrates. Firstly, the excellent SERS activity of noble metal can help to improve the Raman enhancement of composite substrates by introducing the EM and "hot spots" effect into the ZnO nanostructures. Secondly, SERS activity can be combined with other properties of semiconductor ZnO to achieve the multifunctionality of composite substrates. Thirdly, noble metal/ZnO composite substrates can be more environmentally benign and chemically stable than noble metal substrates.

For noble metal/ZnO composite substrates, many detailed parameters can influence the SERS activity; for example, structural configurations of the composite substrate, shapes, sizes, relative location of noble metals and ZnO, etc. With regard to the structural configuration of the composite substrate, noble metal-deposited ZnO substrates and the reversed ZnO-coated noble metal substrates are different [74]. There are different interactions between the metal and ZnO, as well as different Raman enhancement mechanisms in these reversed substrates. For the reversed ZnO-coated noble metal substrate, the charge transfer CT between the semiconductor and molecules dominates the Raman enhancement. However, the noble metal plays a leading role in contributing to the SERS activity of the entire substrate when metallic nanoparticles are deposited on the ZnO nanostructures, and consequently the SERS activity is mainly influenced by shapes, sizes, and aggregation of the deposited metallic nanoparticles [75,76]. The LSPR peak location depends on the morphology, size, and composition of the metallic nanoparticles [76–78], and the number of "hot spots" depends on the aggregation of the metallic nanoparticles, which can be adjusted by distributing noble metals on ZnO substrates [79]. More specifically, for the noble metal Ag and Au, the characteristic LSPR peak of Ag nanoparticles (NPs) is usually centered on 390 nm, and that of Au NPs is usually located at a longer wavelength of 522 nm. Firstly, the LSPR peak position of the mixture of Ag and Au can be influenced by the molar ratios of Ag and Au, which has been well researched by Lakshminarayana Polavarapu et al. [77]. Secondly, the LSPR peak position of Au or Ag can be tuned by changing the morphology and size of the noble metal [78]. For instance, there are transversal and longitudinal LSPR for Au nanorods (NRs). When the Au NRs have a gradually decreased aspect ratio, the longitudinal LSPR will undergo a blue shift, while the transversal LSPR will undergo a red shift until they merge with each other into a single band. A controllable optical property and LSPR peak location can also be achieved by adjusting the size of metallic nanoparticles. "Hot spots" may generate when the metallic nanocrystals are aggregated, and it is critical to control the space between the nanoparticles to create "hot spots". A novel and fantastic study concerning the temperature-controlled formation of "hot spots" has been done through depositing the metallic nanocrystals on a shape-thermoresponsive substrate [79], which can inspire us to creatively design a composite substrate with the ability to control the formation of "hot spots", and consequently modulate the SERS activity.

3. A Unique Advantage of Semiconductor ZnO as SERS Substrate

However, a coin has two sides. It is not the absolute truth that the larger enhancement of the Raman signal means a better application of SERS technique, because the Raman scatterings of bands are usually enhanced selectively and non-uniformly in SERS. In some cases, extremely strong enhancement of a few bands may overshadow some weaker characteristic Raman peaks of "fingerprint regions", which are useful to identify the desired analyte molecules. In spite of a comparatively weaker signal enhancement from the semiconductor ZnO than the noble metal substrates, a moderate enhancement of the Raman signal on the ZnO nanostructures could avoid the occurrence of signal masking and be in favor of the identification of all the spectral components. ZnO has been proven as a potential substrate to reveal some unnoticeable spectral components of the human whole blood with its moderate enhancement (20–30 fold) while ensuring that the original Raman spectrum information is not masked, and the SERS technique based on the ZnO substrate is promising to be a valuable tool to diagnose human diseases with human body fluids [80].

4. Necessity of Reporting a Meaningful Average EF

The enhancement factor is one of the most important parameters used to characterize SERS performance. EF can be largely affected by many test conditions, such as the substrate types, target molecules, and excitation wavelength [24]. Two kinds of frequently used definitions and calculation methods are given below [81]. They respectively are the single molecule enhancement factor (SMEF) and a meaningful spatially average enhancement factor (average EF).

The SMEF is used to calculate the SERS enhancement of a given molecule in a specific position, so it is usually difficult to calculate the SMEF due to its precise definition. The SMEF is only suitable for theoretical estimation because of the difficulty in collecting the signal from just one molecule at a time. The single molecule enhancement factor is defined as the following:

$$\text{SMEF} = \frac{I_{\text{SERS}}^{\text{SM}}}{I_{\text{RS}}^{\text{SM}}} \quad (3)$$

where I_{SERS}^{SM} is the SERS intensity of the single molecule and I_{RS}^{SM} is the average Raman intensity per molecule under the same experimental condition. Only the single target molecule on the "hot spots" can be detected.

In many practical applications, the meaningful average EF is widely used to deal with the average SERS signal. It can be calculated by the following formula:

$$\mathrm{EF} = \frac{I_{\mathrm{SERS}}/N_{\mathrm{SERS}}}{I_{\mathrm{bulk}}/N_{\mathrm{bulk}}} \quad (4)$$

where I_{SERS} and I_{bulk} are the integral intensity of Raman signals at the same peak position under SERS and target molecules conditions, respectively, N_{SERS} and N_{bulk} are respectively the average number of molecules adsorbed on the substrates in the scattering volume for the SERS and Raman (non-SERS) measurement.

At present, it is difficult and controversial to adopt a uniform EF to evaluate the performance of SERS substrates. Both of the above definitions can be used to calculate EF, and there is always a discrepancy between the single molecule EF and the average EF. To be objective, the SMEF is more precise and reasonable, while the averaged EF is more practical. Because the SMEF can greatly change from site to site due to the strong spatial locality of "hot spots", the SMEF is usually used to accurately evaluate the contribution of one single molecule to the Raman enhancement. The meaningful spatially average EF is more practical for evaluating the performance of experimental substrates, while it is usually lower than the maximum SMEF and may underestimate the application of substrates with "hot spots". Moreover, when the average EF is used for actual calculation, the accurate data is often required yet difficult to attain, and a reasonable hypothesis is always needed for the calculation. Thus, it should be realized that the averaged EFs can vary substantially between researchers due to diverse assumptions or inadequate description [82], and an appeal for a much more transparent determination of the performance of a given substrate is needed [63].

5. Synthesis of ZnO Nanostructures as SERS Substrates

5.1. Synthesis of 0-D ZnO Nanostructures

Zero-dimension (0-D) ZnO nanostructures usually include nanocrystals, nanospheres, and nanocages. Template, self-assembly, and thermal decomposition are three primary methods to prepare 0-D ZnO nanostructures.

5.1.1. Synthesis of ZnO Nanospheres

The two-stage thermal decomposition method of the precursor [83–87] and the novel two-stage self-assembly method [88–90] are the most common fabrication methods for 0-D ZnO nanospheres. As one of the liquid phase methods, the thermal decomposition method has the advantage of easy acquisition of highly purified nanoparticles with uniform particle size. The self-assembly method has the superiority of forming a thermodynamically stable and structure-oriented nanostructure easily. In the two-stage thermal decomposition method of the precursor, the first stage is mixing and vigorously stirring the NaOH and zinc acetate ($Zn(Ac)_2$) solution to produce the $Zn(OH)_2$ precipitates, and then the NH_4HCO_3 powder is added to form the precursor of a small crystallite of $Zn_5(CO_3)_2(OH)_6$. In the second stage, the precursor is calcined at a high temperature to obtain the nanospheres. In the novel two-stage self-assembly method, monodisperse ZnO nanospheres can be derived by two-time precipitation at different temperatures. After the first ZnO precipitation is formed through the hydrolysis of zinc acetate dihydrate ($Zn(Ac)_2 \cdot 2H_2O$) with the participation of diethylene glycol (DEG) under a higher temperature (typically 160 °C), the supernatant needs to be preserved by removing the first precipitation. The secondary reaction is performed in a similar way; a certain amount of the preserved supernatant is added to the solution before approaching a relatively low temperature (usually 150 °C), and then the ZnO colloidal spheres can be obtained.

5.1.2. Synthesis of ZnO Nanocages

The ZnO nanocage can be prepared by a template method with a dehydration reaction. In order to synthesize the hollow amorphous ZnO nanocages, Xiaotian Wang [91] used the template method to synthesize the hollow $Zn(OH)_2$ nanocrystals by employing Cu_2O nanocubes as the structure-directing template and adding the stabilizing agent polyvinylpyrrolidone (PVP). Then, the $Zn(OH)_2$ nanocrystals further underwent a dehydration reaction at 250 °C to form hollow amorphous ZnO nanocages (a-ZnO NCs). Crystalline ZnO nanocages (c-ZnO NCs) can be obtained by calcining the a-ZnO NCs. The specific synthetic schematic diagram is shown in Figure 2.

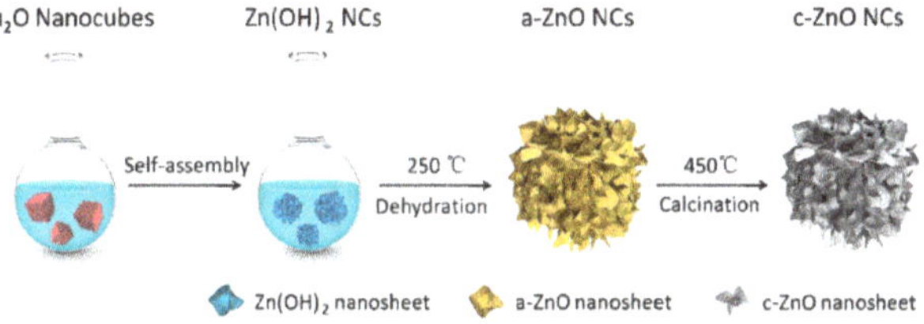

Figure 2. Schematic diagram of the synthesis of a- and c-ZnO NCs (amorphous and crystalline ZnO nanocages). Reproduced with permission from [91]. Copyright Wiley-VCH Verlag GmbH & Co. KGaA, Weinheim, 2017.

5.2. Synthesis of 1-D ZnO Nanostructures

Aqueous chemical growth (ACG) [92] is a simple method to deposit a 1-D ZnO nanostructure such as ZnO nanorods on the substrates. It is conventionally performed at a low temperature with the available inexpensive equipment. Many materials containing Corning 7059, indium tin oxide (ITO)-coated glass and silicon wafer can be used as the substrates for the deposition of ZnO NRs. In general, the $Zn(NO_3)^{2+}$ precursors, hexamethylenetetramine (HMTA) solution, and the substrate are heated at a constant temperature (typically 95 °C) in Pyrex glass bottles for a period of time to obtain ZnO NRs. pH has been adjusted by D. Vernardou [43] to get various ZnO morphologies ranging from the nanorods, tip nanorods, nanoprisms, to flower-like structures under pH of 7, 8, 10, and 12, respectively (Figure 3). Overall, ACG is a simple and practical method to deposit the 1-D ZnO nanostructures on substrates.

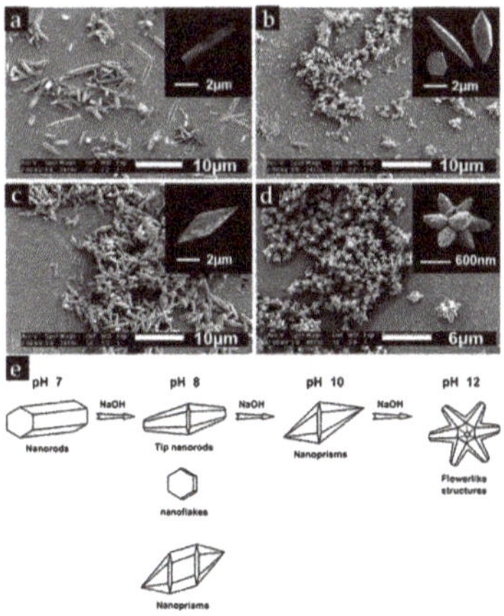

Figure 3. SEM images of ZnO deposited on Corning 7059 for a deposition time of 2 h and (**a**) pH 7, (**b**) pH 8, (**c**) pH 10, (**d**) pH 12, and the individual structures are exhibited in the insets of (**a**–**d**); (**e**) shows a schematic map of the morphology of the nanostructures. Reproduced with permission from [43]. Copyright Elsevier B.V., 2007.

5.3. Synthesis of 2-D ZnO Nanostructures

5.3.1. Synthesis of ZnO Nanosheets

ZnO nanosheet structures are usually prepared by the template method. In a template method, the main body of the template is generally used as a configuration to control the shape and size of synthesized nanostructures. The most common template for porous ZnO nanostructure is the organic additives [93], which are likely to remain when finishing the oriented structure. Layered basic zinc acetate clusters (LBZA-C) as precursor—which can finally transform into the building blocks through refluxing—were creatively used as the template by Qian Liu [94] to assemble porous ZnO nanosheets in one pot (Figure 4). As the refluxing time went on, various evolving morphologies (nanochains, parallelogram frames, semi-filled nanosheets, and porous nanosheets) could be derived.

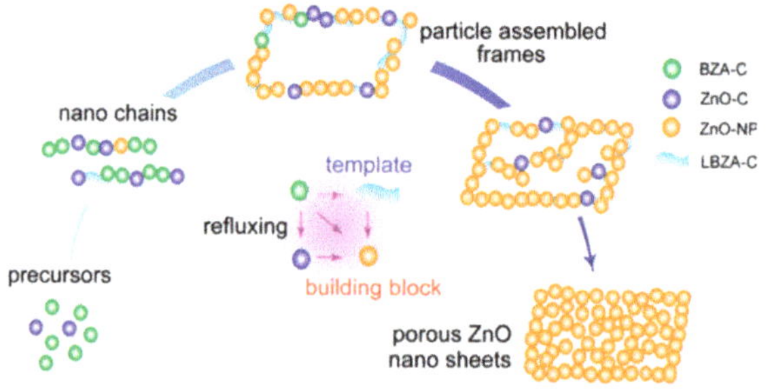

Figure 4. Schematic diagram of the composition and morphology evolution of the porous ZnO nanosheets. Reproduced with permission from [94]. Copyright WILEY-VCH Verlag GmbH & Co. KGaA, Weinheim, 2013. BZA-C: basic zinc acetate clusters; LBZA-C: layered basic zinc acetate clusters; ZnO-C: ZnO clusters; ZnO-NP: ZnO nanoparticles.

5.3.2. Synthesis of ZnO Film

ZnO film is often synthesized on a substrate by the thermal evaporation process [95,96] in a horizontal quartz tube furnace. Evaporation source powder is usually placed in a tungsten, alumina, or quartz boat in the upstream of a flowing inert carrier gas such as argon gas, and the substrate is placed downstream near the evaporation source. After the oxygen in the furnace is thoroughly exhausted, the evaporation source will be heated at a temperature higher than the vaporization temperature, and the substrate is always maintained at a lower temperature for the growth of the film. ZnO film can be grown by this method using ZnO powder as the evaporation source. Naidu Dhanpal Jayram et al. [97] designed a vertical furnace to synthesize a worm-like Ag@ZnO thin film on a glass substrate through a similar thermal evaporation process. The substrate was at the top of the furnace, the ZnO powder and Ag wires in tungsten coil were at the bottom of the furnace as the evaporation source, and the evaporation was performed by setting up a current. A similar atomic layer deposition (ALD) method was used by Yufeng Shan et al. [98] to prepare wheatear-like ZnO nanostructures on Si substrate. Diethyl zinc and deionized (DI) water were respectively employed as the Zn precursor and the oxygen source. This thermal evaporation process was also used in the synthesis of ultra-sharp ZnO nanocone arrays deposited with the Au particle-on-Ag film systems by Youngoh Lee et al. [99]. Varied morphologies (ZnO nanorods, nanonails, and nanocones) (Figure 5) were derived by controlling the axial growth rate and the amount of the ZnO powder. Excessive ZnO powder was used for the growth of ZnO nanocones, and a fast and slow axial growth rate were respectively applied for synthesizing ZnO nanorods and ZnO nanocones.

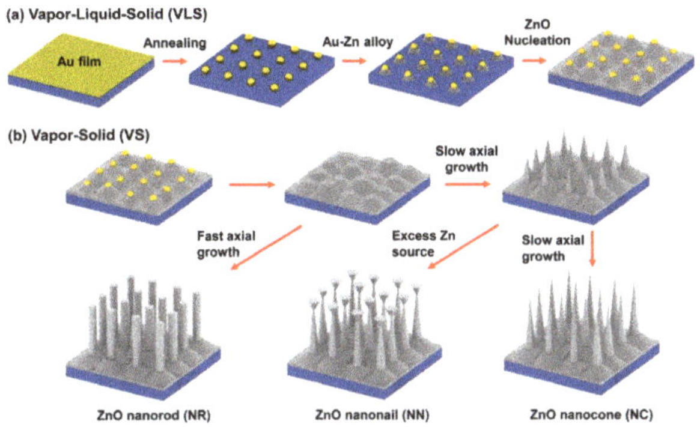

Figure 5. Schematic diagram of the growth mechanism of ZnO nanocones, nanorods, and nanonails. (**a**) Textured ZnO film formed with a vapor–liquid–solid (VLS) mechanism; (**b**) ZnO nanocones, nanorods, and nanonails grown with the vapor–solid (VS) mechanism. Reproduced with permission from [99]. Copyright American Chemical Society, 2015.

5.4. Synthesis of 3-D ZnO Nanostructures

5.4.1. Synthesis of ZnO Nanorod Arrays

The microarray is one of the most representative ZnO nanostructures for SERS application because its distinguished structure can repeatedly scatter and further trap the light. Substrate materials are essential to the deposition of the ZnO microarrays. Electrodeposition [100] can be used to deposit the ZnO microarrays on the surface of the working electrode ITO glass by continuously bubbling oxygen with the electrodeposition solution of KCl and $Zn(Ac)_2$.

The three-step seeded growth process is another frequently used method to grow ZnO nanoarrays [101,102]. Firstly, a well-mixed steady ZnO nanocrystals sol is prepared [103] with KOH and $Zn(Ac)_2$ in methanol or aqueous solution by stirring at 60 °C for 2 h according to a wet chemical method. Secondly, the sol is dropped or spin-cast many times on the surface of the substrate to form a layer of ZnO seed film, and the substrate materials can be ITO, silicon wafers, or plastic substrates. Then, the anneal operation is needed to ensure the adhesion of the ZnO nanocrystals on the substrate. Thirdly, a hydrothermal method is used to grow ZnO NR arrays on the surface of substrates with the addition of zinc nitrate hydrate and diethylenetriamine at a routine temperature of 90 °C or 95 °C. The previous two steps can also be replaced by a magnetron sputtering or atomic layer deposition method to coat a layer of ZnO seeds on the substrates. The zinc nitrate hydrate can also be substituted by zine acetate hexahydrate ($Zn(Ac)_2 \cdot 6H_2O$). HMTA and methenamine can play the same role as the diethylenetriamine. The concrete preparation process of the patterned ZnO nanowire arrays is presented vividly in Figure 6 [18]. In addition, the ZnO nanorod and nanowire arrays can grow not only on plain substrate materials, but also on substrates with various kinds of backbones such as Si nanopillar and carbon-nanotube arrays [104]. Chuawei Cheng et al. [105] have successfully fabricated Si/ZnO nanotrees by growing the ZnO nanorod arrays on the Si nanopillar arrays by using the above method (Figure 7).

The synthesis of noble metal/ ZnO NRs composite nanostructures is primarily based on the above method for growing the ZnO NRs. Firstly, the ZnO NR arrays need to be grown on the substrates by the three-step seeded growth process. Then, the noble metal/ZnO composite will be completed with another two steps. The first step is to prepare the precursor solution for the growth of the noble metal. For various required morphologies of the noble metal, different additives are added to this solution. The next step is immersing the ZnO NR substrates into the precursor solution and then directly irradiating with UV light. Experimental parameters of the morphology-controlled Au/ZnO NRs composite have been studied by Jia-Quan Xu et al. [106]. The precursor solution is $HAuCl_4$

aqueous solution, and the additives for the dendritic, sea-urchin-like, conical, chain-like, sphere-like Au (Figure 8) and Au NPs are, respectively, ammonium hydroxide (28%), phosphate-buffered saline (PBS), 0.15 M ammonium carbonate, 0.15 M *p*-phenylenediamine, 0.1 M HMTA, and saturated melamine (25 °C). Similarly, when decorating the Ag NPs on the ZnO NR arrays, $AgNO_3$ solution is employed and $NaBH_4$ solution usually needs to be added to reduce the adnexed Ag^+ into Ag NPs on the surface of ZnO NRs with the wet chemistry method [107]. In addition, a series of methods including magnetron sputtering, thermal evaporation, or electron beam evaporation and photochemical deposition methods can also be used to deposit the noble metal on the ZnO NRs [18,105].

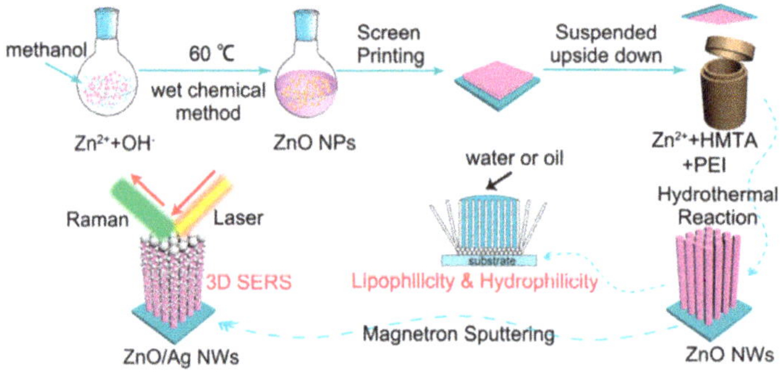

Figure 6. Schematic diagram of the preparation process of patterned ZnO nanowire arrays and the three-dimensional ZnO/Ag nanowire surface-enhanced Raman scattering (SERS) substrate. Reproduced with permission from [18]. Copyright The Royal Society of Chemistry, 2016. HMTA: hexamethylenetetramine; NW: nanowire; PEI: ethylene imine polymer.

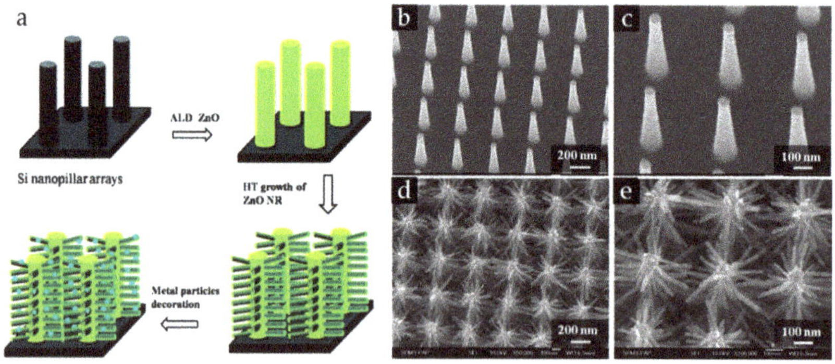

Figure 7. (**a**) Schematic diagram of the fabrication procedures for the three-dimensional ordered Si/ZnO nanotrees decorated by silver nanoparticles; SEM images of the (**b**) Si nanopillar arrays and the (**d**) ordered Si/ZnO nanotrees; as well as (**c**,**e**) the corresponding magnified SEM images. Reproduced with permission from [105]. Copyright American Chemical Society, 2010. ALD: atomic layer; NR: nanorod; HT: hydrothermal.

In recent years, a novel microfluidic technology has been on the rise and is rapidly becoming a new platform for sample preparation. It can manipulate the analyte flexibly and integrate with SERS to allow the instantaneous in-situ detection and investigation of analyte, even a single cell in the future. Yuliang Xie et al. [108] has grown 3-D Ag@ZnO nanostructure clusters by two sequential reactions catalyzed via an optothermal effect within the microfluidic devices. Firstly, ZnO NRs were grown on a gold-coated glass slide when focused by a continuous laser beam in the microfluidic channel containing $Zn(NO_3)_2$ and HMTA solution as precursors. Secondly, Ag NPs were grown on the ZnO

NRs by focusing the laser beam onto the preformed ZnO NRs in the $AgNO_3$ solution. It was very important to control the parameters of the laser (e.g., the heating power and the position of the focused laser spot) to determine the formation of ZnO NRs and Ag NPs.

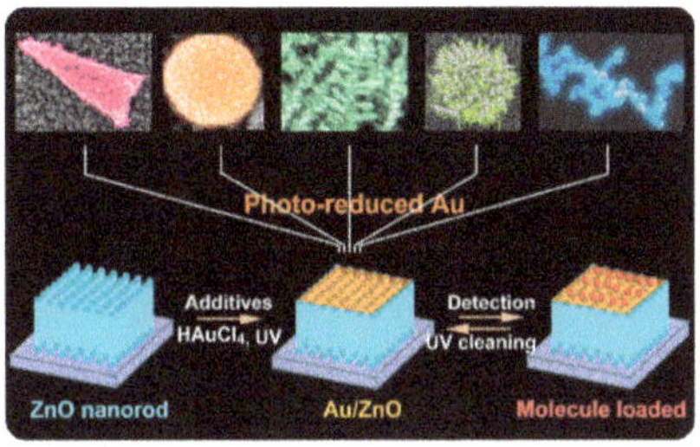

Figure 8. The photoinduced Au nanostructures in the form of conical Au, sphere Au, dendritic Au, sea-urchin-like Au, and Au chain from left to right, respectively. Reproduced with permission from [106]. Copyright American Chemical Society, 2016.

5.4.2. Synthesis of 3-D Sandwich Structure Assembly

The charge transfer in noble metal/molecule/semiconductor assemblies [109] is critical to the exploration of the chemical enhancement mechanism, thus it is necessary to summarize the assembly methods of the 3-D sandwich nanostructures. Here we take the assembly of the representative ZnO/PATP (4-aminothiophenol)/Ag and the reverse Ag/PATP/ZnO (Figure 9) as the example to introduce the fabrication method [110]. For the ZnO/PATP/Ag assembly, ZnO NR film is synthesized on the glass substrate with the above three-step seeded growth process, then the ZnO film is immersed into the PATP ethanol solution at room temperature for some time, and finally the obtained ZnO/PATP substrate is immersed into a silver colloid (which was prepared according to the literature [58]), to derive the ZnO/PATP/Ag assembly. The preparation process of the inverse Ag/PATP/ZnO assembly is similar to the former process. Note, however, that the glass cleaned by a mixed solution of H_2O_2 and H_2SO_4 should be immersed into a poly(diallydimethylammonium chloride) (PDDA) solution for the next self-assembly of Ag.

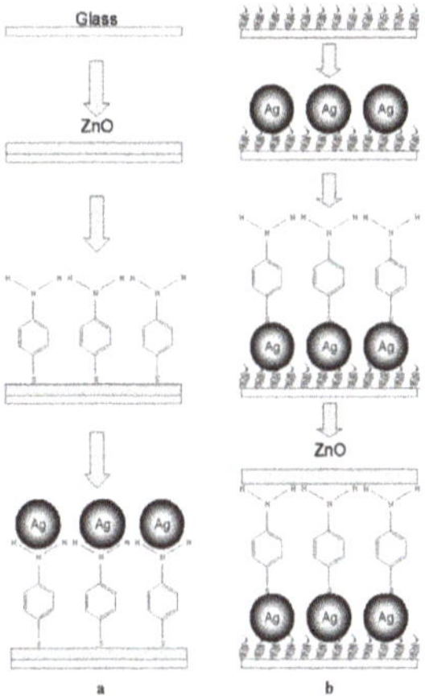

Figure 9. Schematic diagram of (**a**) ZnO/PATP (4-aminothiophenol)/Ag sandwich structure; (**b**) Ag/PATP/ZnO sandwich structure on glass substrates. Reproduced with permission from [110]. Copyright American Chemical Society, 2008.

6. ZnO Nanostructures for Versatile SERS Application

In order to improve the SERS activity of ZnO nanostructures, the pure ZnO nanostructures are usually designed in varied shapes and sizes, doped with different elements and combined with noble metals. In this chapter, in addition to the three ZnO substrates above, three-dimensional sandwich (3-D-sandwich) structure nanomaterials as a distinctive ZnO composite materials are also introduced due to their unique 2-D stacked structures and the addition of typical 2-D materials such as graphene. Therefore, four types of ZnO nanostructures are used as the versatile SERS substrates: pure ZnO nanostructure materials, elemental doped ZnO nanomaterials, noble metal/ZnO composite nanomaterials, and 3-D-sandwich structure nanomaterials. Reported ensemble averaged enhancement factors on different ZnO nanostructure substrates are listed in Table 1, and they serve only as an indicator of SERS substrate performance.

Table 1. Reported ensemble averaged enhancement factors (EFs) on different ZnO nanostructure substrates. 4-MBA: 4-mercaptobenzoic acid; 4-Mpy: 4-mercaptopyridine; CM: chemical mechanism; D266: 1-methyl-1′-propylsulpho-2,2′-cyanine sulphonate; EM: electromagnetic mechanism; a-ZnO: amorphous ZnO; R6G: rhodamine 6G; rGO: reduced graphene oxide.

Substrates	Morphology	Probes	EF/Detection Limits	Mechanism	References
ZnO	Colloids	D266	More than 50	CM	[30]
ZnO	Nanocrystals	4-Mpy	10^3	CM	[72]
ZnO	Nanoparticles	4-MBA & 4-Mpy	-	CM	[85]
ZnO	Nanowires, nanocones	4-Mpy	10^3	"Hot spots" + cavity-like structural resonance	[111]
ZnO	Porous nanosheets	4-MBA	$10^3/10^{-6}$M	CM	[94]
ZnO	ZnO nanorod arrays sheathed with ZnO nanocrystals	4-Mpy	68	CM	[112]
a-ZnO	Nanocages	4-Mpy	6.62×10^5	CM	[91]
Co-doping ZnO	Nanoparticles	4-MBA	-	CM	[83]
Mg-doping ZnO	Nanoparticles	4-MBA	-	CM	[106]
Ag/ZnO	Microspheres	4-Mpy	9×10^4	EM + CM	[88]
Ag/ZnO	Wheatear-like ZnO nanoarrays decorated with Ag nanoparticles	R6G	4.9×10^7	EM + CM	[98]
Ag/ZnO	Worm-like Ag-coated ZnO nanowires	R6G	3.082×10^7 $/10^{-10}$M	EM + CM	[97]
Ag/ZnO	ZnO nanowires deposited on an Ag foil surface	PATP	1.2×10^8 $/10^{-12}$M	EM + CM	[113]
Ag/ZnO	Urchin-like Ag NPs deposited on ZnO hollow nanosphere arrays	R6G	$10^8/10^{-10}$M	CM + "hot spots"	[114]
Ag/ZnO	Ag-nanoparticle-decorated Si/ZnO nanotrees	R6G	1×10^6	EM + CM + structure-induced light trapping	[105]
Ag/ZnO	Ag nanoparticles deposited on ZnO nanowire arrays	Malachite green (MG)/amoxicillin	MG (2.5×10^{10} $/10^{-12}$ M) Amoxicillin (10^{-9}M)	"Hot spots"	[18]
Au/ZnO	Dendritic Au/ZnO composite	R6G	10^{-9}M	EM + CM	[106]
Au/ZnO	Au-coated ZnO nanowires	4-methylbenzenethiol (4-MBT)/1,2-benzendithiol (1,2-BDT)	2.19×10^6/ 4×10^5M	EM + CM + "hot spots"	[50]
Au/ZnO	Flower-shaped ZnO-nanopyramids-coated Au core	PATP	-	CM of ZnO greatly excited by LSPR of Au core	[74]

Table 1. *Cont.*

Substrates	Morphology	Probes	EF/Detection Limits	Mechanism	References
Au/ZnO	Au-coated ZnO nanorods	MB	10^{-12}M	"Hot spots"	[115]
Au/ZnO	Au nano-porous structure electroplated on ZnO nanorods	R6G	2.24×10^6	"Hot spots"	[116]
ZnO/Ag/Au NPs	Ultrasharp nanocones	Benzenethiol (BT), R6G, adenine	10^{10}–10^{11}/BT (10^{-19} M), R6G (10^{-17} M), adenine (10^{-17} M)	EM + CM + "hot spots"	[99]
Au/ZnO/PATP	Layer-by-layer assembly	PATP	-	CM	[117]
ZnO-Ag-graphene nanosheets	Core–shell nanostructure integrated on nanosheets	Acridine orange (AO) dye	-	EM + CM	[118]
Ag/ZnO/rGO	Ag nanoparticles deposited on ZnO/rGO nanocomposite	*E.coli*	10^4 cfu/mL	EM	[119]

6.1. *Pure ZnO Nanostructure Materials*

In 1996, Hao Wen et al. [30] successfully observed the surface-enhanced Raman signal of cyanine dye 1-methyl-1′-propylsulpho-2,2′-cyanine sulphonate (D266) molecules on pure ZnO colloids with an enhancement greater than 50. Semiconductor ZnO can exhibit SERS activity itself, but the enhancement is generally very weak. Many researchers have made a contribution to the design and fabrication of ZnO nanostructures with high SERS activities. Numerous microstructures have been devised and synthesized to realize the improvement of SERS activity of pure ZnO nanostructure materials, such as nanocrystals, nanospheres, nanowires, nanorods, nanoneedles, nanocones, nanosheets, nanocages, etc. SERS activities of these ZnO substrates are shape- or size-dependent due to different numbers of absorption sites for probed molecules, quantum confinement effect, multiple matter–light interactions in photonic microarrays, and optical cavity resonance by architectural configuration. An improved SERS performance has been achieved on a variety of pure ZnO nanostructures by many researchers.

6.1.1. Morphology Optimization Design of Pure ZnO Substrates

Qian Liu et al. [94] synthesized 2-D parallelogram-shaped porous ZnO nanosheets with an enhancement of 10^3 for 4-mercaptobenzoic acid (4-MBA) molecules. The porous morphology was beneficial to improving the SERS performance because it can provide large surface areas, abundant defects, and plentiful surface states, which can promote greater adsorption of probed molecules and favor efficient charge transfer, thus enhancing the SERS activity.

6.1.2. Structure Optimization Design of Pure ZnO Substrates

In addition to the morphology optimization, special nanostructures with the ability to trap the light can also promote the Raman enhancement of pure semiconductors. A giant enhancement of the Raman signal from 4-mercaptopyridine (4-Mpy) adsorbed on 3-D ZnO nanoarray structures (nanowires and nanocones) was observed by Hae-Young Shin et al. [111]. They held the view that the CM dominated by the PICT between the substrates and the adsorbed molecules can only partly explain the enhancement, and that the great enhancement of SERS with EF of 10^3 should be mainly attributed to the cavity-like resonance behavior in the well-constructed ZnO nanostructures, which has been confirmed by the finite-difference time-domain (FDTD) calculation. Their research provided a special way for us to design and employ structure-induced resonance to enhance the SERS activity.

6.1.3. Size Optimization Design of Pure ZnO Substrates

Besides the nanostructures and morphologies of ZnO, particle sizes and measurement conditions also have an important impact on the SERS performance. The size-dependent SERS activity was explored by Zhihua Sun et al. [85], and they prepared ZnO nanocrystals with varied diameters in the range of 18–31 nm. The SERS effect was investigated by using the 4-Mpy and 4-MBA molecules as target molecules. They found that the size-dependence of ZnO nanostructure substrates was subsistent and the optimum particle size for the ZnO nanocrystals was nearly 28 nm. The CM dominated by the charge transfer between substrates and molecules was responsible for the size-dependent SERS activity, and they attributed the size-dependent charge-transfer resonance to the formation of a charge-transfer complex between a surface-bound exciton and the adsorbed molecules.

6.1.4. Effects of Crystallinity on the SERS Activity of Pure ZnO Substrates

Recently, a novel study was carried out by Xiaotian Wang [91] with respect to the relationship between the SERS performance and the crystallinity of ZnO nanocages. It was worth noting that the amorphous ZnO nanocages (a-ZnO NCs) demonstrated a more excellent SERS activity than the crystalline ZnO nanocages (c-ZnO NCs). The difference of the lattice structure and crystallinity between a-ZnO NCs and c-ZnO NCs should account for this interesting finding. c-ZnO NCs had an ordered periodic lattice structure, which may strongly restrict the electrons and limit the escape and transfer of these electrons, while the long-range disordered amorphous lattice structure of a-ZnO NCs can lead the system to a metastable state, and make the charge transfer easier. Accordingly, the PICT process was facilitated and the polarization tensor was expanded, and the SERS activity was enhanced.

6.1.5. Prerequisite for Realizing SERS on Pure ZnO Substrates: PICT

In order to achieve the prerequisite of realizing SERS in the 3-D ZnO NR arrays system, Xiaotian Wang et al. [100] investigated the SERS performance of 4-Mpy and 4-aminothiophenol (PATP) adsorbed on the ZnO NR arrays. They found that the efficient PICT between ZnO NRs and the probed molecules can amplify the probed molecules' polarization tensor and the scattering cross-section, which was vital to the SERS enhancement, and therefore the effective PICT was the prerequisite for improving the SERS activity of semiconductor ZnO NR arrays. In addition, it was encouraging that the logarithmic concentration of pharmaceutical molecules ($(Bu_4N)_2$ [$Ru(dcbpyH)_2$-$(NCS)_2$] (N_{719}) and acetaminophen, as well as the corresponding intensity of detected Raman peaks was linearly dependent. This finding provided feasibility evidence for tracing the photo-induced charges for the dye-sensitized solar cells (DSSCs), and there was promise for the exploitation of the semiconductor SERS substrates as chemosensors for pharmaceutical analysis.

Kwan Kim et al. [101] found a phenomenon that b_2-type bands of 4-aminobenzenethiol (4-ABT) were absent in the 4-ABT Raman spectrum, whereas they were identified when adsorbed onto ZnO NR arrays, which was different from the ever-present a_1-type bands. A similar phenomenon also emerged on 4-ABT derivatives including 4-(methylamino)benzenethiol (4-MABT), 4-(acetamido)benzenethiol (4-AABT), 4-(benzylideneamino)benzenethiol (4-BABT), 4,4′-dimercaptohydrazobenzene (4,4′-DMHAB) and 4,4′-dimercaptoazobenzene (4,4′-DMAB). The above evidence indicated that the b_2-type band was related to the contact and interaction between the ZnO NRs and the probed molecules. To confirm this conjecture, they investigated the effect of the measurement conditions such as pH, excitation wavelength, and electric potential on the SERS signal. The results showed that the b_2-type bands intensity would change with the pH because protonation of the amine group in the acidic solution made an increase of the LUMO level of 4-ABT, and thus the PICT process became harder and the signal became weak. The above discovery implied that b_2-type bands were assigned to the PICT resonance. Similar evidence was also provided by the impact of excitation wavelength and electric potential on a_1- and b_2-type bands.

6.2. *Elemental Doped ZnO Nanomaterials*

Doping is a universal method used to introduce defects into semiconductors and change the lattice constant, the bond energy, and the energy gap of semiconductors. Appropriate doping element and concentration can promote the separation efficiency of the electron and hole, and improve the photocatalytic activity of semiconductors such as TiO_2 and ZnO. Some studies have found that element doping can also be an effective means of enhancing the SERS activity, and the reason has been given as the following. For semiconductor substrates, the CM plays an important role in enhancing the Raman signal. When the probed molecules are chemisorbed on the ZnO nanostructures, surface defects which are introduced by the elemental doping will promote the formation of surface defects energy level, and make the CT process easier to process between the surface defects energy levels and the LUMO of the probed molecules with a relatively lower laser energy. It can be speculated that a higher concentration of dopant will result in a stronger SERS.

6.2.1. Effects of Doping Concentration on SERS Activity of Elemental Doped ZnO Substrates

Xiangxin Xue et al. [83] studied the impact of Co-doping concentration on the SERS intensity with the system of 4-MBA molecules adsorbed on Co-doped ZnO NPs. Because the frequency of the LSPR of the semiconductors was far away from the laser wavelength, CT rather than LSPR was considered to be primarily responsible for the SERS effect. It was interesting that the optimum Co-doping concentration was 1% instead of a higher concentration, which may introduce more defects into the ZnO. It can be suspected that 1% Co-doping ZnO NPs had the largest possibility to generate the CT process. A reasonable explanation was provided that a higher defect concentration may cause the electron–hole recombination, which would compete with the CT from the Co-doping ZnO substrates to the molecules. A similar phenomenon was observed by Limin Chang et al. [84], who observed that there was an optimum Mg doping concentration of 3% when studying the SERS performance of the Mg-doping ZnO nanocrystals by using 4-MBA as probe molecules. Additionally, Szetsen Lee employed hydrogen and oxygen plasmas to introduce defects into hydrothermally synthesized ZnO NRs [92]. SERS activities can be promoted by controlling the concentration of defects with the help of H_2 plasma to reduce the oxygen vacancy and the O_2 plasma to increase the interstitial oxygen. In the meantime, the photoluminescence (PL) intensity of ZnO NRs can also be adjusted by the plasma treatment. A concordant combination of PL and SERS on the semiconductor ZnO nanostructures deserved further research.

6.3. *Noble Metal/ZnO Composite Materials*

Though ZnO nanostructures have many advantages and functionalities, their application as SERS substrates are always plagued by the inferior EF. It is understandable to combine noble metals with semiconductor ZnO nanostructures to achieve multifunctional and sensitive composite SERS substrates. There are two different structural configurations for noble metal/ZnO composite substrates, namely: metal-deposited ZnO substrate and ZnO-coated metal substrates. Ag is well-known for the skyscraping efficiency in enhancing the Raman signal, and is often used to combine with ZnO because the LSPR peak of Ag (390 nm) is adjacent to the UV absorption band (380 nm) of the nanoscale ZnO, which is favorable for a strong interfacial electronic coupling between the Ag and ZnO nanostructures. However, the life span of Ag is limited because it is easily oxidized. Compared with Ag, another common noble metal Au is more stable against oxidation, and the characteristic LSPR peak of Au is located at a longer wavelength of 522 nm [120]. For the biological Raman detection, a longer laser wavelength of 700–1100 nm with low scattering, absorption, and fluorescence is preferred [121,122], and thus the ability of better using the long wavelength laser by tuning LSPR of Au to the NIR region makes Au more biocompatible and extends the application of the SERS to biochemical, biomedical, and biological detection to revel the detailed information about DNA and proteins [108]. Nevertheless, Au NPs are always coated with a layer of chemical substance, which can help to avoid their aggregation and provide them with a functional surface chemistry. In fact, not only the aggregation of the metallic nanoparticles but also the adsorption of target molecules could be prevented by the protective layers. It is advisable to enhance their good points and avoid their shortcomings when using Au and Ag as the SERS substrates.

6.3.1. Ag/ZnO Composite Materials

ZnO nanostructures can not only be deposited on Ag foil, but can also be coated with Ag nanostructures. Wei Song et al. [113] attempted to deposit ZnO nanofibers on a silver foil surface to form a composite substrate. A high SERS intensity of the PATP (Figure 10a) was observed on this substrate, with an EF of 1.2 × 10^8 and a detection limit down to 10^{-12} M. The enhanced scattering could be attributed to the EM arising from the exciton–plasmon interaction between ZnO nanofibers and the silver foil surface, which afforded the localization of the electric field at the gap between ZnO nanofibers and silver foil. It can be further demonstrated by simulating the distribution of the electric field for the system of the silver foil deposited by ZnO nanofibers with the FDTD (Figure 10b). Yufeng Shan et al. [98] fabricated wheatear-like ZnO nanoarrays decorated with Ag NPs, rhodamine 6G (R6G) was used as target molecules, and the EF reached up to 4.9 × 10^7. Firstly, Ag NPs-deposited 3-D ZnO nanoarrays with a large surface area can generate a high-density of "hot spots". Secondly, hydrogenation introduced many defects into ZnO and adjusted the surface energy band structure of the ZnO nanostructures, thus promoting charge transfer between the substrates and target molecules. Therefore, both the EM derived from the "hot spots" and the CM dominated by charge transfer, as well as the target molecules enrichment resulted from the large surface area of the ZnO nanoarrays, contributed to the SERS enhancement. Moreover, the appropriate hydrogenation degree increased the carrier concentration in ZnO nanostructures and evidently enhanced the photocatalytic activities in the meantime.

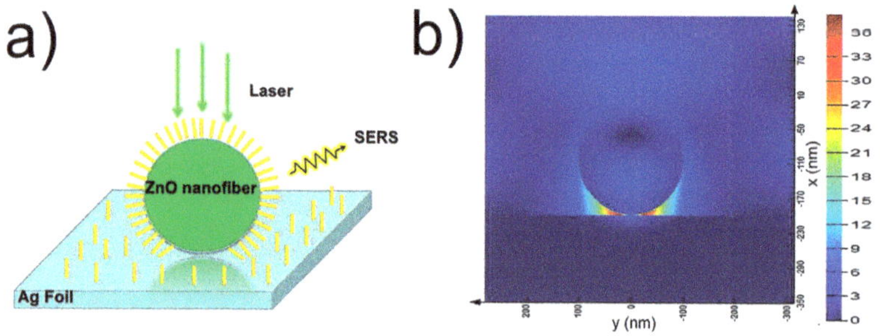

Figure 10. (**a**) Schematic diagram of ZnO nanofibers deposited on the surface of silver foil with probed molecules; (**b**) The distribution of the electric field for ZnO nanofibers deposited on the surface of silver foil calculated with finite-difference time-domain (FDTD) simulation. Reproduced with permission from [113]. Copyright The Royal Society of Chemistry, 2015.

Apart from the photocatalytic performance, the super-hydrophobicity can also be integrated with SERS activity on the ZnO nanostructures. Naidu Dhanpal Jayram et al. [97] successfully prepared super-hydrophobic and high-sensitive worm-like silver-coated ZnO NWs with a contact angle of 163° and an enhancement of 3.082 × 10^7 for detecting 10^{-10} M rhodamine 6G (R6G). The super-hydrophobicity was derived from the great increase of the air/water interface when the large fraction of air was entrapped in the interstices of the rough ZnO NWs. The sensitive SERS activity could be explained by the following two reasons. On the one hand, it was noteworthy that there was a strong interfacial electronic coupling between the neighboring ZnO NWs and Ag NPs. On the other hand, the higher contact angle indicated that the same amount of probed molecules in the droplet could be enriched in a smaller area, and consequently an increased surface coverage can be obtained, leading to an enhanced SERS signal on the substrate with high contact angle.

6.3.2. Au/ZnO Composite Materials

In comparison to the Ag/ZnO composite nanostructures, Au/ZnO composite nanostructures are more biocompatible and can be applied in biological detection. Jiaquan Xu et al. [106] fabricated composite renewable biosensors with various Au nanostructures such as dendritic, spherical, sea-urchin-like, conical, and chain-like structures. Among these composite substrates, the dendritic Au NPs/ZnO composite exhibited the strongest Raman enhancement thanks to the hierarchical structure and the relative high coverage, which could provide abundant gaps for SERS enhancement, and thus even the 10^{-9} M R6G molecules can be detected by the dendritic Au/ZnO composite. The Au NPs/ZnO composite was not limited to the detection of NO released from the cells, it also had self-cleaning functionality, which can allow the substrates to be reused many times for the effective detection of target molecules as a renewable biosensor. Further, the in-situ SERS and electrochemical impedance spectroscopy (EIS) measurements were integrated in the Au nano-porous structure coated ZnO NRs to help us obtain the cellular information and monitor cell response to the environment (Figure 11) [116].

When combining a noble metal with ZnO nanostructures, the metallic nanoparticles are usually deposited on the surface of ZnO nanostructures while few studies pay attention to another reverse structural configuration, which is ZnO-coated metal nanostructures. For the metal-deposited ZnO nanostructures, the interaction between the metals and ZnO is often weak and the contribution of the substrate to SERS is mainly from the metal. Liping Liu et al. [74] have studied the flower-shaped Au-ZnO hybrid nanoparticles, which exhibited a stronger SERS signal of the b_2 modes of PATP molecules because the CT from ZnO to molecules was enhanced by the additional electrons from Au NPs to the ZnO surface. In addition, the ZnO-coated Au substrates had a self-cleaning performance under visible light, and it can be developed into a promising SERS substrate simultaneously accompanied with the biocompatibility and visible-light-induced reproducibility.

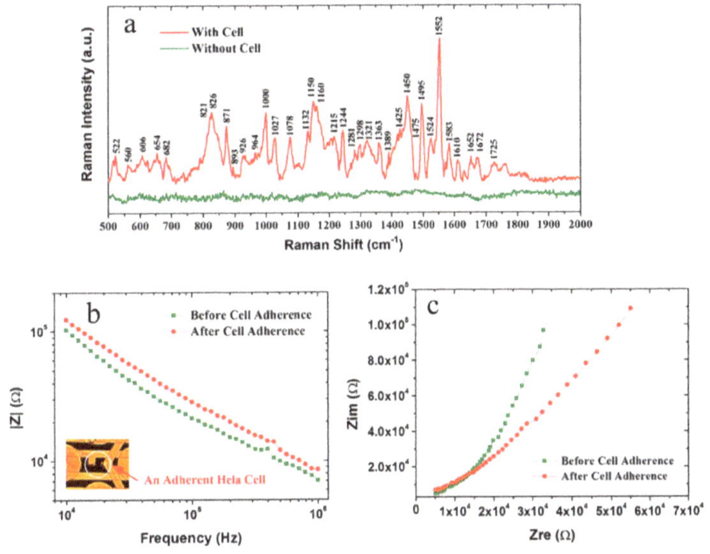

Figure 11. Dual-modal detection of a single Hela cell. (**a**) SERS spectrum of a single Hela cell attaching on the underneath electrode with Au nanostructure; (**b**) Comparison of cellular impedances before and after its attachment. Inset in (b) is the light microscope image of a single Hela cell adhering on the measuring electrodes; (**c**) Cole-Cole plots of the impedance measurements. Reproduced with permission from [116]. Copyright Nature Publishing Group, 2015.

6.3.3. ZnO/Ag/Au Composite Materials

Structural configuration is also crucial for the SERS performance of the 3-D nanostructure. In order to achieve the ultra-high SERS activity with "hot spots" and the light trapping, a 3-D design of ultrasharp ZnO/Ag/Au NPs nanocone array was cleverly constructed by Youngoh Lee et al. [99]. The substrate even can detect target molecules at even zeptomole levels with the help of a large surface area provided by the 3-D ZnO nanostructures for the formation of SERS active sites, as well as the special metal nanostructures, which can act as efficient antennas and make the light absorbance increase. Further, the coupling of Ag film and Au NPs in this metal nanostructure can induce a large electric field at the particle–film gap, which has been theoretically evaluated and calculated by the discrete dipole approximation (DDA) method (Figure 12). The substrate was expected to be used for single-molecule detection at the zeptomole level.

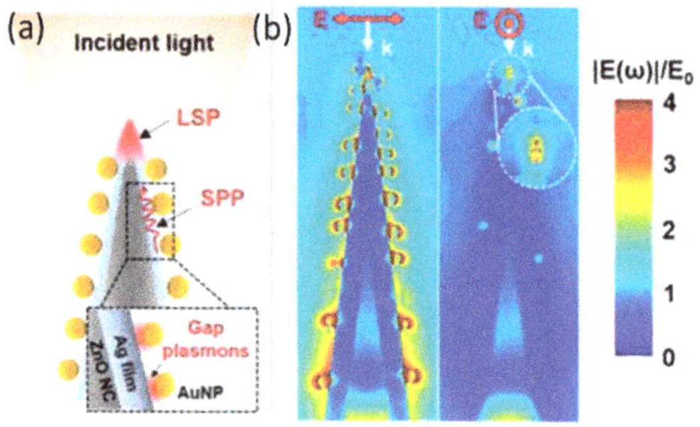

Figure 12. Ultrasharp ZnO nanocone arrays hybridized with the plasmonic systems of Au particle on Ag film (**a**) Schematic showing detection principle of fabricated SERS sensor; (**b**) Normalized electric field distribution for the ZnO/Ag/PDDA/AuNP NCs. Reproduced with permission from [99]. Copyright American Chemical Society, 2015. PDDA: poly(diallydimethylammonium chloride).

6.4. *3-D-Sandwich Structure Nanomaterials*

3-D-sandwich structure nanomaterials with a large contact area can usually exhibit an excellent SERS performance due to their particular structures. A majority of target molecules can be adsorbed by the large contact area of the 3-D-sandwich structure, and their particular structures are beneficial to the formation of "hot spots". In addition, 3-D-sandwich structure can help the substrates to employ the distinctive properties of the constituent materials. The general constituent materials for 3-D-sandwich structures are noble metals, semiconductors, and typical 2-D layered materials such as graphene and MoS_2. The relationship between the SERS performance and the structural configuration of the constituent materials is worthy of further research.

6.4.1. Noble Metal/Molecules/ZnO 3-D-Sandwich Structural Composite Substrates

The ZnO/PATP/Ag sandwich structure and its reverse Ag/PATP/ZnO sandwich structure were prepared by Zhihua Sun et al. [110]. It was discovered that these two substrates with the inverse structures showed the different SERS performance under a laser of 1064 nm due to the diverse connection of functional groups. It can be speculated that CM dominated by the charge transfer process primarily contributed to the SERS performance of these sandwich structures. The conjecture has been further confirmed by the evidence that the non-totally symmetric b_2 modes of PATP were strongly enhanced in the ZnO/PATP/Ag sandwich structure, whereas in the reverse structure, they did not. Their discovery suggests that the charge transfer through the bridge-like interconnecting probed molecules between nanoscale metals and semiconductors may be detected by SERS. Based on the previous research, the relative intensity of non-totally symmetric modes can be considered as an indicator for the contributions of the charge transfer process to the SERS enhancement by Alexander P. Richter et al. [86]. After observing the size- and excitation wavelength-dependent charge transfer in the Ag/PATP/ZnO sandwich structure, they attempted to introduce a quantitative equation, as follows:

$$P_{CT}(k) = \frac{I^k(CT) - I^k(SPR)}{I^k(CT) + I^0(SPR)} \tag{5}$$

where $P_{CT}(k)$ is the degree of charge transfer of a k-line and the k-line may be either a totally symmetric or nontotally symmetric line, $I^k(CT)$ is the intensity of a line where CT induced the increase of the SERS intensity caused by surface plasmon resonances ($I^k(SPR)$), while $I^0(SPR)$ is the intensity of a line in the spectrum where only SPR contributed to the signal.

They also found that for a series of molecules with a low-lying unfilled π^* orbital as an electron acceptor (PATP, 4-Mpy, and 4-MBA), the optimal particle diameter of ZnO was 27.7 nm for all wavelengths, which was consistent with the research results of Zhihua Sun et al. [110]. Through this study, they firmly ensured that the size-dependent SERS effect of the previous study was indeed the charge transfer in nature. Libin Yang et al. [117] considered that the Raman enhancement of PATP adsorbed on the Au/ZnO substrate was attributed to a CT contribution from the metal to molecules instead of the EM by examining the relative enhancement of the non-totally symmetric (b_2) modes. According to the EM model proposed by Creighton [123] and Moskovits [124], the totally symmetric (a_1) modes should have the strongest enhancement regardless of the orientation of the molecules. The selective enhancement of only the b_2 mode among the a_1, b_1, and b_2 modes cannot be explained by the EM mechanism. This series of studies about the relationship between the charge transfer and non-totally symmetric b_2 modes provided us with an important understanding of the role of CM in SERS.

6.4.2. Graphene/Noble Metal/ZnO 3-D-Sandwich Structural Composite Substrates

Graphene is famous for a unique two-dimensional (2-D) layered structure with the exceptional electric, thermal, and optical properties [125,126], and has the potential to be exploited in the 3-D sandwich structure of composite substrates. In recent years, there have been studies about the possibility of using graphene as a SERS substrate [127], and graphene has been reported to have controllable SERS performance [128]. Many scientists have attempted to composite graphene with semiconductors and/or the noble metal nanostructures [129,130] in a layered sandwich form as the SERS substrates. These substrates can exhibit outstanding SERS activity, and the graphene has been confirmed to play an important role for the SERS enhancement. In the meantime, the participation of the graphene is also beneficial to improving the photocatalytic performance of the SERS substrates and making the SRES substrates more environmentally friendly.

Cheng-Chi Kuo et al. [131] have investigated the role that graphene played in the SERS and photocatalytic performance of the graphene–semiconductor (e.g., TiO_2, ZnO) hybrid panel (GHP) substrates. They found that the precise number of graphene layers was critical to the performance of SERS and photocatalysis. Results showed that the hybrid with three layers of graphene (3L-GHP) possessed the maximum SERS performance, with an EF of 10^8 when using R6G as the target molecule. Moreover, it also exhibited excellent photocatalytic activity when photodegrading the methylene blue (MB), due to the rapid electron and hole transfer through the graphene. R. Ajay Rakkesh et al. [118] constructed a ZnO–Ag–graphene nanosheet (ZnO–Ag–GNS) nanoassembly through integrating ZnO–Ag core–shell nanostructures on graphene nanosheets by a wet chemical process (Figure 13). The SERS activity of the ZnO–Ag–GNS nanostructure was enhanced by the LSPR of Ag NPs, as well as the easy interfacial charge transfer process due to the close contact of the graphene nanosheets with the metal Ag. This substrate can rapidly detect organic contaminants such as acridine orange dye (AO dye) and photodegrade the contamination simultaneously. This study gave us illumination that GNS was an excellent candidate for enhancing the SERS effect through accelerating the interfacial charge transfer process and enhancing the photocatalytic activity by means of reducing the recombination rate of the electron–hole pair. The graphene-based substrates have successfully exhibited SERS and photocatalytic activity simultaneously, and have expanded the application of SERS substrates.

Ya Chi Ko et al. [119] have fabricated Ag/ZnO/reduced graphene oxide (rGO) nanocomposite to detect bacteria by SERS and kill them in many ways. It can be realized by combining the photocatalytic property of ZnO NPs, the high specific surface area and NIR photothermal conversion property of rGO, as well as the bacteria-killing capability and SERS property of Ag NPs. *Escherichia coli* (*E. coli*) was successfully detected by this active Ag/ZnO/rGO substrate with a detection limit about 10^4 cfu/mL, and was killed to different degrees by the substrates with or without the NIR or full Xe lamp irradiation.

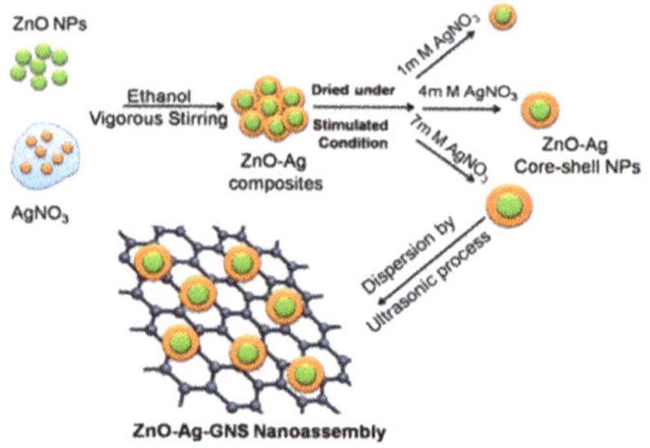

Figure 13. Schematic illustration of ZnO–Ag–graphene nanosheet (ZnO–Ag–GNS) nanoassembly synthesized by a wet chemical process. Reproduced with permission from [118]. Copyright The Royal Society of Chemistry, 2016.

7. Conclusions

In this review, we first discussed the development of SERS substrates from noble metals to semiconductors, and semiconductor ZnO was introduced as one of the potential SERS substrates. In order to improve the SERS activity of ZnO, the primary source of the great disparity between the EF of noble metals and that of ZnO nanostructures was analyzed and clarified, which can be attributed to the LSPR and "hot spots" of noble metals. Then, heavy elemental doping and the combination of noble metals with ZnO were put forward as the major improvement methods. Next, the preparation methods of varied ZnO nanostructures (0–3 dimensions) were summarized. Finally, we presented an overview of ZnO nanostructures for versatile SERS application. For pure ZnO nanostructures, the EF is usually 10^3 due to the predominant chemical enhancement based on the PICT. With regards to the noble metals/ZnO composite substrates, the strongest EF can reach up to 10^{10}–10^{11} with a detection limit at a zeptomole level of 10^{-19} M for the benzenethiol molecules by the ultrasharp ZnO/Ag/Au NPs nanocone substrate, which is promising for application in single molecule detection. In addition to the excellent SERS activity, many of these active ZnO nanostructure substrates are versatile; they not only can be used as chemosensors for detecting the NO released from the cells, but can also integrate EIS to obtain cellular information, identify "fingerprint regions" for disease by investigating human whole blood, and kill bacteria by photocatalysis.

However, there is little breakthrough in the improvement of SERS activity of semiconductors themselves. ZnO nanostructures are more likely to depend on the noble metals rather than themselves as the superior SERS substrates. Additionally, noble metal/ZnO substrates can still be plagued by the imperfection and instability problems during application. We look forward to elevating the SERS activity of ZnO nanostructures with the help of the optimization of shape, structure, and size, as well as sophisticated conformation of "hot spots". A metalloid dielectric property of semiconductors can be expected to be realized in the future to greatly enhance the Raman scattering by tuning the LSPR peak near to the VIS spectra region. Further, the in-situ trace detection can be effectively and harmoniously integrated with other functionalities on the ZnO nanostructure substrates. In my view, there is an urgent need to discover and develop novel versatile semiconductors with ultra-high SERS activity, and this may be the most fundamental way to improve the SERS activity of semiconductors.

Acknowledgments: The authors gratefully acknowledge the financial support of the National Natural Science Foundation of China (No. 51471182), and this work is also supported by Shanghai international science and Technology Cooperation Fund (No. 17520711700).

Author Contributions: Yong Yang and Zhengren Huang provided guidance and advice; Yunfeng Ma, Shuai Li, Yuquan Wei and Nguyen Viet Long provided ideas and help; Lili Yang wrote the paper. All authors contributed to discussion and reviewed the manuscript.

Conflicts of Interest: The authors declare no conflict of interest.

References

1. Griffith Freeman, R.; Grabar, K.C.; Allison, K.J.; Bright, R.M.; Davis, J.A.; Guthrie, A.P.; Hommer, M.B.; Jackson, M.A.; Smith, P.C.; Walter, D.G.; et al. Self-Assembled Metal Colloid Monolayers: An Approach to SERS Substrates. *Science* **1995**, *267*, 1629–1632. [CrossRef] [PubMed]
2. Baker, G.A.; Moore, D.S. Progress in plasmonic engineering of surface-enhanced Raman-scattering substrates toward ultra-trace analysis. *Anal. Bioanal. Chem.* **2005**, *382*, 1751–1770. [CrossRef] [PubMed]
3. Stiles, P.L.; Dieringer, J.A.; Shah, N.C.; Van Duyne, R.P. Surface-enhanced Raman spectroscopy. *Annu. Rev. Anal. Chem.* **2008**, *1*, 601–626. [CrossRef] [PubMed]
4. Fleischmann, M.; Hendra, P.J.; McQuillan, A.J. Raman spectra of pyridzne adsorbed at a silver electrode. *Chem. Phys. Lett.* **1974**, *26*, 163–166. [CrossRef]
5. Albrecht, M.; Creighton, J. Anomalously intense Raman spectra of pyridine at a silver electrode. *J. Am. Chem. Soc.* **1977**, *99*, 5215–5217. [CrossRef]
6. Jeanmaire, D.L.; Duyne, R.P.V. Surface Raman spectroelectrochemistry: Part I. Heterocyclic, aromatic, and aliphatic amines adsorbed on the anodized silver electrode. *J. Electroanal. Chem. Interfacial Electrochem.* **1977**, *84*, 1–20. [CrossRef]
7. Qian, X.M.; Nie, S.M. Single-molecule and single-nanoparticle SERS: From fundamental mechanisms to biomedical applications. *Chem. Soc. Rev.* **2008**, *37*, 912–920. [CrossRef] [PubMed]
8. Yang, Y.; Tanemura, M.; Huang, Z.; Jiang, D.; Li, Z.Y.; Huang, Y.P.; Kawamura, G.; Yamaguchi, K.; Nogami, M. Aligned gold nanoneedle arrays for surface-enhanced Raman scattering. *Nanotechnology* **2010**, *21*, 325701. [CrossRef] [PubMed]
9. Harper, M.M.; McKeating, K.S.; Faulds, K. Recent developments and future directions in SERS for bioanalysis. *Phys. Chem. Chem. Phys.* **2013**, *15*, 5312–5328. [CrossRef] [PubMed]
10. Li, L.; Zhao, A.; Wang, D.; Guo, H.; Sun, H.; He, Q. Fabrication of cube-like Fe_3O_4@SiO_2@Ag nanocomposites with high SERS activity and their application in pesticide detection. *J. Nanopart. Res.* **2016**, *18*, 178. [CrossRef]
11. Hurst, S.J.; Fry, H.C.; Gosztola, D.J.; Rajh, T. Utilizing Chemical Raman Enhancement: A Route for Metal Oxide Support-Based Biodetection. *J. Phys. Chem. C* **2011**, *115*, 620–630. [CrossRef]
12. Brus, L. Noble Metal Nanocrystals: Plasmon Electron Transfer Photochemistry and Single-Molecule Raman Spectroscopy. *Acc. Chem. Res.* **2008**, *41*, 1742–1749. [CrossRef] [PubMed]
13. Lee, S.; Chon, H.; Yoon, S.-Y.; Lee, E.K.; Chang, S.-I.; Lim, D.W.; Choo, J. Fabrication of SERS-fluorescence dual modal nanoprobes and application to multiplex cancer cell imaging. *Nanoscale* **2012**, *4*, 124–129. [CrossRef] [PubMed]
14. Ding, Q.; Zhou, H.; Zhang, H.; Zhang, Y.; Wang, G.; Zhao, H. 3D Fe_3O_4@Au@Ag Nanoflowers Assembled Magnetoplasmonic Chains for in situ SERS Monitoring of Plasmon-assisted Catalytic Reaction. *J. Mater. Chem. A* **2016**, *4*, 8866–8874. [CrossRef]
15. Kundu, S.; Mandal, M.; Ghosh, S.K.; Pal, T. Photochemical deposition of SERS active silver nanoparticles on silica gel and their application as catalysts for the reduction of aromatic nitro compounds. *J. Colloid Interface Sci.* **2004**, *272*, 134–144. [CrossRef] [PubMed]
16. Hakonen, A.; Wang, F.; Andersson, P.O.; Wingfors, H.; Rindzevicius, T.; Schmidt, M.S.; Soma, V.R.; Xu, S.; Li, Y.; Boisen, A.; et al. Hand-Held Femtogram Detection of Hazardous Picric Acid with Hydrophobic Ag Nanopillar SERS Substrates and Mechanism of Elasto-Capillarity. *ACS Sens.* **2017**, *2*, 198–202. [CrossRef] [PubMed]
17. Dasary, S.S.R.; Singh, A.K.; Senapati, D.; Yu, H.; Ray, P.C. Gold Nanoparticle Based Label-Free SERS Probe for Ultrasensitive and Selective Detection of Trinitrotoluene. *J. Am. Chem. Soc.* **2009**, *131*, 13806–13812. [CrossRef] [PubMed]
18. Cui, S.; Dai, Z.; Tian, Q.; Liu, J.; Xiao, X.; Jiang, C.; Wu, W.; Roy, V.A.L. Wetting properties and SERS applications of ZnO/Ag nanowire arrays patterned by a screen printing method. *J. Mater. Chem. C* **2016**, *4*, 6371–6379. [CrossRef]

19. Michaels, A.M.; Nirmal, M.; Brus, L.E. Surface Enhanced Raman Spectroscopy of Individual Rhodamine 6G Molecules on Large Ag Nanocrystals. *J. Am. Chem. Soc.* **1999**, *121*, 9932–9939. [CrossRef]
20. Zeman, E.J.; Schatz, G.C. An accurate electromagnetic theory study of surface enhancement factors for silver, gold, copper, lithium, sodium, aluminum, gallium, indium, zinc, and cadmium. *J. Phys. Chem.* **1987**, *91*, 634–643. [CrossRef]
21. Kneipp, K.; Wang, Y.; Kneipp, H.; Perelman, L.T.; Itzkan, I.; Dasari, R.R.; Feld, M.S. Single Molecule Detection Using Surface-Enhanced Raman Scattering (SERS). *Phys. Rev. Lett.* **1997**, *78*, 1667–1670. [CrossRef]
22. Nie, S. Probing Single Molecules and Single Nanoparticles by Surface-Enhanced Raman Scattering. *Science* **1997**, *275*, 1102–1106. [CrossRef] [PubMed]
23. Xu, H.; Bjerneld, E.; Käll, M.; Börjesson, L. Spectroscopy of Single Hemoglobin Molecules by Surface Enhanced Raman Scattering. *Phys. Rev. Lett.* **1999**, *83*, 4357–4360. [CrossRef]
24. Cao, Y.; Zhang, J.; Yang, Y.; Huang, Z.; Long, N.V.; Fu, C. Engineering of SERS Substrates Based on Noble Metal Nanomaterials for Chemical and Biomedical Applications. *Appl. Spectrosc. Rev.* **2015**, *50*, 499–525. [CrossRef]
25. Yang, Y.; Li, Z.-Y.; Yamaguchi, K.; Tanemura, M.; Huang, Z.; Jiang, D.; Chen, Y.; Zhou, F.; Nogami, M. Controlled fabrication of silver nanoneedles array for SERS and their application in rapid detection of narcotics. *Nanoscale* **2012**, *4*, 2663. [CrossRef] [PubMed]
26. Wang, X.; Shi, W.; She, G.; Mu, L. Surface-Enhanced Raman Scattering (SERS) on transition metal and semiconductor nanostructures. *Phys. Chem. Chem. Phys.* **2012**, *14*, 5891–5901. [CrossRef] [PubMed]
27. Ji, W.; Song, W.; Tanabe, I.; Wang, Y.; Zhao, B.; Ozaki, Y. Semiconductor-enhanced Raman scattering for highly robust SERS sensing: The case of phosphate analysis. *Chem. Commun.* **2015**, *51*, 7641–7644. [CrossRef] [PubMed]
28. Yamada, H.; Yamamoto, Y. Surface enhanced Raman scattering (SERS) of chemisorbed species on various kinds of metals and semiconductors. *Surf. Sci.* **1983**, *134*, 71–90. [CrossRef]
29. Hayashi, S.; Koh, R.; Ichiyama, Y.; Yamamoto, K. Evidence for surface-enhanced Raman scattering on nonmetallic surfaces: Copper phthalocyanine molecules on GaP small particles. *Phys. Rev. Lett.* **1988**, *60*, 1085–1088. [CrossRef] [PubMed]
30. Wen, H.A.O. Surface enhancement of Raman and absorption spectra from cyanine dye D266 adsorbed on ZnO colloids. *Mol. Phys.* **1996**, *88*, 281–290. [CrossRef]
31. Shan, Y.; Zheng, Z.; Liu, J.; Yang, Y.; Li, Z.; Huang, Z.; Jiang, D. Niobium pentoxide: A promising surface-enhanced Raman scattering active semiconductor substrate. *NPJ Comp. Mater.* **2017**, *3*, 11. [CrossRef]
32. Quagliano, L.G. Observation of Molecules Adsorbed on III-V Semiconductor Quantum Dots by Surface-Enhanced Raman Scattering. *J. Am. Chem. Soc.* **2004**, *126*, 7393–7398. [CrossRef] [PubMed]
33. Ji, W.; Zhao, B.; Ozaki, Y. Semiconductor materials in analytical applications of surface-enhanced Raman scattering. *J. Raman Spectrosc.* **2016**, *47*, 51–58. [CrossRef]
34. Tan, X.; Melkersson, J.; Wu, S.; Wang, L.; Zhang, J. Noble-Metal-Free Materials for Surface-Enhanced Raman Spectroscopy Detection. *Chemphyschem* **2016**, *17*, 2630–2639. [CrossRef] [PubMed]
35. Wu, H.; Wang, H.; Li, G. Metal oxide semiconductor SERS-active substrates by defect engineering. *Analyst* **2017**, *142*, 326–335. [CrossRef] [PubMed]
36. Wen, C.; Liao, F.; Liu, S.; Zhao, Y.; Kang, Z.; Zhang, X.; Shao, M. Bi-functional ZnO–RGO–Au substrate: Photocatalysts for degrading pollutants and SERS substrates for real-time monitoring. *Chem. Commun.* **2013**, *49*, 3049. [CrossRef] [PubMed]
37. Shan, Y.; Yang, Y.; Cao, Y.; Huang, Z. Facile solvothermal synthesis of Ag/Fe_3O_4 nanocomposites and their SERS applications in on-line monitoring of pesticide contaminated water. *RCS Adv.* **2015**, *5*, 102610–102618. [CrossRef]
38. Xu, J.; Liu, Y.; Wang, Q.; Duo, H.; Zhang, X.; Li, Y.; Huang, W. Photocatalytically Renewable Micro-electrochemical Sensor for Real-Time Monitoring of Cells. *Angew. Chem. Int. Ed.* **2015**, *54*, 14402–14406. [CrossRef] [PubMed]
39. Willets, K.A.; Van Duyne, R.P. Localized surface plasmon resonance spectroscopy and sensing. *Annu. Rev. Phys. Chem.* **2007**, *58*, 267–297. [CrossRef] [PubMed]
40. Özgür, Ü.; Alivov, Y.I.; Liu, C.; Teke, A.; Reshchikov, M.A.; Doğan, S.; Avrutin, V.; Cho, S.J.; Morkoç, H. A comprehensive review of ZnO materials and devices. *J. Appl. Phys.* **2005**, *98*, 041301. [CrossRef]
41. Fortunato, E.M.C.; Barquinha, P.M.C.; Pimentel, A.C.M.B.G.; Gonçalves, A.M.F.; Marques, A.J.S.; Pereira, L.M.N.; Martins, R.F.P. Fully Transparent ZnO Thin-Film Transistor Produced at Room Temperature. *Adv. Mater.* **2005**, *17*, 590–594. [CrossRef]
42. Wang, Z.L. Nanobelts, Nanowires, and Nanodiskettes of Semiconducting Oxides—From Materials to Nanodevices. *Adv. Mater.* **2003**, *15*, 432–436. [CrossRef]

43. Vernardou, D.; Kenanakis, G.; Couris, S.; Koudoumas, E.; Kymakis, E.; Katsarakis, N. pH effect on the morphology of ZnO nanostructures grown with aqueous chemical growth. *Thin Solid Films* **2007**, *515*, 8764–8767. [CrossRef]
44. Tian, Z.R.; Voigt, J.A.; Liu, J.; McKenzie, B.; McDermott, M.J.; Rodriguez, M.A.; Konishi, H.; Xu, H. Complex and oriented ZnO nanostructures. *Nat. Mater.* **2003**, *2*, 821–826. [CrossRef] [PubMed]
45. Di Mauro, A.; Fragalà, M.E.; Privitera, V.; Impellizzeri, G. ZnO for application in photocatalysis: From thin films to nanostructures. *Mater. Sci. Semicond. Process.* **2017**, *69*, 44–51. [CrossRef]
46. Pearton, S. Recent progress in processing and properties of ZnO. *Prog. Mater. Sci.* **2005**, *50*, 293–340. [CrossRef]
47. Hochbaum, A.; Yang, P. Semiconductor Nanowires for Energy Conversion. *Chem. Rev.* **2010**, *110*, 527–546. [CrossRef] [PubMed]
48. Marimuthu, T.; Anandhan, N.; Thangamuthu, R.; Surya, S. Facile growth of ZnO nanowire arrays and nanoneedle arrays with flower structure on ZnO-TiO_2 seed layer for DSSC applications. *J. Alloys Compd.* **2017**, *693*, 1011–1019. [CrossRef]
49. Wang, L.; Li, J.; Wang, Y.; Yu, K.; Tang, X.; Zhang, Y.; Wang, S.; Wei, C. Construction of 1D SnO_2-coated ZnO nanowire heterojunction for their improved *n*-butylamine sensing performances. *Sci. Rep.* **2016**, *6*, 35079. [CrossRef] [PubMed]
50. Khan, M.A.; Hogan, T.P.; Shanker, B. Gold-coated zinc oxide nanowire-based substrate for surface-enhanced Raman spectroscopy. *J. Raman Spectrosc.* **2009**, *40*, 1539–1545. [CrossRef]
51. Li, Y.; Cai, W.; Duan, G.; Cao, B.; Sun, F.; Lu, F. Superhydrophobicity of 2D ZnO ordered pore arrays formed by solution-dipping template method. *J. Colloid Interface Sci.* **2005**, *287*, 634–639. [CrossRef] [PubMed]
52. Bagra, B.; Pimpliskar, P.; Agrawal, N.K. Bio-compatibility, surface & chemical characterization of glow discharge plasma modified ZnO nanocomposite polycarbonate. *Soild State Phys.* **2014**, 189–191. [CrossRef]
53. Metiu, H. Surface enhanced spectroscopy. *Prog. Surf. Sci.* **1984**, *17*, 153–320. [CrossRef]
54. Lombardi, J.R.; Birke, R.L. A Unified Approach to Surface-Enhanced Raman Spectroscopy. *J. Phys. Chem. C* **2008**, *112*, 5605–5617. [CrossRef]
55. Nitzan, A.; Brus, L.E. Theoretical model for enhanced photochemistry on rough surfaces. *J. Chem. Phys.* **1981**, *75*, 2205–2214. [CrossRef]
56. Shegai, T.; Vaskevich, A.; Rubinstein, I.; Haran, G. Raman Spectroelectrochemistry of Molecules within Individual Electromagnetic Hot Spots. *J. Am. Chem. Soc.* **2009**, *131*, 14390–14398. [CrossRef] [PubMed]
57. Xu, H.; Aizpurua, J.; Kall, M.; Apell, P. Electromagnetic contributions to single-molecule sensitivity in surface-enhanced Raman scattering. *Phys. Rev. E* **2000**, *62*, 4318–4324. [CrossRef]
58. Lee, P.C.; Meise, D. Adsorption and Surface-Enhanced Raman of Dyes on Silver and Gold Sols. *J. Phys. Chem.* **1982**, *86*, 3391–3395. [CrossRef]
59. Campion, A.; Kambhampati, P. Surface-enhanced Raman scattering. *Chem. Sco. Rev.* **1998**, *27*, 241. [CrossRef]
60. Moskovits, M. Surface-enhanced spectroscopy. *Rev. Mod. Phys.* **1985**, *57*, 783–826. [CrossRef]
61. Han, X.X.; Ji, W.; Zhao, B.; Ozaki, Y. Semiconductor-Enhanced Raman Scattering: Active Nanomaterials and Applications. *Nanoscale* **2017**, *9*, 4847–4861. [CrossRef] [PubMed]
62. Lyons, J.L.; Van de Walle, C.G. Computationally predicted energies and properties of defects in GaN. *NPJ Comp. Mater.* **2017**, *3*, 12. [CrossRef]
63. Kleinman, S.L.; Frontiera, R.R.; Henry, A.I.; Dieringer, J.A.; Van Duyne, R.P. Creating, characterizing, and controlling chemistry with SERS hot spots. *Phys. Chem. Chem. Phys.* **2013**, *15*, 21–36. [CrossRef] [PubMed]
64. Le Ru, E.C.; Grand, J.; Sow, I.; Somerville, W.R.C.; Etchegoin, P.G.; Treguer-Delapierre, M.; Charron, G.; Félidj, N.; Lévi, G.; Aubard, J.; et al. A scheme for detecting every single target molecule with surface-enhanced Raman spectroscopy. *Nano Lett.* **2011**, *11*, 5013–5019. [CrossRef] [PubMed]
65. Luther, J.M.; Jain, P.K.; Ewers, T.; Alivisatos, A.P. Localized surface plasmon resonances arising from free carriers in doped quantum dots. *Nat. Mater.* **2011**, *10*, 361–366. [CrossRef] [PubMed]
66. Drude, P. Zur Elektronentheorie der Metalle. *Ann. Phys.* **1990**, *306*, 566–613. [CrossRef]
67. Ma, X.; Dai, Y.; Yu, L.; Huang, B. Noble-metal-free plasmonic photocatalyst: Hydrogen doped semiconductors. *Sci. Rep.* **2014**, *4*, 3986. [CrossRef] [PubMed]
68. Ling, X.; Moura, L.G.; Pimenta, M.A.; Zhang, J. Charge-Transfer Mechanism in Graphene-Enhanced Raman Scattering. *J. Phys. Chem. C* **2012**, *116*, 25112–25118. [CrossRef]
69. Lombardi, J.R.; Birke, R.L. Time-dependent picture of the charge-transfer contributions to surface enhanced Raman spectroscopy. *J. Chem. Phys.* **2007**, *126*, 244709. [CrossRef] [PubMed]

70. Musumeci, A.; Gosztola, D.; Schiller, T.; Dimitrijevic, N.M.; Mujica, V.; Martin, D.; Rajh, T. SERS of Semiconducting Nanoparticles (TiO_2 Hybrid Composites). *J. Am. Chem. Soc.* **2009**, *131*, 6040–6041. [CrossRef] [PubMed]
71. Fonoberov, V.A.; Alim, K.A.; Balandin, A.A.; Xiu, F.; Liu, J. Photoluminescence investigation of the carrier recombination processes in ZnO quantum dots and nanocrystals. *Phys. Rev. B* **2006**, *73*, 165317. [CrossRef]
72. Wang, Y.; Ruan, W.; Zhang, J.; Yang, B.; Xu, W.; Zhao, B.; Lombardi, J.R. Direct observation of surface-enhanced Raman scattering in ZnO nanocrystals. *J. Raman Spectrosc.* **2009**, *40*, 1072–1077. [CrossRef]
73. Brus, L.E. Electron–electron and electron-hole interactions in small semiconductor crystallites: The size dependence of the lowest excited electronic state. *J. Chem. Phys.* **1984**, *80*, 4403–4409. [CrossRef]
74. Liu, L.; Yang, H.; Ren, X.; Tang, J.; Li, Y.; Zhang, X.; Cheng, Z. Au–ZnO hybrid nanoparticles exhibiting strong charge-transfer-induced SERS for recyclable SERS-active substrates. *Nanoscale* **2015**, *7*, 5147–5151. [CrossRef] [PubMed]
75. Polavarapu, L.; Pérez-Juste, J.; Xu, Q.-H.; Liz-Marzán, L.M. Optical sensing of biological, chemical and ionic species through aggregation of plasmonic nanoparticles. *J. Mater. Chem. C* **2014**, *2*, 7460–7476. [CrossRef]
76. Polavarapu, L.; Mourdikoudis, S.; Pastoriza-Santos, I.; Pérez-Juste, J. Nanocrystal engineering of noble metals and metal chalcogenides: Controlling the morphology, composition and crystallinity. *CrystEngComm* **2015**, *17*, 3727–3762. [CrossRef]
77. Polavarapu, L.; Liz-Marzán, L.M. Growth and galvanic replacement of silver nanocubes in organic media. *Nanoscale* **2013**, *5*, 4355. [CrossRef] [PubMed]
78. Gómez-Graña, S.; Fernández-López, C.; Polavarapu, L.; Salmon, J.-B.; Leng, J.; Pastoriza-Santos, I.; Pérez-Juste, J. Gold Nanooctahedra with Tunable Size and Microfluidic-Induced 3D Assembly for Highly Uniform SERS-Active Supercrystals. *Chem. Mater.* **2015**, *27*, 8310–8317. [CrossRef]
79. Fernández-López, C.; Polavarapu, L.; Solís, D.M.; Taboada, J.M.; Obelleiro, F.; Contreras-Cáceres, R.; Pastoriza-Santos, I.; Pérez-Juste, J. Gold Nanorod–pNIPAM Hybrids with Reversible Plasmon Coupling: Synthesis, Modeling, and SERS Properties. *ACS Appl. Mater. Interfaces* **2015**, *7*, 12530–12538. [CrossRef] [PubMed]
80. Gasymov, O.K.; Alekperov, O.Z.; Aydemirova, A.H.; Kamilova, N.; Aslanov, R.B.; Bayramov, A.H.; Kerimova, A. Surface enhanced Raman scattering of whole human blood on nanostructured ZnO surface. *Phys. Status Solidi C* **2017**, *14*, 1600155.
81. Le Ru, E.C.; Blackie, E.; Meyer, M.; Etchegoin, P.G. Surface Enhanced Raman Scattering Enhancement Factors: A Comprehensive Study. *J. Phys. Chem. C* **2007**, *111*, 13794–13803. [CrossRef]
82. Hakonen, A.; Svedendahl, M.; Ogier, R.; Yang, Z.-J.; Lodewijks, K.; Verre, R.; Shegai, T.; Andersson, P.O.; Käll, M. Dimer-on-mirror SERS substrates with attogram sensitivity fabricated by colloidal lithography. *Nanoscale* **2015**, *7*, 9405–9410. [CrossRef] [PubMed]
83. Xue, X.; Ruan, W.; Yang, L.; Ji, W.; Xie, Y.; Chen, L.; Song, W.; Zhao, B.; Lombardi, J.R. Surface-enhanced Raman scattering of molecules adsorbed on Co-doped ZnO nanoparticles. *J. Raman Spectrosc.* **2012**, *43*, 61–64. [CrossRef]
84. Chang, L.; Xu, D.; Xue, X. Photoluminescence and Raman scattering study in ZnO:Mg nanocrystals. *J. Mater. Sci. Mater. Electron.* **2015**, *27*, 1014–1019. [CrossRef]
85. Sun, Z.; Zhao, B.; Lombardi, J.R. ZnO nanoparticle size-dependent excitation of surface Raman signal from adsorbed molecules: Observation of a charge-transfer resonance. *Appl. Phys. Lett.* **2007**, *91*, 221106. [CrossRef]
86. Richter, A.P.; Lombardi, J.R.; Zhao, B. Size and Wavelength Dependence of the Charge-Transfer Contributions to Surface-Enhanced Raman Spectroscopy in Ag/PATP/ZnO Junctions. *J. Phys. Chem. C* **2010**, *114*, 1610–1614. [CrossRef]
87. Jing, L.; Xu, Z.; Shang, J.; Sun, X.; Cai, W.; Guo, H. The preparation and characterization of ZnO ultrafine particles. *Mater. Sci. Eng. A* **2002**, *332*, 356–361. [CrossRef]
88. Song, W.; Wang, Y.; Hu, H.; Zhao, B. Fabrication of surface-enhanced Raman scattering-active ZnO/Ag composite microspheres. *J. Raman Spectrosc.* **2007**, *38*, 1320–1325. [CrossRef]
89. Seelig, E.W.; Tang, B.; Yamilov, A.; Cao, H.; Chang, R.P.H. Self-assembled 3D photonic crystals from ZnO colloidal spheres. *Mater. Chem. Phys.* **2003**, *80*, 257–263. [CrossRef]
90. Jezequel, D.; Guenot, J.; Jouini, N.; Fievet, F. Preparation and Morphological Characterization of Fine, Spherical, Monodisperse Particles of ZnO. *Mater. Sci. Forum* **1994**, *152–153*, 339–342. [CrossRef]
91. Wang, X.; Shi, W.; Jin, Z.; Huang, W.; Lin, J.; Ma, G.; Li, S.; Guo, L. Remarkable SERS Activity Observed from Amorphous ZnO Nanocages. *Angew. Chem. Int. Ed.* **2017**, *56*, 9851–9855. [CrossRef] [PubMed]
92. Lee, S.; Peng, J.-W.; Liu, C.-S. Photoluminescence and SERS investigation of plasma treated ZnO nanorods. *Appl. Surf. Sci.* **2013**, *285*, 748–754. [CrossRef]

93. Xu, F.; Zhang, P.; Navrotsky, A.; Yuan, Z.; Ren, T.; Halasa, M.; Su, B. Hierarchically Assembled Porous ZnO Nanoparticles: Synthesis, Surface Energy, and Photocatalytic Activity. *Chem. Mater.* **2007**, *19*, 5680–5686. [CrossRef]
94. Liu, Q.; Jiang, L.; Guo, L. Precursor-Directed Self-Assembly of Porous ZnO Nanosheets as High-Performance Surface-Enhanced Raman Scattering Substrate. *Small* **2014**, *10*, 48–51. [CrossRef] [PubMed]
95. Zhang, K.; Zhang, Y.; Zhang, T.; Dong, W.; Wei, T.; Sun, Y.; Chen, X.; Shen, G.; Dai, N. Vertically coupled ZnO nanorods on MoS_2 monolayers with enhanced Raman and photoluminescence emission. *Nano Res.* **2014**, *8*, 743–750. [CrossRef]
96. Wu, S.; Huang, C.; Aivazian, G.; Ross, J.S.; Cobden, D.H.; Xu, X. Vapor Solid Growth of High Optical Quality MoS_2 Monolayers with Near-Unity Valley Polarization. *ACS Nano* **2013**, *7*, 2768–2772. [CrossRef] [PubMed]
97. Jayram, N.D.; Sonia, S.; Poongodi, S.; Kumar, P.S.; Masuda, Y.; Mangalaraj, D.; Ponpandian, N.; Viswanathan, C. Superhydrophobic Ag decorated ZnO nanostructured thin film as effective surface enhanced Raman scattering substrates. *Appl. Surf. Sci.* **2015**, *355*, 969–977. [CrossRef]
98. Shan, Y.; Yang, Y.; Cao, Y.; Fu, C.; Huang, Z. Synthesis of wheatear-like ZnO nanoarrays decorated with Ag nanoparticles and its improved SERS performance through hydrogenation. *Nanotechnology* **2016**, *27*, 145502. [CrossRef] [PubMed]
99. Lee, Y.; Lee, J.; Lee, T.K.; Park, J.; Ha, M.; Kwak, S.K.; Ko, H. Particle-on-Film Gap Plasmons on Antireflective ZnO Nanocone Arrays for Molecular-Level Surface-Enhanced Raman Scattering Sensors. *ACS Appl. Mater. Interfaces* **2015**, *7*, 26421–26429. [CrossRef] [PubMed]
100. Wang, X.; She, G.; Xu, H.; Mu, L.; Shi, W. The surface-enhanced Raman scattering from ZnO nanorod arrays and its application for chemosensors. *Sens. Actutator B Chem.* **2014**, *193*, 745–751. [CrossRef]
101. Kim, K.; Kim, K.L.; Shin, K.S. Raman spectral characteristics of 4-aminobenzenethiol adsorbed on ZnO nanorod arrays. *Phys. Chem. Chem. Phys.* **2013**, *15*, 9288–9294. [CrossRef] [PubMed]
102. Greene, L.E.; Law, M.; Goldberger, J.; Kim, F.; Johnson, J.C.; Zhang, Y.; Saykally, R.J.; Yang, P. Low-Temperature Wafer-Scale Production of ZnO Nanowire Arrays. *Angew. Chem. Int. Ed.* **2003**, *42*, 3031–3034. [CrossRef] [PubMed]
103. Pacholski, C.; Kornowski, A.; Weller, H. Self-Assembly of ZnO: From Nanodots to Nanorods. *Angew. Chem. Int. Ed.* **2002**, *41*, 1188–1191. [CrossRef]
104. Zhang, W.-D. Growth of ZnO nanowires on modified well-aligned carbon nanotube arrays. *Nanotechnology* **2006**, *17*, 1036–1040. [CrossRef] [PubMed]
105. Cheng, C.; Yan, B.; Wong, S.M.; Li, X.; Zhou, W.; Yu, T.; Shen, Z.; Yu, H.; Fan, H.J. Fabrication and SERS Performance of Silver-Nanoparticle-Decorated Si/ZnO Nanotrees in Ordered Arrays. *ACS Appl. Mater. Interfaces* **2010**, *2*, 1824–1828. [CrossRef] [PubMed]
106. Xu, J.-Q.; Duo, H.-H.; Zhang, Y.-G.; Zhang, X.-W.; Fang, W.; Liu, Y.-L.; Shen, A.-G.; Hu, J.-M.; Huang, W.-H. Photochemical Synthesis of Shape-Controlled Nanostructured Gold on Zinc Oxide Nanorods as Photocatalytically Renewable Sensors. *Anal. Chem.* **2016**, *88*, 3789–3795. [CrossRef] [PubMed]
107. Huang, C.; Xu, C.; Lu, J.; Li, Z.; Tian, Z. 3D Ag/ZnO hybrids for sensitive surface-enhanced Raman scattering detection. *Appl. Surf. Sci.* **2016**, *365*, 291–295. [CrossRef]
108. Xie, Y.; Yang, S.; Mao, Z.; Li, P.; Zhao, C.; Cohick, Z.; Huang, P.-H.; Huang, T.J. In Situ Fabrication of 3D Ag@ZnO Nanostructures for Microfluidic Surface-Enhanced Raman Scattering Systems. *ACS Nano* **2014**, *8*, 12175–12184. [CrossRef] [PubMed]
109. Mao, Z.; Song, W.; Xue, X.; Ji, W.; Li, Z.; Chen, L.; Mao, H.; Lv, H.; Wang, X.; Lombardi, J.R.; et al. Interfacial Charge-Transfer Effects in Semiconductor–Molecule–Metal Structures: Influence of Contact Variation. *J. Phy. Chem. C* **2012**, *116*, 14701–14710. [CrossRef]
110. Sun, Z.; Wang, C.; Yang, J.; Zhao, B.; Lombardi, J.R. Nanoparticle Metal-Semiconductor Charge Transfer in ZnO/PATP/Ag Assemblies by Surface-Enhanced Raman Spectroscopy. *J. Phy. Chem. C* **2008**, *112*, 6093–6098. [CrossRef]
111. Shin, H.-Y.; Shim, E.-L.; Choi, Y.-J.; Park, J.-H.; Yoon, S. Giant Enhancement of Raman Response Due to OneDimensional ZnO Nanostructures. *Nanoscale* **2014**, *6*, 14622–14626. [CrossRef] [PubMed]
112. Jin, L.; She, G.; Wang, X.; Mu, L.; Shi, W. Enhancing the SERS performance of semiconductor nanostructures through a facile surface engineering strategy. *Appl. Surf. Sci.* **2014**, *320*, 591–595. [CrossRef]

113. Song, W.; Ji, W.; Vantasin, S.; Tanabe, I.; Zhao, B.; Ozaki, Y. Fabrication of a highly sensitive surface-enhanced Raman scattering substrate for monitoring the catalytic degradation of organic pollutants. *J. Mater. Chem. A* **2015**, *3*, 13556–13562. [CrossRef]
114. He, X.; Yue, C.; Zang, Y.; Yin, J.; Sun, S.; Li, J.; Kang, J. Multi-hot spot configuration on urchin-like Ag nanoparticle/ZnO hollow nanosphere arrays for highly sensitive SERS. *J. Mater. Chem. A* **2013**, *1*, 15010. [CrossRef]
115. Sinha, G.; Depero, L.E.; Alessandri, I. Recyclable SERS Substrates Based on Au-Coated ZnO Nanorods. *ACS Appl. Mater. Interfaces* **2011**, *3*, 2557–2563. [CrossRef] [PubMed]
116. Zong, X.; Zhu, R.; Guo, X. Nanostructured gold microelectrodes for SERS and EIS measurements by incorporating ZnO nanorod growth with electroplating. *Sci. Rep.* **2015**, *5*, 16454. [CrossRef] [PubMed]
117. Yang, L.; Ruan, W.; Jiang, X.; Zhao, B.; Xu, W.; Lombardi, J.R. Contribution of ZnO to Charge-Transfer Induced Surface-Enhanced Raman Scattering in Au/ZnO/PATP Assembly. *J. Phys. Chem. C* **2009**, *113*, 117–120. [CrossRef]
118. Rakkesh, R.A.; Durgalakshmi, D.; Balakumar, S. Graphene based nanoassembly for simultaneous detection and degradation of harmful organic contaminants from aqueous solution. *RCS Adv.* **2016**, *6*, 34342–34349. [CrossRef]
119. Ko, Y.Y.; Fang, H.Y.; Chen, D.H. Fabrication of Ag/ZnO/reduced graphene oxide nanocomposite for SERS detection and multiway killing of bacteria. *J. Alloys Compd.* **2017**, *695*, 1145–1153. [CrossRef]
120. Barbillon, G.; Sandana, V.E.; Humbert, C.; Belier, B.; Rogers, D.J.; Teherani, F.H.; Bove, P.; McClintock, R.; Razeghi, M. Study of Au coated ZnO nanoarrays for surface enhanced Raman scattering chemical sensing. *J. Mater. Chem. C* **2017**, *5*, 3528–3535. [CrossRef]
121. Sivapalan, S.T.; DeVetter, B.M.; Yang, T.K.; van Dijk, T.; Schulmerich, M.V.; Carney, P.S.; Bhargava, R.; Murphy, C.J. Off-Resonance Surface-Enhanced Raman Spectroscopy from Gold Nanorod Suspensions as a Function of Aspect Ratio: Not What We Thought. *ACS Nano* **2013**, *7*, 2099–2105. [CrossRef] [PubMed]
122. Zhang, Y.; Lin, J.D.; Vijayaragavan, V.; Bhakoo, K.K.; Tan, T.T.Y. Tuning sub-10 nm single-phase $NaMnF_3$ nanocrystals as ultrasensitive hosts for pure intense fluorescence and excellent T1 magnetic resonance imaging. *Chem. Commun.* **2012**, *48*, 10322. [CrossRef] [PubMed]
123. Creighton, J.A. Surface raman electromagnetic enhancement factors for molecules at the surface of small isolated metal spheres: The determination of adsorbate orientation from sers relative intensities. *Surf. Sci.* **1983**, *124*, 209–219. [CrossRef]
124. Moskovits, M.; Suh, J.S. Surface Selection Rules for Surface-Enhanced Raman Spectroscopy: Calculations and Application to the Surface-Enhanced Raman Spectrum of Phthalazine on Silver J. *Phy. Chem.* **1984**, *88*, 5526–5530. [CrossRef]
125. Kuila, T.; Bose, S.; Mishra, A.K.; Khanra, P.; Kim, N.H.; Lee, J.H. Chemical functionalization of graphene and its applications. *Prog. Mater. Sci.* **2012**, *57*, 1061–1105. [CrossRef]
126. Singh, V.; Joung, D.; Zhai, L.; Das, S.; Khondaker, S.I.; Seal, S. Graphene based materials: Past, present and future. *Prog. Mater. Sci.* **2011**, *56*, 1178–1271. [CrossRef]
127. Ling, X.; Xie, L.; Fang, Y.; Xu, H.; Zhang, H.; Kong, J.; Dresselhaus, M.S.; Zhang, J.; Liu, Z. Can Graphene be used as a Substrate for Raman Enhancement? *Nano Lett.* **2010**, *10*, 553–561. [CrossRef] [PubMed]
128. Yan, T.; Zhang, L.; Jiang, T.; Bai, Z.; Yu, X.; Dai, P.; Wu, M. Controllable SERS performance for the flexible paper-like films of reduced graphene oxide. *Appl. Surf. Sci.* **2017**, *419*, 373–381. [CrossRef]
129. Zhao, Y.; Li, X.; Wang, M.; Zhang, L.; Chu, B.; Yang, C.; Liu, Y.; Zhou, D.; Lu, Y. Constructing sub-10-nm gaps in graphene-metal hybrid system for advanced surface-enhanced Raman scattering detection. *J. Alloys Compd.* **2017**, *720*, 139–146. [CrossRef]
130. Tan, C.; Huang, X.; Zhang, H. Synthesis and applications of graphene-based noble metal nanostructures. *Mater. Today* **2013**, *16*, 29–36. [CrossRef]
131. Kuo, C.C.; Chen, C.H. Graphene thickness-controlled photocatalysis and surface enhanced Raman scattering. *Nanoscale* **2014**, *6*, 12805–12813. [CrossRef] [PubMed]

nanomaterials

MDPI

Review

Engineering the Surface/Interface Structures of Titanium Dioxide Micro and Nano Architectures towards Environmental and Electrochemical Applications

Xiaoliang Wang [1], Yanyan Zhao [2], Kristian Mølhave [3,*] and Hongyu Sun [3,*]

1 College of Science, Hebei University of Science and Technology, Shijiazhuang 050018, China; wxlsr@126.com
2 Department of Chemistry Boston College Merkert Chemistry Center, 2609 Beacon St., Chestnut Hill, MA 02467, USA; zhaogh@bc.edu
3 Department of Micro- and Nanotechnology, Technical University of Denmark, Kongens Lyngby 2800, Denmark
* Correspondence: kristian.molhave@nanotech.dtu.dk (K.M.); hsun@nanotech.dtu.dk (H.S.); Tel.: +45-45-25-68-40 (H.S.)

Received: 30 September 2017; Accepted: 6 November 2017; Published: 9 November 2017

Abstract: Titanium dioxide (TiO_2) materials have been intensively studied in the past years because of many varied applications. This mini review article focuses on TiO_2 micro and nano architectures with the prevalent crystal structures (anatase, rutile, brookite, and TiO_2(B)), and summarizes the major advances in the surface and interface engineering and applications in environmental and electrochemical applications. We analyze the advantages of surface/interface engineered TiO_2 micro and nano structures, and present the principles and growth mechanisms of TiO_2 nanostructures via different strategies, with an emphasis on rational control of the surface and interface structures. We further discuss the applications of TiO_2 micro and nano architectures in photocatalysis, lithium/sodium ion batteries, and Li–S batteries. Throughout the discussion, the relationship between the device performance and the surface/interface structures of TiO_2 micro and nano structures will be highlighted. Then, we discuss the phase transitions of TiO_2 nanostructures and possible strategies of improving the phase stability. The review concludes with a perspective on the current challenges and future research directions.

Keywords: titanium dioxide; crystal structure; surface/interface structure; photocatalysis; lithium/sodium ion batteries; Li–S batteries; phase stability

---

1. Introduction

Environment and energy are important factors, which affect the sustainable development of the society. Clean energy techniques and environmental treatment solutions based on advanced nanomaterials, which are earth abundant and environmentally compatible show the potential to solve the crisis. Titanium dioxide (TiO_2) is such a material that satisfies the criteria [1,2]. As an important and widely used wide bandgap (3.0–3.2 eV) oxide semiconductor, TiO_2 shows unique physical and chemical properties [3]. The applications of TiO_2 materials range from conventional fields (cosmetic, paint, pigment, etc.) to functional devices, such as photo- or electrocatalysis, photoelectrochemical or photovoltaic cells, lithium/sodium ion batteries, Li–S batteries, and biotechnological applications [4–13].

There are at least 11 reported bulk or nanocrystalline phases of TiO_2. In nature, TiO_2 forms four main phases: rutile, anatase, brookite, and TiO_2(B). The crystal models of the four structures are illustrated in Figure 1. All of these TiO_2 phases can be seen as constructed by Ti–O octahedral units. The main structural difference is the connecting ways of the basic Ti–O octahedral repetitive units. For instance, octahedra shares two, three, and four edges in rutile, brookite, and anatase phase, respectively. In TiO_2(B) phase, the Ti–O octahedral connection is similar to the anatase one, but with a different arrangement that shows layer character [14]. Under the condition of normal temperature and atmospheric pressure, the relative stability of bulk phase is rutile > brookite > anatase > TiO_2(B) [15]. However, this stability order can be changed by ambient condition and sample properties (particle size, morphology, surface state, etc.). The four TiO_2 phases can be distinguished by using diffraction, Raman spectroscopy, or electrochemical techniques. Due to the structural difference, these TiO_2 phases each have their specific applications. Therefore, it is important to study the phase transformation among different phases and develop methods to improve the phase stability [16–18].

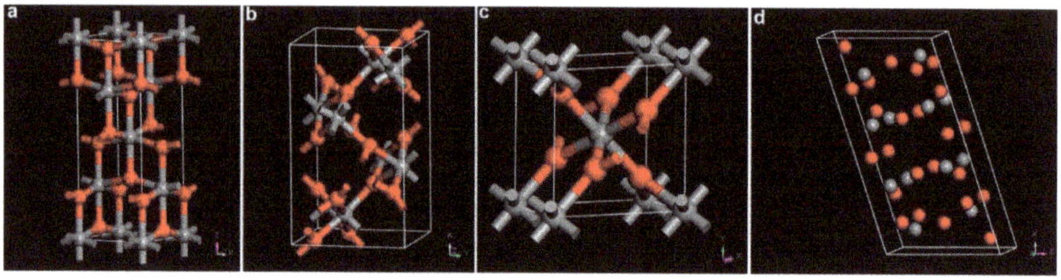

Figure 1. Crystal structures of typical TiO_2 polymorphs: (**a**) rutile; (**b**) brookite; (**c**) anatase; and (**d**) TiO_2(B). Gray and red spheres are Ti^{4+} and O^{2-} ions, respectively.

For a given TiO_2 phase, size and morphology play important roles in the energy conversion and storage. In this regard, TiO_2 nanostructures with well controlled geometric dimension and morphology, such as nanoflowers [19–21], inverse opal- [22–27], urchin- [28–30], and dandelion-like [31–33] structures, have been successfully explored. Besides those geometric parameters, the surface and interface structures are also responsible for the applications mentioned above [12,13,34]. Photo- or electrocatalysis requires the effective adsorption and desorption of reactant molecules/ions and intermediate products on the surface of TiO_2 photocatalysts [35–38]. The ions transportation is occurred across the surface or interface of TiO_2 electrodes during the continuous charging and discharging processes in lithium/sodium ion batteries [39]. Adjusting the interaction between sulfur cathodes and the surface of TiO_2 host is important to improve the cycle stability of Li–S batteries with a higher capacity than those of lithium ion batteries [40]. Therefore, engineering the surface/interface structures of TiO_2 crystals is not only fundamentally important for studying the essential interaction between molecules or ions and TiO_2, but is also valuable to the technical applications [41,42].

In this paper, we summarize the most recent progress in engineering the surface/interface structures of TiO_2 micro and nano structures for the applications in environment and electrochemistry. The article is organized as follows: Section 2 analyzes the benefits of surface/interface engineered TiO_2 micro and nano structures; Section 3 reviews the main strategies used for surface/interface engineering in TiO_2 materials; Section 4 evaluates the advantages and different application of surface/interface engineering in the context of photocatalytic degradation of organic contaminants, water-splitting, CO_2 reduction, antimicrobial and self-cleaning, electrodes for lithium/sodium ion batteries, and Li–S batteries; Section 5 discusses the phase stability of typical TiO_2 structures, and the possible routes to improve the stability; and, finally, we will provide our perspective on the current challenges and important research directions in the future.

2. Advantages of Surface/Interface Engineered TiO_2 Micro and Nano Structures

When compared to the TiO_2 materials in bulk form or other nanostructures, the surface/interface engineered TiO_2 micro and nano structures are promising to transcend the difficulties in photocatalysis and energy storage applications. The benefits of TiO_2 materials with well controlled surface and interface structures are briefly summarized as follows.

(1) Large specific surface area. The surface area of TiO_2 materials plays an important role in their photocatalytic activity and ion storage ability. Firstly, large surface area can increase the contact area with electrolyte, and thus the amount of active reaction sites for photocatalytic applications. Secondly, the high surface area of TiO_2 electrodes is also favorable for the storing more ions.

(2) Tunable band structure and bandgap. The electronic structure of TiO_2 materials can be tuned by engineering surface and interface configurations. Due to the intrinsic limitations of the wide bandgap in bulk form, the practical use of pristine TiO_2 materials in the fields of photocatalysis is hampered. Only ultraviolet (UV) light (<5% of the full solar spectrum) can activate the TiO_2 photocatalysts. By employing surface modification via defect generation, doping, or interface formation, the band structure and the bandgap value of various TiO_2 materials can be adjusted, making it possible to achieve efficient and durable visible light photocatalysis [5–13].

(3) Improved electronic and ionic conductivity. The modulated band structure and bandgap in TiO_2 materials generate additional state within the forbidden band, which facilitates the fast transport of ionic and electronic species, and are important for the rapid migration, transport, and recombination of carriers for catalysis, and high rate battery applications.

(4) Optimized interaction between reactant molecules/ions, intermediate products, and the surface of TiO_2 materials. The binding of species on the engineered TiO_2 surface can be adjusted. It is important to improve the catalytic activity and selectivity, and promote electrochemical performance for novel energy storage device, such as Li–S batteries.

3. Strategies in Surface/Interface Engineering of TiO_2 Micro and Nano Structures

The above discussion shows that surface and interface structures in TiO_2 materials are related to the electronic/optical properties and thus diverse applications ranging from energy to environment. So far, different methods have been proposed to control the surface and interface configurations for TiO_2 micro and nano structures [43–45]. Among the methods, a primary classification can be made by distinguishing physical and chemical methods, which are based on top-down and bottom-up approaches, respectively. There are several excellent reviews describing the specific synthesis methods (such as self-assembly, template, hydrothermal, solvothermal, annealing, electrochemical method, etc.) to control the surface/interface structures [5,34,46]. In this paper, we avoid describing the different synthesis methods, but discuss fundamental strategies, including one-step (sometimes called in-situ) methods, post treatment, and theoretical guidance, those are used to engineer the surface/interface structures.

3.1. One-Step Approach

In order to modify the surface/interface structures via the one-step approach, understanding the nucleation and further growth is essential. Up to now, solution-based and vapor-based approaches have been developed to control the nucleation and growth, and different mechanisms including vapor−liquid−solid, orientation attachment, Ostwald ripening, surfactant-controlled, and growth by surface reaction limitation have been proposed, which have been reviewed elsewhere [5,34,46].

Richter et al. [47] fabricated aligned TiO_2 nanotube arrays by the oxidation of a titanium foil in hydrofluoric acid solution (0.5–3.5 wt %). Electron microscopy images showed that the tubes were open on the tops and were closed on the bottoms. The average tube diameter grew with the increasing of voltage, while the length was independent on reaction time. Field-enhanced void structure was responsible for the tube formation. By suitable choice of the pH value, electrolytes and the Ti sources, the geometry and composition of the nanotube arrays can be controlled more precisely (Figure 2).

Penn et al. [48] proposed that some TiO_2 nanostructures could be formed in solution through the route of oriented attachment, where the merger of nanocrystals is based on orientations of each nanoscale crystal to form single crystalline structure. Experiment and simulations showed that the driving force of an oriented attachment was the reduction of the total surface energy contributed by the removal of certain crystal facets with a high surface energy. The kinetic behaviors of the oriented attachment growth was directly related to the solution properties and reaction temperature. Therefore, it is possible to control the surface/interface properties of the final TiO_2 nanostructures by modifying the crystal facets of the pristine nanocrystals, as well as solution viscosity and others.

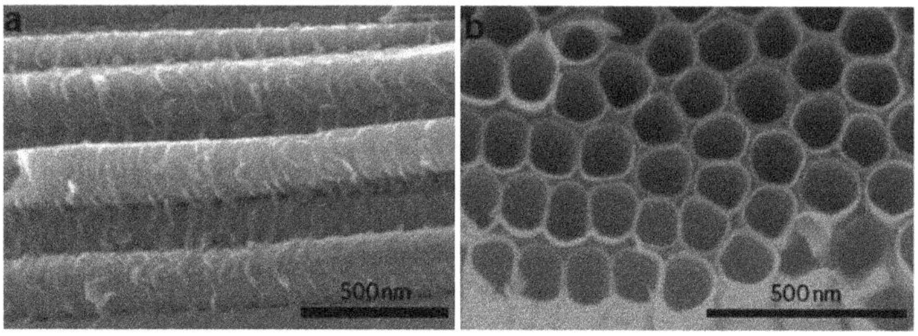

Figure 2. Engineering the surface/interface structures in TiO_2 materials via one step approach. (**a**) Cross section and (**b**) front view scanning electron microscopy (SEM) images of amorphous TiO_2 nanotube arrays fabricated by anodic oxidation. Reproduced with permission from [47], Copyright Nature Publishing Group, 2010.

3.2. *Post Treatment Routes*

Based on the well-established top-down and bottom-up strategies, the synthesis of TiO_2 micro and nano structures with controllable parameters, such as size, morphology, composition, as well as assembly, can be achieved. Those TiO_2 materials with well-defined geometry and chemistry provide abundant possibilities to further tune the atomic scale structures. Therefore, different post-treatment techniques, including thermal annealing, laser irradiation, electrochemical cycling, and solution reaction, have been developed to yield TiO_2 materials with modified surface and interface structures [49–55].

By employing high pressure (~20 bar) hydrogen annealing treatment, Chen et al. [49] successfully converted the pristine white TiO_2 nanoparticles into black hydrogenated particles (Figure 3a–e). The color change indicated that the optical absorption properties had been modified through the treatment. Further structural characterizations showed that the obtained black TiO_2 nanoparticles possessed crystalline core/amorphous shell structure. The surface layer with disordered feature was due to hydrogen dopant, leading to the formation of hydrogen related bonds (such as Ti–H, O–H). Such hydrogen dopant induced surface modification also generates midgap stated, and thus makes the color of the sample as black. Similar to the case of hydrogen treatment, annealing in oxygen deficient atmosphere also results in the effective modification of the surface/interface structures. Huang et al. [50] reported a facile solution reaction, followed by nickel ions assisted ethylene thermolysis to synthesize rutile TiO_2 nanoparticles. The surface of each nanoparticle was etched to form pits with an average size of 2–5 nm (quantum pits). Based on the characterizations, they proposed a possible formation mechanism for the quantum pits. Thanks to the ethylene thermolysis during annealing, a carbon layer was formed on the surface of TiO_2 nanoparticles. The carbon layer then reacted with trace Cl_2 in the chamber, inducing the etching of TiO_2 locally based on the reaction: $TiO_2 + 2C + 2Cl_2 \leftrightarrow TiCl_4 + 2CO$. The microstructure of the rutile TiO_2 nanoparticles is very unique. The abundant quantum-sized pits on the surface generate defect structures and unsaturated bonds, which are important for improving the conductivity and ion storage. Laser irradiation in liquids is also an useful method to modify the surface and interface of different TiO_2 nanostructures [56,57]. During the experiment, laser wavelength, laser energy, irradiation time, and the solution that is employed can be chosen to control the surface structure [58], bandgap, and even phase transformation [59]. In a recent work shown by Filice et al. [58], under-coordinated Ti ions and distorted lattice were formed on the surface of TiO_2 nanoparticles upon laser irradiation, which were important in the modification of the physical and chemical properties. Recently, electrochemical cycling in different mediums (aqueous, organic solution, and ionic liquids) have been used to modify the surface composition, as well as microstructure of TiO_2 materials. The results show that the surface defect structures, especially oxygen vacancies, and their amount can be controlled by adjusting the electrochemical conditions.

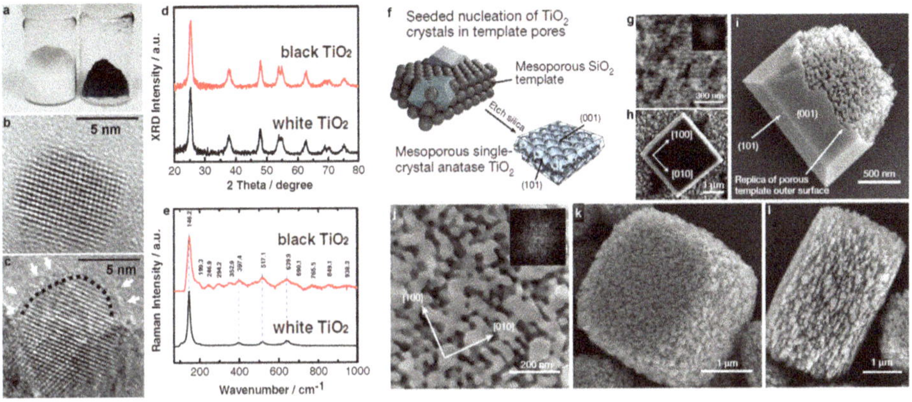

Figure 3. Post treatment route to tune the surface/interface structures in TiO_2 materials. (**a**) A photo comparing unmodified white and disorder-engineered black TiO_2 nanocrystals; (**b**,**c**) High-resolution transmission electron microscopy (HRTEM) images of TiO_2 nanocrystals before and after hydrogenation, respectively. In (**c**), a short dashed curve is applied to outline a portion of the interface between the crystalline core and the disordered outer layer (marked by white arrows) of black TiO_2; (**d**,**e**) X-ray Diffraction (XRD) and Raman spectra of the white and black TiO_2 nanocrystals (reprinted from [49] with permission, Copyright American Association for the Advancement of Science, 2011). (**f**) Schematic and (**g**–**l**) electron microscopy images of mesoporous single-crystal nucleation and growth within a mesoporous template. (**g**) Pristine silica template made up of quasi-close-packed silica beads; (**h**) non-porous truncated bipyramidal TiO_2 crystal; (**i**) template-nucleated variant of the crystal type shown in (**h**); (**j**) replication of the mesoscale pore structure within the templated region; (**k**,**l**) fully mesoporous TiO_2 crystals grown by seeded nucleation in the bulk of the silica template. (Reproduced with permission from [44], Copyright Nature Publishing Group, 2013).

Template assistance is also effective to control the surface/interface of TiO_2 micro and nano configurations. Crossland et al. [44] developed a mesoporous single-crystal anatase TiO_2 based on seed-mediated nucleation and growth inside of a mesoporous template (Figure 3f). In a typical process, silica template was firstly seeded by pre-treatment in a solution of $TiCl_4$ at 70 °C for 60 min. The anatase TiO_2 mesoporous single-crystal was obtained via hydrothermal reaction of TiF_4, with the addition of hydrofluoric acid and pre-treated silica template. The template was then removed by adding aqueous NaOH solution to recover the mesoporous TiO_2 crystals. The final product reveals facet truncated bipyramidal crystals with external symmetry matching that of the homogeneously nucleated bulk crystals, whose mesoscale structure is a negative replica of the silica template. Compared to the conventional TiO_2 nanocrystalline, the TiO_2 mesoporous single-crystal shows a higher conductivity and electron mobility.

3.3. Theoretical Guidance

With the rapid development of modern calculation and simulation, computational material methods based on diverse scale, such as finite element, large scale molecular dynamics (MD) simulation, and density functional theory (DFT) are becoming more and more powerful to provide fundamental insights into experimental results, and more importantly, design and predict the performance of novel functional materials. With the assistance of theoretical methods, it is possible to understand the nucleation, growth, surface properties in liquid and gas environment, which is important to realize controllable synthesis and optimize physical/chemical properties of the nanomaterials [60–62].

The equilibrium morphology of a crystal is given by the standard Wulff construction, which depends on the surface/interface properties. Barnard and Curtiss investigated the effects of surface chemistry on the morphology of TiO_2 nanoparticles by using a thermodynamic model based on surface free energies and surface tensions obtained from DFT calculations. In the condition of hydrated, hydrogen-rich, and hydrogenated surfaces, the shape of anatase and rutile nanoparticles vary little, however, in the case of hydrogen-poor and oxygenated surfaces, the anatase and rutile nanocrystals become elongated. The results show that the exposed facets of the TiO_2 nanocrystals can be controlled through modifying the surface acid-base chemistry.

Besides the acid-base condition, heterogeneous atoms or surfactant adsorption can also affect the surface and interface structures. Based on DFT calculations, Yang et al. [43] systematically studied the adsorption of a wide range of heterogeneous non-metallic atoms X (X = H, B, C, N, O, F, Si, P, S, Cl, Br, or I) on {001} and {101} facets of anatase TiO_2 crystals (Figure 4). The results show that the adsorption of F atoms not only decreases the surface energy for both the (001) and (101) surfaces, but also results in the fact that (001) surfaces are more stable than (101) surfaces, i.e., the F adsorption is favorable for the formation of (001) facets in anatase TiO_2. The theoretical results inspire intense studies on the surface structure control of TiO_2 crystals. Experimentally, a mixture containing titanium tetrafluoride (TiF_4) aqueous solution and hydrofluoric acid was hydrothermally reacted, to generate the truncated anatase bipyramids, and anatase TiO_2 single crystals with a high percentage of {001} facets were obtained.

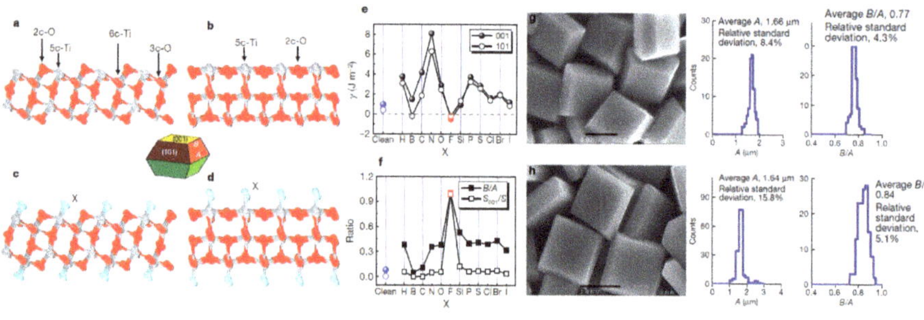

Figure 4. Theoretical calculation guides the modification of surface/interface structures. (**a**–**f**) Slab models and calculated surface energies of anatase TiO_2 (001) and (101) surfaces. (**a**,**b**) Unrelaxed, clean (001) and (101) surfaces; (**c**,**d**) Unrelaxed (001) and (101) surfaces surrounded by adsorbate X atoms; (**e**) Calculated energies of the (001) and (101) surfaces surrounded by X atoms; and, (**f**) Plots of the optimized value of B/A and percentage of {001} facets for anatase single crystals with various adsorbate atoms X. Here, the parameters of A and B are the lengths of the side of the bipyramid and the side of the square {001} "truncation" facets (see the geometric model). The value of B/A describes the area ratio of reactive {001} facets to the total surface. (**g**,**h**) SEM images and statistical data for the size and truncation degree of anatase single crystals (Reproduced with permission from [43], Copyright Nature Publishing Group, 2008).

4. Applications of Surface/Interface Engineered TiO_2 Micro and Nano Structures

Surface and interface structures of TiO_2 materials play important roles in multiple physical/chemical processes. Herein, we will highlight the recent progress in the research activities on the surface/interface engineered TiO_2 micro and nano structures that are used for photocatalysis (including photocatalytic degradation of organic contaminants, photocatalytic hydrogen evolution, photocatalytic CO_2 reduction, antimicrobial, and self-cleaning), lithium/sodium ion batteries, and Li–S batteries.

4.1. Photocatalysis

There are four main steps involved in heterogeneous photocatalysis process (Figure 5a): (1) light absorption; (2) the generation and separation of photoexcited electrons and holes; (3) the migration, transport, and recombination of carriers; and, (4) surface catalytic reduction and oxidation reactions. The overall photocatalysis efficiency is strongly dependent on the cumulative effects of these four consecutive steps. Among different photocatalyst materials, TiO_2 is considered to be a remarkable photocatalyst due to the notable merits such as nontoxicity, biological compatibility, and universality. Since the photocatalytic reaction is a surface or interface sensitive process, control of the surface/interface structures in TiO_2 materials provides a possible way to improve the light absorption and visible light usage, and facilitate the carrier separation, resulting in enhanced photocatalytic properties. Many attempts have been carried out to modify the surface or interface structures of TiO_2 materials, such as exposed crystallographic plane tuning, defect engineering, interface construction, and so on (Figure 5b–d). In the following, we will discuss the effects of those surface/interface modifications on the photocatalytic degradation of organic contaminants, photocatalytic hydrogen evolution, and photocatalytic CO_2 reduction. Other environmental applications such as antimicrobial and self-cleaning are also briefly discussed.

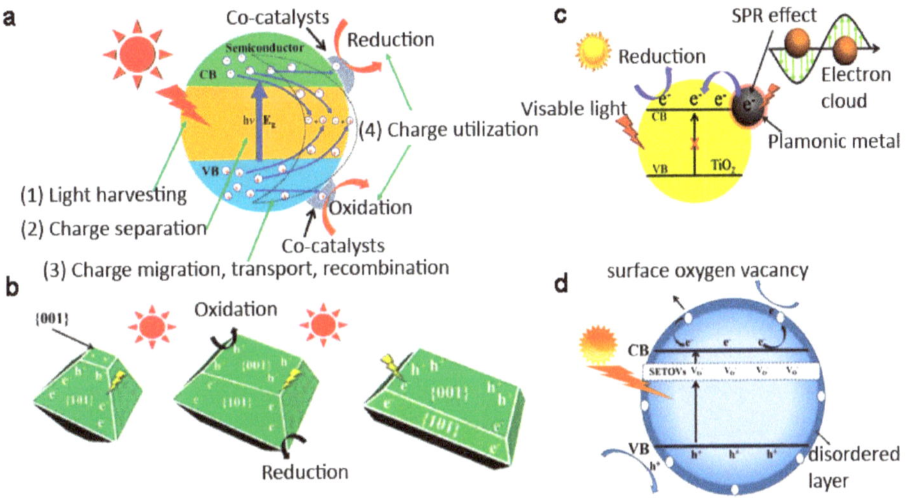

Figure 5. (**a**) Different stages in heterogeneous photocatalysis (Reproduced with permission from [63], Copyright The Royal Society of Chemistry, 2016); surface/interface engineered TiO_2 structures for photocatalytic improvement: (**b**) crystallographic plane tuning (Reproduced with permission from [64], Copyright American Chemical Society, 2014), (**c**) defects engineering (Reproduced with permission from [65], Copyright Elsevier B.V., 2016), and (**d**) creating interfaces in TiO_2 nanostructures (Reproduced with permission from [66], Copyright Elsevier B.V., 2017).

4.1.1. Photocatalytic Degradation of Organic Contaminants

With a rapidly growing world population and expanding industrialization, the development of new materials, techniques, and devices those can provide safe water and air is important to the societal sustainability. Semiconductor photocatalysis has been utilized as an ideal way to degrade various organic contaminants in water and air.

Edy et al. [67] synthesized free-standing TiO_2 nanosheets with different thickness via atomic layer deposition on a dissolvable sacrificial polymer layer. The photocatalytic performance was evaluated for photocatalytic degradation of methyl orange under UV light irradiation. The photocatalytic activity increases with increasing the thickness, which may be due to the existence of Ti^{3+} defect and locally ordered domain structures in the amorphous nanosheets. TiO_2 nanostructures with exposed highly reactive facets, for example, anatase TiO_2 nanosheets with {001} facets, are desirable for the photocatalytic enhancement. Those thin nanosheets are prone to aggregate during the practical usage, which results in the loss of photocatalytic activity. Assembly the individual nanostructure into hierarchical architecture can not only suppress the aggregation of micro/nanoscale building blocks, but also increase specific surface area and the amount of active reaction sites, and reduce the diffusion barrier. We synthesized anatase TiO_2 hollow microspheres assembled with high-energy {001} facets via a facile one-pot hydrothermal method [68]. The percentage of exposed {001} facets on the microspheres was estimated to be about 60%. The photocatalytic ability was evaluated by photodegradation of methylene blue under UV light. The photocatalytic degradation reaction follows pseudo-first-order kinetics among the studied samples. The apparent photochemical degradation rate constant for the hierarchical TiO_2 structures is 4.07×10^{-2} min^{-1}, which is faster than that of control samples (TiO_2 powders (Degussa, P25), 3.11×10^{-2} min^{-1}; porous TiO_2 powders, 2.76×10^{-2} min^{-1}; the etched TiO_2 spheres, 2.17×10^{-2} min^{-1}; the irregular TiO_2 product, 0.86×10^{-2} min^{-1}). The good photocatalytic activity of the hierarchical TiO_2 structures is associated with the hollow structures with bimodal mesopore size distribution and relatively large Brunauer–Emmett–Teller (BET) surface areas. Xiang et al. [69] synthesized a kind of hierarchical flower-like TiO_2 superstructures by alcohothermal treatment method. The superstructures consisted of anatase TiO_2 nanosheets with 87% exposed (001) facets. Photocatalytic oxidative decomposition of acetone was evaluated in air under UV light. The results show that the photocatalytic activity of the flower-like TiO_2 superstructures was better than that of P25 and tubular shaped TiO_2 particles. The synergetic effect of highly exposed (001) facets hierarchically porous structure, and the increased light-harvesting capability is responsible for the enhanced photocatalytic ability.

Besides the exposed high energy facets, the introduction of suitable defect structures in TiO_2 materials can obviously influence the light absorption and the separation of photogenerated electron-hole pairs [70–72]. Cao et al. [70] fabricated mesoporous black TiO_2 spheres with high crystallinity by a facile evaporation-induced self-assembly method combined with mild calcinations after an in-situ hydrogenation under an argon atmosphere. The results indicated that the prepared sample was uniform mesoporous black spheres with Ti^{3+} and N co-doping. The visible-light-driven photocatalytic degradation ratio of methyl orange was up to 96%, which was several times higher than that of pristine TiO_2 material. The excellent photocatalytic activity was due to Ti^{3+} and N doping, which resulted in high visible light utilization and enhanced separation of photogenerated charge carriers, and the mesoporous network structures.

Generating interface structures by depositing plasmonic-metal nanostructures (Ag, Pt, Au, etc.) on TiO_2 materials can increase the generation rate of energetic charge carriers and result in a higher probability of redox reactions [73–76]. By using successive ion layer adsorption and reaction, Shuang et al. [73] synthesized TiO_2 nanopillar arrays with both Au and Pt nanoparticles (~4 nm) decoration. Due to the electron-sink function of Pt and surface plasmon resonance of Au nanoparticles, the charge separation of photoexcited TiO_2 was improved. The obtained Au/Pt nanoparticles decorated TiO_2 nanopillar arrays showed a much higher visible and UV light absorption response, which lead to remarkably enhanced photocatalytic activities in the degradation of methyl orange.

4.1.2. Photocatalytic Hydrogen Evolution

Hydrogen energy is one of the most promising green fuels. Since the first discovery of photoelectrochemical water splitting by Fujishima and Honda in 1972, hydrogen production directly from water and sunlight on semiconductor materials has been intensively investigated [3]. Although numerous semiconductor materials have been explored as photocatalysts to produce hydrogen, TiO_2 remains one of the most studied materials for photocatalytic H_2 evolution due to the main merits of nontoxic and chemical stability.

Highly reactive exposed facets of TiO_2 nanostructures are related to the photocatalytic activity enhancement. For example, Wu et al. [77] synthesized mesoporous rutile TiO_2 single crystal with wholly exposed {111} facets by a seeded-template method. Fluoride ions in the solution played an important role in stabilizing the high energy facet {111} of rutile TiO_2. The ratios of exposed {110} and {111} facets can be controlled by tuning the concentration of fluoride ions. The mesoporous single crystal rutile TiO_2 with wholly exposed {111} reactive facets exhibited a greatly enhanced photocatalytic hydrogen generation. Zhang et al. [78] demonstrated that the TiO_2 single crystal with a novel four-truncated-bipyramid morphology could be synthesized by a facile hydrothermal reaction. The resultant photocatalyst exhibited excellent hydrogen evolution activity from ethanol-water solution. The exposure of both high-energy {001} oxidative and low-energy {101} reductive facets in an optimal ratio are thought to be the key factors for the high photocatalytic activity. In another example, anatase TiO_2 nanoplates with exposed (001) facet were converted from the NH_4TiOF_3 nanoplates [79]. The obtained compact TiO_2 nanoplates exhibited a high H_2-production rate of 13 $mmol \cdot h^{-1} \cdot g^{-1}$ with a H_2-production quantum efficiency of 0.93% at 365 nm.

The influence of defect structures in TiO_2 materials on photocatalytic H_2-evolution is complicated. For one thing, the defects could introduce additional states in the band gap, which cause the recombination of carriers and the weakening of carriers' oxidation and reduction capacities [80–83]. For another, subtly generating specific defects will facilitate the separating of the carriers. Recently, Wu et al. [80] prepared yellow TiO_2 nanoparticles with ultra-small size of ~3 nm. Simulated solar light driven catalytic experiments showed that the evolved H_2 for the yellow TiO_2 was ~48.4 $\mu mol \cdot h^{-1} \cdot g^{-1}$, which was ~3.7 fold when comparing to that of the normal TiO_2 (~13.1 $\mu mol \cdot h^{-1} \cdot g^{-1}$) at the same experimental conditions. It is suggested that the significantly improved H_2-evolution activity can be attributed to the coexistence of titanium vacancies (acceptor) and titanium interstitials (donor) in the TiO_2 materials, which is beneficial for the spontaneous separation of photo-generated charge-carriers. When compared to the complex steps that are required to accurately control of the defects, the passivation of the defect states with elemental doping would be more direct. Recent works show that Mg doping could eliminate the intrinsic deep defect states and weaken the shallow defect states in TiO_2 materials [83]. The result was confirmed by the transient infrared absorption-excitation energy scanning spectroscopic measurement. The photocatalytic over-all water splitting measurements showed the H_2 and O_2 evolution rates can be as high as 850 and 425 $\mu mol \cdot h^{-1} \cdot g^{-1}$ under Air Mass (AM) 1.5 G irradiation and the apparent quantum efficiency of 19.4% was achieved under 350 nm light irradiation.

Rational creating hetero- or homo-interfaces can achieve high-performance photocatalytic hydrogen evolution. When compared to the pure crystalline and amorphous TiO_2 film, high electron concentration and mobility can be concurrently obtained at the homo-interface between crystalline and amorphous layers in a bilayer TiO_2 thin film. Therefore, extraordinary properties could be explored in well-designed interfaces with homogeneous chemical composition. By creating a crystalline Ti^{3+} core/amorphous Ti^{4+} shell structure, Yang et al. [84] successfully activated rutile TiO_2 material with efficient photocatalytic hydrogen evolution properties. The average hydrogen evolution rate was enhanced from 1.7 for pure TiO_2 to 268.3 $\mu mol \cdot h^{-1}$ for TiO_2 with homointerface structures. The origin of the activation was attributed to the regulated the transport behaviors of holes and electrons from the bulk of a particle to the surface by suppressing the transport of electrons in the conduction band and facilitating the transport of holes in the valence band. In addition, hetero-interfaces between TiO_2 materials and other semiconductor or metal nanostructures, including carbon, Si, NiO, ZnS, CdS, MoS_2, MoC_2, layered double hydroxides, and plasmonic metals, has been extensively investigated [85–91]. As an example, Wu et al. [85] reported that anisotropic TiO_2 overgrowth on Au nanorods could be obtained by selective spatial assembly and subsequent hydrolysis. Plasmon-enhanced H_2 evolution under visible/near-infrared light irradiation has been demonstrated. The Au nanorod-TiO_2 interface with the Au nanorod side exposed, as a Schottky junction, can filter out surface plasmon resonance hot electrons from the Au nanorod, which is crucial to boosting the H_2 evolution performance.

4.1.3. Photocatalytic CO_2 Reduction

Due to the increasing consumption of conventional fossil fuels, the concentration of greenhouse gas, especially CO_2, steadily grows over years. Solar-light-driven reduction of CO_2 to useful chemical fuels (such as CH_4, HCO_2H, CH_2O, and CH_3OH) is a promising solution for the serious environmental and energy problems. In the process of photocatalytic CO_2 reduction, typical steps including adsorption of CO_2, generation of electron-hole pair, separation and migration of electron-hole pair, and the reduction of CO_2 are involved. Since CO_2 molecules are highly stable, only the electrons with sufficient reduction potential can be utilized to trigger CO_2 reduction reactions, and suitable photocatalyst is required to decrease the high reaction barrier. Among a wide range of metal and semiconductor photocatalysts for CO_2 reduction, TiO_2 materials has attracted much attention due to the advantageous of high reduction potential, low cost, and high stability. The activity, selectivity, and durability of TiO_2 photocatalysts for CO_2 reduction is related to the efficiency of electron-hole separation and light utilization ability, which are very sensitive to the surface structure, atomic configuration, and chemical composition of the photocatalysts. For example, different kinds of metals (transition, rare, alkali earth metals) have been studied as doping to improve the photocatalytic activity for CO_2 reduction [92–94]. When compared to the metal doping method, which usually suffers from photocorrosion problem, non-metal (carbon, nitrogen, iodine, sulfur, etc.) doping has attracted more attention [95]. However, a large amount of non-intrinsic defects often generated during the doping and created electron-hole recombination centers at the same time. Herein, we mainly focus on surface/interface modification to enhance the performance of TiO_2 photocatalysts towards CO_2 reduction.

Yu et al. [64] investigated the effect of different exposed facets of anatase TiO_2 crystals on the photocatalytic CO_2 reduction activity. By using a simple fluorine-assisted hydrothermal method, they synthesized anatase TiO_2 with different ratios of the exposed {101} and {001} facets. The results showed that the photocatalytic activity of the anatase TiO_2 with the optimized ratio of exposed {001} to {101} facet (55:45) was ~4 times higher than that of P25 powder. They ascribed the enhancement to a concept of "surface heterojunction". Electron and hole are driven to the {101} and {001} facets, inducing the seperation of electron and hole. It is worth mentioning that surface atomic and defect structures on different facets should also contribute the photocatalytic CO_2 reduction processes. Truong et al. [96] synthesized rutile TiO_2 nanocrystals with exposed high-index facets through solvothermal reaction by using a water-soluble titanium-glycolate complex as a precursor. Structural characterizations showed that each branched nanocrystal was bound by four facets of high-index {331} facets, and rutile {101} twinned structures were formed in the boundary of branches. The photocatalytic CO_2 reduction to methanol showed a significantly higher activity was achieved in the synthesized nanostructures due to the abundant surface defects on the high energy facets.

Generating oxygen vacancies is effective to modulate the electronic/optical properties, and thus optimize diverse applications of metal oxides. Generally, bulk oxygen vacancies formed a middle sub-band in the forbidden gap, which made TiO_2 response to the visible light, and those bulk oxygen vacancies also acted as the electron-hole recombination centers. The surface oxygen vacancies not only showed a strong response to the visible light, but also acted as the capture traps to inhibit electrons-holes recombination. By adjusting the concentration ratio of the surface and bulk oxygen vacancies, it is possible to improve the photocatalytic efficiency of TiO_2 nanostructures. Li et al. [66] compared the effects of oxygen vacancies in TiO_2 nanocrystals on the photoreduction of CO_2. By choosing the precursors and post-treatment conditions, they obtained three kinds of TiO_2 materials with different oxygen vacancies, i.e., TiO_2 with surface oxygen vacancies (TiO_2-SO), TiO_2 with bulk single-electron-trapped oxygen vacancies (TiO_2-BO), and TiO_2 with mixed vacancies (TiO_2-SBO).

By analyzing the lifetime and intensity by positron annihilation, the efficiency of photocatalytic CO_2 reduction improved with the increase of the ratio of surface oxygen vacancies to bulk ones. The results revealed the critical role of surface/bulk defects in photocatalytic properties.

Similar to the case of photocatalytic hydrogen evolution, creating metal- or semiconductor- TiO_2 interface via different post-deposition or in-situ forming methods has been demonstrated to be effective to improve the light harvesting and the separation of charged carriers, which are also important for the photoreduction of CO_2. Specifically, Schottky barrier can be formed when the Fermi level of the deposited metals are lower than the conduction band of the TiO_2 materials, which is favorable for the spatial separation of electron-hole pairs. Platinum, which possesses a suitable work function, is one of the most commonly used metal co-catalyst to improve the CO_2 reduction performance of TiO_2 photocatalysts. However, worldwide limited source and the consequent high price of platinum seriously hinder the large scale applications. The deposition of plasmonic nanostructures of metals such as silver and gold on TiO_2 materials has been extensively studied due to the surface plasmon resonance (SPR) effect, which shows important role in improving the photocatalytic activity for CO_2 reduction.

4.1.4. Other Environmental Applications

The essence of antimicrobial by using TiO_2 materials is a photocatalysis process. Therefore, the above surface/interface engineering towards photocatalytic enhancement can also be applied in the antimicrobial studies. Xu and co-workers [97] modified the aligned TiO_2 nanotubes via a thin layer of graphitic C_3N_4 material by a chemical vapor deposition method. Due to the synergetic effect, the bactericidal efficiency against Escherichia coli irradiated by visible-light has been improved. Recently, self-cleaning materials have gained much attention in energy and environmental areas. The self-cleaning properties can be achieved by morphology design to form either hydrophilic or hydrophobic surfaces [98]. Previous works show that the hydrophilic or hydrophobic properties can be controlled by the photocatalytic process [99], making it possible to couple photocatalysis and photoinduced wettability to improve self-cleaning properties in a controllable way. TiO_2 is such a material that shows photocalytic self-cleaning activity. Interface formation via heterojunction or heterostructure [100,101], surface modification [102], and elemental doping [103,104] are typical methods to improve photocatalytic and self-cleaning activities of TiO_2 materials.

4.2. Lithium/Sodium Ion Batteries

Rechargeable lithium ion battery is one of the most important energy storage devices for a wide range of electron devices. The properties of electrode materials play an important role in the final performance of lithium ion batteries. Among the many potential electrode candidates, titanium dioxides with different phases have attracted much attention due to the abundance of raw materials and environmental benignity. Although the theoretical specific capacity of titanium dioxides (335 $mA{\cdot}h{\cdot}g^{-1}$, based on the reaction $TiO_2 + xLi^+ + xe^- \leftrightarrow Li_xTiO_2$, x~0.96) is comparable to that of commercial graphite (372 $mA{\cdot}h{\cdot}g^{-1}$), these materials possess a higher operating voltage platform than that of graphite, which is favorable for inhibiting the formation of lithium dendrite and solid-electrolyte interphase (SEI) layer. Moreover, the minor volume variation during cycling ensures a good cycling stability. It should be noted that the unsatisfied electronic conductivity and sluggish ion diffusion hinder the high-rate applications of these materials. The size, shape, composition, and assembly of TiO_2 anodes are studied to optimize the lithium storage properties.

Recent works also show that nanoscale surface/interface design in TiO_2 nanostructures is beneficial for improving the battery performance (Table 1), which are ascribed to the advantages of micro and nano architectures. For example, theoretical and experimental results demonstrated that lithium insertion was favored on the high-energy {001} facets in anatase phase, because of the open structure, as well as short path for ion diffusion. Since the first synthesis of anatase phase with exposed {001} facets by Yang et al. [43], extensive studies have been reported on the synthesis of TiO_2 anodes with exposed {001} facets. Although the obtained anatase nanostructures possess sheetlike morphology exposed with {001} facets, the samples tend to over-lap to reduce the total surface energy. It is therefore important to prevent the aggregation of anatase nanosheets with exposed {001} facets. By using a simple one-pot solution method, we successfully obtained three-dimensional (3D) anatase TiO_2 hollow microspheres, which were constituted by {001} facets (Figure 6a) [105]. In the synthesis, a mixture containing Ti powder, deionized water, hydrogen peroxide, and hydrofluoric acid was subjected to hydrothermal reaction at a temperature of 180 °C. The addition of hydrofluoric acid and hydrogen peroxide is critical for the formation of {001} facet assembly. The as-prepared sample shows good lithium storage properties. After 50 cycles at a current density of 0.1 C (1 C = 335 mA·h·g^{-1}), a reversible capacity of 157 mA·h·g^{-1} can be retained, which is ~75% retention of the first reversible capacity. Rate performance test show that the discharge capacity reaches about 156 mA·h·g^{-1} after the first 10 cycles at the rate of 1 C, and then it slightly reduces to 135 and 130 mA·h·g^{-1} at the rates of 2 and 5 C, respectively. The electrode can still deliver a reversible capacity of 90 mA·h·g^{-1} even at a high rate of 10 C. The electrode resumes its original capacity of about 150 mA·h·g^{-1} after 10 cycles when the rate returns back to 1 C.

Table 1. Performance comparison of some lithium ion batteries and sodium ion batteries based on typical titanium dioxide (TiO_2) anodes (the voltage is versus Li^+/Li or Na^+/Na).

Material/[Reference]	Capacity (Cycles) (mA·h·g^{-1})	Rate Capability (mA·h·g^{-1})	Voltage (V)
Rutile TiO_2 with quantum pits [50]	145 (80)@168 mA·g^{-1}	102@1675 mA·g^{-1}	1–3/Li
TiO_2 microboxes [106]	187 (300)@170 mA·g^{-1}	63@3400 mA·g^{-1}	1–3/Li
Rutile TiO_2 inverse opals [107]	95 (5000)@450 mA·g^{-1}	-	1–3/Li
Faceted TiO_2 crystals [108]	141.2 (100)@170 mA·g^{-1}	29.9@1700 mA·g^{-1}	1–3/Li
Nanosheet-constructed TiO_2(B) [109]	200 (200)@3350 mA·g^{-1}	216@3350 mA·g^{-1}	1–3/Li
TiO_2 hollow microspheres [105]	157 (50)@170 mA·g^{-1}	90@1700 mA·g^{-1}	1–3/Li
rutile TiO_2 nanostructures [110]	190 (200)@102 mA·g^{-1}	84.5@1700 mA·g^{-1}	1–3/Li
nest-like TiO_2 hollow microspheres [111]	152 (100)@1020 mA·g^{-1}	130@3400 mA·g^{-1}	1–3/Li
Co_3O_4 NPs@TiO_2(B) NSs [112]	677.3 (80)@100 mA·g^{-1}	386@1000 mA·g^{-1}	0.01–3.0/Li
TiO_2(B)@VS_2 nanowire arrays [113]	365.4 (500)@335 mA·g^{-1}	171.2@3350 mA·g^{-1}	0.01–3.0/Li
Nb-doped rutile TiO_2 Mesocrystals [114]	141.9 (600)@850 mA·g^{-1}	96.3@6800 mA·g^{-1}	1–3/Li
TiO_2@defect-rich MoS_2 nanosheets [115]	805.3 (100)@100 mA·g^{-1}	507.6@2000 mA·g^{-1}	0.005–3.0/Li
MoS_2-TiO_2 based composites [116]	648 (400)@1000 mA·g^{-1}	511@2000 mA·g^{-1}	0.005–3.0/Li
macroporous TiO_2 [117]	181 (1000)@1700 mA·g^{-1}	69@12.5 A·g^{-1}	1–3/Li
porous TiO_2 hollow microspheres [118]	216 (100)@170 mA·g^{-1}	112@1700 mA·g^{-1}	1–3/Li
porous TiO_2(B) nanosheets [119]	186 (1000)@1675 mA·g^{-1}	159@6700 mA·g^{-1}	1–3/Li
graphene supported TiO_2(B) sheets [120]	325 (10000)@500 mA·g^{-1}	49@40 A·g^{-1}	1–3/Li
mesoporous TiO_2 coating on carbon [121]	210 (1000)@3400 mA·g^{-1}	150@10.2 A·g^{-1}	1–3/Li
Ti^{3+}-free three-phase $Li_4Ti_5O_{12}$/TiO_2 [122]	136 (1000)@4000 mA·g^{-1}	155.6@8 A·g^{-1}	1.0–2.5/Li
Mesoporous TiO_2 [123]	149 (100)@1000 mA·g^{-1}	104@2000 mA·g^{-1}	1–3/Li
Nanocrystalline brookite TiO_2 [124]	170 (40)@35 mA·g^{-1}	-	1–3/Li
Anatase TiO_2 embedded with TiO_2(B) [125]	190 (1000)@1700 mA·g^{-1}	110@8500 mA·g^{-1}	1–3/Li
TiO_2-Sn@carbon nanofibers [126]	413 (400)@100 mA·g^{-1}	-	0.01–2.0/Na
Double-walled Sb@TiO_{2-x} nanotubes [127]	300 (1000)@2.64 A·g^{-1}	312@13.2 A·g^{-1}	0.1–2.5/Na
Carbon-coated anatase TiO_2 [128]	180 (500)@1675 mA·g^{-1}	134@3.35 A·g^{-1}	0.05–2.0/Na
Nanotube arrays of S-doped TiO_2 [129]	136 (4400)@3350 mA·g^{-1}	167@3350 mA·g^{-1}	0.1–2.5/Na
Amorphous TiO_2 inverse opal [130]	203 (100)@100 mA·g^{-1}	113@5 A·g^{-1}	0.01–3.0/Na
Petal-like rutile TiO_2 [131]	144.4 (1100)@837.5 mA·g^{-1}	59.8@4187 mA·g^{-1}	0.01–3.0/Na
Yolk-like TiO_2 [132]	200.7 (550)@335 mA·g^{-1}	90.6@8375 mA·g^{-1}	0.01–3.0/Na
Blue TiO_2(B) nanobelts [133]	210.5 (5000)@3350 mA·g^{-1}	90.6@5025 mA·g^{-1}	0.01–3.0/Na

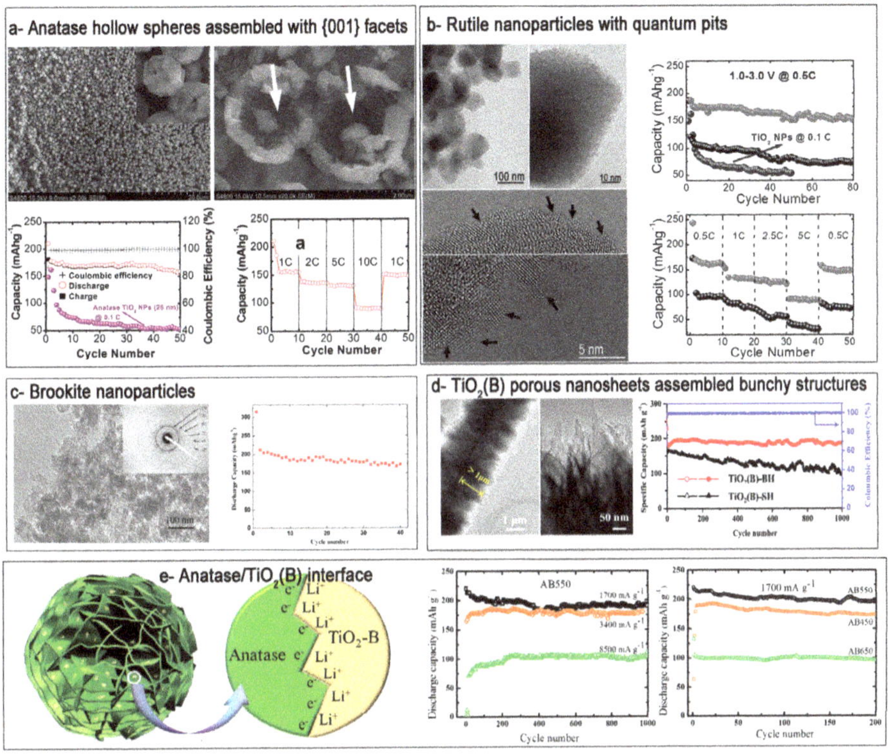

Figure 6. Typical TiO_2 anodes and their lithium storage properties: (**a**) three-dimensional (3D) anatase TiO_2 hollow microspheres assembled with high-energy {001} facets (reprinted from [105] with permission, Copyright The Royal Society of Chemistry, 2012); (**b**) Rutile TiO_2 nanoparticles with quantum pits (reprinted from [50] with permission, Copyright The Royal Society of Chemistry, 2016); (**c**) Brookite TiO_2 nanocrystalline (reprinted from [105] with permission, Copyright The Electrochemical Society, 2007); (**d**) bunchy hierarchical TiO_2(B) structure assembled by porous nanosheets (reprinted from [119] with permission, Copyright Elsevier Ltd., 2017); and (**e**) Ultrathin anatase TiO_2 nanosheets embedded with TiO_2(B) nanodomains (Reproduced with permission from [125], Copyright John Wiley & Sons, 2015).

Rutile TiO_2 is the most stable phase, which can be prepared at elevated temperatures, however, rutile TiO_2 in bulk form is not favorable for the lithium ions intercalation. When the size decreased to nanoscale, rutile TiO_2 phase possesses obvious activity towards the insertion of lithium ions even at room temperature. However, some critical problems should be considered when using rutile TiO_2 nanostructures as anodes in lithium ion batteries, for example, particle aggregation and poor rate capacity. To boost the lithium storage of rutile TiO_2 anodes, the synthesis of micro and nano configurations with optimized surface/interface and improved conductivity is an effective method to overcome the above limitations. We synthesized rutile TiO_2 nanoparticles by a simple solution reaction, followed by annealing treatment (Figure 6b) [50]. The surface of each particle was etched to form quantum-sized pits (average size 2–5 nm), which possessed more unsaturated bond and other defect structures (for example steps, terraces, kinks, and others). The defective rutile TiO_2 nanoparticles provided more active sites for the storage of lithium ions and improved the electron conductivity as well. As a consequence, the sample exhibited a specific capacity of ~145 $mA{\cdot}h{\cdot}g^{-1}$ at a current density of 0.5 C with good rate capability (~102 $mA \cdot h \cdot g^{-1}$ at 5 C) and cycling performance, demonstrating a great potential for lithium ion battery applications.

Among the different TiO_2 polymorphs that were investigated, a severe capacity fading was noted for the brookite phase, although it exhibited nearly one mole of reversible lithium insertion/extraction in its nanostructured form [123]. There has not been extensive research focused on developing such an anode. Reddy et al. [124] demonstrated intercalation of lithium into brookite TiO_2 nanoparticles (Figure 6c). Electrochemical test and ex-situ x-ray diffraction (XRD) studies showed that the structure was stable for lithium intercalation and deintercalation although the intercalation/deintercalation mechanism was not clear. Cycling performance of brookite TiO_2 performed at C/10 rate in the voltage window 1.0–3.0 V showed that there is a gradual loss of capacity in the initial 10 cycles, and the capacity is fairly stable at 170 $mA \cdot h \cdot g^{-1}$ on further cycling. In contrast to other TiO_2 polymorphs, the TiO_2(B) phase possesses relatively more open crystal structure, which allows for the facile insertion/extraction of lithium ions. Moreover, a lower operating potential (~1.55 V vs. Li) when compared to the anatase TiO_2 (~1.75 V vs. Li), an improved reversibility, and a high rate capability make TiO_2(B) phase a promising candidate for lithium storage. Li et al. [119] reported on the orderly integration of porous TiO_2(B) nanosheets into bunchy hierarchical structure (TiO_2(B)-BH) via a facile solvothermal process (Figure 6d). Benefiting from the unique structural merits, TiO_2(B)-BH exhibited a high reversible capacity, long-term cycling stability (186.6 $mA \cdot h \cdot g^{-1}$ at 1675 $mA \cdot g^{-1}$ after 1000 cycles), and a desirable rate performance.

Recently, Jamnik and Maier proposed that it was possible to store additional lithium at the interface of nanosized electrodes, which included solid–liquid (electrode-electrolyte) interface and solid–solid interface between the electrodes (Figure 6e) [125,134]. The interfaces can accommodate additional Li ions, leading to a rise of total Li storage. Meanwhile, an additional synergistic storage is favored if the electrode material is made of a lithium ion-accepting phase and an electron-accepting phase, which is beneficial for charge separation ("Job-sharing" mechanism). Along this line, Wu et al. [125] synthesized a new kind of microsphere that was constructed by ultrathin anatase nanosheets embedded with TiO_2(B) nanodomains, which contained a large amount of interfaces between the two phases. The hierarchical nanostructures show capacities of 180 and 110 $mA \cdot h \cdot g^{-1}$ after 1000 cycles at current densities of 3400 and 8500 $mA \cdot g^{-1}$. The ultrathin nanosheet structure, which provides short lithium diffusion length and high electrode/electrolyte contact area also accounts for the high capacity and long-cycle stability. This study highlights the importance of smart design in the interface structures in the nanoelectrodes.

Although the development and commercialization of lithium ion batteries have gained great success in the past years, one severe drawback of lithium ion batteries is the limited lithium resource in the Earth's crust and its uneven geographical distribution. In this regard, sodium ion batteries have attracted particular attention due to the obvious advantages, including high earth-abundance of sodium, and lower cost vs. lithium ion batteries. In addition, the sodium chemistry is similar as the case of lithium, so the previously established surface/interface engineering strategies for titanium dioxides electrode design in lithium ion batteries system can be transferred to and expedite the sodium ion battery studies. Longoni et al. [39] systematically studied the role of different exposed crystal facets of the anatase nanocrystals on the sodium storage properties. By employing a surfactant-assisted solvothermal route, they synthesized anatase TiO_2 nanostructures with three different morphologies (Rhombic elongated (RE), rhombic (R), and nanobar (NB)), which showed obvious differences in crystal face type exposition. Their electrochemical performance results, together with theoretical analysis, showed that an overcoordinated state of Ti atoms on the crystal surface (low energy density (101) facets of NB and R moieties) strongly inhibits the sodium uptake, while a Goldilocks condition seems to occur for crystalline faces with intermediate energy densities, like (100) in RE. Zhang et al. [131] reported a smart design of the assembly and interface of rutile TiO_2, and fewer layer graphene by using carbon dots as designer additives. The resultant graphene-rich petal-like rutile TiO_2 showed outstanding sodium-storage properties. At a rate of 0.25 C (83.75 $mA \cdot g^{-1}$) after 300 cycles, a high capacity of

245.3 mA·h·g^{-1} was obtained, even at a high current density of 12.5 C (4187.5 mA·g^{-1}), a considerable capacity of 59.8 mA·h·g^{-1} can still be maintained. Notably, the reversible capacity up to 1100 cycles at a current density of 2.5 C (837.5 mA·g^{-1}) can still reach 144.4 mA·h·g^{-1}; even after 4000 cycles at 10 C (3350 mA·g^{-1}), a capacity retention of as high as 94.4% is obtained. Zhang et al. [133] demonstrated the positive function of oxygen vacancies in TiO_2(B) nanobelts for the enhancement of sodium storage. The sample displayed the significantly superior sodium-storage properties, including a higher capacity (0.5 C; 210.5 mA·h·g^{-1} vs. 102.7 mA·h·g^{-1}), better rate performance (15 C; 89.8 vs. 36.7 mA·h·g^{-1}), as compared to those of pristine TiO_2(B) electrodes without oxygen vacancies.

4.3. Li–S Batteries

Li-S batteries possess exceptionally high theoretical energy densities ~2600 Wh·kg^{-1} vs. 580 Wh·kg^{-1} of today's best batteries. Li–S batteries contain low cost materials, sulfur is highly abundant, and the anode consists of lithium metal and does not limit the capacity. Today's Li–S technology falls short in energy density and lifetime because of the limited sulfur loading in the cathode, due to the poor conductivity of sulfur deposits, because of the solvation into the electrolyte of the discharge products (i.e., Li_xS_y polysulfides), and finally because of the large volume expansion of sulfur during the battery cycling affecting the cathode integrity.

Cathodes with high surface area and high electronic conductivity are crucial to improve sulfur loading and rate performance of Li–S batteries. The polysulfides "shuttle" phenomena, via the solvation of the polysulfides in the electrolyte, gradually decrease the mass of active material, leading to continuous fading in capacity and must be avoided. Therefore, the candidate cathodes should have a porous and conductive nature, as well as suitable interactions with polysulfides simultaneously. To overcome those obstacles, a wide range of strategies has been developed, including encapsulation or coating of the sulfur electrode, use of impermeable membranes, and/or the use of electrolytes that minimize the solubility and diffusivity of the polysulfides. However, none of these solutions has led to acceptable results, fulfilling all of the requirements. For example, the main disadvantage of widely used porous conductive carbon electrodes lies in weak physical confinement of lithium sulfides, which is insufficient to prevent the diffusion and shuttling of polysulfides during long-term cycling. Therefore, ideal electrodes should not only possess porous and conductive nature, but also suitable interactions with polysulfides.

On a typical carbon support (Figure 7a), elemental sulfur undergoes reduction to form lithium polysulphides that then dissolve into the electrolyte. In the presence of a polar metal oxide as witnessed for titanium oxides, however, the solvation of the polysulfides is significantly affected (Figure 7b). Not only is the concentration of polysulphides in solution that greatly diminished during discharge, but also a slow, controlled deposition of Li_2S is observed. The results are ascribed to the interface-mediated, spatially controlled reduction of the polysulphides. Yu et al. [136] studied the interactions between intermediate polysulphides, final discharge product Li_2S and stable TiO_2 surface (anatase-TiO_2 (101), rutile-TiO_2 (110)) via theoretical simulation (Figure 7c–f). Their results show that the binding strength of the polysulphides to the anatase-TiO_2 (101) surface (2.30 eV) is a little higher than to rutile-TiO_2 (110) surface (2.18 eV), and the binding energy of Li_2S to the anatase-TiO_2 (101) surface (3.59 eV) is almost the same as with the rutile-TiO_2 (110) surface (3.62 eV). The values are larger than the adsorption binding energies for Li–S composites on graphene (<1 eV), highlighting the efficacy of TiO_2 in binding with polysulfide anions via polar–polar interactions.

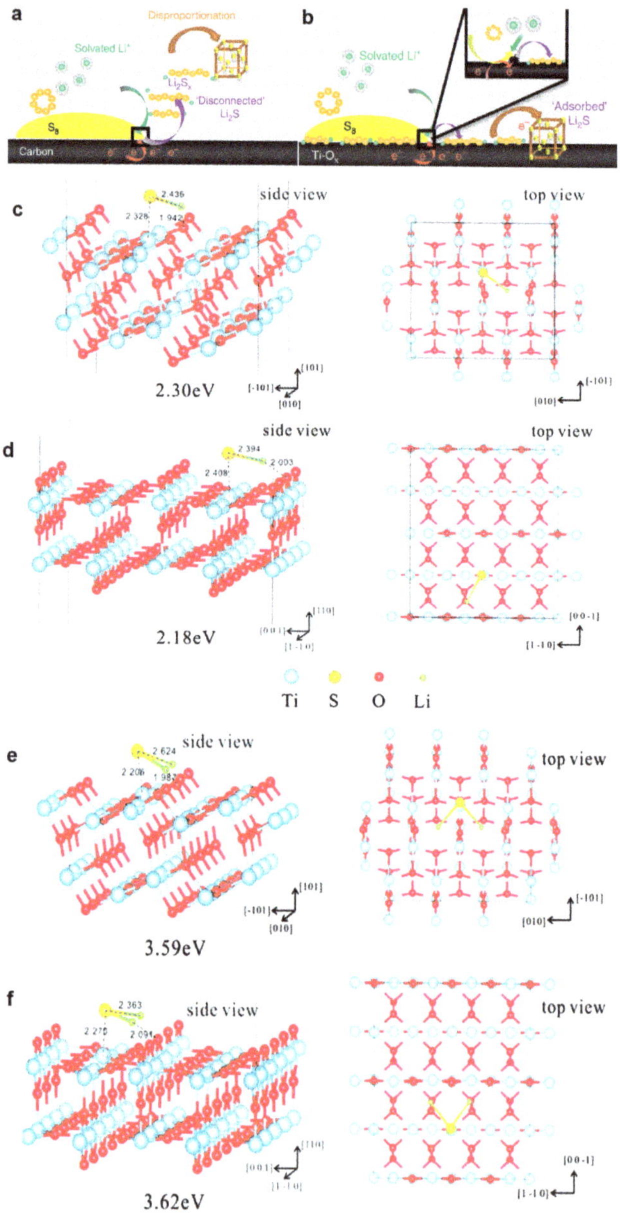

Figure 7. The interaction between sulfur or lithium polysulphides and electrodes. (**a**) On reduction of S_8 on a carbon host, Li_2S_X desorb from the surface and undergo solution-mediated reactions leading to broadly distributed precipitation of Li_2S; (**b**) On reduction of S_8 on the metallic polar Ti_4O_7, Li_2S_X adsorb on the surface and are reduced to Li_2S via surface-mediated reduction at the interface (reprinted from [135] with permission, Copyright Nature Publishing Group, 2014); Adsorption configuration of (**c**,**d**) Li–S* and (**e**,**f**) Li_2S on the (**c**,**e**) anatase-TiO_2 (101) surface and (**d**,**f**) rutile-TiO_2 (110) surface (Reproduced with permission from [136], Copyright The Royal Society of Chemistry, 2016).

Experimentally, Cui et al. [137] designed a unique sulfur-TiO_2 yolk-shell architecture as a sulfur cathode, and obtained an initial specific capacity of 1030 $mA{\cdot}h{\cdot}g^{-1}$ at 0.5 C (1 C = 1673 $mA{\cdot}g^{-1}$) and Coulombic efficiency of 98.4% over 1000 cycles. Impressively, the capacity decay at the end of 1000 cycles is found to be as small as 0.033% per cycle (3.3% per 100 cycles). The excellent properties were ascribed to the yolk–shell morphology, which accommodates the large volumetric expansion of sulfur during cycling, thus preserving the structural integrity of the shell to minimize polysulphide dissolution. Based on the knowledge of chemical interactions between polysulphides and titanium oxides, a wide range of methods have been performed to optimize configuration of sulfur-titanium oxide cathodes. Typical examples include design and synthesis of porous titanium oxides high-surface area, crystalline facts engineering, conductivity enhancement by adding conductive agents (such as carbon fibers, graphene, conductive polymers) into the titanium oxide nanostructures or through annealing in inert/H_2 atmosphere. In this regard, Lou et al. [40] synthesized a sulfur host containing titanium monoxide@carbon hollow nanospheres (TiO@C-HS/S), which possess the key structural elements (i.e., high surface area, conductive, interactions with polysulfides) that are required for high-performance cathodes simultaneously (Figure 8). The TiO@C/S composite cathode delivered high discharge capacities of 41,100 $mA{\cdot}h{\cdot}g^{-1}$ at 0.1 C, and exhibited stable cycle life up to 500 cycles at 0.2 and 0.5 C with a small capacity decay rate of 0.08% per cycle. The Li–S batteries performance based on typical titanium oxides are summarized in Table 2.

Table 2. Comparison of Li–S batteries performance based on typical titanium oxides electrode (the voltage is versus Li^+/Li).

Material/[Reference]	Capacity (Cycles) ($mA{\cdot}h{\cdot}g^{-1}$)	Rate Capability ($mA{\cdot}h{\cdot}g^{-1}$)	Sulfur Loading (%)	Voltage (V)
TiO@carbon [40]	750 (500)@335 $mA{\cdot}g^{-1}$	655 @3.35 $A{\cdot}g^{-1}$	~70	1.9–2.6
Ti_4O_7/S [135]	1070 (500)@3350 $mA{\cdot}g^{-1}$	-	70	1.8–3.0
TiO_2/N-doped graphene [136]	918 (500)@1675 $mA{\cdot}g^{-1}$	833 @6.7 $A{\cdot}g^{-1}$	59	1.7–2.8
S–TiO_2 yolk–shell [137]	1030 (1000)@837 $mA{\cdot}g^{-1}$	630 @3.35 $A{\cdot}g^{-1}$	62	1.7–2.6
TiO_2-porous carbon nanofibers [138]	618 (500)@1675 $mA{\cdot}g^{-1}$	668 @8.375 $A{\cdot}g^{-1}$	55	1.7–2.6
TiO_2-carbon nanofibers [139]	694 (500)@1675 $mA{\cdot}g^{-1}$	540 @3.35 $mA{\cdot}g^{-1}$	68.83	1.7–2.8
TiO_2/graphene [140]	630 (1000)@3350 $mA{\cdot}g^{-1}$	535 @5.025 $A{\cdot}g^{-1}$	51.2	1.6–2.8
Porous Ti_4O_7 particles [141]	989 (300)@167.5 $mA{\cdot}g^{-1}$	873 @1.675 $A{\cdot}g^{-1}$	50-55	1.8–3.0
Polypyrrole/TiO_2 nanotube arrays [142]	1150 (100)@167.5 $mA{\cdot}g^{-1}$	-	61.93	1.8–3.0
Graphene-TiO_2 NPs [143]	663 (100)@1675 $mA{\cdot}g^{-1}$	-	75	1.7–2.8
TiO_2 nanowire/graphene [144]	1053 (200)@335 $mA{\cdot}g^{-1}$	-	60	1.5–2.8
graphene/TiO_2/S [145]	597 (100)@1675 $mA{\cdot}g^{-1}$	-	60	1.5–3.0

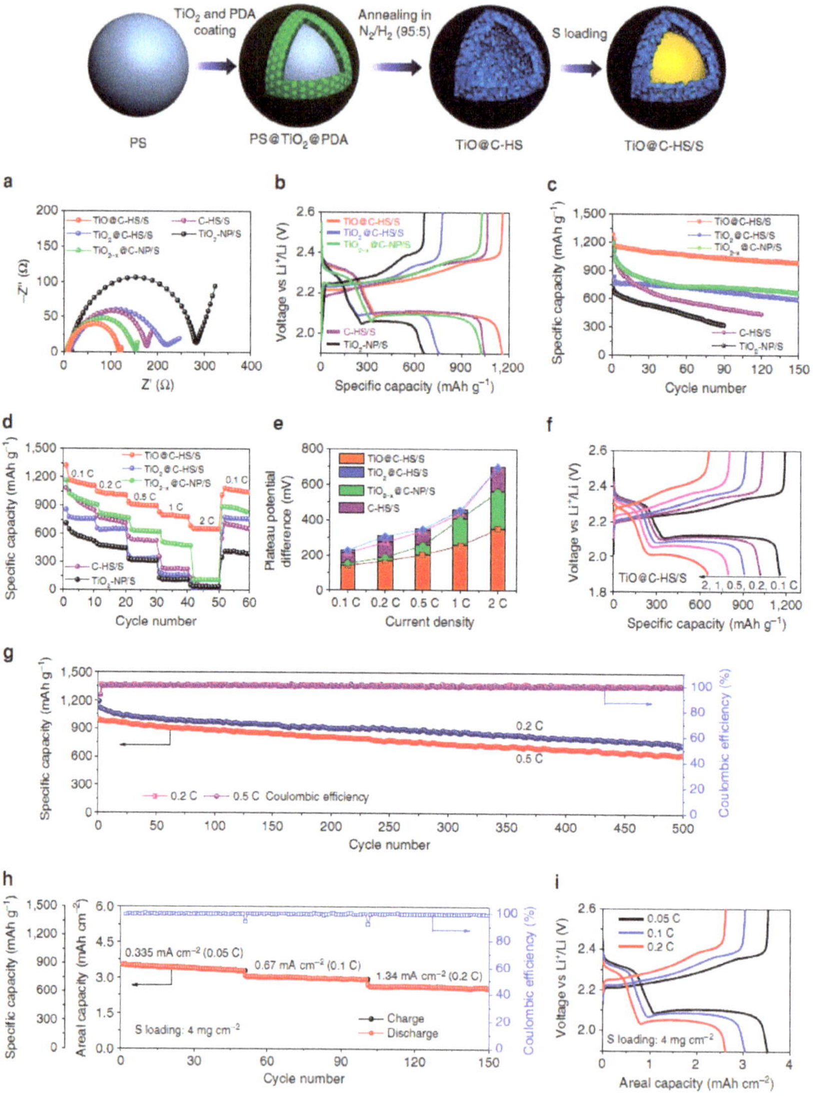

Figure 8. Schematic illustration of the synthesis process and electrochemical properties of TiO@C-HS/S composites. (**a**) Nyquist plots before cycling from 1 MHz to 100 mHz; (**b**) the second-cycle galvanostatic charge/discharge voltage profiles at 0.1 C; (**c**) cycle performances at 0.1 C; (**d**) rate capabilities; and (**e**) the potential differences between the charge and discharge plateaus at various current densities of the TiO@C-HS/S, titanium dioxide@carbon hollow nanospheres/S composite (TiO_2@C-HS/S), carbon coated conductive TiO_{2-x} nanoparticles/S composite (TiO_{2-x}@C-NP/S), pure carbon hollow spheres/S composite (C-HS/S) and TiO_2 nanoparticles/S composite (TiO_2-NP/S) electrodes. (**f**) Voltage profiles at various current densities from 0.1 to 2 C and (**g**) prolonged cycle life and Coulombic efficiency at 0.2 and 0.5C of the TiO@C-HS/S electrode. (**h**) Areal capacities and (**i**) voltage profiles at various current densities from 0.335 (0.05 C) to 1.34 mA·cm^{-2} (0.2 C) of the TiO@C-HS/S electrode with high sulfur mass loading of 4.0 mg·cm^{-2} (reprinted from [40] with permission, Copyright Nature Publishing Group, 2016).

5. Phase Stability of TiO_2 Nanostructures

As a kind of chemically stable and environmentally compatible metal oxides, TiO_2 nanostructures show fantastic physical/chemical properties and find many practical applications, ranging from energy conversion and storage, as mentioned above and others. The properties and applications are determined by the structures of TiO_2 materials, which is related to the external (temperature, pressure, environment, etc.) and internal (composition, stain, etc.) factors. Overall, the relative phase stability in ambient bulk form is TiO_2(B) < anatase < brookite < rutile, and the specific phase shows its unique applications. For example, anatase has been found to be the most active phase in photocatalysis. TiO_2(B) phase is more favorable for the insertion/extraction of lithium ions due to the more open crystal structure when compared to the other TiO_2 phases. Therefore, it is of importance to understand the phase transformation on nanoscale and improve the phase stability of the related TiO_2 nanostructures. General thermodynamic investigation, computational methods (including molecular dynamics simulations and DFT calculations), experimental routes (XRD, calorimetry, electrochemical measurements, etc.) have been successfully employed to study the phase stability and coarsening kinetics of the typical TiO_2 phases under different environment (dry, wet, hydrothermal conditions) [146]. Several excellent reviews describing the topics are available elsewhere, and we do not discuss them in this paper.

With the decreasing of the size or dimension, surface and/or interface will dominate in the nanostructure and play an important role in phase stability. Due to the nature of coordination unsaturation, the atoms at the surface are more active than those within the interior. Therefore, surfaces usually exhibit a lower stability relative to the lattice interior part. For example, the melting point of free-standing nanoparticles is remarkably depressed relative to that of bulk phase (T_0). Stabilizing the surface atoms would be a way to improve the relative phase stability. Typically, when nanoparticles are properly coated by or embedded in a matrix with higher melting point, the melting point of the particles can be elevated above T_0. Herein, we focus on the strategies of surface/interface engineering to tune the phase stability in typical TiO_2 nanostructures.

We systematically studied the crystallization and structural transformation from anatase to rutile phase in the initial amorphous TiO_2 nanowires embedded in anodic aluminum oxide with different diameters (20, 50, and 80 nm, termed as TiO_2-20, TiO_2-50, and TiO_2-80, hereafter) [16]. Electron microscopy analysis and XRD results showed that the crystallization of TiO_2-20, TiO_2-50, and TiO_2-80 from amorphous to anatase occurred at ~600, ~500, and ~400 °C, and the transformation from anatase to rutile phase started at ~900, ~800, and ~750 °C (Figure 9). The results revealed a strong size dependence of the thermal stability of TiO_2 nanowires embedded the template. Control experiments on amorphous TiO_2 powder showed the crystallization and phase transformation temperatures were ~200 and ~600 °C, respectively.

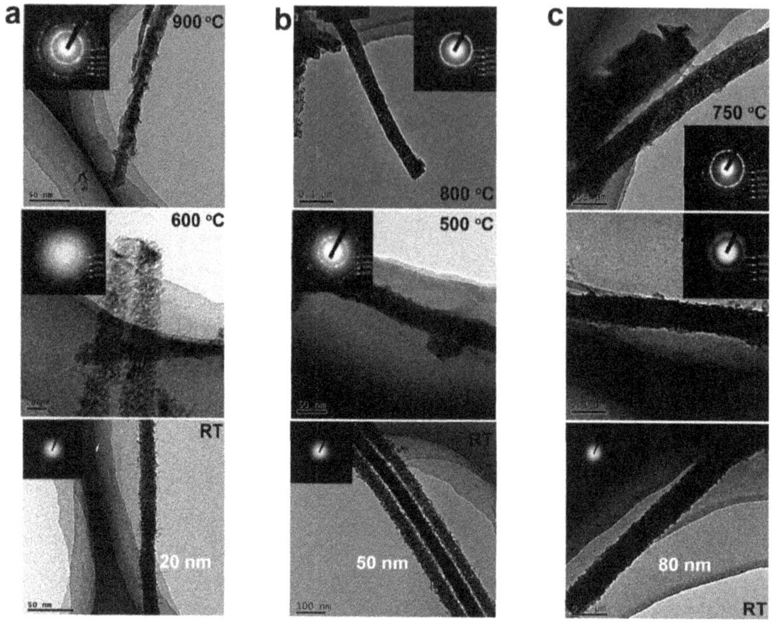

Figure 9. Typical transmission electron microscopy (TEM) images of the as-prepared and annealed TiO_2 nanowires with diameters of (**a**) 20; (**b**) 50; and (**c**) 80 nm. The insets show corresponding selected area electron diffraction (SAED) patterns (Reproduced with permission from [16], Copyright Springer, 2012).

To quantitatively study the nucleation and growth kinetics, in-situ high-temperature X-ray diffraction technique was employed to track the transformation process from anatase to rutile phase. In this method, the position and intensity of diffraction peaks change during the increasing and decreasing temperatures, and thus provide an effective and direct way to trace the phase structure. Taken TiO_2-20 and TiO_2 powder for typical examples, the transformed rutile phase showed an exponential growth versus annealing time t, and the growth of the rutile was a thermally activated process (Figure 10). The rutile growth activation energy (E_g) values of 2.8 ± 0.2 eV and 1.6 ± 0.2 eV were determined in TiO_2-20 and TiO_2 power, respectively. Additionally, no obvious change of the rutile size was observed in the initial stage of the studied temperature range, indicating that the increasing of the rutile volume fraction was induced by the nucleation events. By analyzing the dependence of nucleation rate on the annealing temperature, the rutile nucleation activation energy (E_n) values of 2.7 ± 0.2 eV and 1.9 ± 0.2 eV were yielded for TiO_2-20 and TiO_2 power, respectively. The higher nucleation and growth energy for TiO_2-20 implied that the phase transformation from anatase to rutile was inhibited, i.e. the thermal stability of the anatase phase was improved. Our theoretical work showed the difference of thermal expansion coefficient between the nanoscale channel (aluminum oxide) and the embedded TiO_2 nanowire generated overpressure on the TiO_2/Al_2O_3 interface during annealing. The pressure can be estimated as ~0.13 GPa at 900 °C for TiO_2-20 sample. The pressure compressed the anatase surface and constrained the vibration of surface atoms, which were responsible for the improvement of the anatase phase. By choosing suitable surface layers and other coating techniques (such as Langmuir-Blodgett assembly, atomic layer deposition, etc.), this surface/interface confinement strategy can also be used to improve the phase stability of other TiO_2 polymorphs. For example, Zazpe et al. [15] recently reported on a very obvious enhancement of the phase stability of selforganized TiO_2 nanotubes layers with amorphous structure, which are provided by thin Al_2O_3 coatings of different thicknesses prepared by atomic layer deposition. TiO_2 nanotube layers coated with Al_2O_3 coatings exhibit significantly improved thermal stability, as illustrated by the preservation of the nanotubular structure upon annealing treatment at high temperatures (870 °C). It is worth noting that accompanying by phase transformation during annealing, TiO_2 nanostructures also suffer from the change in size, surface area, bandgap, and morphology [147], which are important parameters that influence the applications and must be considered in the phase stability studies.

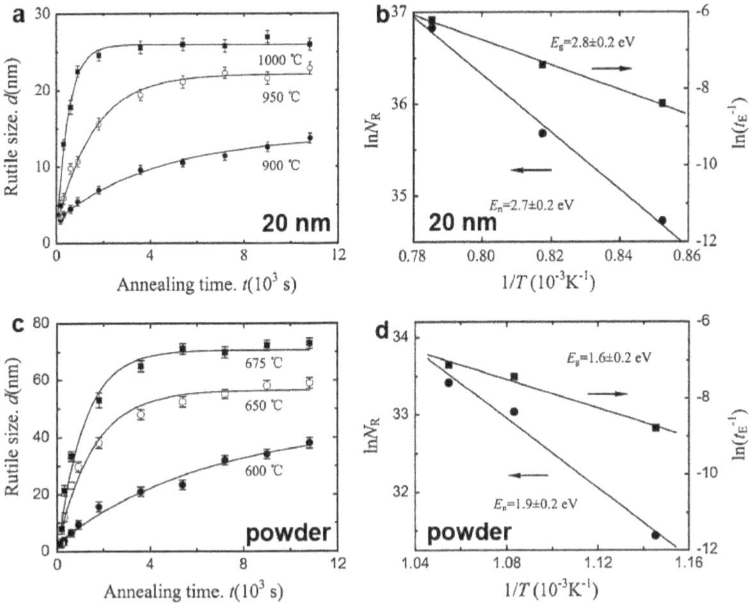

Figure 10. Nucleation and growth kinetics of nanocrystalline anatase to rutile. Annealing time dependence of the size of the rutile in the (**a**) nanowire and (**c**) free-state powders at different temperatures; Annealing temperature variations of the nucleation rate (NR) and the growth saturation rate $t_E{}^{-1}$ for rutile in the (**b**) nanowire and (**d**) free-state powders, respectively (Reproduced with permission from [16], Copyright Springer, 2012).

Besides phase transformation among the different TiO_2 polymorphs, surface atomic rearrangement (reconstruction) also occurs to reach a more stable state at a certain environment (temperature, pressure, atmosphere, humidity, etc.). Remarkably different physical/chemical properties on the surface with respect to the bulk counterpart can be yielded by the reconstruction. The environmental transmission electron microscopy (ETEM) technique allows for the direct imaging of the samples that are placed in a specimen chamber that is high pressures attainable, which can be achieved by either differential pumping systems or delicate TEM holder design [148]. Yuan et al. [149] reported in-situ atomic scale ETEM observations of the formation and evolution of the (1 × 4) reconstruction dynamics on the anatase TiO_2 (001) surface under oxygen atmosphere. They firstly cleaned the wet chemistry synthesized TiO_2 nanosheets with the aid of e-beam irradiation at a temperature of 500 °C in oxygen environment. On the cleaned TiO_2 surface, the real-time dynamics for the transition from metastable (1 × 3) and (1 × 5) to (1 × 4), and the unstable intermediate states were observed and identified (Figure 11). The special reconstruction was driven by the lowly coordinated atoms and surface stress. The results demonstrate the power of in situ real-time technique to study the dynamic formation and evolution of surface structures.

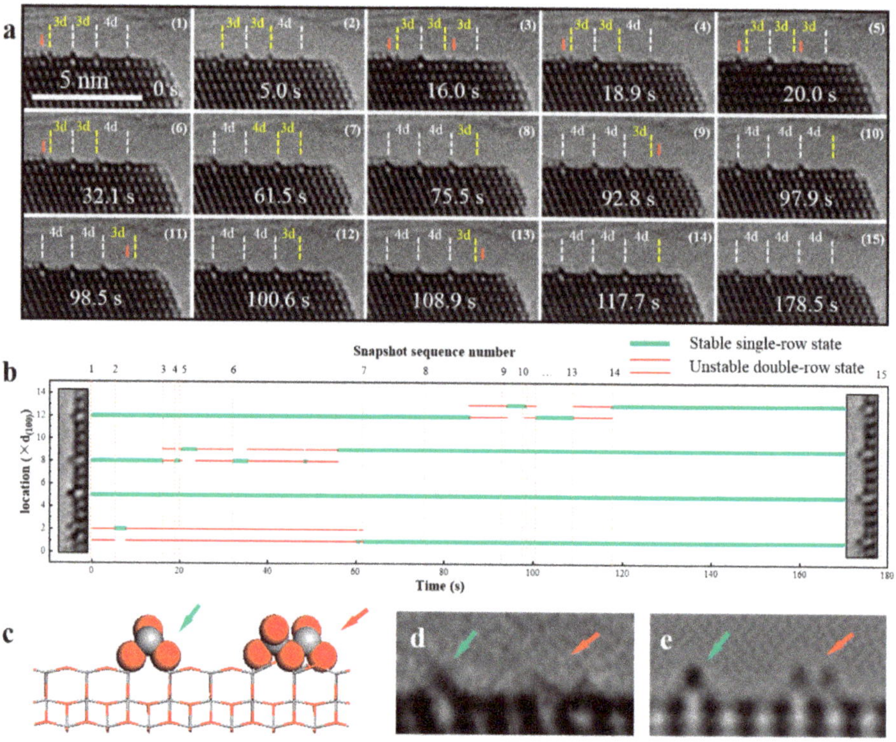

Figure 11. Atomic evolution of the (1 × *n*) reconstructions on anatase TiO_2 (001) surface. (**a**) Sequential HRTEM images of the dynamic structural evolution, viewed from [010] direction, with the red arrows indicating the unstable states; (**b**) The statistical diagram of the locations of the TiO_x rows with green and red lines indicating the stable states and the unstable states; (**c**) Side view of the proposed model for the unstable two-row state with the TiO_x row shown as ball-and-stick (Ti, gray; O, red) on the TiO_2 stick framework. The green and red arrows indicate the stable single-row and instable double-row structures, respectively; (**d**,**e**) Experimental HRTEM image compared with the simulated image based on the model in (**c**). (Reproduced with permission from [149], American Chemical Society, 2016).

6. Conclusions and Perspective

Recent years have witnessed explosive research and development efforts on TiO_2 materials, ranging from controllable synthesis to advanced characterizations and device applications. Although the intrinsic properties, such as wide bandgap, rapid carriers recombination, poor electronic conductivity, and coexistence of multiphases, hampered the practical applications of pristine TiO_2 materials to some extent, the surface/interface modifications have been demonstrated as effective routes to break the limitations, making it possible to be applied in diverse areas. This review article summarized the main progress in engineering the surface/interface structures in TiO_2 micro and nano structures, discussed the effect of surface/interface structures on environmental and electrochemical applications. Specifically, by tuning the exposed crystallographic planes, engineering defect structures, and constructing interface in various TiO_2 materials, the heterogeneous photocatalysis process, including light absorption, the generation and separation of photoexcited carriers, the migration, transport and recombination of carriers, and surface catalytic reactions can be well controlled and optimized. As a result, the photocatalytic properties of TiO_2 materials in the degradation of organic contaminants, hydrogen evolution, CO_2 reduction, antimicrobial, and self-cleaning are greatly improved. For the battery applications, engineering the surface/interface structures of TiO_2 crystal not only increase the sites for ion storage, but also improve the electron and ion conductivity. In Li–S battery system, the interaction between sulfur cathodes and the surface of TiO_2 host can also be adjusted by surface/interface engineering. All of those factors are crucial for improving the specific capacity, rate performance, and cycle durability. In addition, the phase transitions in TiO_2 nanostructures and possible strategies of improving the phase stability have been analyzed. Despite these impressive advances, several challenges still remain.

(1) Developing novel synthesis and treatment methods. Despite great success has been obtained in the controllable synthesis of TiO_2 nanostructures with tailored micro and nano structures, there is still room for improvement in terms of quality of the products. Moreover, the new methods also provide opportunities to further understand the nucleation and growth.
(2) Control of the fine structures. High-index facets and defect sites are chemically active. However, the synthesis of TiO_2 nanocrystals with specific high-index facets is still a challenge. It is highly desirable to synthesize facet-controllable TiO_2 materials and further study the facet effect on energy storage, conversion, and other applications. In addition, selectively generating defect structures and controlling their concentrations in different TiO_2 phases are significant to revel the role of defects in various physical and chemical processes.
(3) In situ/*operando* study the dynamic evolution of the surface/interface. In situ/*operando* spectroscopic or microscopic studies afford the chance to probe the evolution of TiO_2 surface/interface structures in working conditions, which is crucial to study the complex phase transformation and device stability.

Acknowledgments: This work was supported by Chinese National Natural Science Foundation (Grants No. 51401114, 51701063) and Danish Research Council for Technology and Production Case No. 12-126194.

Author Contributions: Hongyu Sun and Xiaoliang Wang wrote the first draft of the manuscript. Editing and revising were carried out by all the authors.

Conflicts of Interest: The authors declare no conflict of interest.

References

1. Sang, L.; Zhao, Y.; Burda, C. TiO_2 Nanoparticles as functional building blocks. *Chem. Rev.* **2014**, *114*, 9283–9318. [CrossRef] [PubMed]
2. Liu, L.; Chen, X. Titanium dioxide nanomaterials: Self-structural modifications. *Chem. Rev.* **2014**, *114*, 9890–9918. [CrossRef] [PubMed]
3. Fujishima, A.; Honda, K. Electrochemical photolysis of water at a semiconductor electrode. *Nature* **1972**, *238*, 37–38. [CrossRef] [PubMed]
4. Nolan, M.; Iwaszuk, A.; Lucid, A.K.; Carey, J.J.; Fronzi, M. Design of novel visible light activse photocatalyst materials: Surface modified TiO_2. *Adv. Mater.* **2016**, *28*, 5425–5446. [CrossRef] [PubMed]
5. Chen, X.; Liu, L.; Huang, F. Black titanium dioxide (TiO_2) nanomaterials. *Chem. Soc. Rev.* **2015**, *44*, 1861–1885. [CrossRef] [PubMed]
6. Cargnello, M.; Gordon, T.R.; Murray, C.B. Solution-phase synthesis of titanium dioxide nanoparticles and nanocrystals. *Chem. Rev.* **2014**, *114*, 9319–9345. [CrossRef] [PubMed]
7. Bai, Y.; Mora-Seró, I.; De Angelis, F.; Bisquert, J.; Wang, P. Titanium dioxide nanomaterials for photovoltaic applications. *Chem. Rev.* **2014**, *114*, 10095–10130. [CrossRef] [PubMed]
8. Ma, Y.; Wang, X.; Jia, Y.; Chen, X.; Han, H.; Li, C. Titanium dioxide-based nanomaterials for photocatalytic fuel generations. *Chem. Rev.* **2014**, *114*, 9987–10043. [CrossRef] [PubMed]
9. Asahi, R.; Morikawa, T.; Irie, H.; Ohwaki, T. Nitrogen-doped titanium dioxide as visible-light-sensitive photocatalyst: Designs, developments, and prospects. *Chem. Rev.* **2014**, *114*, 9824–9852. [CrossRef] [PubMed]
10. Kapilashrami, M.; Zhang, Y.; Liu, Y.-S.; Hagfeldt, A.; Guo, J. Probing the optical property and electronic structure of TiO_2 nanomaterials for renewable energy applications. *Chem. Rev.* **2014**, *114*, 9662–9707. [CrossRef] [PubMed]

11. Schneider, J.; Matsuoka, M.; Takeuchi, M.; Zhang, J.; Horiuchi, Y.; Anpo, M.; Bahnemann, D.W. Understanding TiO_2 photocatalysis: Mechanisms and materials. *Chem. Rev.* **2014**, *114*, 9919–9986. [CrossRef] [PubMed]
12. Ferrighi, L.; Datteo, M.; Fazio, G.; Di Valentin, C. Catalysis under cover: Enhanced reactivity at the interface between (doped) graphene and anatase TiO_2. *J. Am. Chem. Soc.* **2016**, *138*, 7365–7376. [CrossRef] [PubMed]
13. Bourikas, K.; Kordulis, C.; Lycourghiotis, A. Titanium dioxide (anatase and rutile): Surface chemistry, liquid–solid interface chemistry, and scientific synthesis of supported catalysts. *Chem. Rev.* **2014**, *114*, 9754–9823. [CrossRef] [PubMed]
14. Hua, X.; Liu, Z.; Bruce, P.G.; Grey, C.P. The morphology of TiO_2(B) nanoparticles. *J. Am. Chem. Soc.* **2015**, *137*, 13612–13623. [CrossRef] [PubMed]
15. Zazpe, R.; Prikryl, J.; Gärtnerova, V.; Nechvilova, K.; Benes, L.; Strizik, L.; Jäger, A.; Bosund, M.; Sopha, H.; Macak, J.M. Atomic layer deposition Al_2O_3 coatings significantly improve thermal, chemical, and mechanical stability of anodic TiO_2 nanotube layers. *Langmuir* **2017**, *33*, 3208–3216. [CrossRef] [PubMed]
16. Wang, X. Enhancement of thermal stability of TiO_2 nanowires embedded in anodic aluminum oxide template. *J. Mater. Sci.* **2012**, *47*, 739–745. [CrossRef]
17. Zhou, W.; Sun, F.; Pan, K.; Tian, G.; Jiang, B.; Ren, Z.; Tian, C.; Fu, H. Well-ordered large-pore mesoporous anatase TiO_2 with remarkably high thermal stability and improved crystallinity: Preparation, characterization, and photocatalytic performance. *Adv. Funct. Mater.* **2011**, *21*, 1922–1930. [CrossRef]
18. Biswas, D.; Biswas, J.; Ghosh, S.; Wood, B.; Lodha, S. Enhanced thermal stability of Ti/TiO_2/n-Ge contacts through plasma nitridation of TiO_2 interfacial layer. *Appl. Phys. Lett.* **2017**, *110*, 052104. [CrossRef]
19. Zhao, L.; Zhong, C.; Wang, Y.; Wang, S.; Dong, B.; Wan, L. Ag nanoparticle-decorated 3D flower-like TiO_2 hierarchical microstructures composed of ultrathin nanosheets and enhanced photoelectrical conversion properties in dye-sensitized solar cells. *J. Power Sources* **2015**, *292*, 49–57. [CrossRef]
20. Wu, W.-Q.; Xu, Y.-F.; Rao, H.-S.; Su, C.-Y.; Kuang, D.-B. A double layered TiO_2 photoanode consisting of hierarchical flowers and nanoparticles for high-efficiency dye-sensitized solar cells. *Nanoscale* **2013**, *5*, 4362. [CrossRef] [PubMed]
21. Zhang, K.; Zhou, W.; Chi, L.; Zhang, X.; Hu, W.; Jiang, B.; Pan, K.; Tian, G.; Jiang, Z. Black N/H-TiO_2 nanoplates with a flower-Like hierarchical architecture for photocatalytic hydrogen evolution. *ChemSusChem* **2016**, *9*, 2841–2848. [CrossRef] [PubMed]
22. Han, S.-H.; Lee, S.; Shin, H.; Suk Jung, H. A quasi-inverse opal layer based on highly crystalline TiO_2 nanoparticles: A new light-scattering layer in dye-sensitized solar cells. *Adv. Energy Mater.* **2011**, *1*, 546–550. [CrossRef]
23. King, J.S.; Graugnard, E.; Summers, C.J. TiO_2 inverse opals fabricated using low-temperature atomic layer deposition. *Adv. Mater.* **2005**, *17*, 1010–1013. [CrossRef]
24. Kwak, E.S.; Lee, W.; Park, N.-G.; Kim, J.; Lee, H. Compact inverse-opal electrode using non-aggregated TiO_2 nanoparticles for dye-sensitized solar cells. *Adv. Funct. Mater.* **2009**, *19*, 1093–1099. [CrossRef]
25. Seo, Y.G.; Woo, K.; Kim, J.; Lee, H.; Lee, W. Rapid fabrication of an inverse opal TiO_2 photoelectrode for DSSC using a binary mixture of TiO_2 nanoparticles and polymer microspheres. *Adv. Funct. Mater.* **2011**, *21*, 3094–3103. [CrossRef]
26. Cheng, C.; Karuturi, S.K.; Liu, L.; Liu, J.; Li, H.; Su, L.T.; Tok, A.I.Y.; Fan, H.J. Quantum-dot-sensitized TiO_2 inverse opals for photoelectrochemical hydrogen generation. *Small* **2012**, *8*, 37–42. [CrossRef] [PubMed]
27. Cho, C.-Y.; Moon, J.H. Hierarchical twin-scale inverse opal TiO_2 electrodes for dye-sensitized solar cells. *Langmuir* **2012**, *28*, 9372–9377. [CrossRef] [PubMed]
28. Cheng, P.; Du, S.; Cai, Y.; Liu, F.; Sun, P.; Zheng, J.; Lu, G. Tripartite layered photoanode from hierarchical anatase TiO_2 urchin-like spheres and P25: A candidate for enhanced efficiency dye sensitized solar cells. *J. Phys. Chem. C* **2013**, *117*, 24150–24156. [CrossRef]
29. Pan, J.H.; Wang, X.Z.; Huang, Q.; Shen, C.; Koh, Z.Y.; Wang, Q.; Engel, A.; Bahnemann, D.W. Large-scale synthesis of urchin-like mesoporous TiO_2 hollow spheres by targeted etching and their photoelectrochemical properties. *Adv. Funct. Mater.* **2014**, *24*, 95–104. [CrossRef]
30. Chen, J.S.; Liang, Y.N.; Li, Y.; Yan, Q.; Hu, X. H_2O-EG-Assisted synthesis of uniform urchinlike rutile TiO_2 with superior lithium storage properties. *ACS Appl. Mater. Interfaces* **2013**, *5*, 9998–10003. [CrossRef] [PubMed]

31. Bai, X.; Xie, B.; Pan, N.; Wang, X.; Wang, H. Novel three-dimensional dandelion-like TiO_2 structure with high photocatalytic activity. *J. Solid State Chem.* **2008**, *181*, 450–456. [CrossRef]
32. Musavi Gharavi, P.S.; Mohammadi, M.R. The improvement of light scattering of dye-sensitized solar cells aided by a new dandelion-like TiO_2 nanostructures. *Sol. Energy Mater. Sol. Cells* **2015**, *137*, 113–123. [CrossRef]
33. Lan, C.-M.; Liu, S.-E.; Shiu, J.-W.; Hu, J.-Y.; Lin, M.-H.; Diau, E.W.-G. Formation of size-tunable dandelion-like hierarchical rutile titania nanospheres for dye-sensitized solar cells. *RSC Adv.* **2013**, *3*, 559–565. [CrossRef]
34. Liu, G.; Yang, H.G.; Pan, J.; Yang, Y.Q.; Lu, G.Q. (Max); Cheng, H.-M. Titanium dioxide crystals with tailored facets. *Chem. Rev.* **2014**, *114*, 9559–9612. [CrossRef] [PubMed]
35. Wang, Y.; Sun, H.; Tan, S.; Feng, H.; Cheng, Z.; Zhao, J.; Zhao, A.; Wang, B.; Luo, Y.; Yang, J.; Hou, J.G. Role of point defects on the reactivity of reconstructed anatase titanium dioxide (001) surface. *Nat. Commun.* **2013**, *4*, 2214. [CrossRef] [PubMed]
36. Sun, R.; Wang, Z.; Saito, M.; Shibata, N.; Ikuhara, Y. Atomistic mechanisms of nonstoichiometry-induced twin boundary structural transformation in titanium dioxide. *Nat. Commun.* **2015**, *6*. [CrossRef] [PubMed]
37. Selcuk, S.; Selloni, A. Facet-dependent trapping and dynamics of excess electrons at anatase TiO_2 surfaces and aqueous interfaces. *Nat. Mater.* **2016**, *15*, 1107–1112. [CrossRef] [PubMed]
38. Zhang, X.; He, Y.; Sushko, M.L.; Liu, J.; Luo, L.; Yoreo, J.J.D.; Mao, S.X.; Wang, C.; Rosso, K.M. Direction-specific van der Waals attraction between rutile TiO_2 nanocrystals. *Science* **2017**, *356*, 434–437. [CrossRef] [PubMed]
39. Longoni, G.; Pena Cabrera, R.L.; Polizzi, S.; D'Arienzo, M.; Mari, C.M.; Cui, Y.; Ruffo, R. Shape-controlled TiO_2 nanocrystals for Na-ion battery electrodes: The role of different exposed crystal facets on the electrochemical properties. *Nano Lett.* **2017**, *17*, 992–1000. [CrossRef] [PubMed]
40. Li, Z.; Zhang, J.; Guan, B.; Wang, D.; Liu, L.-M.; Lou, X.W. (David). A sulfur host based on titanium monoxide@carbon hollow spheres for advanced lithium–sulfur batteries. *Nat. Commun.* **2016**, *7*, 13065. [CrossRef] [PubMed]
41. Setvín, M.; Aschauer, U.; Scheiber, P.; Li, Y.-F.; Hou, W.; Schmid, M.; Selloni, A.; Diebold, U. Reaction of O_2 with subsurface oxygen vacancies on TiO_2 anatase (101). *Science* **2013**, *341*, 988–991. [CrossRef] [PubMed]
42. Zhou, P.; Zhang, H.; Ji, H.; Ma, W.; Chen, C.; Zhao, J. Modulating the photocatalytic redox preferences between anatase TiO_2 {001} and {101} surfaces. *Chem. Commun.* **2017**, *53*, 787–790. [CrossRef] [PubMed]
43. Yang, H.G.; Sun, C.H.; Qiao, S.Z.; Zou, J.; Liu, G.; Smith, S.C.; Cheng, H.M.; Lu, G.Q. Anatase TiO_2 single crystals with a large percentage of reactive facets. *Nature* **2008**, *453*, 638–641. [CrossRef] [PubMed]
44. Crossland, E.J.W.; Noel, N.; Sivaram, V.; Leijtens, T.; Alexander-Webber, J.A.; Snaith, H.J. Mesoporous TiO_2 single crystals delivering enhanced mobility and optoelectronic device performance. *Nature* **2013**, *495*, 215–219. [CrossRef] [PubMed]
45. Pabón, B.M.; Beltrán, J.I.; Sánchez-Santolino, G.; Palacio, I.; López-Sánchez, J.; Rubio-Zuazo, J.; Rojo, J.M.; Ferrer, P.; Mascaraque, A.; Muñoz, M.C.; et al. Formation of titanium monoxide (001) single-crystalline thin film induced by ion bombardment of titanium dioxide (110). *Nat. Commun.* **2015**, *6*, 6147. [CrossRef] [PubMed]
46. Wang, X.; Li, Z.; Shi, J.; Yu, Y. One-dimensional titanium dioxide nanomaterials: Nanowires, nanorods, and nanobelts. *Chem. Rev.* **2014**, *114*, 9346–9384. [CrossRef] [PubMed]
47. Richter, C.; Schmuttenmaer, C.A. Exciton-like trap states limit electron mobility in TiO_2 nanotubes. *Nat. Nanotechnol.* **2010**, *5*, 769–772. [CrossRef] [PubMed]
48. Penn, R.L.; Banfield, J.F. Morphology development and crystal growth in nanocrystalline aggregates under hydrothermal conditions: Insights from titania. *Geochim. Cosmochim. Acta* **1999**, *63*, 1549–1557. [CrossRef]
49. Chen, X.; Liu, L.; Yu, P.Y.; Mao, S.S. Increasing solar absorption for photocatalysis with black hydrogenated titanium dioxide nanocrystals. *Science* **2011**, *331*, 746–750. [CrossRef] [PubMed]
50. Huang, J.; Fang, F.; Huang, G.; Sun, H.; Zhu, J.; Yu, R. Engineering the surface of rutile TiO_2 nanoparticles with quantum pits towards excellent lithium storage. *RSC Adv.* **2016**, *6*, 66197–66203. [CrossRef]
51. Jiménez, J.M.; Bourret, G.R.; Berger, T.; McKenna, K.P. Modification of charge trapping at particle/particle interfaces by electrochemical hydrogen doping of nanocrystalline TiO_2. *J. Am. Chem. Soc.* **2016**, *138*, 15956–15964. [CrossRef] [PubMed]

52. Giordano, F.; Abate, A.; Correa Baena, J.P.; Saliba, M.; Matsui, T.; Im, S.H.; Zakeeruddin, S.M.; Nazeeruddin, M.K.; Hagfeldt, A.; Graetzel, M. Enhanced electronic properties in mesoporous TiO_2 via lithium doping for high-efficiency perovskite solar cells. *Nat. Commun.* **2016**, *7*, 10379. [CrossRef] [PubMed]
53. Ide, Y.; Inami, N.; Hattori, H.; Saito, K.; Sohmiya, M.; Tsunoji, N.; Komaguchi, K.; Sano, T.; Bando, Y.; Golberg, D.; et al. Remarkable charge separation and photocatalytic efficiency enhancement through interconnection of TiO_2 nanoparticles by hydrothermal treatment. *Angew. Chem. Int. Ed.* **2016**, *55*, 3600–3605. [CrossRef] [PubMed]
54. Liu, J.; Olds, D.; Peng, R.; Yu, L.; Foo, G.S.; Qian, S.; Keum, J.; Guiton, B.S.; Wu, Z.; Page, K. Quantitative analysis of the morphology of {101} and {001} faceted anatase TiO_2 nanocrystals and its implication on photocatalytic activity. *Chem. Mater.* **2017**, *29*, 5591–5604. [CrossRef]
55. Li, W.; Wu, Z.; Wang, J.; Elzatahry, A.A.; Zhao, D. A Perspective on mesoporous TiO_2 materials. *Chem. Mater.* **2014**, *26*, 287–298. [CrossRef]
56. Zhang, D.; Liu, J.; Li, P.; Tian, Z.; Liang, C. Recent advances in surfactant-free, surface-charged, and defect-rich catalysts developed by laser ablation and processing in liquids. *ChemNanoMat* **2017**, *3*, 512–533. [CrossRef]
57. Lau, M.; Straube, T.; Aggarwal, A.V.; Hagemann, U.; de Oliveira Viestel, B.; Hartmann, N.; Textor, T.; Lutz, H.; Gutmann, J.S.; Barcikowski, S. Gradual modification of ITO particle's crystal structure and optical properties by pulsed UV laser irradiation in a free liquid jet. *Dalton Trans.* **2017**, *46*, 6039–6048. [CrossRef] [PubMed]
58. Filice, S.; Compagnini, G.; Fiorenza, R.; Scirè, S.; D'Urso, L.; Fragalà, M.E.; Russo, P.; Fazio, E.; Scalese, S. Laser processing of TiO_2 colloids for an enhanced photocatalytic water splitting activity. *J. Colloid Interface Sci.* **2017**, *489*, 131–137. [CrossRef] [PubMed]
59. Russo, P.; Liang, R.; He, R.X.; Zhou, Y.N. Phase transformation of TiO_2 nanoparticles by femtosecond laser ablation in aqueous solutions and deposition on conductive substrates. *Nanoscale* **2017**, *9*, 6167–6177. [CrossRef] [PubMed]
60. Raghunath, P.; Huang, W.F.; Lin, M.C. Quantum chemical elucidation of the mechanism for hydrogenation of TiO_2 anatase crystals. *J. Chem. Phys.* **2013**, *138*, 154705. [CrossRef] [PubMed]
61. Pan, H.; Zhang, Y.-W.; Shenoy, V.B.; Gao, H. Effects of H-, N-, and (H, N)-doping on the photocatalytic activity of TiO_2. *J. Phys. Chem. C* **2011**, *115*, 12224–12231. [CrossRef]
62. Aschauer, U.; Selloni, A. Hydrogen interaction with the anatase TiO_2 (101) surface. *Phys. Chem. Chem. Phys.* **2012**, *14*, 16595–16602. [CrossRef] [PubMed]
63. Li, X.; Yu, J.; Jaroniec, M. Hierarchical photocatalysts. *Chem. Soc. Rev.* **2016**, *45*, 2603–2636. [CrossRef] [PubMed]
64. Yu, J.; Low, J.; Xiao, W.; Zhou, P.; Jaroniec, M. Enhanced photocatalytic CO_2-reduction activity of Anatase TiO_2 by coexposed {001} and {101} Facets. *J. Am. Chem. Soc.* **2014**, *136*, 8839–8842. [CrossRef] [PubMed]
65. Low, J.; Cheng, B.; Yu, J. Surface modification and enhanced photocatalytic CO_2 reduction performance of TiO_2: A review. *Appl. Surf. Sci.* **2017**, *392*, 658–686. [CrossRef]
66. Li, J.; Zhang, M.; Guan, Z.; Li, Q.; He, C.; Yang, J. Synergistic effect of surface and bulk single-electron-trapped oxygen vacancy of TiO_2 in the photocatalytic reduction of CO_2. *Appl. Catal. B Environ.* **2017**, *206*, 300–307. [CrossRef]
67. Edy, R.; Zhao, Y.; Huang, G.S.; Shi, J.J.; Zhang, J.; Solovev, A.A.; Mei, Y. TiO_2 nanosheets synthesized by atomic layer deposition for photocatalysis. *Prog. Nat. Sci.* **2016**, *26*, 493–497. [CrossRef]
68. Wang, X.; He, H.; Chen, Y.; Zhao, J.; Zhang, X. Anatase TiO_2 hollow microspheres with exposed {001} facets: Facile synthesis and enhanced photocatalysis. *Appl. Surf. Sci.* **2012**, *258*, 5863–5868. [CrossRef]
69. Xiang, Q.; Yu, J. Photocatalytic activity of hierarchical flower-like TiO_2 superstructures with dominant {001} facets. *Chin. J. Catal.* **2011**, *32*, 525–531. [CrossRef]
70. Cao, Y.; Xing, Z.; Shen, Y.; Li, Z.; Wu, X.; Yan, X.; Zou, J.; Yang, S.; Zhou, W. Mesoporous black Ti^{3+}/N-TiO_2 spheres for efficient visible-light-driven photocatalytic performance. *Chem. Eng. J.* **2017**, *325*, 199–207. [CrossRef]
71. An, X.; Zhang, L.; Wen, B.; Gu, Z.; Liu, L.-M.; Qu, J.; Liu, H. Boosting photoelectrochemical activities of heterostructured photoanodes through interfacial modulation of oxygen vacancies. *Nano Energy* **2017**, *35*, 290–298. [CrossRef]
72. Chen, Y.; Li, W.; Wang, J.; Gan, Y.; Liu, L.; Ju, M. Microwave-assisted ionic liquid synthesis of Ti^{3+} self-doped TiO_2 hollow nanocrystals with enhanced visible-light photoactivity. *Appl. Catal. B Environ.* **2016**, *191*, 94–105. [CrossRef]

73. Shuang, S.; Lv, R.; Xie, Z.; Zhang, Z. Surface plasmon enhanced photocatalysis of Au/Pt-decorated TiO_2 nanopillar arrays. *Sci. Rep.* **2016**, *6*, 26670. [CrossRef] [PubMed]
74. Chiu, Y.-H.; Hsu, Y.-J. Au@Cu_7S_4 yolk@shell nanocrystal-decorated TiO_2 nanowires as an all-day-active photocatalyst for environmental purification. *Nano Energy* **2017**, *31*, 286–295. [CrossRef]
75. Jin, J.; Wang, C.; Ren, X.-N.; Huang, S.-Z.; Wu, M.; Chen, L.-H.; Hasan, T.; Wang, B.-J.; Li, Y.; Su, B.-L. Anchoring ultrafine metallic and oxidized Pt nanoclusters on yolk-shell TiO_2 for unprecedentedly high photocatalytic hydrogen production. *Nano Energy* **2017**, *38*, 118–126. [CrossRef]
76. Yu, C.; Yu, Y.; Xu, T.; Wang, X.; Ahmad, M.; Sun, H. Hierarchical nanoflowers assembled with Au nanoparticles decorated ZnO nanosheets toward enhanced photocatalytic properties. *Mater. Lett.* **2017**, *190*, 185–187. [CrossRef]
77. Wu, T.; Kang, X.; Kadi, M.W.; Ismail, I.; Liu, G.; Cheng, H.-M. Enhanced photocatalytic hydrogen generation of mesoporous rutile TiO_2 single crystal with wholly exposed {111} facets. *Chin. J. Catal.* **2015**, *36*, 2103–2108. [CrossRef]
78. Zhang, K.; Liu, Q.; Wang, H.; Zhang, R.; Wu, C.; Gong, J.R. TiO_2 single crystal with four-truncated-bipyramid morphology as an efficient photocatalyst for hydrogen production. *Small* **2013**, *9*, 2452–2459. [CrossRef] [PubMed]
79. Hu, J.; Cao, Y.; Wang, K.; Jia, D. Green solid-state synthesis and photocatalytic hydrogen production activity of anatase TiO_2 nanoplates with super heat-stability. *RSC Adv.* **2017**, *7*, 11827–11833. [CrossRef]
80. Wu, Q.; Huang, F.; Zhao, M.; Xu, J.; Zhou, J.; Wang, Y. Ultra-small yellow defective TiO_2 nanoparticles for co-catalyst free photocatalytic hydrogen production. *Nano Energy* **2016**, *24*, 63–71. [CrossRef]
81. Pei, D.-N.; Gong, L.; Zhang, A.-Y.; Zhang, X.; Chen, J.-J.; Mu, Y.; Yu, H.-Q. Defective titanium dioxide single crystals exposed by high-energy {001} facets for efficient oxygen reduction. *Nat. Commun.* **2015**, *6*, 8696. [CrossRef] [PubMed]
82. Yang, Y.; Gao, P.; Wang, Y.; Sha, L.; Ren, X.; Zhang, J.; Chen, Y.; Wu, T.; Yang, P.; Li, X. A direct charger transfer from interface to surface for the highly efficient spatial separation of electrons and holes: The construction of Ti–C bonded interfaces in TiO_2-C composite as a touchstone for photocatalytic water splitting. *Nano Energy* **2017**, *33*, 29–36. [CrossRef]
83. Gao, L.; Li, Y.; Ren, J.; Wang, S.; Wang, R.; Fu, G.; Hu, Y. Passivation of defect states in anatase TiO_2 hollow spheres with Mg doping: Realizing efficient photocatalytic overall water splitting. *Appl. Catal. B Environ.* **2017**, *202*, 127–133. [CrossRef]
84. Yang, Y.; Liu, G.; Irvine, J.T.S.; Cheng, H.-M. Enhanced photocatalytic H_2 production in core-shell engineered rutile TiO_2. *Adv. Mater.* **2016**, *28*, 5850–5856. [CrossRef] [PubMed]
85. Wu, B.; Liu, D.; Mubeen, S.; Chuong, T.T.; Moskovits, M.; Stucky, G.D. Anisotropic growth of TiO_2 onto gold nanorods for plasmon-enhanced hydrogen production from water reduction. *J. Am. Chem. Soc.* **2016**, *138*, 1114–1117. [CrossRef] [PubMed]
86. Lee, C.-Y.; Park, H.S.; Fontecilla-Camps, J.C.; Reisner, E. Photoelectrochemical H_2 evolution with a hydrogenase immobilized on a TiO_2-protected silicon electrode. *Angew. Chem. Int. Ed. Engl.* **2016**, *55*, 5971–5974. [CrossRef] [PubMed]
87. Valenti, G.; Boni, A.; Melchionna, M.; Cargnello, M.; Nasi, L.; Bertoni, G.; Gorte, R.J.; Marcaccio, M.; Rapino, S.; Bonchio, M.; et al. Co-axial heterostructures integrating palladium/titanium dioxide with carbon nanotubes for efficient electrocatalytic hydrogen evolution. *Nat. Commun.* **2016**, *7*, 13549. [CrossRef] [PubMed]
88. Zhang, R.; Shao, M.; Xu, S.; Ning, F.; Zhou, L.; Wei, M. Photo-assisted synthesis of zinc-iron layered double hydroxides/TiO_2 nanoarrays toward highly-efficient photoelectrochemical water splitting. *Nano Energy* **2017**, *33*, 21–28. [CrossRef]
89. Bendova, M.; Gispert-Guirado, F.; Hassel, A.W.; Llobet, E.; Mozalev, A. Solar water splitting on porous-alumina-assisted TiO_2-doped WO_x nanorod photoanodes: Paradoxes and challenges. *Nano Energy* **2017**, *33*, 72–87. [CrossRef]
90. Yue, X.; Yi, S.; Wang, R.; Zhang, Z.; Qiu, S. A novel architecture of dandelion-like Mo_2C/TiO_2 heterojunction photocatalysts towards high-performance photocatalytic hydrogen production from water splitting. *J. Mater. Chem. A* **2017**, *5*, 10591–10598. [CrossRef]
91. He, H.; Lin, J.; Fu, W.; Wang, X.; Wang, H.; Zeng, Q.; Gu, Q.; Li, Y.; Yan, C.; Tay, B.K.; et al. MoS_2/TiO_2 edge-on heterostructure for efficient photocatalytic hydrogen evolution. *Adv. Energy Mater.* **2016**, *6*, 1600464. [CrossRef]

92. Abdellah, M.; El-Zohry, A.M.; Antila, L.J.; Windle, C.D.; Reisner, E.; Hammarström, L. Time-resolved IR spectroscopy reveals a mechanism with TiO_2 as a reversible electron acceptor in a TiO_2–Re catalyst system for CO_2 photoreduction. *J. Am. Chem. Soc.* **2017**, *139*, 1226–1232. [CrossRef] [PubMed]
93. Matsubu, J.C.; Zhang, S.; DeRita, L.; Marinkovic, N.S.; Chen, J.G.; Graham, G.W.; Pan, X.; Christopher, P. Adsorbate-mediated strong metal–support interactions in oxide-supported Rh catalysts. *Nat. Chem.* **2016**, *9*, 120–127. [CrossRef] [PubMed]
94. Bumajdad, A.; Madkour, M. Understanding the superior photocatalytic activity of noble metals modified titania under UV and visible light irradiation. *Phys. Chem. Chem. Phys.* **2014**, *16*, 7146–7158. [CrossRef] [PubMed]
95. Grigioni, I.; Dozzi, M.V.; Bernareggi, M.; Chiarello, G.L.; Selli, E. Photocatalytic CO_2 reduction vs. H_2 production: The effects of surface carbon-containing impurities on the performance of TiO_2-based photocatalysts. *Catal. Today* **2017**, *281*, 214–220. [CrossRef]
96. Truong, Q.D.; Hoa, H.T.; Le, T.S. Rutile TiO_2 nanocrystals with exposed {331} facets for enhanced photocatalytic CO_2 reduction activity. *J. Colloid Interface Sci.* **2017**, *504*, 223–229. [CrossRef] [PubMed]
97. Xu, J.; Li, Y.; Zhou, X.; Li, Y.; Gao, Z.-D.; Song, Y.-Y.; Schmuki, P. Graphitic C_3N_4-sensitized TiO_2 nanotube layers: A visible-light activated efficient metal-free antimicrobial platform. *Chem. Eur. J.* **2016**, *22*, 3947–3951. [CrossRef] [PubMed]
98. Banerjee, S.; Dionysiou, D.D.; Pillai, S.C. Self-cleaning applications of TiO_2 by photo-induced hydrophilicity and photocatalysis. *Appl. Catal. B Environ.* **2015**, *176–177*, 396–428. [CrossRef]
99. Wang, R.; Hashimoto, K.; Fujishima, A.; Chikuni, M.; Kojima, E.; Kitamura, A.; Shimohigoshi, M.; Watanabe, T. Light-induced amphiphilic surfaces. *Nature* **1997**, *388*, 431–432. [CrossRef]
100. Patrocinio, A.O.T.; Paula, L.F.; Paniago, R.M.; Freitag, J.; Bahnemann, D.W. Layer-by-Layer TiO_2/WO_3 thin films as efficient photocatalytic self-cleaning surfaces. *ACS Appl. Mater. Interfaces* **2014**, *6*, 16859–16866. [CrossRef] [PubMed]
101. Kapridaki, C.; Pinho, L.; Mosquera, M.J.; Maravelaki-Kalaitzaki, P. Producing photoactive, transparent and hydrophobic SiO_2-crystalline TiO_2 nanocomposites at ambient conditions with application as self-cleaning coatings. *Appl. Catal. B Environ.* **2014**, *156–157*, 416–427. [CrossRef]
102. Murakami, A.; Yamaguchi, T.; Hirano, S.; Kikuta, K. Synthesis of porous titania thin films using carbonatation reaction and its hydrophilic property. *Thin Solid Films* **2008**, *516*, 3888–3892. [CrossRef]
103. Nolan, N.T.; Synnott, D.W.; Seery, M.K.; Hinder, S.J.; Van Wassenhoven, A.; Pillai, S.C. Effect of N-doping on the photocatalytic activity of sol–gel TiO_2. *J. Hazard. Mater.* **2012**, *211–212*, 88–94. [CrossRef] [PubMed]
104. Feng, N.; Wang, Q.; Zheng, A.; Zhang, Z.; Fan, J.; Liu, S.-B.; Amoureux, J.-P.; Deng, F. Understanding the high photocatalytic activity of (B, Ag)-codoped TiO_2 under solar-light irradiation with XPS, solid-state NMR, and DFT calculations. *J. Am. Chem. Soc.* **2013**, *135*, 1607–1616. [CrossRef] [PubMed]
105. Yu, Y.; Wang, X.; Sun, H.; Ahmad, M. 3D anatase TiO_2 hollow microspheres assembled with high-energy {001} facets for lithium-ion batteries. *RSC Adv.* **2012**, *2*, 7901–7905. [CrossRef]
106. Gao, X.; Li, G.; Xu, Y.; Hong, Z.; Liang, C.; Lin, Z. TiO_2 Microboxes with controlled internal porosity for high-performance lithium storage. *Angew. Chem. Int. Ed. Engl.* **2015**, *54*, 14331–14335. [CrossRef] [PubMed]
107. McNulty, D.; Carroll, E.; O'Dwyer, C. Rutile TiO_2 inverse opal anodes for li-ion batteries with long cycle life, high-rate capability, and high structural stability. *Adv. Energy Mater.* **2017**, *7*, 1602291. [CrossRef]
108. Liu, G.; Yin, L.-C.; Pan, J.; Li, F.; Wen, L.; Zhen, C.; Cheng, H.-M. Greatly enhanced electronic conduction and lithium storage of faceted TiO_2 crystals supported on metallic substrates by tuning crystallographic orientation of TiO_2. *Adv. Mater.* **2015**, *27*, 3507–3512. [CrossRef] [PubMed]
109. Liu, S.; Jia, H.; Han, L.; Wang, J.; Gao, P.; Xu, D.; Yang, J.; Che, S. Nanosheet-constructed porous TiO_2-B for advanced lithium ion batteries. *Adv. Mater.* **2012**, *24*, 3201–3204. [CrossRef] [PubMed]
110. Gao, R.; Jiao, Z.; Wang, Y.; Xu, L.; Xia, S.; Zhang, H. Eco-friendly synthesis of rutile TiO_2 nanostructures with controlled morphology for efficient lithium-ion batteries. *Chem. Eng. J.* **2016**, *304*, 156–164. [CrossRef]
111. Wang, Z.; Zhang, F.; Xing, H.; Gu, M.; An, J.; Zhai, B.; An, Q.; Yu, C.; Li, G. Fabrication of nest-like TiO_2 hollow microspheres and its application for lithium ion batteries with high-rate performance. *Electrochim. Acta* **2017**, *243*, 112–118. [CrossRef]
112. Mujtaba, J.; Sun, H.; Zhao, Y.; Xiang, G.; Xu, S.; Zhu, J. High-performance lithium storage based on the synergy of atomic-thickness nanosheets of TiO_2(B) and ultrafine Co_3O_4 nanoparticles. *J. Power Sources* **2017**, *363*, 110–116. [CrossRef]

113. Cao, M.; Gao, L.; Lv, X.; Shen, Y. TiO_2-B@VS_2 heterogeneous nanowire arrays as superior anodes for lithium-ion batteries. *J. Power Sources* **2017**, *350*, 87–93. [CrossRef]
114. Lan, T.; Zhang, W.; Wu, N.-L.; Wei, M. Nb-doped rutile TiO_2 mesocrystals with enhanced lithium storage properties for lithium ion battery. *Chem.-Eur. J.* **2017**, *23*, 5059–5065. [CrossRef] [PubMed]
115. Chen, B.; Liu, E.; He, F.; Shi, C.; He, C.; Li, J.; Zhao, N. 2D sandwich-like carbon-coated ultrathin TiO_2@defect-rich MoS_2 hybrid nanosheets: Synergistic-effect-promoted electrochemical performance for lithium ion batteries. *Nano Energy* **2016**, *26*, 541–549. [CrossRef]
116. Chen, B.; Liu, E.; Cao, T.; He, F.; Shi, C.; He, C.; Ma, L.; Li, Q.; Li, J.; Zhao, N. Controllable graphene incorporation and defect engineering in MoS_2-TiO_2 based composites: Towards high-performance lithium-ion batteries anode materials. *Nano Energy* **2017**, *33*, 247–256. [CrossRef]
117. Lui, G.; Li, G.; Wang, X.; Jiang, G.; Lin, E.; Fowler, M.; Yu, A.; Chen, Z. Flexible, three-dimensional ordered macroporous TiO_2 electrode with enhanced electrode–electrolyte interaction in high-power Li-ion batteries. *Nano Energy* **2016**, *24*, 72–77. [CrossRef]
118. Jin, J.; Huang, S.-Z.; Shu, J.; Wang, H.-E.; Li, Y.; Yu, Y.; Chen, L.-H.; Wang, B.-J.; Su, B.-L. Highly porous TiO_2 hollow microspheres constructed by radially oriented nanorods chains for high capacity, high rate and long cycle capability lithium battery. *Nano Energy* **2015**, *16*, 339–349. [CrossRef]
119. Li, X.; Wu, G.; Liu, X.; Li, W.; Li, M. Orderly integration of porous TiO_2(B) nanosheets into bunchy hierarchical structure for high-rate and ultralong-lifespan lithium-ion batteries. *Nano Energy* **2017**, *31*, 1–8. [CrossRef]
120. Ren, G.; Hoque, M.N.F.; Liu, J.; Warzywoda, J.; Fan, Z. Perpendicular edge oriented graphene foam supporting orthogonal TiO_2(B) nanosheets as freestanding electrode for lithium ion battery. *Nano Energy* **2016**, *21*, 162–171. [CrossRef]
121. Liu, Y.; Elzatahry, A.A.; Luo, W.; Lan, K.; Zhang, P.; Fan, J.; Wei, Y.; Wang, C.; Deng, Y.; Zheng, G.; et al. Surfactant-templating strategy for ultrathin mesoporous TiO_2 coating on flexible graphitized carbon supports for high-performance lithium-ion battery. *Nano Energy* **2016**, *25*, 80–90. [CrossRef]
122. Wang, S.; Yang, Y.; Quan, W.; Hong, Y.; Zhang, Z.; Tang, Z.; Li, J. Ti^{3+}-free three-phase $Li_4Ti_5O_{12}/TiO_2$ for high-rate lithium ion batteries: Capacity and conductivity enhancement by phase boundaries. *Nano Energy* **2017**, *32*, 294–301. [CrossRef]
123. Chu, S.; Zhong, Y.; Cai, R.; Zhang, Z.; Wei, S.; Shao, Z. Mesoporous and nanostructured TiO_2 layer with ultra-high loading on nitrogen-doped carbon foams as flexible and free-standing electrodes for lithium-ion batteries. *Small* **2016**, *12*, 6724–6734. [CrossRef] [PubMed]
124. Reddy, M.A.; Kishore, M.S.; Pralong, V.; Varadaraju, U.V.; Raveau, B. Lithium Intercalation into Nanocrystalline Brookite TiO_2. *Electrochem. Solid-State Lett.* **2007**, *10*, A29–A31. [CrossRef]
125. Wu, Q.; Xu, J.; Yang, X.; Lu, F.; He, S.; Yang, J.; Fan, H.J.; Wu, M. Ultrathin anatase TiO_2 nanosheets embedded with TiO_2-B nanodomains for lithium-ion storage: Capacity enhancement by phase boundaries. *Adv. Energy Mater.* **2015**, *5*, 1401756. [CrossRef]
126. Mao, M.; Yan, F.; Cui, C.; Ma, J.; Zhang, M.; Wang, T.; Wang, C. Pipe-wire TiO_2–Sn@carbon nanofibers paper anodes for lithium and sodium ion batteries. *Nano Lett.* **2017**, *17*, 3830–3836. [CrossRef] [PubMed]
127. Wang, N.; Bai, Z.; Qian, Y.; Yang, J. Double-Walled Sb@TiO_{2-x} Nanotubes as a superior high-rate and ultralong-lifespan anode material for Na-ion and Li-ion batteries. *Adv. Mater.* **2016**, *28*, 4126–4133. [CrossRef] [PubMed]
128. Tahir, M.N.; Oschmann, B.; Buchholz, D.; Dou, X.; Lieberwirth, I.; Panthöfer, M.; Tremel, W.; Zentel, R.; Passerini, S. Extraordinary performance of carbon-coated anatase TiO_2 as sodium-ion anode. *Adv. Energy Mater.* **2016**, *6*, 1501489. [CrossRef] [PubMed]
129. Ni, J.; Fu, S.; Wu, C.; Maier, J.; Yu, Y.; Li, L. Self-supported nanotube arrays of sulfur-doped TiO_2 enabling ultrastable and robust sodium storage. *Adv. Mater.* **2016**, *28*, 2259–2265. [CrossRef] [PubMed]
130. Zhou, M.; Xu, Y.; Wang, C.; Li, Q.; Xiang, J.; Liang, L.; Wu, M.; Zhao, H.; Lei, Y. Amorphous TiO_2 inverse opal anode for high-rate sodium ion batteries. *Nano Energy* **2017**, *31*, 514–524. [CrossRef]
131. Zhang, Y.; Foster, C.W.; Banks, C.E.; Shao, L.; Hou, H.; Zou, G.; Chen, J.; Huang, Z.; Ji, X. Graphene-rich wrapped petal-like rutile TiO_2 tuned by carbon dots for high-performance sodium storage. *Adv. Mater.* **2016**, *28*, 9391–9399. [CrossRef] [PubMed]
132. Zhang, Y.; Wang, C.; Hou, H.; Zou, G.; Ji, X. Nitrogen doped/carbon tuning yolk-like TiO_2 and its remarkable impact on sodium storage performances. *Adv. Energy Mater.* **2017**, *7*, 1600173. [CrossRef]

133. Zhang, Y.; Ding, Z.; Foster, C.W.; Banks, C.E.; Qiu, X.; Ji, X. Oxygen vacancies evoked blue TiO_2(B) nanobelts with efficiency enhancement in sodium storage behaviors. *Adv. Funct. Mater.* **2017**, *27*, 1700856. [CrossRef]
134. Jamnik, J.; Maier, J. Nanocrystallinity effects in lithium battery materials Part IV. *Phys. Chem. Chem. Phys.* **2003**, *5*, 5215–5220. [CrossRef]
135. Pang, Q.; Kundu, D.; Cuisinier, M.; Nazar, L.F. Surface-enhanced redox chemistry of polysulphides on a metallic and polar host for lithium-sulphur batteries. *Nat. Commun.* **2014**, *5*, 4759. [CrossRef] [PubMed]
136. Yu, M.; Ma, J.; Song, H.; Wang, A.; Tian, F.; Wang, Y.; Qiu, H.; Wang, R. Atomic layer deposited TiO_2 on a nitrogen-doped graphene/sulfur electrode for high performance lithium–sulfur batteries. *Energy Environ. Sci.* **2016**, *9*, 1495–1503. [CrossRef]
137. Wei Seh, Z.; Li, W.; Cha, J.J.; Zheng, G.; Yang, Y.; McDowell, M.T.; Hsu, P.-C.; Cui, Y. Sulphur–TiO_2 yolk–shell nanoarchitecture with internal void space for long-cycle lithium–sulphur batteries. *Nat. Commun.* **2013**, *4*, 1331. [CrossRef] [PubMed]
138. Song, X.; Gao, T.; Wang, S.; Bao, Y.; Chen, G.; Ding, L.-X.; Wang, H. Free-standing sulfur host based on titanium-dioxide-modified porous-carbon nanofibers for lithium-sulfur batteries. *J. Power Sources* **2017**, *356*, 172–180. [CrossRef]
139. Liang, G.; Wu, J.; Qin, X.; Liu, M.; Li, Q.; He, Y.-B.; Kim, J.-K.; Li, B.; Kang, F. Ultrafine TiO_2 decorated carbon nanofibers as multifunctional interlayer for high-performance lithium–sulfur battery. *ACS Appl. Mater. Interfaces* **2016**, *8*, 23105–23113. [CrossRef] [PubMed]
140. Xiao, Z.; Yang, Z.; Wang, L.; Nie, H.; Zhong, M.; Lai, Q.; Xu, X.; Zhang, L.; Huang, S. A lightweight TiO_2/graphene interlayer, applied as a highly effective polysulfide absorbent for fast, long-life lithium-sulfur batteries. *Adv. Mater.* **2015**, *27*, 2891–2898. [CrossRef] [PubMed]
141. Mei, S.; Jafta, C.J.; Lauermann, I.; Ran, Q.; Kärgell, M.; Ballauff, M.; Lu, Y. Porous Ti_4O_7 particles with interconnected-pore structure as a high-efficiency polysulfide mediator for lithium-sulfur batteries. *Adv. Funct. Mater.* **2017**, *27*, 1701176. [CrossRef]
142. Zhao, Y.; Zhu, W.; Chen, G.Z.; Cairns, E.J. Polypyrrole/TiO_2 nanotube arrays with coaxial heterogeneous structure as sulfur hosts for lithium sulfur batteries. *J. Power Sources* **2016**, *327*, 447–456. [CrossRef]
143. Huang, J.-Q.; Wang, Z.; Xu, Z.-L.; Chong, W.G.; Qin, X.; Wang, X.; Kim, J.-K. Three-dimensional porous graphene aerogel cathode with high sulfur loading and embedded TiO_2 nanoparticles for advanced lithium–sulfur batteries. *ACS Appl. Mater. Interfaces* **2016**, *8*, 28663–28670. [CrossRef] [PubMed]
144. Zhou, G.; Zhao, Y.; Zu, C.; Manthiram, A. Free-standing TiO_2 nanowire-embedded graphene hybrid membrane for advanced Li/dissolved polysulfide batteries. *Nano Energy* **2015**, *12*, 240–249. [CrossRef]
145. Ding, B.; Xu, G.; Shen, L.; Nie, P.; Hu, P.; Dou, H.; Zhang, X. Fabrication of a sandwich structured electrode for high-performance lithium–sulfur batteries. *J. Mater. Chem. A* **2013**, *1*, 14280–14285. [CrossRef]
146. Zhang, H.; Banfield, J.F. Structural Characteristics and mechanical and thermodynamic properties of nanocrystalline TiO_2. *Chem. Rev.* **2014**, *114*, 9613–9644. [CrossRef] [PubMed]
147. Tayade, R.J.; Kulkarni, R.G.; Jasra, R.V. Photocatalytic degradation of aqueous nitrobenzene by nanocrystalline TiO_2. *Ind. Eng. Chem. Res.* **2006**, *45*, 922–927. [CrossRef]
148. Wagner, J.B.; Cavalca, F.C.; Damsgaard, C.D.; Duchstein, L.D.L.; Hansen, T.W.; Renu Sharma, P.A.C. Exploring the environmental transmission electron microscope. *Micron* **2012**, *43*, 1169–1175. [CrossRef] [PubMed]
149. Yuan, W.; Wang, Y.; Li, H.; Wu, H.; Zhang, Z.; Selloni, A.; Sun, C. Real-time observation of reconstruction Dynamics on TiO_2(001) surface under oxygen via an environmental transmission electron microscope. *Nano Lett.* **2016**, *16*, 132–137. [CrossRef] [PubMed]

nanomaterials

Review

Wet-Chemical Preparation of TiO_2-Based Composites with Different Morphologies and Photocatalytic Properties

Liqin Xiang and Xiaopeng Zhao *

Smart Materials Laboratory, Department of Applied Physics, Northwestern Polytechnical University, Xi'an 710129, China; lqxiang@nwpu.edu.cn
* Correspondence: xpzhao@nwpu.edu.cn

Received: 7 September 2017; Accepted: 2 October 2017; Published: 9 October 2017

Abstract: TiO_2-based composites have been paid significant attention in the photocatalysis field. The size, crystallinity and nanomorphology of TiO_2 materials have an important effect on the photocatalytic efficiency. The synthesis and photocatalytic activity of TiO_2-based materials have been widely investigated in past decades. Based on our group's research works on TiO_2 materials, this review introduces several methods for the fabrication of TiO_2, rare-earth-doped TiO_2 and noble-metal-decorated TiO_2 particles with different morphologies. We focused on the preparation and the formation mechanism of TiO_2-based materials with unique structures including spheres, hollow spheres, porous spheres, hollow porous spheres and urchin-like spheres. The photocatalytical activity of urchin-like TiO_2, noble metal nanoparticle-decorated 3D (three-dimensional) urchin-like TiO_2 and bimetallic core/shell nanoparticle-decorated urchin-like hierarchical TiO_2 are briefly discussed.

Keywords: TiO_2-based materials; photocatalysis; nanomorphology; preparation

1. Introduction

Based on its unique chemical and physical characteristics, titanium dioxide (TiO_2) has attracted much attention in many fields including paint pigments, photocatalysis, solar cells, antibacterial agents, electrical energy storage and some advanced functional materials [1–3]. The performance in these applications strongly depends on the microstructure, crystallinity and nanomorphology of TiO_2 [1]. In particular, in the photocatalysis field, although a new type of polymeric photocatalyst—that is, graphitic carbon nitride—has been intensively investigated recently due to its huge advantages—including metal-free contents, visible light absorption ability, suitable band gap for water splitting, stability, and being environmentally benign [4–7]—TiO_2 is still regarded as one of most ideal candidates for photocatalysis because of its strong oxidization, harmlessness to surroundings, chemical inactivity, good stability and low cost [1,3,8,9]. There are three important processes including photo excitation, bulk diffusion and the surface transfer of photoinduced charge carriers in photocatalysis [8]. Thus, the performance of photocatalysis depends strongly upon the charge transfer at the material surface and the light-response range of materials [1,10,11]. The processes of light harvesting and charge transfer efficiencies are affected mainly by the size, crystallinity and nanomorphology of TiO_2 materials [1,8,10–14]. The preparation and photocatalytic properties of TiO_2 with different morphologies including zero-dimensional (micro/nanospheres), one-dimensional (rods, tubes, and nanowires), two-dimensional (films, layers and sheets), and three-dimensional (porous spheres, urchin-like spheres) TiO_2 structures have been widely investigated in the past decade [15–21]. Different ways have been developed for preparing TiO_2 materials with different nanostructures. The general synthesis approaches for the fabrication of TiO_2 materials include sol-gel, hydrothermal and solvothermal techniques [1]. Controlling the microscopic structures of TiO_2 is still a challenge because TiO_2 precursors are highly reactive.

Pure TiO_2 is not a perfect photocatalyst due to the disadvantages of low photocatalytic efficiency and the narrow light-response region [8]. Doping metal ions or introducing noble metal nanoparticles onto the surface of TiO_2 was demonstrated to be one of the effective ways to enhance the photocatalytic efficiency because these TiO_2-based composites can combine the functions of TiO_2 and metal ions or noble metals [22–25]. Furthermore, the properties of the composites can be adjusted by controlling the ingredients and the microstructures of the TiO_2.

In the past two decades, our group has focused on the synthesis, electrorheological (ER) properties, luminescence properties and photocatalytic activities of TiO_2-based materials [26–32]. A series of TiO_2-based materials with different compositions, crystallinity and interior microstructures have been synthesized by different methods [33–36]. The TiO_2 particles with a familiar microstructure, such as solid spheres [37], hollow spheres [29], porous spheres [28,38], hollow porous spheres [39] and urchin-like spheres [30], were synthesized and characterized in detail. In addition, some TiO_2 composites with a special interior microstructure were also designed and synthesized [40–43]. According to the dielectric design, rare-earth-doped TiO_2 particles were synthesized by sol-gel methods [44–49]. Inspired by the structure of biological surfaces, a kind of composite particle possessing both nano- and micro-scale structures was prepared via a hydrothermal method [50,51]. TiO_2 particles with a cell-like structure were also synthesized [52]. It is noteworthy that the TiO_2-based materials described above show excellent properties in different applications. For example, the rare-earth-doped TiO_2, and the micro- or nano-structured composites with TiO_2 have been demonstrated to show a distinct enhancement in their ER properties [44,50–52]. The hollow Sm^{3+}-doped TiO_2 and the monodisperse mesoporous Eu-doped TiO_2 spheres have shown good luminescent performance [28,29]. The urchin-like TiO_2 and urchin-like TiO_2 decorated with Au, Ag, Co@Au or Co@Ag nanoparticles have shown significant improvement in photocatalytic activities [30–32].

Until now, there have been many review articles introducing the progress made in the field of TiO_2-based materials [1–3,8]. Based on our group's research work on TiO_2 materials, this review is primarily concentrated on the preparation of TiO_2 composites with different morphologies and the photocatalytic activities of urchin-like TiO_2 composites.

2. Preparation of TiO_2 and TiO_2-Based Composites with Different Morphologies

2.1. *Micro- and Nano-Spheres*

2.1.1. Solid Spheres

Spherical particles with a specific size can be used in many fields, such as photonic crystals, pigments, and so on [15]. In order to obtain monodisperse spherical TiO_2 particles, many methods have been developed. However, it is still a challenge to control the morphology and size of TiO_2 microspheres because of the high reactivity of precursors. Increasing the charge of the particle surface and the steric repulsion of the particles are effective methods of controlling the stability of TiO_2 microspheres [53]. We have reported a simple and reproducible sol-gel method for synthesizing well-defined spherical TiO_2 particles with diameters within 200–800 nm. In this method, polymers including polyethylene glycol (PEG), poly(ethylene oxide)$_{106}$-poly(propylene oxide)$_{70}$-poly(ethylene oxide)$_{106}$ (F127) copolymer, octadecylamine (ODA), and surfactant Span-80 were used to control the size of TiO_2 particles [37]. For example, quasi-monodisperse TiO_2 submicron spheres were synthesized by controlling the hydrolysis of tetrabutyl titanate in ethanol containing the above polymers and small amounts of deionized water. During this process, depending on the used polymer, the transmission time from the transparent solution into white suspension changed from several seconds to minutes. As soon as the transparent solution changed into white suspension, the stirring had to be stopped and the suspension was further aged for 8 h to form quasi-monodisperse TiO_2 submicron spheres.

After high temperature annealing, the spheres were crystallized into the anatase phase. Figure 1 shows the quasi-monodisperse TiO_2 submicron spheres with different diameters within 200–800 nm, synthesized with different polymers.

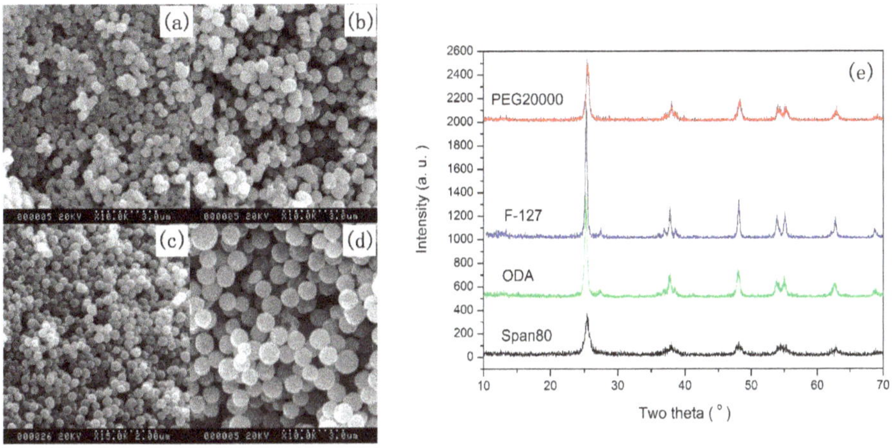

Figure 1. SEM (Scanning electron microscopy) images ((**a**) PEG (polyethylene glycol) 20000, (**b**) ODA (octadecylamine), (**c**) F-127, (**d**) Span80) and the XRD (X-ray diffraction) patterns (**e**) of TiO_2 spheres synthesized by adding different polymers [37].

2.1.2. Hollow Spheres

Due to high specific surface area and low density, hollow structured materials have been widely used in many fields [54]. TiO_2 hollow spheres with a well-defined crystal phase are highly desirable for photocatalysis use [55–57]. Hollow structured TiO_2 can be feasibly synthesized by hard template and soft template methods. Compared to the soft template method, the hard template method is simpler, and so it is frequently used.

An et al. have used polystyrene (PS) spheres as a hard template to prepare hollow Sm^{3+}-doped TiO_2 spheres [29]. The schematic illustration of the formation mechanism is shown in Figure 2. Since the surface of PS spheres obtained by surfactant-free microemulsion polymerization is negatively charged, no additional surface modification of PS spheres is needed for the next coating of TiO_2. In an ethanol/acetonitrile mixed solvent, a small amount of ammonia was used to induce the hydrolysis of tetrobutyl titanate to form the amorphous Sm^{3+}-doped TiO_2 coating layer on the surface of the PS spheres. After washing with ethanol, drying, and annealing, hollow TiO_2:Sm^{3+} spheres, as shown in Figure 3, could be obtained.

Figure 2. Schematic illustration of the formation mechanism for hollow TiO_2:Sm^{3+} spheres [29].

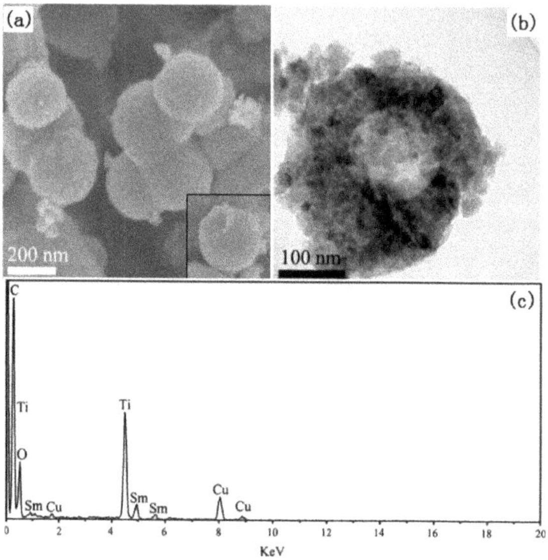

Figure 3. (**a**) SEM image, (**b**) TEM (Transmission Electron Microscopy) image and (**c**) EDS (Energy Dispersive Spectrum) of TiO_2:Sm^{3+} hollow spheres [29].

2.1.3. Porous Spheres

Due to their high surface area, porous materials are very popular for different applications including energy storage, solar cells and catalyzers [58–60]. Mesoporous TiO_2-based materials have attracted much attention for their enhanced reactivity and light harvesting [60]. The macrochannels in mesoporous TiO_2 particles have served as a light-transfer path that can introduce incident photon flux to the interior surface of the TiO_2 particles [58]. A mesoporous structure gives light waves more chances to penetrate deep inside the photocatalyst and more light waves are captured. The crystallinity, pore size and composition are important for tuning the properties of mesoporous TiO_2 spheres [1,8,60]. There has been intensive research concentrated on the design and preparation of porous TiO_2 materials with unique porosities and tunable pore sizes [61–65].

We have synthesized mesoporous Ce-doped TiO_2 spheres by a low-temperature hydrothermal method by using neutral dodecylamine (DDA) as a surfactant and tetrabutyl titanate as a Ti source [64]. To control the rate of hydrolysis of tetrabutyl titanate, a solvent mixture of ethanol and propanol (2:1, *v*/*v*) was used. No additional water was used to initialize the hydrolysis and condensation reaction of the tetrabutyl titanate, due to the used $CeCl_3{\cdot}7H_2O$ containing structured water. After the $CeCl_3{\cdot}7H_2O$ was dissolved, the structured water was released. DDA was able to make the solution alkaline and this made it easy to increase the rate of hydrolysis of the tetrabutyl titanate. However, the dissolution of $CeCl_3{\cdot}7H_2O$ also could result in a decrease of the pH value of solution. Thus, $CeCl_3{\cdot}7H_2O$ could service as not only as a dopant but also as an initiator and buffer. After reaction, a precipitate was formed and it was further refluxed for 2 h at 80 °C in an acid solution to get rid of the template and obtain sphere-like mesoporous Ce-doped TiO_2 particles with a diameter of 100–1000 nm as shown in Figure 4. The XRD (X-ray diffraction) patterns in Figure 4 show that the TiO_2 particles are semi-crystalline. The formed anatase crystalline size is very small, about 2–3 nm. In addition, from the TEM (Transmission Electron Microscopy) image, it could be found that the pore structure was worm-like, with a size of 2–3 nm. The Brunauer-Emmett-Teller (BET) surface area of the mesoporous Ce-doped TiO_2 was 118 m^2/g, which is much higher than the 9.6 m^2/g of single-doped TiO_2 obtained without a surfactant.

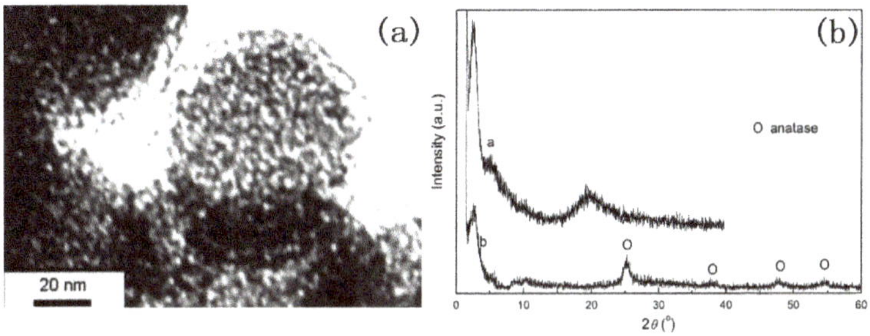

Figure 4. TEM photograph of mesoporous-doped TiO_2 and XRD patterns (**a**) before hydrothermal and acid treatment, (**b**) after hydrothermal and acid treatment of mesoporous-doped TiO_2 [64].

The mesoporous Eu-doped TiO_2 spheres have also been developed by the hydrolysis of tetrabutyl titanate [28]. To increase the thickness of the pore wall, nonionic copolymer Pluronic F-127 was used as a template agent. Europium ethoxide was specially prepared as a doping agent in order to increase the rate of hydrolysis and condensation of the tetrabutyl titanate. Small amounts of water were added to initiate the reaction under stirring. As soon as the solution became slightly white after several minutes, the stirring was stopped. Meanwhile, in order to control the water content, the preparation was conducted under the protection of flowing N_2. After aging for 24–48 h at 35–40 °C, the suspended particles were filtered and washed with ethanol several times. The final products were obtained after calcination at 400 °C for 4 h. The synthesized Eu-doped TiO_2 particles have a spherical morphology and a mesoporous structure, with a pore size of 7–10 nm. The special surface area of the phosphor particles is 158 m^2/g. The high resolution TEM images in Figure 5c show that the pore wall is semi-crystalline that many anatase nanocrystallites are dispersed in the amorphous TiO_2. The XRD patterns showed in Figure 5d have indicated that no peaks corresponding to the europium compound was detected and no shift of the anatase peaks was observed after doping with Eu^{3+}. It can be concluded that Eu^{3+} ions are mainly dispersed in the amorphous TiO_2 region.

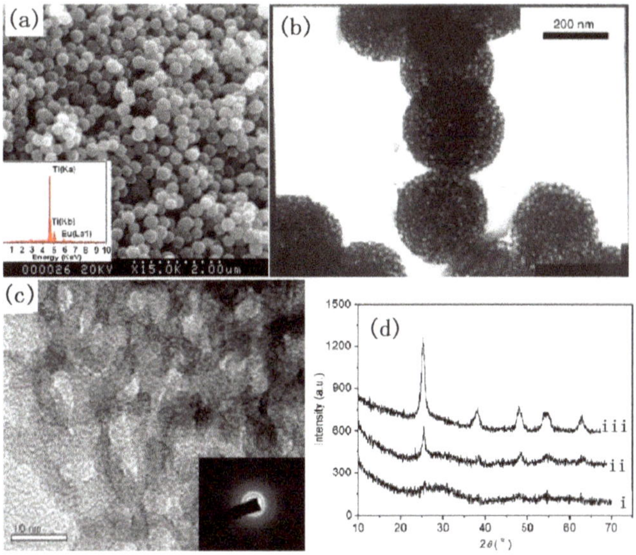

Figure 5. (**a**) SEM image and EDS spectra, (**b**) TEM image, (**c**) high resolution TEM image and corresponding electron diffraction pattern of monodisperse mesoporous after 400 °C calcinations, (**d**) XRD patterns: (i) before calcinations, (ii) after 400 °C calcinations, and (iii) after 500 °C calcinations [28].

2.1.4. Hollow and Porous Spheres

Compared to single hollow spheres, hollow TiO_2 spheres with a porous shell are more interesting in photocatalysis because they can provide a higher surface area and active site points, decreased diffusion resistance, and increased accessibility to reactants [66]. Several methods have been reported to synthesize hollow TiO_2 spheres with a porous shell [66,67]. Among these methods, the template method is the most popular. Different sacrificial templates can be used for controlling the size and morphology of such a hollow nanostructure. The template method followed by a hydrothermal or calcination treatment is often used to synthesize hollow TiO_2 spheres with a crystalline shell [39]. Figure 6 shows a typical process of preparing hollow TiO_2 spheres with a crystalline shell. In this process, amorphous TiO_2 was firstly coated onto the surface of SiO_2 spheres by the sol-gel method in an alkaline solution. Then, the composite microspheres were subjected to a hydrothermal or calcination treatment. Meanwhile, the amorphous TiO_2 was crystallized into nanocrystals and the mesoporous structure was formed by nanocrystal stacking. Finally, the SiO_2 core was removed by etching in an alkaline solution. As shown in Figure 7, the sample prepared by hydrothermal treatment had a mean diameter of 620 nm with a 180 nm thick mesoporous TiO_2 shell. The BET surface area was 231.1 m^2/g and the pore size was 6.5 nm. However, the sample prepared by calcination had a mean diameter of 440 nm with a 90 nm thick mesoporous TiO_2 shell. The BET surface area was 158.3 m^2/g.

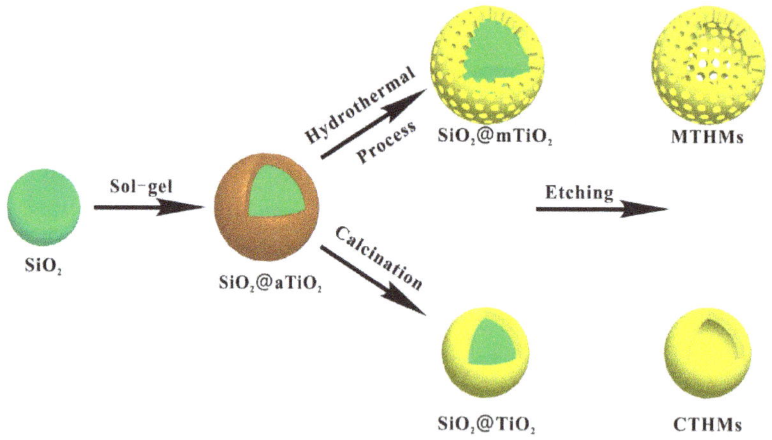

Figure 6. Schematic illustration of the process for the fabrication of the mesoporous TiO_2 hollow microspheres [39].

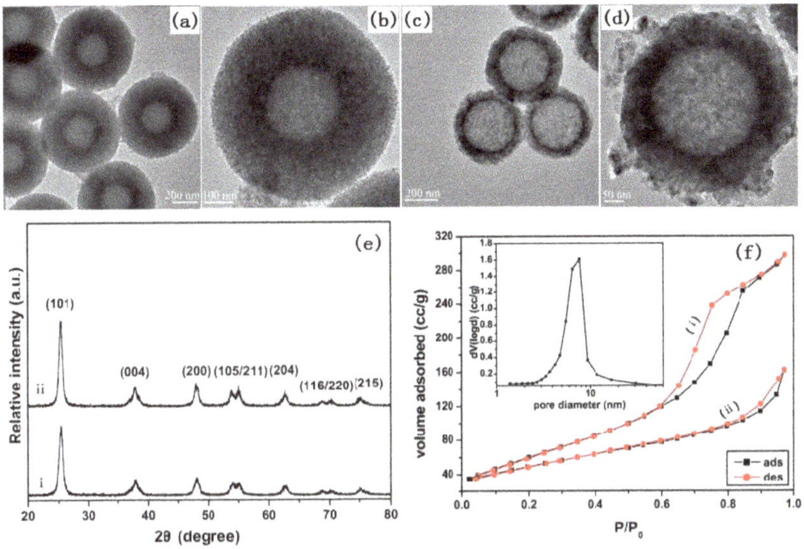

Figure 7. TEM images (**a**,**b**), XRD patterns (**e**-i) and nitrogen adsorption-desorption isotherms (**f**-i) of the sample prepared by the hydrothermal process; TEM images (**c**,**d**), XRD patterns (**e**-ii) and nitrogen adsorption-desorption isotherms (**f**-ii) of the sample prepared by the calcination process [39].

2.2. *3D Urchin-Like Hierarchical Particles*

Urchin-like microspheres possess an epitaxial multilevel structure. The unique micro/nano hierarchical structure has two obvious advantages over single nano-scale or micro-scale structures when they are used as photocatalysts [54,68]. One is that urchin-like TiO_2 is more efficient at absorbing incidental light because of the increase of multiple-reflection of the hierarchical microspheres [54]. The other is that urchin-like hierarchical TiO_2 is easy to separate from waste water by the filtration or sedimentation methods. The template-assisted method is a familiar approach to prepare the hierarchical materials. However, it is troublesome to remove templates from products. Impurities are easily introduced into products in the process of utilization and removal of templates. The template-free method is accepted as an ideal strategy which can overcome the drawbacks. Recently, TiO_2 particles with different hierarchical structures have been successfully fabricated via the template-free method [69–73].

2.2.1. Urchin-Like Hierarchical TiO_2 Particles

We have developed a synthesis of a kind of 3D urchin-like TiO_2 microspheres via a solvothermal method without adding any surfactant or template [30]. Tetrabutyl titanate and titanium tetrachloride ($TiCl_4$) aqueous solution were used as the reactant, and toluene was used as the solvent. The solvothermal reaction took place in a Teflon-lined autoclave at 150 °C for 24 h. Sea-urchin-like 3D hierarchical TiO_2 microspheres with a uniform size of 2.5–3.0 μm were obtained (Figure 8). The 3D hierarchical microspheres were made of single crystalline rutile nanoneedles with diameters about 20–40 nm, which grew radially from the core of the microspheres. The morphology and crystal phase of the 3D hierarchical TiO_2 microspheres could be influenced by some factors, such as the ratio of tetrabutyl titanate to $TiCl_4$, the solvothermal temperature, and so on. By tracing the particle morphology change by SEM and XRD techniques, we concluded that the formation of 3D hierarchical TiO_2 microspheres mainly concerned three steps, i.e., nucleation, self-assembly, dissolution and recrystallization, as depicted in Figure 9. In the nucleation stage, nanoparticles were formed. Then, the nanoparticles assembled into microspheres. Finally, the microspheres gradually changed into the urchin-like hierarchical structure by dissolution and recrystallization.

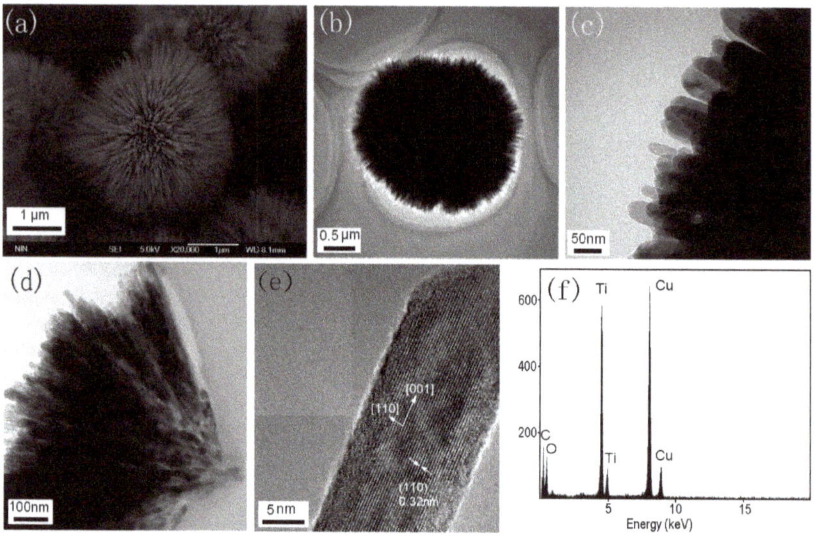

Figure 8. SEM image (**a**), TEM images (**b–e**), and EDS (**f**) of the urchin-like hierarchical TiO_2 [30].

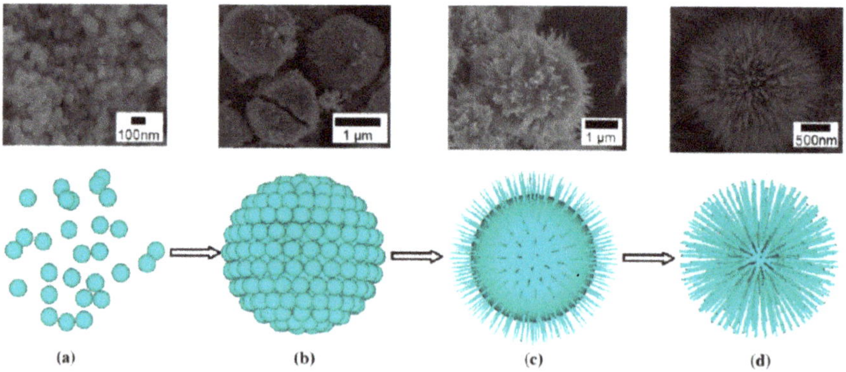

Figure 9. Schematic illustration of the formation process of 3D urchin-like hierarchical TiO_2: (**a**) nanoparticle; (**b**) microsphere; (**c**) similar urchin-like sphere; (**d**) urchin-like sphere [30].

2.2.2. Cr Doped Urchin-Like Hierarchical TiO_2 Particles

Urchin-like Cr-doped TiO_2 particles could be also synthesized by the same solvothermal method described above in a solution of titanium tetrabutyl titanate dissolving $CrCl_3$ [73]. The morphology of Cr-doped TiO_2 particles is characterized by SEM images shown in Figure 10. The mean particle size of the hierarchical microspheres can be adjusted within 1–5 μm and the diameter of the nanorods is about 20–30 nm. The EDS results showed that the content of the Cr element in the Cr-doped TiO_2 particles was ~2.9 mol %.

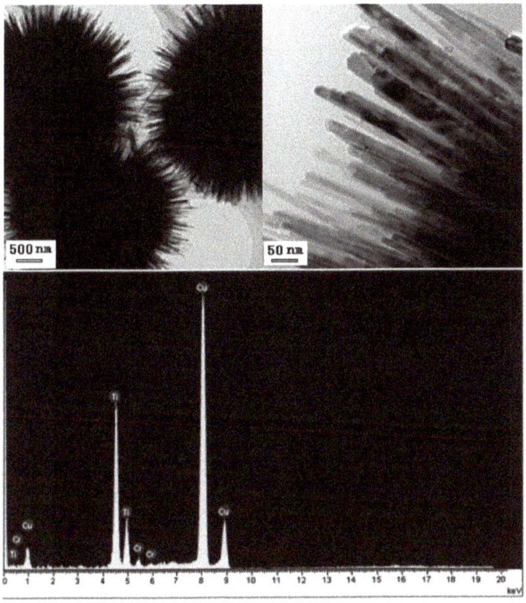

Figure 10. TEM images and EDS spectra of Cr-doped urchin-like TiO2 particles [73].

2.2.3. Noble Metal Nanoparticle-Decorated 3D Urchin-Like TiO_2 Particles

Noble metal nanoparticle-decorated semiconductors are interesting for photocatalysis because of their combined properties [74]. Decorating the noble metal nanoparticles (e.g., Au, Ag and Pt) onto the surface of TiO_2 is an effective method to improve the photocatalytic activity because not only light-harvesting efficiency can be enhanced due to the surface plasmon resonance of noble metal nanoparticles, but the recombination of surface radicals can also be slowed down by capturing photogenerated electrons of noble metal nanoparticles [75–79]. Figure 11 shows a schematic illustration of 3D urchin-like hierarchical TiO_2 microspheres decorated with Au nanoparticles via a two-step wet-chemical process [31]. In the first step, the surface of the urchin-like TiO_2 microspheres was modified with APTES (3-aminopropyl-triethoxysilane) that possess amidocyanogen. Then, the modified particles were decorated with Au nanoparticles in $HAuCl_4$ aqueous solution by the reduction of $NaBH_4$. Since the amidocyanogen could interact with Au nanoparticles by a weak covalent bond, Au nanoparticles were closely attached to the surface of the TiO_2 nanostructures, as shown in Figure 12. It was seen that the Au nanoparticles with diameters of about 2–10 nm mainly adhered to the surface of the needles uniformly. Most of the Au nanoparticles possess a rhombic dodecahedra structure. The UV-Vis (ultraviolet-visible) spectra show an absorption band located at the wavelength of about 530 nm due to the surface plasmon resonance of Au nanoparticles.

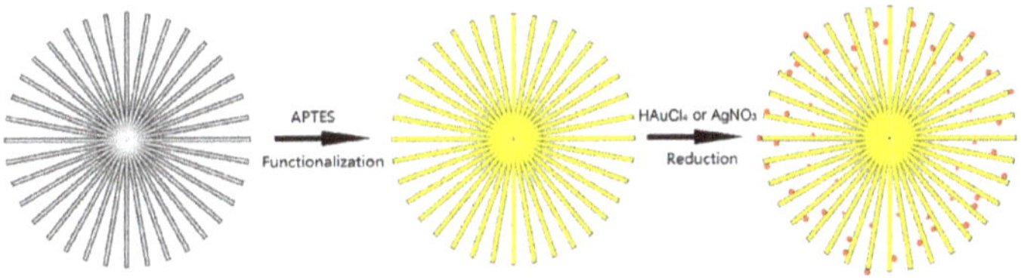

Figure 11. Schematic illustration of the synthesis process of urchin-like TiO_2 decorated with Au or Ag nanoparticles [31].

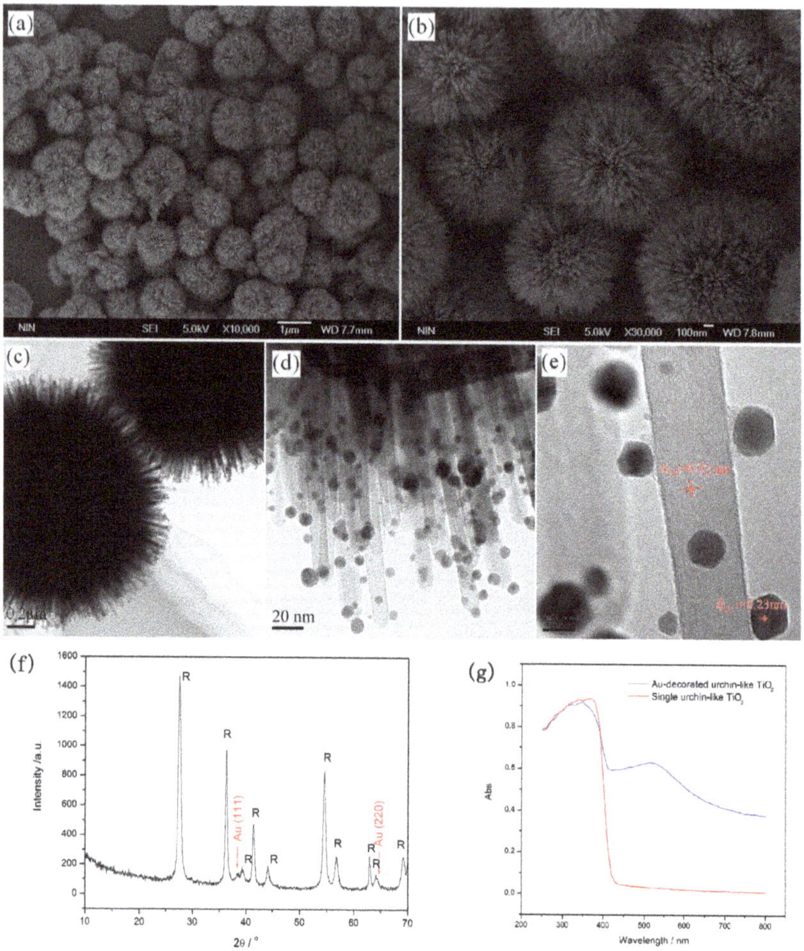

Figure 12. SEM images (**a**,**b**), TEM images (**c**–**e**), XRD pattern (**f**) and UV-Vis absorption spectra (**g**) of Au-decorated 3D urchin-like TiO_2 nanostructures [31].

The urchin-like TiO_2 microspheres decorated with Ag nanoparticles could be also prepared by the similar process. As shown in Figure 13, the Ag nanoparticles with diameters of about 2 nm are decorated on the TiO_2 nanoneedles homogeneously. A broad absorption band at around 500 nm, corresponding to the surface plasmon resonance of the Ag nanoparticles, appears in the UV-Vis absorption spectrum (Figure 13f).

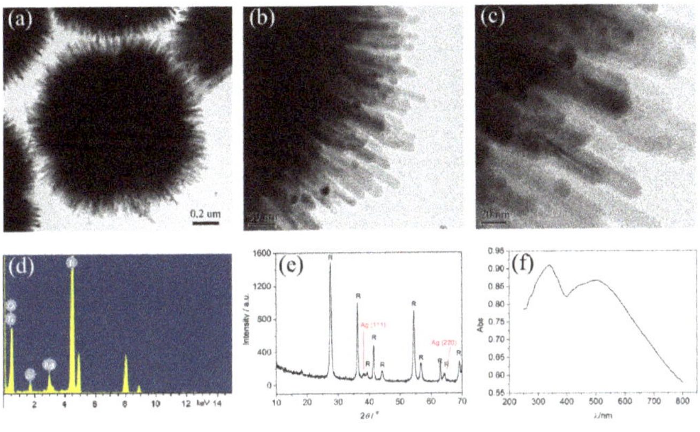

Figure 13. TEM images (**a–c**), EDS (**d**), XRD (**e**), and UV-Vis absorption spectrum (**f**) of Ag-decorated 3D urchin-like TiO_2 nanostructures [31].

2.2.4. Core/Shell-Structured Bimetallic Nanoparticle-Decorated 3D Urchin-Like Hierarchical TiO_2 Particles

Bimetallic nanostructures often show a more excellent comprehensive performance over their monometallic counterpart [80]. Especially core/shell bimetallic nanostructures with a magnetic core and a noble-metallic shell have aroused researchers' interest [81–85]. The magnetic core can offer a drive force for the recycling of samples, while the noble-metal shell can improve the optical properties [85]. Figure 14 shows a typical preparation process of a kind of 3D urchin-like hierarchical TiO_2 decorated with a bimetallic core/shell nanoparticle (Co@Au and Co@Ag). The preparation includes three steps, i.e., surface activation, electroless plating and in-situ replacement [32]. First, the surface of the urchin-like TiO_2 microspheres was activated by implanting Pd nanodots. Then, Co nanoparticles were formed and adhered to the nanoneedle surface of the urchin-like TiO_2 microspheres by electroless plating. Finally, Ag or Au were further formed and coated on the surface of the Co nanoparticles by an in-situ replacement process.

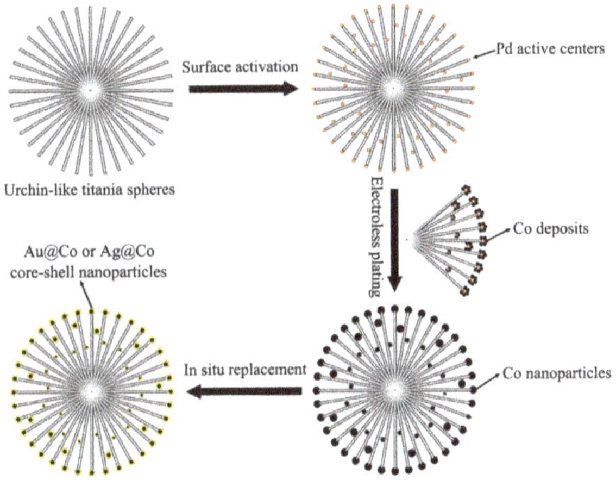

Figure 14. A schematic synthesis process of urchin-like TiO_2 decorated with core/shell-structured Co@Au or Co@Ag bimetallic nanoparticles [32].

The SEM and TEM images in Figure 15 show the morphology of Co@Au/TiO_2 composites. It can be seen that many core/shell nanostructured nanoparticles with diameters of 10–80 nm are attached to the surface of TiO_2 nanorods. The images of the elemental mapping of core/shell nanoparticles further identify that the core is Co and the shell is Au. The size and distribution of the bimetallic particles can be adjusted by controlling the ratio of Co to TiO_2 during the electroless plating process. The thickness of the Au or Ag shell could be controlled by adjusting the concentration of $HAuCl_4$ or $AgNO_3$ in the solution and the reaction time. Both the Co@Au/TiO_2 and Co@Ag/TiO_2 particles showed good response to an applied external magnetic field [32].

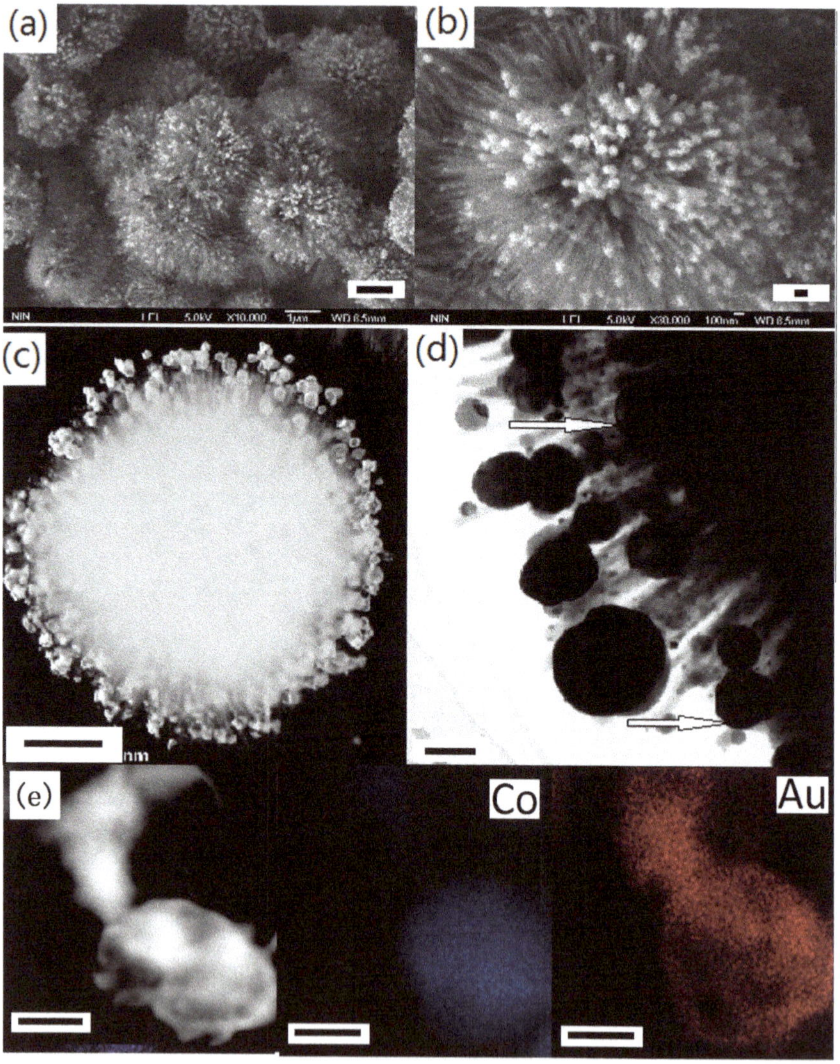

Figure 15. The SEM and TEM images of Co@Au/TiO_2 composites: (**a**,**b**) SEM images, (**c**,**d**) TEM images, (**e**) high-resolution TEM images; (**e**) the local elemental mapping of Co and Au (Scale bar = 1 μm for (**a**); scale bar = 100 nm for (**b**); scale bar = 500 nm for (**c**); scale bar = 100 nm for (**d**); scale bar = 50 nm for (**e**)) [32].

2.2.5. Photocatalytic Activity of Urchin-Like Hierarchical TiO_2 and Their Composites

Although TiO_2 is an ideal candidate for photocatalysis because of its strong oxidization, harmlessness to surroundings, chemical inactivity, good stability and low cost, the main weakness of TiO_2 is the lack of visible light response due to the large band gap. Therefore, the question of how to increase the efficiency of visible light harvesting is an important research topic in this field [1]. Although single controlling the morphology of TiO_2 materials cannot increase the efficiency of visible light harvesting, it is possible to enhance the visible light harvesting of TiO_2 composites by combining ion doping or noble metal decoration with morphology control. Ion doping or noble metal decoration can induce or increase the visible light absorption of TiO_2, while the absorption effect can be further enhanced by other effects from material morphology, such as decreased light scattering or increased multiple reflection, etc.

The photocatalytic efficiency of commercial P25, urchin-like TiO_2 and Au or Ag-decorated urchin-like TiO_2 was evaluated by degrading MB (methyl blue) under UV-Vis light irradiation. It was found that the photocatalytic degradation efficiency under the same conditions followed the order: $Ag/TiO_2 > Au/TiO_2 > TiO_2 >$ P25 (Figure 16). By the photoluminescence spectra, Au or Ag nanoparticles decorated on the surface of TiO_2 were demonstrated to be able to effectively capture photogenerated electrons and prevent electron/hole recombining (Figure 16). In addition, the urchin-like micro/nano hierarchical structure also may increase the visible light harvesting efficiency by the multiple-reflection of nanoneedles. These should be responsible for the enhanced photocatalytic efficiency of urchin-like TiO_2 microspheres after decoration with Au or Ag nanoparticles. Similarly, the light-harvesting efficiency could also be enhanced by decorating with Co@Au or Co@Ag bimetallic nanoparticles, as shown in Figure 17. As a result, the photocatalytical efficiency of urchin-like TiO_2 was enhanced obviously, as shown in Figure 17, by the experiment of decolorizing methyl blue (MB) solution.

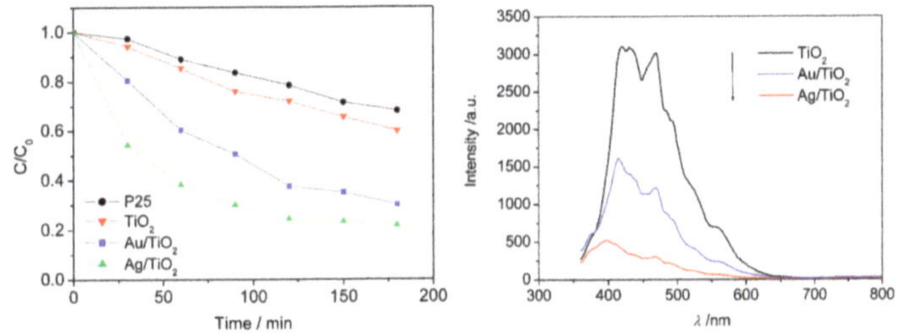

Figure 16. Photodegradation curves of MB (methyl blue) in the presence of P25, TiO_2, Au/TiO_2 and Ag/TiO_2 (**left**); Photoluminence spectra (λ_{ex} = 215 nm) of pure urchin-like TiO_2, Au/TiO_2 and Ag/TiO_2 (**right**) [31].

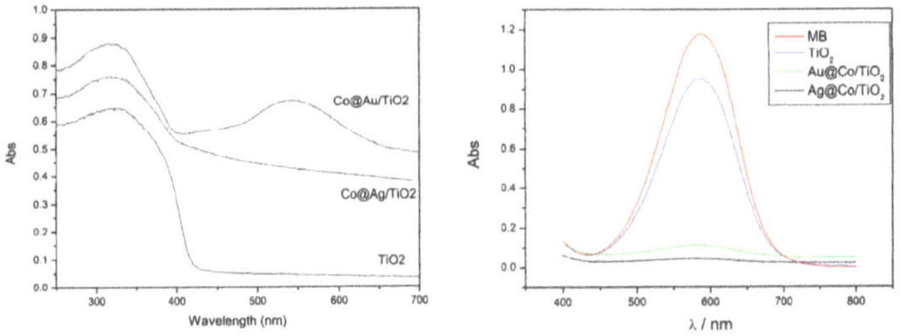

Figure 17. UV-Vis absorption spectra of TiO_2, $Co@Au/TiO_2$ and $Co@Ag/TiO_2$ (**left**); UV-Vis absorption spectra of MB before and after degradation with TiO_2, $Co@Au/TiO_2$ and $Co@Ag/TiO_2$ for 15 min at room temperature (**right**) [32].

3. Summary

Based on our group's research work, we provided a brief review of the synthesis of TiO_2 with different morphologies and the photocatalytic properties of urchin-like TiO_2, noble metal nanoparticle-decorated 3D urchin-like TiO_2 and core/shell-structured bimetallic nanoparticle-decorated 3D urchin-like hierarchical TiO_2. The examples of the fabrication of solid spheres, hollow spheres, porous spheres, and porous and hollow microspheres of anatase TiO_2-based materials were introduced. The synthesis and photocatalytic efficiency of urchin-like rutile TiO_2, urchin-like rutile TiO_2 nanostructures decorated with Au or Ag nanoparticles and core/shell-structured bimetallic nanoparticles (Co@Au and Co@Ag) were especially introduced. The results of photocatalytic tests show that 3D urchin-like hierarchical structures have unique merits in the efficient harvesting of solar light, and decorating Au, Ag or bimetallic nanoparticles on the surface of 3D urchin-like TiO_2 can promote photoinduced charge-carrier separation.

Acknowledgments: This work was supported by the National Natural Science Foundation of China (Grant Nos. 51502247 and 11674267)

Author Contributions: Liqin Xiang and Xiaopeng Zhao wrote the paper.

Conflicts of Interest: The authors declare no conflict of interest.

References

1. Fattakhova-Rohlfing, D.; Zaleska, A.; Bein, T. Three-Dimensional Titanium Dioxide Nanomaterials. *Chem. Rev.* **2014**, *114*, 9487–9558. [CrossRef] [PubMed]
2. Abdullah, N.; Kamarudin, S.K. Titanium dioxide in fuel cell technology: An overview. *J. Power Sources* **2015**, *278*, 109–118. [CrossRef]
3. Chen, X.B.; Mao, S.S. Titanium dioxide nanomaterials: Synthesis, properties, modifications, and applications. *Chem. Rev.* **2007**, *107*, 2891–2959. [CrossRef] [PubMed]
4. Zhang, G.; Lan, Z.A.; Wang, X. Conjugated Polymers: Catalysts for Photocatalytic Hydrogen Evolution. *Angew. Chem. Int. Ed.* **2016**, *55*, 15712–15727. [CrossRef] [PubMed]
5. Kessler, F.K.; Zheng, Y.; Schwarz, D.; Merschjann, C.; Schnick, W.; Wang, X.; Bojdys, M.J. Functional carbon nitride materials-design strategies for electrochemical devices. *Nat. Rev. Mater.* **2017**, *2*, 17030. [CrossRef]
6. Zhang, G.; Lan, Z.A.; Wang, X. Surface engineering of graphitic carbon nitride polymers with cocatalysts for photocatalytic overall water splitting. *Chem. Sci.* **2017**, *8*, 5261–5274. [CrossRef] [PubMed]
7. Ong, W.J.; Tan, L.L.; Ng, Y.H.; Yong, S.T.; Chai, S.P. Graphitic carbon nitride (g-C_3N_4)-based photocatalysts for artificial photosynthesis and environmental remediation: Are we a step closer to achieving sustainability? *Chem. Rev.* **2016**, *116*, 7159–7329. [CrossRef] [PubMed]
8. Liu, G.; Wang, L.; Yang, H.G.; Cheng, H.M.; Lu, G.Q. TiO_2-based photocatalysts-crystal growth, doping and heterostructuring. *J. Mater. Chem.* **2010**, *20*, 831–843. [CrossRef]
9. Ruokolainen, M.; Ollikainen, E.; Sikanen, T.; Kotiaho, T.; Kostiainen, R. Oxidation of tyrosine-phosphopeptides by titanium dioxide photocatalysis. *J. Am. Chem. Soc.* **2016**, *138*, 7452–7455. [CrossRef] [PubMed]
10. Kisch, H. Semiconductor photocatalysis-mechanistic and synthetic aspects. *Angew. Chem. Int. Ed.* **2013**, *52*, 812–847. [CrossRef] [PubMed]
11. Du, J.; Lai, X.; Yang, N.; Zhai, J.; Kisailus, D.; Su, F.; Wang, D.; Jiang, L. Hierarchically Ordered Macro-Mesoporous TiO_2-Graphene Composite Films: Improved Mass Transfer, Reduced Charge Recombination, and Their Enhanced Photocatalytic Activities. *ACS Nano* **2011**, *5*, 590–596. [CrossRef] [PubMed]
12. Li, Y.F.; Liu, Z.P. Particle size, shape and activity for photocatalysis on TiO_2 anatase nanoparticles in aqueous surroundings. *J. Am. Chem. Soc.* **2011**, *133*, 15743–15752. [CrossRef] [PubMed]
13. Meulen, T.; Mattson, A.; Oesterlund, L. A comparative study of the photocatalytic oxidation of propane on anatase, rutile, and mixed-phase anatase-rutile TiO_2 nanoparticles: Role of surface intermediates. *J. Catal.* **2007**, *25*, 131–138.
14. Kang, X.W.; Chen, S.W. Photocatalytic reduction of methylene blue by TiO_2 nanotube arrays: Effects of TiO_2 crystalline phase. *J. Mater. Sci.* **2010**, *45*, 2696–2702. [CrossRef]

15. Tanaka, S.; Nogami, D.; Tsuda, N.; Miyake, Y. Synthesis of highly-monodisperse spherical TiO_2 particles with diameters in the submicron range. *J. Colloid Interface Sci.* **2009**, *334*, 188–194. [CrossRef] [PubMed]
16. Momeni, M.; Ghayeb, Y. Fabrication, characterization and photoelectrochemical performance of chromium-sensitized TiO_2 nanotubes as efficient photoanodes for solar water splitting. *J. Solid State Electrochem.* **2016**, *20*, 683–689. [CrossRef]
17. Ghicov, A.; Schmuki, P. Self-ordering electrochemistry: A review on growth and functionality of TiO_2 nanotubes and other self-aligned MO_x structures. *Chem. Commun.* **2009**, 2791–2808. [CrossRef] [PubMed]
18. Zhang, J.; Zhu, Z.; Tang, Y.; Müllen, K.; Feng, X. TiO_2 Nanosheet-Mediated Construction of a Two-Dimensional TiO_2/Cadmium Sulfide Heterostructure for High Hydrogen Evolution Activity. *Adv. Mater.* **2014**, *26*, 734–738. [CrossRef] [PubMed]
19. Bartl, M.H.; Boettcher, S.W.; Frindell, K.L.; Stucky, G.D. 3-D molecular assembly of function in TiO_2-based composite material systems. *Acc. Chem. Res.* **2005**, *38*, 263–271. [CrossRef] [PubMed]
20. Zhu, T.; Li, J.; Wu, Q. Construction of TiO_2 hierarchical nanostructures from nanocrystals and their photocatalytic properties. *ACS Appl. Mater. Interfaces* **2011**, *3*, 3448–3453. [CrossRef] [PubMed]
21. Li, H.; Bian, Z.; Zhu, J.; Zhang, D.; Li, G.; Huo, Y.; Li, H.; Lu, Y. Mesoporous TiO_2 spheres with tunable chamber lasmon and enhanced photocatalytic activity. *J. Am. Chem. Soc.* **2007**, *129*, 8406–8407. [CrossRef] [PubMed]
22. Choi, W.; Termin, A.; Hoffman, M.R. The role of metal ion dopants in quantum-sized TiO_2: Correlation between photoreactivity and charge carrier recombination dynamics. *J. Phys. Chem.* **1994**, *98*, 13669–13679. [CrossRef]
23. Su, R.; Tiruvalam, R.; He, Q.; Dimitratos, N.; Kesavan, L.; Hammond, C.; Lopez-Sanchez, J.A.; Bechstein, R.; Kiely, C.J.; Hutchings, G.J.; et al. Promotion of phenol photodecomposition over TiO_2 using Au, Pd, and AuPd nanoparticles. *ACS Nano* **2012**, *6*, 6284–6292. [CrossRef] [PubMed]
24. Smirnova, N.; Vorobets, V.; Linnik, O.; Manuilov, E.; Kolbasovb, G.; Eremenkoa, A. Photoelectrochemical and photocatalytic properties of mesoporous TiO_2 films modified with silver and gold nanoparticles. *Surf. Interface Anal.* **2010**, *42*, 1205–1208. [CrossRef]
25. Arabtzis, I.M.; Stergiopoulos, T.; Andreeva, D.; Kitova, S.; Neophytides, S.G.; Falaras, P. Charaterizarion and photocatalytic activity of Au/TiO_2 thin films for azo-dye degradation. *J. Catal.* **2003**, *220*, 127–135. [CrossRef]
26. Yin, J.B.; Zhao, X.P. Preparation and Enhanced Electrorheological Activity of TiO_2 Doped with Chromium Ion. *Chem. Mater.* **2004**, *16*, 321–328. [CrossRef]
27. Zhao, X.P.; Duan, X. In situ sol-gel preparation of polysaccharide/titanium oxide hybrid colloids and their electrorheological effect. *J. Colloid Interface Sci.* **2002**, *251*, 376–380. [CrossRef] [PubMed]
28. Yin, J.B.; Xiang, L.Q.; Zhao, X.P. Monodisperse spherical mesoporous Eu-doped TiO_2 phosphor particles and the luminescence properties. *Appl. Phys. Lett.* **2007**, *90*, 113112. [CrossRef]
29. An, G.F.; Yang, C.S.; Jin, S.H.; Chen, G.W.; Zhao, X.P. Hollow TiO_2:Sm^{3+} spheres with enhanced photoluminescence fabricated by a facile method using polystyrene as template. *J. Mater. Sci.* **2013**, *48*, 5483–5488. [CrossRef]
30. Xiang, L.Q.; Zhao, X.P.; Yin, J.B.; Fan, B.L. Well-organized 3-D urchin-like multilevel TiO_2 microspheres with high photocatalytic activity. *J. Mater. Sci.* **2012**, *47*, 1436–1445. [CrossRef]
31. Xiang, L.Q.; Zhao, X.P.; Shang, C.H.; Yin, J.B. Au or Ag nanoparticle-decorated 3D urchin-like TiO_2 nanostructures: Synthesis, characterization, and enhanced photocatalytic activity. *J. Colloid Interface Sci.* **2013**, *403*, 22–28. [CrossRef] [PubMed]
32. Xiang, L.Q.; Liu, S.; Yin, J.B.; Zhao, X.P. Bimetallic core/shell nanoparticle-decorated 3D urchin-like hierarchical TiO_2 nanostructures with magneto-responsive and decolorization characteristics. *Nanoscale Res. Lett.* **2015**, *10*, 84. [CrossRef] [PubMed]
33. Yin, J.B.; Zhao, X.P. Titanate nano-whisker electrorheological fluid with high suspended stability and ER activity. *Nanotechnology* **2006**, *17*, 192–196. [CrossRef]
34. Guo, H.X.; Zhao, X.P.; Ning, G.H.; Liu, G.Q. Synthesis of Ni/Polystyrene/TiO_2 multiply coated microsphere. *Langmuir* **2003**, *19*, 4884–4888. [CrossRef]
35. Guo, H.X.; Zhao, X.P.; Guo, H.L.; Zhao, Q. Preparation of porous SiO_2/Ni/TiO_2 multiply microsphere responsive to electric and magnetic fields. *Langmuir* **2003**, *19*, 9799–9803. [CrossRef]
36. Yin, J.B.; Zhao, X.P. Electrorheological fluids based on glycerol-activated TiO_2 gel and silicone oil with high yield strength. *J. Colloid Interface Sci.* **2003**, *257*, 228–236. [CrossRef]

37. Xiang, L.Q.; Yin, J.B.; Gao, W.S.; Zhao, X.P. Controllable Preparation of Quasi-monodispersed Spherical TiO_2 Particles. *J. Inorg. Mater. Chin.* **2007**, *22*, 253–258.
38. Ma, L.M.; Zheng, F.X.; Zhao, X.P. Sedimentation behaviour of hierarchical porous TiO_2 microspheres electrorheological fluids. *J. Intel. Mater. Syst. Struct.* **2015**, *26*, 1936–1944. [CrossRef]
39. Wang, J.H.; Chen, G.W.; Yin, J.B.; Luo, C.R.; Zhao, X.P. Enhanced electrorheological performance and antisedimentation property of mesoporous anatase TiO_2 shell prepared by hydrothermal process. *Smart Mater. Struct.* **2017**, *26*, 035036. [CrossRef]
40. Wang, B.X.; Zhao, X.P. Core/Shell Nanocomposite Based on the Local Polarization and Its Electrorheological Behavior. *Langmuir* **2005**, *21*, 6553–6559. [CrossRef] [PubMed]
41. Yin, J.B.; Xia, X.; Wang, X.X.; Zhao, X.P. The electrorheological effect and dielectric properties of suspensions containing polyaniline@TiO_2 nanocable-like particles. *Soft Matter* **2011**, *7*, 10978–10986. [CrossRef]
42. Xiang, L.Q.; Zhao, X.P. Electrorheological activity of a composite of TiO_2-Coated Montmorillonite. *J. Mater. Chem.* **2003**, *13*, 1529–1532. [CrossRef]
43. Wang, B.X.; Zhao, X.P. Preparation of kaolinite/TiO_2 coated nanocomposite and their electro-rheological properties. *J. Mater. Chem.* **2003**, *13*, 2248–2253. [CrossRef]
44. Zhao, X.P.; Yin, J.B. Preparation and electrorheological characteristics of rare-earth-doped TiO_2 suspensions. *Chem. Mater.* **2002**, *14*, 2258–2263. [CrossRef]
45. Zhao, X.P.; Yin, J.B.; Xiang, L.Q.; Zhao, Q. Electrorheological fluids containing Ce-doped TiO_2. *J. Mater. Sci.* **2002**, *37*, 2569–2573. [CrossRef]
46. Zhao, X.P.; Yin, J.B.; Xiang, L.Q.; Zhao, Q. Effect of rare earth substitution on electrorheological properties of TiO_2. *Int. J. Mod. Phys. B* **2002**, *16*, 2371–2377. [CrossRef]
47. Yin, J.B.; Zhao, X.P. Temperature effect of rare earth-doped TiO_2 Electrorheological fluids. *J. Phys. D Appl. Phys.* **2001**, *34*, 2063–2067. [CrossRef]
48. Yin, J.B.; Zhao, X.P. Electrorheological effect of Cerium-Doped TiO_2. *Chin. Phys. Lett.* **2001**, *18*, 1144–1146.
49. Yin, J.B.; Zhao, X.P. Enhanced electrorheological activity of mesoporous Cr-doped TiO_2 from activated pore wall and high surface area. *J. Phys. Chem. B* **2006**, *110*, 12916–12925. [CrossRef] [PubMed]
50. Wang, B.X.; Zhao, X.P. A bionic nano-papilla particle and its super-oleophilic ability. *Adv. Funct. Mater.* **2005**, *15*, 1815–18212. [CrossRef]
51. Xiang, L.Q.; Zhao, X.P.; Yin, J.B. Micro/nano-structured montmorillonite/TiO_2 particles with high electrorheological activity. *Rheol. Acta* **2011**, *50*, 87–95. [CrossRef]
52. Qiao, Y.P.; Zhao, X.P.; Su, Y.Y. Dielectric metamaterial particles with enhanced efficiency of mechanical/electrical energy transformation. *J. Mater. Chem.* **2011**, *21*, 394–399. [CrossRef]
53. Eiden-Assmann, S.; Widoniak, J.; Maret, G. Synthesis and Characterization of Porous and Nonporous Monodisperse Colloidal TiO_2 Particles. *Chem. Mater.* **2004**, *16*, 6–11. [CrossRef]
54. Zhao, Y.; Jiang, L. Hollow Micro/Nanomaterials with Multilevel Interior Structures. *Adv. Mater.* **2009**, *21*, 3621–3638. [CrossRef]
55. Réti, B.; Kiss, G.I.; Gyulavári, T.; Baan, K.; Magyari, K.; Hernadi, K. Carbon sphere templates for TiO_2 hollow structures: Preparation, characterization and photocatalytic activity. *Catal. Today* **2017**, *284*, 160–168. [CrossRef]
56. Orellana, M.; Osiglio, L.; Arnal, P.M.; Pizzio, L.R. TiO_2 hollow spheres modified with tungstophosphoric acid with enhanced visible light absorption for the photodegradation of 4-chlorophenol. *Photochem. Photobiol. Sci.* **2017**, *16*, 46–52. [CrossRef] [PubMed]
57. Liu, Y.; Lan, K.; Bagabas, A.A.; Zhang, P.; Gao, W.; Wang, J.; Sun, Z.; Fan, J.; Elzatahry, A.A.; Zhao, D. Ordered Macro/Mesoporous TiO_2 Hollow Microspheres with Highly Crystalline Thin Shells for High-Efficiency Photoconversion. *Small* **2016**, *12*, 860–867. [CrossRef] [PubMed]
58. Li, Y.; Fu, Z.Y.; Su, B.L. Hierarchically Structured Porous Materials for Energy Conversion and Storage. *Adv. Funct. Mater.* **2012**, *22*, 4634–4667. [CrossRef]
59. Kim, Y.J.; Lee, M.H.; Kim, H.J.; Lim, G.; Choi, Y.S.; Park, N.G.; Kim, K.; Lee, W.I. Formation of Highly Efficient Dye-Sensitized Solar Cells by Hierarchical Pore Generation with Nanoporous TiO_2 Spheres. *Adv. Mater.* **2009**, *21*, 3668–3673. [CrossRef]
60. Ismail, A.A.; Bahnemann, D.W. mesoporous TiO_2 photocatalysts: Preparation, characterization and reaction mechanisms. *J. Mater. Chem.* **2011**, *21*, 11686–11707. [CrossRef]

61. Li, S.; Zheng, J.; Zhao, Y.; Liu, Y. Preparation of a three-dimensional ordered macroporous titanium dioxide material with polystyrene colloid crystal as a template. *J. Appl. Polym. Sci.* **2008**, *107*, 3903–3908. [CrossRef]
62. Hegazy, A.; Prouzet, E. Room temperature synthesis and thermal evolution of porous nanocrystalline TiO_2 anatase. *Chem. Mater.* **2012**, *24*, 245–254. [CrossRef]
63. Zheng, X.; Lv, Y.; Kuang, Q.; Zhu, Z.; Long, X.; Yang, S. Close-Packed Colloidal SiO_2 as a Nanoreactor: Generalized Synthesis of Metal Oxide Mesoporous Single Crystals and Mesocrystals. *Chem. Mater.* **2014**, *26*, 5700–5709. [CrossRef]
64. Yin, J.B.; Zhao, X.P. Preparation and Electrorheological Activity of Mesoporous Rare-Earth-Doped TiO_2. *Chem. Mater.* **2002**, *14*, 4633–4640. [CrossRef]
65. Yin, J.B.; Zhao, X.P. Wormhole-like mesoporous Ce-doped TiO_2: A new electrorheological material with high activity. *J. Mater. Chem.* **2003**, *13*, 689–695. [CrossRef]
66. Joo, J.B.; Lee, I.; Dahl, M.; Moon, G.D.; Zaera, F.; Yin, Y. Controllable Synthesis of Mesoporous TiO_2 Hollow Shells: Toward an Efficient Photocatalyst. *Adv. Funct. Mater.* **2013**, *23*, 4246–4254. [CrossRef]
67. Bian, Z.F.; Zhu, J.; Cao, F.L.; Huo, Y.N.; Lu, Y.F.; Li, H.X. Solvothermal synthesis of well-defined TiO_2 mesoporous nanotubes with enhanced photocatalytic activity. *Chem. Commun.* **2010**, *46*, 8451–8453. [CrossRef] [PubMed]
68. Seisenbaeva, G.A.; Moloney, M.P.; Tekoriute, R.; Hardy-Dessources, A.; Nedelec, J.M.; Gun'ko, Y.K.; Kessler, V.G. Biomimetic Synthesis of Hierarchically Porous Nanostructured Metal Oxide Microparticles-Potential Scaffolds for Drug Delivery and Catalysis. *Langmuir* **2010**, *26*, 9809–9817. [CrossRef] [PubMed]
69. Wang, C.; Yin, L.; Zhang, L.; Qi, Y.; Lun, N.; Liu, N. Large scale synthesis and gas-sensing properties of anatase TiO_2 Three-dimensional hierarchical nanostructures. *Langmuir* **2010**, *26*, 12841–12848. [CrossRef] [PubMed]
70. Rui, Y.; Wang, L.; Zhao, J.; Wang, H.; Li, Y.; Zhang, Q.; Xu, J. Template-free synthesis of hierarchical TiO_2 hollow microspheres as scattering layer for dye-sensitized solar cells. *Appl. Surf. Sci.* **2016**, *369*, 170–177. [CrossRef]
71. Zhang, D.; Li, G.; Wang, F.; Yu, J.C. Green synthesis of a self-assembled rutile mesocrystalline photocatalyst. *CrystEngComm* **2010**, *12*, 1759–1763. [CrossRef]
72. Liu, G.; Yang, H.G.; Sun, C.; Cheng, L.; Wang, L.; Lu, G.Q.; Cheng, H.M. Titania polymorphs derived from crystalline titanium diboride. *CrystEngComm* **2009**, *11*, 2677–2682. [CrossRef]
73. Yin, J.B.; Zhao, X.P.; Xiang, L.Q.; Xia, X.; Zhang, Z.S. Enhanced electrorheology of suspensions containing sea-urchin-like hierarchical Cr-doped TiO_2 particles. *Soft Matter* **2009**, *5*, 4687–4697. [CrossRef]
74. Vaneski, A.; Susha, A.S.; Rodríguez-Fernández, J.; Berr, M.; Jäckel, F.; Feldmann, J.; Rogach, A.L. Hybrid colloidal heterostructures of anisotropic semiconductor nanocrystals decorated with noble metals: Synthesis and function. *Adv. Funct. Mater.* **2011**, *21*, 1547–1556. [CrossRef]
75. Kochuveedu, S.T.; Kim, D.P.; Kim, D.H. Surface lasmon induced visible light photocatalytic activity of TiO_2 nanospheres decorated by Au nanoparticles with controlled configuration. *J. Phys. Chem. C* **2012**, *116*, 2500–2506. [CrossRef]
76. Grigorieva, A.V.; Goodilin, E.A.; Derlyukova, L.E.; Anufrieva, T.A.; Tarasov, A.B.; Dobrovolskii, Y.A.; Tretyakov, Y.D. TiO_2 nanotubes supported platinum catalyst in CO oxidation process. *Appl. Catal. A Gen.* **2009**, *362*, 20–25. [CrossRef]
77. Kafizas, A.; Dunnil, C.W.; Parkin, I.P. The relationship between photocatalytic activity and Photochromic state of nanoparticulate silver surface loaded titanium dioxide thin-films. *Phys. Chem. Chem. Phys.* **2011**, *13*, 13827–13838. [CrossRef] [PubMed]
78. Wodka, D.; Bielanska, E.; Socha, R.P.; Elzbieciak-Wodka, M.; Gurgul, J.; Nowak, P.; Warszynski, P.; Kumakiri, I. Photocatalytic activity of titanium dioxide modified by silver nanoparticles. *ACS Appl. Mater. Interfaces* **2010**, *2*, 1945–1953. [CrossRef] [PubMed]
79. Zhang, N.; Liu, S.; Fu, X.; Xu, Y.J. Selective Pt deposition onto the face (110) of TiO_2 assembled microspheres that substantially enhances the photocatalytic properties. *J. Phys. Chem. C* **2011**, *115*, 9136–9145. [CrossRef]
80. Feng, L.L.; Wu, X.C.; Ren, L.R.; Xiang, Y.J.; He, W.W.; Zhang, K.; Zhou, W.Y.; Xie, S.S. Well-controlled synthesis of Au@Pt nanostructures by gold-nanorod-seeded growth. *Chem. Eur. J.* **2008**, *14*, 9764–9771. [CrossRef] [PubMed]

81. Levin, C.S.; Hofmann, C.; Ali, T.A.; Kelly, A.T.; Morosan, E.; Nordlander, P.; Whitmire, K.H.; Halas, N.J. Magnetic-plasmonic core-shell nanoparticles. *J. Am. Chem. Soc.* **2009**, *3*, 1379–1388. [CrossRef] [PubMed]
82. Chen, L.Y.; Fujita, T.; Ding, Y.; Chen, M.W. A Three-dimensional gold-decorated nanoporous copper core-shell composite for electrocatalysis and nonenzymatic biosensing. *Adv. Funct. Mater.* **2010**, *20*, 2279–2285. [CrossRef]
83. Xuan, S.; Xiang, Y.; Wang, J.; Yu, J.C.; Leung, K.C. Preparation, Characterization, and Catalytic Activity of Core/Shell Fe_3O_4@Polyaniline@Au Nanocomposites. *Langmuir* **2009**, *25*, 11835–11843. [CrossRef] [PubMed]
84. Wang, L.; Clavero, C.; Huba, Z.; Carroll, K.J.; Carpenter, E.E.; Gu, D.F.; Lukaszew, R.A. Plasmonics and Enhanced Magneto-Optics in Core-Shell Co-Ag Nanoparticles. *Nano Lett.* **2011**, *11*, 1237–1240. [CrossRef] [PubMed]
85. Lu, Y.; Zhao, Y.; Yu, L.; Dong, L.; Shi, C.; Hu, M.J.; Xu, Y.J.; Wen, L.P.; Yu, S.H. Hydrophilic Co@Au Yolk/Shell Nanospheres: Synthesis, Assembly, and Application to Gene Delivery. *Adv. Mater.* **2010**, *22*, 1407–1411. [CrossRef] [PubMed]

nanomaterials

Article

Oxygen Partial Pressure Impact on Characteristics of Indium Titanium Zinc Oxide Thin Film Transistor Fabricated via RF Sputtering

Ming-Hung Hsu, Sheng-Po Chang *, Shoou-Jinn Chang, Wei-Ting Wu and Jyun-Yi Li

Institute of Microelectronics & Department of Electrical Engineering Center for Micro/Nano Science and Technology Advanced Optoelectronic Technology Center, National Cheng Kung University, Tainan 701, Taiwan; hsuminghung0121@gmail.com (M.-H.H.); changsj@mail.ncku.edu.tw (S.-J.C.); waiting31317@gmail.com (W.-T.W.); z823040@gmail.com (J.-Y.L.)

* Correspondence: changsp@mail.ncku.edu.tw; Tel.: +886-6-275-7575 (ext. 62400-1208)

Academic Editor: Andrea Lamberti

Received: 21 April 2017; Accepted: 23 June 2017; Published: 26 June 2017

Abstract: Indium titanium zinc oxide (InTiZnO) as the channel layer in thin film transistor (TFT) grown by RF sputtering system is proposed in this work. Optical and electrical properties were investigated. By changing the oxygen flow ratio, we can suppress excess and undesirable oxygen-related defects to some extent, making it possible to fabricate the optimized device. XPS patterns for O 1s of InTiZnO thin films indicated that the amount of oxygen vacancy was apparently declined with the increasing oxygen flow ratio. The fabricated TFTs showed a threshold voltage of −0.9 V, mobility of 0.884 cm^2/Vs, on-off ratio of 5.5×10^5, and subthreshold swing of 0.41 V/dec.

Keywords: indium titanium zinc oxide; thin film transistor; oxygen partial pressure

1. Introduction

Oxide semiconductor, as the active layer for TFTs, has drawn much interest recently owing to their advantageously high carrier mobility, high optical transparency in the visible light region, high thermal/environmental stability, and low process temperature [1–7]. Among them, ZnO has been regarded as a promising material and is ubiquitous in optoelectronic devices as a result of its low toxicity, abundance on Earth, wide energy bandgap of 3.37 eV, and large exciton-binding energy of 60 meV at room temperature [8,9]. Many research groups have endeavored to develop TFTs with high electrical and optical properties, which can be utilized in organic light emitting diodes (OLEDs) and active matrix liquid crystal displays (AMLCDs). The resistivity of the thin film cannot be too high since it will cause low output current, while it cannot be too low as it will bring about distastefully high leakage current and surface current. Conventional TFT material, such as hydrogenated amorphous silicon (a-Si:H) usually benefit from ambient deposition conditions and ease of fabrication. Nonetheless, such devices exhibit poor mobility [10], which is deemed undesirable for high-speed devices or fast switching circuits.

One way to enhance the mobility is addition of appropriate metal element [11]. Hosono et al. reported that amorphous oxides comprising heavy metal cations with an ns^0 ($n \geq 4$) electronic configuration can accomplish elevation of mobility due to high overlapping, large diameter, and high spherical symmetry of ns^0 orbitals [12,13]. Consequently, indium-doped or tin-doped ZnO materials become attractive and popular. For example, indium zinc oxide (IZO) [14], zinc tin oxide (ZTO) [15], indium zinc tin oxide (IZTO) [16], and indium gallium zinc oxide (IGZO) [17] have been reported in the literature. Compared to the previous studies, obviously there is little research related to incorporation of titanium (Ti). Although IGZO holds the edge and is commonly-seen in TFT fabrication, it is known that scientists have tried to seek alternatives to replace gallium because the weak Ga–O bond may result in instability issues. Like magnesium (Mg), hafnium (Hf), and zirconium (Zr), titanium has the ability to suppress excess carrier generation and make devices more stable under bias and illumination stress [18–20]. Furthermore, titanium is non-noxious, and has a lower electronegativity (1.54) as well as a lower standard electrode potential (−1.63 V) compared to those of Zn (1.65 and −0.76 V) [21], which means it is more likely to oxidize than zinc and can be used as a carrier suppressor in ZnO-based TFTs. Accordingly, InTiZnO, as a novel material, is expected to exhibit high optical and electrical performance.

In this paper, we proposed the fabrication of indium titanium zinc oxide (InTiZnO) TFT by utilizing Radio-Frequency (RF) sputtering system. Our goal is to seek whether InTiZnO has possibility of being a promising semiconductor material that can be used for optoelectronic component fabrication. It is known thin films can be grown through spray pyrolysis, chemical vapor deposition, sol–gel, and sputtering methods. Sputtering is preferable because of its large-area deposition, stable growth rate, and good film quality [22]. We focused on the impact of oxygen-related defects on the electrical performance of the device. The compensation level of oxygen-related defects was investigated and discussed.

2. Materials and Methods

The schematic diagram of the fabricated bottom-gate InTiZnO TFT was shown in Figure 1. First, quartz substrates were chemically cleaned by an ultrasonicator. A 70-nm-thick aluminum was subsequently deposited on the as-cleaned substrate by thermal evaporation with a metal mask. Silica of 200 nm was then grown by plasma-enhanced chemical vapor deposition (PECVD, SAMCO PD-220NA, Kyoto, Japan). The channel layer InTiZnO of 50 nm was sputtered with various oxygen flow ratios. It is noted an InTiZnO target (In:Ti:Zn = 99:1:99 in molar ratio) was utilized, the RF power was 80 W, and the working pressure was 5 mTorr. Sputtering was performed in argon and oxygen ambience at room temperature with manipulated gas flows. Argon flow was from 49 sccm to 45 sccm, while oxygen flow was from 1 sccm to 5 sccm. Next, aluminum of 70 nm was thermally grown to act as source and drain, whose patterns were defined by another metal mask. All processes were done without intentional heating or post-annealing. No etching, lift-off, or other photolithography technique was included. The gate length and gate width of the fabricated TFTs were 100 μm and 1000 μm, respectively. For thin film analysis, we prepared 100-nm-thick InTiZnO on cleaned substrate. Surface morphology of the films was examined via atomic force microscope (AFM, NT-MDT Solver P47-PRO, Moscow, Russia). Transmittance spectrum was recorded via a UV-3101 UV–Vis–NIR spectrophotometer (Shimadzu UV-3101PC, SHIMADZU Corp., Kyoto, Japan). Crystallinity of the films was investigated by X-ray Diffraction (XRD, Rigaku ATX-E, Tokyo, Japan) with a Cu Kα radiation source (λ = 1.54056 Å). X-ray Photoelectron Spectroscopy (XPS, VG ESCALAB220i-XL, Thermo Scientific, Waltham, MA, USA) was applied to analyze chemical state of oxygen. The TFTs were subject to current–voltage (*I*–*V*) characteristics measurement in dark at room temperature by a semiconductor parameter analyzer (Agilent B1500, Palo Alto, CA, USA).

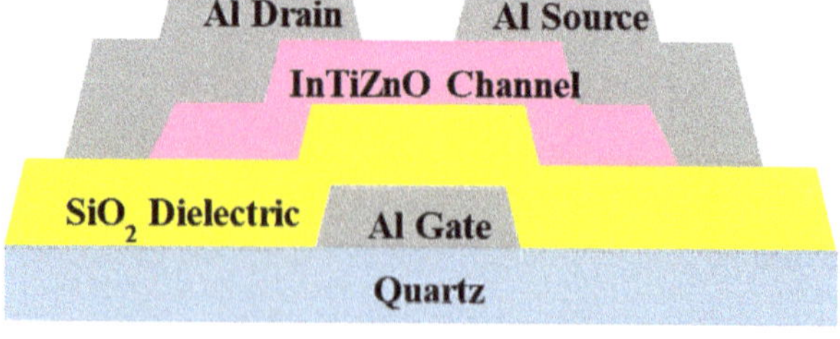

Figure 1. Schematic diagram of the sputtered InTiZnO thin film transistor.

3. Results and Discussion

The measured XRD patterns are demonstrated in Figure 2. It is notable that we prepared thin films of 100 nm on quartz substrate for film analyses. Five respective samples were named in light of their oxygen flow ratios during the sputtering process. In other words, Sample A indicated the device with oxygen flow ratio of 2%, Sample B was for 4%, Sample C was for 6%, Sample D was for 8%, and Sample E was for 10%. No evident diffraction peak of crystalline phase was observed from the spectrum, except for the peak at 21.4° to 21.5° attributed to quartz substrate. The result was similar to the report of A. Liu et al. in 2014 [23]. There is a weak peak located at around 32°, which may be attributed to ZnO crystal (JCPDS #890510). However, sputtered quaternary compounds are complex and usually amorphous owing to the nature of deposition method. Consequently, we tend to believe the characterization results revealed that InTiZnO films were amorphous. Figure 3a shows the transmittance of InTiZnO thin films sputtered under various oxygen flow conditions. The variation of oxygen flow ratio had little influence on the absorption edge. It was observed that the average transmittance of each sample was over 80% in the visible region. To go further, the relation between absorption coefficient (α) and the incident photon energy (hν) was plotted, as shown in Figure 3b. It is known the calculation of the corresponding optical bandgap of a certain semiconductor material is described by Tauc's Law and is given as

$$(\alpha h\nu)^2 = B(h\nu - Eg),$$

where B a constant and *Eg* the bandgap [24]. By estimating the tangent lines from the curves it turned out the calculated optical bandgap of InTiZnO was 3.90–4.06 eV, implicating InTiZnO is a wide-gap material having potential for further photo-sensing applications. It is notable that optical absorption by defects appears at energies lower than the optical gap. Band tailing is speculated to be attributed to the combing effects of impurity disorder and other defects [25,26].

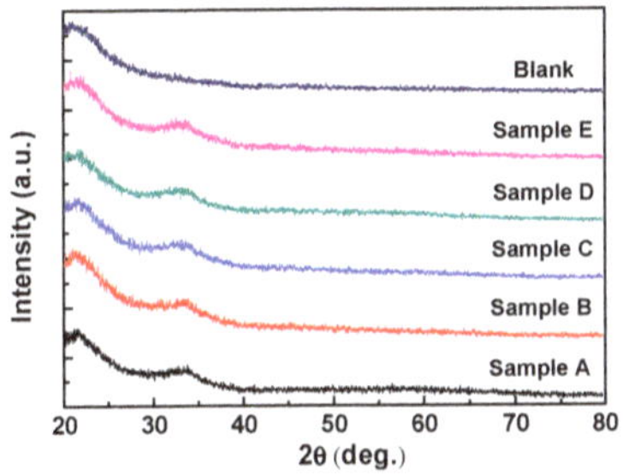

Figure 2. XRD spectrum measured from a blank quartz glass and Samples A, B, C, D, and E.

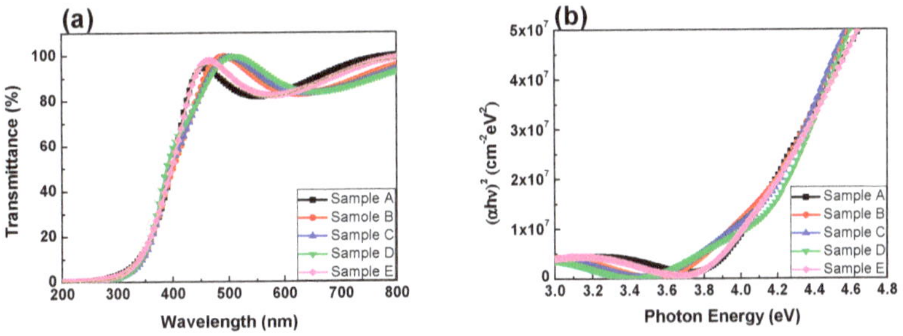

Figure 3. (**a**) Transmittance spectrum measured from Samples A, B, C, D, and E. (**b**) Absorption coefficient versus photon energy for these five samples.

Figure 4a–e shows the output (I_D-V_D) curves of InTiZnO TFTs measured in dark at room temperature. The drain voltage was swept from 0 V to 15 V with a step of 0.1 V, while the gate voltage was from 0 V to 12 V with a step of 3 V. The obvious pinch-off and drain-current saturation suggested the electron transport in the active layers was well controlled by the gate and drain bias. Compared to other samples, Sample E seemed to be a bit worse, which was speculated to stem from excess incorporation of oxygen. It will be discussed later. The transfer curves of InTiZnO TFTs derived in dark at room temperature is shown in Figure 5. The drain voltage was fixed at 3 V while the gate voltage was from −2 V to 12 V with a step of 0.1 V. It can be clearly seen that five samples exhibited typical n-type characteristics. For comparison, the threshold voltage (V_T), mobility (μ_{eff}), on-off current ratio, subthreshold swing (SS), and total equivalent trap states (N_t) of each sample are listed in Table 1 (Values with error for Samples A, B, and C are shown in Table S1 in Supplementary Materials). By drawing the $I_D{}^{1/2}$ versus V_G figure, V_T could be determined by setting $I_D{}^{1/2}$ = 0. The device mobility and subthreshold swing are defined as

$$I_D = \frac{W}{2L} C_{ox} \mu_{eff} (V_G - V_T)^2,$$

$$SS = \frac{\partial V_G}{\partial \log I_D},$$

where I_D the drain current, $\frac{W}{L}$ the dimension of the device, C_{ox} the gate capacitance per unit area, V_G is the gate voltage, and V_T the threshold voltage. It is known that SS is related to the interface and bulk trap density of the TFTs. The total equivalent trap states can be acquired by using the following equation [27],

$$N_t = [\frac{SS \log(e)}{kT/q} - 1] \frac{C_{ox}}{q},$$

where q is the charge of the electron, k the Boltzmann constant, T the temperature, SS the subthreshold swing, and C_{ox} the gate capacitance per unit area. The dielectric constant of SiO_2 is 3.9, and the calculated capacitance was about 17 nF/cm^2. It is noteworthy that when sweeping from 12 V to −2 V during I_D-V_D measurement, another curve is expected to appear, which is related to a reliability issue.

Table 1. Electronic parameters of each unannealed InTiZnO TFT measured in dark at room temperature.

Samples	Oxygen Flow Ratio	V_T (V)	μ_{eff} (cm^2/Vs)	On-Off Current Ratio	SS (V/dec)	N_t (cm^{-2})
Sample A	2%	−0.35	1.625	1.5×10^5	0.32	4.6×10^{11}
Sample B	4%	−0.9	0.884	5.5×10^5	0.41	6.2×10^{11}
Sample C	6%	−0.5	0.235	1.1×10^3	1.62	2.8×10^{12}
Sample D	8%	−1.4	0.006	1.4×10^2	2.46	4.3×10^{12}
Sample E	10%	−4	0.004	1.1×10^1	8.57	1.5×10^{13}

InTiZnO films prepared under various gas mixture revealed a homogeneous surface (images shown in Figure S1 in Supplementary Materials). The grains are shaped as round cones uniformly distributed on the 2 × 2 µm area. From AFM analysis, surface roughness varied as a function of oxygen flow ratio. It was gradually increased from 1.003 nm to 1.354 nm by increasing the oxygen flow. The rougher surface was due to the more drastic ion crashing on the film during sputtering. Hence, higher oxygen flow ratio may cause objectionable surface scattering, leading to mobility reduction. A rougher surface should also be responsible for the variation of N_t.

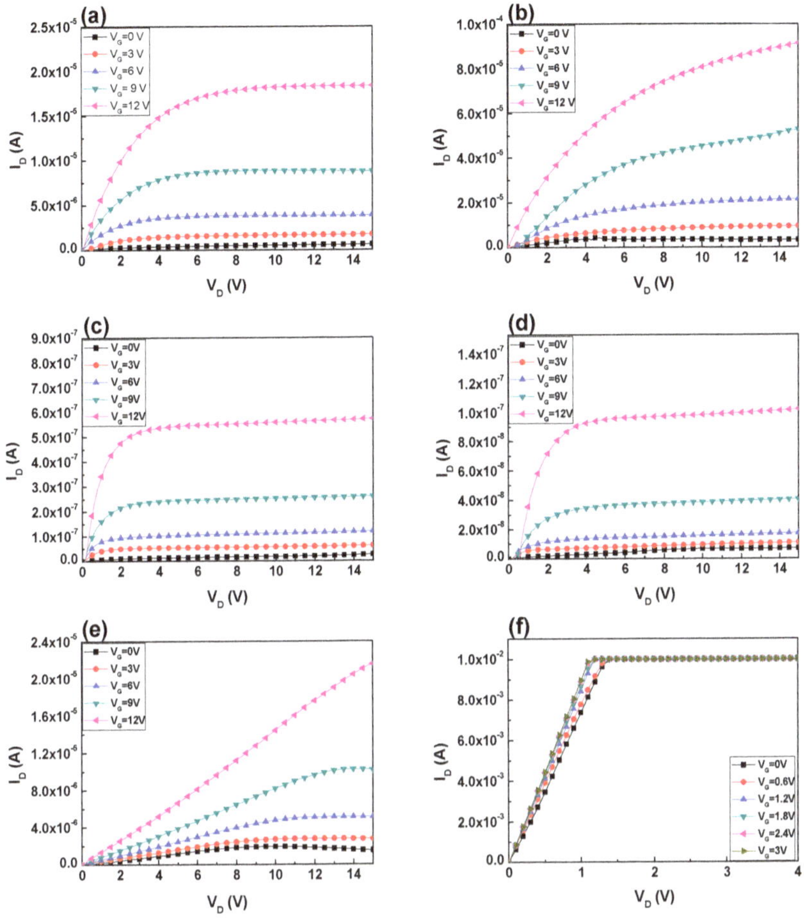

Figure 4. I_D-V_D of InTiZnO TFTs as a function of oxygen flow ratios: (**a**) Sample A, (**b**) Sample B, (**c**) Sample C, (**d**) Sample D, and (**e**) Sample E, and (**f**) Sample F with 0% oxygen flow ratio.

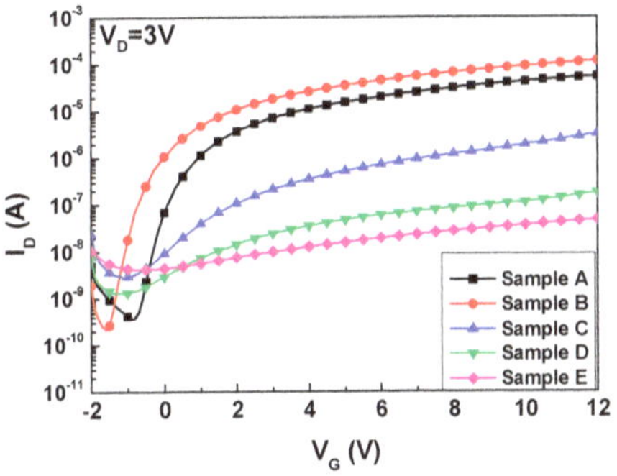

Figure 5. Transfer curve of InTiZnO TFTs with different oxygen flow ratios.

On the other hand, it is known that oxygen vacancy acts as a carrier provider in ZnO-based materials. When oxygen flow ratio increased, the amount of oxygen vacancies in the thin film reduced. It can be seen that free carriers became less from the *I–V* measurement. In I_D-V_D measurement, samples with a higher oxygen flow ratio would reach a lower saturation current. The saturation current: Sample D < Sample C < Sample B. In the I_D-V_G transfer curve, the on current at 12 V: Sample E < Sample D < Sample C < Sample B. Accordingly, it is confirmed that concentration of carriers was suppressed. However, participation of excess oxygen during film growth could compromise the performance of devices. Namely, it would bring about undesirable oxygen-related deficiencies and make TFTs difficult to operate desirably. Augmentation in trap states due to increasing oxygen flow ratio could hinder the carriers in the channel layer from smooth transport. Therefore, the mobility decreased from 1.625 cm^2/Vs to 0.004 cm^2/Vs and subthreshold swing increased from 0.32 V/dec to 8.57 V/dec. It is notable that our devices with a negative threshold voltage were in depletion mode (D-mode). For a depletion-mode transistor, the device is normally-on at zero gate–source voltage. It indicates that there may be a power consumption issue, which could be solved by increasing the thickness of dielectric layer or replacing high-k material with the original gate oxide.

Figure 6a–e is XPS spectra of O 1s of InTiZnO films grown with different oxygen flow ratios. The spectra were deconvoluted into two peaks by Gaussian fitting: O_I occurred at 529.6 eV was assigned to O^{2-} species in the lattice, and O_{II} located at 531.3 eV was associated with oxygen vacancies or defects, or O^{2-} ion in oxygen-deficient region [8,28]. By evaluating peak area O_{II} over (O_I + O_{II}) it can be seen the proportion decreased from 78.8% to 65.3% as the oxygen flow ratio increased from 2% to 10%. Assuming small contributions of the film surfaces, the result indicated that concentration of oxygen vacancy in the films was reduced in terms of proportion with the increasing oxygen flow ratio, and it was consistent with the device performance. That is, manipulation of oxygen flow and argon flow is a performance trade-off issue. On the basis of the electric properties, we deduced that Sample A and Sample B were superior to other samples, whereas on-off current ratio of Sample B was higher than that of Sample A. It implied the gate control ability of Sample B was better, as a result of well-manipulated oxygen flow ratio of 4%. Sample B was considered to be the optimal device in this work, exhibiting a mobility of 0.884 cm^2/Vs, on-off ratio of 5.5×10^5, and subthreshold swing of 0.41. Table 2 lists previous studies similar to InTiZnO. It can be seen that our devices are competent.

Table 2. Electrical performance of ZnO-based TFTs reported in the literature

Materials	Deposition Method	V_T (V)	μ_{eff} (cm^2/Vs)	On-Off Current Ratio	*SS* (V/dec)	N_t (cm^{-2})
InZnO [29]	Sol–gel	0.18	0.15	10^5	0.86	N.A.
InMgZnO [30]	Sol–gel	N.A.	0.56	$<10^5$	2.2	N.A.
InTiZnO [31]	Sol–gel	8.49	0.04	10^4	1.06	N.A
InTiZnO [24]	PLD	7.89	2.58	10^8	0.76	1.5×10^{12}
InTiZnO, Sample A (this work)	sputter	−0.35	1.625	1.5×10^5	0.32	5.7×10^{11}
InTiZnO, Sample B (this work)	sputter	−0.9	0.884	5.5×10^5	0.41	7.3×10^{11}

On the other hand, sample without oxygen (oxygen flow ratio = 0%, only pure argon) was also prepared. During *I–V* measurement, the applied drain voltage swept from 0 V to 4 V with an increment of 0.1 V, while the gate voltage was from 0 V to 3 V with an increment of 0.6 V. In Figure 4f, it was observed the output current reached the compliance current (10 mA) of the semiconductor parameter analyzer. The result indicated that the InTiZnO 0% TFT was over-conductive. The innate defects from ZnO-based material were not suitably compensated, lacking feasibility of device operation.

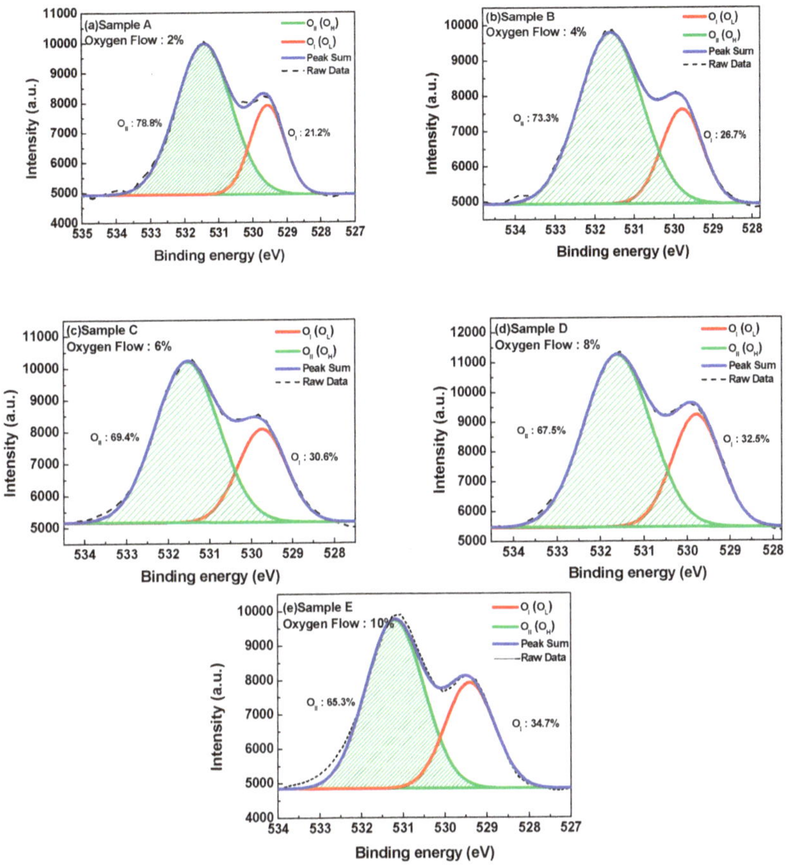

Figure 6. XPS O 1s spectra of InTiZnO thin films with various oxygen flow ratios: (**a**) Sample A, (**b**) Sample B, (**c**) Sample C, (**d**) Sample D, and (**e**) Sample E.

4. Conclusions

In summary, we reported the fabrication of InTiZnO TFTs. The transmittance of InTiZnO was more than 80% in the visible region. The energy bandgap of InTiZnO was derived to be 3.90–4.06 eV. The results showed InTiZnO had great potential for UV sensors. By manipulating the oxygen flow ratio during sputtering, oxygen vacancies could be filled as confirmed by XPS measurement, making it possible to prepare the optimal InTiZnO TFTs that exhibited high electrical performance. If only argon took part in growth of active layers, excess carriers would exist, resulting in high leakage currents. We found that 4% oxygen flow ratio was preferable. Under such conditions, Sample B showed a threshold voltage of −0.9 V, mobility of 0.884 cm^2/Vs, on-off ratio of 5.5×10^5, and subthreshold swing of 0.41 V/dec. We believe the findings reveal a great step toward knowledge of the quaternary semiconductor material InTiZnO. The combination of transparency, ease of fabrication under room temperature, and high on-off current ratio makes InTiZnO TFT very promising for the next-generation optoelectronic device.

Supplementary Materials: The following are available online at www.mdpi.com/2079-4991/7/7/156/s1, Figure S1: AFM images of each InTiZnO samples, Table S1: Electronic parameters for five Samples, with errors for Samples A, B, and C.

Acknowledgments: This work was supported by the Ministry of Science and Technology under contract number MOST 103-2221-E-006-098 and 105-2221-E-006-118. This work was also supported in part by the Center for Frontier Materials and Micro/Nano Science and Technology, National Cheng Kung University, Taiwan, and by the Advanced Optoelectronic Technology Center, National Cheng Kung University, for projects from the Ministry of Education.

Author Contributions: Ming-Hung Hsu, Sheng-Po Chang, Shoou-Jinn Chang, Wei-Ting Wu, and Jyun-Yi Li conceived and designed the experiment; Ming-Hung Hsu, Sheng-Po Chang, Shoou-Jinn Chang, Wei-Ting Wu, and Jyun-Yi Li performed the experiment; Ming-Hung Hsu, Sheng-Po Chang, Shoou-Jinn Chang, Wei-Ting Wu, and Jyun-Yi Li analyzed the data; Ming-Hung Hsu, Sheng-Po Chang, Wei-Ting Wu, and Jyun-Yi Li wrote the paper.

Conflicts of Interest: The authors declare no conflict of interest.

References

1. Jeong, Y.; Song, K.; Jun, T.; Jeong, S.; Moon, J. Effect of gallium content on bias stress stability of solution-deposited Ga–Sn–Zn–O semiconductor transistors. *Thin Solid Films* **2011**, *519*, 6164–6168. [CrossRef]
2. Rim, Y.S.; Kim, D.L.; Jeong, W.H.; Kim, H.J. Effect of Zr addition on ZnSnO thin-film transistors using a solution process. *Appl. Phys. Lett.* **2010**, *97*, 233502. [CrossRef]
3. Fortunato, E.M.; Barquinha, P.M.; Pimentel, A.C.; Gonçalves, A.M.; Marques, A.J.; Martins, R.F.; Pereira, L.M. Wide-bandgap high-mobility ZnO thin-film transistors produced at room temperature. *Appl. Phys. Lett.* **2004**, *85*, 2541–2543. [CrossRef]
4. Fortunato, E.; Barquinha, P.; Martins, R. Oxide Semiconductor Thin-Film Transistors: A Review of Recent Advances. *Adv. Mater.* **2012**, *24*, 2945–2986. [CrossRef] [PubMed]
5. Yu, X.; Marks, T.J.; Facchetti, A. Metal oxides for optoelectronic applications. *Nat. Mater.* **2016**, *15*, 383–396. [CrossRef] [PubMed]
6. Tian, J.; Cao, G. Design, fabrication and modification of metal oxide semiconductor for improving conversion efficiency of excitonic solar cells. *Coord. Chem. Rev.* **2016**, *320*, 193–215. [CrossRef]
7. Petti, L.; Münzenrieder, N.; Vogt, C.; Faber, H.; Büthe, L.; Cantarella, G.; Bottacchi, F.; Anthopoulos, T.D.; Tröster, G. Metal oxide semiconductor thin-film transistors for flexible electronics. *Appl. Phys. Rev.* **2016**, *3*, 021303. [CrossRef]
8. Zhang, X.; Qin, J.; Xue, Y.; Yu, P.; Zhang, B.; Wang, L.; Liu, R. Effect of aspect ratio and surface defects on the photocatalytic activity of ZnO nanorods. *Sci. Rep.* **2014**, *4*, 4596. [CrossRef] [PubMed]
9. Li, J.Y.; Chang, S.P.; Hsu, M.H.; Chang, S.J. High Responsivity MgZnO Ultraviolet Thin-Film Phototransistor Developed Using Radio Frequency Sputtering. *Materials* **2017**, *10*, 126. [CrossRef]
10. Wrench, J.S.; Brunell, I.F.; Chalker, P.R.; Jin, J.D.; Shaw, A.; Mitrovic, I.Z.; Hall, S. Compositional tuning of atomic layer deposited MgZnO for thin film transistors. *Appl. Phys. Lett.* **2014**, *105*, 202109. [CrossRef]
11. Yue, H.Y.; Wu, A.M.; Hu, J.; Zhang, X.Y.; Li, T.J. Relationship between Structure and Functional Properties of the ZnO:Al Thin Films. *Mater. Sci. Forum* **2011**, *675*, 1275–1278. [CrossRef]
12. Ma, T.Y.; Choi, M.H. Optical and electrical properties of Mg-doped zinc tin oxide films prepared by radio frequency magnetron sputtering. *Appl. Surf. Sci.* **2013**, *286*, 131–136. [CrossRef]
13. Hosono, H.; Kikuchi, N.; Ueda, N.; Kawazoe, H. Working hypothesis to explore novel wide band gap electrically conducting amorphous oxides and examples. *J. Non-Cryst. Solids* **1996**, *198*, 165–169. [CrossRef]
14. Lee, S.; Park, H.; Paine, D.C. A study of the specific contact resistance and channel resistivity of amorphous IZO thin film transistors with IZO source-drain metallization. *J. Appl. Phys.* **2011**, *109*, 063702. [CrossRef]
15. Avis, C.; Kim, Y.G.; Jang, J. Solution processed hafnium oxide as a gate insulator for low-voltage oxide thin-film transistors. *J. Mater. Chem.* **2012**, *22*, 17415–17420. [CrossRef]
16. Fuh, C.S.; Liu, P.T.; Huang, W.H.; Sze, S.M. Effect of annealing on defect elimination for high mobility amorphous indium-zinc-tin-oxide thin-film transistor. *IEEE Electron Device Lett.* **2014**, *35*, 1103–1105. [CrossRef]
17. Yao, J.; Xu, N.; Deng, S.; Chen, J.; She, J.; Shieh, H.P.D.; Liu, P.T.; Huang, Y.P. Electrical and photosensitive characteristics of a-IGZO TFTs related to oxygen vacancy. *IEEE Trans. Electron Devices* **2011**, *58*, 1121–1126.
18. Jeong, Y.; Bae, C.; Kim, D.; Song, K.; Woo, K.; Shin, H.; Cao, G.; Moon, J. Bias-stress-stable solution-processed oxide thin film transistors. *ACS Appl. Mater. Interfaces* **2010**, *2*, 611–615. [CrossRef] [PubMed]

19. Huang, C.X.; Li, J.; Fu, Y.Z.; Zhang, J.H.; Jiang, X.Y.; Zhang, Z.L.; Yang, Q.H. Characterization of dual-target co-sputtered novel Hf-doped ZnSnO semiconductors and the enhanced stability of its associated thin film transistors. *J. Alloys Compd.* **2016**, *681*, 81–87. [CrossRef]
20. Choi, W.S.; Jo, H.; Kwon, M.S.; Jung, B.J. Control of electrical properties and gate bias stress stability in solution-processed a-IZO TFTs by Zr doping. *Curr. Appl. Phys.* **2014**, *14*, 1831–1836. [CrossRef]
21. Bard, A.J.; Parsons, R.; Jordan, J. *Standard Potentials in Aqueous Solution*; CRC Press: Boca Raton, FL, USA, 1985; Volume 6.
22. Satoh, K.; Kakehi, Y.; Okamoto, A.; Murakami, S.; Moriwaki, K.; Yotsuya, T. Electrical and optical properties of Al-doped ZnO–SnO_2 thin films deposited by RF magnetron sputtering. *Thin Solid Films* **2008**, *516*, 5814–5817. [CrossRef]
23. Liu, A.; Zhang, Q.; Liu, G.X.; Shan, F.K.; Liu, J.Q.; Lee, W.J.; Shin, B.C.; Bae, J.S. Oxygen pressure dependence of Ti-doped In-Zn-O thin film transistors. *J. Electroceram.* **2014**, *33*, 31–36. [CrossRef]
24. Sarma, D.; Das, T.M.; Baruah, S. Bandgap Engineering of ZnO Nanostructures through Hydrothermal Growth. *ADBU J. Eng. Technol.* **2016**, *4*, 216–218.
25. Lemlikchi, S.; Abdelli-Messaci, S.; Lafane, S.; Kerdja, T.; Guittoum, A.; Saad, M. Study of structural and optical properties of ZnO films grown by pulsed laser deposition. *Appl. Surf. Sci.* **2010**, *256*, 5650–5655. [CrossRef]
26. Chen, Y.; Xu, X.L.; Zhang, G.H.; Xue, H.; Ma, S.Y. A comparative study of the microstructures and optical properties of Cu-and Ag-doped ZnO thin films. *Phys. B Condens. Matter* **2009**, *404*, 3645–3649. [CrossRef]
27. Jeong, W.H.; Kim, G.H.; Shin, H.S.; Du Ahn, B.; Kim, H.J.; Ryu, M.K.; Park, K.B.; Seon, J.B.; Lee, S.Y. Investigating addition effect of hafnium in InZnO thin film transistors using a solution process. *Appl. Phys. Lett.* **2010**, *96*, 093503. [CrossRef]
28. Leelavathi, A.; Madras, G.; Ravishankar, N. Origin of enhanced photocatalytic activity and photoconduction in high aspect ratio ZnO nanorods. *Phys. Chem. Chem. Phys.* **2013**, *15*, 10795–10802. [CrossRef] [PubMed]
29. Koo, C.Y.; Song, K.; Jun, T.; Kim, D.; Jeong, Y.; Kim, S.H.; Ha, J.; Moon, J. Low temperature solution-processed InZnO thin-film transistors. *J. Electrochem. Soc.* **2010**, *157*, J111–J115. [CrossRef]
30. Hu, C.F.; Feng, J.Y.; Zhou, J.; Qu, X.P. Investigation of oxygen and argon plasma treatment on Mg-doped InZnO thin film transistors. *Appl. Phys. A* **2016**, *122*, 941. [CrossRef]
31. Yong Chong, H.; Wan Han, K.; Soo No, Y.; Whan Kim, T. Effect of the Ti molar ratio on the electrical characteristics of titanium-indium-zinc-oxide thin-film transistors fabricated by using a solution process. *Appl. Phys. Lett.* **2011**, *99*, 161908. [CrossRef]

nanomaterials

MDPI

Article

Growth Mechanism Studies of ZnO Nanowires: Experimental Observations and Short-Circuit Diffusion Analysis

Po-Hsun Shih and Sheng Yun Wu *

Department of Physics, National Dong Hwa University, Hualien 97401, Taiwan; libra.kevin.t@gmail.com
* Correspondence: sywu@mail.ndhu.edu.tw; Tel.: +886-3-863-3717

Academic Editor: Andrea Lamberti
Received: 3 July 2017; Accepted: 18 July 2017; Published: 21 July 2017

Abstract: Plenty of studies have been performed to probe the diverse properties of ZnO nanowires, but only a few have focused on the physical properties of a single nanowire since analyzing the growth mechanism along a single nanowire is difficult. In this study, a single ZnO nanowire was synthesized using a Ti-assisted chemical vapor deposition (CVD) method to avoid the appearance of catalytic contamination. Two-dimensional energy dispersive spectroscopy (EDS) mapping with a diffusion model was used to obtain the diffusion length and the activation energy ratio. The ratio value is close to 0.3, revealing that the growth of ZnO nanowires was attributed to the short-circuit diffusion.

Keywords: nanocrystalline materials; zinc oxide; nanowire; EDS mapping; short-circuit diffusion

1. Introduction

In the nanometer era, one-dimensional semiconductor nanomaterials with different morphologies have been fabricated by various methods and techniques. The properties of nanomaterials are usually quite distinct from these of bulk. Among the various semiconductors, zinc oxide (ZnO) has many charming properties that include a direct and wide band gap, large exaction energy, a large piezoelectric constant, strong ultraviolet emissions, stable structure, high penetrability, and good conductivity [1]. One-dimensional ZnO nanostructures have some potential applications such as nano-lasers, nano-detectors, and nano-sensors [2]. The ZnO nanowires can be fabricated by the vapor transport method [3,4], molecular beam epitaxy method [5], laser ablation method [6], simple thermal method [7,8], and so on. In recent years, the morphology, lengths, diameters, and growing directions of ZnO nanostructures could be roughly controlled by adjusting the parameters in the manufacturing process. Moreover, one-dimensional nanowire arrays [9], 3D network nanowires [10,11], and coaxial core–shell nanowires [12] have been synthesized using special methods with various catalysts and auxiliaries have been used in many of these methods. However, despite plenty of studies having been performed to probe the various properties of ZnO nanowires, only a few have specialized in investigating the optical properties of a single nanowire. Recent studies on ZnO nanowires have slightly shifted the focus to different aspects that include the growth mechanism, structural transformation, electron mobility, and phonon transmission. After understanding the disguised reaction mechanism, it could be possible to control and modify the electronic properties of nanowires. The mechanical characterization of single ZnO nanowire was reported by Argawal et al. [13,14], the size effect of Young's modulus of ZnO nanowires was investigated experimentally, as well as computationally. Experimentally, ZnO nanowires with diameters ranging from 20.4 to 412.9 nm were tested under a uniaxial tensile load using a nanoscale materials testing system inside a transmission electron microscope, revealing the Young's modulus of ZnO nanowires monotonically decreases from 160 to 140 GPa as the nanowires diameter increases from 20 to ~80 nm. Therefore, to date, how to synthesize high-crystalline nanowires and how to establish an ordered investigation are very important scientific and technical issues. However, some challenges have to be overcome. Firstly, researchers have to develop various fabrication methods to investigate the properties of ZnO nanowires and to make large-area ZnO nanowire arrays for the purpose of potential applications. In most of these various synthesis methods, various catalysts or auxiliaries were utilized in these fabrication processes. Residual contaminations could be found on the surface or at the edge of the nanomaterials with the result that it is difficult to observe the essential properties of ZnO. Therefore, the synthesis of nanowires without a catalyst or using a one-step growth method would be of interest. These syntheses would be useful to directly characterize the physical properties. Moreover, the 3D network ZnO nanowires with high surface-to-volume ratio have attracted a great deal of attention, while how to control the dimensionality and size of ZnO nanowires has also become popular. Secondly, compared to the developed growth mechanisms for zero- and two-dimensional nano systems, the growth mechanism of nanowires deserves to be investigated in-depth. The growth mechanism of metal-oxide nanowires using catalytic methods is often attributed to the vapor-liquid-solid mechanism, but without using catalysts it is not clear. Some reporters have proposed that the metal-oxide nanowires grown below the melting temperatures are attributed to the short-circuit diffusion mechanism. Investigating the atomic diffusion in metal-oxide films is becoming an increasingly important issue. Lastly, plenty of studies have been performed to probe the diverse properties of ZnO nanowires, but only a few have focused on the physical properties of a single nanowire since analyzing the phonon confinement along a single nanowire is difficult. In addition to this, previous studies of size effects were usually investigated by fabricating various sized nanoparticles and examining these properties. Nevertheless, the multi-contributions of nanoparticle shapes, size distributions, thermal effects, surface effects, and strain have resulted in a complicated situation in which the experimental pieces of information could not be compared with one another [15]. The investigation of a single ZnO nanowire, therefore, provides a reasonable possibility for probing the size effect.

In this study, we report the syntheses of a single ZnO nanowire using a Ti-assisted chemical vapor deposition (CVD) method to avoid the appearance of catalytic contamination. The dimensionality and size of ZnO nanowires can be controlled through fabrication time. A two-dimensional energy dispersive spectrum is a conventional technique to examine the diffusion situation in nanostructures. A short-circuit diffusion model was presented.

2. Materials and Methods

One-dimensional ZnO nanostructures can be fabricated by various methods. Two typical methods are introduced as follows. The chemical vapor deposition (CVD) method with metal catalysts is a common method to fabricate semiconductor nanowires, such as TiO_2, ZnO, GaN, and so on [4,16,17]. A typical example of ZnO nanowires synthesized by chemical vapor transport and a condensation system has been reported by Yang et al. [17]. During the process of nanowire growth, the zinc powder was heated to generate zinc vapor and then flowed to the substrate. The zinc vapor reacted with the gold solvent on the substrate at a lower temperature region to form alloy droplets. When the alloy droplet became supersaturated, crystalline ZnO nanowires were grown on the substrate surface. The oxygen atom source originated from the reaction between zinc and CO/CO_2 vapor. The growth mechanism could be attributed to the VLS (vapor-liquid-solid) crystal growth mechanism which is widely used for explaining the growth mechanism of oxide nanowires.

Compared with the previous methods, the thermal evaporation method without using catalysts [18,19] is another simple method for the production of ZnO nanowires. The fabrication of ZnO nanowires under ambient air and collected products on the surface of the sample stage has been reported by Wang et al. [19]. No distinguishable suboxide (ZnO_{1-x}) and impurity phase can be observed on the XRD patterns. In addition, 3D network ZnO structures can be fabricated by this method [20,21]. Chang et al. [20] fabricated ZnO nanowires using a chemical vapor deposition method at 700 °C under an argon gas flow in a quartz tube. The ZnO samples with different morphology were collected in various regions on the substrate. The authors proposed that the tuning pressure effect results in a different nucleation rate which further affects the morphology.

In this study, we used the CVD method to fabricate ZnO nanowires without an auxiliary template or any catalyst. Only a pure titanium grid (melting point: ~1941 K) was used as an auxiliary and substrate in this fabrication process. No other catalyst and auxiliary were used. The titanium is a good candidate for being a substrate due to both zinc and titanium having a hexagonal structure with a space group of $P6_3/mmc$, while ZnO has a hexagonal structure with a space group of $P6_3mc$, as shown in Figure 1a–c, respectively. As shown in the supplementary materials of Table S1 [22], the lattice constants are close with each other. In addition to this, no TiO_x nanowires were grown on the surface of the Ti grid during the annealing process at a temperature range of 300–800 °C. Based on this, the titanium grid was selected as the substrate. Compared with other methods, this approach is a simple, convenient, and reliable method for preparing ZnO nanowires. The complete synthesis process is as follows: (1) a porcelain boat, a cut zinc ingot (0.2 g), and a pure silicon plate with a side length of 0.5 cm were cleaned with ethanol and then were washed in a low-energy ultrasonic cleaning bath for five minutes, respectively; (2) the zinc ingot was mounted on a pure titanium grid with a mesh number of 200; (3) the grid was put on the cleaned silicon plate and then all were placed on a porcelain boat as shown in Figure 1d; (4) the boat was placed in a quartz tube in the middle region of a heated oven as shown in Figure 1e; (5) the pressure of the quartz tube was reduced to less than 1×10^{-2} Torr by a mechanical pump; (6) the heating temperature in the quartz tube was set for various samples in a temperature range of 300–800 °C, respectively; (7) these temperatures were automatically adjusted by a current controller; (8) after the temperature was stabilized, a mixed gas of oxygen (20 *sccm*) and argon (80 *sccm*) was introduced into the tube and the pressure was kept at 760 Torr by a flux controller; (9) the boat was heated at a set temperature for two hours; (10) after heating, the samples were cooled to room temperature naturally after the heating; and (11) these as-grown samples were saved in a low-pressure container to avoid further oxidation. The zinc ingot on the grid melted in the heating process and then the liquid zinc uniformly covered the grid. The zinc atoms combined with oxygen atoms after which ZnO nanowires grew on the ZnO film. The optical images of various annealing temperatures samples were shown in the supplementary materials of Figure S1 [22]. As can be seen in Figure S1, the heated titanium grids were usually curly and fragile. These samples were characterized by various techniques and are discussed in further sections. In order to achieve the research purpose, we used several measuring instruments to characterize the properties of ZnO nanowires. The morphological appearance and elemental composition were characterized by field-emission scanning electron microscopy and an energy dispersion spectrometer, respectively. The atomic image and crystal structure were obtained by analytical transmission electron microscopy.

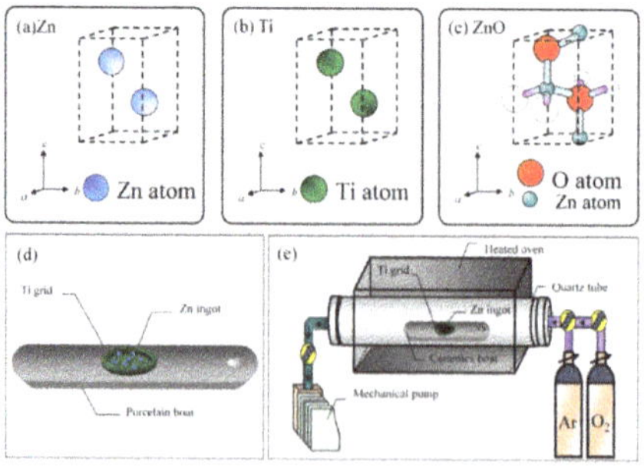

Figure 1. Schematic figures of unit cells of (**a**) zinc; (**b**) titanium and (**c**) ZnO; (**d**) the schematic figure for a porcelain boat, in which a high-purity zinc ingot (~0.2 g, 99.99%) on a cleaned Ti grid wafer was mounted on a cut silicon wafer; (**e**) the schematic figure for the chemical vapor deposition (CVD) instrument.

3. Results and Discussion

3.1. Morphological Analysis of ZnO Nanowire

An electron microscope is a precision electro-optical instrument for observing the scattering between incident electrons and samples to investigate material morphology and fine structure. In this study, field-emission scanning electron microscopy (FE-SEM, JEOL, JSM 6500F, Peabody, MA, USA) was used to characterize the morphology of ZnO nanowires. In this study, the elemental component and elemental mapping of ZnO nanowires was observed by energy-dispersive X-ray spectroscopy (Inca X-sight model 7557 Oxford Instrument, Abingdon, Oxfordshire, UK) that was equipped with the above-mentioned scanning electron microscope. The energy resolution is less than 133 eV. The detectable range of elements is from boron to uranium. The maximum detectable energy of X-rays is related to the incident electron beam energy. Following the SEM process, 15 keV is the limit. The sample preparation and experimental process are similar to the steps for SEM measurement. Figure 2a–i displays these SEM images with a magnification of 50,000 for various samples synthesized at annealing temperatures of 300–800 °C, respectively. In the further sections, the sample synthesized at 300 °C is labeled as T300, and other samples are tagged in the same way. It can be seen that, the ZnO nanowires only can be seen on the samples of T400–T700. The one-dimensional ZnO nanowires were observed on the surface of T400 and T450 and 3D network ZnO nanowires were obtained in a temperature range of 500–700 °C. In the one-dimensional ZnO samples as shown in Figure 2b,c, the ZnO nanowires were individually separated and straight. The diameters and lengths of the ZnO nanowires were in the ranges of a few tens of nanometers and several μm, respectively. In the three-dimensional (3D) ZnO samples, as shown in Figure 2d–h, the branching ZnO nanowires were found on the grid surface. The branches grew randomly on the trunk and were not perpendicular to the truck. With the annealing temperature increasing, the branch numbers and structural complexity clearly increased. As the annealing temperatures were lower than 400 °C or higher than 700 °C, no nanowire can be seen on the grid surface. With the high temperature range shown in Figure 2i, the surfaces become very rough as a result of grain growth. The grain size is in a range from 100 nanometers to 5 μm. The diameter distribution of ZnO nanowires could be described using a log-normal function. We measured the diameters for various samples and fitted the curves using a log-normal function, respectively. The log-normal function is defined as follows: $f(d) = \frac{1}{(2\pi)^{0.5} d\sigma} \exp[-\frac{(\ln d - \ln <d>)^2}{2\sigma^2}]$, where d is the diameter, <d> is the mean value and σ is the standard deviation. The experimental data and fitting curves were plotted by black hollow bars and red solid lines in Figure 3a–g, respectively. The corresponding fitting parameters are shown in Table 1. For the sample of T400, as an example, the mean value is 33.2 ± 0.3 nm and the standard deviation is 0.24 ± 0.01. The low standard deviation (<0.5) means that the nanowire size is confined in a small range. The regression constant R^2 of 0.99 indicates the fitting curve is very close to the size distribution. From the overall results, the mean diameter of ZnO nanowires reveals an obvious dependence on annealing temperature as shown in Figure 3h; the mean diameter increases with increasing annealing temperatures. In the temperature range of 400–650 °C, the temperature dependence of growth mean diameter of ZnO nanowires can be described by a parabolic law (solid curve). The standard deviations as shown in Figure 3i are less than 0.5, revealing that the distribution of mean diameters is uniform.

Table 1. Summary of the fitting parameters obtained from the log normal function.

T_A (°C)	Sample	<d> (nm)	δ (nm)	Area	R^2
400	T400	33.2 ± 0.3	0.24 ± 0.01	3219 ± 113	0.99
450	T450	33.4 ± 0.4	0.39 ± 0.01	2789 ± 78	0.99
500	T500	36.6 ± 0.4	0.35 ± 0.01	1834 ± 204	0.99
550	T550	38.5 ± 1.6	0.31 ± 0.04	1834 ± 204	0.92
600	T600	37.4 ± 1.9	0.35 ± 0.04	2817 ± 331	0.88
650	T650	57.5 ± 2.5	0.49 ± 0.04	3348 ± 441	0.97
700	T700	191.5 ± 8.2	0.34 ± 0.04	4177 ± 442	0.93

Figure 2. Scanning electron microscopy (SEM) images for a series of samples synthesized at (**a**) 300 °C; (**b**) 400 °C; (**c**) 450 °C; (**d**) 500 °C; (**e**) 550 °C; (**f**) 600 °C; (**g**) 650 °C; (**h**) 700 °C and (**i**) 800 °C, respectively.

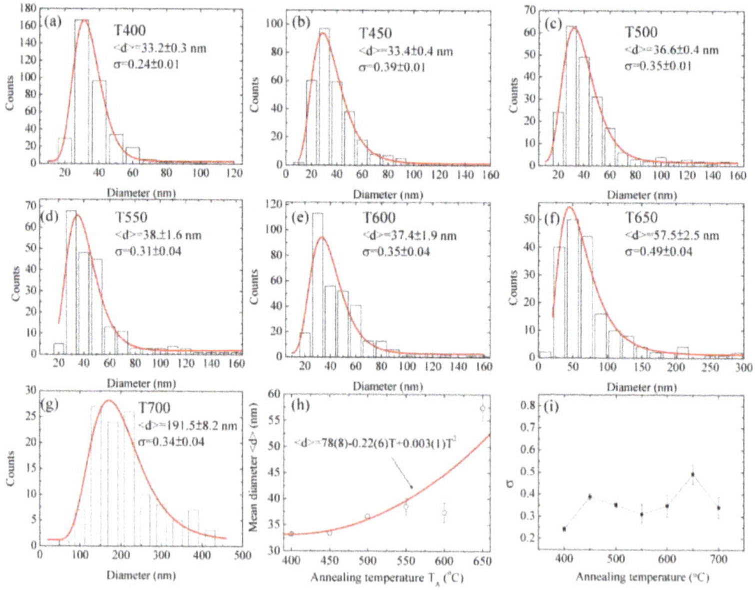

Figure 3. Size distributions for a series of samples synthesized at (**a**) 400 °C; (**b**) 450 °C; (**c**) 500 °C; (**d**) 550 °C; (**e**) 600 °C; (**f**) 650 °C and (**g**) 700 °C, respectively; and (**h**,**i**) show the mean diameter and the standard deviation versus annealing temperatures.

3.2. Crystal Structure Analysis of ZnO Nanowires

Transmission electron microscopy (TEM) is an important microscopic technique for cancer research, material science, and nanotechnology. In the present study, the high-resolution (HR) images and the selected area electron diffraction (SAED) patterns were obtained by an analytical transmission electron microscope (JEOL 3010, Peabody, MA, USA) and a field emission transmission electron microscope (JEOL JEM-2100, Peabody, MA, USA). Figure 4a shows an example of the TEM result of a straight ZnO nanowire of the T450 sample. The diameter of the ZnO nanowire is about 32 nm. The single crystalline nature is revealed. The diffraction pattern along the (001) direction shown in Figure 4b can be indexed as hexagonal ZnO in structure with a space group of $P6_3mc$. A high-magnification enlargement of a selected area of the high-resolution TEM image is shown in Figure 4c. It can be seen that the normal direction of planes is not parallel to the growth direction of the ZnO nanowire. A gray-level analysis was used to extract the scattering intensity which can be fit using a multi-Gaussian function to obtain the average plane spacing. Figure 4d shows the fitting result, in which the asterisk and the solid line show the experimental data and fitting curve, respectively. The fitted distance is 0.279 (4) nm, corresponding to the *d*-spacing of (100) planes of ZnO hexagonal structure. The lattice parameter of the hexagonal structure can be calculated using the relationship: $\frac{1}{{d_{hkl}}^2} = \frac{4}{3}\frac{h^2+hk+k^2}{a^2} + \frac{l^2}{c^2}$, where d_{hkl} is a spacing between two planes of (*hkl*), *a* and *c* are lattice parameters, and *h*, *k*, and *l* are the Miller indices. The result shows the lattice parameter *a* of 0.32(3) nm. Since the *c*-axis of the hexagonal structure is perpendicular to the scattering plane, we could not measure the lattice constant *c*. Based on this analysis, a schematic crystal structure of a single ZnO nanowire is shown in Figure 4e. The growth direction of ZnO nanowires is along the (110) direction. As the annealing temperatures was higher than 500 °C, dendritic ZnO nanowires could be easily seen in the TEM images of Figure S2a (see supplementary materials of Figure S2) [22], showing the morphology of a typical dendritic ZnO nanowire. The Bragg spots as shown in Figure 2b correspond to the zone axis (001) reflection of the ZnO wurtzite structure.

The detailed structure of the ZnO nanowire was investigated using high-resolution images. Two high-magnification enlargements of selected regions, marked in Figure S2a, are shown in Figure S2c,d, respectively. It can be seen that there is an obvious atomic arrangement of hexagonal symmetry and the atomic spacing can be obtained by the above gray-level analysis. The values obtained from the fittings, as shown in Figure S2e,f were, respectively, 0.33 (3) and 0.33 (1) nm, corresponding to the lattice constant *a* of the ZnO wurtzite structure. This result of lattice parameters is consistent with that obtained by SAED observations. The growth directions of these branches are indicated by the (110) direction. It can be explained that the two nanowires were inclined towards each other by 60 degrees. In our experience, the branching nanowires appear randomly on the surface of dendritic ZnO nanowires and the angle between the branches and trunks is closed to multiples of 60 degrees and with no other angle able to be found [23–25]. Details of the corresponding lattice parameters for various samples are summarized into Table 2. As can be seen in Table 2, the lattice parameters of various samples are slightly smaller than that of bulk [26], assumed that the strain effect is responsible for the lattice contraction [27].

The ZnO nanowires have a growth direction of (110), which is in contrast to the common (001) growth direction [28–31]. Only a few papers [32,33] have reported that the ZnO nanowires grew along the (110) direction. Similar analysis method of the structure and lattice parameters for a single nanowire has been reported by Barriga et al. [34]. Four different cylindrical nanowire systems (Ni, Co and $Co_{58}Ni_{42}/Co_{83}Ni_{17}$ nanowires), grown by standard electrodeposition techniques in the nanometer size channels of porous alumina templates, were investigated using TEM and SAED. In their comprehensive analysis, these results can be explained by considering the characteristics of the measurement technique and the confined template-assisted growth, which force the atoms to be accommodated in a cylindrical volume with nanoscale dimensions. The growth mechanism of ZnO nanowires could be attributed to the short-circuit diffusion [33,35], the high zinc vapor pressure [36] and the diffusion-limited supersaturated environment [23,25]. Rackauskas et al. [33] assumed that the growth of ZnO nanowires is related to the diffusion through grain boundaries in the ZnO layer and the crystal defects in ZnO nanowires. Fan's group [36] proposed that the growths of the trunk and branches go through a self-catalytic liquid-solid and vapor-solid process, respectively. Zinc atoms were heated to form a vapor at 600 °C and then nucleated on the nanowire surface to form branches. Complementarily, Park et al. [25] assumed that the supersaturated reactant vapors play an important role in forming the dendritic side branches. In this point of view, a catalyst is not necessary in the formation of branch growth and the growth process of ZnO nanowires should be dependent on a surface diffusion, with respect to the annealing temperatures and growth times. The formation of ZnO nanowires was carried out by a two-dimensional EDS mapping investigation.

Table 2. A summary of lattice parameters for various samples.

Sample	Trunk		Branch	
	Diameter (nm)	Lattice Constant *a* (nm)	Diameter (nm)	Lattice Constant *a* (nm)
T400	13.0	0.32 (2)		
T450	32.1	0.32 (3)		
T500	54.1	0.33 (1)	19.1	0.33 (3)
T550	37.1	0.32 (3)	25.8	0.32 (3)
T600	65.2	0.32 (2)	56.4	0.32 (6)
T650	83.3	0.32 (3)	33.3	0.32 (3)

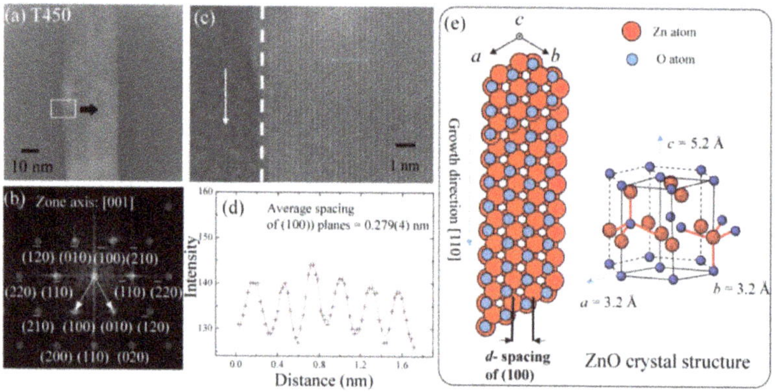

Figure 4. (**a**) Transmission electron microscopy (TEM) image; (**b**) corresponding selected-area electron pattern; (**c**) high-resolution image of the selection region (marked in (**a**)) of ZnO nanowires for T450; (**d**) height-position intensity along the line taken from high resolution (HR)-TEM (marked in (**b**)); and (**e**) schematic figure of the crystal structure for a single ZnO nanowire.

3.3. *Two-Dimensional EDS Mapping*

To observe element and component distribution is important for investigating the diffusion in nanomaterials. There are numerous experimental methods, such as secondary ion mass spectrometry, electron microprobe analysis, auger electron spectroscopy, nuclear reaction analysis, nuclear magnetic relaxation, confocal Raman spectroscopy, transmission electron microscopy, and energy dispersive spectroscopy, for studying diffusion in solids, in which the energy dispersive spectrometer is a convenient and useful tool to analyze spatial element distributions, especially EDS mapping images that can offer direct evidence of element distributions. Figure 5a shows the schematic illustration of EDS mapping of the cross-sections of ZnO samples. The broken samples were fixed on cleaned silicon plates by carbon tape and then were mounted on a special sample holder for taking cross section images. An EDS detector scans an edge portion near the sample surface and then depicts a corresponding element mapping of a cross-section of the ZnO samples. Along the normal direction of the sample surface, we assumed that the bottom part is titanium substrate, followed by pure Zn film, ZnO_x film, and ZnO nanowires as shown in Figure 5b.

We assumed that the zinc atoms were diffused from the bottom of the ZnO films to the surface, and then formed the ZnO nanowires. In this view, the diffusion length is related to the thickness of layers on the grid surface. We can estimate the ZnO thickness to understand the growth mechanism through EDS mapping technique. Figure 6a shows an example of the EDS mapping result for T500 sample, in which the corresponding SEM image is used as a background. The size of the scanning area is about 60 × 54.2 μm, covering a cross-sectional area of a tube of titanium grid. The number of scanning times is 5 in order to improve the measurement accuracy. The elements of zinc (blue), oxygen (red), and titanium (green) were, respectively, indicated using the lock-in energy of Zn–L_α (0.8–1.2 keV), O–K_α (0.4–0.6 keV), and Ti–K_α (4.3–4.7 keV). As shown in the Figure 6a, the center of the tube (point A) shows the blue color, revealing the existence of the zinc component.

We assume that, due to thermal heating and grid expansion, zinc flows into the hollow tube from the surface through crevices and fills the tube. On the upper surface of the grid (point B), the red dots and blue dots are distributed uniformly, revealing that the grid has been covered in a thin zinc oxide layer. On the other side of the grid surface (point C), the fewer oxygen signals were attributed to the smaller contact area with oxygen during the annealing process. At the edge of the grid surface (point D), a green ring showed clearly the position of the grid section. Incidentally, the number of signal points in the upper part of the Figure 6a is more than that in the lower part due to the detection angle and the focal length. The three element distribution mappings for Ti, O and Zn are, respectively, shown in Figure 6b, in which the purple curves show the intensities of each element along the vertical line (yellow), respectively. Line profile EDS analysis clearly shows the presence of O in the sample. It can be seen that the intensity of oxygen signals deceases suddenly near the wall of the Ti and the decreasing curve shown in Figure 6c can be used to obtain the thickness of ZnO layers. The curve can be obtained by an exponential function [37]. The line width of the distribution can be used to define the mean diffusion length $<\xi_d>$. The obtained diffusion length ξ_d is near 3.40 μm for T500 sample. A series of EDS mapping were examined and obtained on various samples. The corresponding diffusion lengths of the ZnO layer versus the annealing temperatures are shown in Figure 7. As seen in Figure 7, the estimated values of ξ_d versus the annealing temperature T_A are plotted, revealing an increase with the increase in the diffusion length ξ_d. The red solid curve indicates the fit of the data to the theoretical curve for an exponential decay function, namely $\xi_d = \xi_{do} + \beta T_A$, where ξ_{do} = 0.41 (1) nm and β = 0.073 (2) nm/K represents the initial constant and the fitted parameters, respectively.

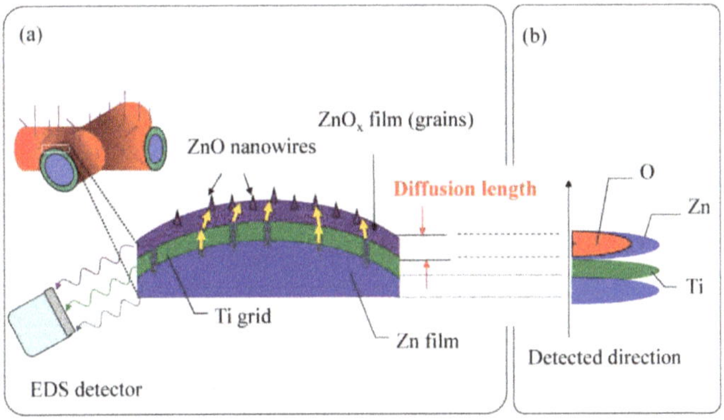

Figure 5. (**a**) Schematic diagram of EDS mapping; and (**b**) hypothetical results of the corresponding EDS spectra along the line scan.

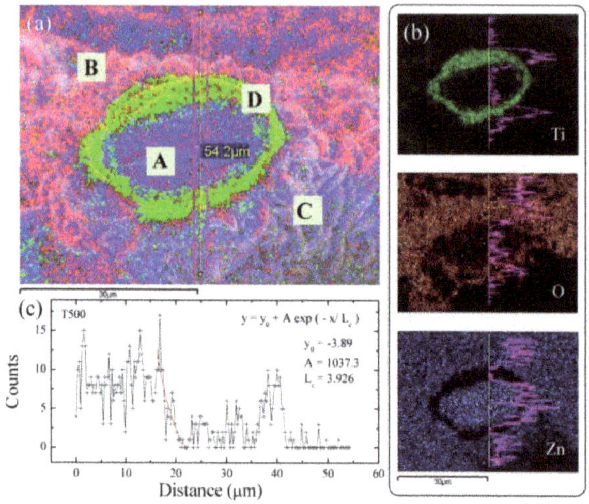

Figure 6. (**a**) A selected area EDS mapping of T500; (**b**) corresponding EDS mappings for a single element; and (**c**) the counts of oxygen versus the positions.

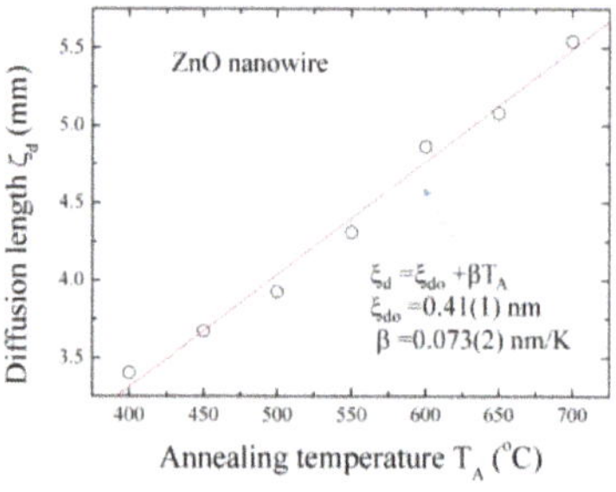

Figure 7. Diffusion length is dependent on the annealing temperature, revealing a growth rate of 0.0073 μm/°C^{-1}.

In general, the short-circuit diffusion plays an important role below 0.5 times the melting temperature [38]. The various samples in this study were fabricated below 0.5 times the melting temperature of ZnO (1975 °C) [26], in which recrystallization and grain growth proceed slowly and polycrystallinity provides effective short-circuit diffusion paths [35,39]. The short-circuit mechanism was used to describe the formation of one-dimensional nanostructures [40]. For example, Lu et al. [41] fabricatedα–Fe_3O_4 nanowires by oxidizing iron in pure oxygen between 400 and 600 °C. The oxide layer can be controlled by varying the oxidation temperatures to form grains. They pointed out that the iron ions diffuse from the Fe_2O_3/Fe_3O_4 interface to the free surface via grain boundary diffusion. Subsequently, the Fe ions diffuse from the grain boundary to the nanowire root via surface diffusion to form Fe_3O_4 nanowires on the top of Fe_3O_4 grains.

Xu et al. [42] fabricated CuO nanowires on Cu foils in wet air at a temperature range of 400–700 °C, with diameters between 50 to 400 nm and lengths between 1 and 15 micrometers. The authors assumed that a high density of sub-boundaries in the surface layer enhances the formation of nanowires. They emphasized that the short-circuit diffusion dominates in the middle temperatures, while the lattice diffusion would be important at high temperatures. Yuan et al. [43] proposed a method to enhance the nanowire growth density and length by increasing the surface roughness. It was found that the increased surface roughness (smaller grains) results in more short diffusion paths and surface sites that contribute to the diffusive transport of copper atoms along grain boundaries and the nucleation of CuO in order to improve the nanowire density and length. We purposed that the growth mechanism of ZnO nanowires was attributed to the short-circuit diffusion. A diffusion model [44] was used to interpret the diffusion length, where a parameter of γ is taken as the percentage of the lattice activation energy of zinc ions to clarify the contribution from short-circuit- or lattice-diffusion. The values γ of 1/3 and 1 indicate that the diffusion prefers short-circuit- and lattice-diffusion, respectively [35]. The equation of diffusion length is shown as follows: $\Delta L = \sqrt{D_L \cdot \tau} = [\frac{\beta}{2}\alpha^2 v_D \exp(-\frac{\gamma Q}{RT}) \cdot \tau]^{1/2}$, where ΔL is the diffusion length, D_L is the diffusion coefficient, τ is the growth time (approximately 7200 s), β is the number of atoms jumping along (110), α (= 0.162 nm) is the *d*-spacing of (110) related to the diffusion direction, v_D (= 1.73×10^{11} s^{-1}) is the vibrational frequency, Q (= 318 kJ/mol) is the activation energy of ZnO [45], R (= 1.987 cal·mol·K^{-1}) is the gas constant, and T is the growth temperature. The vibrational frequency can be calculated by the equation: $v_D = \frac{1}{2^{1/2}}\left(\frac{Q}{m\alpha^2}\right)$, where m (= 65.4 g/mol) is the zinc molar weight. Figure 8a shows the diffusion ratio γ versus annealing temperatures. The colors indicate the different diffusion lengths as shown on the left, in which the gray color denotes a diffusion length of less than 1 μm and the red color represents that of more than 200 μm. The obtained diffusion lengths are marked in the figure by white solid circles. It can be seen that the diffusion lengths corresponding to the growth temperatures of ZnO nanowires are located in the region of 0.26–0.35 times the activation energy of ZnO lattice diffusion, revealing that short-circuit diffusion dominates the diffusion process. Moreover, the value of activation energy at T400 sample is ~83 kJ/mol, close to the migration energies of zinc interstitials (77 kJ/mol) and vacancies (88 kJ/mol) [46]. Figure 8b and Table 3 show the diffusion ratio γ versus the annealing temperatures, in which the experimental data can be described by a linear function.

In the high-temperature range, the diffusion ratio is high, meaning that the lattice diffusion plays an important role in the transport process. The ZnO nanowires were found in a γ range of 0.261–0.353 where these γ values are close to 0.3, revealing that the short-circuit diffusion mechanism dominates the diffusion. The dimensionality of ZnO nanowires can be controlled by adjusting the annealing temperatures and the diffusion ratio. Figure 9 shows the schematic illustration of the growth process. In the first step, the zinc ingot melted to form a zinc film in a reduced oxygen environment, as shown in Figure 9a. Then, the zinc film reacted with the introduced oxygen gas to construct a thin ZnO film on the top of the zinc film, as shown in Figure 9b. The thickness of the ZnO layer could be obtained by probing the oxygen distribution in EDS mapping, as shown in Figure 9c. Finally, zinc atoms could diffuse through the ZnO film to form various structures on the surface at temperature regions of $T_A < 400$ °C (Figure 9d), $400 < T_A < 700$ °C (Figure 9e), and $T_A > 700$ °C (Figure 9f), respectively, in which the ZnO nanowires were found at middle temperatures and ZnO film was obtained both at high and low temperatures. The contribution of oxygen migration is not considered due to the large atomic size and high migration energies [32]. According to the previous report [33], the migration energies of oxygen interstitial and vacancy are 118 and 124 kJ/mol, respectively. ZnO nanowires cannot grow at high temperatures, which could be explained by the grain sizes and thickness of ZnO_x. It is well known that the contribution from boundary diffusion decreases with increasing nanocrystal size [47,48]. Brass and Chanfreau [48] proposed that a large density of grain boundaries would provide fast atomic diffusion paths along the boundaries. Based on this, atoms that diffuse through small grains to the surface would be faster than those in large grains. The grain size increases from tens to hundreds of nanometers with increasing annealing temperatures. Besides this, the thickness of ZnO_x films also increased with annealing temperatures that were obtained in EDS mapping. Both the large grains and thick films fabricated at high temperatures would prevent zinc atomic diffusion along grain boundaries from the zinc film to the surface to form ZnO nanowires, in which the lattice diffusion would be the dominant transport mechanism.

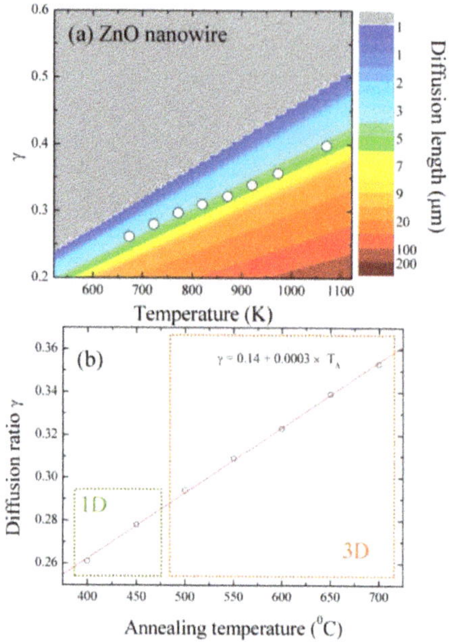

Figure 8. (**a**) Plots of the diffusion length and diffusion ratio γ versus the annealing temperatures; and (**b**) the diffusion ratio γ versus the annealing temperatures.

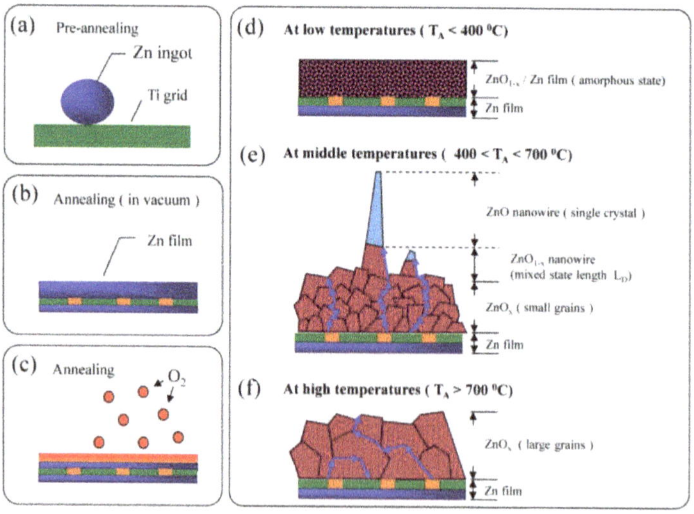

Figure 9. A schematic plot of growth process of ZnO structures: (**a**) sample preparation before annealing; (**b**) the formation of zinc film; (**c**) the oxidation on the film surface and the formation of various ZnO structures at (**d**) $T_A < 400$ °C, (**e**) $400 < T_A < 700$ °C, and (**f**) $T_A > 700$ °C.

Table 3. Diffusion lengths along with simulated results for a diffusion theory. The value γ is the ratio of the activation energy.

Dimensionality	Sample	Diffusion Length (μm)	Diffusion Ratio γ
1-D	T400	3.404	0.261
1-D	T450	3.674	0.278
3-D	T500	3.927	0.294
3-D	T550	4.310	0.309
3-D	T600	4.861	0.323
3-D	T650	5.080	0.339
3-D	T700	5.542	0.353
Large grains	T800	4.462	0.402

4. Conclusions

The growth mechanism and the phonon confinement effect in 1D and 3D network ZnO nanowires are investigated. The ZnO nanowires were fabricated by a Ti-assisted chemical vapor deposition method without any catalyst in a temperature range of 400–700 °C. The mean diameter ranging from 33.2 to 191.5 nm increased with the increasing temperature. The dimensionality can be controlled by adjusting the annealing temperatures. Below 500 °C, only one-dimensional ZnO nanowires can be found on the sample surface, and, above it, three-dimensional ZnO nanowires were able to be grown. ZnO nanowires have a hexagonal structure with a space group of $P6_3mc$ and the (110) growth direction. The formation of ZnO nanowires was attributed to the short-circuit diffusion. Energy dispersive X-ray spectroscopic mapping technique was used to depict the diffusion of zinc atoms through ZnO_x film from the zinc base to the film surface. A diffusion model was utilized to calculate the activation energy of the diffusion for various samples. The result shows that the activation energy is 0.26–0.35 times the activation energy of ZnO lattice diffusion, revealing that the growth of ZnO nanowires was related to the diffusion goes through the grain boundaries or sub-boundaries and then forms ZnO nanowires.

Supplementary Materials: The following are available online at http://www.mdpi.com/2079-4991/7/7/188/s1, Table S1. A list of space groups and lattice parameters of Zn, Ti and ZnO; Figure S1. Optical images for various annealing temperaure of samples (a) an unheated Ti grid and a series of samples synthesized at (b) 300 °C; (c) 350 °C; (d) 400 °C; (e) 450 °C; (f) 500 °C; (g) 550 °C; (h) 600 °C; (i) 650 °C; (j) 700 °C; (k) 750 °C; and (l) 800 °C, respectively; Figure S2. (a) TEM image; (b) corresponding selected-area electron pattern; (c,d) high-resolution images of selected regions (marked in (a)) of ZnO nanowires for T500; (e) and (f) show the height-position intensity along the lines taken from HR-TEM marked in (c) and (d), respectively.

Acknowledgments: We would like to thank the Ministry of Science and Technology (MOST) of the Republic of China for their financial support of this research through project numbers MOST-105-2112-M-259-003 and MOST-106-2112-M-259-001.

Author Contributions: Sheng Yun Wu wrote, conceived, and designed the experiments. Po-Hsun Shih grew the samples and analyzed the data. All authors discussed the results, contributed to the manuscript text, commented on the manuscript, and approved its final version.

Conflicts of Interest: The authors declare no conflict of interest.

References and Note

1. Janotti, A.; Van de Walle, C.G. Fundamentals of zinc oxide as a semiconductor. *Rep. Prog. Phys.* **2009**, *72*, 126501. [CrossRef]
2. Morkoç, H.; Özgür, Ü. *Zinc Oxide: Fundamentals, Materials and Device Technology*, 1st ed.; Wiley-VCH: Weinheim, Germany, 2009; p. 477.
3. Shih, H.Y.; Chen, Y.T.; Huang, N.H.; Wei, C.M.; Chen, Y.F. Size-dependent photoelastic effect in ZnO nanorods. *Appl. Phys. Lett.* **2009**, *94*, 021908. [CrossRef]
4. Huang, M.H.; Mao, S.; Feick, H.; Yan, H.; Wu, Y.; Kind, H.; Weber, E.; Russo, R.; Yang, P. Room-Temperature Ultraviolet Nanowire Nanolasers. *Science* **2001**, *292*, 1897–1899. [CrossRef] [PubMed]
5. Heo, Y.W.; Norton, D.P.; Tien, L.C.; Kwon, Y.; Kang, B.S.; Ren, F.; Pearton, S.J.; LaRoche, J.R. ZnO nanowire growth and devices. *Mater. Sci. Eng. R Rep.* **2004**, *47*, 1–47. [CrossRef]

6. Wang, R.P.; Xu, G.; Jin, P. Size dependence of electron-phonon coupling in ZnO nanowires. *Phys. Lett. B* **2004**, *69*, 113303. [CrossRef]
7. Dang, H.Y.; Wang, J.; Fan, S.S. The synthesis of metal oxide nanowires by directly heating metal samples in appropriate oxygen atmospheres. *Nanotechnology* **2003**, *14*, 738–741. [CrossRef]
8. Zhang, B.; Zhou, S.M.; Wang, H.W.; Du, Z.L. Raman scattering and photoluminescence of Fe-doped ZnO nanocantilever arrays. *Chin. Sci. Bull.* **2008**, *53*, 1639–1643.
9. Kar, S.; Pal, B.N.; Chaudhuri, S.; Chakravorty, D. One-Dimensional ZnO Nanostructure Arrays: Synthesis and Characterization. *J. Phys. Chem. B* **2006**, *110*, 4605–4611. [CrossRef] [PubMed]
10. Gao, P.X.; Wang, Z.L. Mesoporous Polyhedral Cages and Shells Formed by Textured Self-Assembly of ZnO Nanocrystals. *J. Am. Chem. Soc.* **2003**, *125*, 11299–11305. [CrossRef] [PubMed]
11. Wang, Z.-L. Nanostructures of zinc oxide. *Materialstoday* **2004**, *7*, 26. [CrossRef]
12. Hu, J.Q.; Li, Q.; Meng, X.M.; Lee, C.S.; Lee, S.T. Thermal reduction route to the fabrication of coaxial Zn/ZnO nanocables and ZnO nanotubes. *Chem. Mater.* **2003**, *15*, 305–308. [CrossRef]
13. Agrawal, R.; Peng, B.; Espinosa, H.D. Elasticity Size Effects in ZnO Nanowires—A Combined Experimental-Computational Approach. *Nano Lett.* **2009**, *8*, 3668. [CrossRef] [PubMed]
14. Agrawal, R.; Peng, B.; Espinosa, H.D. Experimental-Computational Investigation of ZnO nanowires Strength and Fracture. *Nano Lett.* **2009**, *9*, 4177. [CrossRef] [PubMed]
15. Chou, M.H.; Liu, S.B.; Huang, C.Y.; Wu, S.Y.; Cheng, C.L. Confocal Raman spectroscopic mapping studies on a single CuO nanowire. *Appl. Surf. Sci.* **2008**, *254*, 7539–7543. [CrossRef]
16. Ng, H.T.; Chen, B.; Li, J.; Han, J.; Meyyappan, M. Optical properties of single-crystalline ZnO nanowires on m-sapphire. *Appl. Phys. Lett.* **2003**, *82*, 2023. [CrossRef]
17. Yamg, P.; Yan, H.; Mao, S.; Russo, R.; Johnson, T.; Saykally, R.; Morris, N.; Pham, J.; He, R.; Choi, H.-J. Controlled growth of ZnO nanowires and their optical properties. *Adv. Funct. Mater.* **2002**, *12*, 323. [CrossRef]
18. Wan, Q.; Lin, C.L.; Yu, X.B.; Wang, T.H. Room-temperature hydrogen storage characteristics of ZnO nanowires. *Appl. Phys. Lett.* **2004**, *84*, 124. [CrossRef]
19. Wang, X.; Zhang, J.; Zhu, Z. Ammonia sensing characteristics of ZnO nanowires studied by quartz crystal microbalance. *Appl. Surf. Sci.* **2006**, *15*, 2404. [CrossRef]
20. Chang, P.C.; Fan, Z.; Wang, D.; Tseng, W.Y.; Chiou, W.A.; Hong, J.; Lu, J.G. ZnO Nanowires Synthesized by Vapor Trapping CVD Method. *Chem. Mater.* **2004**, *16*, 5133. [CrossRef]
21. Fan, D.H.; Shen, W.Z.; Zheng, M.J.; Zhu, Y.F.; Lu, J.J. Integration of ZnO Nanotubes with Well-Ordered Nanorods through Two-Step Thermal Evaporation Approach. *J. Phys. Chem. C* **2007**, *111*, 9116. [CrossRef]
22. Supplementary materials at http://www.mdpi.com/2079-4991/7/7/188/s1 for the detail of a list of space groups and lattice parameters of Zn, Ti and ZnO (Table S1); Optical images for various annealing temperature of samples (Figure S1); TEM image and corresponding selected-area electron pattern. (Figure S2).
23. Leung, Y.H.; Djurišić, A.B.; Gao, J.; Xie, M.H.; Wei, Z.F.; Xu, S.J.; Chan, W.K. Zinc oxide ribbon and comb structures: Synthesis and optical properties. *Chem. Phys. Lett.* **2004**, *394*, 452–457. [CrossRef]
24. Huang, H.; Yang, S.; Gong, J.; Liu, H.; Duan, J.; Zhao, X.; Zhang, R. Controllable Assembly of Aligned ZnO Nanowires/Belts Arrays. *J. Phys. Chem. B* **2005**, *109*, 20746–20750. [CrossRef] [PubMed]
25. Park, J.H.; Choi, H.J.; Choi, Y.J.; Sohn, S.H.; Park, J.G. Ultrawide ZnO nanosheets. *J. Mater. Chem.* **2003**, *14*, 35–36. [CrossRef]
26. Xu, J.; Pan, Q.; Shun, Y.; Tian, Z. Grain size control and gas sensing properties of ZnO gas sensor. *Sens. Actuators* **2000**, *66*, 277–279. [CrossRef]
27. Giri, P.K.; Bhattacharyya, S.; Singh, D.K.; Kesavamoorthy, R.; Panigrahi, B.K.; Nair, K.G.M. Correlation between microstructure and optical properties of ZnO nanoparticles synthesized by ball milling. *J. Appl. Phys.* **2007**, *102*, 093515. [CrossRef]
28. Zhou, S.M.; Zhang, X.H.; Meng, X.M.; Fan, X.; Wu, S.K.; Lee, S.T. Preparation and photoluminescence of Sc-doped ZnO nanowires. *Phys. E Low-Dimens. Syst. Nanostruct.* **2005**, *25*, 587–591. [CrossRef]
29. Vayssieres, L. Growth of Arrayed Nanorods and Nanowires of ZnO from Aqueous Solutions. *Adv. Mater.* **2003**, *15*, 464–466. [CrossRef]
30. Geng, C.; Jiang, Y.; Yao, Y.; Meng, X.; Zapien, J.A.; Lee, C.S.; Lifshitz, Y.; Lee, S.T. Well-Aligned ZnO Nanowire Arrays Fabricated on Silicon Substrates. *Adv. Funct. Mater.* **2004**, *14*, 589–594. [CrossRef]

31. Zheng, M.J.; Zhang, L.D.; Li, G.H.; Shen, W.Z. Fabrication and optical properties of large-scaleuniform zinc oxide nanowire arrays by one-step electrochemical deposition technique. *Chem. Phys. Lett.* **2002**, *363*, 123–128. [CrossRef]
32. Ren, S.; Bai, Y.F.; Chen, J.; Deng, S.Z.; Xu, N.S.; Wu, Q.B.; Yang, S. Catalyst-free synthesis of ZnO nanowire arrays on zinc substrate by low temperature thermal oxidation. *Mater. Lett.* **2007**, *61*, 666–670. [CrossRef]
33. Rackauskas, S.; Nasibulin, A.G.; Jiang, H.; Tian, Y.; Statkute, G.; Shandakov, S.D.; Lipsanen, H.; Kauppinen, E.I. Mechanistic investigation of ZnO nanowire growth. *Appl. Phys. Lett.* **2009**, *95*, 183114. [CrossRef]
34. Barriga-Castro, E.D.; Mendoza-Resendéz, R.; García, J.; Pridab, V.M.; Luna, C. Pseudo-monocrystalline properties of cylindrical nanowires confinedly grown by electrodeposition in nanoporous alumina templates. *RSC Adv.* **2017**, *7*, 13817. [CrossRef]
35. Shih, P.H.; Hung, H.J.; Ma, Y.R.; Wu, S.Y. Tuning the dimensionality of ZnO nanowires through thermal treatment: An investigation of growth mechanism. *Nanoscale Res. Lett.* **2012**, *7*, 354. [CrossRef] [PubMed]
36. Fan, H.J.; Scholz, R.; Kolb, F.M.; Zacharias, M. Two-dimensional dendritic ZnO nanowires from oxidation of Zn microcrystals. *Appl. Phys. Lett.* **2004**, *85*, 4142. [CrossRef]
37. Gandhi, A.C.; Hung, H.J.; Shih, P.H.; Cheng, C.L.; Ma, Y.R.; Wu, S.Y. In Situ Confocal Raman Mapping Study of a Single Ti-Assisted ZnO Nanowire. *Nanoscale Res. Lett.* **2010**, *5*, 581–586. [CrossRef] [PubMed]
38. Cheng, C.L.; Ma, Y.R.; Chou, M.H.; Huang, C.Y.; Yeh, V.; Wu, S.Y. Direct observation of short-circuit diffusion during the formation of a single cupric oxide nanowire. *Nanotechnology* **2007**, *18*, 245604. [CrossRef]
39. Herchl, R.; Khoi, N.N.; Homma, T.; Smeltzer, W.W. Short-circuit diffusion in the growth of nickel oxide scales on nickel crystal faces. *Oxid. Met.* **1972**, *4*, 35–49. [CrossRef]
40. Li, S.B.; Bei, G.P.; Zhai, H.X.; Zhang, Z.L.; Zhou, Y.; Li, C.W. The origin of driving force for the formation of Sn whiskers at room temperature. *J. Mater. Res.* **2007**, *22*, 3226. [CrossRef]
41. Lu, Y.; Wang, Y.; Cai, R.; Jiang, Q.; Wang, J.; Li, B.; Sharma, A.; Zhou, G. The origin of hematite nanowire growth during the thermal oxidation of iron. *Mater. Sci. Eng. B* **2012**, *177*, 327.
42. Xu, C.H.; Woo, C.H.; Shi, S.Q. Formation of CuO nanowires on Cu foil. *Chem. Phys. Lett.* **2004**, *399*, 62. [CrossRef]
43. Lu, Y.; Zhou, G. Enhanced CuO Nanowire Formation by Thermal Oxidation of Roughened Copper. *J. Electrochem. Soc.* **2012**, *159*, C205.
44. Metselaar, R. Diffusion in solids. *Part I: Introduction to the theory of diffusion. J. Mater. Educ.* **1984**, *6*, 229.
45. Gao, F.; Chino, N.; Naik, S.P.; Sasaki, Y.; Okubo, T. Photoelectric properties of nano-ZnO fabricated in mesoporous silica film. *Mater. Lett.* **2007**, *61*, 3179–3184. [CrossRef]
46. Tomlins, G.W.; Routbort, J.L.; Mason, T.O. Zinc self-diffusion, electrical properties, and defect structure of undoped, single crystal zinc oxide. *J. Appl. Phys.* **2000**, *87*, 117. [CrossRef]
47. Kofstad, P. *Nonstoichiometry, Diffusion, and Electrical Conductivity in Binary Metal Oxides*; Wiley: New York, NY, USA, 1972.
48. Brass, A.M.; Chanfreau, A. Accelerated diffusion of hydrogen along grain boundaries in nickel. *Acta Mater.* **1996**, *44*, 3823–3831. [CrossRef]

nanomaterials

MDPI

Article

The Critical Role of Thioacetamide Concentration in the Formation of ZnO/ZnS Heterostructures by Sol-Gel Process

Eloísa Berbel Manaia [1,2,*], Renata Cristina Kiatkoski Kaminski [3], Bruno Leonardo Caetano [1], Marina Magnani [4], Florian Meneau [5], Amélie Rochet [5], Celso Valentim Santilli [4], Valérie Briois [6], Claudie Bourgaux [2] and Leila Aparecida Chiavacci [1,*]

1 Department of Drugs and Medicines, School of Pharmaceutical Sciences, São Paulo State University (UNESP), Araraquara, São Paulo 14800-903, Brazil; caetano@fcfar.unesp.br
2 Institut Galien, University Paris-Sud, The National Center for Scientific Research (CNRS), UMR 8612, 92296 Châtenay-Malabry, France; claudie.bourgaux@u-psud.fr
3 Department of Chemistry, Sergipe Federal University—Campus Itabaiana, Av. Vereador Olimpio Grande, s/n—Itabaiana, SE 49506-036, Brazil; re_kaminski@hotmail.com
4 Chemistry Institute of São Paulo State University—UNESP, Prof. Francisco Degni Street, 55, Araraquara, São Paulo 14800-060, Brazil; marina@iq.unesp.br (M.M.); santilli@iq.unesp.br (C.V.S.)
5 Brazilian Synchrotron Light Laboratory (LNLS), Brazilian Center for Research in Energy and Materials (CNPEM), São Paulo 13083-970, Brazil; florian.meneau@lnls.br (F.M.); amelie.rochet@lnls.br (A.R.)
6 Synchrotron Optimized Light Source of Intermediate Energy to LURE (SOLEIL), L'Orme des Merisiers, BP48, Saint Aubin, 91192 Gif-sur Yvette, France; valerie.briois@synchrotron-soleil.fr
* Correspondence: manaiaeb@fcfar.unesp.br (E.B.M.); leila@fcfar.unesp.br (L.A.C.); Tel.: +55-16-3301-4686 (E.B.M.); +55-16-3301-6966 (L.A.C.)

Received: 30 September 2017; Accepted: 6 November 2017; Published: 23 January 2018

Abstract: ZnO/ZnS heterostructures have emerged as an attractive approach for tailoring the properties of particles comprising these semiconductors. They can be synthesized using low temperature sol-gel routes. The present work yields insight into the mechanisms involved in the formation of ZnO/ZnS nanostructures. ZnO colloidal suspensions, prepared by hydrolysis and condensation of a Zn acetate precursor solution, were allowed to react with an ethanolic thioacetamide solution (TAA) as sulfur source. The reactions were monitored in situ by Small Angle X-ray Scattering (SAXS) and UV-vis spectroscopy, and the final colloidal suspensions were characterized by High Resolution Transmission Electron Microscopy (HRTEM). The powders extracted at the end of the reactions were analyzed by X-ray Absorption spectroscopy (XAS) and X-ray diffraction (XRD). Depending on TAA concentration, different nanostructures were revealed. ZnO and ZnS phases were mainly obtained at low and high TAA concentrations, respectively. At intermediate TAA concentrations, we evidenced the formation of ZnO/ZnS heterostructures. ZnS formation could take place via direct crystal growth involving Zn ions remaining in solution and S ions provided by TAA and/or chemical conversion of ZnO to ZnS. The combination of all the characterization techniques was crucial to elucidate the reaction steps and the nature of the final products.

Keywords: ZnO; ZnS; quantum dots; heterostructure; sol-gel; Small Angle X-ray Scattering; X-ray Absorption Spectroscopy; UV-vis spectroscopy

1. Introduction

ZnO and ZnS are wide band gap semiconductors with outstanding electronic and optical properties. They are used in a wide range of applications such as photocatalysts, sensors, electroluminescent devices and lasers [1–4]. More specifically, ZnO quantum dots (QDs) are promising luminescent probes for bioimaging due to their biodegradability and very low toxicity in vivo [5–9]. In recent years, ZnO/ZnS heterostructures have emerged as an attractive approach for tailoring the particle characteristics and properties [10,11]. When core–shell nanoparticles are formed, the shell acts as a barrier between the core and the surrounding medium, and can change the chemical reactivity and colloidal stability of the core. It is also a strategy for improving the photoluminescence properties of semiconductor nanoparticles (NPs) [12]. Coating NPs with a higher band gap semiconductor can passivate the core surface, eliminating surface-related defect states that induce non-radiative recombination of photogenerated electron–hole pairs (excitons), thereby lowering fluorescence quantum yield and giving rise to "blinking" [13]. ZnO/ZnS core–shell nanocables [14], nanorods [15,16], nanotubes [17], nanowires [18] and nanospheres or powders [19–23], with dimensions in the 25–200 nm range, were obtained. They were synthesized in solution at moderate temperatures via a partial chemical conversion of ZnO to ZnS in the presence of a sulfur source or via addition of both a sulfur source and a Zn precursor to preformed ZnO NPs. Most reports deal with sulfidation processes performed in an aqueous medium using Na_2S as the sulfur source. Different kinds of shells were obtained, from porous shells formed of small nanoparticles to compact shells ensuring a full coverage and, accordingly, different luminescent properties of the ZnO/ZnS structures were described.

The aim of our study is to yield insight into the detailed mechanisms of the heterostructure formation, important for a controllable and reproducible synthesis. We have investigated the formation of ZnO/ZnS heterostructures or the conversion of ZnO QDs to ZnS QDs using a straightforward, low-temperature (60 °C) route. ZnO colloidal suspensions were prepared by a sol-gel process, involving hydrolysis and condensation of a Zn acetate precursor solution, as described in our previous reports [24–26]. These suspensions were then allowed to react with an ethanolic thioacetamide (TAA) solution, as sulfur source. TAA concentration was varied to reveal the influence of this parameter on the so-formed particles. The formation of ZnO/ZnS nanostructures was monitored by Small Angle X-ray Scattering (SAXS) and UV-vis spectroscopy techniques. Nanoparticles were further characterized by X-ray diffraction (XRD), X-ray Absorption Spectroscopy (XAS) and High Resolution Transmission Electron Microscopy (HRTEM). Different ZnO/ZnS nanostructures were identified, depending on the synthesis conditions. The combination of all the characterization techniques was crucial to elucidate the reaction steps and the final products.

2. Materials and Methods

Zinc acetate dehydrate, $ZnAc_2{\cdot}2H_2O$ (Qhemis, Indaiatuba, Brazil, 98%), thioacetamide, TAA (Sigma-Aldrich, St. Louis, MO, USA, 99.0%), lithium hydroxide monohydrate, $LiOH{\cdot}H_2O$ (Vetec, Speyer, Germany, 98%), ethanol (Qhemis, Indaiatuba, Brazil, 99.5%) and heptane (Synth, Diadema, Brazil, 99%), were used as received, without further purification. Zinc oxide (Alfa Aesar, Haverhill, MA, USA) and zinc sulfide (Prolabo, Fontenay-sous-Bois, France) were used as standard.

2.1. Synthesis

ZnO colloidal suspensions (ZnO Susp) were synthesized according to the sol-gel route proposed by Spanhel and Anderson [27]. The $Zn_4O(Ac)_6$ tetrameric precursor (labeled herein ZnAc precursor) was first prepared by refluxing an absolute ethanol solution containing 0.05 M $ZnAc_2{\cdot}2H_2O$ over a period of ~2 h at 80 °C. The thus obtained transparent precursor solution was stored at ~4 °C before to be used for the ZnO QDs preparation. Hydrolysis and condensation reactions leading to the ZnO colloidal suspensions were done by adding under continuous stirring a $LiOH{\cdot}H_2O$ absolute ethanol solution (0.5 M). Reactions were carried out at 40 °C and, after 5 s of stirring, the ZnO colloidal suspension was immediately cooled and stored at ~4 °C to prevent any further nucleation and growth process. A nominal molar ratio of [OH]/[Zn] = 0.5 was used. The thus obtained suspension contained about 20% ZnO QDs and 80% remaining ZnAc precursor [24].

To investigate the sulfidation process, 10 mL of ZnO colloidal suspension were allowed to react with 10 mL of TAA ethanolic solution (as sulfur source) under continuous magnetic stirring at 60 °C for 40 min. The suspension was then cooled and stored at ~4 °C. Different concentrations of TAA were used while the other parameters remained unchanged. The final products prepared with different amounts of TAA (1.5, 5 and 50 mM) were designed as TAA1.5, TAA5 and TAA50, respectively. The corresponding [Zn]/[S] nominal molar ratios were 33.33, 10, and 1, respectively.

A ZnS colloidal suspension was synthesized by mixing 10 mL of ZnAc precursor with 10 mL of TAA 5 mM ethanolic solution under continuous stirring at 60 °C for 10 min. The [Zn]/[S] nominal molar ratio was 10. The suspension was then cooled and stored at ~4 °C. This sample, designed as ZnS QDs, was used for comparison with QDs prepared as described above.

2.2. Powder Extraction

The as-synthesized colloidal suspensions (ZnO QDs, TAA1.5, TAA5, TAA50 and ZnS QDs) were mixed with a "nonsolvent" heptane [28] (1:4) to induce the precipitation of the QDs, and then centrifuged at 20 °C for 10 min (10,000 rpm). The supernatant solution was discarded and the washed powder was dried under vacuum at room temperature. This method allows extracting the QDs without modifying their size and structure. The dried powders were characterized by XRD and XAS.

2.3. Characterization

2.3.1. X-ray Diffraction (XRD) of powders

XRD analysis of the powders was performed on a Bruker D2 PHASER diffractometer (Karlsruhe, Germany) using the Cu Kα radiation, λ = 1.5418 Å, selected by a curved graphite monochromator and a fixed divergence slit of 1/8 deg. in a Bragg–Brentano configuration. The diffraction patterns were measured in the 2θ range 5–70° by the step counting method (0.1 step and 3 s counting time).

2.3.2. High Resolution Transmission Electron Microscopy (HRTEM)

HRTEM investigations were performed with a Philips microscope model CM 200 (FEI Compay, Hillsboro, OR, USA) operating at 200 keV. A drop of the dilute colloidal suspension of QDs was deposited on a copper grid carbon film and dried. Image analysis was carried out with ImageJ (National Institutes of Health, Bethesda, MD, USA) and Digital Micrograph software (Gatan Inc., Pleasanton, CA, USA) packages.

2.3.3. X-ray Absorption Spectroscopy (XAS)

X-ray absorption experiments were performed at the Spectroscopy Applied to Material Based on Absorption (SAMBA) beamline at the French synchrotron source Optimized Light Source of Intermediate Energy to LURE (SOLEIL) (Saint-Aubin, France). A fixed exit sagitally focusing Si (220) double crystal monochromator was used. The grazing incidence of the white and monochromatic beams on both Pd-coated collimating and focusing mirrors was set at 4 mrad, ensuring an efficient harmonic rejection. The beam size was about 2 mm (H) × 0.5 mm (V) at the sample position. The ionization chambers were filled with a N_2/He gas mixture and the transmission mode was used to record the data.

The software packages Athena (Washington, DC, USA) and Artemis (Washington, DC, USA) [29] were used to analyze the XAS data. To calibrate each data set in energy the maximum of the first derivative of the zinc reference foil, recorded simultaneously with the data, was used. A linear background was fitted to the pre-edge region and subtracted from the spectra. A post-edge background using the AUTOBK algorithm was applied with a cutoff Rbkg (distance (in Å) for χ(R) above which the signal is ignored) = 1.15 and k-weight = 3 in order to isolate the extended X-ray absorption fine structure (EXAFS) oscillations $\chi(k)$. Then, the Fourier transformations of EXAFS data were carried out between 3.7 and 12 Å^{-1} using a k^3-weighting Kaiser–Bessel window with a dk (FFT window parameter) = 2 apodization window.

2.3.4. Small-Angle X-ray Scattering (SAXS)

The SAXS measurements were performed at SAXS1 beamline of the Brazilian Synchrotron Light Laboratory (LNLS, Campinas, Brazil). The beamline was equipped with a monochromator ($\lambda = 1.550$ Å), a two-dimensional (2D) detector, Pilatus 300 K, localized at 934.934 mm from the sample to record the scattering intensity I(q) as a function of the scattering vector, q. The resulting scattering vector, q, ranged from 0.11 to 4.15 nm^{-1}. Silver behenate standard was used to calibrate the q range.

The nucleation and growth of nanoparticles have been followed in situ, injecting the freshly prepared reactional solution into the thermostated sample holder (Campinas, Brazil) which was set to 60 °C. Time-resolved SAXS patterns could be recorded from the beginning of the reaction. The curves were collected within an interval of 1 min.

Data were normalized taking into account the beam decay, acquisition time and sample transmission. The scattered intensity of the sample holders, and the solvent (ethanol) were subtracted from the total intensity. The analysis of the SAXS data was carried out using the software package SASfit (Villigen, Switzerland) [30].

2.3.5. UV-vis Spectroscopy

The absorption spectra were measured using a Cary Win 4000 UV-vis spectrophotometer (Santa Clara, California, United States) with a cuvette of 1mm optical path. Five aliquots were collected at different times (1, 5, 10, 20 and 40 min) for each reaction (TAA1.5, TAA5 and TAA50) and diluted 8 times in ethanol to be recorded in the linear absorption range. The reaction TAA50, as well as ZnS QDs, was also monitored in situ using a Cary 60 UV-vis (Santa Clara, CA, USA) with an immersion probe with optical path of 2 mm.

All spectra of the QDs colloidal suspensions were recorded between 200 and 400 nm, with a wavelength step of 1 nm, and an average counting time of 0.2 s per point. The UV-vis spectra were corrected from the absorption spectrum of ethanol. The size of ZnO QDs was determined from the absorption spectra using the effective mass model derived by Brus [31].

3. Results and Discussion

ZnO/ZnS QDs were prepared via a simple sol-gel synthesis. The base-catalyzed hydrolysis and condensation reactions, leading from the ZnAc precursor to ZnO QDs, were stopped before completion and an ethanolic solution of TAA, used as sulfur source, was then added to the suspension containing both ZnO NPs and remaining ZnAc precursor; sulfur ions released upon TAA hydrolyze could react with Zn ions to form ZnS. Insights into the nature and size of crystallites could be obtained from HRTEM and XRD, the later technique being more representative of the whole sample. XAS was recorded at the Zn K-edge to probe the local environment around Zn atoms and evidence ZnO or ZnS structures hardly detectable by XRD because of their very small size and/or lattice disorder. Finally, SAXS and UV-vis spectroscopy allowed monitoring in situ the formation of nanoparticles through either the nanoparticle size measurement or the ZnO and ZnS excitonic peak growth.

3.1. Structural Features of Powders

Diffraction patterns of powders collected at the end of the synthesis are presented in Figure 1a. The positions of peaks characteristic of the ZnO hexagonal wurtzite structure are marked with dashed black lines while those indicative of the ZnS cubic zinc blende structure are marked with dashed red lines. These phases are the most stable for ZnO and ZnS, respectively [32,33]. The TAA1.5 sample exhibits peaks characteristic of ZnO. Figure 1b displays a zoom of the diffraction curves between $2\theta = 23°$ and $2\theta = 40°$ for ZnO QDs, ZnS QDs, and TAA5 sample. A careful observation shows the presence of small broad peaks characteristic of ZnO and ZnS structures in TAA5 sample pattern, pointing out the coexistence of both phases. The TAA50 pattern is similar to that of ZnS QDs, suggesting the formation of the cubic zinc blende structure. However, because of the broadening of the diffraction peaks in the [42°, 65°] 2θ range (Figure 1c), the wurtzite and zinc blende phases could not be distinguished based on XRD patterns only [34]. HRTEM images further supported the formation of the cubic zinc blende structure in TAA50 sample (results not shown).

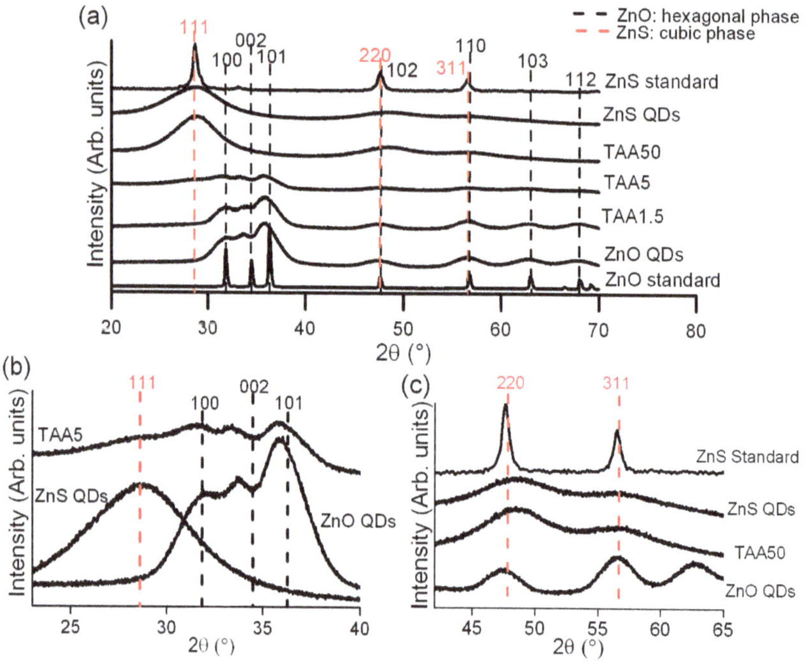

Figure 1. (**a**) X-ray diffraction (XRD) patterns of ZnO and ZnS standards, ZnO and ZnS quantum dots (QDs), and samples prepared with different concentration of the sulfur source (thioacetamide (TAA)), where the vertical lines indicate the hexagonal wurtzite phase (black) and cubic zinc blende phase (red); (**b**) Zoom of the peaks (111), (100), (002) and (101) in the 2θ range from 23° to 40° of the ZnO QDs, ZnS QDs, and TAA5; and (**c**) Zoom of the peaks (220) and (311) in the 2θ range from 42° to 65° of the ZnO QDs, ZnS QDs, ZnS Standard and TAA50.

The average size of ZnO and ZnS nanocrystals could be estimated using the Debye–Scherrer relation [35] applied to the (100), (002) and (101) reflections of ZnO and (111) reflection of ZnS, respectively:

$$D = \frac{k\lambda}{\beta \cos\theta}$$

where D is the average crystallite size; k is a constant (shape factor, 0.89 for spherical nanoparticles), λ is the X-ray wavelength, β is the Full-Width-at-Half-Maximum (FWHM) of the diffraction peak and 2θ is the diffraction angle. The crystallite sizes of ZnO and ZnS QDs were about 5.3 and 1.4 nm, respectively. Regarding TAA5 sample, the crystallite size could not be accurately determined because of the weak broad pattern.

TAA5 sample imaged by HRTEM with low (Figure 2a) and high (Figure 2b) magnification is presented in Figure 2. Figure 2a shows that the sample is crystalline. In Figure 2b, the lattice fringes of two attached ZnO and ZnS crystals are clearly observed. The 0.26 nm spacing arises from the 002 lattice planes of the wurtzite ZnO phase [36], while the 0.31 nm spacing results from the (111) planes of the blende ZnS phase [37]. The HRTEM study corroborates XRD data, revealing the presence of wurtzite ZnO and zinc blende ZnS phases; the morphology evidences the coexistence of ZnS and ZnO nanocrystals.

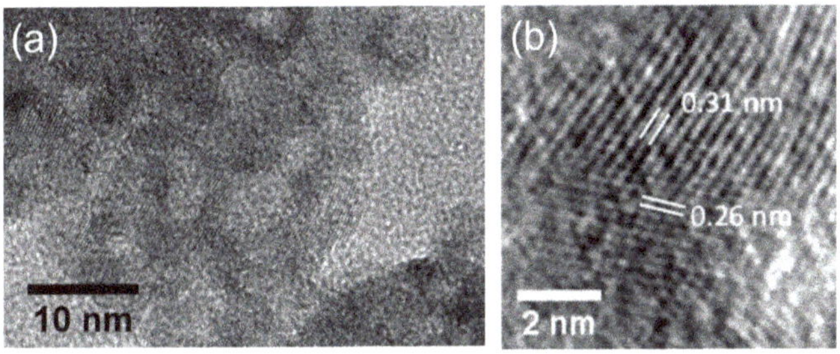

Figure 2. (**a**) High Resolution Transmission Electron Microscopy (HRTEM) image of TAA5 showing in (**b**) the interplanar spacing of ZnO and ZnS phases.

Figure 3a,b displays the X-ray absorption near edge structure (XANES) spectra and Fourier Transforms (FT) of EXAFS spectra recorded for the different samples and compared to the standard references. Significant differences in XANES shape and white line positions can be observed between the standard ZnO and ZnS references. The FT of ZnO EXAFS spectrum is characterized by two main contributions, the first one corresponding to the oxygen tetrahedral coordination shell at 1.96 Å and a second one related to the zinc second next neighbors at 3.23 Å [38], whereas ZnS presents essentially a first tetrahedral sulfur coordination shell at 2.34 Å and a broad contribution with low intensity compared to ZnO corresponding to the zinc second next neighbors at 3.82 Å [39].

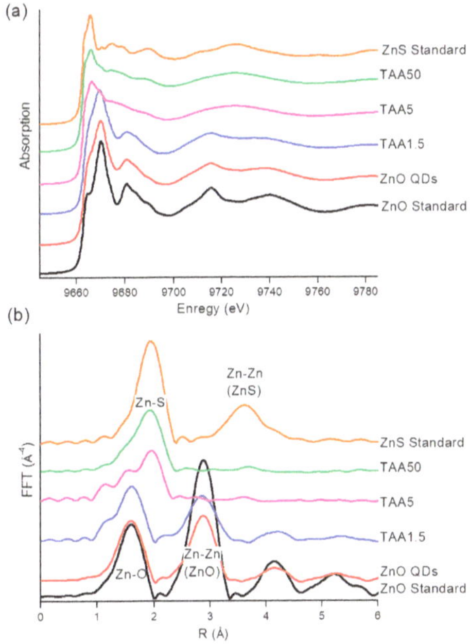

Figure 3. (**a**) X-ray absorption near edge structure (XANES) spectra; and (**b**) Fourier Transforms of extended X-ray absorption fine structure (EXAFS) spectra recorded for the different samples and ZnO and ZnS standard references.

TAA1.5 sample displays a XANES spectrum very similar to those recorded for ZnO QDs and ZnO standard. This feature indicates that this sample consists mainly of ZnO nanoparticles. Conversely, the sample prepared with the highest TAA amount (50 mM) displays a XANES spectrum closer to that of ZnS standard. This finding evidences that the large amount of sulfur ions in the medium favored the formation of nanoparticles presenting local order arrangement comparable to ZnS.

FT spectra fully support the conclusions drawn from the XANES spectra. Samples with ZnO QD characteristics (TAA1.5) and those with ZnS features (TAA50) show FT peaks located at the same positions as the respective standards. The FT spectrum of the sample prepared with intermediate TAA concentration (5 mM) is characterized by the double maximum of the first contribution, revealing the simultaneous presence of ZnO and ZnS, in agreement with XRD and HRTEM results.

Further insights into the structural features of the samples were obtained by the least-square fitting procedures of the first coordination shells. The so obtained structural parameters are gathered in Table 1. As expected, TAA1.5 is characterized by 3.6 ± 0.2 oxygen atoms at 1.99 ± 0.01 Å. This confirms that only ZnO QDs are observed for the lowest TAA concentration. TAA50 can be described by a coordination shell made of 3.0 ± 0.2 sulfur atoms at 2.34 ± 0.01 Å and 1.2 ± 0.1 oxygen atoms at 2.02 ± 0.01 Å. Of note, no contribution at larger distances is observed. The XANES spectrum of TAA50 is satisfactorily fitted by a linear combination of ZnO QD and ZnS standard spectra, in the ~25:75% proportion. This composition is in perfect agreement with the oxygen and sulfur coordination numbers reported in Table 1, suggesting that this sample is a mixture of both QDs: ~25% of ZnO QDs and 75% of ZnS QDs. The lack of medium range order can be explained by the small size of ZnS QDs (~1.4 nm), compared to that of ZnO QDs (~5.3 nm), resulting in no detectable contribution of Zn-Zn second neighbors to the FT spectra.

Table 1. Extended X-ray absorption fine structure (EXAFS) structural parameters for the first Zinc coordination sphere of ZnO and ZnS, and the percentages of ZnO and ZnS deduced from the Linear Combination Fittings (LCF) of X-ray absorption near edge structure (XANES) spectra.

Sample	$N_{Zn\text{-}O}$	$R_{Zn\text{-}O}$	$\sigma^2_{Zn\text{-}O}$ (Å^2)	$N_{Zn\text{-}S}$	$R_{Zn\text{-}S}$	$\sigma^2_{Zn\text{-}S}$ (Å^2)	R_{factor}	LCF	
								% ZnO	% ZnS
ZnO Standard	4	1.97	0.004 ± 0.002				0.0003		
ZnO QDs	3.7 ± 0.4	1.97	0.005 ± 0.001				0.0014		
TAA1.5	3.6 ± 0.2	1.99	0.006 ± 0.001				0.0003	98	2
TAA5	2.0 ± 0.6	2	0.007 ± 0.005	2.2 ± 1.0	2.34	0.007 ± 0.005	0.0027	37	63
TAA50	1.2 ± 0.1	2.02	0.008 ± 0.001	3.0 ± 0.2	2.34	0.008 ± 0.001	0.0001	23	77
ZnS Standard				4	2.34	0.006 ± 0.001	0.0004		

In the case of TAA5, the relative proportions of ZnO QDs and ZnS QDs deduced from LCF of the XANES spectrum are 37% and 63%, respectively, which should lead to about 1.5 oxygen atoms and 2.5 sulfur atoms in the first coordination shell of Zn. These values are very close from those retrieved from the least-square fitting procedures, suggesting a first coordination shell around Zn composed of 2.0 ± 0.6 oxygen atoms at 2.00 ± 0.02 Å and 2.2 ± 1.0 sulfur atoms at 2.34 ± Å. The slight increase of the number of O atoms, 2.0 instead of 1.5, observed could be associated to the formation of the core–shell structure.

3.2. *Time-Resolved Study of the Nanoparticle Synthesis*

The formation of the QDs was first studied by UV-Vis absorption spectroscopy. Figure S1a in the Supplementary Materials shows the UV-vis spectra of ZnO QDs, ZnS QDs and a mixture of ZnO and ZnS QDs and Figure S1b presents the UV-vis spectrum of the ethanolic solution of TAA. The excitonic peaks at about 290 nm and 350 nm are characteristic of ZnS QDs [40] and ZnO QDs [26], respectively, and the band around 266 nm is the fingerprint of TAA. The decrease of the TAA absorption band reflects the release of S^{-2} ions into the solution. QDs are characterized by a decreasing band gap, and thus a red-shift of their excitonic absorption, with increasing QD size. The radius of ZnO QDs

($R_{UV\text{-}vis}$) could therefore be determined from the absorption spectra using the effective mass model derived by Brus.

UV-vis absorption spectra measured at 1, 5, 10, 20 and 40 min of reaction for the different TAA concentrations are shown in Figure 4. ZnS QDs formation is also show in Figure 4a for comparison to the other reactions. The inset in Figure 4a shows the spectrum of the ZnS washed and redispersed in ethanol, which exhibited the excitonic peak at about 290 nm characteristic of ZnS QDs. For TAA50, the reaction was also monitored in situ and time-resolved UV-vis spectra are shown in Figure 5.

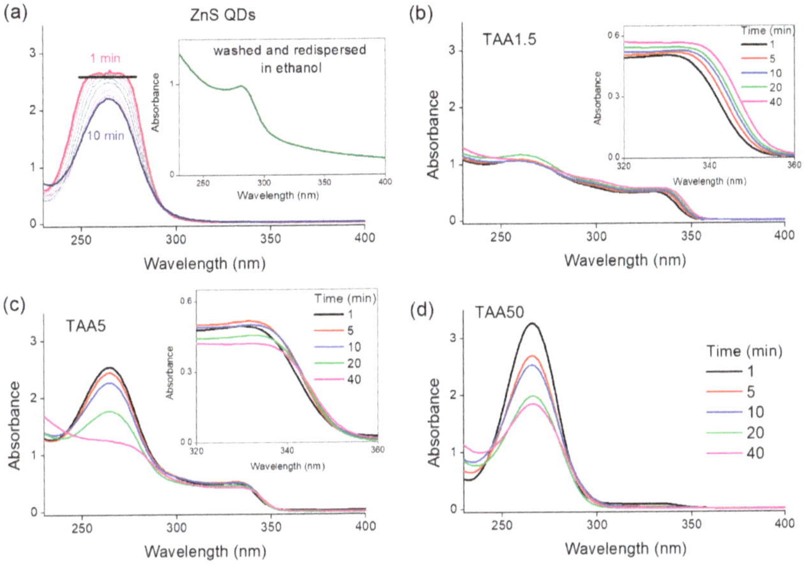

Figure 4. Selected UV-vis spectra measured at the indicated reaction times (min) for: (**a**) ZnS quantum dots (QDs); and different thioacetamide (TAA) concentrations: (**b**) 1.5 mM; (**c**) 5 mM; and (**d**) 50 mM. The inset in (**a**) shows the spectrum of the ZnS QDs washed and redispersed in ethanol. The insets in (b) and (c) show a zoom of the ZnO excitonic peaks at the indicated reaction times (min) for TAA1.5 and TAA5, respectively.

At the lower TAA concentration, we observe a red-shift of the excitonic peak of ZnO corresponding to a growth of about 0.3 nm. Moreover, the absence of a ZnS excitonic peak confirms the sole presence of ZnO nanocrystals.

For TAA5 we do not observe any displacement of the excitonic peak of ZnO as a function of time, indicating that the size of the ZnO nanoparticles remains constant. This behavior might be explained by the formation of a ZnS shell around the ZnO core, preventing the growth of the latter. On the other hand, the absorbance varied along the time. It increased slightly in the first minutes of the reaction, reflecting an increase in the number of ZnO QDs (inset of Figure 4c). It can be assumed that the nucleation of new ZnO QDs resulted from the hydrolysis and condensation of the precursor still present in the suspension. After about 5 min, the opposite trend was observed: the absorbance decreased along time. We can suggest that some of the ZnO QDs could be converted into ZnS. At the end of the reaction, both excitonic peaks of ZnO and ZnS were observed.

For TAA50, the ZnO peak present at the beginning of the reaction slowly decreases to finally vanish as the reaction proceeds. This reaction was monitored in situ and the Figure 5 presents selected UV-vis spectra of different reaction times (min) (Figure 5a) 1–5, (Figure 5b) 5–21, and (Figure 5c) 21–40, as well as (Figure 5d) the spectrum of the final product washed and redispersed in ethanol. From 1 to 5 min (Figure 5a), a red-shift accompanied by the increase of absorbance intensity is observed, demonstrating a slight increase of the size of ZnO QDs and the increase of the number of ZnO nanoparticles, as observed for TAA5. From 5 to 21 min of reaction (Figure 5b), the absorbance intensity of the ZnO excitonic peak decreases. This can be interpreted as the consumption of the ZnO nanoparticles with the simultaneous formation of ZnS. From 23 to 40 min (Figure 5c), the ZnO excitonic peak is absent and a red-shift of the peak around 320 nm is observed demonstrating the growth of ZnS QDs. To confirm the formation of ZnS nanoparticles, the TAA50 reaction product was washed in order to remove the unreacted TAA, which dominates the absorption spectrum in this wavelength range (Figure 4c), and re-suspended in ethanol before UV spectroscopy analysis. As expected, the peak around 290 nm fully confirms the formation of ZnS nanoparticles as the main phase (Figure 5d).

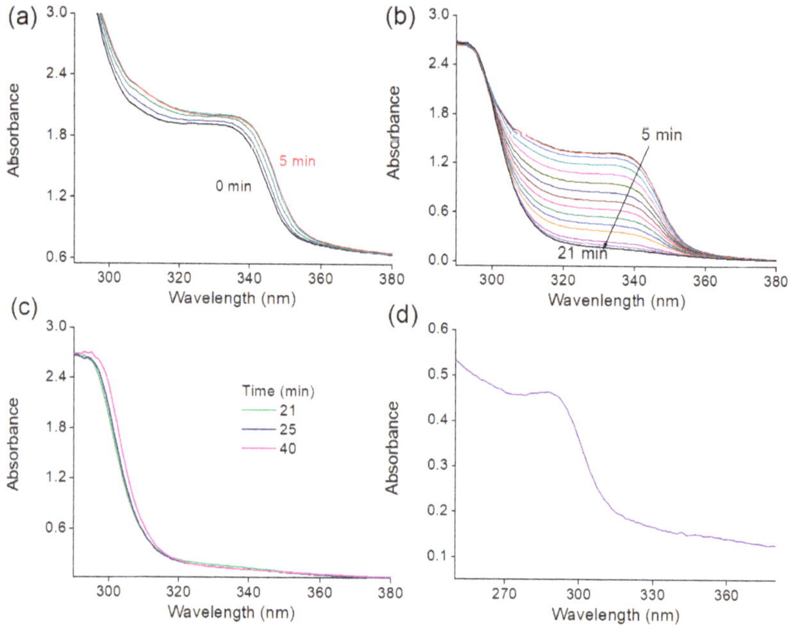

Figure 5. Selected UV-vis spectra of TAA50 reaction measured at the indicated reaction times (min): (**a**) 1–5; (**b**) 5–21; and (**c**) 21–40; and (**d**) the spectrum of the final product washed and redispersed in ethanol.

The growth of the different populations of nanoparticles was further evidenced by time-resolved SAXS patterns recorded from the beginning of the reactions. The reactional solution was injected in the sample holder and kept at 60 °C and the reaction was monitored during 40 min. Figure 6 presents the log–log plot of selected SAXS curves measured at increasing times of reaction for the different concentrations of TAA. The SAXS curves of ZnO QDs ([TAA] = 0, Figure 6a) display a plateau at low q-range (Guinier region) and an asymptotic linear decrease in the high-q range (Porod region), characteristic of a dilute suspension of nanoparticles. In the Guinier region, the scattered intensity can be approximated by $I(q) = I(0)\exp(-Rg^2q^2/3)$ where Rg is the radius of gyration (Guinier radius) of the particles or aggregates and $I(0)$ is the limit of $I(q)$ when $q \to 0$, given by $I(0) = N \times (\rho_p - \rho_s)2 \times V^2$, where N is the particle number density, ρ_p and ρ_s are the average electron densities of the particles and the solution, respectively, and V is the volume of the particle [41]. The gradual shift of curves toward lower q values reflects the increase of the mean nanoparticle size while the nucleation of new nanoparticles induces an increase in $I(0)$. A similar behavior was observed for TAA1.5 and TAA5.

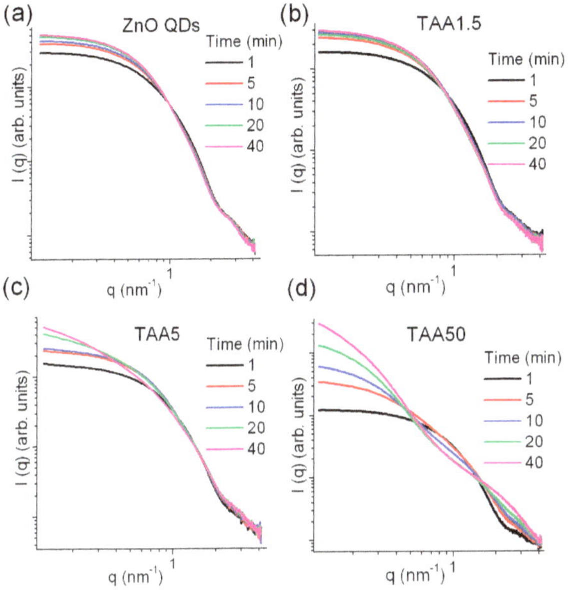

Figure 6. Selected small angle X-ray scattering (SAXS) profiles recorded in situ at the indicated reaction times (min) for: (**a**) ZnO quantum dots (QDs); and for different thioacetamide (TAA) concentrations: (**b**) 1.5 mM; (**c**) 5 mM; and (**d**) 50 mM.

Figure 7 gathers the time evolution of the QD average radii (R_{SAXS}) deduced from Rg values ($R_{SAXS} = (5/3)^{1/2}$ Rg for a spherical NP) of colloidal particles formed using 0 (ZnO QDs), 1.5 and 5 mM TAA concentrations. For TAA concentration equal to 1.5 mM the evolution of R_{SAXS} is almost identical to that of ZnO QD R_{SAXS} ([TAA] = 0.0 mM). TAA had a negligible effect on the growth of colloids. R_{SAXS} of TAA5 displays a similar evolution until 10 min. After that, R_{SAXS} increases from 3.5 to 4.8 nm at the end of the reaction. The comparison of $R_{UV\text{-}vis}$, which remains constant in the course of the reaction, and R_{SAXS} suggest that the R_{SAXS} increase is related to aggregation of nanoparticles and/or to the growth of ZnO/ZnS heterostructures: ZnS shells could form on ZnO, yielding larger scattering objects.

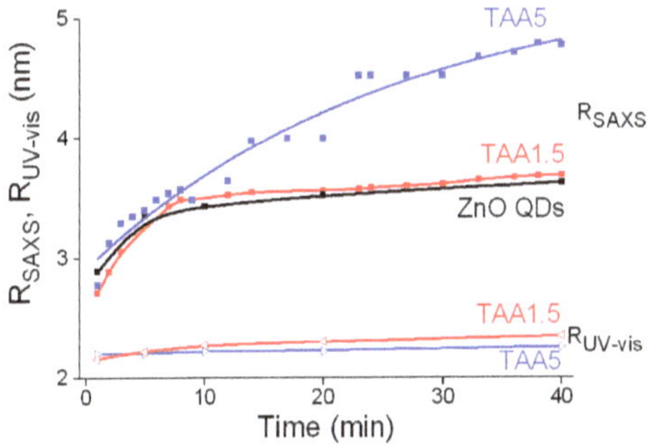

Figure 7. Time evolution of the quantum dots (QDs) average radii (R_{SAXS}) deduced from Rg values ($R_{SAXS} = (5/3)^{1/2}$ Rg for a spherical NP) of colloidal particles formed using 0 (ZnO QDs), 1.5 and 5 mM thioacetamide (TAA) concentrations and the $R_{UV\text{-}vis}$ calculated by Brus equation from UV spectra of TAA1.5 and TAA5. Solid lines are guides for the eye.

The SAXS profiles recorded during synthesis of TAA50 samples are markedly different (Figure 6d): the first curves, characterized by a Guinier regime and an asymptotic linear decrease at high q, are typical of a single population of diluted nanoparticles, whereas the curves recorded later show two Gaussian decays, suggesting the presence of two populations of nanoparticles having different average sizes.

We highlight in Figure 8 four time intervals characteristic of the different evolution steps of TAA50 SAXS curves between: (a) 1–5 min; (b) 5–9 min; (c) 9–23 min; and (d) 24–40 min.

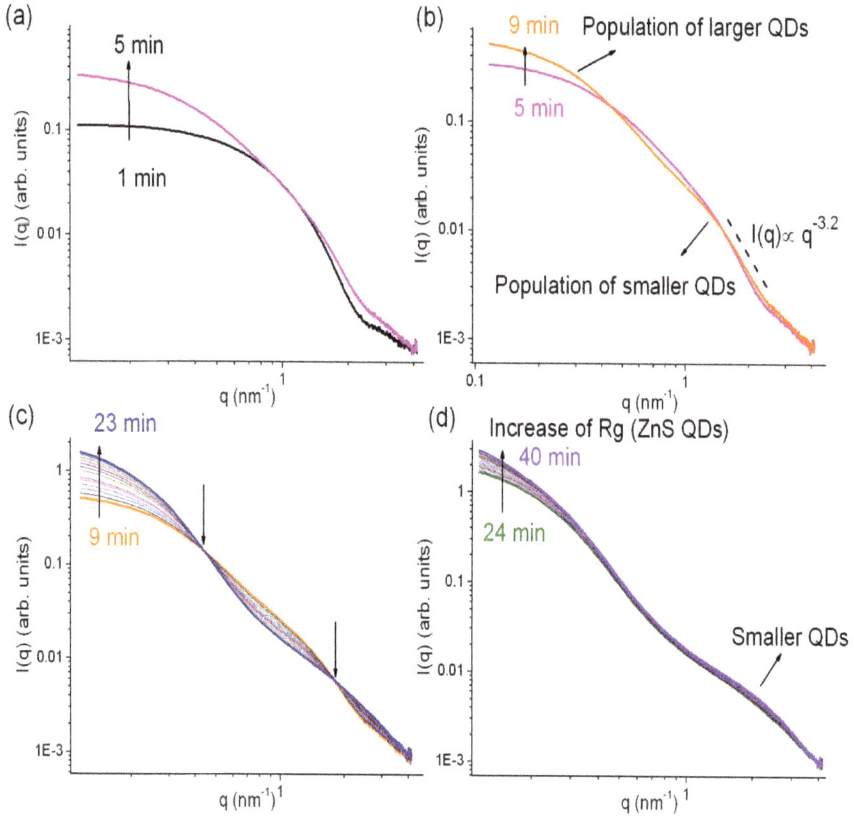

Figure 8. Small-angle X-ray scattering (SAXS) curves recorded during TAA50 reaction at different time intervals: (**a**) 1–5 min; (**b**) 5–9 min; (**c**) 9–23 min; and (**d**) 24–40 min. In (**c**) the two arrows at $q1 = 0.44$ nm^{-1} and at $q2 = 1.78$ nm^{-1} indicate the two isobestic points.

At the early reaction time (Figure 8a), the mean QD size increases, in agreement with the red-shift of the ZnO excitonic peak. The main modification in the shape of the SAXS curves observed between ~5 and 9 min (Figure 8b) is then the emergence of the second Gaussian decays. The slope at high q range (−3.2) corresponds to a diffuse interface, possibly due to the formation of a shell of variable composition or to localized etching of the nanoparticles resulting in a rough surface. After this step, all the SAXS curves clearly display two Gaussian decays. In the intermediate period (Figure 8c: 9 to 23 min), the curves present a double crossover at $q1 = 0.44$ nm^{-1} and at $q2 = 1.78$ nm^{-1}, characterizing two isobestic points. The existence of these isobestic points implies that at least two equilibrium states describe the nanostructural transformation occurring from 9 to 23 min. One population of scatters grows at the expense of the other one. In the last step (Figure 8d), the largest nanoparticles continue to grow.

The analysis of the SAXS experimental results for TAA50 was therefore carried out considering two populations (denoted as 1 and 2) of nanoparticles; SAXS data were satisfactorily fitted with the sum of two distributions, characterized by the mean average radii of gyration *Rg*1 and *Rg*2 (see Supplementary Materials). Figure 9a gathers the time evolution of R_{SAXS1} and R_{SAXS2} deduced from *Rg*1 and *Rg*2 values, respectively ($R_{SAXS} = (5/3)^{1/2}$ *Rg* for a spherical NP), Figure 9b the variation of *I*1(0) versus $Rg1^6$ and Figure 9c *I*2(0) versus $Rg2^6$. The increase in *I*1(0) in the first minutes while $Rg1^6$ remains almost constant highlights the nucleation of new nanoparticles. Then, as the reaction progresses, the evolution of *I*1(0) suggests the growth and aggregation of existing nanoparticles (Figure 9b). On the other hand, the size *Rg*2 of the nanoparticles belonging to the second population slightly decreases with time until about 27 min (Figure 9c).

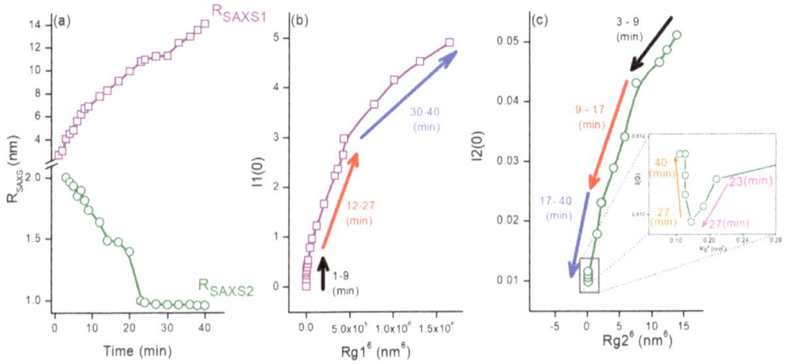

Figure 9. (**a**) Time evolution of the average radii R_{SAXS1} and R_{SAXS2} deduced from *Rg* values ($R_{SAXS} = (5/3)^{1/2}$ *Rg* for a spherical NP); (**b**) *I*1(0) versus $Rg1^6$ plot; and (**c**) *I*2(0) versus $Rg2^6$ plot of TAA50, where *I*(0) is the limit of *I*(*q*) when $q \rightarrow 0$.

The large objects seen in the low *q*-range at the end of the synthesis are likely aggregates of small ZnS crystals. However, it is less clear whether the second population corresponds to precursor clusters, to ZnO QDs, that progressively dissolve, or to ZnS QDs that nucleate.

The above results can be summarized as follows:

TAA1.5: ZnO QDs were almost exclusively formed, as shown by UV-vis absorption, XAS and XRD. ZnS did not form, likely because the available amount of sulfur was too small.

TAA5: All of the characterization techniques evidenced the coexistence of ZnO and ZnS species and/or the formation of heterostructures. Taken together, UV-vis absorption spectroscopy and SAXS experiments indicated that ZnO nanoparticle growth or dissolution might be hindered by the formation of a ZnS shell. Although thin shells cannot easily be evidenced by HRTEM, some HRTEM images displayed a ZnS crystal attached on a ZnO nanoparticle. The weak UV emission and the reduced green emission, as compared to the photoluminescence of ZnO QDs and TAA1.5, also supported the formation of ZnO/ZnS heterostructures (see Supplementary Materials). However, the full coverage of the ZnO core with a dense ZnS shell would enhance the UV luminescence more strongly. This suggests the formation of a discontinuous shell, likely made of small ZnS nanocrystallites. Regarding the visible green emission, originating mainly from surface defects such as oxygen vacancies, it could be quenched by even a small amount of sulfur ions on the surface [10,17,20,21,42].

TAA50: ZnS QDs were mainly formed as shown by XRD, XAS and UV-vis spectroscopy. Small QDs formed large aggregates. In addition to Zn precursor remaining in solution at the beginning of the reaction (~80%), Zn ions from the surface of the ZnO QDs could easily react with the abundant S ions to form ZnS.

The concentration of TAA, used as sulfur source, is the main parameter controlling the conversion of ZnO to ZnS. However, the reaction medium is complex and many chemical reactions can take place. Indeed, some water is brought into the ethanolic precursor solution by the dissolution of $ZnAc_2{\cdot}2H_2O$ and $LiOH{\cdot}H_2O$. When the precursor is prepared, acetate moieties are released, which are able to form acetic acid or to react with ethanol to yield ester and additional water [38,43]. In the presence of water, TAA can decompose to release acetamide and S ions, leading to the formation of ZnS. The progressive hydrolysis of TAA with time could sustain the growth of ZnS in the TAA5 and TAA50 systems. Of note, for the TAA50 system, the [Zn]/[S] ratio is 1. Therefore, a full conversion of ZnO into ZnS could be expected according to their respective solubility constants. However, XAS unambiguously demonstrated the presence of O atoms near Zn atoms. As a tentative explanation, we might suggest that completion of the reaction would have needed more time. Indeed, when ZnS formation is initiated on the ZnO surface, further conversion reaction requires diffusion of ionic species (inward S diffusion and outward O diffusion).

4. Conclusions

Depending on synthesis conditions, ZnS formation could take place via direct crystal growth involving Zn ions remaining in solution and S ions provided by TAA and/or interfacial sulfidation of ZnO. The influence of TAA concentration was evidenced. ZnO and ZnS phases were mainly obtained at low and high TAA concentrations, respectively. At intermediate TAA concentrations, we observed the formation of ZnO/ZnS heterostructures. The reaction steps were monitored in situ by SAXS, and UV-vis spectroscopy; the final products were further characterized by XRD, HR-TEM and XAS. This detailed study contributes to the understanding of the mechanisms underlying the formation of ZnO/ZnS heterostructures via a sol-gel route.

Supplementary Materials: The following are available online at www.mdpi.com/2079-4991/8/2/55/s1, Figure S1: UV-vis spectra of: (a) ZnO QDs, ZnS QDs and the mixture of ZnO and ZnS QDs as-synthesized colloidal suspensions; and (b) ethanolic solution of TAA, Figure S2: Example of SAXS curve of TAA50 at the end of the reaction (about 36 min) fitted with the form factors of two populations of homogeneous spheres displaying Lognormal radius distribution using SASfit software, Figure S3: Photoluminescence emission spectra of ZnO QDs, TAA1.5 and TAA5 excited at 353 nm, and TAA50 and ZnS QDs excited at 302 nm.

Acknowledgments: The authors thank LNLS for providing beamtime at SAXS1 beamline (project number: SAXS1-14321) and SOLEIL for providing beamtime at the SAMBA beamline (project number: 20131299) and S. Blanchandin for DRX measurements. We would like to thank the Electron Microscopy Laboratory of the Chemistry Institute (LME-IQ) for the transmission electron microscopy (TEM) facilities. The authors thank the reviewers for helpful suggestions. This work was also partially supported by the São Paulo Research Foundation (FAPESP) (project number 2012/07570-4), Coordination for Improvement of Higher Education Personnel (CAPES)/French Committee for the Evaluation of Academic and Scientific Cooperation with Brazil (COFECUB) cooperation program (ME 767-13) and the Support Program for Scientific Development/School of Pharmaceutical Sciences-São Paulo State University (PADC/FCF-UNESP)

Author Contributions: Renata Cristina Kiatkoski Kaminski, Bruno Leonardo Caetano, Valérie Briois, Claudie Bourgaux and Leila Aparecida Chiavacci conceived and designed the experiments; Eloísa Berbel Manaia, Renata Cristina Kiatkoski Kaminski, Bruno Leonardo Caetano and Marina Magnani performed the experiments; Eloísa Berbel Manaia, Amélie Rochet, Florian Meneau, Celso Valentim Santilli and Claudie Bourgaux analyzed the data; and Eloísa Berbel Manaia, Claudie Bourgaux, Valérie Briois, Celso Valentim Santilli and Leila Aparecida Chiavacci wrote the paper.

Conflicts of Interest: The authors declare no conflict of interest. The founding sponsors had no role in the design of the study; in the collection, analyses, or interpretation of data; in the writing of the manuscript, and in the decision to publish the results.

References

1. Liu, K.-K.; Shan, C.-X.; He, G.-H.; Wang, R.-Q.; Sun, Z.-P.; Liu, Q.; Dong, L.; Shen, D.-Z. Advanced encryption based on fluorescence quenching of ZnO nanoparticles. *J. Mater. Chem. C* **2017**, *5*, 7167–7173. [CrossRef]
2. Wang, G.; Huang, B.; Li, Z.; Lou, Z.; Wang, Z.; Dai, Y.; Whangbo, M.-H. Synthesis and characterization of ZnS with controlled amount of S vacancies for photocatalytic H_2 production under visible light. *Sci. Rep.* **2015**, *5*, 8544. [CrossRef] [PubMed]

3. Wang, L.; Kang, Y.; Liu, X.; Zhang, S.; Huang, W.; Wang, S. ZnO nanorod gas sensor for ethanol detection. *Sens. Actuators B-Chem.* **2012**, *162*, 237–243. [CrossRef]
4. Lim, J.H.; Kang, C.K.; Kim, K.K.; Park, I.K.; Hwang, D.K.; Park, S.J. UV Electroluminescence Emission from ZnO Light-Emitting Diodes Grown by High-Temperature Radiofrequency Sputtering. *Adv. Mater.* **2006**, *18*, 2720–2724. [CrossRef]
5. Xiong, H.-M. ZnO Nanoparticles Applied to Bioimaging and Drug Delivery. *Adv. Mater.* **2013**, *37*, 5329–5335. [CrossRef] [PubMed]
6. Matsuyama, K.; Ihsan, N.; Irie, K.; Mishima, K.; Okuyamam, T.; Mutom, H. Bioimaging application of highly luminescent silica-coated ZnO-nanoparticle quantum dots with biotin. *J. Colloid Interface Sci.* **2013**, *399*, 19–25. [CrossRef] [PubMed]
7. Moussodia, R.-O.; Balan, L.; Merlin, C.; Mustin, C.; Schneider, R. Biocompatible and stable ZnO quantum dots generated by functionalization with siloxane-core PAMAM dendrons. *J. Mater. Chem.* **2010**, *20*, 1147–1155. [CrossRef]
8. Manaia, E.B.; Kaminski, R.C.K.; Caetano, B.L.; Briois, V.; Chiavacci, L.A.; Bourgaux, C. Surface modified Mg-doped ZnO QDs for biological imaging. *Eur. J. Nanomed.* **2015**, *7*, 109–120. [CrossRef]
9. Zhao, H.; Lv, P.; Huo, D.; Zhang, C.; Ding, Y.; Xu, P.; Hu, Y. Doxorubicin loaded chitosan-ZnO hybrid nanospheres combining cell imaging and cancer therapy. *RSC Adv.* **2015**, *5*, 60549–60551. [CrossRef]
10. Sharma, S.; Chawla, S. Enhanced UV emission in ZnO/ZnS core shell nanoparticles prepared by epitaxial growth in solution. *Electron. Mater. Lett.* **2013**, *9*, 267–271. [CrossRef]
11. Luo, J.; Zhao, S.; Wu, P.; Zhang, K.; Peng, C.; Zheng, S. Synthesis and characterization of new Cd-doped ZnO/ZnS core-shell quantum dots with tunable and highly visible photoluminescence. *J. Mater. Chem. C* **2015**, *3*, 3391–3398. [CrossRef]
12. Reiss, P.; Protière, M.; Li, L. Core/Shell Semiconductor Nanocrystals. *Small* **2009**, *5*, 154–168. [CrossRef] [PubMed]
13. Isnaeni; Kim, K.H.; Nguyen, D.L.; Lim, H.; Nga, P.T.; Cho, Y.-H. Shell layer dependence of photoblinking in CdSe/ZnSe/ZnS quantum dots. *Appl. Phys. Lett.* **2011**, *98*, 012109. [CrossRef]
14. Wang, X.; Ren, X.; Kahen, K.; Hahn, M.A.; Rajeswaran, M.; Maccagnano-Zacher, S.; Silcox, J.; Cragg, G.E.; Efros, A.L.; Krauss, T.D. Non-blinking semiconductor nanocrystals. *Nature* **2009**, *459*, 686–689. [CrossRef] [PubMed]
15. Panda, S.K.; Dev, A.; Chaudhuri, S. Fabrication and luminescent properties of *c*-axis oriented ZnO-ZnS core-shell and ZnS nanorod arrays by sulfidation of aligned ZnO nanorod arrays. *J. Phys. Chem. C* **2007**, *111*, 5039–5043. [CrossRef]
16. Wu, D.; Jiang, Y.; Yuan, Y.; Wu, J.; Jiang, K. ZnO–ZnS heterostructures with enhanced optical and photocatalytic properties. *J. Nanopart. Res.* **2011**, *13*, 2875–2886. [CrossRef]
17. Shuai, X.M.; Shen, W.Z. A Facile Chemical Conversion Synthesis of ZnO/ZnS Core/Shell Nanorods and Diverse Metal Sulfide Nanotubes. *J. Phys. Chem. C* **2011**, *115*, 6415–6422. [CrossRef]
18. Liu, L.; Chen, Y.; Guo, T.; Zhu, Y.; Su, Y.; Jia, C.; Wei, M.; Cheng, Y. Chemical Conversion Synthesis of ZnS Shell on ZnO Nanowire Arrays: Morphology Evolution and Its Effect on Dye-Sensitized Solar Cell. *ACS Appl. Mater. Interfaces* **2011**, *4*, 17–23. [CrossRef] [PubMed]
19. Geng, J.; Liu, B.; Xu, L.; Hu, F.-N.; Zhu, J.-J. Facile Route to Zn-Based II−VI Semiconductor Spheres, Hollow Spheres, and Core/Shell Nanocrystals and Their Optical Properties. *Langmuir* **2007**, *23*, 10286–10293. [CrossRef] [PubMed]
20. Verma, P.; Pandey, A.C.; Bhargava, R.N. Synthesis and characterisation: Zinc oxide–sulfide nanocomposites. *Physica B* **2009**, *404*, 3894–3897. [CrossRef]
21. Nam, W.; Lim, Y.; Seo, W.-S.; Cho, H.; Lee, J. Control of the shell structure of ZnO–ZnS core-shell structure. *J. Nanopart. Res.* **2011**, *13*, 5825–5831. [CrossRef]
22. Sadollahkhani, A.; Kazeminezhad, I.; Lu, J.; Nur, O.; Hultman, L.; Willander, M. Synthesis, structural characterization and photocatalytic application of ZnO@ZnS core-shell nanoparticles. *RSC Adv.* **2014**, *4*, 36940–36950. [CrossRef]
23. Sookhakian, M.; Amin, Y.M.; Basirun, W.J.; Tajabadi, M.T.; Kamarulzaman, N. Synthesis, structural, and optical properties of type-II ZnO–ZnS core–shell nanostructure. *J. Lumin.* **2014**, *145*, 244–252. [CrossRef]

24. Caetano, B.L.; Silva, M.N.; Santilli, C.V.; Briois, V.; Pulcinelli, S.H. Unified ZnO Q-dot growth mechanism from simultaneous UV-Vis and EXAFS monitoring of sol-gel reactions induced by different alkali base. *Opt. Mater.* **2016**, *61*, 92–97. [CrossRef]
25. Caetano, B.L.; Briois, V.; Pulcinellim, S.H.; Meneau, F.; Santilli, C.V. Revisiting the ZnO Q-dot Formation Toward an Integrated Growth Model: From Coupled Time Resolved UV−Vis/SAXS/XAS Data to Multivariate Analysis. *J. Phys. Chem. C* **2017**, *121*, 886–895. [CrossRef]
26. Caetano, B.L.; Santilli, C.V.; Meneau, F.; Briois, V.; Pulcinellim, S.H. In Situ and Simultaneous UV−vis/SAXS and UV−vis/XAFS Time-Resolved Monitoring of ZnO Quantum Dots Formation and Growth. *J. Phys. Chem. C* **2011**, *115*, 4404–4412. [CrossRef]
27. Spanhel, L.; Anderson, M.A. Semiconductor clusters in the sol-gel process: Quantized aggregation, gelation, and crystal growth in concentrated zinc oxide colloids. *J. Am. Chem. Soc.* **1991**, *113*, 2826–2833. [CrossRef]
28. Meulenkamp, E.A. Synthesis and Growth of ZnO Nanoparticles. *J. Phys. Chem. B* **1998**, *102*, 5566–5572. [CrossRef]
29. Ravel, B.; Newville, M. ATHENA, ARTEMIS, HEPHAESTUS: Data analysis for X-ray absorption spectroscopy using IFEFFIT. *J. Synchrotron Radiat.* **2005**, *12*, 537–541. [CrossRef] [PubMed]
30. Kohlbrecher, J.; Bressler, I. *SASfit*; Paul Scherrer Institut: Villigen, Switzerland, 2014.
31. Brus, L.E. Electron–electron and electron-hole interactions in small semiconductor crystallites: The size dependence of the lowest excited electronic state. *J. Chem. Phys.* **1984**, *80*, 4403–4409. [CrossRef]
32. Baranov, A.N.; Sokolov, P.S.; Tafeenko, V.A.; Lathe, C.; Zubavichus, Y.V.; Veligzhanin, A.A.; Chukichev, M.V.; Solozhenko, V.L. Nanocrystallinity as a Route to Metastable Phases: Rock Salt ZnO. *Chem. Mater.* **2013**, *25*, 1775–1782. [CrossRef]
33. Lin, P.-C.; Hua, C.C.; Lee, T.-C. Low-temperature phase transition of ZnS: The critical role of ZnO. *J. Solid State Chem.* **2012**, *194*, 282–285. [CrossRef]
34. La Porta, F.A.; Andres, J.; Li, M.S.; Sambrano, J.R.; Varela, J.A.; Longo, E. Zinc blende versus wurtzite ZnS nanoparticles: Control of the phase and optical properties by tetrabutylammonium hydroxide. *Phys. Chem. Chem. Phys.* **2014**, *16*, 20127–20137. [CrossRef] [PubMed]
35. West, A.R. *Solid State Chemistry and Its Applications*, 2nd ed.; John Wiley and Sons: New York, NY, USA, 1992; ISBN 978-1-119-94294-8.
36. Xu, X.; Xu, C.; Wang, X.; Lin, Y.; Dai, J.; Hu, J. Control mechanism behind broad fluorescence from violet to orange in ZnO quantum dots. *CrystEngComm* **2013**, *15*, 977–981. [CrossRef]
37. Liu, C.; Ji, Y.; Tan, T. One-pot hydrothermal synthesis of water-dispersible ZnS quantum dots modified with mercaptoacetic acid. *J. Alloys Compd.* **2013**, *570*, 23–27. [CrossRef]
38. Briois, V.; Giorgetti, C.; Baudelet, F.; Blanchandin, S.; Tokumoto, M.S.; Pulcinelli, S.H.; Santilli, C.V. Dynamical Study of ZnO Nanocrystal and Zn-HDS Layered Basic Zinc Acetate Formation from Sol−Gel Route. *J. Phys. Chem. C* **2007**, *111*, 3253–3258. [CrossRef]
39. Curcio, A.L.; Bernardi, M.I.B.; Mesquita, A. Local structure and photoluminescence properties of nanostructured $Zn_{1-x}Mn_xS$ material. *Phys. Status Solidi Rapid Res. Lett.* **2015**, *12*, 1367–1371. [CrossRef]
40. Mehta, S.K.; Kumar, S.; Chaudhary, S.; Bhasin, K.K.; Gradzielski, M. Evolution of ZnS Nanoparticles via Facile CTAB Aqueous Micellar Solution Route: A Study on Controlling Parameters. *Nanoscale Res. Lett.* **2008**, *4*, 17–28. [CrossRef] [PubMed]
41. Guinier, A.; Fournet, G. *Small Angle Scattering of X-rays*, 1st ed.; Wiley: New York, NY, USA, 1955; pp. 1–268.
42. Rabani, J. Sandwich colloids of zinc oxide and zinc sulfide in aqueous solutions. *J. Phys. Chem.* **1989**, *93*, 7707–7713. [CrossRef]
43. Spanhel, L. Colloidal ZnO nanostructures and functional coatings: A survey. *J. Sol-Gel Sci. Technol.* **2006**, *39*, 7–24. [CrossRef]

nanomaterials

MDPI

Article

Preparation and Characterization of ZnO Nanoparticles Supported on Amorphous SiO_2

Ying Chen, Hao Ding * and Sijia Sun

Beijing Key Laboratory of Materials Utilization of Nonmetallic Minerals and Solid Wastes, National Laboratory of Mineral Materials, School of Materials Science and Technology, China University of Geosciences (Beijing), Beijing 100083, China; chenying@cugb.edu.cn (Y.C.); 1012122105@cugb.edu.cn (S.S.)

* Correspondence: dinghao113@126.com; Tel.: +86-010-8232-2982

Received: 30 June 2017; Accepted: 24 July 2017; Published: 10 August 2017

Abstract: In order to reduce the primary particle size of zinc oxide (ZnO) and eliminate the agglomeration phenomenon to form a monodisperse state, Zn^{2+} was loaded on the surface of amorphous silica (SiO_2) by the hydrogen bond association between hydroxyl groups in the hydrothermal process. After calcining the precursors, dehydration condensation among hydroxyl groups occurred and ZnO nanoparticles supported on amorphous SiO_2 (ZnO–SiO_2) were prepared. Furthermore, the SEM and TEM observations showed that ZnO nanoparticles with a particle size of 3–8 nm were uniformly and dispersedly loaded on the surface of amorphous SiO_2. Compared with pure ZnO, ZnO–SiO_2 showed a much better antibacterial performance in the minimum inhibitory concentration (MIC) test and the antibacterial properties of the paint adding ZnO–SiO_2 composite.

Keywords: amorphous SiO_2; load; monodisperse; ZnO nanoparticle; antibacterial

1. Introduction

Antimicrobial tests and environmental toxicity tests have been widely explored in order to improve health, safety, and the environment [1–3]. Zinc oxide (ZnO), as a semiconductor material with a band gap of 3.3 eV at room temperature [4], has high chemical stability, strong photosensitivity and non-toxicity property and is widely used in antibacterial materials [5]. Compared with ordinary ZnO powder, ZnO nanoparticles have a large specific surface area and small size effect, and show wide application potential in microbial inhibition and mildew removal [6,7].

However, like most of the nanoparticles, ZnO nanoparticles are prone to forming serious agglomeration, including hard agglomeration among the particles formed via the chemical reaction of the surface groups and soft agglomeration formed by other physical effects [8]. It is difficult to depolymerize the particles involved in hard agglomeration. Therefore, the apparent grain size of the primary ZnO particles tends to increase to the micron scale and the normal performance of ZnO nanoparticles is inhibited. In the preparation process of ZnO nanoparticles, in addition to the control of the ZnO morphology and primary particle size, the agglomeration phenomenon of ZnO particles should be suppressed to obtain dispersed nanoparticles. Wang et al. synthesized the doped ZnO nanoparticles with the mixture of alcohol and water as the solvent according to a precipitation method [9]. Chen et al. prepared ZnO nanocrystals via the reaction of zinc stearate with excessive alcohol in the hydrocarbon solvent [10]. Weller et al. used the low-temperature solvent thermal method to synthesize dispersible spherical ZnO nanoparticles and nano-rods with zinc acetate as precursors in methanol [11]. However, these methods have low synthesis performance and limited control ability. Especially, the solvent thermal process [12,13] is required to deal with organic solvents and it is difficult to realize industrial production. Therefore, some nanoparticles (such as titanium dioxide, TiO_2) [14,15] are supported on the surface or pores of the inorganic carrier. In this way, the strong interaction between the carrier surface and nanoparticles, and the forced isolation among the carrier particles efficiently prevent the agglomeration among the nanoparticles and improve the dispersion effects and functions.

Amorphous SiO_2, commercially known as white carbon black, is an aggregate of SiO_2 particles ($SiO_2 \cdot nH_2O$) and commonly used as a rubber reinforcing additive [16,17]. The primary particle size of SiO_2 particles is generally 10–100 nm. SiO_2 particles containing rich Si–OH groups, which can form a strong interaction between the SiO_2 carrier surface and Zn–OH (precursors of ZnO). This interaction reduces the combination between Zn–OH and prevents its aggregation, thus contributing to the formation of monodisperse ZnO nanoparticles. In addition, small amorphous SiO_2 particles have high dispensability and can prevent further agglomeration of ZnO–SiO_2 composite. Therefore, amorphous SiO_2 was selected as the carrier of supported ZnO nanoparticles.

Based on the above results, in this paper, the environmentally friendly hydrothermal method was adopted to prepare ZnO–SiO_2 by loading Zn^{2+} on the surface of amorphous SiO_2 and calcining active products at high temperatures. Moreover, the structures and antibacterial properties of as-prepared ZnO–SiO_2 were explored.

2. Experimental Procedure

2.1. Materials

In this study, amorphous SiO_2 was purchased from Henan Jiaozuo Fluoride New Energy Technology Co., Ltd (Jiaozuo, Henan, China). The properties of amorphous SiO_2 are described as follows: SiO_2 content of 96.63%, whiteness of 96.76%, average aggregate size of 20 μm, primary particle size of 20–30 nm, and specific surface area of 59.54 m^2/g. Zinc nitrate ($Zn(NO_3)_2 \cdot 6H_2O$) as the source of Zn^{2+} was from Beijing Yili Fine Chemical Co., Ltd (Beijing, China). Sodium polyacrylate (PAAS) as a dispersant was supplied by Changzhou Run Yang Chemical Co., Ltd (Changzhou, Jiangsu, China). Pure ZnO, as an antibacterial agent, was compared with the ZnO–SiO_2 composite in antibacterial performance. It was produced by the Xi Long Chemical Co., Ltd (Guangzhou, Guangdong, China) and the size of the particles was about 200 nm. Figure 1 shows SEM images of amorphous SiO_2 and pure ZnO.

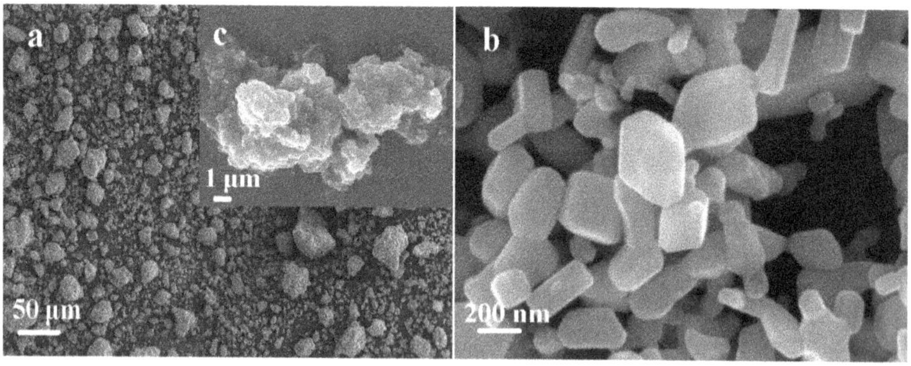

Figure 1. Micrographs of (**a**,**c**) amorphous SiO_2 and (**b**) pure ZnO.

2.2. ZnO–SiO_2 Precursor

Amorphous SiO_2, sodium polyacrylate (1% of the weight of SiO_2) and H_2O were mixed and stirred to prepare the suspension with solid content of 18%. Ceramic polishing balls (diameter: 1–3 mm) were added into the suspension according to the proportion of 50% of the solid content and then stirred at a speed of 1000 r/min for 1 h to prepare the depolymerized amorphous SiO_2 slurry. A zinc nitrate solution (0.09 wt %) was added into the slurry and the pH of the mixture was respectively adjusted to 5.0 and 7.0 by adding 6 mol/L NaOH and 6 mol/L HNO_3. The mixture was stirred at 60 °C for 1 h. The precursors were obtained after suction filtration, washing, and drying and denoted as Zn–SiO_2-pH5.0 (the precursor was prepared with pH value at 5.0) and Zn–SiO_2-pH7.0 (the precursor was prepared with pH value at 7.0) respectively. The preparation process is shown in Figure 2.

2.3. *Preparation of ZnO–SiO₂*

The precursors Zn–SiO_2-pH5.0 and Zn-SiO_2-pH7.0 were calcined at 400 °C for 1 h to obtain the composite particles of ZnO and SiO_2 and denoted as ZnO–SiO_2-pH5.0 (the composite obtained by calcining the precursor which was prepared with pH value at 5.0) and ZnO–SiO_2-pH7.0 (the composite obtained by calcining the precursor which was prepared with pH value at 7.0). The loads of ZnO were 4.51 and 11.26%, respectively.

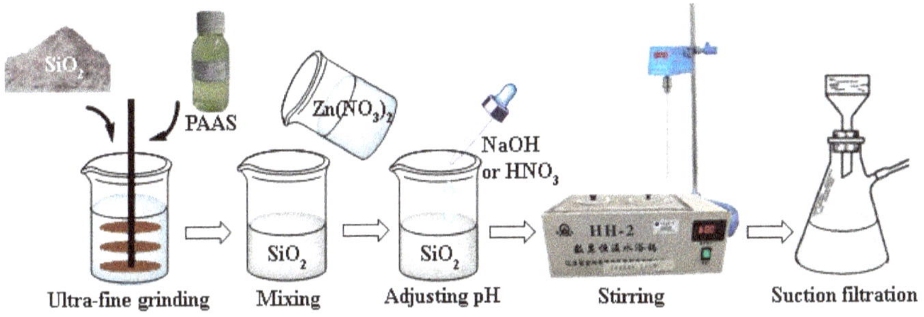

Figure 2. Preparation of composite particles of ZnO–SiO_2 precursor.

2.4. *Characterization*

The X-ray diffraction (XRD) was measured by using a D/max-Ra X-ray diffractometer (Ouyatu, Japan, Cu Kα radiation = 1.54 Å) in an angular range of 10–80° (2θ) with a step of 0.02° (2θ). Scherer Equation [18] is used to calculate the average grain size of ZnO nanoparticles:

$$\beta = \frac{K\lambda}{Dcos\theta''} \quad (1)$$

where K is the shape factor constant (0.94); λ is X-ray wavelength; D is the grain size; θ is the diffraction angle; β is the diffraction peak half width.

An X-ray fluorescence spectrometer (XRF Shimadzu-1800, Kyoto, Japan) was used to analyze the oxide content of samples. The particle size and size distribution of the composite particles of ZnO and SiO_2 were characterized by transmission electron microscopy (TEM FEI Tecnai G220, Portland, OR, USA). Scanning electron microscopy (SEM) was used to explore the morphology of ZnO–SiO_2 by a Hitachi field emission scanning electron microscope (Hitachi S4800, Tokyo, Japan) under the voltage of 10 kV. The Fourier transform infrared spectroscopy (FTIR, Madison, WI, USA) measurement was carried out to explore the changes in functional groups of ZnO-SiO_2 by Nicolet IS50. The samples were finely pulverized and then diluted in dried KBr to form a homogeneous mixture according to the sample-KBr ratio of 1/200. The X-ray photoelectron spectroscopy (XPS, Manchester, UK) measurement was conducted on an Axis Ultra spectrometer with monochromatic Mg Kα (1253.6 eV) radiation to investigate the valence state of Zn.

2.5. *Antimicrobial Test*

The antimicrobial ability of ZnO–SiO_2 under dark conditions was investigated through antibacterial tests [19–21]. Different concentrations of ZnO–SiO_2-pH5.0, ZnO–SiO_2-pH7.0, and pure

ZnO were added to the agar medium, and then *E. coli* (CGMCC 1.2385) was inoculated on the medium to observe the growth of bacteria and determine the minimum inhibitory concentration (MIC) [22].

The antimicrobial coating was obtained by mixing 12 wt % styrene-acrylic emulsion, 34 wt % of H_2O, 50 wt % of the filler (0–8 wt % of ZnO–SiO_2-pH7.0), and 4 wt % of paint additive. The antibacterial property of ZnO–SiO_2 was evaluated by testing the antibacterial property of the coating. The antibacterial rate of the coating was tested according to Chinese national standard GB/T21866-2008 [23]. The antibacterial rate (R) is calculated as:

$$R = 100\% \times (A - B)/A, \quad (2)$$

where A and B are the average number of colonies of the blank control plate and antibacterial coating plate after 24 h.

3. Results and Discussion

3.1. Structure and Characterization of ZnO–SiO_2

3.1.1. Phase and Chemical Constitution of ZnO–SiO_2

Figure 3 shows the XRD patterns of ZnO–SiO_2. Table 1 shows the XRF results of each sample. The XRD pattern of the SiO_2 carrier shows a strong bread peak near 2θ of 23°, indicating that the main phase is an amorphous phase corresponding to amorphous SiO_2. In the XRD patterns of ZnO–SiO_2-pH5.0 and ZnO–SiO_2-pH7.0, in addition to the above-mentioned peak reflecting the amorphous phase, the peaks at 31.8°, 34.5°, 36.3°, and 47.5° correspond to the ZnO diffraction peak [24,25], indicating that Zn^{2+} has been transformed into ZnO after the thermal reaction with the SiO_2 carrier and calcination. The ZnO diffraction intensity of ZnO–SiO_2-pH7.0 was significantly larger than that of ZnO–SiO_2-pH5.0 due to the different loadings of ZnO. The contents of ZnO in ZnO–SiO_2-pH5.0 and ZnO–SiO_2-pH7.0 are respectively 4.51 and 11.26% (Table 1). The SiO_2 content in ZnO–SiO_2-pH5.0 is lower than that in ZnO–SiO_2-pH7.0. The results are consistent with the XRD results.

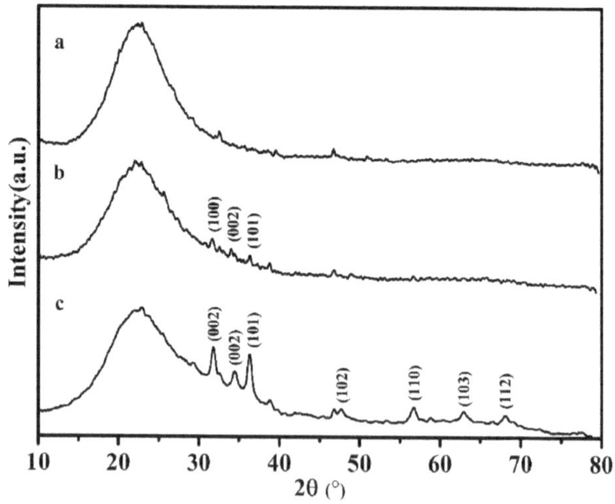

Figure 3. XRD of (**a**) amorphous SiO_2; (**b**) ZnO–SiO_2-pH5.0; and (**c**) ZnO–SiO_2-pH7.0.

Table 1. XRF of amorphous SiO_2, ZnO–SiO_2-pH5.0 and ZnO–SiO_2-pH7.0.

Samples	SiO_2/%	ZnO/%	Na_2O/%
amorphous SiO_2	96.63	0	0.98
ZnO–SiO_2-pH5.0	92.78	4.51	1.05
ZnO–SiO_2-pH7.0	84.22	11.26	2.68

According to the XRD data in Figure 3, the grain size of ZnO–SiO_2-pH7.0 was calculated to be 3.63 nm according to the Scherer Equation.

3.1.2. Microstructure of ZnO–SiO_2

Figure 4 shows the distribution of three elements (O, Si, and Zn) in ZnO–SiO_2-pH5.0 and ZnO–SiO_2-pH7.0. The distributions of these three elements are consistent with the distribution of ZnO–SiO_2 particles. The distribution densities of O and Si are larger than that of Zn. The Zn density in the elemental distribution of ZnO–SiO_2-pH7.0 is greater than that of ZnO–SiO_2-pH5.0. The results indicate that the main components of ZnO–SiO_2 are SiO_2. The content of ZnO is low, but evenly distributed on the surface of SiO_2 particles.

Figure 5 shows the TEM images of the amorphous SiO_2 carrier, ZnO–SiO_2-pH5.0, and ZnO–SiO_2-pH7.0. At a small scale, all the samples are regular particle aggregates. The particle size is about 20–30 nm. Although these particles overlap each other, the overall dispersion effect is good. These unit particles are obviously amorphous SiO_2 particles. At the scale of 10 nm, the surface morphology of SiO_2 particles in the SiO_2 carrier is uniform, indicating that no other material is loaded. At the scale of 2 nm, only homogeneous non-crystal phase particles are observed. Dark spots with a size of 3–8 nm are uniformly distributed in ZnO–SiO_2-PH5.0 and ZnO–SiO_2-pH7.0 at the scale of 10 nm. These dark spots are crystal phase particles at the larger magnification. The stripe spacing can reflect the lattice size. The stripe spacing of ZnO–SiO_2-pH5.0 and ZnO–SiO_2-pH7.0 are respectively measured to be 2.45 and 2.59 nm. ZnO (101) plane spacing and ZnO (002) plane spacing are respectively measured to be 2.45 and 2.59 Å, which are almost consistent with standard ZnO (101) plane spacing of 2.47 Å and ZnO (002) plane spacing of 2.60 Å (ICDD card # 89-7102). These data indicate that these ZnO nanoparticles were monodispersedly loaded on the SiO_2 surface. The size of ZnO particles is 3–8 nm, which is consistent with the average particle size of 3.63 nm obtained in the XRD test [18].

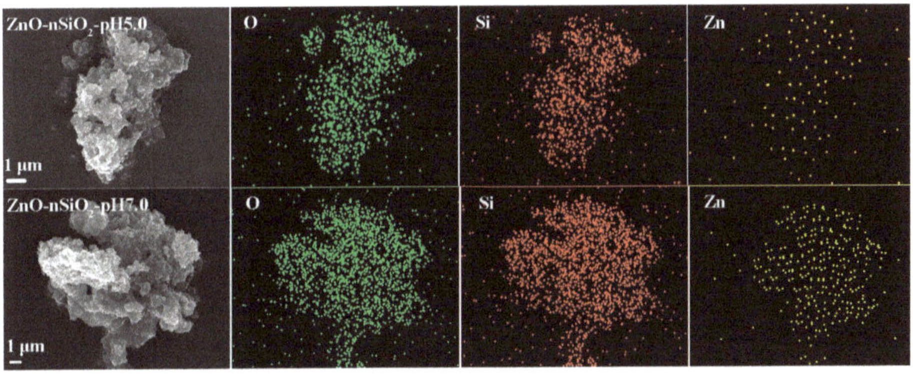

Figure 4. SEM images of ZnO–SiO_2 and corresponding mapping results.

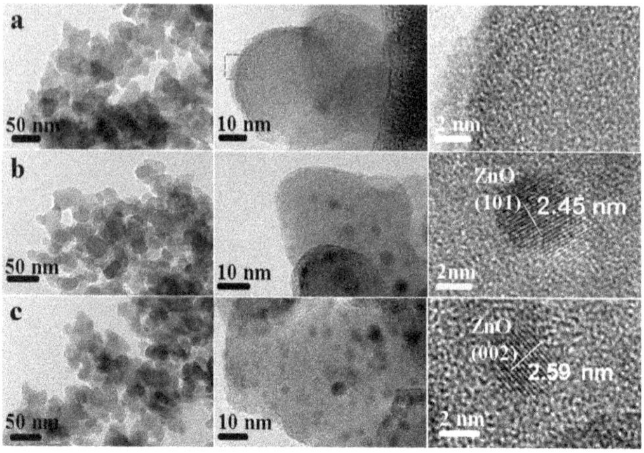

Figure 5. TEM maps of (**a**) amorphous SiO_2; (**b**) ZnO–SiO_2-pH5.0; (**c**) ZnO–SiO_2-pH7.0 at different scales.

3.1.3. Formation Mechanism of ZnO–SiO_2

Figure 6 shows the XPS pattern between 1015 and 1050 eV of the ZnO–SiO_2. The peaks of ZnO–SiO_2-pH5.0 and ZnO–SiO_2-pH7.0 at 1046.0 and 1045.5 eV correspond to Zn2$p_{1/2}$ orbital; the peaks at 1023.0 and 1022.5 eV correspond to the Zn2$p_{3/2}$ orbital [26]. These peaks are equivalent to the 2$p_{1/2}$ and 2$p_{3/2}$ energy peaks (1044.2 and 1021.2 eV) of ZnO [27]. Moreover, the energy difference between Zn2$p_{1/2}$ and Zn2$p_{3/2}$ orbitals is 23 eV, which is the same as that of ZnO. Therefore, it can be determined that the valence of Zn in ZnO–SiO_2-pH5.0 and ZnO–SiO_2-pH7.0 is +2, which is consistent with the results of XRD, XRF, and TEM.

Table 2 shows the percentages of the amorphous SiO_2 carrier, composite precursors (pH 5.0 and 7.0) and ZnO–SiO_2 based on XPS. Compared with the amorphous SiO_2 carrier, the Zn^{2+} composite precursors show the increasing ratio of O/Si with the increase in the Zn content. The change may be interpreted as follows. Hydroxyl groups generated by the hydrolysis of Zn form hydrogen bonds on the SiO_2 surface, thus resulting in an increase in the amount of O. Compared with the precursors, ZnO–SiO_2 products show a decreased O/Si ratio because the amount of oxygen is decreased by the dehydration condensation reaction among the –OH bonds on the surface of precursors during calcination. The above analysis suggests that ZnO nanoparticles was loaded on the surface of amorphous SiO_2.

Figure 7 shows the infrared spectra of the amorphous SiO_2 carrier, Zn^{2+} composite precursor (Zn–SiO_2-pH5.0 and Zn–SiO_2-pH7.0), and the final products (ZnO–SiO_2-pH5.0 and ZnO–SiO_2-pH7.0). The absorption peaks of each sample at 450 and 1062 cm^{-1} are respectively ascribed to symmetrical and antisymmetric stretching vibration of Si–O–Si. The absorption peak at 799 cm^{-1} corresponds to bending vibration, reflecting the characteristics of SiO_2 [28]. As shown in Figure 7, the infrared spectra of Zn^{2+} composite precursors (Zn–SiO_2-pH5.0 and Zn–SiO_2-pH7.0) at 3317 and 3319 cm^{-1} correspond to Zn–OH stretching vibration. The absorption peaks of hydroxyl groups in zinc hydroxide ($Zn(OH)_2$) at 1345 and 1347 cm^{-1} reflect the bridging effect of the hydroxyl group in the product, and the Si–OH bending vibration peak in the two products moves from 958 cm^{-1} (the vibration peak of the raw material SiO_2) to 952 and 954 cm^{-1}, respectively. The changes indicate that in the hydrothermal reaction during the preparation process of Zn–SiO_2-pH5.0, Zn^{2+} forms a complex of $Zn(OH)_2$, which yields hydrogen bonds with Si–OH on the surface of amorphous SiO_2 [29–31]. In addition, the –OH bending vibration peaks of water adsorbed on the surface of Zn–SiO_2-pH5.0 and Zn–SiO_2-pH7.0 occur at 1413 and 1411 cm^{-1}, respectively. The –OH shear vibration peaks occur at 1580 and 1579 cm^{-1} [32]. The Zn–OH and Si–OH of calcined products and the –OH bond of adsorbed water disappeared after the calcination of the precursors. Therefore, the high-temperature calcination resulted in the evaporation of SiO_2 surface water and the dehydration condensation reaction between Si–OH and Zn–OH, and yielded Si–O–Zn chemical bond.

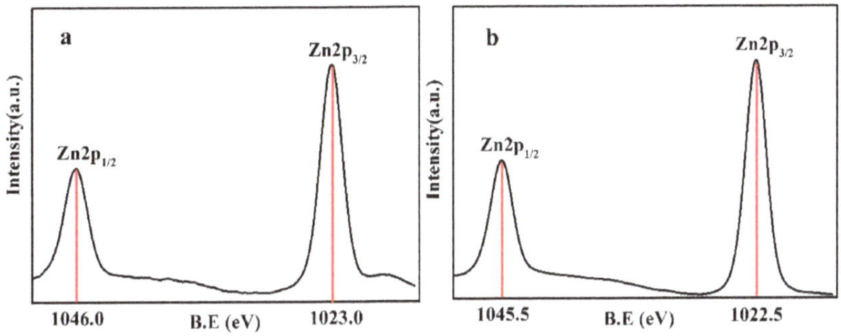

Figure 6. XPS of (**a**) ZnO–SiO_2-pH5.0; (**b**) ZnO–SiO_2-pH7.0.

Table 2. Element analysis based on XPS results

Samples	C1s (%)	Zn2p (%)	Si2p (%)	O1s (%)	O/Si
Amorphous SiO_2	2.74	0	31.27	65.84	2.11
Zn–SiO_2-pH5.0	6.18	2.24	27.55	62.04	2.25
ZnO–SiO_2-pH5.0	3.48	2.47	30.03	63.68	2.12
Zn–SiO_2-pH7.0	6.90	6.24	24.38	60.84	2.50
ZnO–SiO_2-pH7.0	5.96	10.55	25.18	58.39	2.31

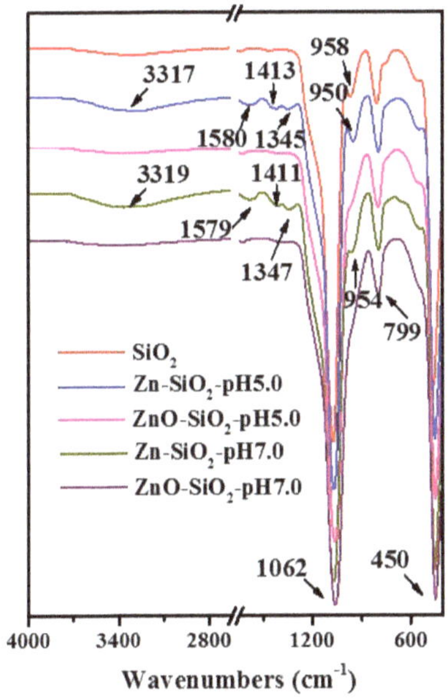

Figure 7. FTIR spectra of amorphous SiO_2, Zn–SiO_2-pH5.0, ZnO–SiO_2-pH5.0, Zn–SiO_2-pH7.0, and ZnO–SiO_2-pH7.0.

Figure 8 shows a schematic diagram of the synthesis process of ZnO–SiO_2 by loading Zn^{2+} on the amorphous SiO_2 carrier via the hydrothermal reaction and composite precursor calcination. Due to the large number of Si–OH bonds on the surface of SiO_2 and the strong activity of Si–OH bonds, the interaction between Si–OH and Zn–OH is greater than the interaction in Zn–OH (multinuclear ions). Therefore, Zn^{2+} is immobilized on the surface of SiO_2 and dispersed ZnO nanoparticles are formed after the calcination of precursors.

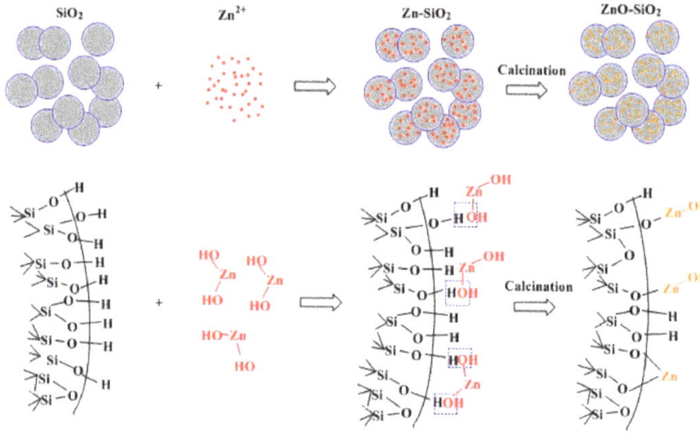

Figure 8. Synthesis of the ZnO–nSiO_2 Complex

3.2. *Antibacterial Properties of ZnO–SiO_2*

ZnO has antibacterial activity under light and dark conditions and is mostly applied under dark conditions. In order to investigate the antibacterial properties of ZnO–SiO_2 under dark conditions, ZnO–SiO_2-pH5.0, ZnO–SiO_2-pH7.0, and pure ZnO were respectively prepared. Figure 9 shows the bacterial growth profiles obtained by the plate test. Obvious colonies were formed in the blank control without the antimicrobial material (Figure 9a). When the concentration of ZnO–SiO_2-pH5.0 was 10 mg/mL, obvious colonies were observed on the culture plate; when the concentration of ZnO–SiO_2-pH5.0 was 20 mg/mL, the number of colonies decreased but colonies did not completely disappear; when the concentration of ZnO–SiO_2-pH5.0 was increased to 36 mg/mL, no colony was formed (Figure 9b). The concentrations of ZnO–SiO_2-pH7.0 and pure ZnO required for colony-free results were respectively 19 mg/mL and 20 mg/mL (Figure 9c,d). Based on the above results, the minimum inhibitory concentration (MIC) of each sample was determined and converted into the minimum inhibitory concentration of ZnO based on the content of ZnO in the composite (Table 3). The MIC values of ZnO–SiO_2-pH5.0 and ZnO–SiO_2-pH7.0 were respectively 1.60 and 2.14 mg/mL, which were equivalent to 10% of the MIC of pure ZnO (20 mg/mL), indicating that the antimicrobial ability of ZnO nanoparticles loaded on the SiO_2 surface was about 10 times that of pure ZnO. Obviously, the formation of dispersed nanoparticles (3–8 nm) loaded on amorphous SiO_2 greatly improved its antimicrobial performance.

Table 3. MIC of ZnO–SiO_2-pH5.0, ZnO–SiO_2-pH7.0, and pure ZnO.

MIC (mg/mL)	ZnO–SiO_2-pH5.0	ZnO–SiO_2-pH7.0	ZnO
E. coli	36	19	20
E. coli (ZnO)	1.60	2.14	20

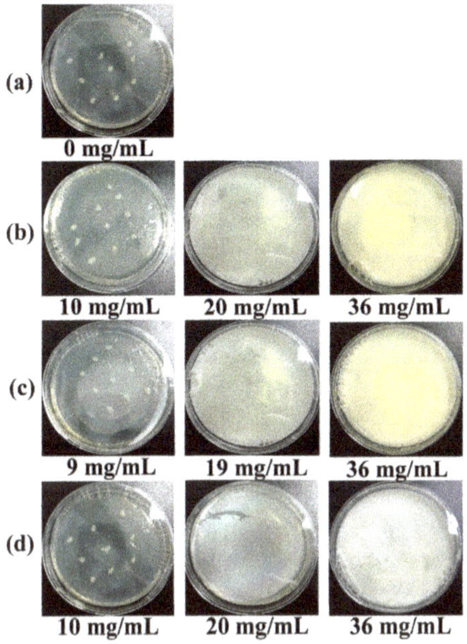

Figure 9. Antimicrobial tests of (**a**) blank control, (**b**) different concentrations of ZnO–SiO_2-pH5.0 in agar medium, (**c**) different concentrations of ZnO–SiO_2-pH7.0 in agar medium, (**d**) different concentrations of pure ZnO in agar medium.

Figure 10 shows the antibacterial rate and colony growth conditions of *E. coli* on the plates added with different amounts of ZnO–SiO_2-pH7.0 coating. The antibacterial rate of the coating without ZnO–SiO_2-pH7.0 was 0 and a large number of colonies were formed on the plate, indicating that the coating showed no antibacterial property. When the addition of ZnO–SiO_2-pH7.0 in the coating was only 2%, the antibacterial rate was increased above 70%, showing a good antibacterial effect; when the dosage was gradually increased to 8%, the antibacterial rate of the coating to *E. coli* was 90.48%, which met the requirements of the antibacterial effect of antibacterial coating in Chinese national standard GBT21866-2008. The increasing antibacterial rate of the paint indicated less colonies and a better antibacterial effect.

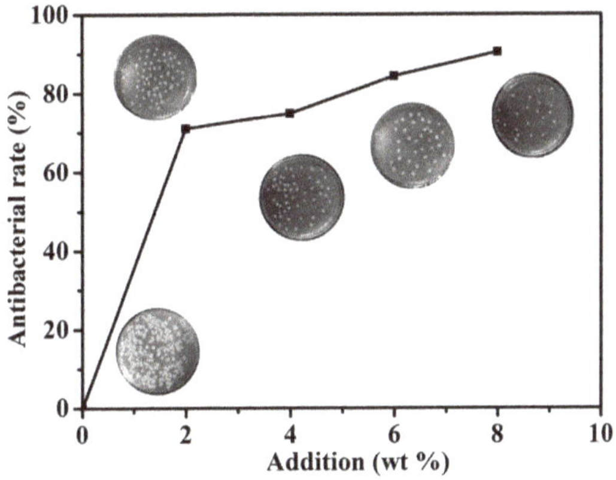

Figure 10. Effects of different additives on antibacterial rate (*E. coli*).

The antimicrobial properties of the ZnO nanoparticles supported uniformly and dispersedly on the surface of SiO_2 were greatly improved, due to the large specific surface area and surface activity of ZnO nanoparticles compared with the pure ZnO of large particles, and the contact and inhibition with microbes is stronger. This should be considered as one of the means to enhance the function of ZnO.

4. Conclusions

ZnO–SiO_2 composite was prepared by an environmentally friendly hydrothermal method and high-temperature calcination. In this composite, ZnO nanoparticles with a particle size of 3–8 nm were uniformly and dispersedly loaded on the surface of amorphous SiO_2. The size of the ZnO particles used in the industry is about 500 nm, and there is a certain degree of agglomeration among particles. According to the analysis of relevant tests, the strong interaction between the SiO_2 carrier surface and Zn–OH (precursors of ZnO) reduced the combination between Zn–OH, prevented its aggregation and formed monodispersed nanoparticles. Compared with pure ZnO, ZnO–SiO_2 showed much better antibacterial performance in the MIC test and the characterization test of paint properties.

In general, ZnO nanoparticles loaded uniformly and depressively on the surface of amorphous SiO_2 greatly enhanced its antibacterial function.

Acknowledgments: This work was supported by the Fundamental Research Funds for the Central Universities of China (No. 2652016160) and the Project Commissioned by Shandong Private Enterprises (No. 2015-KY19-139 20151224).

Author Contributions: Ying Chen and Hao Ding conceived and designed the experiments; Ying Chen performed the experiments; Ying Chen, Hao Ding and Sijia Sun anamyzed the date; Hao Ding contributed reagents/materials/analysis tools; Ying Chen wrote the paper.

Conflicts of Interest: The authors declare no competing financial interest.

References

1. Ficociello, G.; Zanni, E.; Cialfi, S.; Aurizi, C.; Biolcati, G.; Palleschi, C.; Talora, C.; Uccelletti, D. Glutathione S-transferase θ-subunit as a phenotypic suppressor of *pmr1*Δ strain, the *Kluyveromyces lactis* model for Hailey-Hailey disease. *Biochim. Biophys. Acta* **2016**, *1863*, 2650–2657. [CrossRef] [PubMed]
2. Zanni, E.; Laudenzi, C.; Schifano, E.; Palleschi, C.; Perozzi, G.; Uccelletti, D.; Devirgiliis, C. Impact of a Complex Food Microbiota on Energy Metabolism in the Model Organism *Caenorhabditis elegans*. *BioMed Res. Int.* **2015**, *2015*, 621709. [CrossRef] [PubMed]
3. Uccelletti, D.; Zanni, E.; Guerisoli, M.; Palleschi, C. Caenorhabditis elegans: an emerging animal model for biomonitoring environmental toxicity. *J. Biotechnol.* **2010**, *150*, 257. [CrossRef]
4. Malachová, K.; Praus, P.; Rybková, Z.; Kozák, O. Antibacterial and antifungal activities of silver, copper and zinc montmorillonites. *Appl. Clay Sci.* **2011**, *53*, 642–645. [CrossRef]
5. Zanni, E.; Chandraiahgari, C.R.; Bellis, G.D.; Montereali, M.R.; Armiento, G.; Ballirano, P.; Polimeni, A.; Sarto, M.S.; Uccelletti, D. Zinc Oxide Nanorods-Decorated Graphene Nanoplatelets: A Promising Antimicrobial Agent against the Cariogenic Bacterium *Streptococcus mutans*. *Nanomaterials* **2016**, *6*, 179. [CrossRef] [PubMed]
6. Li, Y.; Zhang, W.; Niu, J.; Chen, Y. Mechanism of Photogenerated Reactive Oxygen Species and Correlation with the Antibacterial Properties of Engineered Metal-Oxide Nanoparticles. *ACS Nano* **2012**, *6*, 5164–5173. [CrossRef] [PubMed]
7. Cha, S.H.; Hong, J.; Mcguffie, M.; Yeom, B.; Vanepps, J.S.; Kotov, N.A. Shape-Dependent Biomimetic Inhibition of Enzyme by Nanoparticles and Their Antibacterial Activity. *ACS Nano* **2015**, *9*, 9097–9105. [CrossRef] [PubMed]
8. Vincenzini, P. *Ceramic Powders*; Elsevier Scientific Pub. Co: New York, NY, USA, 1983; Volume 16, pp. 843–849.
9. Wang, M.-H.; Ma, X.-Y.; Jiang, W.; Zhou, F. Synthesis of doped ZnO nanopowders in alcohol–water solvent for varistors applications. *Mater. Lett.* **2014**, *121*, 149–151. [CrossRef]
10. Chen, Y.; Kim, M.; Lian, G.; Johnson, M.B.; Peng, X. Side reactions in controlling the quality, yield, and stability of high quality colloidal nanocrystals. *J. Am. Chem. Soc.* **2005**, *127*, 13331–13337. [CrossRef] [PubMed]

11. Pacholski, C.; Kornowski, A.; Weller, H. Self-Assembly of ZnO: From Nanodots to Nanorods. *Angew. Chem. Int. Ed.* **2002**, *41*, 1188–1191. [CrossRef]
12. Choi, S.H.; Kim, E.G.; Park, J.; An, K.; Lee, N.; And, S.C.K.; Hyeon, T. Large-Scale Synthesis of Hexagonal Pyramid-Shaped ZnO Nanocrystals from Thermolysis of Zn-Oleate Complex. *J. Phys. Chem. B* **2005**, *109*, 14792–14794. [CrossRef] [PubMed]
13. Pacholski, C.; Kornowski, A.; Weller, H. ZnO nanorods: Growth mechanism and anisotropic functionalization. In Proceedings of the SPIE 49th Annual Meeting on Optical Science and Technology, Denver, CO, USA, 2–6 August 2004.
14. Ito, M.; Fukahori, S.; Fujiwara, T. Adsorptive removal and photocatalytic decomposition of sulfamethazine in secondary effluent using TiO_2-zeolite composites. *Environ. Sci. Pollut. Res. Int.* **2014**, *21*, 834–842. [CrossRef] [PubMed]
15. Hassani, A.; Khataee, A.; Karaca, S.; Karaca, C.; Gholami, P. Sonocatalytic degradation of ciprofloxacin using synthesized TiO_2 nanoparticles on montmorillonite. *Ultrason. Sonochem.* **2017**, *35*, 251–262. [CrossRef] [PubMed]
16. Xu, K.; Sun, Q.; Guo, Y.; Zhang, Y.; Dong, S. Preparation of super-hydrophobic white carbon black from nano-rice husk ash. *Res. Chem. Intermed.* **2013**, *40*, 1965–1973. [CrossRef]
17. Chen, Y.; Peng, Z.; Kong, L.X.; Huang, M.F.; Li, P.W. Natural rubber nanocomposite reinforced with nano silica. *Polym. Eng. Sci.* **2008**, *48*, 1674–1677. [CrossRef]
18. Khorsand Zak, A.; Majid, W.H.A.; Ebrahimizadeh Abrishami, M.; Yousefi, R.; Parvizi, R. Synthesis, magnetic properties and X-ray analysis of $Zn_{0.97}X_{0.03}O$ nanoparticles (X = Mn, Ni, and Co) using Scherrer and size-strain plot methods. *Solid State Sci.* **2012**, *14*, 488–494. [CrossRef]
19. Jones, N.; Ray, B.; Ranjit, K.T.; Manna, A.C. Antibacterial activity of ZnO nanoparticle suspensions on a broad spectrum of microorganisms. *FEMS Microbiol. Lett.* **2008**, *279*, 71–76. [CrossRef] [PubMed]
20. Tam, K.H.; Djurišić, A.B.; Chan, C.M.N.; Xi, Y.Y.; Tse, C.W.; Leung, Y.H.; Chan, W.K.; Leung, F.C.C.; Au, D.W.T. Antibacterial activity of ZnO nanorods prepared by a hydrothermal method. *Thin Solid Films* **2008**, *516*, 6167–6174. [CrossRef]
21. Zhong, Z.; Zhe, X.; Sheng, T.; Yao, J.; Xing, W.; Yong, W. Unusual Air Filters with Ultrahigh Efficiency and Antibacterial Functionality Enabled by ZnO Nanorods. *ACS App. Mater. Interfaces* **2015**, *7*, 21538–21544. [CrossRef] [PubMed]
22. Schwalbe, R.; Steele-Moore, L.; Goodwin, A.C. *Antimicrobial Susceptibility Testing Protocols*; CRC Press: Boca Raton, FL, USA, 2007.
23. Test Method and Effect for Antibacterial Capability of Paints Film. Available online: http://c.gb688.cn/bzgk/gb/showGb?type=online&hcno=BD23209B12F6B07650FD542B0EB1456A (accessed on 1 October 2008).
24. Zuo, Z.; Liao, R.; Zhao, X.; Song, X.; Qiao, Z.; Guo, C.; Zhuang, A.; Yuan, Y. Anti-frosting performance of superhydrophobic surface with ZnO nanorods. *Appl. Therm. Eng.* **2017**, *110*, 39–48. [CrossRef]
25. Perillo, P.M.; Atia, M.N.; Rodríguez, D.F. Effect of the reaction conditions on the formation of the ZnO nanostructures. *Phys. E Low-Dimens. Syst. Nanostruct.* **2017**, *85*, 185–192. [CrossRef]
26. Al-Gaashani, R.; Radiman, S.; Daud, A.R.; Tabet, N.; Al-Douri, Y. XPS and optical studies of different morphologies of ZnO nanostructures prepared by microwave methods. *Ceram. Int.* **2013**, *39*, 2283–2292. [CrossRef]
27. Gallegos, M.V.; Peluso, M.A.; Thomas, H.; Damonte, L.C.; Sambeth, J.E. Structural and optical properties of ZnO and manganese-doped ZnO. *J. Alloys Compd.* **2016**, *689*, 416–424. [CrossRef]
28. Gao, H.; Song, Z.; Yang, L.; Wu, H. Synthesis Method of White Carbon Black Utilizing Water-Quenching Blast Furnace Slag. *Energy Fuels* **2016**, *30*, 9645–9651. [CrossRef]
29. Ghotbi, M.Y. Synthesis and characterization of nano-sized ε-$Zn(OH)_2$ and its decomposed product, nano-zinc oxide. *J. Alloys Compd.* **2010**, *491*, 420–422. [CrossRef]
30. Hubert, C.; Naghavi, N.; Canava, B.; Etcheberry, A. Zinc Sulfide Based Chemically Deposited Buffer Layers for Electrodeposited CIS Solar Cells. In Proceedings of the Conference Record of the 2006 IEEE 4th World Conference on Photovoltaic Energy Conversion, Waikoloa, HI, USA, 7–12 May 2006.

31. Nie, H.; He, A.; Zheng, J.; Xu, S.; Li, J.; Han, C.C. Effects of Chain Conformation and Entanglement on the Electrospinning of Pure Alginate. *Biomacromolecules* **2008**, *9*, 1362–1365. [CrossRef] [PubMed]
32. Tai, Y.; Qian, J.; Zhang, Y.; Huang, J. Study of surface modification of nano-SiO_2 with macromolecular coupling agent (LMPB-g-MAH). *Chem. Eng. J.* **2008**, *141*, 354–361. [CrossRef]

nanomaterials

MDPI

Article

Growth Method-Dependent and Defect Density-Oriented Structural, Optical, Conductive, and Physical Properties of Solution-Grown ZnO Nanostructures

Abu ul Hassan Sarwar Rana, Ji Young Lee, Areej Shahid and Hyun-Seok Kim *

Division of Electronics and Electrical Engineering, Dongguk University-Seoul, Seoul 04620, Korea; a.hassan.rana@gmail.com (A.u.H.S.R.); ljy010425@naver.com (J.Y.L.); areejshahid.146@gmail.com (A.S.)

* Correspondence: hyunseokk@dongguk.edu; Tel.: +82-2-2260-3996; Fax: +82-2-2277-8735

Received: 1 August 2017; Accepted: 7 September 2017; Published: 10 September 2017

Abstract: It is time for industry to pay a serious heed to the application and quality-dependent research on the most important solution growth methods for ZnO, namely, aqueous chemical growth (ACG) and microwave-assisted growth (MAG) methods. This study proffers a critical analysis on how the defect density and formation behavior of ZnO nanostructures (ZNSs) are growth method-dependent. Both antithetical and facile methods are exploited to control the ZnO defect density and the growth mechanism. In this context, the growth of ZnO nanorods (ZNRs), nanoflowers, and nanotubes (ZNTs) are considered. The aforementioned growth methods directly stimulate the nanostructure crystal growth and, depending upon the defect density, ZNSs show different trends in structural, optical, etching, and conductive properties. The defect density of MAG ZNRs is the least because of an ample amount of thermal energy catered by high-power microwaves to the atoms to grow on appropriate crystallographic planes, which is not the case in faulty convective ACG ZNSs. Defect-centric etching of ZNRs into ZNTs is also probed and methodological constraints are proposed. ZNS optical properties are different in the visible region, which are quite peculiar, but outstanding for ZNRs. Hall effect measurements illustrate incongruent conductive trends in both samples.

Keywords: ZnO; defects; structural properties; convection; microwave; nanostructures; hydrothermal

1. Introduction

There is wide interest and research into oxide nanomaterials, including binary oxides, for example, ZnO, CuO, MgO, TiO_2, and SnO_2; ternary oxides, for example, $BaTiO_3$, $PbTiO_3$, $BiFeO_3$, and $KNbO_3$; and complex compounds, for example, $Ba_{1-x}Sr_xTiO_3$, $La_{0.325}Pr_{0.300}Ca_{0.375}MnO_3$, and $La_{0.5}Ca_{0.5}MnO_3$, due to their distinct geometries and cutting-edge physical and chemical properties [1]. ZnO has attracted significant attention because of its wide bandgap, and particular electrical, optical, structural, physical, chemical, piezo, and thermoelectric properties [2–5]. The occurrence of ZnO in polymorphic nanoscale shapes, such as nanorods (ZNRs), nanowires, nanoflowers (ZNFs), nanostars, nanoparticles, nanotubes (ZNTs), tetrapods, and polypods, further enhances its ambit under the wide canvas of myriad applications, such as field effect transistors, light-emitting diodes, ultraviolet lasers, photodetectors, thermo- and piezo-nanogenerators, solar cells, sensors, and so forth [6–12].

Optimized growth methods have occupied a foreground in the fabrication of ZnO nanostructures (ZNSs). The most eminent methods for ZNS synthesis are gas phase reaction, liquid phase deposition, metal organic chemical vapor deposition (MOCVD), vapor liquid solid, thermal evaporation, pulsed laser deposition, molecular beam epitaxy (MBE), thermal evaporation, and aqueous chemical growth (ACG) [13–17]. ACG, also called the hydrothermal method, is one of the most effective methods for

ZNS growth due to simple setup, low cost, and green chemistry aspects [18]. However, the dilemma envisaged with the process are long heating times and temperature gradients in the solution autoclave, which can affect the structural and crystalline properties of ZNSs.

Microwave-assisted growth (MAG) methods have been recently proposed to address these issues [19]. MAG exploits the benefits of ACG, while addressing heat transfer problems and shortening fabrication time. Heat is produced by absorbance of longer wavelength and lower energy electromagnetic microwaves within the material rather than via convection. The advantages associated with MAG method are homogeneous heating profiles, higher heating rates, shorter fabrication time, precise control of the reaction mixture, selective heating with different microwave power, higher yields, and energy savings. Some disadvantages are high equipment cost; short penetration depth, which limits large-scale growth; arduous in situ monitoring; and very high homogeneous nucleation rates, which results in growth stoppage. However, the overall advantages outweigh the disadvantages, and MAG is usually recommended over the ACG method.

In this study, we present how to control the ZNSs crystal defect formation using convective ACG and irradiation MAG methods. Depending upon the crystal growth phenomenon, an in-depth comparison of growth, material, structural, optical, and conduction properties of ZNSs is performed. The experimental parameters, such as chemical constituents, molar concentrations of the precursor solutions, solution pH, ultimate growth temperature, and aspect ratio of the grown ZNSs were synchronized to allow fair comparison between ZNS properties grown with ACG and MAG methods. It was found that the ZnO morphology and defect density could be controlled by judiciously opting for the ZNS growth method. Furthermore, the structural, optical, etching, and conductive properties had a direct relation with ZNS defect density. The results are critically analyzed and all the antithetical trends have been propounded.

2. Materials and Methods

2.1. Sample Preparation

Commercially available chemicals of analytical grade were purchased for the experiments. Aqueous solutions employed 18 MΩ de-ionized (DI) water. P-Si (100, 1–10 Ω·cm) was used as a test substrate for ZNS growth. The P-Si substrates were cut into 1 × 1 inch segments and exposed to buffered oxide etchant to remove the insulating SiO_2 layer at ambient conditions. After 2 min of immersion time, the substrates were cleaned with DI water and dried with N_2 gas.

2.2. Thin Film Seeds

A buffer layer for ZNS growth is essential because of the large lattice mismatch between ZnO and Si (2.19 Å) [20]. Hence, a buffer layer of ZnO seeds was used to catalyze the ZnO growth species. Seeds were fabricated by mixing 0.022 M zinc acetate dihydrate [$Zn(CH_3COO)_2{\cdot}2H_2O$] ($M_W$ 219.51 g/mol) in 10 mL n-propanol [C_3H_8O] (M_W 60.10 g/mol). The mixture was well stirred or sonicated for 30 min. The solution was ready to be deposited upon the substrate when the color changed from transparent to milky and back to transparent. The seed solution was then spin coated twice onto the substrate surface at 3000 RPM for 30 s. The spin coated samples were annealed at 110 °C for 2 min for the first coating, and at 300 °C for 60 min for the second coating to provide proper bondage between the seeds and substrate surface.

2.3. ZnO Nanorods Using ACG and MAG Methods

ZNRs were grown using ACG and MAG methods. Separate growth solutions were made for the two methods: 50 mM zinc nitride hexahydrate [$Zn(NO_3)_2{\cdot}6H_2O$] (M_W 297.48 g/mol) was mixed with methenamine [$C_6H_{12}N_4$] (M_W 140.186 g/mol) in DI water. Once the solutions were prepared, the seeded samples were immersed at the top of the solution to maximize heating effects within the solution. The solution bottles, with samples attached, were placed on a hot plate or in a 800 watt domestic microwave oven for the ACG and MAG methods, respectively. In ACG, a magnetic stirrer was placed at the base of the solution container, which was used to mobilize the reactants at the base, while the revolving disk in the microwave oven served the purpose in MAG method. The sensing probe of a wired thermometer was immersed in the solution autoclave to monitor the in situ temperature profile of the solutions. For synchronization, the ACG solution temperature was set above 100 °C to provide a perfect comparison between growth conditions and final products of both methods. After six hours, the ACG sample was removed from the solution, rinsed in DI water, and dried with N_2 gas. For the MAG method, we used the solution replacement method to address growth stoppage. The solution was replaced four times after each 5-min microwave exposure. The sample was then rinsed in DI water and dried with N_2.

2.4. ZnO Nanoflowers Using ACG and MAG Methods

ZNF growth was promoted using ammonium hydroxide [NH_4OH] (M_W 35.05 g/mol) as a pH buffer, with 15–20 mL dissolved in 50 mM zinc nitride hexahydrate and methenamine solution, then stirred for 1 h to produce a homogeneous solution. The seeded substrates were immersed in the solutions under the ACG and MAG conditions described above. To ensure fair comparison between methods, the temperature and pH of the solutions were synchronized. The samples were removed after 6 h and 10 min for the ACG and MAG methods, respectively, rinsed in DI water, and dried with N_2.

2.5. ZnO Nanotubes Using ACG and MAG Methods

Various methods have been proposed for ZNT fabrication [21–25]. However, to allow fair comparison between ACG and MAG methods, we opted for ZNT growth using potassium chloride [KCl] (74.55 g/mol). This is one of the most effective, simplest, and safest metamorphosis methods, where formed ZNRs are etched into ZNTs. ACG and MAG ZNR samples were immersed in a 3–5 M KCl etching solution at 95 °C for 6 h. The samples were then removed, cleaned with DI water, and dried with N_2.

2.6. Characterization Tools

To probe ZNS morphology, samples were imaged with scanning electron microscopy (SEM: Hitachi S-4800, Suwon, Gyeonggi-do, South Korea) operating at an emission energy of 25 KeV. Purity and crystalline quality were assessed via X-ray diffraction (XRD: Rigaku Ultima IV, Dongguk University, Seoul, South Korea) with Cu Kα radiation (λ = 0.15418 nm). The 2θ rage was taken from 20 to 50 degrees. Defect-centric optical properties were recorded with photoluminescence (PL: Accent RPM 2000, Suwon, Gyeonggi-do, South Korea) spectroscopy at scan rate of 15 pts/s with a laser excitation wavelength of 230–260 nm and power of 2.09 mW at room temperature and pressure (RTP). The PL range was set from 300 to 700 nm. The electrical characteristics and the carrier concentration were determined using Hall effect measurement system (ECOPIA AHT55T5, Dongguk University, Seoul, South Korea) at RTP in dark conditions. The in situ solution temperature and pH were monitored with a wired digital pH and thermometer. A Samsung domestic 850 watt microwave oven (Dongguk University, Seoul, South Korea) was used for MAG method. The photolithographic patterns were defined on the substrate with the help of Karl Suss MA 6 mask aligner (Dongguk University, Seoul, South Korea), and the metal contacts were made with E-beam metal evaporator (Dongguk University, Seoul, South Korea).

3. Results and Discussion

3.1. Growth Mechanism and Internal Chemistry

ZNS growth depends upon growth temperature to facilitate the chemical reactions,

$$C_6H_{12}N_4 + 6H_2O \overset{\text{Heat}}{\rightarrow} 4NH_3 + 6HCHO,$$

$$NH_3 + H_2O \overset{\text{Heat}}{\rightarrow} NH_4{}^+ + OH^-,$$

$$Zn(NO_3)_2{\cdot}6H_2O \overset{\text{Heat}}{\rightarrow} Zn^{2+} + 2NO_3{}^-,$$

and

$$Zn^{2+} + 2OH^- \overset{\text{Heat}}{\rightarrow} Zn(OH)_2 \overset{\text{Heat}}{\rightarrow} ZnO + H_2O,$$

where the reaction rates depend strongly on the heat transfer method [18,19].

Figure 1 demonstrates ZNS growth for ACG and MAG methods, with solution temperature shown by darker (hotter) and paler (cooler) red. The samples were studiously attached at the topmost possible position to scrutinize the distribution of heat inside the solution. Convective heating (ACG method, Figure 1a) produces a temperature gradient between the bottom and top of the solution. Black body radiation is convectively conducted into the solution, facilitating the reaction, and the reaction vessel only intercedes between energy transfer from the hotplate into the solution. The temperature gradient results in higher and lower reaction rates near the bottom and top surfaces, respectively, which leads to inhomogeneous ZNS growth. The stirrer at the base mobilizes the homogeneously nucleated large ZNRs at the bottom and the small ZNRs at the top surface, which are nucleated on the seeded substrate. A thin layer of ZnO seeds, with crystal orientation towards 0001, lowers the surface energy at the ZnO-Si interface and improves heterogeneous nucleation of the crystal nuclei on the seeded substrate rather than homogeneous nucleation in the solution [26]. The resulted orientation of the formed ZNRs is also towards the 0001 direction. Furthermore, temperature rise is also relatively slow, taking approximately 50 min for the solution to reach the required nucleation temperature. It is well established that ZnO crystal quality is growth temperature-dependent [27]. ZnO growth requires a particular amount of thermal energy for the derivation of chemical reactions and crystal growth. The prevalent temperature gradients in ACG provide bases for the difference in thermal energies provided to the reactants at different catalytic temperatures [28]. The said phenomenon affects the crystalline quality of the formed nanostructures.

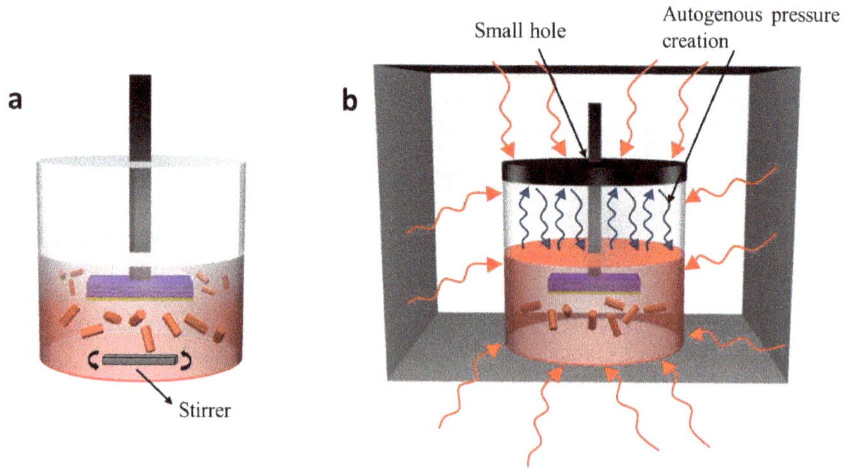

Figure 1. Heating profile for (**a**) aqueous chemical growth (ACG) and (**b**) microwave-assisted growth (MAG) methods.

We also believe that the said difference in the ACG ZNR dimensions is because of the formation, growth, and implosion of sono-chemical acoustic cavitations in the solution [29]. The stirrer at the base provides sound waves and shear kinetic energy to the reaction solution. The impinging sound waves, with a wavelength longer than the bond length of the reactants, do not have the ability to affect the formation energies of the reactants and cannot influence the chemical reaction directly [30]. However, the stirrer forms acoustic cavitations in the form of bubbles, which act as packets of high energy, temperature, and pressures inside the solution [31,32]. The revolving packets act as carriers of very high energy, which is converted into heat upon implosion and speeds up the chemical activity in the vicinity, while the reaction rate remains the same elsewhere. The phenomenon results in the formation and nucleation of disproportionate dimension ZNSs in the solution.

In contrast, the MAG heating profile is quite smooth and enticing, as shown in Figure 1b. The 2.85 GHz microwaves bombard the reaction flask, and hence the solution, from all sides. The solution is heated by ionic conduction of the dissolved chemicals and dipolar polarization of water molecules. Microwaves are relatively evenly distributed within the solution, and the solution experiences a steep and homogeneous temperature rise in approximately 2 min, which stimulates both nucleation and crystal growth [33]. It is noteworthy that the ionic conduction has more profound heat generating capacity than dipolar polarization, which has considerable implications on nanomaterial growth in ionic liquids. Furthermore, operational parameters, such as reaction temperature, irradiation power, and vessel pressure, must be precisely controlled to ensure a smooth microwave interaction with the material. Hence, a specially designed reaction container that facilitates autogenous pressure creation within the reaction chamber expedites the chemical reactions.

3.2. Defect Density and Conductive Properties of ZNRs Grown with ACG and MAG Methods

ZnO is naturally an n-type semiconductor, but the longstanding controversy to discern the unintentional n-type ZNS characteristics is yet a moot point in research and development. Non-stoichiometry during ZNS growth is cited as the main reason for this dilemma, and is the origin of the prevalent stalemate. Intrinsic defects within the ZnO crystal structure are pivotal for feasible theories of the origin of n-type characteristics. Oxygen vacancies (V_o) and zinc interstitials (Zn_i) are considered as potential donors in some literature, with hydrogen (H) also sometimes considered important [34–36]. Mollwo, Thomas, and Lander first proposed the theory of H donor in intrinsic ZnO, which was later substantiated by Van de Walle [37,38]. It is believed that H replaces V_o via four-fold coordination with neighboring atoms in ZnO crystal structure, and acts as a donor. Recently, many groups have countered the concept of H being an intrinsic donor impurity with experimentation and hypotheses [39]. Despite the enduring controversy, the V_o concept is central to all the proposed theories on the origin of unintentional n-type conductivity in intrinsic ZnO.

Table 1 shows the measured electrical characteristics; carrier concentration, resistivity, and mobility; of samples grown with ACG and MAG methods. The characterization was done with the Hall effect measurement system. The ZNRs were grown on an insulating glass substrate and ohmic indium contacts were fabricated on the four corners of the samples to provide optimal results. The In dots were fabricated directly on ZNRs by defining a four-spot window via shadow mask photolithography and deposition via e-beam metal evaporation. We did not use any top layer for the deposition of metal contacts and the contacts were made directly upon the ZNRs. The four probes were attached on the contacts with alternating rotations for optimization. Both samples show n-type conductivity, but carrier concentration in ACG samples is very high compared to MAG samples because of the plethora of donor defects. The high-power microwave irradiations provide sufficient thermal energy to the atoms to nucleate on the proper crystal lattice points, reducing the probability of V_o production. Although, the crystalline quality of MAG samples is superior, a steep descent in carrier concentration is the reason for high resistivity. On the contrary, the inefficient convective heating profile of ACG provides a platform for the production of V_o in the crystal lattice and hence enhanced n-type conductivity. The following equations were used to govern the relation between Hall voltage, carrier concentration, and mobility of the samples. The relation for Hall voltage (V_H) is:

$$V_H = \frac{B_z I_x}{ned} \quad (1)$$

Hence, the majority carrier concentration (n) would be:

$$n = \frac{B_z I_x}{V_H e d} \quad (2)$$

where B_z is the magnetic field towards z-direction, I_x is drift current, d is film thickness, and e is electronic charge magnitude. The relation between majority carrier concentration (n) and mobility (μ_n) of the samples was governed by:

$$\mu_n = \frac{L I_x}{e n V_x W d}, \quad (3)$$

where L is the length and W is the width of the sample under test, and V_x is the drift voltage. It is evident from (3) that the mobility of the samples is inversely proportional to the majority carrier concentration, which is the primary reason of very high mobility in MAG ZNRs as compared to ACG ZNRs, as seen in Table 1. Hence, it is proved that ZnO carrier concentration is growth method-dependent and can also be controlled by an optimal use of a growth method rather than doping. The conductive properties of both samples are in accordance with XRD and PL data. Furthermore, the defect density influenced the ZnO structural, optical, and etching behaviors, as shown in the next.

Table 1. Electrical and conductive properties of samples grown with ACG and MAG methods.

Growth Method	Carrier Concentration (cm^{-3})	Resistivity (Ω·cm)	Mobility (cm^2/V-s)
ACG	1.8×10^{17}	31.8	1.05
MAG	1.08×10^{14}	1010	56.8

3.3. Methodological Constraints for Nanorod Growth Via ACG and MAG Methods

Figure 2 shows SEM images of ZNRs grown with ACG and MAG methods. Figure 2a,b shows top and the cross-sectional SEM images of the buffer layer of ZnO seeds, respectively. Seeds are necessary to provide the nucleation energy required for ZNR growth, lower interfacial energy at the ZnO substrate interface, and minimize the large lattice mismatch and ultimately the stress and the strain between both materials. The 0.022 M zinc acetate seeds were spin coated twice upon the surface to provide a 15-nm seed layer. Double seed coating was chosen to reduce fabrication time.

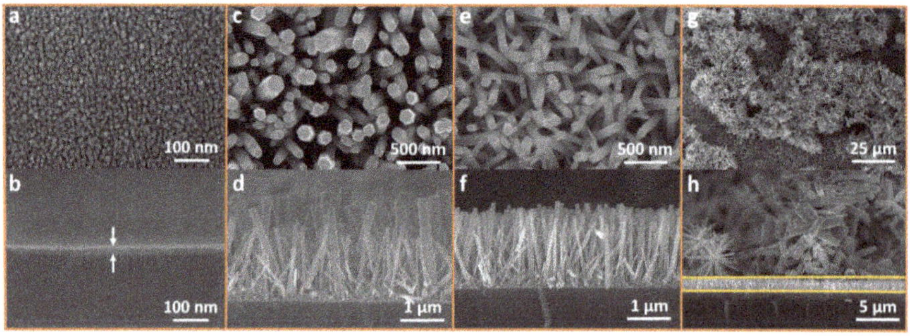

Figure 2. Scanning electron microscope (SEM) (top and side views, respectively): (**a**,**b**) Seed; (**c**,**d**) ZnO nanorods (ZNRs) fabricated using the ACG method; (**e**,**f**) ZNRs fabricated using the MAG method; (**g**,**h**) ZnO debris on the surface of vertical ZNRs.

Figure 2c,d shows top and cross-sectional images, respectively, of ZNRs grown with the ACG method at temperatures above 100 °C. The resultant ZNRs are hexagonal and densely populated. However, ZNR dimensions are eccentric and inhomogeneous, due to inept convective heating process and stirring. In contrast, ZNRs grown by the MAG method are also densely populated, but relatively homogeneous and immutable, as shown in Figure 2e,f. To allow fair comparison, the MAG ZNR lengths were synchronized with ACG ZNR lengths to approximately 2 μm. In this regard, the fabrication of ZNRs via a facile MAG solution-replacement method is an ideal method to adopt because it is easy to control the length of ZNRs within the solution [19]. Using 25 mM solution, ZNR length increased 250 nm per solution-replacement cycle on average, which doubled to 500 nm per solution-replacement cycle for a molar concentration 50 mM. Hence, replacing the solution four times produced 2 μm ZNRs. Above all, the ZNRs were grown in an amazingly short fabrication time of 20 min, in contrast to 6 h using the ACG method.

Another hitch associated with the ACG method is deposition of debris ZNSs on the surface of the active ZNR layer. Some ZNRs were homogeneously nucleated inside the solution, mobilized by the stirrer, and deposited on the substrate surface. Longer ACG fabrication time exacerbates debris deposition. As shown in Figure 2g, the debris is quite dense and the active area is covered with debris. The cross-sectional view of Figure 2h shows the havoc played by the debris, where the debris layer thickness is quite high compared to the 2-μm vertical ZNRs enclosed in the yellow strip. The debris could be proved devastating and offers an utter mess to measure the material, optical, and electrical properties of vertically aligned ZNRs, and requires multiple measurements to acquire meaningful outcomes for useful ZNRs. Because of very short fabrication time, such debris was not deposited in ZNRs grown with MAG method, as shown in the background of Figure 2f. A short-term solution to address debris is to clean the sample with de-ionized (DI) water immediately after removing it from the growth solution. This removes much of the debris from the surface, since it is poorly attached compared to the desired vertical ZNRs.

3.4. Methodological Constraints for Nanoflower Growth via ACG and MAG Methods

ZNFs were fabricated via NH_4OH treatment, one of the most effective growth methods [19]. The name of the structure reflects the geometrical similarity to a flower. Most research groups use NaOH for ZNF growth, but we find NH_4OH treatment more beneficial and efficient. The detailed growth mechanism is portrayed in Figure 3. Before the addition of NH_4OH, methanemine is decomposed to provide OH^- ions, which then react to form $Zn(OH)_4{}^{2-}$ growth units and $Zn(OH)_2$ nuclei. However, at low pH, the nuclei are more than the growth units, which halts ZNF growth, and only ZNRs are formed. With the addition of NH_4OH as a mineralizer and pH buffer, OH^- ion population is significantly increased in the solution. The initial solution pH was approximately 6.8, and increased to as much as 12 after NH_4OH addition. Zn^{2+} and OH^- ions react to form ZnO nuclei and growth units, as shown in Figure 3a,b. When the pH was elevated to 12, the growth units were more than ZnO nuclei and aligned via electrostatic attractive forces between the ZNF nuclei, as shown in Figure 3c. The growth units and nuclei start amalgamating through Ostwald ripening, consuming smaller crystals into larger crystals. Due to very low surface energy under the influence of homogeneous nucleation, a number of ZnO nuclei amalgamate to form ZnO crystallites, as shown in Figure 3d. All the active sites circumambulate at the outer edges of the crystallite to provide a base for petals, and are ready to be grown into ZNFs at high temperatures. The process flow for both ACG and MAG methods remain the same until this point.

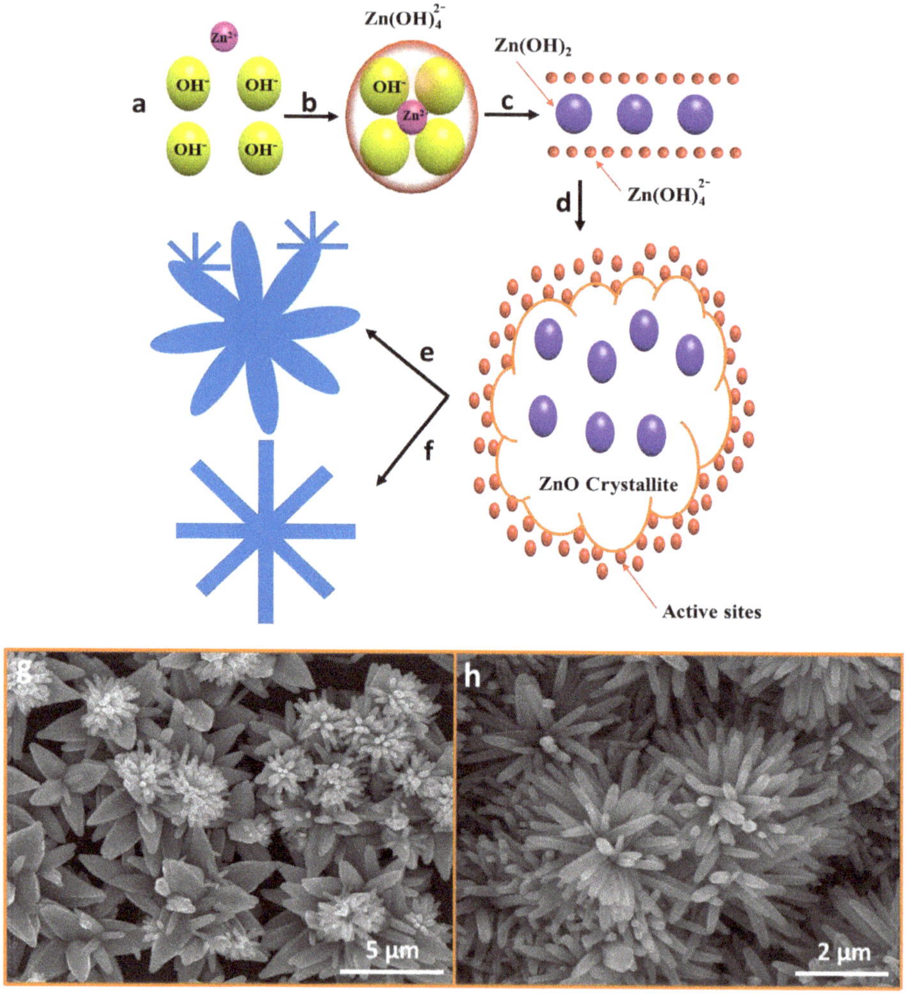

Figure 3. (**a**–**f**) Process flow for ZnO nanoflowers (ZNFs) growth using the ACG and MAG methods, (**g**) SEM image of ZNF using the ACG method, and (**h**) SEM image of ZNF using the MAG method.

Figure 3e,g shows the schematic for a single ZNF and SEM image, respectively, grown with the ACG method. As discussed above, the homogeneous nucleation rate is higher and lower near bottom and top surfaces of the reaction flask, respectively, which leads to inhomogeneous ZNF growth. Thus, the SEM image shows that ZNFs are quite large, but small ZNFs also start nucleating at the petals and at different centers of the already grown ZNFs. Relatively large ZNFs are formed at the bottom of the flask, which are mobilized by the stirrer and adsorbed at the substrate surface at the top edge of the solution flask. Simultaneously, small ZNFs are homogeneously nucleated and adsorbed upon the petals of already formed ZNFs, which produces inhomogeneous ZNF growth for the ACG method. Another point to ponder is the difference in the diameter of the top (0001) and bottom $(000\bar{1})$ surfaces of the ZNF petals. The underlying reason for this phenomena is the difference in growth rate of various ZNF planes: $V(0001) > V(\bar{1}0\bar{1}\bar{1}) > V(\bar{1}010) > V(\bar{1}011) > V(000\bar{1})$ [40]. The bottom $000\bar{1}$ surface, with the lowest growth rate, becomes stable, and the top 0001, with the highest growth rate, erodes at the growth stoppage point because of the presence of the sample inside the solution for such a long time at high temperatures in the ACG method. Hence, pointed-tip ZNFs are formed (see Figure 3e,g).

In contrast, ZNF growth via the MAG method is relatively homogeneous and quick, as shown in the schematic and SEM image of Figure 3f,h, respectively. All the flower petals have a similar length and diameter, and the ZNFs are quite dense and homogeneously distributed across the substrate surface. The homogeneous growth is because of MAG efficient heat transfer. Erosion at the top 0001 surface is not evident, because of the very short fabrication time: the samples were ready to be removed from the solution within just 10 min for the MAG method. It is evident that the erosion would be seen at the top 0001 surface in MAG if the sample is left under MAG growth conditions beyond 10 min. However, that is not required because the precursor solution has already been saturated by MAG in 10 min.

3.5. Defect-Centric Etching of Nanorods into Nanotubes

ZNTs have paramount importance because of their distinct chemical properties, hollow structure, high surface-to-volume ratio, high anisotropy, and current carrying capability. However, ZNT growth is quite challenging. We propose ZNT growth via defect-centric etching of ZNRs grown with the ACG and MAG methods. Defect-centric etching is the only known feasible method for ZNT formation. Hence, it is important to compare the practical application of this method on ZNRs grown with the ACG and MAG methods [41].

Figure 4a shows the SEM image of ZNTs formed via defect-centric etching of ZNRs grown with the ACG method in a KCl solution. The inset shows a corresponding higher magnification portion. Almost all the ZNRs are etched at the core towards the bottom, and a hollow tubular structure is formed. The tube walls are quite thick and the hole is formed only at the core. The tube shape remains hexagonal after etching, which means the etching was only performed in the unstable ZNR core. Hence, it is inferred that the formation of ZNTs from ZNRs grown with the ACG method is feasible because of the polycrystalline structure and the profusion of defects created within the ZNR crystal structure during the growth process. The absence of etching for a few of the inhomogeneous, smaller-diameter ZNRs are highlighted in the inset of Figure 4a. Thus, etching does not only depend upon defect density, but also diameter and area available for etching. Etching appears to be proportional to defect density and ZNR diameter. Trimming of such a small diameter ZNRs into ZNTs, which is quite arduous, has not been propounded previously, and is successfully performed here for the first time.

Figure 4b shows the SEM for MAG method ZNRs etched into ZNTs in a KCl solution with the corresponding high magnification image in the inset. None or only very few ZNRs are etched, but the etching stops at the surface. Furthermore, in the case of partial etching, it is found that the core is etched with the most stable ZNR lateral walls. The absence of defect-centric etching is because of the pure crystalline structure of ZNRs grown with the MAG method. Because of the immaculate nature of the MAG method, only a few defects are incorporated in the crystalline structure, which makes it difficult for the etchant to trim the formed ZNRs into hollow ZNTs. Furthermore, the few existing defects tend to be at the ZNR surface, which results in a skimpy etching profile.

Figure 4c further clarifies how the Cl^- ions are adsorbed on the polar surface of the formed ZNRs and etch the defect-rich ACG ZNRs from core to base. A hollow surface towards the bottom is seen with quite thick and stable lateral walls. This also shows how MAG ZNR was either not etched or etched only at the surface with the dissolution of most stable lateral walls. These visualizations are in accordance with X-ray diffraction (XRD), photoluminescence (PL), and Hall effect measurements of the samples.

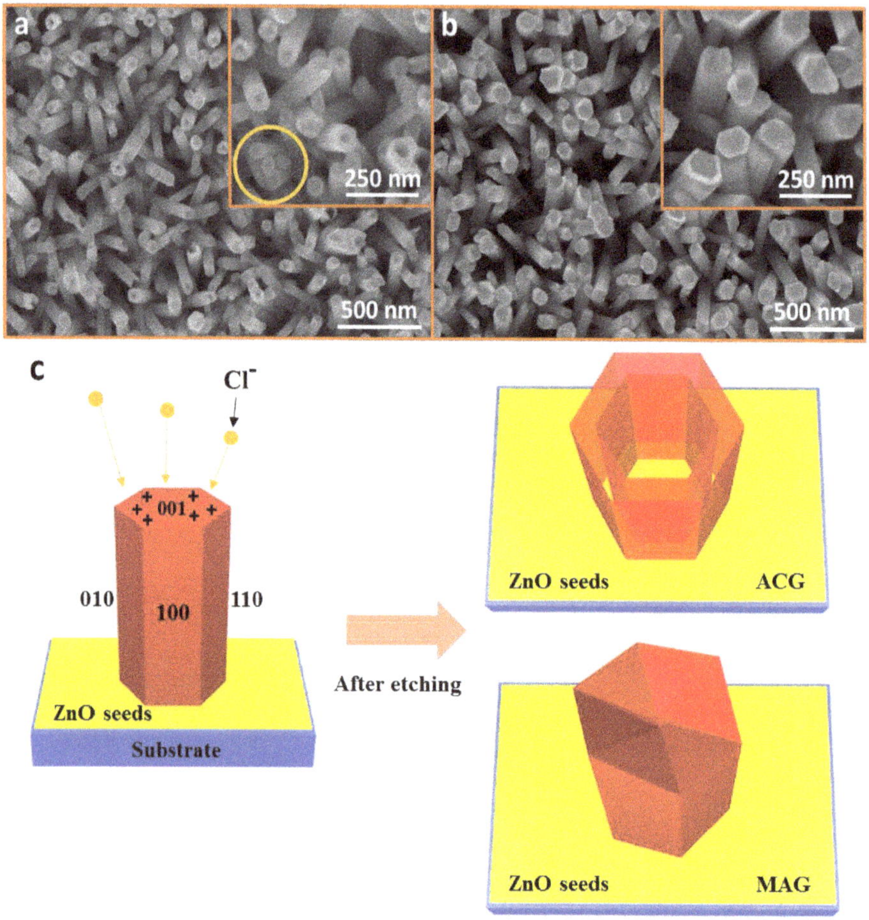

Figure 4. SEM images of ZnO nanotubes (ZNTs) (inset: magnified image), formed via defect-centric etching of ZNRs grown with (**a**) ACG and (**b**) MAG methods; (**c**) Etching mechanism for ACG and MAG ZNRs.

3.6. XRD for ZNRs Grown with ACG and MAG Methods

The structural and crystalline properties of ZNRs grown with ACG and MAG methods were investigated using XRD, as shown in Figure 5. The multiple peaks across 100, 002, 101, and 102 in Figure 5a and a single peak across 002 in Figure 5b correspond to the hexagonal phases in ZNRs, confirming the wurtzite nature of both samples. The incorporation of multiple peaks in ACG ZNRs imply the deterioration of ZnO 002 texture. However, the highest peak in both samples were across 002, which confirms ZNR vertical alignment along the *c*-axis perpendicular to the substrate, in accordance with the SEM data. The multiple peaks and a single peak also show that ACG and MAG ZNRs are poly and single crystalline, respectively.

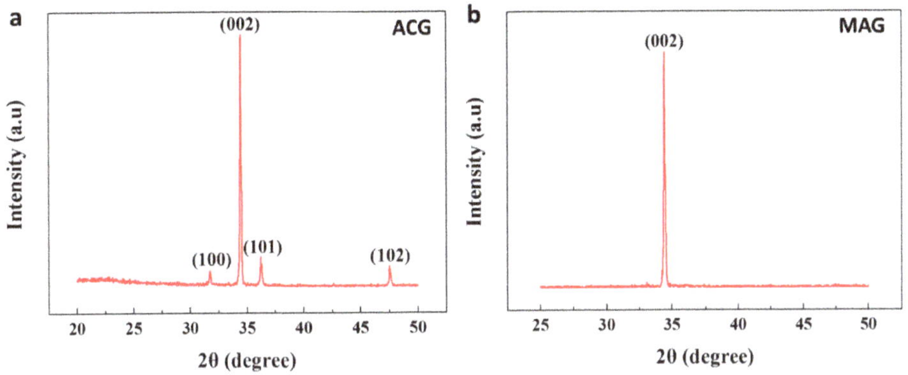

Figure 5. ZNR X-ray diffraction (XRD) profiles: (**a**) the ACG and (**b**) the MAG method.

Stress and the strain in the crystal structures depend upon multiple factors, such as the lattice mismatch, 2θ position of the 002 peak, and the lattice constant (C_o) of the grown crystals. Extrinsic stress/strain was not considered because both samples were grown on p-Si with a ZnO buffer layer, but an in-depth structural analysis has been performed for intrinsic stress/strain levels depending upon the growth method used. The lattice constants, a and c, of ZnO were calculated via Braggs law:

$$2dsin\theta = n\lambda \tag{4}$$

where d is the spacing between the lattice planes of Miller indices, n is order of diffraction which is normally taken as 1, λ is X-ray wavelength (1.54 Å), and θ implies Bragg's angle. The lattice constant for (100) plane is calculated by the relation:

$$a = \frac{\lambda}{2sin\theta}\sqrt{\frac{4}{3}(h^2 + hk + k^2) + \left(\frac{a}{c}\right)^2 l^2} \tag{5}$$

$$a = \frac{\lambda}{\sqrt{3}sin\theta} \tag{6}$$

where θ is the diffraction angle responding 100 peak. The lattice constant c for 002 plane is calculated by the relation:

$$c = \frac{\lambda}{2sin\theta}\sqrt{\frac{4}{3}\left(\frac{c}{a}\right)^2 (h^2 + hk + k^2) + l^2} \tag{7}$$

$$c = \frac{\lambda}{sin\theta} \tag{8}$$

where θ corresponds to 002 peak. The 2θ value of the 002 peak and C_o for stress-free bulk ZnO are 34.42 and 0.5205 nm, respectively [42]. The 2θ of samples grown with the ACG and MAG methods were 34.31 and 34.43, respectively. The 0.1 degree left-shift in ACG samples was caused by stress [43]. The 2θ value of MAG samples are more near to the bulk ZnO, which is the direct result of intrinsic stress/strain relaxation during MAG process. The process minimizes the surface energy of the film and adds value to the crystallinity of the sample. Furthermore, the lattice constant C_{ACG} is 0.5238 and C_{MAG} is 0.5201 nm. ACG samples show tensile stress ($C_{ACG} > C_o$), whereas the MAG samples show compressive stress ($C_{MAG} < C_o$), so the strain along the c-axis, $[(C - C_o)/C_o] \times 100$, was 63% (compressive) and 7% (tensile) for ACG and MAG samples, respectively. Thus, the stress and strain levels in MAG samples are almost negligible, which validates the methodological efficacy of the MAG method.

The full width at half maximum (FWHM) of the 002 peak and grain size (D) are indicative of the crystalline quality of the ZNRs. The Scherrer formula was used to calculate D [44,45],

$$D = (0.89\lambda)/(B\ cos\theta), \tag{9}$$

where λ is the X-ray wavelength (0.15418 nm), B is the FWHM in radians, and θ is the diffraction angle. Using the FWHM shown in Figure 5a,b, D = 27 nm and 68 nm, for ACG and MAG samples, respectively. D is indicative of crystal quality and yield stress in the structure. Larger D implies that less driving force is required to move a dislocation pileup near the edges from one crystallite to another. MAG process stimulates the grain boundary migration, which results in facile coalescence of crystallites and ponder large grain growth. In short, unlike ACG, MAG provides sufficient energy to the atoms to expeditiously occupy the legitimate crystal sites on the crystal lattice. Hence, material with smaller grains (ACG sample) exhibits higher yield stress than material with larger grains (MAG sample). Yield stress can also be governed by the Hall-Petch equation,

$$\sigma y = \sigma_0 + \frac{ky}{\sqrt{d}}, \tag{10}$$

where σ_y is the yield stress; σ_o is the material constant; k_y is the strength coefficient specific to each material; and d is the average grain diameter. Grain diameter is inversely proportional to the yield stress, which implies that ACG samples will have more yield stress than MAG samples. Hence, it is proved that stress/strain in the films are growth process-dependent rather than of thermal origin. Moreover, the crystalline quality of MAG samples is comparable to samples grown with expensive and sophisticated equipment, such as MOCVD and MBE [46,47]. The bond length (L) of ZnO nanostructures grown with ACG and MAG methods were calculated via:

$$L = \sqrt{\frac{a^2}{3} + \left(\frac{1}{2} - u\right)^2 c^2} \tag{11}$$

where u parameter, which depends upon a/c ratio in wurtzite structures, can be calculated as:

$$u = \frac{a^2}{3c^2} + 0.25 \tag{12}$$

The calculated values of L, along with other structural parameters, are given in Table 2.

Table 2. Structural parameters of samples grown with ACG and MAG growth methods. Full width at half maximum (FWHM); grain size (D); bond length (L).

Growth Method	c (Å)	2θ (Degree)	FWHM (Degree)	D (nm)	L (Å)	Strain (%)
ACG	5.238	34.38	0.311	26.86	1.9751	63
MAG	5.201	34.43	0.122	68	1.9769	07

3.7. *ZNR Optical Properties from ACG and MAG Methods*

ZNR absorption and emission spectroscopy are important research and development parameters due to the unique optical properties of ZnO. The optical properties were measured with PL Accent RPM 2000 at room temperature. The nanostructure luminescent properties were strongly dependent upon the growth method used and the nanostructure crystal growth. Figure 6a shows the PL response of ACG ZNRs, displaying an orthodox high-intensity UV peak at 377 nm (3.27 eV) and a broad peak in the visible range. The high-intensity UV peak, with FWHM ~16.5 nm, is caused by free exciton recombination, and the broad visible emission is due to defects formed within the crystal during fabrication [48]. Various contradictory theories have been proposed, including V_o, oxygen interstitials, zinc vacancies, and zinc interstitials Zn_i, for the luminescent recombination centers. However, the presence of V_o is the most pertinent theory for the broad luminescence band in the visible region. The most plausible charge states for V_o in the crystal are neutral ($V^{\circ}{}_o$), singly ionized ($V^{+}{}_o$), and doubly ionized ($V^{2+}{}_o$) [49]. Singly ionized vacancies ($V^{+}{}_o$) are the most susceptible to act as a green emission and electron recombination centers [50]. A peculiar large and sharp kink at 528 nm was also observed in the visible region.

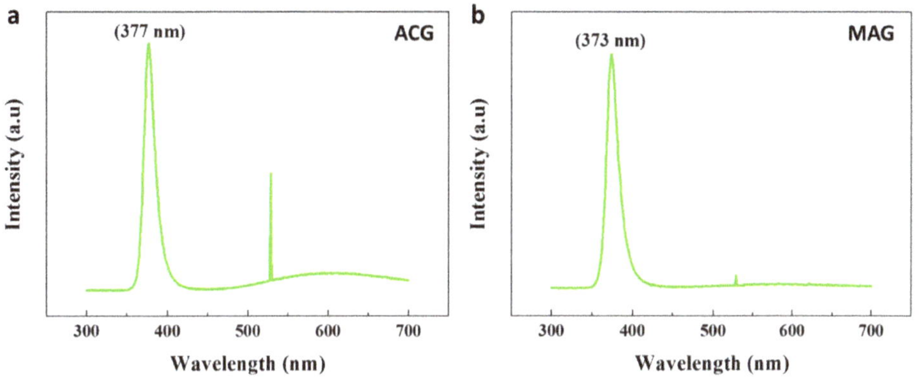

Figure 6. ZNR Photoluminescence spectra: (**a**) the ACG and (**b**) the MAG method.

Similarly, Figure 6b shows PL spectra for MAG ZNRs, which is quite distinct from ACG ZNRs. The UV peak was 373 nm (3.30 eV) with high-intensity and low FWHM (~15 nm), displaying significantly better optical quality compared to ACG ZNRs. Another important criterion to judge optical performance is the ultraviolet-to-visible emission ratio, and MAG ZNRs have ~3 times the value of ACG ZNRs, due to their higher intensity UV peak and very low peak intensity in the visible region [28]. MAG ZNRs also show a flat band in the visible region because of the absence of defect-centric recombination centers in the crystal structure. The authenticity of the results could be checked by matching the structural and optical performance of the same ZNRs [51]. It is found that the optical performance of the samples, such as the defect density, crystalline quality, and optical structure of ZNRs, is in accordance with the findings of the structural properties of the same ZNRs in Figure 5.

Another point to ponder in the PL data is the deviation of ACG ZNR UV peak from 373 nm in MAG ZNRs to 377 nm. We believe that the observed red shift and the reduction in near band edge emission in ACG ZNRs are because of band gap renormalization (BGR) effect. It has already been established that the free electron density of ACG ZNRs is far superior to MAG ZNRs (Table 1). This high electron density results in BGR via many body effect and free carrier screening inside the structure and is the primary reason for red shift in the sample [52]. Furthermore, slightly large diameter and high lattice stress in ACG ZNRs could be secondary reasons for this red shift [53]. Additionally, the sharp peaks at 528–530 nm in both ACG and MAG ZNRs are quite unusual for ZnO. Previously, we believed that the sharp peaks were because of the presence of confined defects (V_o) in ZnO lattice. However, the sharp peaks are actually the PL laser response imbibed during the data acquisition with PL: Accent RPM 2000.

4. Conclusions

We present method-dependent and crystallization-oriented growth, material, structural, optical, and conductive properties of ZNSs grown using the classic ACG and emerging MAG methods. The effects of convective (ACG) and radiative (MAG) processes were discussed in the context of ZNS crystal growth. The two methods could be exploited to control the ZnO defect density. The ACG ZNRs and ZNFs showed inhomogeneous growth because of convective temperature gradients, and sono-chemical acoustic cavitations in the solution and MAG ZNRs led to homogeneous growth trends because of immutable irradiative heating. ACG and MAG ZNSs were poly and single crystalline, respectively, which provided further verification of the optical and electrical properties. Conversion of MAG ZNRs to ZNTs was quite difficult because of their defect-free structure. Furthermore, MAG ZNRs showed superior optical profile and flat PL response in the visible region. Crystalline and growth properties provided the explanation for high and low n-type intrinsic conductivity in ACG and MAG samples, respectively. Further research on growth and methodological constraints of ACG and MAG methods is required to address basic problems for ZNS growth, and illuminate fresh opportunities for application-oriented experimental and theoretical studies on ZNS growth via ACG and MAG methods.

Acknowledgments: This work was supported by the Korea Institute of Energy Technology Evaluation and Planning (KETEP) and the Ministry of Trade, Industry & Energy (MOTIE) of the Republic of Korea (No. 20174030201520) and the Basic Science Research Program through the National Research Foundation of Korea (NRF) funded by the Ministry of Education (No. 2017R1D1A1A09000823).

Author Contributions: A.u.H.S.R. fabricated the nanostructures, designed and conducted all the experiments, and analyzed the data. J.Y.L. and A.S. helped in experimentation and data collection. H.-S.K. planned and supervised the project. All authors contributed to discussing the results and writing the manuscript.

Conflicts of Interest: The authors declare no competing financial interests.

References

1. Patzke, G.R.; Zhou, Y.; Kontic, R.; Conrad, F. Oxide Nanomaterials: Synthetic Developments, Mechanistic Studies, and Technological Innovations. *Angew. Chem. Int. Ed.* **2011**, *50*, 826–859. [CrossRef] [PubMed]
2. Bagnall, D.M.; Chen, Y.F.; Zhu, Z.; Yao, T.; Koyama, S.; Shen, M.Y.; Goto, T. Optically Pumped Lasing of ZnO at Room Temperature. *Appl. Phys. Lett.* **1997**, *70*, 2230–2232. [CrossRef]
3. Keis, K.; Vayssieres, L.; Rensmo, H.; Lindquist, S.E.; Hagfeldt, A. Photoelectrochemical Properties of Nano-to Microstructured ZnO Electrodes. *J. Electrochem. Soc.* **2001**, *148*, A149–A155. [CrossRef]
4. Amin, M.; Manzoor, U.; Islam, M.; Bhatti, A.S.; Shah, N.A. Synthesis of ZnO nanostructures for low temperature CO and UV sensing. *Sensors* **2012**, *12*, 13842–13851. [CrossRef] [PubMed]
5. Kong, X.Y.; Wang, Z.L. Spontaneous Polarization-Induced Nanohelixes, Nanosprings, and Nanorings of Piezoelectric Nanobelts. *Nano Lett.* **2003**, *3*, 1625–1631. [CrossRef]
6. Hjiri, M.; El Mir, L.; Leonardi, S.G.; Donato, N.; Neri, G. CO and NO_2 selective monitoring by ZnO-based sensors. *Nanomaterials* **2013**, *3*, 357–369. [CrossRef] [PubMed]
7. Konenkamp, R.; Word, R.C.; Schlegel, C. Vertical Nanowire Light-Emitting Diode. *Appl. Phys. Lett.* **2004**, *85*, 6004–6006. [CrossRef]
8. Goldberger, J.; Sirbuly, D.J.; Law, M.; Yang, P. ZnO Nanowire Transistors. *J. Phys. Chem. B* **2005**, *109*, 9–14. [CrossRef] [PubMed]
9. Huang, M.H.; Mao, S.; Feick, H.; Yan, H.; Wu, Y.; Kind, H.; Weber, E.; Russo, R.; Yang, P. Room-z Temperature Ultraviolet Nanowire Nanolasers. *Science* **2001**, *292*, 1897–1899. [CrossRef] [PubMed]
10. Suehiro, J.; Nakagawa, N.; Hidaka, S.I.; Ueda, M.; Imasaka, K.; Higashihata, M.; Okada, T.; Hara, M. Dielectrophoretic Fabrication and Characterization of a ZnO Nanowire-Based UV Photosensor. *Nanotechnology* **2006**, *17*, 2567. [CrossRef] [PubMed]
11. Zhang, F.; Chao, D.; Cui, H.; Zhang, W.; Zhang, W. Electronic structure and magnetism of Mn-Doped ZnO nanowires. *Nanomaterials* **2015**, *5*, 885–894. [CrossRef] [PubMed]
12. Law, M.; Greene, L.E.; Johnson, J.C.; Saykally, R.; Yang, P. Nanowire Dye-Sensitized Solar Cells. *Nat. Mater.* **2005**, *4*, 455–459. [CrossRef] [PubMed]
13. Al-Salman, H.S.; Abdullah, M.J. RF sputtering enhanced the morphology and photoluminescence of multi-oriented ZnO nanostructure produced by chemical vapor deposition. *J. Alloys Compd.* **2013**, *547*, 132–137. [CrossRef]
14. Yang, P.; Yan, H.; Mao, S.; Russo, R.; Johnson, J.; Saykally, R.; Morris, N.; Pham, J.; He, R.; Choi, H.J. Controlled Growth of ZnO Nanowires and Their Optical Properties. *Adv. Funct. Mater.* **2002**, *12*, 323. [CrossRef]

15. Park, W.I.; Yi, G.C.; Kim, M.; Pennycook, S.J. ZnO Nanoneedles Grown Vertically on Si Substrates by Non-Catalytic Vapor-Phase Epitaxy. *Adv. Mater.* **2002**, *14*, 1841–1843. [CrossRef]
16. Sun, Y.; Fuge, G.M.; Ashfold, M.N. Growth of Aligned ZnO Nanorod Arrays by Catalyst-Free Pulsed Laser Deposition Methods. *Chem. Phys. Lett.* **2004**, *396*, 21–26. [CrossRef]
17. Yao, B.D.; Chan, Y.F.; Wang, N. Formation of ZnO Nanostructures by a Simple Way of Thermal Evaporation. *Appl. Phys. Lett.* **2002**, *81*, 757–759. [CrossRef]
18. Rana, A.S.; Ko, K.; Hong, S.; Kang, M.; Kim, H.S. Fabrication and Characterization of ZnO Nanorods on Multiple Substrates. *J. Nanosci. Nanotechnol.* **2015**, *15*, 8375–8380. [CrossRef] [PubMed]
19. Rana, A.S.; Kang, M.; Kim, H.S. Microwave-Assisted Facile and Ultrafast Growth of ZnO Nanostructures and Proposition of Alternative Microwave-Assisted Methods to Address Growth Stoppage. *Sci. Rep.* **2016**, *6*, 24870. [CrossRef] [PubMed]
20. Tao, Y.; Fu, M.; Zhao, A.; He, D.; Wang, Y. The effect of seed layer on morphology of ZnO nanorod arrays grown by hydrothermal method. *J. Alloys Compd.* **2010**, *489*, 99–102. [CrossRef]
21. Sun, Y.; Fuge, G.M.; Fox, N.A.; Riley, D.J.; Ashfold, M.N. Synthesis of Aligned Arrays of Ultrathin ZnO Nanotubes on a Si Wafer Coated with a Thin ZnO Film. *Adv. Mater.* **2005**, *17*, 2477–2481. [CrossRef]
22. Yu, H.; Zhang, Z.; Han, M.; Hao, X.; Zhu, F.A. General Low-Temperature Route for Large-Scale Fabrication of Highly Oriented ZnO Nanorod/Nanotube Arrays. *J. Am. Chem. Soc.* **2005**, *127*, 2378–2379. [CrossRef] [PubMed]
23. Cheng, C.L.; Lin, J.S.; Chen, Y.F. A simple approach for the growth of highly ordered ZnO nanotube arrays. *J. Alloys Compd.* **2009**, *476*, 903–907. [CrossRef]
24. Tang, Y.; Luo, L.; Chen, Z.; Jiang, Y.; Li, B.; Jia, Z.; Xu, L. Electrodeposition of ZnO Nanotube Arrays on TCO Glass Substrates. *Electrochem. Commun.* **2007**, *9*, 289–292. [CrossRef]
25. Xu, L.; Liao, Q.; Zhang, J.; Ai, X.; Xu, D. Single-Rystalline ZnO Nanotube Arrays on Conductive Glass Substrates by Selective Dissolution of Electrodeposited ZnO Nanorods. *J. Phys. Chem. C* **2007**, *111*, 4549–4552. [CrossRef]
26. Vayssieres, L.; Keis, K.; Lindquist, S.E.; Hagfeldt, A. Purpose-built anisotropic metal oxide material: 3D highly oriented microrod array of ZnO. *J. Phys. Chem. B* **2001**, *105*, 3350–3352. [CrossRef]
27. Cheng, Q.; Ostrikov, K.K. Temperature-dependent growth mechanisms of low-dimensional ZnO nanostructures. *CrystEngComm* **2011**, *13*, 3455–3461. [CrossRef]
28. Zhao, J.; Hu, L.; Wang, Z.; Wang, Z.; Zhang, H.; Zhao, Y.; Liang, X. Epitaxial growth of ZnO thin films on Si substrates by PLD technique. *J. Cryst. Growth* **2005**, *280*, 455–461. [CrossRef]
29. Suslick, K.S.; Flannigan, D.J. Inside a Collapsing Bubble, Sonoluminescence and Conditions during Cavitation. *Annu. Rev. Phys. Chem.* **2008**, *59*, 659–683. [CrossRef] [PubMed]
30. Suslick, K.S. Sonochemistry. *Science* **1990**, *247*, 1439–1445. [CrossRef] [PubMed]
31. Suslick, K.S.; Hammerton, D.A.; Cline, R.E. Sonochemical hot spot. *J. Am. Chem. Soc.* **1986**, *108*, 5641–5642. [CrossRef]
32. Flint, E.B.; Suslick, K.S. The temperature of cavitation. *Science* **1991**, *253*, 1397. [CrossRef] [PubMed]
33. Jhung, S.H.; Jin, T.; Hwang, Y.K.; Chang, J.S. Microwave Effect in the Fast Synthesis of Microporous Materials: Which Stage between Nucleation and Crystal Growth is Accelerated by Microwave Irradiation? *Chem. Eur. J.* **2007**, *13*, 4410–4417. [CrossRef] [PubMed]
34. Lany, S.; Zunger, A. Anion Vacancies as a Source of Persistent Photoconductivity in II-VI and Chalcopyrite Semiconductors. *Phys. Rev. B* **2005**, *72*, 035215. [CrossRef]
35. Look, D.C.; Hemsky, J.W.; Sizelove, J.R. Residual Native Shallow Donor in ZnO. *Phys. Rev. Lett.* **1999**, *82*, 2552. [CrossRef]
36. Janotti, A.; Van de Walle, C.G. Hydrogen Multicentre Bonds. *Nat. Mater.* **2007**, *6*, 44–47. [CrossRef] [PubMed]
37. Thomas, D.G.; Lander, J.J. Hydrogen as a Donor in Zinc Oxide. *J. Chem. Phys.* **1956**, *25*, 1136–1142. [CrossRef]
38. Van de Walle, C.G. Hydrogen as a Cause of Doping in Zinc Oxide. *Phys. Rev. Lett.* **2000**, *85*, 1012. [CrossRef] [PubMed]
39. Sun, F. On the Origin of Intrinsic Donors in ZnO. *Appl. Surf. Sci.* **2010**, *256*, 3390–3393. [CrossRef]
40. Zhang, Z.; Mu, J. Hydrothermal synthesis of ZnO nanobundles controlled by PEO–PPO–PEO block copolymers. *J. Colloid Interface Sci.* **2007**, *307*, 79–82. [CrossRef] [PubMed]

41. Israr, M.Q.; Sadaf, J.R.; Yang, L.L.; Nur, O.; Willander, M.; Palisaitis, J.; Persson, P.A. Trimming of Aqueous Chemically Grown ZnO Nanorods into ZnO Nanotubes and Their Comparative Optical Properties. *Appl. Phys. Lett.* **2009**, *95*, 073114. [CrossRef]
42. McMurdie, H.F.; Morris, M.C.; Evans, E.H.; Paretzkin, B.; Wong-Ng, W.; Ettlinger, L.; Hubbard, C.R. Standard X-ray Diffraction Powder Patterns from the JCPDS Research Associateship. *Powder Diffr.* **1986**, *1*, 64–77. [CrossRef]
43. Gupta, V.; Mansingh, A. Influence of Postdeposition Annealing on the Structural and Optical Properties of Sputtered Zinc Oxide Film. *J. Appl. Phys.* **1996**, *80*, 1063–1073. [CrossRef]
44. Zhang, Y.; Liu, Y.; Wu, L.; Li, H.; Han, L.; Wang, B.; Xie, E. Effect of Annealing Atmosphere on the Photoluminescence of ZnO Nanospheres. *Appl. Surf. Sci.* **2009**, *255*, 4801–4805. [CrossRef]
45. Keskenler, E.F.; Tomakin, M.; Dogan, S.; Turgut, G.; Aydın, S.; Duman, S.; Gurbulak, B. Growth and characterization of Ag/n-ZnO/p-Si/Al heterojunction diode by sol-gel spin technique. *J. Alloys Compd.* **2013**, *550*, 129–132. [CrossRef]
46. Park, D.J.; Kim, D.C.; Lee, J.Y.; Cho, H.K. Epitaxial Growth of ZnO Layers Using Nanorods with High Crystalline Quality. *Nanotechnology* **2007**, *18*, 395605. [CrossRef] [PubMed]
47. Ting, S.Y.; Chen, P.Y.; Wang, H.C.; Liao, C.H.; Chang, M.W.; Hsieh, Y.P.; Yang, C.C. Crystallinity Improvement of ZnO Thin Film on Different Buffer Layers Grown by MBE. *J. Nanomater.* **2012**, *2012*, 6. [CrossRef]
48. Fonoberov, V.A.; Alim, K.A.; Balandin, A.A.; Xiu, F.; Liu, J. Photoluminescence Investigation of the Carrier Recombination Processes in ZnO Quantum Dots and Nanocrystals. *Phys. Rev. B* **2006**, *73*, 165317. [CrossRef]
49. Pal, U.; Serrano, J.G.; Santiago, P.; Xiong, G.; Ucer, K.B.; Williams, R.T. Synthesis and Optical Properties of ZnO Nanostructures with Different Morphologies. *Opt. Mater.* **2006**, *29*, 65–69. [CrossRef]
50. Vanheusden, K.; Warren, W.L.; Seager, C.H.; Tallant, D.R.; Voigt, J.A.; Gnade, B.E. Mechanisms Behind Green Photoluminescence in ZnO Phosphor Powders. *J. Appl. Phys.* **1996**, *79*, 7983–7990. [CrossRef]
51. Sardari, S.E.; Iliadis, A.A.; Stamataki, M.; Tsamakis, D.; Konofaos, N. Crystal quality and conductivity type of (002) ZnO films on (100) Si substrates for device applications. *Solid State Electron.* **2010**, *54*, 1150–1154. [CrossRef]
52. Reynolds, D.C.; Look, D.C.; Jogai, B. Combined effects of screening and band gap renormalization on the energy of optical transitions in ZnO and GaN. *J. Appl. Phys.* **2000**, *88*, 5760–5763. [CrossRef]
53. Fair, R.B. The effect of strain-induced band-gap narrowing on high concentration phosphorus diffusion in silicon. *J. Appl. Phys.* **1979**, *50*, 860–868. [CrossRef]

nanomaterials

MDPI

Article

Tuning the Electronic Conductivity in Hydrothermally Grown Rutile TiO_2 Nanowires: Effect of Heat Treatment in Different Environments

Alena Folger [1], Julian Kalb [2], Lukas Schmidt-Mende [2] and Christina Scheu [1,3,*]

1 Max-Planck-Institut für Eisenforschung GmbH, Max-Planck-Str. 1, 40237 Düsseldorf, Germany; a.folger@mpie.de
2 Department of Physics, University of Konstanz, POB 680, 78457 Konstanz, Germany; julian.kalb@uni-konstanz.de (J.K.); lukas.schmidt-mende@uni-konstanz.de (L.S.-M.)
3 Materials Analytics, RWTH Aachen University, Kopernikusstr. 10, 52074 Aachen, Germany
* Correspondence: c.scheu@mpie.de; Tel.: +49-211-6792-720

Received: 28 August 2017; Accepted: 19 September 2017; Published: 23 September 2017

Abstract: Hydrothermally grown rutile TiO_2 nanowires are intrinsically full of lattice defects, especially oxygen vacancies. These vacancies have a significant influence on the structural and electronic properties of the nanowires. In this study, we report a post-growth heat treatment in different environments that allows control of the distribution of these defects inside the nanowire, and thus gives direct access to tuning of the properties of rutile TiO_2 nanowires. A detailed transmission electron microscopy study is used to analyze the structural changes inside the nanowires which are correlated to the measured optical and electrical properties. The highly defective as-grown nanowire arrays have a white appearance and show typical semiconducting properties with n-type conductivity, which is related to the high density of oxygen vacancies. Heat treatment in air atmosphere leads to a vacancy condensation and results in nanowires which possess insulating properties, whereas heat treatment in N_2 atmosphere leads to nanowire arrays that appear black and show almost metal-like conductivity. We link this high conductivity to a TiO_{2-x} shell which forms during the annealing process due to the slightly reducing N_2 environment.

Keywords: black TiO_2; nanowire; conductivity; electron energy loss spectroscopy; oxygen vacancy; defects

1. Introduction

Nanostructured titanium dioxide (TiO_2) is a promising material in the field of energy conversion and storage [1]. In most TiO_2 applications, the efficiency of the device is determined by three consecutive processes: light absorption, charge separation, and electron transport. Although TiO_2 is widely used for energy applications, the efficiency of bare TiO_2 is limited by a wide band gap of around 3 eV [2] and a relatively low electron conductivity [3,4]. To overcome these limitations, defect engineering can be used to optimize the optical band gap and the electrical properties. In combination with an optimized geometry, which can be derived from theoretical calculations [5], defect engineering enables the fabrication of highly active devices.

Defects can be introduced in TiO_2 by metal [6] and nonmetal [7,8] impurities or dopants. However, this approach has the drawback that the dopants, especially d-block transition metals, also act as recombination centers for the generated electron hole pairs [9], which in turn lowers the efficiency of the device. Other approaches to produce defective TiO_2 without doping are mediated by the incorporation of Ti^{3+} and oxygen vacancies (O_{vac}) via reduction [10–12], which might introduce surface disorder in addition [13–15]. Most approaches use hydrogen environment and elevated temperatures [14,16,17] or a hydrogen plasma [18,19] to produce defective TiO_2. Similar types of defective TiO_2 can be obtained if active metals, such as Zn [10,20], Al [21], or Mg [22], are used as a reductant. However, these harsh reductive conditions are not mandatory to obtain defective TiO_2. Instead, black TiO_2 nanoparticles with surface disorder can be obtained by annealing amorphous nanoparticles in Ar gas [13]. Concededly, there are no reports which show that crystalline TiO_2, e.g., rutile TiO_2 nanowires (NWs), can be reduced in an oxygen-deficient atmosphere, such as vacuum, Ar, or N_2.

The effect of defect engineering on the optical band gap and the apparent color, which can be tuned from yellow over blue to black, has been studied extensively [16,23,24]. Apart from that, reports about how structural changes, such as the introduction of O_{vac} or surface disorder, influence the electrical properties are rare. So far, Nowotny and co-workers studied the influence of defect disorder on the semiconducting properties of rutile TiO_2 and found a strong effect on the electrical properties [25–27]. Especially, a high amount of O_{vac}, which is intrinsically found in rutile TiO_2, leads to strong n-type characteristics. In addition, the influence of O_{vac} on the conducting properties of TiO_2 NWs was evaluated by intensity-modulated photocurrent spectroscopy. For oxygen-deficient NWs, two electron-transport modes, a trap-free mode in the core and a trap-limited mode near the surface, were detected [28]. Recently, Lü et al. [29] investigated the effect of the surface disorder on the electrical properties. On a 40 nm thick bilayer structure of crystalline anatase ($\approx$20 nm) and amorphous TiO_2 (20 nm), which serves as a model system, they found a metallic conductivity at the interface between the crystalline and the amorphous part. These results give a first hint on the electrical properties of the defective, black TiO_2. Admittedly, in this model system the amorphous layer does not represent the surface disorder found in black TiO_2 adequately. It is much thicker and does not show any ordering phenomena [29].

In this work, we present how the electrical properties of TiO_2 NW arrays, incorporating rutile TiO_2 NWs with different defect states, can be changed. A detailed analysis of the nanostructure and the local chemical environment of three differently treated NW arrays, in combination with our results from ultraviolet-visible (UV–Vis) and current-voltage (IV) measurements, leads to a better understanding of the underlying mechanism that are responsible for the electronic properties of defective TiO_2. The results show how TiO_2 NWs, which are intrinsically n-type semiconductors in the as-grown state, can be converted to almost insulating TiO_2 NWs or NWs with a metal-like conductivity simply by using an appropriate atmosphere for the post-growth annealing.

2. Results

2.1. (Internal) Nanostructure and Local Chemical Environment

Scanning electron microscopy (SEM) investigations (Figure 1) reveal that the NW arrays of the three samples consist of NWs which grow almost perpendicular to the fluorine tin oxide (FTO) substrate and are of similar size (diameter of as-grown: 164 $\pm$ 31 nm, annealed in air: 172 $\pm$ 16 nm, annealed in N_2: 157 $\pm$ 28 nm). The high magnification SEM images in the insets of Figure 1 disclose slight morphological changes at the tip of the NWs. The as-grown NWs (Figure 1a) possess a rough tip, which is built by a bundle of nanofibers, as shown before by Wisnet et al. [30]. This structure is removed for the NWs annealed in air (Figure 1b), which have a much smoother surface. The tip of the NWs annealed in N_2 looks like an intermediate state between the as-grown NWs and the NWs annealed in air, although it was annealed at the same temperature for the same time. For the NWs annealed in N_2, the nanofiber bundle is still visible at the tip (Figure 1c), but not as prevalent as in the as-grown NWs.

The high-angle annular dark-field scanning transmission electron microscopy ((S)TEM) images in Figure 2, all taken from the central part of appropriate NWs, show more significant changes inside the NWs due to the annealing. While the as-grown NW is built by a bundle of nanofibers, as indicated from the SEM image, the annealed NWs are a single-crystalline material, which is interspersed with voids [31,32]. Nevertheless, SEM showed that even for the annealed NWs there are still residuals of the former nanofiber bundle at the tip (Figure 1b,c). The NW annealed in air does not show any further changes besides the voids, whereas the NW annealed in N_2 has internal voids and in addition a distinct core-shell like structure with an approximately 10 nm thick shell. A similar shell can be detected for as-grown NWs and NWs annealed in air but it is only 1–3 nm thick. Although the nanostructures of the three NWs differ, no changes in crystallography can be detected. The diffraction patterns in the insets of Figure 2 correspond to rutile TiO_2 acquired in the $[1\bar{1}0]$ zone axis and deviate only by the streaking in the diffraction peaks in [110] direction, which is visible for the as-grown NW. This streaking arises from the nanofiber bundle and the high defect density in the as-grown NW [30]. Thus, neither the heat treatment in air nor in N_2 leads to a phase transformation.

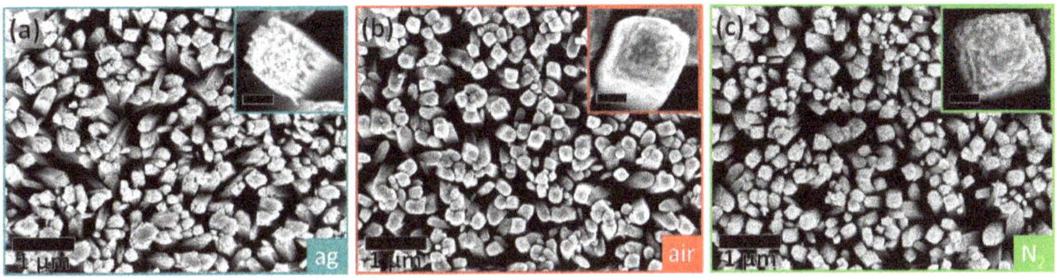

Figure 1. Scanning electron microscopy (SEM) images of nanowire (NW) arrays, which are (**a**) as-grown (ag), (**b**) annealed in air, and (**c**) annealed in N_2. The insets show a high magnification SEM image of a single NW from the respective NW array. The scale bar of the inset is 50 nm.

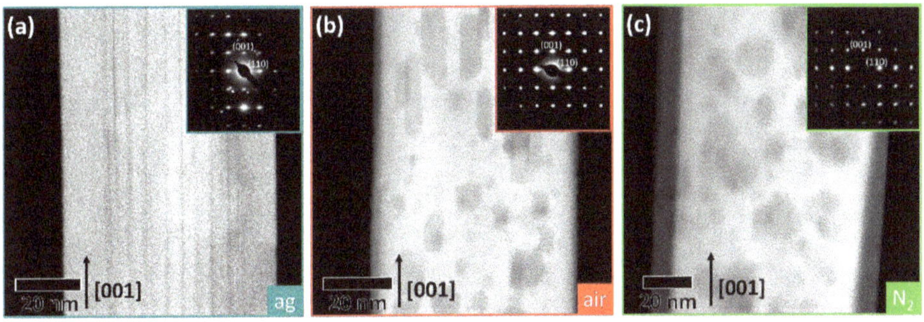

Figure 2. High-angle annular dark-field scanning transmission electron microscopy (STEM) image and a corresponding electron diffraction pattern (inset) for (**a**) an as-grown NW, (**b**) a NW annealed in air, and (**c**) a NW annealed in N_2. All images show a representative area in the center of its respective NW and the diffraction patterns are taken from entire NWs.

Despite the changes in the nanostructure, there are also differences in the local chemical environment of the three different NWs close to the surface. Figure 3a–c shows electron energy-loss (EEL) spectra of the Ti-$L_{2,3}$ edge with different distances to the surface. Close to the surface (yellow lines), the Ti-$L_{2,3}$ edge is shifted to lower energies by around 1 eV and the energy loss near edge fine structure (ELNES) shows that the splitting of the L_2 and L_3 peaks into a doublet is not resolved. This t_{2g}-e_g splitting is typical for rutile TiO_2 and results from a distorted octahedral surrounding of Ti by oxygen ions [33], but cannot be detected for Ti close to the surface. Instead, the Ti-$L_{2,3}$-edge is formed by broad peaks. Depending on the heat treatment, the typical ELNES of rutile TiO_2 occurs closer or more far away from the surface. For the NWs annealed in air, the ELNES shows the typical shape of rutile TiO_2 with a pronounced t_{2g}-e_g splitting after moving 1.8 nm towards the center (orange line in Figure 3b). For the as-grown NW, the broad L_2 and L_3 peaks in the ELNES are observed in the first

2.9 nm of the surface region (red line in Figure 3b). The NW annealed in N_2 has the largest region (up to 4.8 nm, dark red line in Figure 3c), where one can find an ELNES without pronounced t_{2g}-e_g splitting. Moving farer away from the surface, the ELNES of the as-grown NW and the NW annealed in air does not change anymore, but for the NW annealed in N_2 one can see that 10.4 nm away from the surface (light cyan line in Figure 3c), the ELNES changes again. Following the method described by Stroyanov et al. [34], the Ti-$L_{2,3}$ edge is used to calculate the amount of Ti^{4+} relative to the total amount of Ti, which is mainly a sum of Ti^{4+} and Ti^{3+}. Figure 3d–f are overlays of the resulting Ti^{4+} gradients with a STEM image of the analyzed NW area. The shift of the Ti-$L_{2,3}$ edge towards lower energies close to the surface is related to a lower amount of Ti^{4+} in this area. Thus, close to the surface, the NWs are not fully oxidized. Inside the NW, the as-grown NW and the NW annealed in air have a constant amount of Ti^{4+} of around 80%. For the NW annealed in N_2, the changes of the ELNES around 10.4 nm are also linked to a lower amount of Ti^{4+} and the overlay in Figure 3f shows that this decrease of Ti^{4+} is closely related to the core-shell interface. The lack of Ti^{4+} results in an off-stoichiometric TiO_{2-x}. In the following, the shell material will be denoted as TiO_{2-x} to account for the high oxygen deficiency.

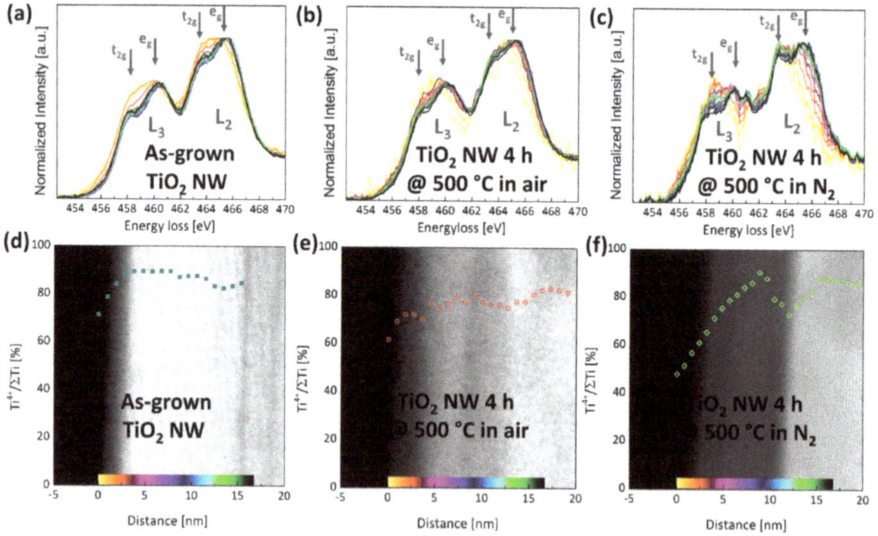

Figure 3. (**a**–**c**) Position resolved electron energy loss (EEL) spectra of the Ti-$L_{2,3}$ edge, for (**a**) an as-grown NW, (**b**) a NW annealed in air, and (**c**) a NW annealed in N_2. The positions of the spectra are marked in the STEM images of (**d**–**f**) with a specific color, which is the same for the respective Ti-$L_{2,3}$ edge (the color changes from the NW surface to the center (left to right) in the following order: yellow, orange, red, pink, purple, blue, cyan, green, black). In (**d**–**f**), the Ti^{4+} gradient is overlaid with the STEM image.

To study the core-shell structure in more detail, Figure 4a shows a high resolution (HR) TEM image of a NW annealed in N_2. This NW has a comparable thick shell to facilitate the analysis. One can see that the NW consists not only of a core and a shell, but of four distinctive areas. The rutile TiO_2 core and the crystalline TiO_{2-x} shell are separated by a defective interface area and the shell is covered with a disordered surface layer. Around 80% of the NW volume can be assigned to the core, which is rutile. The shell is also crystalline and covers around 20% of the NW volume. The high resolution annular bright-field STEM image in Figure 4b shows no differences in the crystal structure of the rutile core and the shell, except a small change in the d-spacing between {110} planes (core: d_{110} = 3.33 Å, shell: d_{110} = 3.29 Å). Although the electron energy-loss spectroscopy (EELS) analysis shows that the shell consists of off-stoichiometric TiO_{2-x}, no inhomogeneity in the oxygen distribution can be detected in the annular bright field STEM image (Figure 4b). Thus, an ordering of a significant amount of O_{vac} in this part of the NW is unlikely, as it would lead to periodic changes in the atomic columns, which should be visible in annular bright-field STEM. However, in all imaging conditions, this shell appears in a different contrast compared to the core. In the shell area, a sample thickness of 110 nm is derived using the low-loss EEL spectrum and assuming an inelastic mean free path of 276 nm for rutile TiO_2 [35]. Considering an error of around 10% for the thickness determination by EELS [36], this thickness estimation is in good agreement with the total thickness of the analyzed NW, which is also shown in Figure 2c (around 100 nm). Thus, the changes in contrast cannot be related to a thickness effect but might be related to a change in the density of the material. The contrast changes might also be affected by the incorporation of nitrogen, but EELS measurements in the shell area show no incorporation of nitrogen within our detection limits of ≈1 at%. It is noteworthy that this observation cannot be confirmed by methods other than EELS with high lateral resolution because the nitrogen and the titanium signal overlap in other spectroscopic techniques, such as Auger and wavelength dispersive X-ray spectroscopy. The defective area, which can be seen in the HR TEM image of Figure 4a between the TiO_2 core and the TiO_{2-x} shell is around 1.9 ± 0.3 nm thick. The disordered surface layer of the NW has a thickness of 2.2 ± 0.3 nm and is not completely amorphous, but shows some periodicity perpendicular to the [001] direction. Figure 4c is an intensity profile of Figure 4a in the first 4 nm next to the vacuum and perpendicular to the NW surface. This profile shows two periodic areas, but with different periodicity. The periodicity of the TiO_{2-x} shell corresponds to the lattice spacing of {110} planes in rutile TiO_2. Closer to the vacuum, there is a second material, which is also periodic to a certain extent, but the related lattice distances are much bigger (≈5 Å). This in-plane ordering in an amorphous phase is due to the underlying substrate periodicity and has been observed for other systems as well [37].

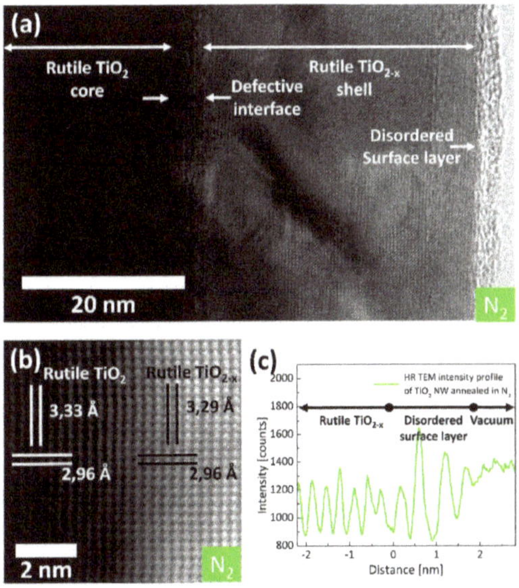

Figure 4. (**a**) High resolution (HR) transmission electron microscopy (TEM) image of a NW annealed in N_2 showing the surface near region. (**b**) Annular bright field STEM image of the interface between the rutile TiO_2 core and the TiO_{2-x} shell. (**c**) Intensity line scan of (**a**) showing an out-of-plane periodicity in the disordered surface layer parallel to the [001] growth direction.

2.2. Optical and Electrical Properties

The changes in the internal nanostructure, which are induced by annealing in different environments, influence the optical properties of the NWs. While the NW arrays incorporating as-grown NWs or NWs annealed in air appear white, the NW arrays annealed in N_2 are black. This color change indicates more light absorption in the visible range for the NWs annealed in N_2. Figure 5a shows Tauc plots for direct allowed band gap transitions of the three different NW arrays. The as-grown NWs have a band gap of 2.98 ± 0.06 eV, which is in good agreement with previous measurements [32]. The band gap of NWs annealed in air is significantly reduced to 2.59 ± 0.04 eV, but for the NWs annealed in N_2, the obtained direct band gap is again 2.96 ± 0.03 eV. In addition, there is an indirect transition for the NWs annealed in N_2 with an indirect band gap of around 2.57 ± 0.02 eV (inset in Figure 5a). In contrast, no strong indirect transition can be detected for the as-grown NWs and the NWs annealed in air. Absorption spectra allow not only the determination of the band gap but are also suitable to measure the so-called Urbach energy, which is a measure of the disorder in materials and leads to additional states within the band gap [38]. The Urbach energy of the as-grown NWs and the NWs annealed in air and in N_2 is 0.61 ± 0.01 eV, 0.55 ± 0.01 eV, and 1.65 ± 0.01 eV, respectively. Since UV–Vis can only probe the band gap on a large scale and as the results might be influenced by the periodicity of the NW array, and the resulting interference effects, additional band gap measurements were performed using EELS. Figure 5b shows the corresponding zero-loss subtracted low-loss EEL spectra of the different NWs. The band gap values derived from the EELS measurements (as-grown: 2.93 ± 0.12 eV, annealed in air 2.41 ± 0.06 eV, annealed in N_2 2.66 ± 0.14 eV) are in good agreement with the values obtained by UV-Vis, considering the indirect transition for the NW annealed in N_2.

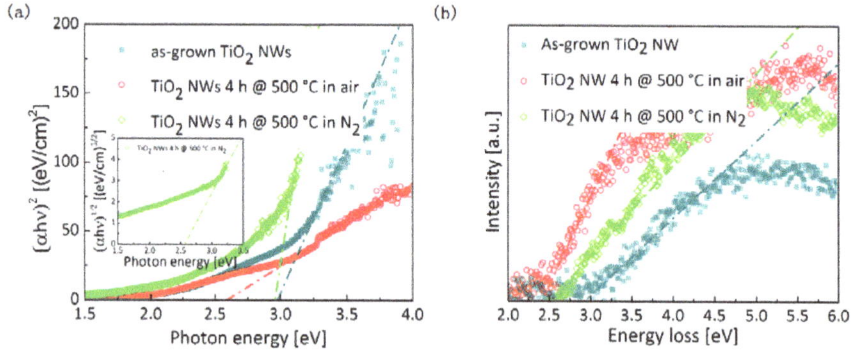

Figure 5. (**a**) Tauc plot for direct band gap and (**b**) zero-loss subtracted low-loss EEL spectra for NWs which are as-grown (petrol squares), annealed in air (red circles), and annealed in N_2 (green lozenges). The inset in (**a**) shows the Tauc plot for an indirect band gap for the NW array annealed in N_2.

Besides the optical properties of the NWs, the electronic properties are affected by the heat treatments. Figure 6a shows the IV-characteristics of the as-grown NWs, the NWs annealed in air, and the NWs annealed in N_2. Significant differences in the electronic properties of the three devices regarding the conduction limiting mechanisms can be observed.

The as-grown NWs block the transient current for electrical fields between 0 and 12 kV/cm (petrol line, Figure 6a). At higher electrical fields, the transient current is increasing exponentially and is hence affected by Schottky emission (petrol line, Figure 6b). For an increasing negative bias, the IV-characteristic turns quickly from an exponential increase into an increase that is proportional to the squared electric field arising from a space-charge-limited current (Figure 6d) [39]. Thus, the Schottky barrier at the PtIr/TiO_2 interface is smaller than the one at the FTO/TiO_2 interface.

TiO_2 NWs annealed in air block the transient current for electrical fields between −25 kV/cm and at least 100 kV/cm (red line, Figure 6a,b), which corresponds to the highest applicable bias in the employed setup. The IV-characteristics of the PtIr/TiO_2 interface become completely exponential and hence the transient current is limited by a Schottky emission across the whole measured bias range (red line, Figure 6b) [39].

In contrast, the transient current of the NWs annealed in N_2 is not blocked at any bias, which indicates an almost complete vanishing of both Schottky barriers (green line, Figure 6b). Only at very low, negative fields up to roughly −2 kV/cm, we found a Fowler–Nordheim tunneling behavior for the electrons passing from the PtIr tip to the TiO_2 NW (Figure 6c). For larger field amplitudes, the transient current becomes linear, showing a relatively large ohmic resistance (green line, Figure 6a). However, the slope and thus the absolute ohmic resistance depends on the applied voltage.

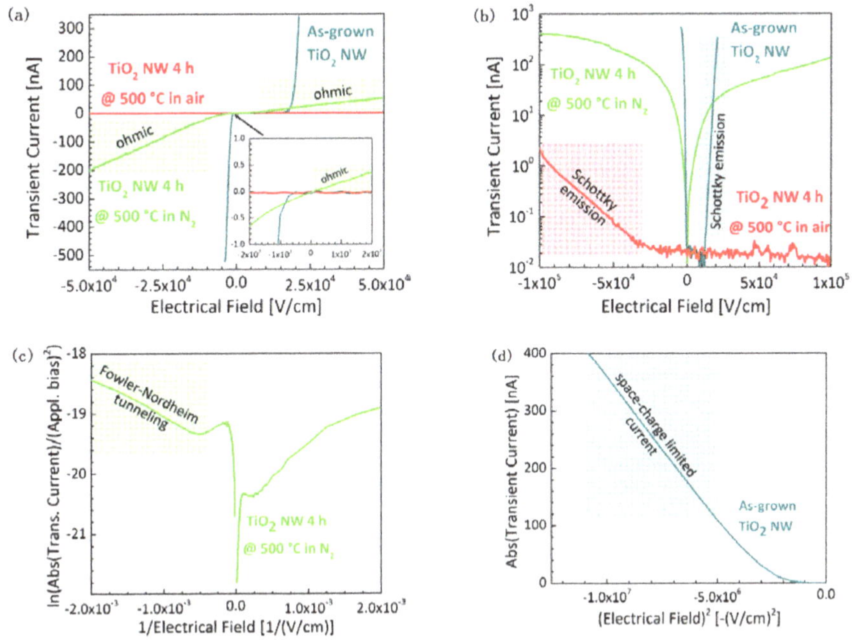

Figure 6. Transient current characteristics through as-grown NWs, NWs annealed in air, or in N_2 that is measured between a PtIr top and an FTO bottom electrode. The different plots emphasize several conduction-limiting mechanisms: (**a**) Linear plot showing ohmic behavior and the inset is a zoom in on the point of origin, (**b**) Schottky plot, (**c**) Fowler–Nordheim plot and (**d**) space–charge-limited current plot [39].

3. Discussion

Our results show that different heat treatments change the nanostructure and the properties of hydrothermally grown rutile TiO_2 NWs significantly. In the following, the interaction of the structural changes on the properties will be discussed.

As-grown NWs are intensively studied and used in many application and thus serve as a reference in this work. The detailed electron microscopic analysis showed that these NWs are rutile TiO_2, but contrary to many reports [40], they are not single-crystalline [30,32]. Instead, they show a meso-crystalline structure that is built by a bundle of nanofibers and incorporate many crystal defects [30], especially a high amount of O_{vac} [32]. The optical band gap of around 3 eV is in accordance with literature values for rutile TiO_2 [2], but the Urbach energy of 0.61 eV is much larger than reported for single-crystalline rutile TiO_2 nanoparticles [41], and can be assigned to the high defect disorder of the O_{vac} and other structural defects. In addition, the O_{vac} influences the electronic properties, as they are prominent electron donors that tune TiO_2 into an n-type semiconductor [42–44]. The O_{vac} in as-grown TiO_2 NWs have two effects on the electronic properties. In the first instance, the local donor density moves the Fermi level upward, closer to the conduction band minimum. As a consequence, the summit of the Schottky barrier between the metallic cathode and the TiO_2 drops with increasing electron donor density close to the interface. In addition, an increased number of O_{vac} lowers the resistivity of TiO_2 by increasing the number of mobile electrons in the conduction band [45,46] and thus the as-grown rutile TiO_2 NWs show n-type conductivity.

As shown in a previous study, TiO_2 NWs annealed in air have a significantly reduced density of O_{vac} in the crystal structure, as vacancy condensation takes place during the heat treatment [32]. The NWs are single-crystalline and the rutile crystal structure of the NWs annealed in air is almost O_{vac} free. In addition, the vacancies close to the NW surface are vanished due to the oxygen atmosphere during the heat treatment, resulting in NWs that have only a 1.8 nm thick surface layer, which deviates from the perfect rutile TiO_2 environment, as shown by changes in the ELNES. These changes during the heat treatment influence the optical and electronic properties of the NW array, as both the amount of trap states and electron donors incorporated in the crystal structure are significantly reduced. This deduction is verified by the UV-Vis measurements, which show that the band gap as well as the Urbach energy shrink. The reduced band gap can be assigned to less O_{vac} in the crystalline rutile TiO_2 [47] and a high Ti^{3+} concentration in the defective area surrounding each void [32]. It is noteworthy that these NWs appear white although the band gap indicates absorption in the visible blue regime. This effect is related to a strong light scattering, which is caused by the high refractive index of TiO_2 [48]. Furthermore, a reduced Urbach energy indicates less disorder. However, Urbach energy is still higher than expected for a single-crystalline rutile nanoparticle [41]. This deviation results from the 1.8 nm thick surface layer covering the NWs and a defective, Ti^{3+} rich area surrounding each void [32]. Concurrently, the transient current is blocked over a broad range of electrical fields. Only for highly negative electrical fields a Schottky emission-limited current can be detected. These results are in good accordance with the discussion above. As the density of O_{vac} is significantly reduced in the rutile crystal structure of NWs annealed in air, the Schottky barrier heights and the bulk resistance are expected to increase. Nevertheless, at high negative electrical fields the Schottky barrier can still be passed. We assume a constant work function for the PtIr tip and the FTO substrate for all experiments, so the Schottky barrier is mainly influenced by the Fermi level of the TiO_2 NWs. Structural inhomogeneity at the TiO_2 NW tip surface might influence the Schottky barrier, but SEM analysis showed that the surface of the NWs annealed in air is the smoothest, so we assume only a minor contribution of surface inhomogeneity on the height of the Schottky barrier.

Annealing in N_2 changes the distribution of O_{vac} as well. According to the TEM results presented in this work, the NWs annealed in N_2 have a complex core-shell structure. From these results, it is reasonable to assume that the defect density in the core, which is riddled by voids, is similar to the defect density of the NWs annealed in air. Consequently, the electronic properties of the core, possessing a low defect density, are similar to the electronic properties of NWs annealed in air. However, the IV-characteristics measured for the NWs annealed in N_2 differ strongly from those, measured for the NWs annealed in air. Thus, the core of the NW annealed in N_2 has no significant influence on the conductivity in these NWs. The Fowler–Nordheim tunneling behavior that occurs at low electrical fields is supposed to be a result of a disordered surface layer (Figure 4a,c) covering the metal-like TiO_{2-x} shell. For strong electric fields, the influence of this ultra-thin layer is negligible. Without this metallization, the Schottky barrier is much thicker and Schottky emission, as observed for the as-grown NWs, instead of tunneling dominates. The metal-like behavior of the shell is in good agreement with the black color of the NW array, as absorption throughout the entire spectral range is common for metals. Several observations indicate that the metallization takes place in a confined volume. Firstly, the optical measurements are still dominated by the properties known for white

TiO_2. It is known that the transmittance of light of thin metal films drops below 20% for films being thicker than about 10–20 nm [49]. As our NW arrays show a high transmittance, it is reasonable to assume that the metallic part in the NWs annealed in N_2 does not exceed 20 nm. In addition, the ohmic resistance measured for these NWs is relatively large. Such large ohmic resistances stem from the tiny cross-sections of the highly conductive part of the NWs annealed in N_2. According to the TEM and EELS results, NWs annealed in N_2 are covered by a TiO_{2-x} shell that contains a very high amount of O_{vac}, as the vacancies cannot be removed at the surface due to the slightly reducing environment of the N_2 atmosphere. Although an incorporation of N cannot be ruled out completely due to the EELS detection limit of around 1%, we assume no influence of a potential N doping (which would be below 1 at% of N) on the electrical properties. This assumption is based on the fact that changes in the electronic properties for $TiO_{2-2x}N_x$ were only detected for N incorporation higher than 5 at% N [50]. Such high concentrations can be excluded due to the absence of an N K-edge in the EEL spectra throughout the NW, although it is not possible to confirm this result with other methods due to signal overlap. Nevertheless, even undoped but strongly reduced TiO_{2-x}, as found in the shell of the NW annealed in N_2, is highly conductive [26,44]. Hence, it is reasonable to assume that the TiO_{2-x} shell is responsible for the unusual properties of these NWs, but due to the small dimensions it is difficult to localize the origin of these effects within the shell. According to the TEM results, the shell can be divided in three parts, namely the disordered surface layer (2.2 nm), the crystalline TiO_{2-x} shell (8–20 nm) and a defective interfacial area between the TiO_{2-x} shell and the TiO_2 core (1.9 nm). The high Urbach energy measured for these NWs originates from the high degree of disorder in the surface layer. Similar surface layers were found in various black TiO_2 nanomaterials and seem to be the origin of the black color [16]. This change in color is mainly related to the presence of a big Urbach tail at the upper part of the valence band [16]. These results are in good accordance with the high Urbach energy which was measured for NWs annealed in N_2. The metallization and the high transient current might result from the entire shell but there are some indications that it is confined on the defective interface between the TiO_{2-x} shell and the TiO_2 core. The EELS analysis showed a higher concentration of Ti^{3+} at this interface, which might arise from a great amount of O_{vac} confined at this interface. Both are electron donor type defects and can lead to high conductivity. Lü et al. found a similar conducting interface at the homojunction of a bilayer thin film. This homojunction is formed between an oxygen-deficient, amorphous TiO_{2-x} layer with around 20 nm thickness and a comparable thick layer of anatase TiO_2 [29]. Our experimental setup does not allow direct proof of this assumption, but the results obtained in this study give evidence that not the entire shell, but a conductive interface might be responsible for the highly conducting properties of the black NWs annealed in N_2. In addition, there is a certain hysteresis of the IV characteristics, which indicates that the O_{vac} are able to drift through the TiO_{2-x} shell. This effect is well known from resistive switching [51–53] and might be the reason why the O_{vac} cannot be detected by annular bright field STEM. Due to the high mobility of the O_{vac}, their density at the PtIr/TiO_2 and FTO/TiO_2 interfaces differ slightly, resulting in the observed asymmetry of the IV characteristics for positive and negative applied bias.

4. Materials and Methods

4.1. Synthesis Procedure

TiO_2 NW arrays were synthesized by a hydrothermal procedure adapted from Liu et al. [40]. All chemicals were used as supplied without further purification. In a typical synthesis, 250 μL titanium butoxide (Ti(nOBu)$_4$), Sigma-Aldrich, St. Louis, MI, USA) was dropped into a mixture of 5 mL concentrated hydrochloric acid (37 wt%, analytical grade, Sigma-Aldrich) and 5 mL deionized water under vigorous stirring. Ultrasonically cleaned (isopropyl alcohol, acetone, ethanol) FTO substrates were placed vertically in a Teflon liner, which was filled with the growth solution and placed into a steel autoclave. The hydrothermal reaction was performed at 150 °C for 4.5 h. Afterwards, the autoclave was cooled down to room temperature. The FTO substrates, covered with TiO_2 NW arrays, were rinsed with deionized water and dried with compressed air. Heat treatment of the samples was performed at 500 °C (50 °C/min ramp up to 500 °C) on an Anton Paar DHS 1100 (Anton Paar, Graz, Austria) heating stage. For the TiO_2 NWs annealed in N_2, a constant N_2 atmosphere of 1.35 bar was applied during the experiment, whereas the other sample was annealed in air.

4.2. Characterization

SEM analysis: The morphology of the NW arrays in top-view was investigated using a Zeiss AURIGA Modular CrossBeam workstation (Zeiss, Oberkochen, Germany) equipped with an in-lens detector. All measurements were carried out at 4 kV.

TEM analysis: TEM was applied for the morphological and crystallographic analysis. The TiO_2 nanowires were scraped off the FTO substrate and the resulting powder was dispersed on a copper grid with a holy carbon film. A Philipps CM20 (FEI, Hillsboro, OR, USA) and a Jeol JEM-2200FS field emission gun instrument (Jeol, Akishima, Japan), both operated at 200 kV, were used for conventional bright field TEM, selected area electron diffraction, and HR TEM.

STEM images and EELS data were acquired at 300 kV with a FEI Titan Themis 60–300 (FEI, Hillsboro, OR, USA) equipped with a high brightness field emission (XFEG™) source, a monochromator, an aberration-corrector for the probe-forming lens system, a BRUKER EDS Super X detector, and a high-resolution energy filter (post-column Quantum ERS energy filter). EEL spectra were acquired in STEM mode with a dispersion of 0.1 eV per channel. To measure the band gap on a local scale, low-loss spectra in monochromated STEM mode were acquired with a dispersion of 0.01 eV. An energy resolution of 0.3 eV, as determined by the full-width at half maximum of the zero-loss peak, was obtained. Using a power-law fit, the tail of the zero-loss peak was removed and the band gap was extracted according to the linear fit method [54]. For all EELS measurements, the convergence semi angle was 23.8 mrad and the collection semi-angle was 35 mrad. All EELS data were taken using the dual-channel acquisition technique [55] and the spectra were corrected for dark current and channel-to-channel gain variation [56]. The background was removed using a standard power law fit [56].

EEL spectra of the Ti-$L_{2,3}$ edge were used to determine the Ti^{3+}/Ti^{4+} ratio with high lateral resolution. Therefore, a calibration technique of Stoyanov et al. [34]. was used. It is based on the position and intensities of the Ti L_2 and L_3 white lines.

UV-Vis spectroscopy: A PerkinElmer Lambda 800 spectrometer (PerkinElmer, Waltham, MA, USA) in transmission mode was utilized to measure the absorption spectra in a wavelength range of 350–850 nm. The step size was 1 nm. The detected UV–Vis data were used to determine direct and indirect band gaps using Tauc plots [57] and to calculate the Urbach energy [38] of the different samples.

IV characteristics: Qualitative information about the electronic properties of the investigated TiO_2 NWs was obtained by IV-measurements. A platinum-iridium (PtIr) (4:1) tip served as a top electrode. The IV-characteristics of the transient current through the PtIr/TiO_2/FTO sandwich were determined. The tip was taken because typical deposition techniques used for flat metal electrodes would infiltrate the interspace between the NWs and cause shorts. The tip was placed manually and softly on the NW array and pushed by its own weight (0.1 g) on a bunch of NWs during the measurement. The bottom contact was established by removing the NWs using a diamond writer and connecting the uncovered FTO with a thin insulated copper wire and a drop of silver paste. The sample holder was transferred into a vacuum chamber, where the humid air was replaced by dry nitrogen during several pumping and purging steps. A Keithley 2401 (Ketihly Instruments, Cleveland, OH, USA) was used as a voltage source and to measure the transient current. In the presented graphic, a positive electric field is pointing from the PtIr tip towards the FTO and the IV-curves were obtained by changing the field from negative to positive values.

5. Conclusions

In this study, we propose heat treatments in different environments in order to manipulate the structure of hydrothermally grown rutile TiO_2 NWs in such a way that their optical and electrical properties can be tailored. The as-grown NWs incorporate a high amount of defects, especially O_{vac}, which are responsible for the n-type conductivity in these NWs. Independent of the environment, the heat treatment leads to a condensation of these vacancies and to the formation of single-crystalline, lattice defect free, rutile TiO_2 NWs that incorporate voids. The absence of O_{vac} results in a blocking of the transient current and concurrently improves the optical properties by decreasing the band gap and Urbach energy. For an oxidizing environment, such as air, the resulting NWs are almost insulating. Although NWs annealed in N_2 contain up to around 80% of an insulating rutile TiO_2 core, their properties are completely different. They possess a black color and an almost metal-like conductivity. These properties are related to the slightly reducing atmosphere of N_2 during the heat treatment. It inhibits the vanishing of the surface-near O_{vac} and thus a core-shell structure with a highly oxygen deficient shell is formed.

Acknowledgments: This work was funded by the German Research Foundation (DFG). The authors thank Christian Liebscher and Siyuan Zhang for their help during EELS band gap measurements and Benjamin Breitbach for performing the heating at the Anton Paar Heating device.

Author Contributions: Alena Folger, Julian Kalb, Lukas Schmidt-Mende and Christina Scheu conceived and designed the experiments; Alena Folger did the synthesis and performed the SEM, TEM, and UV measurements and Julian Kalb performed the IV measurements; Alena Folger and Julian Kalb analyzed the data of the respective experiments; Alena Folger, Julian Kalb, Lukas Schmidt-Mende, and Christina Scheu interpreted and discussed the data, Alena Folger wrote the paper with contributions from Julian Kalb, all authors have given approval to the final version of the manuscript.

Conflicts of Interest: The authors declare no conflict of interest.

References

1. Weng, Z.; Guo, H.; Liu, X.; Wu, S.; Yeung, K.W.K.; Chu, P.K. Nanostructured TiO_2 for energy conversion and storage. *RSC Adv.* **2013**, *3*, 24758. [CrossRef]
2. Scanlon, D.O.; Dunnill, C.W.; Buckeridge, J.; Shevlin, S.A.; Logsdail, A.J.; Woodley, S.M.; Catlow, C.R.A.; Powell, M.J.; Palgrave, R.G.; Parkin, I.P.; et al. Band alignment of rutile and anatase TiO_2. *Nat. Mater.* **2013**, *12*, 798–801. [CrossRef] [PubMed]
3. Yukio, K.; Ryozo, A.; Kichinosuke, Y. Electrical properties of rutile (TiO_2) thin film. *Jpn. J. Appl. Phys.* **1971**, *10*, 976.
4. Tang, H.; Prasad, K.; Sanjinès, R.; Schmid, P.E.; Lévy, F. Electrical and optical properties of TiO_2 anatase thin films. *J. Appl. Phys.* **1994**, *75*, 2042–2047. [CrossRef]
5. Liu, B.; Nakata, K.; Liu, S.; Sakai, M.; Ochiai, T.; Murakami, T.; Takagi, K.; Fujishima, A. Theoretical kinetic analysis of heterogeneous photocatalysis by TiO_2 nanotube arrays: The effects of nanotube geometry on photocatalytic activity. *J. Phys. Chem. C* **2012**, *116*, 7471–7479. [CrossRef]
6. Liu, B.; Chen, H.M.; Liu, C.; Andrews, S.C.; Hahn, C.; Yang, P. Large-scale synthesis of transition-metal-doped TiO_2 nanowires with controllable overpotential. *J. Am. Chem. Soc.* **2013**, *135*, 9995–9998. [CrossRef] [PubMed]
7. Ansari, S.A.; Khan, M.M.; Ansari, M.O.; Cho, M.H. Nitrogen-doped titanium dioxide (n-doped TiO_2) for visible light photocatalysis. *New J. Chem.* **2016**, *40*, 3000–3009. [CrossRef]
8. Lin, T.; Yang, C.; Wang, Z.; Yin, H.; Lü, X.; Huang, F.; Lin, J.; Xie, X.; Jiang, M. Effective nonmetal incorporation in black titania with enhanced solar energy utilization. *Energy Environ. Sci.* **2014**, *7*, 967. [CrossRef]
9. Choi, W.; Termin, A.; Hoffmann, M.R. The role of metal ion dopants in quantum-sized TiO_2: Correlation between photoreactivity and charge carrier recombination dynamics. *J. Phys. Chem.* **1994**, *98*, 13669–13679. [CrossRef]
10. Zheng, Z.; Huang, B.; Meng, X.; Wang, J.; Wang, S.; Lou, Z.; Wang, Z.; Qin, X.; Zhang, X.; Dai, Y. Metallic zinc-assisted synthesis of Ti^{3+} self-doped TiO_2 with tunable phase composition and visible-light photocatalytic activity. *Chem. Commun.* **2013**, *49*, 868–870. [CrossRef] [PubMed]

11. Liu, X.; Bi, Y. In situ preparation of oxygen-deficient TiO_2 microspheres with modified {001} facets for enhanced photocatalytic activity. *RSC Adv.* **2017**, *7*, 9902–9907. [CrossRef]
12. Ullattil, S.G.; Periyat, P. A 'one pot' gel combustion strategy towards Ti^{3+} self-doped 'black' anatase TiO_{2-x} solar photocatalyst. *J. Mater. Chem. A* **2016**, *4*, 5854–5858. [CrossRef]
13. Tian, M.; Mahjouri-Samani, M.; Eres, G.; Sachan, R.; Yoon, M.; Chisholm, M.F.; Wang, K.; Puretzky, A.A.; Rouleau, C.M.; Geohegan, D.B.; et al. Structure and formation mechanism of black TiO_2 nanoparticles. *ACS Nano* **2015**, *9*, 10482–10488. [CrossRef] [PubMed]
14. Chen, X.; Liu, L.; Yu, P.Y.; Mao, S.S. Increasing solar absorption for photocatalysis with black hydrogenated titanium dioxide nanocrystals. *Science* **2011**, *331*, 746–750. [CrossRef] [PubMed]
15. Chen, X.; Liu, L.; Liu, Z.; Marcus, M.A.; Wang, W.C.; Oyler, N.A.; Grass, M.E.; Mao, B.; Glans, P.A.; Yu, P.Y.; et al. Properties of disorder-engineered black titanium dioxide nanoparticles through hydrogenation. *Sci. Rep.* **2013**, *3*, 1510. [CrossRef] [PubMed]
16. Naldoni, A.; Allieta, M.; Santangelo, S.; Marelli, M.; Fabbri, F.; Cappelli, S.; Bianchi, C.L.; Psaro, R.; Dal Santo, V. Effect of nature and location of defects on bandgap narrowing in black TiO_2 nanoparticles. *J. Am. Chem. Soc.* **2012**, *134*, 7600–7603. [CrossRef] [PubMed]
17. Liu, N.; Schneider, C.; Freitag, D.; Hartmann, M.; Venkatesan, U.; Muller, J.; Spiecker, E.; Schmuki, P. Black TiO_2 nanotubes: Cocatalyst-free open-circuit hydrogen generation. *Nano Lett.* **2014**, *14*, 3309–3313. [CrossRef] [PubMed]
18. Wang, Z.; Yang, C.; Lin, T.; Yin, H.; Chen, P.; Wan, D.; Xu, F.; Huang, F.; Lin, J.; Xie, X.; et al. H-doped black titania with very high solar absorption and excellent photocatalysis enhanced by localized surface plasmon resonance. *Adv. Funct. Mater.* **2013**, *23*, 5444–5450. [CrossRef]
19. Lepcha, A.; Maccato, C.; Mettenbörger, A.; Andreu, T.; Mayrhofer, L.; Walter, M.; Olthof, S.; Ruoko, T.P.; Klein, A.; Moseler, M.; et al. Electrospun black titania nanofibers: Influence of hydrogen plasma-induced disorder on the electronic structure and photoelectrochemical performance. *J. Phys. Chem. C* **2015**, *119*, 18835–18842. [CrossRef]
20. Zhao, Z.; Tan, H.; Zhao, H.; Lv, Y.; Zhou, L.J.; Song, Y.; Sun, Z. Reduced TiO_2 rutile nanorods with well-defined facets and their visible-light photocatalytic activity. *Chem. Commun.* **2014**, *50*, 2755–2757. [CrossRef] [PubMed]
21. Wang, Z.; Yang, C.; Lin, T.; Yin, H.; Chen, P.; Wan, D.; Xu, F.; Huang, F.; Lin, J.; Xie, X.; et al. Visible-light photocatalytic, solar thermal and photoelectrochemical properties of aluminium-reduced black titania. *Energy Environ. Sci.* **2013**, *6*, 3007. [CrossRef]
22. Sinhamahapatra, A.; Jeon, J.-P.; Yu, J.-S. A new approach to prepare highly active and stable black titania for visible light-assisted hydrogen production. *Energy Environ. Sci.* **2015**, *8*, 3539–3544. [CrossRef]
23. Das, T.K.; Ilaiyaraja, P.; Mocherla, P.S.V.; Bhalerao, G.M.; Sudakar, C. Influence of surface disorder, oxygen defects and bandgap in TiO_2 nanostructures on the photovoltaic properties of dye sensitized solar cells. *Sol. Energy Mater. Sol. Cells* **2016**, *144*, 194–209. [CrossRef]
24. Shah, M.W.; Zhu, Y.; Fan, X.; Zhao, J.; Li, Y.; Asim, S.; Wang, C. Facile synthesis of defective TiO_{2-x} nanocrystals with high surface area and tailoring bandgap for visible-light photocatalysis. *Sci. Rep.* **2015**, *5*. [CrossRef]
25. Nowotny, M.K.; Sheppard, L.R.; Bak, T.; Nowotny, J. Defect chemistry of titanium dioxide. Application of defect engineering in processing of TiO_2-based photocatalysts. *J. Phys. Chem. C* **2008**, *112*, 5275–5300. [CrossRef]
26. Bak, T.; Bogdanoff, P.; Fiechter, S.; Nowotny, J. Defect engineering of titanium dioxide: Full defect disorder. *Adv. Appl. Ceram.* **2013**, *111*, 62–71. [CrossRef]
27. Nowotny, J.; Bak, T.; Burg, T. Electrical properties of polycrystalline TiO_2 at elevated temperatures. Electrical conductivity. *Phys. Status Solidi B* **2007**, *244*, 2037–2054. [CrossRef]
28. Chen, H.; Wei, Z.; Yan, K.; Bai, Y.; Yang, S. Unveiling two electron-transport modes in oxygen-deficient TiO_2 nanowires and their influence on photoelectrochemical operation. *J. Phys. Chem. Lett.* **2014**, *5*, 2890–2896. [CrossRef] [PubMed]
29. Lu, X.; Chen, A.; Luo, Y.; Lu, P.; Dai, Y.; Enriquez, E.; Dowden, P.; Xu, H.; Kotula, P.G.; Azad, A.K.; et al. Conducting interface in oxide homojunction: Understanding of superior properties in black TiO_2. *Nano Lett.* **2016**, *16*, 5751–5755. [CrossRef] [PubMed]

30. Wisnet, A.; Betzler, S.B.; Zucker, R.V.; Dorman, J.A.; Wagatha, P.; Matich, S.; Okunishi, E.; Schmidt-Mende, L.; Scheu, C. Model for hydrothermal growth of rutile wires and the associated development of defect structures. *Cryst. Growth Des.* **2014**, *14*, 4658–4663. [CrossRef]
31. Wisnet, A.; Bader, K.; Betzler, S.B.; Handloser, M.; Ehrenreich, P.; Pfadler, T.; Weickert, J.; Hartschuh, A.; Schmidt-Mende, L.; Scheu, C.; et al. Defeating loss mechanisms in 1D TiO_2-based hybrid solar cells. *Adv. Funct. Mater.* **2015**, *25*, 2601–2608. [CrossRef]
32. Folger, A.; Ebbinghaus, P.; Erbe, A.; Scheu, C. Role of vacancy condensation in the formation of voids in rutile TiO_2 nanowires. *ACS Appl. Mater. Interfaces* **2017**, *9*, 13471–13479. [CrossRef] [PubMed]
33. Brydson, R.; Sauer, H.; Engel, W.; Thomass, J.M.; Zeitler, E.; Kosugi, N.; Kuroda, H. Electron energy loss and X-ray absorption spectroscopy of rutile and anatase: A test of structural sensitivity. *J. Phys. Condens. Matter* **1989**, *1*, 797–812. [CrossRef]
34. Stoyanov, E.; Langenhorst, F.; Steinle-Neumann, G. The effect of valence state and site geometry on Ti $L_{3,2}$ and O K electron energy-loss spectra of Ti_xO_y phases. *Am. Mineral.* **2007**, *92*, 577–586. [CrossRef]
35. Tanuma, S.; Powell, C.J.; Penn, D.R. Calculations of electron inelastic mean free paths (IMFPs). IV. Evaluation of calculated IMFPs and of the predictive IMFP formula TPP-2 for electron energies between 50 and 2000 eV. *Surf. Interface Anal.* **1993**, *20*, 77–89. [CrossRef]
36. Egerton, R.F.; Cheng, S.C. Measurement of local thickness by electron energy-loss spectroscopy. *Ultramicroscopy* **1987**, *21*, 231–244. [CrossRef]
37. Oh, S.H.; Kauffmann, Y.; Scheu, C.; Kaplan, W.D.; Ruhle, M. Ordered liquid aluminum at the interface with sapphire. *Science* **2005**, *310*, 661–663. [CrossRef] [PubMed]
38. Dow, J.D.; Redfield, D. Toward a unified theory of urbach's rule and exponential absorption edges. *Phys. Rev. B* **1972**, *5*, 594–610. [CrossRef]
39. Chiu, F.-C. A review on conduction mechanisms in dielectric films. *Adv. Mater. Sci. Eng.* **2014**, *2014*, 1–18. [CrossRef]
40. Liu, B.; Aydil, E.S. Growth of oriented single-crystalline rutile TiO_2 nanorods on transparent conducting substrates for dye-sensitized solar cells. *J. Am. Chem. Soc.* **2009**, *131*, 3985–3990. [CrossRef] [PubMed]
41. Choudhury, B.; Choudhury, A. Oxygen defect dependent variation of band gap, urbach energy and luminescence property of anatase, anatase—Rutile mixed phase and of rutile phases of TiO_2 nanoparticles. *Phys. E Low Dimens. Syst. Nanostruct.* **2014**, *56*, 364–371. [CrossRef]
42. Kofstad, P. Nonstoichiometry, diffusion, and electrical conductivity in binary metal oxides. *Mater. Corros. Werkst. Korros.* **1974**, *25*, 801–802.
43. Soerensen, O.T. *Nonstoichiometric Oxides*; Elsevier Science: Amsterdam, The Netherlands, 2012.
44. Nowotny, M.K.; Bak, T.; Nowotny, J. Electrical properties and defect chemistry of TiO_2 single crystal. I. Electrical conductivity. *J. Phys. Chem. B* **2006**, *110*, 16270–16282. [CrossRef] [PubMed]
45. Austin, I.G.; Mott, N.F. Metallic and nonmetallic behavior in transition metal oxides. *Science* **1970**, *168*, 71–77. [CrossRef] [PubMed]
46. Klusek, Z.; Pierzgalski, S.; Datta, S. Insulator–metal transition on heavily reduced TiO_2 (110) surface studied by high temperature-scanning tunnelling spectroscopy (HT-STS). *Appl. Surf. Sci.* **2004**, *221*, 120–128. [CrossRef]
47. Lei, Y.; Leng, Y.; Yang, P.; Wan, G.; Huang, N. Theoretical calculation and experimental study of influence of oxygen vacancy on the electronic structure and hemocompatibility of rutile TiO_2. *Sci. China Ser. E Technol. Sci.* **2009**, *52*, 2742–2748. [CrossRef]
48. Bhatt, P.J.; Tomar, L.J.; Desai, R.K.; Chakrabarty, B.S. Pure single crystallographic form of TiO_2 nanoparticles: Preparation and characterization. *AIP Conf. Proc.* **2015**, *1667*. [CrossRef]
49. Axelevitch, A.; Gorenstein, B.; Golan, G. Investigation of optical transmission in thin metal films. *Phys. Procedia* **2012**, *32*, 1–13. [CrossRef]
50. Koslowski, U.; Ellmer, K.; Bogdanoff, P.; Dittrich, T.; Guminskaya, T.; Tributsch, H. Structural, electrical, optical, and photoelectrochemical properties of thin titanium oxinitride films ($TiO_{2-2x}N_x$ with $0 \leq x \leq 1$). *J. Vac. Sci. Technol. A Vac. Surf. Films* **2006**, *24*, 2199–2205. [CrossRef]
51. Kwon, D.H.; Kim, K.M.; Jang, J.H.; Jeon, J.M.; Lee, M.H.; Kim, G.H.; Li, X.S.; Park, G.S.; Lee, B.; Han, S.; et al. Atomic structure of conducting nanofilaments in TiO_2 resistive switching memory. *Nat. Nanotechnol.* **2010**, *5*, 148–153. [CrossRef] [PubMed]

52. Kim, K.M.; Jeong, D.S.; Hwang, C.S. Nanofilamentary resistive switching in binary oxide system: A review on the present status and outlook. *Nanotechnology* **2011**, *22*, 254002. [CrossRef] [PubMed]
53. Zhang, F.; Gan, X.; Li, X.; Wu, L.; Gao, X.; Zheng, R.; He, Y.; Liu, X.; Yang, R. Realization of rectifying and resistive switching behaviors of TiO_2 nanorod arrays for nonvolatile memory. *Electrochem. Solid State Lett.* **2011**, *14*, H422. [CrossRef]
54. Park, J.; Heo, S.; Chung, J.; Kim, H.; Park, G.S. Bandgap measurement of dielectric thin films by using monochromated STEM-EELS. *Microsc. Microanal.* **2007**, *13*, 1306–1307. [CrossRef]
55. Scott, J.; Thomas, P.J.; Mackenzie, M.; McFadzean, S.; Wilbrink, J.; Craven, A.J.; Nicholson, W.A. Near-simultaneous dual energy range EELS spectrum imaging. *Ultramicroscopy* **2008**, *108*, 1586–1594. [CrossRef] [PubMed]
56. Egerton, R.F. Electron energy-loss spectroscopy in the TEM. *Rep. Prog. Phys.* **2009**, *72*, 16502–16525. [CrossRef]
57. Tauc, J.; Grigorovici, R.; Vancu, A. Optical properties and electronic structure of amorphous germanium. *Phys. Status Solidi B* **1966**, *15*, 627–637. [CrossRef]

nanomaterials

MDPI

Article

TiO_2 Nanowire Networks Prepared by Titanium Corrosion and Their Application to Bendable Dye-Sensitized Solar Cells

Saera Jin †, Eunhye Shin † and Jongin Hong *

Department of Chemistry, Chung-Ang University, 84 Heukseok-ro, Dongjak-gu, Seoul 06974, Korea; saera0907@gmail.com (S.J.); swh0904@gmail.com (E.S.)

* Correspondence: hongj@cau.ac.kr; Tel.: +82-2-820-5196

† Equal contribution to this work.

Received: 10 August 2017; Accepted: 25 September 2017; Published: 12 October 2017

Abstract: TiO_2 nanowire networks were prepared, using the corrosion of Ti foils in alkaline (potassium hydroxide, KOH) solution at different temperatures, and then a further ion-exchange process. The prepared nanostructures were characterized by field emission scanning electron microscopy, Raman spectroscopy, and X-ray photoelectron spectroscopy. The wet corroded foils were utilized as the photoanodes of bendable dye-sensitized solar cells (DSSCs), which exhibited a power conversion efficiency of 1.11% under back illumination.

Keywords: TiO_2; wet corrosion; dye-sensitized solar cells

1. Introduction

The dye-sensitized solar cell (DSSC, or Grätzel cell) was first presented in 1991 [1], and has since been considered a low-cost alternative to conventional silicon solar cells, and a promising construction element for building-integrated photovoltaics (BIPV). This is due to its transparency and diverse colors [2]. Recently, cost-effective (Ti and stainless steel) metal foils have been proposed for the fabrication of bendable DSSCs for flat and curved building skins instead of plastic materials. This is primarily because of the lack of limitations with respect to high-temperature processing [3–6]. In particular, Ti substrates decreased the series resistance of DSSCs, and thus allowed for a better fill factor (FF) and power conversion efficiency, as compared with F-doped SnO_2 (FTO)-conducting glasses [5,6]. The DSSC consists of a dye-sensitized porous photoanode, a redox electrolyte, and a platinized counter electrode. Photoanodes are generally prepared using TiO_2 nanoparticles with a size of 5–20 nm. However, such mesoporous nanoparticle films are hampered by the limited electron transport caused by particle-to-particle hopping and charge recombination at interfaces [7,8]. With the aim of improving the DSSC performance, one-dimensional (1D) TiO_2 nanostructures, such as nanotubes, nanorods, and nanowires, have been synthesized by various methods including anodization [9–11], electrospinning [12] and hydrothermal alkali treatment of titania nanoparticles [13,14]. It is suggested that one-dimensional (1D) nanostructures offer better transport pathways for photogenerated electrons than nanoparticles because of their longer carrier diffusion lengths [8,15]. Recently, 1D titania arrays have been directly prepared on a Ti foil via the anodic oxidation of Ti in different electrolytes containing fluoride ions [16], and have been used as the photoanode of DSSCs [9,11,17]. The 1D titania arrays could also be detached and transferred onto the FTO glass for the fabrication of DSSCs [17]. However, it is not easy to reliably produce such nanostructures over a large area, since the anodization process is sensitive to electrochemical reaction conditions. Direct oxidation of Ti foils can also produce nanorods or nanowires directly grown on Ti substrates, but they require a long growth time [18] or high temperatures [19]. In this study, we report a facile approach for fabricating TiO_2 nanowire networks on Ti foil using a Ti corrosion reaction in KOH aqueous solutions at different temperatures, followed by a further ion-exchange process. We also investigate the efficacy of the TiO_2 nanowire networks as photoanodes for bendable DSSCs.

2. Results and Discussion

Figure 1 shows field emission scanning electron microscopy (FE-SEM) images of the nanostructures prepared using Ti wet corrosion in 5 M KOH aqueous solution at different temperatures. Porous nanowire networks were formed on the Ti surface at all corrosion temperatures. Details of the chemical reactions occurring between Ti and alkaline solution can be found elsewhere [20,21]. Ti reacts with hydroxide ions through hydration reactions, which results in hydrated TiO_2. Meanwhile, hydrated TiO_2 can be also dissolved as negatively charged hydrates by hydroxyl attack. Notably, a higher corrosion temperature yielded thicker nanowires, networks that were more aggregated, and a deeper corrosion depth. Figure 2 shows FE-SEM images of nanowire networks formed in 5 M KOH at 50 °C, and compares various corrosion time periods. Interestingly, compact nanowire networks were formed near the metal/oxide interface, whereas porous networks were produced near the surface. This result might be related to the local concentration gradient of hydroxide ions. The average thicknesses of the nanostructures formed for corrosion times of 6, 12, 24, and 48 h were 1.4 ± 0.1, 1.6 ± 0.2, 2.6 ± 0.2, and 3.3 ± 0.1 μm, respectively.

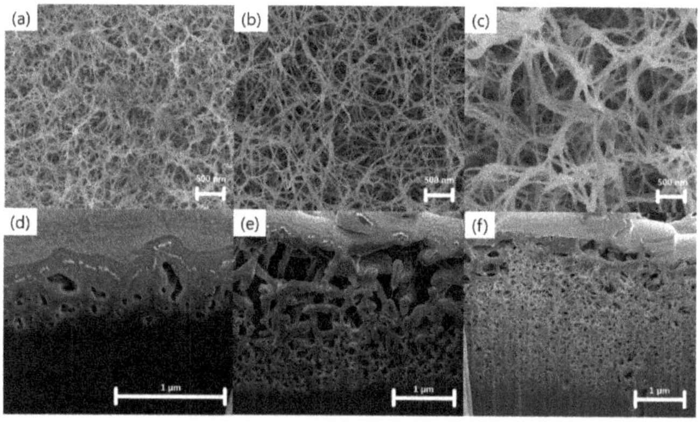

Figure 1. Field emission scanning electron microscopy (FE-SEM) images of the surface and cross-section of wet-corroded Ti foil at various temperatures in 5 M KOH aqueous solution: (**a**,**d**) 20 °C, (**b**,**e**) 50 °C, and (**c**,**f**) 80 °C.

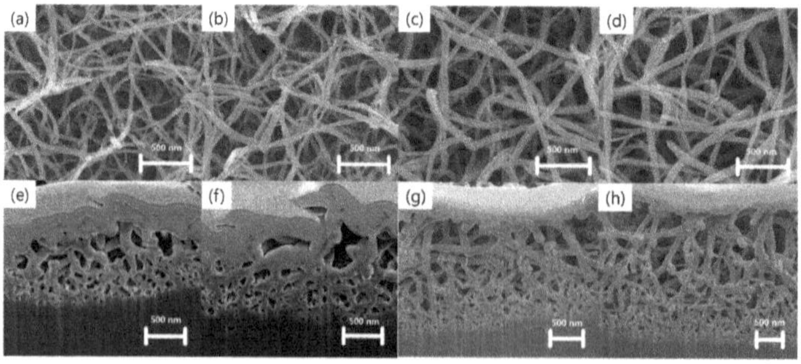

Figure 2. FE-SEM images of the surface and cross-section of wet-corroded Ti foil at 50 °C in 5 M KOH aqueous solution for various time periods: (**a**,**e**) 6 h, (**b**,**f**) 12 h, (**c**,**g**) 24 h, and (**d**,**h**) 48 h.

As reported previously, the resulting nanostructures could be K-incorporated TiO_2 nanowires containing K–Ti–O bonds [20–24]. Figure 3 shows the Raman scattering spectra of all of the wet-corroded samples under 531 nm excitation to identify Ti–O bonds of the anatase and rutile phases and K–Ti–O bonds of potassium titanates. Anatase has six Raman active modes (i.e., 144 cm^{-1} (E_g), 197 cm^{-1} (E_g), 399 cm^{-1} (B_{1g}), 513 cm^{-1} (A_{1g}), and 639 cm^{-1} (E_g)) and rutile has four Raman active modes (i.e., 143 cm^{-1} (B_{1g}), 447 cm^{-1} (E_g), 612 cm^{-1} (A_{1g}), and 826 cm^{-1} (B_{2g})) [25]. All samples showed typical Raman peaks originating from the anatase and rutile phases of TiO_2 (i.e., 197 cm^{-1}, 448 cm^{-1}, 640 cm^{-1}, and 826 cm^{-1}). Interestingly, prominent Raman peaks resulting from the potassium-doped TiO_2 (K-doped TiO_2; i.e., 285 cm^{-1} and 660 cm^{-1}) were also observed [26,27]. As the reaction temperature increased, all of the peaks became more intense and sharper, as a result of which the crystallinity of the nanostructures improved. Similarly, the corrosion time also had an influence on the degree of crystallinity. The X-ray diffraction (XRD) patterns of the wet-corroded Ti foils were also obtained (Figure S1 in Supplementary Materials). The intensity of the crystalline Ti peaks decreased with the increase in the reaction temperature. Unfortunately, prominent crystalline TiO_2 peaks were not observed using XRD.

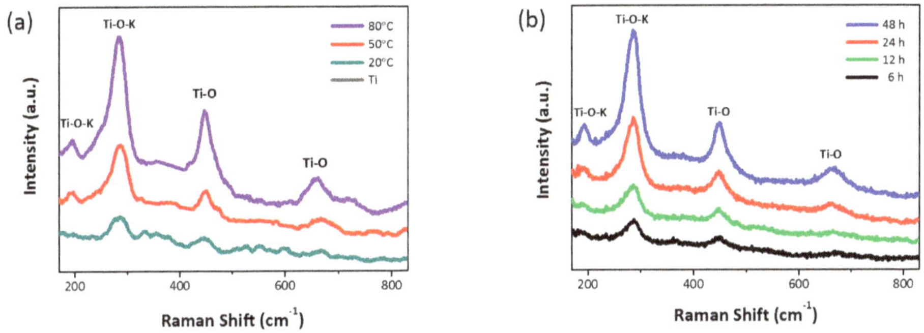

Figure 3. Raman spectra of Ti foils corroded (**a**) in 5 M KOH aqueous solution for 24 h at different temperatures and (**b**) in 5 M KOH aqueous solution at 50 °C for different time periods.

The chemical identity of the wet-corroded samples was confirmed by X-ray photoelectron spectroscopy (XPS). To replace K^+ with H^+, the wet-corroded Ti foil was immersed in HCl solution and the efficacy of the ion-exchange (hereafter referred to as "ion-exchange" and abbreviated as "IE") was investigated as well. According to the survey XPS scans, the samples contained K, Ti, and O; no other elements were detected, except for carbon (Figure S2 in Supplementary Materials). Figure 4 shows the X-ray photoelectron narrow scan spectra of the K 2p, Ti 2p, and O 1s levels in the nanostructures prepared at 50 °C for 48 h. The observed binding energies of the K 2p, Ti 2p, and O 1s levels are summarized in Table 1. As can be seen from Figure 4a, the binding energy of the K $2p_{3/2}$ level before IE treatment was in fairly good agreement with that of the K-incorporated titanates reported in the literature. However, the K 2p peak disappeared completely after the IE treatment. As shown in Figure 4b, the Ti 2p doublet peaks of Ti $2p_{1/2}$ and Ti $2p_{3/2}$ were observed at 464.5 eV and 458.8 eV, respectively, and this is ascribed to the Ti–Ti bond. In Figure 4c, the binding energy of the O 1s level corresponds mainly to the Ti–O bond (bulk O^{2-}) in TiO_2 (530.0 eV and 530.5 eV for anatase and rutile, respectively). The small peak at the higher binding energy was a result of OH groups belonging to hydroxyl groups and adsorbed H_2O, and its intensity became slightly higher after the IE treatment. On the basis of the observed binding energies, the as-received nanostructures were concluded to be K-doped TiO_2. Unfortunately, the K-doped TiO_2 exhibited p-type characteristics [24], and thus the removal of potassium dopant was crucial in order to use the nanostructures as the DSSC photoanode. It should be noted that the IE process was highly effective in removing potassium from the nanostructures without affecting the Ti–Ti and Ti–O bonds.

Table 1. Binding energies of K 2p, Ti 2p, and O 1s levels in X-ray photoelectron spectroscopy (XPS) fitting. IE: ion exchange.

(Unit: eV)	K 2p		Ti 2p		O 1s	
	$2p_{1/2}$	$2p_{3/2}$	$2p_{1/2}$	$2p_{3/2}$	O–H	Ti–O
48 h	295.6	292.9	464.5	458.8	532.1	530.4
48 h (IE)	-	-	464.5	458.8	531.9	530.4

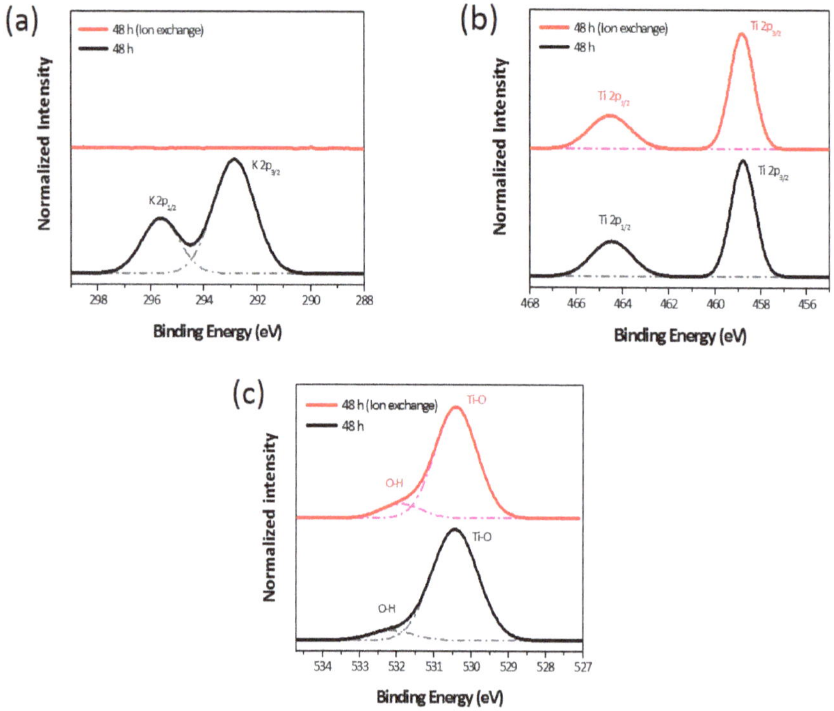

Figure 4. Normalized intensity of the XPS narrow scan spectra of (**a**) K 2p, (**b**) Ti 2p, and (**c**) O 1s levels. The dotted lines below the XPS spectra represent the Lorentzian-fitted curves.

The fabricated nanowire networks were sensitized with N719 dye on the Ti foil and then used as the photoanode of a DSSC. However, because the substrate was metallic, the DSSC was required to be illuminated from the Pt counter electrode side (i.e., back illumination). The main drawback of this configuration relates to the transmission losses due to the Pt-based catalyst and the I^-/I_3^- liquid electrolyte. Figure 5a shows the current density-voltage (J-V) characteristics of the DSSCs under back illumination. Table 2 summarizes the photovoltaic parameters. The photovoltaic performance improved with an increase in the reaction time: open-circuit voltage (V_{oc}) increased from 0.63 V to 0.69 V; photocurrent density (J_{sc}) increased from 0.60 mA/cm^2 to 2.08 mA/cm^2; and power conversion efficiency (η) increased from 0.27% to 1.03%. Notably, the IE treatment resulted in better photovoltaic performance: η increased from 1.03% to 1.11%. Electrochemical impedance spectroscopy (EIS) can offer valuable insights into interfacial charge-transfer processes of DSSCs. Figure 5b shows the Nyquist plots of the DSSCs under back illumination with applied open-circuit voltage. The semicircle in the intermediate frequency region reflects the charge-transfer resistance at the

TiO_2/photosensitizer/electrolyte interface. As the photoanode thickness increased, the charge-transfer resistance decreased; thus, this coincides with the resultant DSSC performance.

Table 2. Photovoltaic characteristics of wet-corroded Ti foil dye-sensitized solar cells (DSSCs) at 50 °C for different corrosion time periods. IE: ion exchange; FF: fill factor; V_{oc}: open-circuit voltage; η: power conversion efficiency; J_{sc}: photocurrent density.

Wet Corrosion Time (h)	V_{oc} (V)	J_{sc} (mA/cm^2)	*FF* (%)	η (%)
6	0.63	0.60	72.0	0.27
12	0.64	0.92	71.7	0.42
24	0.67	1.76	68.3	0.80
48	0.69	2.08	71.5	1.03
48 (IE)	0.69	2.28	70.1	1.11

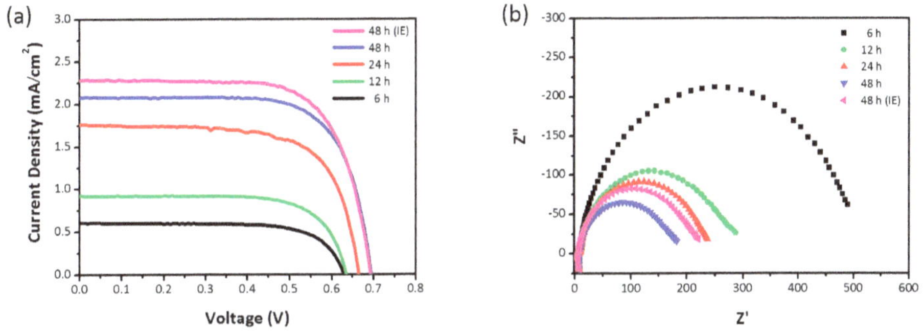

Figure 5. (**a**) J-V characteristics and (**b**) Nyquist plots of DSSCs.

3. Material and Methods

3.1. Three-Dimensional TiO_2 Nanowire Networks

A pure titanium foil (Ti > 99.5%, Nilaco Co., Tokyo, Japan) with a thickness of 1 mm was used as the starting material for wet corrosion. TiO_2 nanostructures could be prepared through a Ti corrosion reaction in KOH aqueous solution [20]. A Ti substrate 15 mm × 30 mm in size was polished with a SiC sheet (No. 1000) and subsequently cleaned by ultrasonication in acetone, isopropanol, and deionized (DI) water. The cleaned substrate was immersed in 5 M KOH (95%) for 24 h at different temperatures (20, 50, and 80 °C). The wet-corroded Ti substrate was thoroughly rinsed with DI water. The 3D morphology of the TiO_2 nanostructures was investigated by field emission scanning electron microscopy (FE-SEM, S-4800, Hitachi, Tokyo, Japan). A Focused Ion Beam (FIB, Thermo Fisher Scientific, Waltham, MA, USA) was used to prepare the cross-sectional samples. Micro-Raman spectroscopy was performed in a back-scattering geometry by using a laser operating at a wavelength of approximately 531 nm and with a spectral resolution of 1.4 cm^{-1} (FEX, NOST, Seongnam, Korea). The Raman signals were detected using a charge-coupled-device (CCD) camera (iDus DV401A, Andor, Concord, MA, USA). The XRD patterns were collected on a D/max250/PC (Rigaku, Tokyo, Japan) using Cu radiation at 40 kV and 200 mA at room temperature. X-ray photoelectron spectroscopy (XPS) was performed with the K-Alpha XPS system (Thermo Fischer Scientific, Waltham, MA, USA) using a monochromated Al Kα X-ray source with an energy of 1486.6 eV. The spectra of Ti 2p and O 1s energy levels were calibrated with respect to the C 1s peak of the adventitious carbon on the sample surface at 285.0 eV.

3.2. DSSCs

The wet-corroded foil was immersed in 0.1 M HCl aqueous solution for 24 h at room temperature (RT) to replace K^+ with H^+, after which it was rinsed with deionized (DI) water and dried under N_2 flow. The titanium tetrachloride ($TiCl_4$) treatment was performed by soaking the foil in 0.04 M $TiCl_4$ aqueous solution at 75 °C for 30 min. It was then rinsed with DI water and sintered at 500 °C for 30 min. The foil was exposed to O_2 plasma and then immersed in 0.1 M HNO_3 solution for 30 min to facilitate dye adsorption. The final foil was immersed in a 0.5 mM N719 (Solaronix) ethanol solution for 12 h. A Pt counter electrode was prepared on fluorine-doped SnO_2 (FTO)-coated conducting glass (TEC 8, Pilkington; thickness: 2.2 mm, sheet resistance: 8 Ω/sq) by spin-coating of 0.04 M chloroplatinic acid (H_2PtCl_6) solution and post-annealing at 400 °C for 1 h. Both the dye-sensitized foil and the Pt counter electrode were sealed with a 25-μm-thick layer of Surlyn (Solaronix, Aubonne, Switzerland). An iodide based redox electrolyte (Iodolyte AN-50, Solaronix, Aubonne, Switzerland) was injected into the cell. The photovoltaic characteristics of the cell were measured using a solar cell I–V measurement system (K3000 LAB, McScience Inc., Suwon, Korea) under air mass 1.5 (AM 1.5) global, one-sun illumination (100 mW/cm^2). The effective area of the fabricated solar cell was 1 cm × 0.7 cm. The open-circuit voltage (V_{oc}), photocurrent density (J_{sc}), fill factor (FF), and power conversion efficiency (η) were recorded simultaneously. EIS experiments were performed using a frequency response analyzer (Solartron 1260, AMETEK. Inc., Berwyn, PA, USA). A sinusoidal potential perturbation with an amplitude of 10 mV was applied over a frequency range from 100 kHz to 0.1 Hz.

4. Conclusions

TiO_2 nanowire networks were easily prepared with Ti corrosion in strong basic solutions at different temperatures and then a further IE process. Importantly, the prepared nanostructures on Ti foils were utilized as the photoanodes of bendable DSSCs, and consequently, the DSSCs exhibited a power conversion efficiency of 1.11%, even under back illumination. Our work towards further developments (e.g., fabrication optimization and transfer of the TiO_2 nanowire networks to various substrates [11,17,28] for front illumination) will be explored and published in due course.

Supplementary Materials: The following are available online at http://www.mdpi.com/2079-4991/7/10/315/s1, Figure S1: XRD patterns of the wet-corroded Ti foil samples at various temperatures in 5 M KOH aqueous solution and normal Ti foil, Figure S2: Survey XPS spectrum of wet-corroded Ti foil sample at corrosion temperature of 50 °C and corrosion time of 48 h.

Acknowledgments: This work was supported by a grant (17CTAP-C129910-01) from the Technology Advancement Research Program (TARP) funded by the Ministry of Land, Infrastructures, and Transport (MOLIT) of Korea and the Chung-Ang University Graduate Research Scholarship (2016).

Author Contributions: Saera Jin, Eunhye Shin and Jongin Hong conceived and designed the experiments; Saera Jin and Eunhye Shin performed the experiments; Saera Jin and Eunhye Shin analyzed the data; Saera Jin, Eunhye Shin and Jongin Hong wrote the paper; Saera Jin and Eunhye Shin significantly contributed to manuscript preparation.

Conflicts of Interest: The authors declare no conflict of interest.

References

1. O'Regan, B.; Gratzel, M. A low-cost, high-efficiency solar-cell based on dye-sensitized colloidal TiO_2 films. *Nature* **1991**, *353*, 737–740. [CrossRef]
2. Hagfeldt, A.; Boschloo, G.; Sun, L.; Kloo, L.; Pettersson, H. Dye-sensitized solar cells. *Chem. Rev.* **2010**, *110*, 6595–6663. [CrossRef] [PubMed]
3. Kang, M.G.; Park, N.-G.; Ryu, K.S.; Chang, S.H.; Kim, K.-J. A 4.2% efficient flexible dye-sensitized TiO_2 solar cells using stainless steel substrate. *Sol. Energy Mater. Sol. Cell* **2006**, *90*, 574–581. [CrossRef]
4. Park, J.H.; Jun, Y.; Yun, H.-G.; Lee, S.-Y.; Kang, M.G. Fabrication of an Efficient Dye-Sensitized Solar Cell with Stainless Steel Substrate. *J. Electrochem. Soc.* **2008**, *155*, F145–F149. [CrossRef]

5. Onoda, K.; Ngamsinlapasathian, S.; Fujieda, T.; Yoshikawa, S. The superiority of Ti plate as the substrate of dye-sensitized solar cells. *Sol. Energy Mater. Sol. Cell* **2007**, *91*, 1176–1181. [CrossRef]
6. Chen, H.W.; Huang, K.-C.; Hsu, C.-Y.; Lin, C.-Y.; Chen, J.-G.; Lee, C.-P.; Lin, L.-Y.; Vittal, R.; Ho, K.-C. Electrophoretic deposition of TiO_2 film on titanium foil for a flexible dye-sensitized solar cell. *Electrochim. Acta* **2011**, *56*, 7991–7998. [CrossRef]
7. Frank, A.J.; Kopidakis, N.; van de Lagemaat, J. Electrons in nanostructured TiO_2 solar cells: Transport, recombination and photovoltaic properties. *Coord. Chem. Rev.* **2004**, *248*, 1165–1179. [CrossRef]
8. Zhang, Q.; Cao, G. Nanostructured photoelectrodes for dye-sensitized solar cells. *Nano Today* **2011**, *6*, 91–109. [CrossRef]
9. Mor, G.K.; Varghese, O.K.; Paulose, M.; Shankar, K.; Grimes, C.A. A review on highly ordered, vertically oriented TiO_2 nanotube arrays: Fabrication, material properties, and solar energy applications. *Sol. Energy Mater. Sol. Cell* **2006**, *90*, 2011–2075. [CrossRef]
10. Wang, J.; Lin, Z. Anodic formation of ordered TiO_2 nanotube arrays: Effects of electrolyte temperature and anodization potential. *J. Phys. Chem.* **2009**, *113*, 4026–4030. [CrossRef]
11. Yip, C.T.; Huang, H.; Zhou, L.; Xie, K.; Wang, Y.; Feng, T.; Li, J.; Tam, W.Y. Direct and seamless coupling of TiO_2 nanotube photonic crystal to dye-sensitized solar cell: A single-step approach. *Adv. Mater.* **2011**, *23*, 5624–5628. [CrossRef] [PubMed]
12. Song, M.Y.; Kim, D.K.; Ihn, K.J.; Jo, S.M.; Kim, D.Y. Electrospun TiO_2 electrodes for dye-sensitized solar cells. *Nanotechnology* **2004**, *15*, 1861–1865. [CrossRef]
13. Ohsaki, Y.; Masaki, N.; Kitamura, T.; Wada, Y.; Okamoto, T. Dye-sensitized TiO_2 nanotube solar cells: Fabrication and electronic characterization. *Phys. Chem. Chem. Phys.* **2005**, *7*, 4157–4163. [CrossRef] [PubMed]
14. Wu, W.-Q.; Rao, H.-S.; Xu, Y.-F.; Wang, Y.-F.; Su, C.-Y.; Kuang, D.-B. Hierarchical oriented anatase TiO_2 nanostructure arrays on flexible substrate for efficient dye-sensitized solar cells. *Sci. Rep.* **2013**, *3*, 1892. [CrossRef] [PubMed]
15. Wu, W.-Q.; Xu, Y.-F.; Su, C.-Y.; Kuang, D.-B. Ultra-long anatase TiO_2 nanowire arrays with multi-layered configuration on FTO glass for high-efficiency dye-sensitized solar cells. *Energy Environ. Sci.* **2014**, *7*, 644–649. [CrossRef]
16. Gong, D.; Grimes, C.A.; Varghese, O.K.; Hu, W.C.; Singh, R.S.; Chen, Z.; Dickey, E.C. Titanium oxide nanotube arrays prepared by anodic oxidation. *J. Mater. Res.* **2001**, *16*, 3331–3334. [CrossRef]
17. Park, J.; Lee, T.-W.; Kang, M.G. Growth, detachment and transfer of highly-ordered TiO_2 nanotube arrays: Use in dye-sensitized solar cells. *Chem. Commun.* **2008**, 2867–2869. [CrossRef] [PubMed]
18. Wu, J.M. Low-temperature preparation of titania nanorods through direct oxidation of titanium with hydrogen peroxide. *J. Cryst. Growth* **2004**, *269*, 347–355. [CrossRef]
19. Peng, X.; Chen, A. Aligned TiO_2 nanorod arrays synthesized by oxidizing titanium with acetone. *J. Mater. Chem.* **2004**, *14*, 2542–2548. [CrossRef]
20. Shin, E.; Jin, S.; Kim, J.; Chang, S.-J.; Jun, B.-H.; Park, K.-W.; Hong, J. Preparation of K-doped TiO_2 nanostructures by wet corrosion and their sunlight-driven photocatalytic performance. *Appl. Surf. Sci.* **2016**, *379*, 33–38. [CrossRef]
21. Lee, S.Y.; Lee, C.H.; Kim, D.Y.; Locquet, J.-P.; Seo, J.W. Preparation and photocatalytic activity of potassium-incorporated titanium oxide nanostructures produced by the wet corrosion process using various titanium alloys. *Nanomaterials* **2015**, *5*, 1397–1417. [CrossRef] [PubMed]
22. Kim, J.-I.; Lee, S.-Y.; Pyun, J.-C. Characterization of photocatalytic activity of TiO_2 nanowire synthesized from Ti-plate by wet corrosion process. *Curr. Appl. Phys.* **2009**, *9*, e252–e255. [CrossRef]
23. Lee, S.-Y.; Takai, M.; Kim, H.-M.; Ishihara, K. Preparation of nano-structured titanium oxide film for biosensor substrate by wet corrosion process. *Curr. Appl. Phys.* **2009**, *9*, e266–e269. [CrossRef]
24. Lee, S.-Y.; Matsuno, R.; Ishihara, K.; Takai, M. Electrical transport ability of nanostructured potassium-doped titanium oxide film. *Appl. Phys. Express* **2011**, *4*, 407–416. [CrossRef]
25. Zhang, J.; Li, M.; Feng, Z.; Chen, J.; Li, C. UV Raman spectroscopic study on TiO_2. I. phase transformation at the surface and in the bulk. *J. Phys. Chem. B* **2006**, *110*, 927–935. [CrossRef] [PubMed]
26. Liu, C.; Lu, X.; Yu, G.; Feng, X.; Zhang, Q.; Xu, Z. Role of an intermediate phase in solid state reaction of hydrous titanium oxide with potassium carbonate. *Mater. Chem. Phys.* **2005**, *94*, 401–407. [CrossRef]

27. Chen, L.-C.; Huang, C.-M.; Tsai, F.-R. Characterization and photocatalytic activity of K^+-doped TiO_2 photocatalysts. *J. Mol. Catal. A Chem.* **2007**, *265*, 133–140. [CrossRef]
28. Ke, S.; Chen, C.; Fu, N.; Zhou, H.; Ye, M.; Lin, P.; Yuan, W.; Zeng, X.; Chen, L.; Huang, H. Transparent indium tin oxide electrodes on muscovite mica for high-temperature-processed flexible optoelectronic devices. *ACS Appl. Mater. Interfaces* **2016**, *8*, 28406–28411. [CrossRef] [PubMed]

nanomaterials

MDPI

Article

Periodic TiO_2 Nanostructures with Improved Aspect and Line/Space Ratio Realized by Colloidal Photolithography Technique

Loïc Berthod [1], Olga Shavdina [1,2], Isabelle Verrier [1], Thomas Kämpfe [1,*], Olivier Dellea [2], Francis Vocanson [1], Maxime Bichotte [1], Damien Jamon [1] and Yves Jourlin [1]

1 Lyon, UJM-Saint-Etienne, Laboratoire Hubert Curien UMR 5516, CNRS, Institut d'Optique Graduate School, F-42023 Saint-Etienne, France; loic.berthod@univ-st-etienne.fr (L.B.); olga.shavdina@univ-st-etienne.fr (O.S.); isabelle.verrier@univ-st-etienne.fr (I.V.); francis.vocanson@univ-st-etienne.fr (F.V.); maxime.bichotte@univ-st-etienne.fr (M.B.); damien.jamon@univ-st-etienne.fr (D.J.); Yves.Jourlin@univ-st-etienne.fr (Y.J.)

2 Laboratoire des Composants pour le Conversion de l'Energie (L2CE), Laboratoire d'Innovation pour les Technologies des Energies Nouvelles et des nanomatériaux (CEA/LITEN), F-38054 Grenoble, France; olivier.dellea@cea.fr

* Correspondence: thomas.kampfe@univ-st-etienne.fr; Tel.: +33-4-7791-5819

Received: 11 September 2017; Accepted: 5 October 2017; Published: 12 October 2017

Abstract: This paper presents substantial improvements of the colloidal photolithography technique (also called microsphere lithography) with the goal of better controlling the geometry of the fabricated nano-scale structures—in this case, hexagonally arranged nanopillars—printed in a layer of directly photopatternable sol-gel TiO_2. Firstly, to increase the achievable structure height the photosensitive layer underneath the microspheres is deposited on a reflective layer instead of the usual transparent substrate. Secondly, an increased width of the pillars is achieved by tilting the incident wave and using multiple exposures or substrate rotation, additionally allowing to better control the shape of the pillar's cross section. The theoretical analysis is carried out by rigorous modelling of the photonics nanojet underneath the microspheres and by optimizing the experimental conditions. Aspect ratios (structure height/lateral structure size) greater than 2 are predicted and demonstrated experimentally for structure dimensions in the sub micrometer range, as well as line/space ratios (lateral pillar size/distance between pillars) greater than 1. These nanostructures could lead for example to materials exhibiting efficient light trapping in the visible and near-infrared range, as well as improved hydrophobic or photocatalytic properties for numerous applications in environmental and photovoltaic systems.

Keywords: sol-gel; TiO_2; sub-wavelength structures; colloidal photolithography

1. Introduction

Colloidal photolithography [1,2] has several advantages, the most important one being its ability to periodically nano-structure large surfaces which can be planar or non-planar (curved of cylinder based shape). The method uses microspheres arranged in a regular grid to focus light into a photosensitive material. It is based on a 2D hexagonal self-arrangement of the microspheres in a monolayer. The concentration of the optical field underneath the microspheres called 'photonic nanojet' can illuminate the photosensitive layer locally, leading to a latent image according to the arrangement of the microspheres, which is then chemically developed.

Among the photosensitive materials, TiO_2 sol-gel material is attractive because of its optical and chemical properties [3], especially when it is nanostructured [4]. TiO_2 is well known for its high refractive index (up to 2.2 in its anatase phase), for its high mechanical and chemical stability, as well as for its photocatalytic properties. Association of the colloidal photolithography with TiO_2 material leads to innovative components that could be used for example in attractive environmental applications [5,6] as well as in the domain of solar and photovoltaic energy [7–9]. Combining colloidal lithography and direct photopatternable sol-gel TiO_2 material leads to a unique and powerful technology allowing to perform microstructuring in only one technological step, without etching process, while being compatible with standard and non-standard large substrates.

When TiO_2 sol-gel material, being a negative photoresist, is periodically structured by colloidal lithography, the nanojets issued from each microsphere will lead to the origination of periodic nanopillars [10]. One limit of this process according to the state of the art is a low aspect ratio (height/lateral size) of the achieved nanopillars (or rods) due to the shape of the nanojet inside the TiO_2. This is unfortunate since very high nanopillars have interesting properties, for instance, with regards to hydrophobicity. In the present study this limit was exceeded, with nanopillars of up to several hundreds of nanometers in height compared to the commonly achieved several tens of nanometers. This was achieved by depositing the sol-gel material on a reflecting substrate, like aluminum, and exploiting standing wave effects between the different materials. This approach of using metal layers to confine and form the electric field has been successfully applied for other lithographic techniques (e.g., two-beam lithography [11]) but, to the best of the author's knowledge, this approach is new to colloidal lithography.

A second optimization of the form of the nanojets promises to result in wider than usual nanopillars, increasing the line/space ratio (size of the pillar/period of the grating). Wide nanopillars can, for example, increase the absorption of light in the UV region and can lead to useful photo-catalysis phenomena as well as higher efficiency in solar cells. The idea is to apply a tilt to the incident wave focused into the TiO_2 material. Some authors have already considered microsphere photolithography under oblique incidence to produce arbitrary nano-patterns by projection of a pixelated optical mask into the photoresist [12]. In comparison, our approach is simpler because it is based on exposure of the TiO_2 sol-gel directly through a mask of a SiO_2 microspheres monolayer without any other intermediary optical system.

In the following, the results of rigorous simulations of the optical field behind the microspheres are presented, followed by experimental demonstrations of the two mentioned approaches for achieving wider nanopillars with complex shapes, as well as very high aspect ratio columns.

2. Materials and Methods

2.1. Rigorous Optical Simulation of Nanojets Created by Microsphere Arrays

In order to predict the distribution of the electric field from the array of microspheres, the geometry of the elementary cell of a hexagonal grating of silica microspheres was defined using a MATLAB routine for later use in an rigorous coupled-wave analysis (RCWA) [13] based optical propagation code ("MCGratings" [14]), which allows the calculation of the electromagnetic field distribution behind the microspheres. Each microsphere of 1 μm diameter was longitudinally (i.e., in the direction of light propagation) discretized in 35 layers, which is a good compromise for determining a sufficiently exact representation of the electromagnetic field in an acceptable calculation time. The number of considered Fourier-orders was determined by repeatedly calculating several representative structures with an increasing number of Fourier-orders and verifying the convergence of the results towards a solution that is sufficiently stable. 15×25 Fourier-orders were used for the calculation in this paper, as they proved to be a good compromise between calculation time and the required precision of the result.

2.2. Direct Photolithography of the Sol-Gel TiO_2

In order to prepare TiO_2-based photoresist, specific sol-gel formulations were used. The final sol is prepared from titanium isopropoxide orthotitanate (TIPT) complexed by benzoyl acetone (BzAc), using a mixture of two different primary sols as detailed in [4]. The so-obtained final sol can be coated on glass substrates by spin-coating and is thus compatible with large sized substrates. The deposited sol-gel TiO_2 film is coated by a microspheres monolayer using the Langmuir Blodgett method that leads to hexagonal self-organization of the particles, as detailed in [10]. Further chemical and optical properties of the sol-gel and details about its preparation, as well as required parameters and details about the photolithographic patterning process can be found in previous works by the authors [4,10].

3. Results and Discussion

3.1. Simulation and Optimization of the Microsphere-Created Nanojets

The result of the optical simulations are presented in Figure 1, showing the obtained mapping of the component E_y of the electric field for a linearly polarized incident wave at the wavelength $\lambda = 365$ nm (corresponding to the i-line of the used narrow-band gas discharge mercury vapor lamp) propagating along the z-axis from air (top) to the sol-gel layer (down) with varying tilting angle, using optical borosilicate-crown glass (BK7) as substrate. The influence of the polarization direction of the incident light was found to be negligible for the overall structure of the field distribution, it was therefore fixed to having the E-Field along the y direction. In the simulation, the sol-gel layer thickness is supposed to be infinite underneath the microspheres that are arranged hexagonally with a period of 1 μm in x-direction, corresponding to their diameter. The incidence angle in x-direction of the incident wave varies from 0° (Figure 1a) to 30° (Figure 1f). The refractive index and the absorption of the sol-gel depend weakly on the illumination parameters and are thus difficult to fix in the modeling. As those changes are minor [10], we restricted for the scope of this paper the simulation to a non-absorbing sol-gel with a fixed refractive index of $n = 1.63$, measured by ellipsometry.

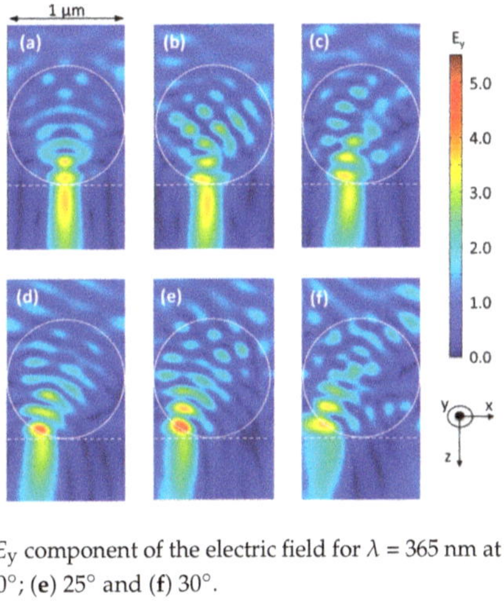

Figure 1. Mapping of the E_y component of the electric field for $\lambda = 365$ nm at different incidence angles: (**a**) 0°; (**b**) 5°; (**c**) 12°; (**d**) 20°; (**e**) 25° and (**f**) 30°.

Having calculated the electric field, several parameters were studied to fully analyze and exploit the properties of the resulting nanojet. These parameters include the distance from the microsphere's output face to the maximum intensity inside the nanojet, the nanojet's length and diameter as well as the ratio between them, and the energy of the nanojet outside of the microsphere. The analysis of these parameters permits to choose the best exposure angle in order to obtain a required geometry (the length/width ratio and the shift of the nanojet along x-axis) with sufficient field concentration to expose the TiO_2 in a reasonable time. It turns out that the overall shape and the length/width ratio of the nanojet remains stable up to tilting angles of about 25°, allowing for a flexible use of tilted incidence waves to change the shape of the developed structures. As expected, increasing the exposure angle shifts the lateral position of the maximal field amplitude laterally (Figure 2).

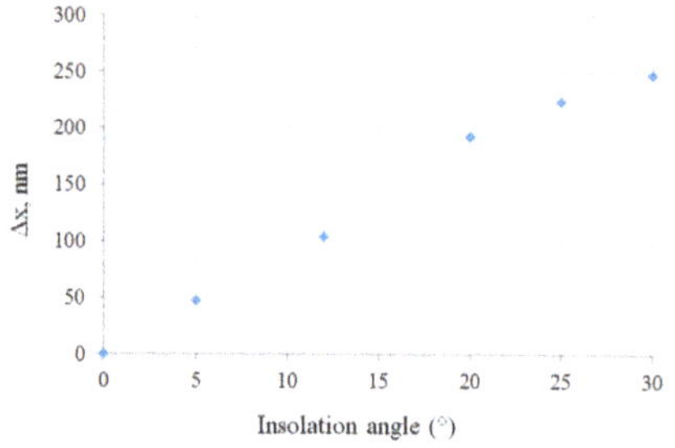

Figure 2. Lateral shift of the maximum intensity of the nanojet inside the TiO_2 film versus exposure angle.

3.2. *First Optimization: Increasing the Nanopillars Width and Creating Variable Shapes*

Figure 3 shows the different investigated optical setups and the resulting structures, comprising two exposures (Figure 3a) and four exposures (Figure 3b) under 20° exposure angle, as well as a continuous exposure at 20° (Figure 3c) and 25° (Figure 3d) inclination coupled to a continuous rotation of the substrate. According to Figure 2b for an angle of illumination of 20°, the maximum intensity shifts about 200 nm in the lateral direction x leading to different shapes on the illuminated area according to the three configurations shown in the middle row of Figure 3. After development in ethanol, each condition of illumination leads to different shapes of the resulting hexagonally periodic structure illustrated by the SEM photographs at the bottom row of Figure 3. For the two-beam case of Figure 3a a bow-tie pattern appears, the four exposures case of Figure 3b creates a clover-shaped pattern, whereas the rotation under an angle of 20° leads to large nanopillars with approximately 500 nm diameter (Figure 3c), and close to 600 nm diameter for 25° exposure angle, corresponding to a line/space ratio greater than 1 (Figure 3d).

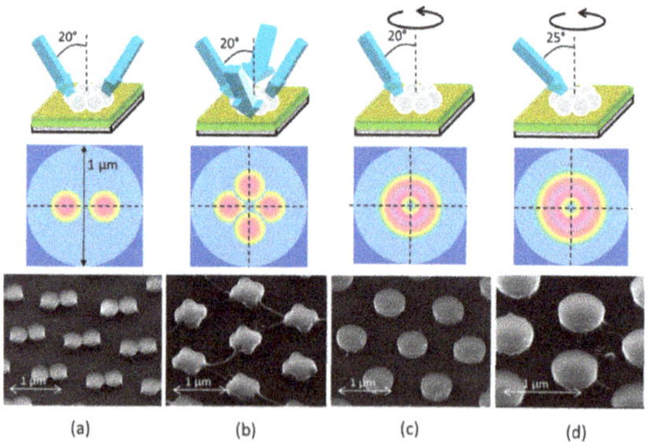

Figure 3. Illustration of the illumination conditions (top row), top view the expected nanojet arrangement in the TiO_2 layer (middle row, red indicates increased intensity) and SEM photographs of the resulting TiO_2 pattern (bottom row) of hexagonally periodic structures on BK7. (**a**) two opposite exposures of angle 20° leading to a bow-tie structure, (**b**) four exposures of angle 20° leading to clover leaf structure, (**c**) one exposure under an angle of 20° with substrate rotation leading to nanopillars of 500 nm diameter and (**d**) one exposure under an angle of 25° with substrate rotation leading to nanopillars of 600 nm diameter.

3.3. Second Optimization: Increasing the Nanopillar's Height

The second optimization of the geometry of the nanopillars, where the goal was to increase the pillar's height, consists in using a reflecting substrate like, in the present case, a BK7 glass coated with a thin film (170 nm) of aluminum. As previously mentioned, the microspheres are deposited on the sol-gel TiO_2 film. At normal incidence, the nanojet coming from the microspheres is reflected back, leading to a standing wave pattern which is notably longer than the nanojet without reflection, thus allowing higher nanopillars to be created. However, the standing wave creates intensity minima and maxima along the z-axis, which can be detrimental to the formation of smooth nanopillars if the sidewalls follow this modulation. Furthermore, at the interface TiO_2-Al, the electric field value must be 0 in order to satisfy the continuity condition of its transverse component. This condition is not favorable for the stability of the structure because exposure of the TiO_2 at the interface will be particularly ineffective. The distance L between two successive minima is related to the real part of the refractive index n of the film and to the insolating wavelength λ by $L = \lambda/2n$. For TiO_2 of refractive index $n = 1.84$ at the insolating wavelength $\lambda = 365$ nm, L takes the value of 100 nm. The number of nodes is given by the film thickness divided by the distance L. For example, for a TiO_2 film of thickness 600 nm, the nodes number is 6, as confirmed in the simulation shown in Figure 4b (top row).

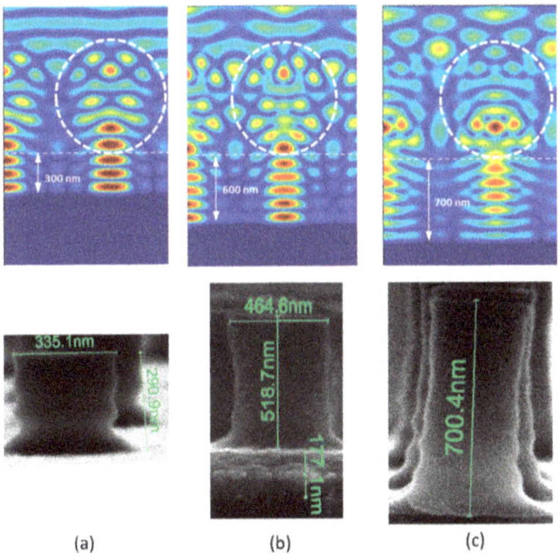

Figure 4. Simulation mapping of the electric field and SEM of the corresponding nanopillar with TiO_2 initial film thickness of: (**a**) 300 nm; (**b**) 600 nm; and (**c**) 700 nm.

Using the same experimental process as in the first part of this section but without any tilt, the nanopillar created by the nanojet reveals a shape in Figure 4b (bottom row) that is not so far from that predicted by the simulation. However, a smoothing effect of the exposed image is present, consolidating the pillars at their basis and also avoiding a too important modulation of the sidewalls. The smoothing can be attributed to the spectral linewidth of the illumination (which is neglected in the simulation), leading to a minor position change of the maxima and minima of the intensity and thus slightly blurring the interference pattern. Furthermore, a change of the refractive index and the absorption of the sol-gel can occur during exposure (bleaching effect, densification [10]) which will dynamically alter the field distribution and thus also create blurring. A detailed analysis of the importance of those effects is beyond the scope of this paper, the good agreement between simulation and results however confirm the eligibility of the modelling. The obtained pattern is 465 nm wide and 520 nm high with an under layer of thickness 177 nm. The addition of the height of the nanopillar and of the under-layer's thickness corresponds to the initial TiO_2 thickness deposited onto the substrate.

The aspect ratio (height/width) is in this case equal to 1.2, but it has been improved with other samples when using a thicker film of TiO_2 as shown below. Figure 4 shows the simulation results and the SEM images for different thicknesses of the TiO_2 layer. For a thickness of 300 nm (Figure 4a), we can notice undulations on the nanopillars' edge that follows the three nodes of the standing wave. For the higher nanopillars, the effect is less visible because of a reduced contrast between minima and maxima of the standing wave. In the case of a 300 nm thickness TiO_2 film, the aspect ratio is 0.87 and is, as expected, smaller than the one obtained previously for a 600 nm thick film.

However, using even thicker TiO_2 film, the height of nanopillars can reach 700 nm (Figure 4c) and the aspect ratio is 2.18. The SEM photographs of the hexagonal arrangement of the nanopillar are presented in Figure 5a. The diffraction effect shown in Figure 5b confirms the presence of the structuration on the whole surface. It also shows that the size of the perfectly crystalline regions of the surface is limited, in our case from several hundreds of µm to some mm, which is a known effect for Langmuir–Blodgett type monolayers [15].

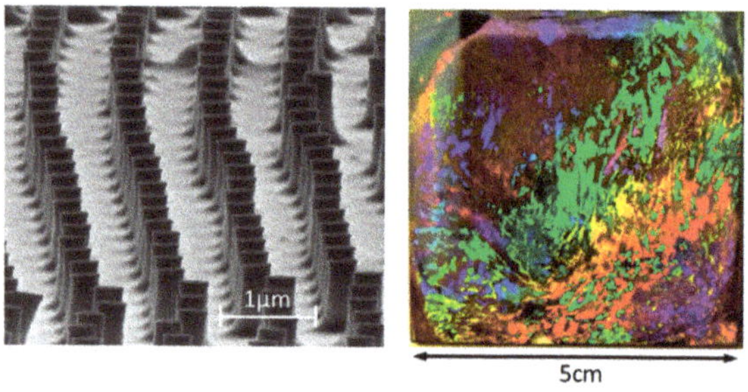

(a) (b)

Figure 5. Global view of the hexagonally arranged nanostructures with high aspect ratio: (**a**) SEM of the array of nanopillars; (**b**) macroscopic view of the structure under white light illumination, showing the typical rainbow effect of periodic surface structures.

4. Conclusions

In conclusion, the colloidal lithography technique is used to create original TiO_2 nano-structures with high aspect ratio or with large line/space ratio, opening up its use for new applications. This study has demonstrated both theoretically and experimentally that the shape of the photonic nanojet limiting the nano-plots height can be modified using field enhancement by wave reflection and tilted illumination. The aspect ratio of the nanopillars was improved by 147% when using reflecting aluminum substrates compared to the case of transparent substrates. In this case, the shape of the pillars is additionally no longer trapezoidal, which is characteristic for transparent substrates, but approximates a square shape. Furthermore, it has been shown that the nanopillar's width can be increased by employing multiple illumination technique and substrate rotation under oblique incidence, resulting in line/width ratios of the nanostructure grating larger than 1, strongly increased in comparison to the case of normal incidence. These improvements provide much flexibility to the colloidal lithography technique regarding the geometry of the microscopic structures, paving the way for a more widespread application to large scale planar and non-planar substrates.

Acknowledgments: The authors would like to thank the French Region Rhône-Alpes for its support, in the framework of an ARC 4 Energies, and Stéphanie Reynaud from the Hubert Curien Laboratory for her contribution to the SEM images.

Author Contributions: Yves Jourlin and Loïc Berthod conceived and designed the lithographic setups; Olga Shavdina and Thomas Kämpfe prepared the numerical simulations; Olivier Dellea and Loïc Berthod prepared the microsphere arrays; Francis Vocanson prepared and optimized the TiO_2 Solgel solution; Isabelle Verrier, Loïc Berthod, Damien Jamon, and Maxime Bichotte carried out the lithographic patterning; Loïc Berthod, Isabelle Verrier, and Thomas Kämpfe wrote the paper.

Conflicts of Interest: The authors declare no conflict of interest.

References

1. Wu, W.; Dey, D.; Memis, O.G.; Katsnelson, A.; Mohseni, H. A Novel Self-aligned and Maskless Process for Formation of Highly Uniform Arrays of Nanoholes and Nanopillars. *Nanoscale Res. Lett.* **2008**, *3*, 123–127. [CrossRef]
2. Hsu, C.-M.; Connor, S.T.; Tang, M.X.; Cui, Y. Wafer-scale silicon nanopillars and nanocones by Langmuir-Blodgett assembly and etching. *Appl. Phys. Lett.* **2008**, *93*, 133109. [CrossRef]
3. Fallet, M.; Permpoon, S.; Deschanvres, J.L.; Langlet, M. Influence of physico-structural properties on the photocatalytic activity of sol-gel derived TiO_2 thin films. *J. Mater. Sci.* **2006**, *41*, 2915–2927. [CrossRef]
4. Briche, S.; Tebby, Z.; Riassetto, D.; Messaoud, M.; Gamet, E.; Pernot, E.; Roussel, H.; Dellea, O.; Jourlin, Y.; Langlet, M. New insights in photo-patterned sol-gel-derived TiO_2 films. *J. Mater. Sci.* **2011**, *46*, 1474–1486. [CrossRef]
5. Lazar, M.; Varghese, S.; Nair, S. Photocatalytic Water Treatment by Titanium Dioxide: Recent Updates. *Catalysts* **2012**, *2*, 572–601. [CrossRef]
6. Pelaez, M.; Nolan, N.T.; Pillai, S.C.; Seery, M.K.; Falaras, P.; Kontos, A.G.; Dunlop, P.S.M.; Hamilton, J.W.J.; Byrne, J.A.; O'Shea, K.; et al. A review on the visible light active titanium dioxide photocatalysts for environmental applications. *Appl. Catal. B Environ.* **2012**, *125*, 331–349. [CrossRef]
7. Mor, G.K.; Shankar, K.; Paulose, M.; Varghese, O.K.; Grimes, C.A. Use of Highly-Ordered TiO_2 Nanotube Arrays in Dye-Sensitized Solar Cells. *Nano Lett.* **2006**, *6*, 215–218. [CrossRef] [PubMed]
8. Zukalová, M.; Zukal, A.; Kavan, L.; Nazeeruddin, M.K.; Liska, P.; Grätzel, M. Organized Mesoporous TiO_2 Films Exhibiting Greatly Enhanced Performance in Dye-Sensitized Solar Cells. *Nano Lett.* **2005**, *5*, 1789–1792. [CrossRef] [PubMed]
9. Spinelli, P.; Macco, B.; Verschuuren, M.A.; Kessels, W.M.M.; Polman, A. Al_2O_3/TiO_2 nano-pattern antireflection coating with ultralow surface recombination. *Appl. Phys. Lett.* **2013**, *102*, 233902. [CrossRef]
10. Shavdina, O.; Berthod, L.; Kämpfe, T.; Reynaud, S.; Veillas, C.; Verrier, I.; Langlet, M.; Vocanson, F.; Fugier, P.; Jourlin, Y.; et al. Large Area Fabrication of Periodic TiO_2 Nanopillars Using Microsphere Photolithography on a Photopatternable Sol-Gel Film. *Langmuir* **2015**, *31*, 7877–7884. [CrossRef] [PubMed]
11. Kusaka, K.; Kurosawa, H.; Ohno, S.; Sakaki, Y.; Nakayama, K.; Moritake, Y.; Ishihara, T. Waveguide-mode interference lithography technique for high contrast subwavelength structures in the visible region. *Opt. Express* **2014**, *22*, 18748–18756. [CrossRef] [PubMed]
12. Bonakdar, A.; Rezaei, M.; Brown, R.L.; Fathipour, V.; Dexheimer, E.; Jang, S.J.; Mohseni, H. Deep-UV microsphere projection lithography. *Opt. Lett.* **2015**, *40*, 2537. [CrossRef] [PubMed]
13. Moharam, M.G.; Gaylord, T.K.; Pommet, D.A.; Grann, E.B. Stable implementation of the rigorous coupled-wave analysis for surface-relief gratings: Enhanced transmittance matrix approach. *J. Opt. Soc. Am. A* **1995**, *12*, 1077–1086. [CrossRef]
14. MC Grating Software. Available online: https://mcgrating.com/ (accessed on 11 October 2017).
15. Bardosova, M.; Pemble, M.E.; Povey, I.M.; Tredgold, R.H. The Langmuir-Blodgett Approach to Making Colloidal Photonic Crystals from Silica Spheres. *Adv. Mater.* **2010**, *22*, 3104–3124. [CrossRef] [PubMed]

nanomaterials

MDPI

Article

Organozinc Precursor-Derived Crystalline ZnO Nanoparticles: Synthesis, Characterization and Their Spectroscopic Properties

Yucang Liang [1,*], Susanne Wicker [1], Xiao Wang [2], Egil Severin Erichsen [3] and Feng Fu [4,*]

1 Institut für Anorganische Chemie, Eberhard Karls Universität Tübingen, Auf der Morgenstelle 18, 72076 Tübingen, Germany; susanne.wicker@ipc.uni-tuebingen.de
2 School of Physics and Electronics, Hunan University, Changsha 410082, China; xiao_wang@hnu.edu.cn
3 Laboratory for Electron Microscopy, University of Bergen, Allégaten 41, 5007 Bergen, Norway; Egil.Erichsen@mnfa.uib.no
4 College of Chemistry & Chemical Engineering, Yan'an University, Shaanxi Key Laboratory of Chemical Reaction Engineering, Yan'an 716000, China
* Correspondence: yucang.liang@uni-tuebingen.de (Y.L.); yadxfufeng@126.com (F.F.); Tel.: +49-07071-29-76216 (Y.L.); +86-911-2332037 (F.F.)

Received: 5 November 2017; Accepted: 21 December 2017; Published: 4 January 2018

Abstract: Crystalline $ZnO_{\text{-ROH}}$ and $ZnO_{\text{-OR}}$ (R = Me, Et, *i*Pr, *n*Bu) nanoparticles (NPs) have been successfully synthesized by the thermal decomposition of in-situ-formed organozinc complexes $Zn(OR)_2$ deriving from the reaction of $Zn[N(SiMe_3)_2]_2$ with ROH and of the freshly prepared $Zn(OR)_2$ under an identical condition, respectively. With increasing carbon chain length of alkyl alcohol, the thermal decomposition temperature and dispersibility of in-situ-formed intermediate zinc alkoxides in oleylamine markedly influenced the particle sizes of $ZnO_{\text{-ROH}}$ and its shape (sphere, plate-like aggregations), while a strong diffraction peak-broadening effect is observed with decreasing particle size. For $ZnO_{\text{-OR}}$ NPs, different particle sizes and various morphologies (hollow sphere or cuboid-like rod, solid sphere) are also observed. As a comparison, the calcination of the fresh-prepared $Zn(OR)_2$ generated $ZnO_{\text{-R}}$ NPs possessing the particle sizes of 5.4~34.1 nm. All crystalline ZnO nanoparticles are characterized using X-ray diffraction analysis, electron microscopy and solid-state ^{1}H and ^{13}C nuclear magnetic resonance (NMR) spectroscopy. The size effect caused by confinement of electrons' movement and the defect centres caused by unpaired electrons on oxygen vacancies or ionized impurity heteroatoms in the crystal lattices are monitored by UV-visible spectroscopy, electron paramagnetic resonance (EPR) and photoluminescent (PL) spectroscopy, respectively. Based on the types of defects determined by EPR signals and correspondingly defect-induced probably appeared PL peak position compared to actual obtained PL spectra, we find that it is difficult to establish a direct relationship between defect types and PL peak position, revealing the complication of the formation of defect types and photoluminescence properties.

Keywords: organozinc precursor; thermal decomposition; zinc oxide; nanoparticle; size effect; spectroscopic properties

1. Introduction

Nanostructured metal oxide nanoparticles have attracted increasing attention due to their specific physical and chemical properties in optic, magnetism, conductivity and reactivity. A lots of metal oxides have been applied to industrial products in sensors, in cosmetics, in medical diagnosis, or as new devices for optical and electronic applications. These metal oxides were often prepared by top down from physical approach and bottom up from chemical method. The physical method is very difficult to control uniform shape and obtain very small grain size less than 10 nm [1]. Nowadays, chemical methods (such as co-precipitation, microemulsion, solvothermal synthesis, thermal decomposition, sol-gel technique, sonochemical route, microwave-assisted technique and electrochemical deposition) are becoming a popular strategy for fabricating nanostructured metal oxide or metal hydroxide with controllable morphology and uniform particle size at molecular level. Among these metal oxides (SnO_2, ZnO, NiO, TiO_2, Fe_xO_y, Co_xO_y, In_2O_3, WO_3, $CoFe_2O_4$ etc.) [2–11], nanostructured ZnO materials were extensively investigated due to several advantages in friendly environment, high electron mobility, flexible synthesis methods, various morphologies, the most interesting hot research targets for building field-effect transistors and energy harvesting (piezoelectric nanogenerators and photovoltaics) [12–17], for bioimaging and drug delivery [18] and sensors [1,19].

Zinc oxide with a large direct wide band gap of about 3.37 eV at room temperature and a large exciton binding energy of 60 meV [20,21] has attracted significant attention because of its special electronic and photonic properties and its broad applications in electronics, optoelectronics, electrochemistry, fabricating piezoelectric nanodevices, light emitting diodes, solar cell, nanolasers, sensors and catalysis [22]. More recently, transition-metal-doped p-ZnO NPs-based sensory array can be used for instant discrimination of explosive vapours [23]. Up to now, 1D nanostructured ZnO materials have been prepared by using various synthetic approaches including wet chemical method [22,24–26], physical or chemical vapour deposition [22,27], pulsed electro-chemical deposition [22,28–32], pulsed laser deposition [33,34], molecular beam epitaxy [35], electrospinning [36] and microwave heating technique [37–40]. Wet chemical approach is a facile, cost-effective and convenient method using cheap inorganic zinc salts as precursors in an alkaline medium to fabricate nanostructured ZnO materials with uniform morphologies such as nanospheres, nanotubes, nanorods, nanoprisms, nanobelts and nanowires and pure or transition-metal-doped ultrathin ZnO nanosheets [22–26]. For example, the monodisperse morphology-controlled ZnO troughs could be prepared at the air-water interface under mild conditions [41] and ZnO microspheres could be fabricated by the chemical conversion of ZnSe [42]. But its shortage is that the ZnO nanoparticles obtained often show polydispersity or poor uniformity or aggregations in solution. To overcome these shortcomings, a thermal decomposition method was developed in order to control the dispersibility or hinder self-assemble of single nanoparticle in the presence of stabilizer [40,43]. The formation mechanism of ZnO nanocrystals can be monitored by in-situ IR spectra [44,45]. Moreover, the improvement of the dispersibility of ZnO NPs can also be carried out by microemulsion method [46]. However, in general, thermal decomposition of inorganic zinc salts such as $ZnC_2O_4 \cdot 2H_2O$, $Zn_5(OH)_6(CO_3)_2$, $Zn(CH_3COO)_2 \cdot 2H_2O$ and $Zn_3(OH)_4(NO_3)_2$ often occurs at a high temperature (300–600 °C) with a long reaction time to generate crystalline ZnO NPs [47,48]. Compared to inorganic zinc salts, organozinc complexes as a potential candidate can readily decompose at a low temperature. For the past ten years, some organozinc complexes have been chosen as the zinc source for the preparation of nanostructured zinc oxide or porous zinc oxide materials. In these complexes, reaction temperature and the use of different organozinc precursors markedly influenced the quality and properties of the obtained nanostructured ZnO, such as mono/polydispersibility, nucleation and growth rate and optical properties. For example, the thermal decomposition of diethylzinc under O_2-rich environment produced würtzite ZnO nanocrystals via hot-injection method in the presence of trioctylphosphine oxide or alkylamines [49]. The alcoholization of diethylzinc generated zinc alkoxide, which could be hydrolysed to form crystalline zinc oxide with a particle size of 3–5 nm and the resulting ZnO particles could further aggregate together to form spherical particles that have a large surface area and enhanced reactivity [50]. Metal zinc NPs prepared by bis(cyclohexyl) zinc complex in the presence of different solvents and stabilizers under argon protection were exposed to moisture air to be slowly oxidized and form size and shape-controlled crystalline ZnO NPs [51,52].

In 1994, the heteroleptic zinc complexes MeZn(O*i*Pr) or MeZn(O*t*Bu) were used as a single precursor for growth of ZnO film by metal-organic chemical vapour deposition(CVD) [53]. In 2005, Driess and co-workers further investigated the CVD process of heterocubane precursor $[MeZn(OiPr)]_4$ for the formation of size-selected ZnO NPs and proposed the reaction mechanism of the gas-phase decomposition of [MeZn(O*i*Pr)]$_4$ [54,55]. Moreover, the pyrolysis of EtZnO*i*Pr could also form monodisperse spherical ZnO NPs with an average size of 3.1 ± 0.3 nm in the presence of trioctylphosphine oxide; the resulting ZnO NPs indicated a blue-shifted phenomenon of excitonic absorption peak, revealing a quantum confinement effect of nanostructured ZnO [56]. In 2007, Polarz and co-workers explored in detail the reaction of $ZnMe_2$ with polyethylene glycol in toluene and the formed [$MeZnOPEG_{400}$] gels could be used as a ZnO precursor for the preparation of mesoporous ZnO materials [57]. Except for the above-mentioned, these organozinc complexes—[$MeZn(OCH_2CH_2OMe)$] and RZn(OH)-type such as [*t*BuZn(μ-OH)]$_n$ and its derivatives—could also be converted into ZnO NPs [58–60]. In addition to these zinc complexes, the directly thermal decomposition of zinc-organic framework $Zn_4O(BDC)_3$ (MOF-5, BDC represents benzene-1,4-dicarboxylate) at above 400 °C formed amorphous carbon-covered ZnO NPs [61]. Decomposition of zinc acetylacetonate in oleylamine yielded ZnO NPs of 7–10 nm with increasing reaction temperature [62]. Furthermore, the doping or surface modification of heterometals can also effectively adjust the electrical, optical and magnetic properties of obtained ZnO materials.

In addition, the photoluminescent property of bulk or nanostructured zinc oxide particles were also extensively investigated. Especially, zinc oxide materials with different morphologies showed a variety of optical properties. ZnO nanowires synthesized with a vapour phase transport process via catalysed epitaxial crystal growth on the substrate indicated a band gap at 377 nm (3.29 eV) [63]. ZnO nanowires prepared by vapour transport had a strong emission at 380 nm (3.26 eV) [64]. ZnO materials with spheres, triangular prisms and rods prepared by thermal decomposition method displayed various UV emission ranging from 3.19 eV (spheres) to 3.30 eV (triangular prisms), implying that the band gaps depended on the morphology of nanostructured materials [65]. Hydrothermally synthesized ZnO nanorods had a UV emission at 390 nm, a broad shoulder (400–425 nm) and weak peaks at 417, 446 and 465 nm from the photoluminescent spectrum [24]. For ZnO quantum rods, different photoluminescent properties were also observed due to quantum confinement effects [66]. However, the photoluminescence spectrum of ZnO prepared by physical method such as thermal evaporation deposition showed different emission at various temperatures, for example, bound exciton (3.354 eV), free excitons (3.375 and 3.421 eV), the first/second/third longitudinal optical phonon order replicas (3.315/3.243/3.171 eV) of free exciton (3.375 eV) and donor acceptor pairs (3.188 eV) at low temperature (6 K), the first longitudinal optical phonon replica (3.315 eV) of free exciton at room temperature [67]. The photoluminescence characteristics of catalyst free ZnO nanowires at different temperatures and excitation intensities were explored by Mohanta and Thareja [68]. Moreover, ZnO NPs in vapour phase showed 42 meV shift in peak position of PL spectrum compared to that of bulk ZnO [69]. As an extensive research the exciton-exciton scattering in vapour phase ZnO NPs was also investigated [70]. In addition, CdO-modified ZnO tailored the band gap of ZnO to achieve luminescence from ultraviolet to the blue and green spectral region [71]. With increasing Cd concentration in ZnCdO, the band gap gradually decreased due to a larger ionic radius of Cd^{2+} [72]. Note that the photoluminescence spectrum of ZnCdO with 50 wt % Cd showed an abnormal red-blue-redshift with increasing temperature [73]. Based on these investigations, it is easier to find that the photoluminescent behaviour of nanostructured ZnO materials depends on size, morphology, surface defect types and doping of surface impurity as well as preparation methods.

In this study, we investigated in detail the thermal decomposition of the in-situ formed zinc alkoxide, $Zn(OR)_2$ [74] (R = Me, Et, *i*Pr and *n*Bu), which originated from the reaction between $Zn[N(SiMe_3)_2]_2$ [75] and alkyl alcohol (MeOH, EtOH, *i*PrOH and *n*BuOH) in the presence of oleylamine, to fabricate the nanostructured ZnO particles and explored the influence of the type of alkyl alcohol, reaction temperature and the capped-stabilizer on morphology, size, the degree of crystallinity, spin paramagnetic resonance spectra and optical property of ZnO NPs obtained. As a comparison, the direct pyrolysis of zinc alkoxide was also investigated in the presence or absence of high boiling-point solvent oleylamine or a mixture of oleic acid and 1-octadecene.

2. Results and Discussion

2.1. *Structural Characterization of a Series of Nano-Structured Zinc Oxide Particles $ZnO_{\text{-}ROH}$, $ZnO_{\text{-}OR}$ and $ZnO_{\text{-}R}$ (R = Me, Et, iPr, nBu, Gc)*

Homoleptic organozinc complexes including alkyl zinc ($ZnMe_2$ or $Zn(C_6H_{11})_2$ etc.) and zinc amide complexes ($Zn[N\textit{i}Pr_2]_2$ or $Zn[N(SiMe_3)_2]_2$ etc.) have been used as a zinc precursor for the preparation of metal zinc NPs due to low thermodynamic stability of Zn–N and Zn–C bonding [51,52]. But the direct decomposition of these complexes cannot result in the formation of zinc oxide NPs due to the absence of oxygen source. In present work, $ZnO_{\text{-}ROH}$ NPs were prepared via thermal decomposition of an in-situ formed zinc alkoxide which derived from the reaction of $Zn[N(SiMe_3)_2]_2$ with alkyl alcohol ROH (R = Me, Et, *i*Pr and *n*Bu) in the presence of high boiling-point solvent oleylamine. Under an identically synthetic condition, reaction procedure is proposed as follows:

As viewed in Scheme 1, zinc silylamido complex first reacts with alkyl alcohol to form zinc alkoxide complex and eliminate silylamide ligand [74], follows by in-situ thermal decomposition to generate ZnO NPs. The thermal decomposition temperature mainly depends on the nature of formed intermediate zinc alkoxide. Moreover, volatile compositions including by-product and unreacted alkyl alcohol can be removed completely under vacuum.

$$Zn[N(SiMe_3)_2]_2 + \underset{(R=Me,\ Et,\ iPr,\ nBu)}{2\,ROH} \xrightarrow[-\,2HN(SiMe_3)_2]{\text{hexane, oleylamine}} Zn(OR)_2 \quad (1)$$

$$Zn(OR)_2 \xrightarrow[240\ ^\circ C,\ 2\ h]{\text{oleylamine}} ZnO \quad (2)$$

Scheme 1. Proposed reaction process for the formation of nanostructured ZnO particles using homoleptic zinc silylamido complex $Zn[N(SiMe_3)_2]_2$ as a zinc precursor and alkyl alcohol as a reactant in the presence of high boiling-point solvent as a stabilizer.

2.1.1. Crystalline $ZnO_{\text{-}MeOH}$ NPs

According to synthetic procedure (Scheme 1), when methanol was used as a reactant and solvent in the presence of oleylamine, $ZnO_{\text{-}MeOH}$ particles were obtained. The powder X-ray diffraction (PXRD) pattern of $ZnO_{\text{-}MeOH}$ clearly shows several well-resolution diffraction peaks, which can be indexed as (100), (002), (101), (102), (110), (103) and (112) reflections (Figure 1), revealing a characteristic würtzite structure of ZnO (a = 3.25 Å, c = 5.21 Å, $P6_3mc$, Powder Diffraction File Database (PDF-2), entry: JCPDS 36-1451). For the synthesis of $ZnO_{\text{-}MeOH}$ NPs, we observed that the dropwise addition of hexane solution of $Zn[N(SiMe_3)_2]_2$ into reaction system quickly formed the sphere-shaped aggregations. Part of the aggregations was separated, washed several times with methanol under argon protection and dried under vacuum to get white powder. Elemental analysis confirmed that the white powder was $Zn(OMe)_2$ (found wt % C: 19.38, H: 4.39, N: 0.09; calcd. C: 18.84, H: 4.74). IR spectrum of white powder clearly shows characteristic C–O vibration at 1050 cm^{-1}, C–H stretching vibrations in the range of 2820~2930 cm^{-1} and bending vibration at 1450 cm^{-1} from $-CH_3$ group and Zn–O vibration at 482 and 560 cm^{-1} (Figure S1, Electronic Supplementary Information (ESI)), further verifying the formation of pure intermediate $Zn(OMe)_2$. It is noted that the sphere-shaped aggregations $\{Zn(OMe)_2\}_n$ after removal of volatile compositions could not effectively disperse in oleylamine solvent even if it was heated to 240 °C. The thermal decomposition of in-situ formed $\{Zn(OMe)_2\}_n$ generated $ZnO_{\text{-}MeOH}$ NPs still preserved sphere-like aggregations with average particle size of 234.5 ± 6.1 nm (Figure 2, left). These agglomerations consist of small nanoparticles with an average particle size of 5.9 ± 0.3 nm and the lattice fringes are clearly visible with spacing of 0.281 and 0.510 nm and the lattice spacing of 0.281 nm corresponds to (100) planes of ZnO with $P6_3mc$ symmetry (Figure 2, right). Moreover, the visual grain boundaries clearly confirm the connection of small particles each other. In addition, according to the Scherrer formula, $D_{hkl} = K\lambda/(\beta\cos\theta)$, where D_{hkl} is the mean size of the ordered crystalline grain size, K is a dimensionless shape factor (0.89), λ is the X-ray wavelength (0.15406 nm), β is the full width at half the maximum intensity (in radians), θ is the Bragg angle (in degrees), the calculated mean particle size of ZnO_{-MeOH} is 4.6 nm from the (100) plane and 4.9 nm from the (101) reflection plane. These results are in good agreement with the result measured by transmission electron microscopy (TEM) images. Detailed particle sizes determined by different methods are listed in Table 1.

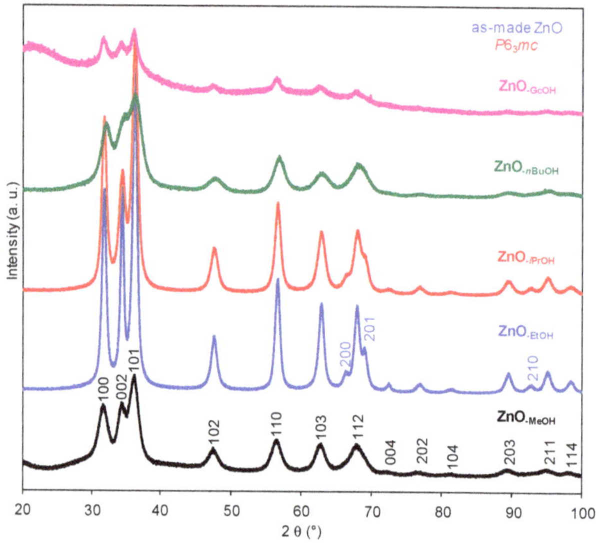

Figure 1. Wide-angle PXRD patterns of as-synthesized ZnO_{-ROH} NPs by using $Zn[N(SiMe_3)_2]_2$ as a zinc precursor and alkyl alcohol as a reactant and solvent in the presence of stabilizer.

Table 1. Particle size determined by TEM and XRD data and absorption peak from UV spectra.

Sample	Particle Size/nm [a]	Particle Size/nm [b]	Electron *g*-Factor	Absorption Edge/nm
ZnO_{-MeOH}	5.9 ± 0.3	4.55; 4.86	2.00	352
ZnO_{-EtOH}	11.7 ± 0.6	10.7; 10.2	2.02, 2.00, 1.99, 1.96	358
ZnO_{-iPrOH}	7.5 ± 0.2	9.0; 8.3	2.00	356
ZnO_{-nBuOH}	-	4.3	2.12, 2.07, 2.00, 1.96	351
ZnO_{-GcOH}	-	3.8	-	-
ZnO_{-OMe}	different sizes	5.3	2.01, 2.00, 1.99	364
ZnO_{-OEt}	5.4 ± 0.1	4.8	2.12~1.90	361
ZnO_{-OiPr}	4.2 ± 0.6	4.8	2.12~1.90	347
ZnO_{-OnBu}	4.5 ± 0.2	5.1	-	346
ZnO_{-OGc}	-	-	2.00	348
ZnO_{-Me}	-	34.1	1.96	373
ZnO_{-Et}	-	7.5	1.96	358
ZnO_{-iPr}	-	28.4	1.96	365
ZnO_{-nBu}	-	5.4	1.96	358
ZnO_{-Gc}	-	21.3	2.00, 1.96	373

Note: [a] Particle size was measured using TEM images; [b] Particle size was calculated using the Scherrer formula from (100) diffraction peak of XRD patterns.

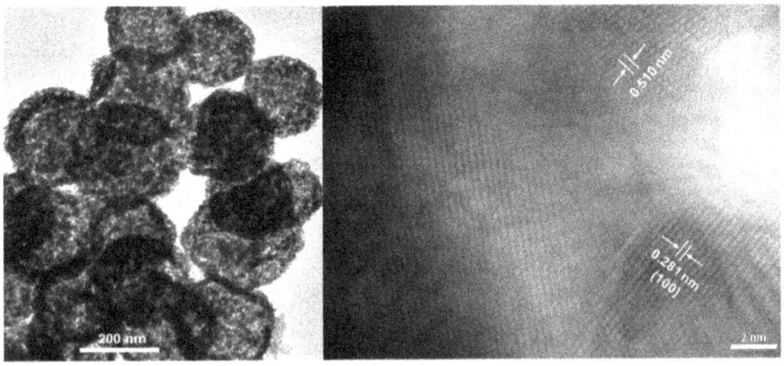

Figure 2. (High-resolution) TEM images of the crystalline ZnO_{-MeOH} NPs with visual lattice fringes (correspond to crystal planes).

2.1.2. Alkyl Alcohol Effect

When EtOH, *i*PrOH and *n*BuOH were respectively used to replace MeOH in the preparation of ZnO_{-MeOH}, the corresponding nanoscaled ZnO_{-ROH} (R = Et, *i*Pr, *n*Bu) particles were obtained. All ZnO_{-ROH} NPs have a same crystal structure (space group: $P6_3mc$) verified by PXRD patterns (Figure 1) as that of ZnO_{-MeOH}, but the ZnO_{-EtOH} and ZnO_{-iPrOH} show a relatively narrow and strong diffraction peaks compared to that of ZnO_{-MeOH} and ZnO_{-nBuOH}, implying that use of different alkyl alcohols as a reactant and a reaction solvent can effectively influences the crystalline particle size of zinc oxides obtained and the results show a strong diffraction peak-broadening/weakening effect with decreased grain size [76]. This phenomenon is probably caused by the change of growth rate of ZnO particles influencing by decomposition temperature, dispersibility and solubility of formed intermediate $Zn(OR)_2$ in oleylamine during the thermal decomposition.

Moreover, for ZnO_{-EtOH} NPs, TEM images and the selected area electron diffraction (SAED) pattern clearly revealed a crystalline structure with $P6_3mc$ symmetry (Figure 3a,b). For the sphere-like ZnO_{-iPrOH} and plate-like-aggregated ZnO_{-nBuOH} particles (Figure 3c,e), the lattice fringes can also be observed (Figure 3d,f), although ZnO_{-nBuOH} has a series of slightly broad and weak diffraction peaks (similar to that of ZnO_{-MeOH}). The particle size calculated by Scherrer formula is listed in Table 1: 10.7 nm from the (100) plane and 10.2 nm from the (101) reflection, respectively, for the spherical ZnO_{-EtOH}; 9.0 nm from the (100) plane and 8.3 nm from the (101) reflection plane, respectively, for sphere-like ZnO_{-iPrOH}; 4.3 nm from (100) plane and 4.0 nm from the (101) reflection plane, respectively, for ZnO_{-nBuOH}. These results are quite close to corresponding average particle size measured by TEM images (Table 1, 11.7 $\pm$ 0.6 nm for ZnO_{-EtOH}; 7.5 $\pm$ 0.2 nm for ZnO_{-iPrOH}; for the multiply plate-like-aggregated ZnO_{-nBuOH} spheres (Figure 3e), it is very difficult to accurately measure their particle sizes). Based on these results, it is obvious to observe size-induced the weakening of diffraction peak intensity and the broadening of diffraction peaks (grain size: ZnO_{-EtOH} > ZnO_{-iPrOH} > ZnO_{-MeOH} > ZnO_{-nBuOH}) and alkyl alcohol-induced the change of morphologies. As a special example, when high boiling-point glycerol (GcOH) was used as a reactant, the formed intermediate zinc glycerolate $Zn[OCH_2CH(OH)CH_2O]$ [77–79] (monoclinic $P2_1/c$, Powder Diffraction File Database (PDF-2), entry: JCPDS 23-1975) could not be thermally decomposed at 240 °C for 2 h under an identical condition to yield crystalline ZnO_{-GcOH} NPs. However, ZnO_{-GcOH} NPs can be prepared at 320 °C for 2 h in the presence of oleic acid and 1-octadecene. Obviously, this thermal decomposition temperature is markedly less than 400–500 °C which has been previously used for the preparation of ZnO NPs by choosing zinc glycerolate as a molecular precursor [77,79]. Compared to ZnO_{-ROH} (R = Me, Et, *i*Pr and *n*Bu), several weak diffraction peaks are observed for ZnO_{-GcOH} (Figure 1). The particle size calculated by Scherrer formula is 3.8 nm from the (100) plane and 3.6 nm from the (101) reflection plane, respectively.

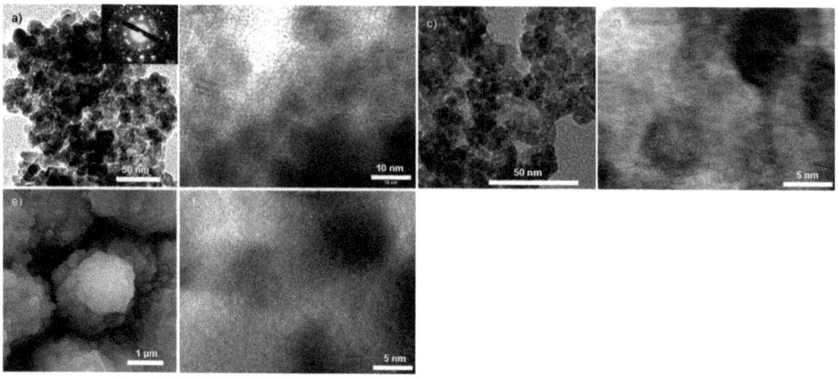

Figure 3. (High-resolution) TEM images of crystalline (**a**,**b**) ZnO_{-EtOH} and (**c**,**d**) ZnO_{-iPrOH}, and SEM and TEM image of (**e**,**f**) ZnO_{-nBuOH}. The inset in (**a**) is the SAED pattern.

As a comparison, thermal decomposition of intermediate $Zn(OiPr)_2$ forms ZnO NPs composing of spherical agglomerations of crystallites. This result is quite similar to small-sized ZnO prepared previously via a hydrolysis of $Zn(OiPr)_2$ [50]. In the present research, it is very difficult to perform the synthesis of strictly size-controlled and monodisperse ZnO NPs in the presence of stabilizer due to poor solubility and high concentration of intermediate $Zn(OR)_2$ formed in reaction system. This reaction system completely differs from that previously reported by Chaudret and co-workers for the synthesis of size- and shape-controlled crystalline ZnO NPs at room-temperature [51,52].

To further corroborate alkyl alcohol effect, thermal gravimetric analysis of intermediates $Zn(OR)_2$ (R = Me, Et, *i*Pr and *n*Bu) was carried out to monitor their thermal decomposition temperature (TDT). As viewed in Figure S2 (ESI), the starting TDT markedly depends upon alkoxyl group of $Zn(OR)_2$. The ordering of decomposition from low to high temperature is $Zn(OiPr)_2 < Zn(OnBu)_2 < Zn(OEt)_2 < Zn(OMe)_2 < Zn(OGc)$. High TDT of Zn(OGc) has been reported previously [78]. For complexes $Zn(OMe)_2$ and Zn(OGc), high TDT should be attributed to the polymeric structure of $Zn(OMe)_2$ and the quite stable chelated structure of Zn(OGc), respectively. This result is in agreement with the experimental phenomena observed. More important is that high TDT probably influences the rate of growth of ZnO NPs and thereby results in the formation of different particle sizes and various morphologies.

Moreover, in order to confirm the formation of intermediate $Zn(OR)_2$ and their compositions before thermal decomposition, the separated intermediates were dried and characterized by elemental analysis and IR spectrum. The results show that IR spectra (Figure S1, ESI) and the composition of the corresponding intermediate are quite similar to that of correspondingly fresh prepared zinc alkoxide molecule. For example, intermediate $Zn(OEt)_2$, the contents of C, H and N are 30.57, 6.59 and 0.07 wt %, respectively. It is in good agreement with fresh $Zn(OEt)_2$ prepared by the reaction of $Zn[N(SiMe_3)_2]_2$ with anhydrous ethyl alcohol in hexane. For the synthesis of ZnO_{-OR} (R = *i*Pr, *n*Bu and Gc), the contents of C, H and N of intermediate are also in good agreement with the corresponding organozinc molecular precursor $Zn(OR)_2$ (R = Et, *i*Pr, *n*Bu) and Zn(OGc), demonstrating the formation of intermediate zinc alkoxide. Note that intermediate did not contain any stabilizer or organic solvent after washing with the corresponding alkyl alcohol several times and drying.

Furthermore, the chemical composition of as-made ZnO NPs was determined using the solid-state ^1H and ^{13}C NMR spectra, elemental analysis and infrared resonance spectra. All crystalline ZnO_{-ROH} NPs still contain high carbon contents ranging from 6 to 30 wt % (not shown). Due to low resolution of solid-state ^1H NMR spectra, only chemical shifts of hydrogen atoms at δ = 0.6–5.5 ppm attributed to $-CH_3$, $-CH_2$, –CH=CH– groups from unreacted zinc precursor and stabilizer were observed for as-made ZnO_{-ROH} (R = Me, Et, *i*Pr and Gc) NPs (Figure 4a). In addition, for ZnO_{-GcOH} NPs, ^1H NMR spectrum showed a very weak signal at δ = 11 ppm, confirming the presence of –COOH group from stabilizer oleic acid. However, high-resolution ^{13}C spectra are markedly different (Figure 4b). For material ZnO$_{-MeOH}$, solid-state ^{13}C NMR spectrum clearly showed a single characteristic and strong signal at 56.3 ppm attributed to the undecomposed $Zn(OMe)_2$ precursor but no characteristic peak of carbon from stabilizer oleylamine was observed. For crystalline ZnO$_{-EtOH}$ and ZnO$_{-iPrOH}$ NPs, the ^{13}C NMR spectra indicate a series of signals at δ = 169.8 (C=O), 127.6 (CH=CH), 47.6–15.5 (CH_2, CH_3) and 0.7 ppm (trapped $HN(SiMe_3)_2$), revealing the presence of oleylamine stabilizer and by-products containing carbonyl group and the trapped by-product $HN(SiMe_3)_2$ and the absence of undecomposed zinc alkoxides. The formation of carbonyl group is relevant to intermediate of thermal decomposition of $Zn(OR)_2$ (R = Et, *i*Pr) [54,80]. Furthermore, the appearance of C–H, C=C and C=O vibrations in IR spectra also further demonstrate the existence of stabilizer and incompletely decomposed by-product (such as Zn-carbonyl intermediate) from zinc precursor (Figure S3, ESI). Similarly, for ZnO$_{-GcOH}$ material, the carbon signals at δ = 65.7 and 72.1 ppm are attributed to the incompletely decomposed zinc complex $Zn[OCH_2CH(OH)CH_2O]$. The characteristic carbon signals at δ = 184.8 (–CO_2H), 139 and 118 ($CH=CH_2$) and 130 (CH=CH) as well as δ = 14–35 ppm (CH_3, CH_2) belong to oleic acid and 1-octadecene.

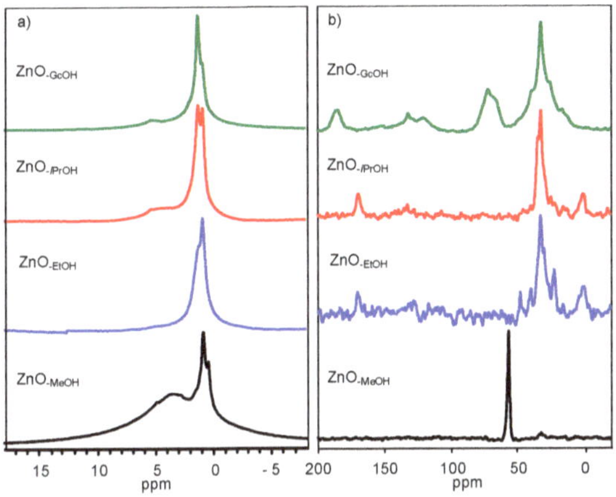

Figure 4. Solid-state (**a**) ^1H and (**b**) ^{13}C NMR spectra of as-made ZnO NPs obtained by using different alkyl alcohols as reactants under an identical condition except for ZnO$_{-GcOH}$.

Unsurprisingly, all organic composites in as-made ZnO$_{-ROH}$ NPs can be completely removed by calcinations at 500 °C for 4 h. Representative TEM images for the calcined ZnO$_{-ROH}$ (R = Me, Et and *i*Pr) are shown in Figure 5. A typically spherical morphology was preserved and an average particle size is 57.7 ± 1.5 nm for calcined ZnO$_{-MeOH}$, 17.6 ± 0.6 nm for calcined ZnO$_{-EtOH}$ and 12.3 ± 0.4 nm for calcined ZnO$_{-iPrOH}$, respectively. Note that particle sizes markedly increased after high-temperature calcinations due to aggregation and growth of small particles. In addition, high crystalline structures with grain boundaries can be observed anywhere (Figure 5b,d,f). The selected area electron diffraction pattern of calcined ZnO$_{-iPrOH}$ NPs reveals a typical würtzite-structured ZnO with $P6_3mc$ symmetry (Figure 5e, inset). Furthermore, highly well-resolved diffraction peaks in Figure S4 (ESI) also confirmed that hexagonal structure of crystalline ZnO$_{-ROH}$ (R= Me, Et, *i*Pr, *n*Bu) were preserved after high-temperature calcinations but the intensity of diffraction peaks were much higher than the parent materials, implying that increasing crystalline particle size enhanced the intensity of diffraction peaks. The crystalline particle size calculated by Scherrer formula is 55.3 nm from the (100) plane for the spherical ZnO$_{-MeOH}$, 19.1 nm from the (100) plane for the spherical ZnO$_{-EtOH}$, 14.6 nm from the

(100) plane for sphere-like ZnO_{-iPrOH} and 16.3 nm from the (100) plane for ZnO_{-nBuOH}. These results are in good agreement with data measured by TEM images.

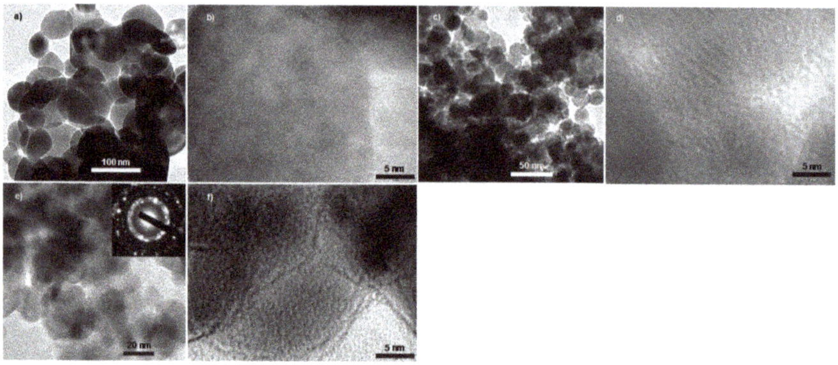

Figure 5. TEM images of organic composite-removed crystalline (**a**,**b**) ZnO_{-MeOH}, (**c**,**d**) ZnO_{-EtOH} and (**e**,**f**) ZnO_{-iPrOH}. The inset in (**e**) is corresponding SAED pattern.

2.1.3. Zinc Precursor Effect

As above-mentioned, the thermal decomposition of in-situ formed intermediates $Zn(OR)_2$ (R = Me, Et, *i*Pr and *n*Bu) derived from the reaction of alkyl alcohol with $Zn[N(SiMe_3)_2]_2$ markedly affect size and morphology of as-made ZnO_{-ROH} NPs in the presence of oleylamine. In order to corroborate this hypothesis, the directly thermal decomposition of fresh-prepared zinc precursors $Zn(OR)_2$ (R = Me, Et, *i*Pr and *n*Bu) was performed in the presence of oleylamine to fabricate corresponding ZnO_{-OR} (R = Me, Et, *i*Pr and *n*Bu) NPs. The high crystalline würtzite-structure ZnO with $P6_3mc$ symmetry was corroborated by PXRD patterns (Figure 6), indicative of same structure as ZnO_{-ROH} NPs prepared by using thermal decomposition of in-situ formed zinc alkoxide. But TEM analyses confirm that particle size and morphology of ZnO_{-OR} NPs are quite different compared to corresponding ZnO_{-ROH} materials.

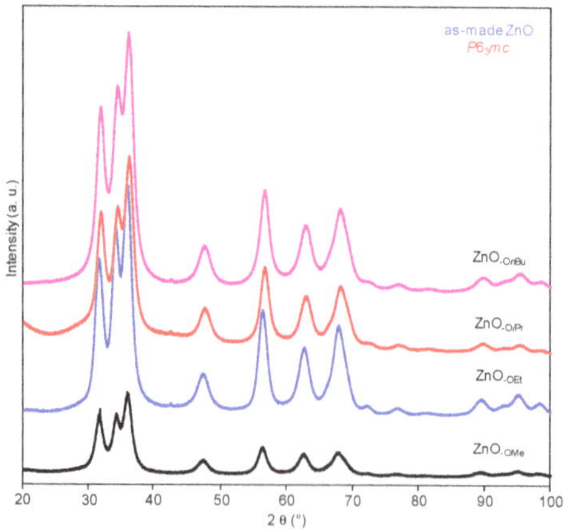

Figure 6. Wide-angle PXRD patterns of as-made ZnO_{-OR} NPs derived from directly thermal decomposition of fresh-prepared $Zn(OR)_2$ (R = Me, Et, *i*Pr and *n*Bu) in the presence of oleylamine at 240 °C for 2 h.

As viewed in Figure 7, all ZnO_{-OR} NPs with sphere-like or cuboid-like morphology show different sizes. Obviously, ZnO_{-OMe} particles are composed of hollow spheres with average diameter of 113.8 ± 12.2 nm and cuboid-like particles with different sizes in length and width (Figure 7a) to compare with that of ZnO_{-MeOH} (Figure 2). In fact, hollow spheres consist of very small particles (Table 1, the calculated particle size is 5.3 nm from the (100) reflection of XRD pattern). A representatively characteristic TEM image of ZnO_{-OMe} nanocrystals is shown in Figure 7b, revealing a typical crystalline structure. As viewed in Table 1, for crystalline ZnO_{-OEt} nanocrystals, the average particle size measured by TEM image is 5.4 ± 0.1 nm (Figure 7c,d). This result is close to 4.8 nm calculated by the Scherrer formula. For ZnO_{-OiPr}, the average size of spherical aggregation is 56.2 ± 1.8 nm (Figure 7e). These aggregations consist of crystalline nanoparticle with an average size of 4.2 ± 0.6 nm (Figure 7f, particle size calculated by the Scherrer formula is 4.8 nm from the (100) plane). For ZnO_{-OnBu} NPs, the average particle size is 4.5 ± 0.2 nm and the lattice fringes can be observed (Figure 7g,h). Moreover, for ZnO_{-OMe} and ZnO_{-OiPr} NPs, the shapes observed by SEM images (Figure 8a,b) are in accordance with that of TEM images observed under a low magnification.

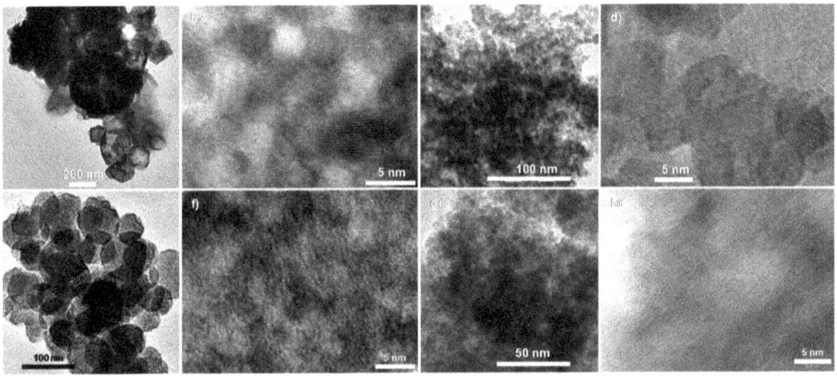

Figure 7. TEM images of ZnO_{-OR} NPs, for (**a**,**b**) ZnO_{-OMe}, (**c**,**d**) ZnO_{-OEt}, (**e**,**f**) ZnO_{-OiPr} and (**g**,**h**) ZnO_{-OnBu}.

Figure 8. The representative SEM images of obtained ZnO NPs by the directly thermal decomposition and calcinations of $Zn(OMe)_2$ and $Zn(OiPr)_2$; (**a**) ZnO_{-OMe}; (**b**) ZnO_{-OiPr}; (**c**) ZnO_{-Me}; and (**d**) ZnO_{-iPr}.

On the basis of above-mentioned characterizations, we found that use of two different approaches based on thermal decomposition of in-situ-synthesized zinc alkoxides and directly thermal decomposition of fresh-prepared zinc alkoxides under an identical condition led to different results for preparing ZnO nanocrystals. With increasing carbon chain, the results trend to identical, such as ZnO_{-nBuOH} and ZnO_{-OnBu}. For ZnO_{-ROH} (R = Et, *i*Pr, *n*Bu) series, the calculated and measured average particle sizes are uniform and gradually decrease with increasing carbon atoms of alkoxyl group but ZnO_{-MeOH} makes an exception. For ZnO_{-OR} (R = Me, Et, *i*Pr, *n*Bu) series, no remarkable tendency is observed on the change of particle size and shape. For ZnO_{-MeOH} and ZnO_{-OMe}, and ZnO_{-iPrOH} and ZnO_{-OiPr}, two approaches indicate a completely opposite phenomenon on the size of spherical aggregations. These differences can be contributed to two causes, one is that the in-situ formed zinc alkoxide can better disperse in oleylamine solvent except for $\{Zn(OMe)_2\}_n$ and the fresh-prepared zinc alkoxides cannot better disperse in oleylamine solvent. Second is the change of different TDT-induced growth rate of zinc oxide from zinc alkoxides. Synergism of two aspects affects particle sizes and morphologies of as-made ZnO NPs. For example, $Zn(OMe)_2$ showed a complicatedly thermal decomposition steps that thereby induced the formation of particles with different growth rate and aggregated together with different shapes and poor dispersion in oleylamine also enabled them to readily form aggregations.

As a comparison, the direct calcination of $Zn(OR)_2$ (R = Me, Et, *i*Pr, *n*Bu and Gc) at 500 °C for 4 h generated ZnO_{-R} NPs showed crystalline ZnO structures with 3D hexagonal symmetry, which were confirmed by PXRD analysis (Figure 9) and TEM images (Figure S5, ESI). The results reveal that the high-temperature calcination is obviously beneficial to the formation of crystalline ZnO NPs with the large particle size. According to Scherrer formula, the calculated particle size is 34.1 nm for ZnO_{-Me}, 7.5 nm for ZnO_{-Et}, 28.4 nm for ZnO_{-iPr}, 5.4 nm for ZnO_{-nBu} and 21.3 nm for ZnO_{-Gc} from the (100) plane of XRD patterns (Table 1). Moreover, as a representative comparison, SEM images (Figure 8) between ZnO_{-OMe} and ZnO_{-Me} and ZnO_{-OiPr} and ZnO_{-iPr} displayed various morphologies for ZnO_{-R} series. As viewed in Figure 8, large particles of ZnO_{-Me} are composed of unnumbered small spherical particles. This result is quite similar to that of as-made ZnO_{-MeOH} NPs, indirect indicating poor dispersibility of precursor $[Zn(OMe)_2]_n$ in solvents. Hence, three different approaches in the present study were used to prepare ZnO NPs, clearly revealing that their morphologies, crystalline particle sizes and the aggregation behaviour strongly depend on the zinc precursors and synthetic methods. These investigations are in agreement with the results reported previously [37]. Thermal decomposition and CVD method often lead to the formation of aggregation of ZnO with various particle sizes and with or without regular shapes from different organozinc precursors [34,54,59,60,81].

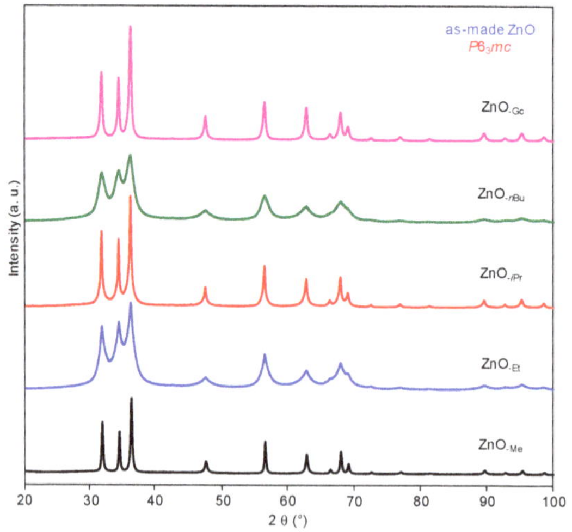

Figure 9. PXRD patterns of ZnO_{-R} NPs obtained by the direct calcinations of precursors $Zn(OR)_2$ (R = Me, Et, *i*Pr, *n*Bu and Gc) at 500 °C for 4 h.

2.2. Electron Paramagnetic Resonance (EPR) Spectra of a Series of Nano-Structured Zinc Oxide Particles ZnO_{-ROH}, ZnO_{-OR} and ZnO_{-R} (R = Me, Et, iPr, nBu, Gc)

Different synthesis methods and various zinc precursors often induce the change of physical and chemical properties of obtained nanostructured materials due to different defects. In general, for ZnO NPs, the types of defect centres include zinc vacancies (V_{Zn}, V_{Zn}^{+}, V_{Zn}^{2+}), zinc on interstitial sites

(Zn_i, Zn_i^+, Zn_i^{2+}), oxygen vacancies (V_O, V_O^-, V_O^{2-}, V_O^+, V_O^{2+}) and oxygen on interstitial sites (O_i, O_i^-, O_i^{2-}). In these intrinsic defect centres only V_{Zn}, V_{Zn}^+, Zn_i^+, O_i, O_i^-, V_O, V_O^-, V_O^{2-} and V_O^+ defect centres can be monitored by EPR spectra [82], especially, oxygen vacancies [80,81,83–88]. Hence, the EPR investigation of pure ZnO materials are not surprising. Herein, as a standard field marker, polycrystalline DPPH with electron spin g-factor ($g = 2.0036$) was used for the exact determination of the magnetic field offset. For ZnO materials, the previous reports have discussed in detail on that different defects generated the appearance of different signals at $g = 1.96$ and 2.00 in EPR spectra, although these assignments of the EPR signals have caused much controversy [80,81,89,90]. At present, the low-field signal at $g = 2.00$ is often assigned to an unpaired electron on an oxygen vacancy site [81,90,91] and high-field signal with $g = 1.96$ is either attributed to shallow donor centres such as ionized impurity atoms in the crystal lattices of ZnO [91–98], or to unpaired electrons on oxygen vacancies in some cases. When tetrameric methylzinc *tert*-butoxide $[MeZnO\textit{t}Bu]_4$ or $[MeZnO\textit{i}Pr]_4$ complex was used as a molecular precursor to prepare ZnO NPs, Driess and co-workers emphasized that EPR signals with $g \sim 2.000$ and 1.96 were attributed to oxygen vacancies with a single trapped electron (V_O^+) and impurity atoms in the ZnO lattices, respectively [80,81]. It is worth noting that with increasing calcination temperature EPR signal with $g \sim 2.000$ decreased until completely disappeared.

In the present study, the X-band EPR properties of ZnO NPs prepared by three different approaches are investigated and their spectra are shown in Figure 10. All electron spin g-factors of all ZnO particles are listed in Table 1. For $ZnO_{\text{-ROH}}$ series, EPR spectra of $ZnO_{\text{-MeOH}}$ and $ZnO_{\text{-}i\text{PrOH}}$ NPs only show a signal with an electron spin g-factor around 2.00, suggesting an unpaired electron trapped on an oxygen vacancy site [80,81,90]. However, a series of weak or strong signals at $g = 2.02$, 2.00, 1.99 and 1.96 for $ZnO_{\text{-EtOH}}$ and at $g = 2.12$, 2.07, 2.03, 2.00, 1.98, 1.96 and 1.94 for $ZnO_{\text{-}n\text{BuOH}}$ NPs, are identified, respectively, revealing the co-existence of unpaired electrons on oxygen vacancies and impurity atoms in the ZnO lattices (Figure 10a). This phenomenon has been reported by Driess and co-workers due to oxygen vacancies and the type and concentration of impurity heteroatoms trapped in the ZnO lattices during the growth of ZnO particles [81]. In fact, the above-mentioned ^{13}C NMR spectra have clearly confirmed doping of impurity heteroatoms in ZnO in our ZnO NPs. Additionally, note that when the fresh-prepared $Zn(OR)_3$ (R = Me, Et and *i*Pr) was used as the zinc precursor and other conditions are identical for the fabrication of $ZnO_{\text{-OR}}$ NPs, the X-band EPR spectra clearly verified the formation of different defect centres with g-factors of 2.01, 2.00, 1.99 for $ZnO_{\text{-OMe}}$, of the ranging from 2.12 to 1.90 for $ZnO_{\text{-OR}}$ (R = Et, *i*Pr), respectively. Weak or strong EPR signals revealed the ratio of predominant and secondary defect centres (Figure 10b) and their relevance to use of organozinc molecular precursor, for example, starting precursor or intermediate is $Zn(OMe)_2$, no EPR signal with $g \sim 1.96$ appeared, if with $Zn(OEt)_2$ two signals with $g \sim 2.00$ and 1.96 appeared in EPR spectra.

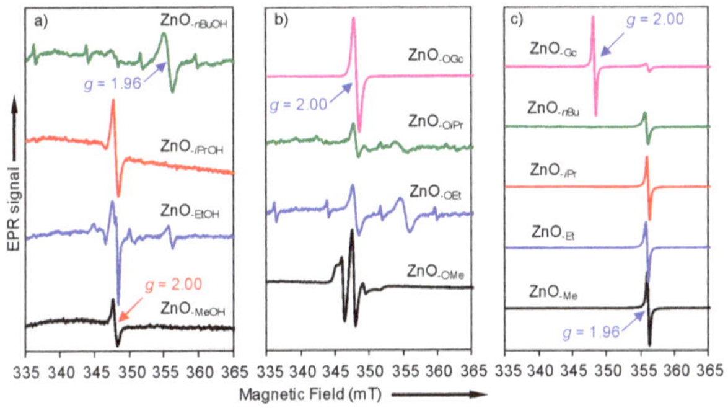

Figure 10. Room temperature EPR spectra of ZnO NPs derived from different synthetic approaches, for (**a**) $ZnO_{\text{-ROH}}$; (**b**) $ZnO_{\text{-OR}}$; and (**c**) $ZnO_{\text{-R}}$ (R = Me, Et, *i*Pr, *n*Bu and Gc).

As a comparison, a series of ZnO_{-R} materials prepared by the direct calcination of $Zn(OR)_3$ (R = Me, Et, *i*Pr and *n*Bu) at 500 °C for 4 h only exhibited one signal with *g*-factor of 1.96 in X-band EPR spectra (Figure 10c), except ZnO_{-GC} that indicated two signals with *g*-factor of 2.00 and 1.96. The previous researches have carefully investigated the case of the appearance of the EPR signals with $g = 2.00$ and 1.96 in ZnO nanocrystals. The former is attributed to surface defects (a singly ionized Zn vacancy or an unpaired electron trapped in an oxygen vacancy site or stable O^{2-} vacancy) due to the formation of core-shell structured ZnO nanocrystals during the milling process [82,88] but this attribution is still controversial and the latter is assigned to a singly ionized oxygen vacancy (V_O^+) which has been confirmed by UV-light irradiation that produced an enhanced EPR signal at $g = 1.96$ [99] or is relevant to oxygen vacancy concentration or impurity atoms effect [81,86]. In some cases, low-field paramagnetic signal ($g = 2.00$) in ZnO materials can also be attributed to oxygen vacancies with a single trapped electron (V_O^+) [81]. Moreover, ZnO materials maybe have interstitial zinc, V_O^-, V_{Zn} and O_i defects but V_{Zn} and O_i-type defect are thermodynamically stable in the ZnO crystal lattices at higher oxygen partial pressure. In our ZnO NPs, no clear core-shell structure is observed and no special controlled condition except argon environment was performed during the preparation of ZnO NPs, hence we suggest that EPR signal at $g = 1.96$ is related to oxygen vacancy and impurity atoms (C, Si or N) in the ZnO lattices. Note that the calcinations of ZnO_{-R} (R = Me, Et, *i*Pr, *n*Bu) at 500 °C for 4 h formed ZnO particles did not show EPR signal at $g = 2.00$. This result is in better agreement with that the signal at $g = 2.00$ appears only at lower temperatures [81]. But the EPR spectra of ZnO_{-GC} NPs indicates two signals at $g = 2.00$ and 1.96, clearly confirming that the types of defects are probably influenced by the coordination modes and the structure of coordination ligand of zinc precursor during calcinations. On the basis of above-mentioned discussions, the formation of defect centres or defect chemistry in ZnO NPs depends not only on the use of the original alkyl alcohols but also on the synthesis conditions [37,100,101]. Physical-chemical property of materials reflects the alteration of structure of materials in composition and defect chemistry.

2.3. *Optical Properties of a Series of Nano-Structured Zinc Oxide Particles ZnO_{-ROH}, ZnO_{-OR} and ZnO_{-R} (R = Me, Et, iPr, nBu, Gc)*

2.3.1. UV-Vis Spectra

For a series of ZnO NPs, size-dependent ultraviolet-visible (UV-Vis) absorption spectra are shown in Figure 11. The band centred at 352, 358, 356 and 351 nm for nanostructured ZnO_{-MeOH}, ZnO_{-EtOH}, ZnO_{-iPrOH} and ZnO_{-nBuOH} were respectively observed, indicating the occurrence of blue shift in the ZnO_{-ROH} NPs (Figure 11a) compared to bulk ZnO (380 nm) due to size quantization effect caused by the confinement of the movement of electrons [102]. As listed in Table 1, position of band from UV-Vis absorption spectra indirectly reflects particle sizes of ZnO NPs. The results are in agreement with particle sizes measured by TEM microscopy and calculated by Scherrer formula ($ZnO_{-EtOH} > ZnO_{-iPrOH} > ZnO_{-MeOH} > ZnO_{-nBuOH}$). Similar blue shift phenomena were also observed for ZnO_{-OR} and ZnO_{-R} (R = Me, Et, *i*Pr, *n*Bu and Gc) (Figure 11b,c) but UV-Vis adsorption spectra of ZnO_{-Me}, ZnO_{-iPr} and ZnO_{-Gc} are very similar to that of bulk ZnO due to large particles under high temperature condition. It is worth noting that the occurrence of blue shift depends on not only synthesis approaches but also nature and structure of precursors in the present reaction system.

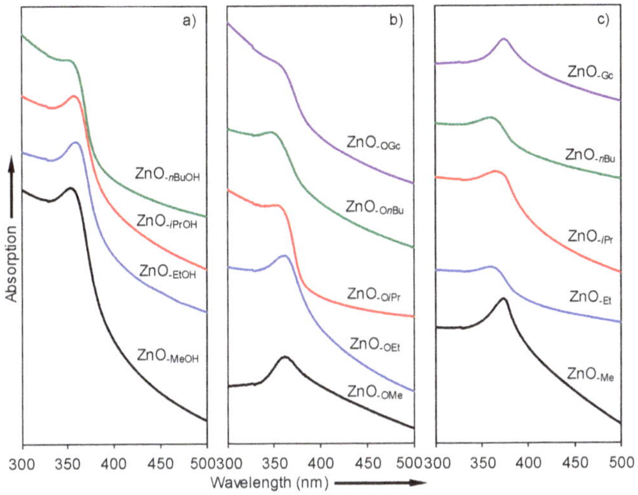

Figure 11. Size-dependent UV-Vis absorption spectra of (**a**) ZnO_{-ROH}; (**b**) ZnO_{-OR}; and (**c**) ZnO_{-R} (R = Me, Et, *i*Pr, *n*Bu and Gc) NPs.

2.3.2. Photoluminescence (PL) Property

To further characterize the optical natures of a series of ZnO-ROH, ZnO-OR and ZnO-R (R = Me, Et, *i*Pr, *n*Bu and Gc) NPs, photoluminescent spectra were recorded (Figure 12). All ZnO NPs clearly show a strong peak nearly at 394 nm (3.14 eV) with excitation wavelength of 290 nm. Note that owing to use of different organozinc precursors, PL spectra of ZnO_{-ROH} (Figure 12a) and ZnO_{-OR} (R = Me, Et, *i*Pr and *n*Bu) NPs (Figure 12b) also show a shoulder peak at about 378 nm (3.28 eV) and the intensity of shoulder peak almost keeps same for the ZnO_{-ROH} (R = Me, Et) NPs, generally increases for the ZnO_{-ROH} (R = *i*Pr, *n*Bu) and ZnO_{-OR} (R = Me, Et, *i*Pr and *n*Bu) NPs. After carefully checked our experimental data and compared with those previously reported spectra [103–106], we concluded that the PL peaks in the UV range from 370 to 400 nm are due to near band gap emissions and can be considered as exciton in origin, although it is slightly less than 3.37 eV at room temperature [1,2]. In general, the band gap emission at room temperature is dominated by phonon replica of free exciton due to strong exciton-phonon coupling [1,2,67]. This phenomenon was also observed in our experiments. As an example, the spectral analyses of a series of ZnO_{-ROH} (R = Me, Et, *i*Pr and *n*Bu) NPs are shown in Figure 13. After fitting the PL spectra in the range from 2.9 to 3.5 eV, we find three peaks located at 3.139 eV (394 nm), 3.211 eV (386 nm) and 3.283 eV (378 nm). The energy difference of these peaks is about 72 meV, which corresponds to the longitudinal optical (LO) phonon energy in ZnO. Therefore, we can assign the peak at 3.283, 3.211 and 3.139 eV as the free exciton (FX) emission, the first order phono-assisted emission (free exciton-1 LO, FX-1LO) and the second order phono-assisted emission (free exciton-2 LO, FX-2LO), respectively. As shown in Figure 13b,c, for different ZnO NPs, the intensity ratio of these contributions is different.

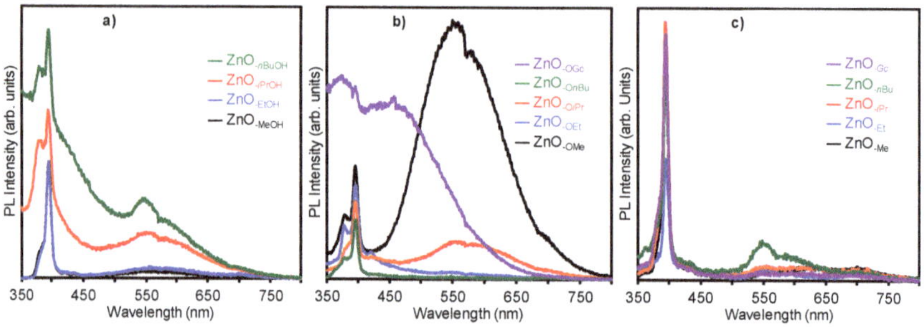

Figure 12. Photoluminescence spectra of (**a**) ZnO_{-ROH}; (**b**) ZnO_{-OR}; and (**c**) ZnO_{-R} (R = Me, Et, *i*Pr, *n*Bu and Gc) NPs derived from different synthetic approaches.

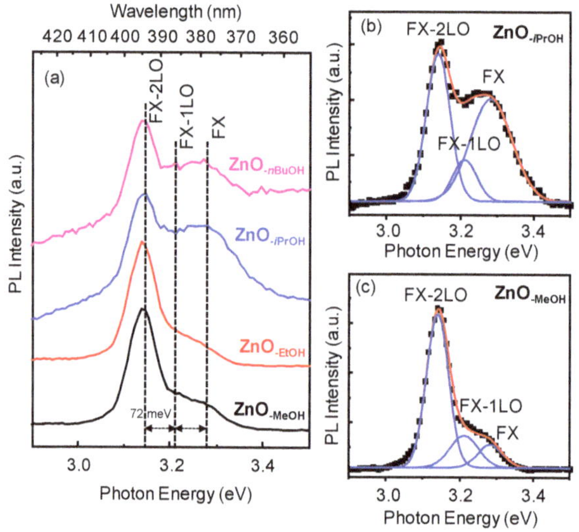

Figure 13. (**a**) PL spectra of ZnO_{-ROH} (R = Me, Et, *i*Pr and *n*Bu) NPs in the UV range; (**b**,**c**) spectral fitting of the peaks for ZnO_{-iPrOH} and ZnO_{-MeOH}, respectively.

Indeed, the observed PL peaks show less photon energy compared to the bandgap of the bulk ZnO. The PL of ZnO nanostructures previously reported at room temperature showed that near band gap emission was much less energy than the band gap (3.37 eV) of bulk ZnO. For instance, Wang and co-workers early reported that the ultrathin ZnO nanobelts (6 nm) exhibited a near band edge emission at 373 nm (3.32 eV) [107], while the normal ZnO nanobelts showed a peak at 387 nm (3.2 eV). In addition, one review paper also elucidated the PL of ZnO nanostructures varying from 372 to 390 with different shapes [2]. These below the bulk band gap (3.37 eV) of ZnO are probably caused by different morphologies [65], surface defects and impurity [108], as well as size effect [109], which can affect the near band edge emission. Hence the red shift of the near band edge emission of ZnO NPs in the present study can be attributed to the influences of these factors, such as ZnO_{-ROH}, ZnO_{-OR} and ZnO_{-R} series prepared by using zinc precursors with increasing carbon chain length of alkoxyl groups, possessing different defects caused by the preparation process. These results are probably related to that the presence of stabilizer and the incomplete decomposition of zinc alkoxides led to the doping of impurity into the ZnO lattices and surface defects (Figure 4). Based on the above-mentioned analysis of the alteration of the origin of the UV peak in these nanocrystalline ZnO particles, the change of near band gap emission can be readily understood by free excitonic or defect-related emission, surface impurity and defects depending on the synthesis method [5,64–66]. When chelated zinc glycerate was used as the zinc precursor, the obtained ZnO_{-OGc} NPs showed a strong photoluminscent peak at 378 nm and a broad peak centred at 450 nm (Figure 12b) and the peak at 450 nm is unclear. However, besides the fundamental ZnO emission, some weak or strong peaks with subgap energies at 550 nm (2.25 eV, "green" luminescence) [110] and 600 nm (2.06 eV) are detected in the photoluminescent spectra (Figure 12), especially, for ZnO_{-ROH} and ZnO_{-OR} (R = Me, *i*Pr) and ZnO_{-R} (R = *i*Pr, *n*Bu) NPs. The luminescent peaks can be attributed to oxygen vacancies at 550 nm and surface interstitial oxygen at about 600 nm [87]. This result is in good agreement with the energy level of oxygen vacancies calculated by Fu and co-workers [111]. Based on above-mentioned analysis, the intensity of luminescent peaks of ZnO NPs mainly depends on the ratio or concentration of predominant and secondary defect centres, such as zinc on interstitial sites (Zn_i), oxygen vacancies, oxygen on interstitial sites (O_i) and impurity heteroatoms. In fact, the formation of these defect centres are often affected by zinc precursor used, the presence of stabilizer and synthesis conditions [81,86,100,101].

Based on above-mentioned EPR investigations and PL spectra, we try to explain the relationship between EPR signals (g = 2.00, 1.96) and photoluminescence spectra of ZnO (λ = 378, 550, 600 nm) by having obtained luminescence property. In general, the types of defects or vacancies in a material often influence the emission spectrum of material. However, for the present ZnO NPs, we find that it is difficult to establish a direct relation between EPR signals and luminescence behaviour, for example, ZnO_{-MeOH} NPs showed a strong EPR signal at g = 2.00 which designed to an unpaired electron trapped on an oxygen vacancy site but PL spectrum did not show any strong photo-luminescent peak at λ = 550 nm as "green" luminescence, ZnO_{-EtOH} NPs showed two EPR signals at g = 2.00 and 1.96 but its luminescence behaviour is quite similar to that of ZnO_{-MeOH}. Note that ZnO_{-iPrOH} and ZnO_{-nBuOH} NPs showed a completely different surface defect type (oxygen vacancy (g = 2.00) and doping of impurity or surface interstitial oxygen (g = 1.96), Figure 10) but they showed a quite similar photoluminescence property reflected by "green" luminescence peak at λ = 550 nm and impurity-doping or surface interstitial oxygen-induced luminescence peak at λ = 600 nm from the PL spectrum (Figure 12). For ZnO_{-OR} and ZnO_{-R} series, a similar situation are also observed (Figure 12). These results confirm that for our ZnO NPs no unified regulation can be found between EPR signals and photoluminescence behaviours. Hence, we think that these results clearly reveal the complication of defect chemistry and photoluminescence properties. In fact, our study is in good accordance with results previously reported [81,86,88].

3. Materials and Methods

3.1. General Consideration

For synthesis of organozinc compounds, all operations were performed with rigorous exclusion of air and moisture, using glovebox techniques (MB Braun MB150B-G-II; <1 ppm O_2, <1 ppm H_2O, argon) or the standard Schlenk technique. Anhydrous methanol (99.8%), anhydrous isopropanol (99.5%) and oleylamine (tech. 70%) were purchased from Sigma-Aldrich (St. Louis, MO, USA) and used as received. Absolute ethanol (99.5%, Sigma-Aldrich, St. Louis, MO, USA) and anhydrous *n*-butanol (99.8%, Sigma-Aldrich, St. Louis, MO, USA) were used after drying with 5 Å molecular sieves and distillation. Glycerol (ACS reagent, ≥99.5%, Sigma-Aldrich, St. Louis, MO, USA) was used after drying with 5 Å molecular sieves. Hexane solvent was purified by using Grubbs columns (MBraun SPS, solvent purification system) before using. $Zn[N(SiMe_3)_2]_2$ were synthesized according to previously reported method with slight modification [75]. $Zn(OR)_2$ (R = Me, Et, *i*Pr and *n*Bu) was prepared by the reaction of one mole of $Zn[N(SiMe_3)_2]_2$ and two moles of alkyl alcohol in hexane [74].

3.2. *Synthesis of Crystalline Zinc Oxide NPs*

3.2.1. The Preparation of ZnO_{-ROH} (R = Me, Et, *i*Pr, *n*Bu, Gc) Nanoparticles

ZnO nanoparticles were synthesized by thermal decomposition method. The typical preparation procedure is presented as follows. Synthesis of spherical crystalline ZnO_{-MeOH} nanoparticles: Oleylamine (4 mL) was added into a two-neck flask and degassed at 100 °C for 30 min, followed by the filling of argon. 4 mL of anhydrous methanol was then added under argon protection. After being stirred for 10 min, the solution of $Zn[N(SiMe_3)_2]_2$ (3.00 g) in 12 mL of hexane was added dropwise by syringe rejection under vigorous stirring. White flocculent turbidity formed slowly. The following the mixture was refluxed at 100 °C for 1.5 h and then naturally cooled down to room temperature. The volatile matters were removed under vacuum and the final suspension was heated at a rate of 5 °C per minute to 240 °C and kept this temperature for 2 h. The resulting suspension was cooled down to room temperature, then 10 mL of methanol was added and stirred for several minutes. White precipitate was collected by centrifugation at a rate of 6000 rotations per minute for 10 min and washed twice with 5 mL of hexane. The precipitate was then dispersed in hexane. Part of separated precipitates was dried at 100 °C and used for analysis, denoted as ZnO_{-MeOH}.

Similarly, for materials ZnO_{-EtOH}, ZnO_{-iPrOH} and ZnO_{-nBuOH}, the detailed synthesis procedures are completely similar to that of ZnO_{-MeOH}, except that different alkyl alcohol was used as a reaction solvent. But for the preparation of ZnO_{-GcOH} (GcOH = glycyl alcohol), the mixture of 1-octadecene and oleic acid was used as a mixed solvent, the decomposition temperature was 320 °C and other conditions are identical.

3.2.2. The Preparation of ZnO_{-OR} (R = Me, Et, *i*Pr, *n*Bu, Gc) Nanoparticles

The fresh-prepared $Zn(OR)_2$ (R = Me, Et, *i*Pr and *n*Bu) was respectively used as a zinc molecular precursor and oleylamine was used as a reaction solvent. Reaction temperature and time are same as the preparation of ZnO_{-MeOH} NPs. The obtained ZnO material was denoted as ZnO_{-OR}. For the ZnO_{-OGC} preparation, the mixed oleic acid and 1-octadecene was used as a high boiling-point solvent and heating temperature was 320 °C for 2 h.

3.2.3. The Preparation of ZnO_{-R} (R = Me, Et, *i*Pr, *n*Bu, Gc) Nanoparticles

As a comparison, a series of zinc oxide materials denoted as ZnO_{-R} (R = Me, Et, *i*Pr, *n*Bu and Gc) were obtained by the direct calcination of fresh-prepared $Zn(OR)_2$ (R = Me, Et, *i*Pr, *n*Bu and Gc) at 500 °C for 4 h (Temperature-controlled programme: (i) from room temperature to 500 °C with a heating rate of 5 °C per minute; (ii) 500 °C, 4 h; (iii) naturally cool down to room temperature).

3.3. *Characterization*

Powder X-ray diffraction (PXRD) analysis of the as-synthesized and stabilizer-free ZnO nanostructured materials were recorded on a Bruker Advance D8 instrument in the step/scan mode using monochromatic CuK_{α} radiation (λ = 1.5406 Å). The wide-angle diffractograms were collected in the 2θ range of 10–100 °C with a scanspeed of 5. The crystallite size of the NPs is evaluated from X-ray powder diffraction data using Scherrer formula $D_{hkl} = K\lambda/(\beta\cos\theta)$, where D_{hkl} is the mean size of the ordered crystalline grain size, K is a dimensionless shape factor (0.89), λ is the X-ray wavelength of the Cu target (0.15406 nm), β is the full width at half the maximum (FWHM) intensity (in radians), θ is the Bragg angle (in degrees). Transmission electron microscopy (TEM) images were obtained on a JEOL JEM2100 microscopy (Akishima, Tokyo, Japan) with an operating voltage of 160 kV. For TEM measurement, a drop of fine powdery material suspended in ethanol (99.9%) under the ultrasonic vibration was loaded onto a holey carbon film on a square 400 mesh copper 3.05 mm grid. Scanning electron microscopy (SEM) images were recorded on a JEOL JSM-5900LV microscope (JEOL Ltd., Tokyo, Japan) accompanying with EDX system operated at an accelerating voltage of 15 kV. All SEM images reported here are representative of the corresponding materials.

Diffuse Reflectance Infrared Fourier-Transform (DRIFT) spectra of ZnO nanostructured materials were recorded on a Thermo Scientific Nicolet 6700 FTIR spectrometer (Waltham, MA, USA) with KBr reference (256 scans, a resolution of 4 cm^{-1}) in the range of 4000–400 cm^{-1}. ^{1}H and ^{13}C CP MAS NMR spectra were obtained at room temperature on a Bruker DSX 200 instrument (Billerica, MA, USA) equipped with magic angle spinning (MAS) hardware and using ZrO_2 rotor with an inside diameter of 3 mm. Adamantine was used as a reference (^{1}H: 1.76 and 1.87 ppm; ^{13}C: 28.46 and 37.85 ppm). Ultraviolet-visible (UV-Vis) absorption spectra were taken by using a Perkin Elmer Lambda 35 spectrometer (Waltham, MA, USA). Photoluminescence (PL) studies were conducted using a Varian Cary Eclipse fluorescence spectrophotometer (Palo Alto, CA, USA) with a Xe lamp at room temperature with excitation wavelength of 290 nm. X-band (10 GHz) electron paramagnetic resonance (EPR) spectra were obtained on a Bruker ESR 300E spectrometer (Billerica, MA, USA) at ambient temperature. The magnetic field was determined using a nuclear magnetic resonance gaussmeter (Villebon-Sur-Yvette, France). As a standard field marker, polycrystalline DPPH with electron spin g-factor (g = 2.0036) was used for the exact determination of the magnetic field offset. Elemental analyses of C, H and N were performed on an Elementar Vario EL III (Langenselbold, Germany). Thermogravimetric analyses (TGA) were obtained on a Netzsch STA 449F3 instrument (Selb, Germany) equipped with a quartz crucible at a heating rate of 2 K min^{-1} under $Ar/O_2/Ar$ atmosphere.

4. Conclusions

Crystalline ZnO nanoparticles have been successfully synthesized by using three different approaches—thermal decomposition of in-situ formed zinc alkoxide from the reaction of $Zn[N(SiMe_3)_2]_2$ and various alkyl alcohol, thermal decomposition of fresh-prepared $Zn(OR)_2$ (R = Me, Et, *i*Pr, *n*Bu, Gc) in the presence of stabilizer and the direct calcination of $Zn(OR)_2$ (R = Me, Et, *i*Pr, *n*Bu, Gc) at 500 °C. The alteration of morphology, crystalline particle sizes, aggregation behaviour and defect chemistry of obtained ZnO NPs depend on not only organozinc precursor but also on synthesis condition. The change of adsorption peaks in UV-Vis spectra demonstrated particle size effect. Various EPR and photoluminescent spectra confirmed defect behaviour of zinc or oxygen vacancies or impurity heteroatoms in crystalline ZnO NPs. These results demonstrated the relationship between structure and property and the complication of the formation of defect chemistry and photoluminescence behaviour in ZnO NPs.

Supplementary Materials: The following are available online at http://www.mdpi.com/2079-4991/8/1/22/s1, Figure S1: DRIFT spectra of the separated and dried intermediate zinc alkoxides before thermal decomposition, Figure S2: Thermogravimetric analysis curve of zinc alkoxides, $Zn(OR)_2$ (R = Me, Et, *i*Pr and *n*Bu), Figure S3: DRIFT spectra of as-made ZnO_{-MeOH}, ZnO_{-EtOH}, ZnO_{-iPrOH}, ZnO_{-BuOH} and ZnO_{-GCOH} NPs (Left) and zinc alkoxides (Right), Figure S4: PXRD patterns of calcined ZnO_{-ROH} (R = Me, Et, *i*Pr and *n*Bu), Figure S5: High-resolution TEM images of the crystalline ZnO_{-Me} (Left) and ZnO_{-iPr} (Right) NPs obtained by direct calcinations of $Zn(OMe)_2$ and $Zn(OiPr)_2$.

Acknowledgments: Yucang Liang is grateful to Reiner Anwander for normal financial support and to the University of Tübingen for funding within the excellent EXPAND program as well as Markus Ströbele for measuring TGA, Paul Schuler for recording the EPR spectra and Kristina Strohmaier for measuring the solid-state NMR spectra. Feng Fu gratefully acknowledges the support of the National Natural Science Foundation of China (No. 21663030).

Author Contributions: Yucang Liang conceived, designed and performed the experiments and characterization of materials, analysed the data and wrote the paper; Susanne Wicker performed part of the experiments and collected data; Xiao Wang completed the spectral fitting analysis; Egil Severin Erichsen performed TEM and SEM measurements; and Feng Fu designed part of experiments, analysed the data and discussed in detail about organizing the paper.

Conflicts of Interest: The authors declare no conflict of interest.

References

1. Comini, E.; Sberveglieri, G. Metal Oxides Nanowires as Chemical Sensors. *Mater. Today* **2014**, *13*, 36–44. [CrossRef]
2. Wu, W.-Q.; Chen, D.; Caruso, R.A.; Cheng, Y.-B. Recent Progress in Hybrid Perovskite Solar Cells Based on n-Type Materials. *J. Mater. Chem. A* **2017**, *5*, 10092–10109. [CrossRef]
3. Xue, N.; Zhang, Q.; Zhang, S.; Zong, P.; Yang, F. Highly Sensitive and Selective Hydrogen Gas Sensor Using the Mesoporous SnO_2 Modified Layers. *Sensors* **2017**, *17*, 2351. [CrossRef] [PubMed]
4. Cheng, L.; Ma, S.Y.; Li, X.B.; Luo, J.; Li, W.Q.; Li, F.M.; Mao, Y.Z.; Wang, T.T.; Li, Y.F. Highly Sensitive Acetone Sensors Based on Y-Doped SnO_2 Prismatic Hollow Nanofibers Synthesized by Electrospinning. *Sens. Actuators B Chem.* **2014**, *200*, 181–190. [CrossRef]
5. Zhai, T.; Li, L.; Wang, X.; Fang, X.; Bando, Y.; Golberg, D. Recent Developments in One-Dimensional Inorganic Nanostructures for Photodetectors. *Adv. Funct. Mater.* **2010**, *20*, 4233–4248. [CrossRef]
6. Yoshida, T.; Zhang, J.; Komatsu, D.; Sawatani, S.; Minoura, H.; Pauporté, T.; Lincot, D.; Oekermann, T.; Schlettwein, D.; Tada, H.; et al. Electrodeposition of Inorganic/Organic Hybrid Thin Films. *Adv. Funct. Mater.* **2009**, *19*, 17–43. [CrossRef]
7. Nikoobakht, B.; Wang, X.; Herzing, A.; Shi, J. Scalable Synthesis and Ddevice Integration of Self-Registered One-Dimensional Zinc Oxide Nanostructures and Related Materials. *Chem. Soc. Rev.* **2013**, *42*, 342–365. [CrossRef] [PubMed]
8. Lee, J.; Zhang, S.; Sun, S. High-Temperature Solution-Phase Syntheses of Metal-Oxide Nanocrystals. *Chem. Mater.* **2013**, *25*, 1293–1304. [CrossRef]
9. Bonomo, M.; Naponiello, G.; Venditti, I.; Zardetto, V.; Carlo, A.D.; Dini, D. Electrochemical and Photoelectrochemical Properties of Screen-Printed Nickel Oxide Thin Films Obtained from Precursor Pastes with Different Compositions. *J. Electrochem. Soc.* **2017**, *164*, H137–H147. [CrossRef]
10. Naponiello, G.; Venditti, I.; Zardetto, V.; Saccone, D.; Carlo, A.D.; Fratoddi, I.; Barolo, C.; Dini, D. Photoelectrochemical Characterization of Squaraine-Sensitized Nickel Oxide Cathodes Deposited via Screen-Printing for p-Type Dye-Sensitized Solar Cells. *Appl. Surf. Sci.* **2015**, *356*, 911–920. [CrossRef]
11. Lu, L.T.; Dung, N.T.; Tung, L.D.; Thanh, C.T.; Quy, O.K.; Chuc, N.V.; Maenosono, S.; Thanh, N.T.K. Synthesis of Magnetic Cobalt Ferrite Nanoparticles with Controlled Morphology, Monodispersity and Composition: the Influence of Solvent, Surfactant, Reductant and Synthetic Conditions. *Nanoscale* **2015**, *7*, 19596–19610. [CrossRef] [PubMed]
12. Li, L.; Zhai, T.; Bando, Y.; Golberg, D. Recent Progress of One-Dimensional ZnO Nanostructures Solar Cells. *Nano Energy* **2012**, *1*, 91–96. [CrossRef]
13. Venditti, I.; Barbero, N.; Russo, M.V.; Carlo, A.D.; Decker, F.; Fratoddi, I.; Barolo, C.; Dini, D. Electrodeposited ZnO with Squaraine Sentisizers as Photoactive Anode of DSCs. *Mater. Res. Express* **2014**, *1*, 015040. [CrossRef]
14. Wang, X. Piezoelectric Nanogenerators—Harvesting Ambient Mechanical Energy at the Nanometer Scale. *Nano Energy* **2012**, *1*, 13–24. [CrossRef]
15. Briscoe, J.; Dunnn, S. Piezoelectric Nanogenerators—A Review of Nanostructured Piezoelectric Energy Harvesters. *Nano Energy* **2015**, *14*, 15–29. [CrossRef]
16. Lu, M.-P.; Lu, M.-Y.; Chen, L.-J. p-Type ZnO Nanowires: From Synthesis to Nanoenergy. *Nano Energy* **2012**, *1*, 247–258. [CrossRef]
17. Kumar, B.; Kim, S.-W. Energy Harvesting Based on Semiconducting Piezoelectric ZnO Nanostructures. *Nano Energy* **2012**, *1*, 342–355. [CrossRef]
18. Xiong, H.-M. ZnO Nanoparticles Applied to Bioimaging and Drug Delivery. *Adv. Mater.* **2013**, *25*, 5329–5335. [CrossRef] [PubMed]
19. Dhahri, R.; Leonardi, S.G.; Hjiri, M.; El Mir, L.; Bonavita, A.; Donato, N.; Iannazzo, D.; Neri, G. Enhanced Performance of Novel Calcium/Aluminum co-Doped Zinc Oxide for CO_2 Sensors. *Sens. Actuators B Chem.* **2017**, *239*, 36–44. [CrossRef]
20. Özhür, Ü.; Alivov, Y.I.; Liu, C.; Teke, A.; Reshchikov, M.A.; Doğan, S.; Avrutin, V.; Cho, S.-J.; Morkoç, H. A Comprehensive Review of ZnO Materials and Devices. *J. Appl. Phys.* **2005**, *98*, 041301.
21. Djurišić, A.B.; Leung, Y.H. Optical Properties of ZnO Nanostructures. *Small* **2006**, *2*, 944–961. [CrossRef] [PubMed]

22. Xu, S.; Wang, Z.L. One-Dimensional ZnO Nanostructures: Solution Growth and Functional Properties. *Nano Res.* **2011**, *4*, 1013–1098. [CrossRef]
23. Qu, J.; Ge, Y.; Zu, B.; Li, Y.; Dou, X. Transition-Metal-Doped p-Type ZnO Nanoparticle-Based Sensory Array for Instant Discrimination of Explosive Vapors. *Small* **2016**, *12*, 1369–1377. [CrossRef] [PubMed]
24. De la Rosa, E.; Sepúlveda-Guzman, S.; Reeja-Jayan, B.; Torres, A.; Salas, P.; Elizondo, N.; Jose Yacaman, M. Controlling the Growth and Luminescence Properties of Well-Faceted ZnO Nanorods. *J. Phys. Chem. C* **2007**, *111*, 8489–8495. [CrossRef]
25. Wu, X.; Zheng, L.; Wu, D. Fabrication of Superhydrophobic Surfaces from Microstructured ZnO-Based Surfaces via a Wet-Chemical Route. *Langmuir* **2005**, *21*, 2665–2667. [CrossRef] [PubMed]
26. Demir, M.M.; Muoz-Esp, R.; Lieberwirth, I.; Wegner, G. Precipitation of Monodisperse ZnO Nanocrystals via Acid-Catalyzed Esterification of Zinc Acetate. *J. Mater. Chem.* **2006**, *16*, 2940–2947. [CrossRef]
27. Wu, J.-J.; Liu, S.-C. Low-Temperature Growth of Well-Aligned ZnO Nanorods by Chemical Vapor Deposition. *Adv. Mater.* **2002**, *14*, 215–218. [CrossRef]
28. Izaki, M.; Omi, T. Transparent Zinc Oxide Films Prepared by Electrochemical Reaction. *Appl. Phys. Lett.* **1996**, *68*, 2439–2440. [CrossRef]
29. Wu, K.; Sun, Z.; Cui, J. Unique Approach toward ZnO Growth with Tunable Properties: Influence of Methanol in an Electrochemical Process. *Cryst. Growth Des.* **2012**, *12*, 2864–2871. [CrossRef]
30. Elias, J.; Lvy-Clment, C.; Bechelany, M.; Michler, J.; Wang, G.-Y.; Wang, Z.; Philippe, L. Hollow Urchin-like ZnO Thin Films by Electrochemical Deposition. *Adv. Mater.* **2010**, *22*, 1607–1612. [CrossRef] [PubMed]
31. Manzano, C.V.; Caballero-Calero, O.; Hormeño, S.; Penedo, M.; Luna, M.; Martín-González, M.S. ZnO Morphology Control by Pulsed Electrodeposition. *J. Phys. Chem. C* **2013**, *117*, 1502–1508. [CrossRef]
32. Minch, R.; Es-Souni, M. A Versatile Approach to Processing of High Active Area Pillar Coral- and Sponge-Like Pt-Nanostructures. Application to Electrocatalysis. *J. Mater. Chem.* **2011**, *21*, 4182–4188. [CrossRef]
33. Sun, Y.; Fuge, G.M.; Ashfold, M.N.R. Growth of Aligned ZnO Nanorod Arrays by Catalyst-Free Pulsed Laser Deposition Methods. *Chem. Phys. Lett.* **2004**, *396*, 21–26. [CrossRef]
34. Hong, J.I.; Bae, J.; Wang, Z.L.; Snyder, R.L. Room-Temperature, Texture-Controlled Growth of ZnO Thin Ffilms and Their Application for Growing Aligned ZnO Nanowire Arrays. *Nanotechnology* **2009**, *20*, 085609. [CrossRef] [PubMed]
35. Heo, Y.W.; Varadarajan, V.; Kaufman, M.; Kim, K.; Norton, D.P.; Ren, F.; Fleming, P.H. Site-Specific Growth of Zno Nanorods Using Catalysis-Driven Molecular-Beam Epitaxy. *Appl. Phys. Lett.* **2002**, *81*, 3046–3048. [CrossRef]
36. Lin, D.; Wu, H.; Pan, W. Photoswitches and Memories Assembled by Electrospinning Aluminum-Doped Zinc Oxide Single Nanowires. *Adv. Mater.* **2007**, *19*, 3968–3972. [CrossRef]
37. Schneider, J.J.; Hoffmann, R.C.; Engstler, J.; Klyszcz, A.; Erdem, E.; Jakes, P.; Eichel, R.-A.; Pitta-Bauermann, L.; Bill, J. Synthesis, Characterization, Defect Chemistry, and FET Properties of Microwave-Derived Nanoscaled Zinc Oxide. *Chem. Mater.* **2010**, *22*, 2203–2212. [CrossRef]
38. Bilecka, I.; Djerdj, I.; Niederberger, M. One-Minute Synthesis of Crystalline Binary and Ternary Metal Oxide Nanoparticles. *Chem. Commun.* **2008**, 886–888. [CrossRef] [PubMed]
39. Bilecka, I.; Elser, P.; Niederberger, M. Kinetic and Thermodynamic Aspects in the Microwave-Assisted Synthesis of ZnO Nanoparticles in Benzyl Alcohol. *ACS Nano* **2009**, *3*, 467–477. [CrossRef] [PubMed]
40. Hu, X.; Gong, J.; Zhang, L.; Yu, J.C. Continuous Size Tuning of Monodisperse ZnO Colloidal Nanocrystal Clusters by a Microwave-Polyol Process and Their Application for Humidity Sensing. *Adv. Mater.* **2008**, *20*, 4845–4850. [CrossRef]
41. Tang, J.; Cui, X.; Liu, Y.; Yang, X. Morphology-Controlled Synthesis of Monodisperse ZnO Troughs at the Air-Water Interface under Mild Conditions. *J. Phys. Chem. B* **2005**, *109*, 22244–22249. [CrossRef] [PubMed]
42. Peng, Q.; Xu, S.; Zhuang, Z.; Wang, X.; Li, Y. A General Chemical Conversion Method to Various Semiconductor Hollow Structures. *Small* **2005**, *1*, 216–221. [CrossRef] [PubMed]
43. Du, Y.-P.; Zhang, Y.-W.; Sun, L.-D.; Yan, C.-H. Efficient Energy Transfer in Monodisperse Eu-Doped ZnO Nanocrystals Synthesized from Metal Acetylacetonates in High-Boiling Solvents. *J. Phys. Chem. C* **2008**, *112*, 12234–12241. [CrossRef]
44. Joo, J.; Kwon, S.G.; Yu, J.H.; Hyeon, T. Synthesis of ZnO Nanocrystals with Cone, Hexagonal Cone, and Rod Shapes via Non-Hydrolytic Ester Elimination Sol-Gel Reactions. *Adv. Mater.* **2005**, *17*, 1873–1877. [CrossRef]

45. Kwon, S.G.; Hyeon, T. Colloidal Chemical Synthesis and Formation Kinetics of Uniformly Sized Nanocrystals of Metals, Oxides, and Chalcogenides. *Acc. Chem. Res.* **2008**, *41*, 1696–1709. [CrossRef] [PubMed]
46. Sarkara, D.; Tikkub, S.; Thaparb, V.; Srinivasac, R.S.; Khilar, K.C. Formation of Zinc Oxide Nanoparticles of Different Shapes in Water-in-Oil Microemulsion. *Colloids Surf. A Physicochem. Eng. Asp.* **2011**, *381*, 123–129. [CrossRef]
47. Lin, C.-C.; Li, Y.-Y. Synthesis of ZnO Nanowires by Thermal Decomposition of Zinc Acetate Dihydrate. *Mater. Chem. Phys.* **2009**, *113*, 334–337. [CrossRef]
48. Audebrand, N.; Auffrédic, J.-P.; Louër, D. X-ray Diffraction Study of the Early Stages of the Growth of Nanoscale Zinc Oxide Crystallites Obtained from Thermal Decomposition of Four Precursors. General Concepts on Precursor-Dependent Microstructural Properties. *Chem. Mater.* **1998**, *10*, 2450–2461. [CrossRef]
49. Shim, M.; Guyot-Sionnest, P. Organic-Capped ZnO Nanocrystals: Synthesis and n-Type Character. *J. Am. Chem. Soc.* **2001**, *123*, 11651–11654. [CrossRef] [PubMed]
50. Carnes, C.L.; Klabunde, K.J. Synthesis, Isolation, and Chemical Reactivity Studies of Nanocrystalline Zinc Oxide. *Langmuir* **2000**, *16*, 3764–3772. [CrossRef]
51. Monge, M.; Kahn, M.L.; Maisonnat, A.; Chaudret, B. Room-Temperature Organometallic Synthesis of Soluble and Crystalline ZnO Nanoparticles of Controlled Size and Shape. *Angew. Chem. Int. Ed.* **2003**, *42*, 5321–5324. [CrossRef] [PubMed]
52. Kahn, M.L.; Monge, M.; Collière, V.; Senocq, F.; Maisonnat, A.; Chaudret, B. Size- and Shape-Control of Crystalline Zinc Oxide Nanoparticles: A New Organometallic Synthetic Method. *Adv. Funct. Mater.* **2005**, *15*, 458–468. [CrossRef]
53. Auld, J.; Houlton, D.J.; Jones, A.C.; Rushworth, S.A.; Malik, M.A.; O'Brien, P.; Critchlow, G.W. Growth of ZnO by MOCVD using Alkylzinc Alkoxides as Singlesource Precursors. *J. Mater. Chem.* **1994**, *4*, 1249–1253. [CrossRef]
54. Polarz, S.; Roy, A.; Merz, M.; Halm, S.; Schröder, D.; Schneider, L.; Bacher, G.; Kruis, F.E.; Driess, M. Chemical Vapor Synthesis of Size-Selected Zinc Oxide Nanoparticles. *Small* **2005**, *1*, 540–552. [CrossRef] [PubMed]
55. Polarz, S.; Orlov, A.; Hoffmann, A.; Wagner, M.R.; Rauch, C.; Kirste, R.; Gehlhoff, W.; Aksu, Y.; Driess, M.; van den Berg, M.W.E.; et al. A Systematic Study on Zinc Oxide Materials Containing Group I Metals (Li, Na, K)-Synthesis from Organometallic Precursors, Characterization, and Properties. *Chem. Mater.* **2009**, *21*, 3889–3897. [CrossRef]
56. Kim, C.G.; Sung, K.; Chung, T.-M.; Jung, D.Y.; Kim, Y. Monodispersed ZnO Nanoparticles from a Single Molecular Precursor. *Chem. Commun.* **2003**, 2068–2069. [CrossRef]
57. Polarz, S.; Regenspurger, R.; Hartmann, J. Self-Assembly of Methylzinc-Polyethylene Glycol Amphiphiles and Their Application to Materials Synthesis. *Angew. Chem. Int. Ed.* **2007**, *46*, 2426–2430. [CrossRef] [PubMed]
58. Polarz, S.; Neues, F.; van den Berg, M.W.E.; Grünert, W.; Khodeir, L. Mesosynthesis of ZnO-Silica Composites for Methanol Nanocatalysis. *J. Am. Chem. Soc.* **2005**, *127*, 12028–12034. [CrossRef] [PubMed]
59. Bury, W.; Krajewska, E.; Dutkiewicz, M.; Sokołowski, K.; Justyniak, I.; Kaszkur, Z.; Kurzydłowski, K.J.; Płocińskic, T.; Lewiński, J. *tert*-Butylzinc Hydroxide as an Efficient Predesigned Precursor of ZnO Nanoparticles. *Chem. Commun.* **2011**, *47*, 5467–5469. [CrossRef] [PubMed]
60. Sokołowski, K.; Justyniak, I.; Bury, W.; Grzonka, J.; Kaszkur, Z.; Mąkolski, Ł.; Dutkiewicz, M.; Lewalska, A.; Krajewska, E.; Kubicki, D.; et al. *tert*-Butyl(tert-butoxy)zinc Hydroxides: Hybrid Models for Single-Source Precursors of ZnO Nanocrystals. *Chem. Eur. J.* **2015**, *21*, 5488–5495. [CrossRef] [PubMed]
61. Zhang, L.; Hu, Y.H. A Systematic Investigation of Decomposition of Nano $Zn_4O(C_8H_4O_4)_3$ Metal-Organic Framework. *J. Phys. Chem. C* **2010**, *114*, 2566–2572. [CrossRef]
62. Liu, J.F.; Bei, Y.Y.; Wu, H.P.; Shen, D.; Gong, J.Z.; Li, X.G.; Wang, Y.W.; Jiang, N.P.; Jiang, J.Z. Synthesis of Relatively Monodisperse ZnO Nanocrystals from a Precursor Zinc 2,4-Pentanedionate. *Mater. Lett.* **2007**, *61*, 2837–2840. [CrossRef]
63. Huang, M.H.; Mao, S.; Feick, H.; Yan, H.; Wu, Y.; Kind, H.; Weber, E.; Russo, R.; Yang, P. Room-Temperature Ultraviolet Nanowire Nanolasers. *Science* **2001**, *292*, 1897–1899. [CrossRef] [PubMed]
64. Huang, M.H.; Wu, Y.; Feick, H.; Tran, N.; Weber, E.; Yang, P. Catalytic Growth of Zinc Oxide Nanowires by Vapor Transport. *Adv. Mater.* **2001**, *13*, 113–116. [CrossRef]

65. Andelman, T.; Gong, Y.; Polking, M.; Yin, M.; Kuskovsky, I.; Neumark, G.; O'Brien, S. Morphological Control and Photoluminescence of Zinc Oxide Nanocrystals. *J. Phys. Chem. B* **2005**, *109*, 14314–14318. [CrossRef] [PubMed]
66. Yin, M.; Gu, Y.; Kuskovsky, I.L.; Andelman, T.; Zhu, Y.; Neumark, G.F.; O'Brien, S. Zinc Oxide Quantum Rods. *J. Am. Chem. Soc.* **2004**, *126*, 6206–6207. [CrossRef] [PubMed]
67. Mohanta, A.; Thareja, R.K. Photoluminescence Study of ZnO Nanowires Grown by Thermal Evaporation on Pulsed Laser Deposited ZnO Buffer Layer. *J. Appl. Phys.* **2008**, *104*, 044906. [CrossRef]
68. Mohanta, A.; Thareja, R.K. Photoluminescence Characteristics of Catalyst Free ZnO Nanowires. *Mater. Res. Express* **2014**, *1*, 015023. [CrossRef]
69. Mohanta, A.; Singh, V.; Thareja, R.K. Photoluminescence from ZnO Nanoparticles in Vapor Phase. *J. Appl. Phys.* **2008**, *104*, 064903. [CrossRef]
70. Mohanta, A.; Kung, P.; Thareja, R.K. Exciton-Exciton Scattering in Vapor Phase ZnO Nanoparticles. *Appl. Phys. Lett.* **2015**, *106*, 013108. [CrossRef]
71. Ma, D.; Huang, J.; Ye, Z.; Wang, L.; Zhao, B. Relationship between Photoluminescence and Structural Properties of the Sputtered $Zn_{1-x}Cd_xO$ Films on Si Substrates. *Opt. Mater.* **2004**, *25*, 367–371. [CrossRef]
72. Mohanta, A.; Thareja, R.K. Photoluminescence Study of ZnCdO Alloy. *J. Appl. Phys.* **2008**, *103*, 024901. [CrossRef]
73. Mohanta, A.; Thareja, R.K. Temperature-Dependent S-Shaped Photoluminescence in ZnCdO Alloy. *J. Appl. Phys.* **2010**, *107*, 084904. [CrossRef]
74. Boulmaâz, S.; Hubert-Pfalzgraf, L.G. The Quest for Mixed-Metal Alkoxides Based on Zinc: Synthesis and Characterization of Zinc-Tantalum Oxoisopropoxides. *J. Sol-Gel Sci. Technol.* **1994**, *2*, 11–15. [CrossRef]
75. Bochmann, M.; Bwembya, G.; Webb, K.J. Arene Chalrogenolato Complexes of Zinc And Cadmium. *Inorg. Synth.* **1997**, *31*, 19–24.
76. Wang, Y.; Zhang, J.; Zhao, Y. Strength Weakening by Nanocrystals in Ceramic Materials. *Nano Lett.* **2007**, *7*, 3196–3199. [CrossRef] [PubMed]
77. Das, J.; Khushalani, D. Nonhydrolytic Route for Synthesis of ZnO and Its Use as a Recyclable Photocatalyst. *J. Phys. Chem. C* **2010**, *114*, 2544–2550. [CrossRef]
78. Dong, H.; Feldmann, C. Porous ZnO Platelets via Controlled Thermal Decomposition of Zinc Glycerolate. *J. Alloys Compd.* **2012**, *513*, 125–129. [CrossRef]
79. Reinoso, D.M.; Damiani, D.E.; Tonetto, G.M. Zinc Glycerolate as a Novel Heterogeneous Catalyst for the Synthesis Offatty Acid Methyl Esters. *Appl. Catal. B Environ.* **2014**, *144*, 308–316. [CrossRef]
80. Polarz, S.; Strunk, J.; Ischenko, V.; van den Berg, M.W.E.; Hinrichsen, O.; Muhler, M.; Driess, M. On the Role of Oxygen Defects in the Catalytic Performance of Zinc Oxide. *Angew. Chem. Int. Ed.* **2006**, *45*, 2965–2969. [CrossRef] [PubMed]
81. Ischenko, V.; Polarz, S.; Grote, D.; Stavarache, V.; Fink, K.; Driess, M. Zinc Oxide Nanoparticles with Defects. *Adv. Funct. Mater.* **2005**, *15*, 1945–1954. [CrossRef]
82. Parashar, S.K.S.; Murty, B.S.; Repp, S.; Weber, S.; Erdem, E. Investigation of Intrinsic Defects in Core-Shell Structured ZnO Nanocrystals. *J. Appl. Phys.* **2012**, *111*, 113712. [CrossRef]
83. Vanheusden, K.; Seager, C.H.; Warren, W.L.; Tallant, D.R.; Voigt, J.A. Correlation Between Photoluminescence and Oxygen Vacancies in ZnO Phosphors. *Appl. Phys. Lett.* **1996**, *68*, 403–405. [CrossRef]
84. Wang, X.Y.; Vlasenko, L.S.; Pearton, S.J.; Chen, W.M.; Buyanova, I.A. Oxygen and Zinc Vacancies in As-Grown ZnO Single Crystals. *J. Phys. D Appl. Phys.* **2009**, *42*, 175411. [CrossRef]
85. Janotti, A.; Van de Walle, C.G. Oxygen Vacancies in ZnO. *Appl. Phys. Lett.* **2005**, *87*, 122102. [CrossRef]
86. Drouilly, C.; Krafft, J.-M.; Averseng, F.; Casale, S.; Bazer-Bachi, D.; Chizallet, C.; Lecocq, V.; Vezin, H.; Lauron-Pernot, H.; Costentin, G. ZnO Oxygen Vacancies Formation and Filling Followed by in Situ Photoluminescence and in Situ EPR. *J. Phys. Chem. C* **2012**, *116*, 21297–21307. [CrossRef]
87. Stavale, F.; Nilius, N.; Freund, H.-J. STM Luminescence Spectroscopy of Intrinsic Defects in ZnO(0001) Thin Films. *J. Phys. Chem. Lett.* **2013**, *4*, 3972–3976. [CrossRef]
88. Kaftelen, H.; Ocakoglu, K.; Thomann, R.; Tu, S.; Weber, S.; Erdem, E. EPR and Photoluminescence Spectroscopy Studies on the Defect Structure of ZnO Nanocrystals. *Phys. Rev. B* **2012**, *86*, 014113. [CrossRef]
89. Galland, D.; Herve, A. ESR Spectra of the Zinc Vacancy in ZnO. *Phys. Lett. A* **1970**, *33*, 1–2. [CrossRef]
90. Taylor, A.L.; Filipovi, G.; Lindeber, G. Electron Paramagnetic Resonance Associated with Zn Vacancies in Neutron-Irradiated ZnO. *Solid State Commun.* **1970**, *8*, 1359–1361. [CrossRef]

91. Yu, B.; Zhu, C.; Gan, F.; Huang, Y. Electron Spin Resonance Properties of ZnO Microcrystallites. *Mater. Lett.* **1998**, *33*, 247–250. [CrossRef]
92. Jing, L.Q.; Xu, Z.L.; Shang, J.; Sun, X.J.; Cai, W.M.; Guo, H.C. The Preparation and Characterization of ZnO Ultrafine Particles. *Mater. Sci. Eng. A* **2002**, *332*, 356–361. [CrossRef]
93. Zhang, L.Y.; Yin, L.W.; Wang, C.X.; Lun, N.; Qi, Y.X.; Xiang, D. Origin of Visible Photoluminescence of ZnO Quantum Dots: Defect-Dependent and Size-Dependent. *J. Phys. Chem. C* **2010**, *114*, 9651–9658. [CrossRef]
94. Kakazev, N.G.; Sreckovic, T.V.; Ristic, M.M. Electronic Paramagnetic Resonance Investigation of the Evolution of Defects in Zinc Oxide During Tribophysical Activation. *J. Mater. Sci.* **1997**, *32*, 4619–4622. [CrossRef]
95. Schulz, M. ESR Experiments on Ga Donors in ZnO Crystals. *Phys. Status Solidi A* **1975**, *27*, K5–K8. [CrossRef]
96. Block, D.; Hereve, A.; Cox, R.T. Optically Detected Magnetic Resonance and Optically Detected ENDOR of Shallow Indium Donors in ZnO. *Phys. Rev. B* **1982**, *25*, 6049. [CrossRef]
97. Gonzales, C.; Block, D.; Cox, R.T.; Herve, A. Magnetic Resonance Studies of Shallow Donors in Zinc Oxide. *J. Cryst. Growth* **1982**, *59*, 357–362. [CrossRef]
98. Garces, N.Y.; Wang, L.; Bai, L.; Giles, N.C.; Halliburton, L.E.; Cantwell, G. Role of Copper in the Green Luminescence from ZnO Crystals. *Appl. Phys. Lett.* **2002**, *81*, 622–624. [CrossRef]
99. La Porta, F.A.; Andrés, J.; Vismara, M.V.G.; Graeff, C.F.O.; Sambrano, J.R.; Li, M.S.; Varela, J.A.; Longo, E. Correlation between Structural and Electronic Order-Disorder Effects and Optical Properties in ZnO Nanocrystals. *J. Mater. Chem. C* **2014**, *2*, 10164–10174. [CrossRef]
100. Mondal, O.; Pal, M. Strong and Unusual Violet-Blue Emission in Ring Shaped ZnO Nanocrystals. *J. Mater. Chem.* **2011**, *21*, 18354–18358. [CrossRef]
101. Zhang, Y.; Xu, J.; Xiang, Q.; Li, H.; Pan, Q.; Xu, P. Brush-Like Hierarchical ZnO Nanostructures: Synthesis, Photoluminescence and Gas Sensor Properties. *J. Phys. Chem. C* **2009**, *113*, 3430–3435. [CrossRef]
102. Koch, U.; Fojtik, A.; Weller, H.; Henglein, A. Photochemistry of Semiconductor Colloids. Preparation of Extremely Small ZnO Particles, Fluorescence Phenomena and Size Quantization Effects. *Chem. Phys Lett.* **1982**, *122*, 507–510. [CrossRef]
103. Wang, L.; Giles, N.C. Temperature Dependence of the Free-Exciton Transition Energy in Zinc Oxide by Photoluminescence Excitation Spectroscopy. *J. Appl. Phys.* **2003**, *94*, 973–978. [CrossRef]
104. Hsu, H.-C.; Hsieh, W.-F. Excitonic Polaron and Phonon Assisted Photoluminescence of ZnO Nanowires. *Solid State Commun.* **2004**, *131*, 371–375. [CrossRef]
105. Liu, R.; Pan, A.; Fan, H.; Wang, F.; Shen, Z.; Yang, G.; Xie, S.; Zou, B. Phonon-Assisted Stimulated Emission in Mn-Doped ZnO Nanowires. *J. Phys. Condens. Matter* **2007**, *19*, 136206. [CrossRef]
106. Yang, S.; Tian, X.; Wang, L.; Wei, J.; Qi, K.; Li, X.; Xu, Z.; Wang, W.; Zhao, J.; Bai, X.; et al. In-situ Optical Transmission Electron Microscope Study of Exciton Phonon Replicas in ZnO Nanowires by Cathodoluminescence. *Appl. Phys. Lett.* **2014**, *105*, 071901. [CrossRef]
107. Wang, X.; Ding, Y.; Summers, C.J.; Wang, Z.L. Large-Scale Synthesis of Six-Nanometer-Wide ZnO Nanobelts. *J. Phys. Chem. B* **2004**, *108*, 8773–8777. [CrossRef]
108. Fonoberov, V.A.; Balandin, A.A. Origin of Ultraviolet Photoluminescence in ZnO Quantum Dots: Confined Excitons Versus Surface-Bound Impurity Exciton Complexes. *Appl. Phys. Lett.* **2004**, *85*, 5971–5973. [CrossRef]
109. Li, J.W.; Yang, L.W.; Zhou, Z.F.; Chu, P.K.; Wang, X.H.; Zhou, J.; Li, L.T.; Sun, C.Q. Bandgap Modulation in ZnO by Size, Pressure, and Temperature. *J. Phys. Chem. C* **2010**, *114*, 13370–13374. [CrossRef]
110. Kohan, A.F.; Ceder, G.; Morgan, D.; Van de Walle, C.G. First-Principles Study of Native Point Defects in ZnO. *Phys. Rev. B* **2000**, *61*, 15019. [CrossRef]
111. Lin, B.; Fu, Z.; Jia, Y. Green Luminescent Center in Undoped Zinc Oxide Films Deposited on Silicon Substrates. *Appl. Phys. Lett.* **2001**, *79*, 943–945. [CrossRef]

nanomaterials

MDPI

Article

Enhanced UV-Visible Light Photocatalytic Activity by Constructing Appropriate Heterostructures between Mesopore TiO_2 Nanospheres and Sn_3O_4 Nanoparticles

Jianling Hu, Jianhai Tu, Xingyang Li, Ziya Wang, Yan Li, Quanshui Li and Fengping Wang *

Department of Physics, School of Mathematics and Physics, University of Science and Technology Beijing, Beijing 10083, China; jlhu_ustb@163.com (J.H.); jianhaituustb@163.com (J.T.); xl99936@uga.edu (X.L.); wzywolf@163.com (Z.W.); liyan000998@163.com (Y.L.); qsli@ustb.edu.cn (Q.L.)

* Correspondence: fpwang@ustb.edu.cn; Tel.:+86-10-6233-2587; Fax: +86-10-6233-2993

Received: 25 September 2017; Accepted: 13 October 2017; Published: 19 October 2017

Abstract: Novel TiO_2/Sn_3O_4 heterostructure photocatalysts were ingeniously synthesized via a scalable two-step method. The impressive photocatalytic abilities of the TiO_2/Sn_3O_4 sphere nanocomposites were validated by the degradation test of methyl orange and •OH trapping photoluminescence experiments under ultraviolet (UV) and visible light irradiation, respectively. Especially under the visible light, the TiO_2/Sn_3O_4 nanocomposites demonstrated a superb photocatalytic activity, with 81.2% of methyl orange (MO) decomposed at 30 min after irradiation, which greatly exceeded that of the P25 (13.4%), TiO_2 (0.5%) and pure Sn_3O_4 (59.1%) nanostructures. This enhanced photocatalytic performance could be attributed to the mesopore induced by the monodispersed TiO_2 cores that supply sufficient surface areas and accessibility to reactant molecules. This exquisite hetero-architecture facilitates extended UV-visible absorption and efficient photoexcited charge carrier separation.

Keywords: photocatalyst; heterostructures; TiO_2; Sn_3O_4

1. Introduction

As a stable, low-cost and environmentally benign material, nanoscaled titanium dioxide (TiO_2) with unique structural and functional properties has become a widely used semiconductor photocatalyst for various solar-driven clean energy technologies [1]. Tailoring the morphology of TiO_2 photoanode is a preferred route to achieving high performance in solar cells due to its enhanced properties, such as high surface area, faster electron transport, lower electron-hole recombination rate and good light-harvesting features [2,3]. Nevertheless, the wide optical bandgap of TiO_2, which seriously limits its light harvesting capability, leaving about 96% of the solar light energy wasted [4]. Compared with the solution of generating donor or acceptor states in the band gap by adding impurities, rationally designing and constructing the surface heterostructures would be a more efficient strategy for achieving an excellent photocatalyst [5].

Recently, Sn_3O_4, a novel non-stoichiometric oxide, has raised particular interest in the field of photocatalysis—especially in terms of its catalytic behavior under visible light irradiation—due to a suitable band-gap inside visible light (2.2–2.9 eV) and a distinct surface structure composed of both valences of tin [6,7]. Several studies have shown its great potential as an auspicious photocatalyst under visible light, both for generating hydrogen and degrading dyes [8]. However, some drawbacks have hindered its performance in practice. As a semiconductor with a relatively narrow band gap, pure Sn_3O_4 generally leads to a fast recombination of photoexcited electron-hole pairs, which ultimately decreases its degradation rate [9].

Discussing these problems together, it proposes an intriguing idea that Sn_3O_4, as the second component, attaches to the surface of TiO_2 nanostructures, for an exquisite TiO_2/Sn_3O_4 heterostructure. On the one hand, a theoretical analysis indicates that the interface between TiO_2 and Sn_3O_4 is to be a perfect type-II heterojunction (both the potentials of valence band (VB) and conduction band (CB) of Sn_3O_4 are higher than that of TiO_2) [10], which is actually conducive to the separation of photoexcited electron-hole pairs. Furthermore, latest reports have exhibited the superiority of the heterogeneous composite of this kind [11,12]. On the other hand, increased photoactive facets can effectively facilitate the efficiency of photo-absorption and oxygen chemisorption, and bring about a fast rate of surface reactions [13]. Therefore, highly dispersive anatase TiO_2 mesopore nanospheres, which possess a large number of active surfaces, would likely be an amazing matrix in TiO_2/Sn_3O_4 nanocomposites.

The two-step self-assembly approach is a feasible strategy for the refined design of hierarchical nanostructures with complex morphologies, and has been proven to be an effective way to design multiscale nanostructures, since the morphology and composition obtained from the first step can be further tuned and adjusted by a subsequent second process. Moreover, this approach also allows the combination of multiple synthetic techniques, and the synthesis of complex nanostructures with hierarchical multiscale structures compared with the conventional one-step self-assembling method [14]. Recently, a lot of complex nanostructures with high photocatalytic performance for both visible light and ultraviolet has been acquired by the two-step synthesis method. Usually, these methods can be classified into two categories: (1) synthesis under two continuous identical methods, such as in [9,15,16]; and (2) synthesis under two different methods, such as in [17,18]. In 2015, TiO_2/Sn_3O_4 nanobelts [9] were successfully produced by first synthesizing the TiO_2 nanobelts, and then assembling Sn_3O_4 onto the TiO_2 nanobelts in a subsequent hydrothermal procedure; in the same year, hierarchical Sn_3O_4/N-TiO_2 nanotubes [17] were synthesized by first weaving N-doped TiO_2 nanotube via electro-spinning, and then modifying them with Sn_3O_4 via hydrothermal reaction. In 2017, a range of heterojunction WO_3/TiO_2 thin films were deposited via a two-step process using chemical vapor deposition (CVD) methods [15]. Generally, electro-spinning and chemical vapor deposition is associated with at least one of the following factors: expensive equipment, high voltage, hazardous by-products, or toxic chemicals, rendering the method less environmental friendly and much more complicated than the hydrothermal method or sol-gel synthesis. Hence, synthesizing complex nanostructures via a combination of sol-gel and hydrothermal methods would be a low-cost, scalable, easy to control, and eco-friendly strategy in terms of preparing high-quality, uniform, catalysts.

Herein, we developed a scalable two-step route, combining the sol-gel method and hydrothermal progress to achieve excellent visible and ultraviolet photocatalytic activity by uniformly synthesizing the Sn_3O_4 nanoparticles on the surface of TiO_2 nanospheres. As expected, an enormous enhancement of photocatalytic efficiency was achieved by the distinctive TiO_2/Sn_3O_4 nanocomposites.

2. Experimental Section

2.1. Chemicals

The chemicals used in this study were of analytic grade, and were used without further purification. Tetrabutyl titanate was purchased from Beijing Xingjin Chemical Factory, Beijing, China. Methyl orange (MO) was obtained from Tianjin Jinke Fine Chemical Industry Research Institute, Tianjin, China. Tin dichloride dihydrate ($SnCl_2 \cdot 2H_2O$) and trisodium citrate dihydrate ($Na_3C_6H_5O_7 \cdot 2H_2O$) were purchased from Xilong Chemical Industry Co., Ltd., Guangdong, China. Terephthalic acid was purchased from Alfa Aesar (Tianjin, China). All the other organic solvents and salts, including ethylene glycol, acetone, NaOH, were purchased from Sinopharm Chemical Reagent Beijing Co. P25 (nanoscale TiO_2 powder, surface area 50 $m^2 \cdot g^{-1}$) was purchased from Degussa AG (Hanau, Germany).

2.2. Synthesis of Samples

In this research, we propose a two-step synthesis method to obtain TiO_2/Sn_3O_4 nanocomposites by preparing TiO_2 core via sol-gel route first and then synthesizing Sn_3O_4 on the surface of TiO_2. (1) Synthesis of core TiO_2 nanospheres [19]: 3.5 mL tetrabutyl titanate was dissolved in the 50 mL ethylene glycol while stirring vigorously for 10 h. Then, the mixture was immediately poured into a solution containing 170 mL acetone and 2.7 mL deionized water under constant stirring, until white precipitation appeared. The acquired precipitate was calcined in air at 500 °C for 1 h to produce the TiO_2 powders; (2) Coating TiO_2 with Sn_3O_4: 0.2 g of the TiO_2 product described above and 5.0 mmol $SnCl_2{\cdot}2H_2O$ were mixed with 25 mL deionized water, followed by the addition of 12.5 mmol $Na_3C_6H_5O_7{\cdot}2H_2O$ and 2.5 mmol NaOH under magnetic stirring. During this process, Sn(II) ions were attached to the surface of hydroxyl-rich TiO_2 spherical colloids through inorganic grafting. The resulting precursor was then transferred to a 50 mL Teflon-lined stainless autoclave and maintained at 180 °C for 12 h. Finally, the collected powder was washed several times with deionized water and ethanol, and dried at 60 °C for 12 h.

2.3. Characterization of Samples

In order to obtain the physical and chemical properties of as-prepared samples, several characterizations were conducted. X-ray diffraction (XRD) patterns were recorded by a Rigaku D/MAX-2500 diffractometer with Cu Kα radiation (Rigaku, Tokyo, Japan). Raman spectra were obtained using a HORIBA HR800 spectrometer with an Nd:YAG laser at a wavelength of 532 nm (Horiba Yvon, Paris, France). Scanning electron microscopy (SEM) and transmission electron microscopy (TEM) images were acquired from the ZEISS SUPRA 55 (Zeiss, Oberkochen, Germany)and JEOL JEM-2010 (JEOL, Tokyo, Japan), respectively. X-ray photoelectron spectra (XPS) were recorded on a scanning X-ray microprobe PHI Quantera II (Ulvac-PHI, Chigasaki, Japan). The nitrogen adsorption-desorption isotherm was measured at 77 K on an Autosorb-iQ2-MP analyzer (Quantachrome Instruments, Boynton Beach, FL, USA). The absorption spectra were carried out by UV-visible spectrophotometer (Lambda 950, Perkin-Elmer, Shelton, WA, USA) and the hydroxyl radicals (•OH) trapping photoluminescence spectra were examined by a fluorescence spectrophotometer (Hitachi F-4500, Hitachi, Tokyo, Japan) using excitation a wavelength of 315 nm.

2.4. Photocatalytic Experiments

The photocatalytic activity of TiO_2/Sn_3O_4 nanocomposites was evaluated via methyl orange (MO) degradation rate. 80 mL aqueous suspension of MO (20 mg/L) and 80 mg of photocatalyst powder were placed in a 100 mL beaker. Prior to irradiating, the suspensions were magnetically stirred in the dark for 40 min to establish adsorption-desorption equilibrium. A 250 W mercury lamp and a 500 W Halogen lamp with a 420 nm cut-off filter were used as the UV and visible light sources, respectively. After given irradiation time intervals, aliquots of the mixed solution were collected and centrifuged to remove the catalyst particulates for analysis. Four consecutive cycles were tested. The samples were washed thoroughly with water and dried after each cycle.

Using terephthalic acid as a probe molecule, the hydroxyl radicals (•OH) at the photo-illuminated sample/water interface were examined by a special photoluminescence (PL) technique. Terephthalic acid reacts readily with •OH, producing 2-hydroxyterephthalic acid, a great fluorescent material with a unique photoluminescence peak at 426 nm [20], which makes it easy to detect by fluorescence spectrum (excitation wavelength: 315 nm, fluorescence peak: 426 nm). In a typical experiment, 80 mL 0.5 mM terephthalic acid and 2 mM NaOH aqueous solution were completely mixed and then transferred into a 100 mL beaker. The rest of steps are the same as for the degradation of MO.

3. Results and Discussion

Figure 1a illustrates the XRD pattern of the TiO_2/Sn_3O_4 nanocomposites. All the diffraction peaks can be indexed to the anatase TiO_2 (JCPDS 21-1272, marked with black •) and triclinic Sn_3O_4 (JCPDS 16-0737, marked with red ☆), validating the high purity of the synthesized TiO_2/Sn_3O_4 composite phase. Compared with the single-phase TiO_2 and Sn_3O_4, all the diffraction peaks are broad and weak, which indicates that the crystallinities are slightly reduced [21]. This result may be attributed to lattice distortion induced by interfacial strain because of different lattice parameters between Sn_3O_4 and TiO_2 [22]. In addition, Raman spectroscopy results (Figure 1b) further confirm the purity of the synthesized TiO_2/Sn_3O_4 composite phase. Specifically, Raman activities of 144, 196, 396, 520 and 638 cm^{-1} were assigned to the anatase TiO_2 [23], and the 133, 143, 170 and 238 cm^{-1} Raman peaks could be attributed to the Sn_3O_4, in accordance with previous reports [7,24]. The textures of the as-synthesized TiO_2/Sn_3O_4, TiO_2, Sn_3O_4 and P25 were characterized by N_2 physisorption experiments. The Brunauer–Emmett–Teller (BET) surface area data of samples are provided in Table 1. The N_2 adsorption-desorption isotherm and pore-size distribution of TiO_2/Sn_3O_4 nanocomposites are shown in Figure 1c. The results display that the TiO_2/Sn_3O_4 nanocomposites possess an average pore diameter of 2.733 nm and a larger surface area of 68.1 m^2/g than the as-prepared TiO_2 (0.04 m^2/g), Sn_3O_4 (35.2 m^2/g), and the reported TiO_2/Sn_3O_4 nanobelt heterostructure (51.5 m^2/g) [9]. Such a high surface-to-volume ratio for the TiO_2/Sn_3O_4 nanocomposites might be of extreme good value in photocatalytic processes, as they would provide more active sites for the adsorption of reactant molecules, and their optical absorbance would increase at visible wavelengths [25]. Figure 1d shows the UV-visible diffusion reflectance spectra (DRS) and plots of $(F(R)h\nu)^{1/2}$ versus photo energy ($h\nu$) of the TiO_2/Sn_3O_4 along with spectra of the pristine TiO_2 and Sn_3O_4 for comparison. The absorption spectra of the TiO_2/Sn_3O_4 nanocomposites exhibit the mixed absorption properties of both the components. In particular, the absorption edge for TiO_2/Sn_3O_4 nanocomposites is clearly shifted towards visible region (near 505 nm). The optical band gap determined from the plot of the Kubelka-Munk function was found to be 2.46 eV, compared to the observed values of 3.22 eV and 2.61 eV for TiO_2 and Sn_3O_4, respectively. These data reveal the Sn_3O_4/TiO_2 nanocomposites have a lower band gap than the pure Sn_3O_4 and TiO_2 nanoparticles, which is consistent with the published literature [9], and can be explained by the reduced crystallinity of both materials [1], as shown by XRD analysis.

Table 1. The specific surface area and apparent reaction rate constants (κ) of TiO_2, P25, Sn_3O_4 TiO_2/Sn_3O_4 samples.

Photocatalyst	TiO_2	P25	Sn_3O_4	TiO_2/Sn_3O_4
	κ (min^{-1})	κ (min^{-1})	κ (min^{-1})	κ (min^{-1})
UV irradiation	0.028	0.24	0.064	0.24
Visible light	0.0010	0.0023	0.024	0.052
Surface Area ($m^2 \cdot g^{-1}$)	0.04	50	35.2	68.1

The chemical composition and valence state were characterized by X-ray photoelectron spectroscopy (XPS). The full range of XPS spectra, ranging from 0 to 1000 eV, of TiO_2/Sn_3O_4 nanocomposites are shown in Figure 2a. No impurities were observed in the spectra, which is consistent with the results of XRD and Raman. Figure 2b shows the curve fitting data of the Sn 3d core-level spectra. Moreover, the Sn 3d doublet characterized by Sn $3d_{3/2}$–Sn $3d_{5/2}$ splitting peak can be clearly observed. The prominent peak of Sn $3d_{5/2}$ level is dissolved into two peaks centered at 486.77 and 486.15 eV, which can be attributed to Sn(IV) and Sn(II) configurations [26], respectively. The Sn $3d_{3/2}$ spectra exhibit two peaks at 495.14 and 494.51 eV, which are assigned to Sn^{2+} and Sn^{4+} [26]. As shown in Figure 2c, the binding energies (BE) of Ti $2p_{3/2}$ and Ti $2p_{1/2}$ are 458.5 and 464.2 eV respectively, which are ascribed to the Ti^{4+} oxidation states [27]. On the basis of the above discussion, it can be concluded that the TiO_2/Sn_3O_4 sample is composed of Ti(IV), Sn(II and IV), and O, which is in good agreement with the XRD and Raman results. In addition, the calculated Ti/Sn ratio is 0.20, indicating that most of the surface of the TiO_2 nanocrystals is covered by Sn_3O_4 nanocrystals (SEM and TEM experiments further confirm this result, and will be discussed later).

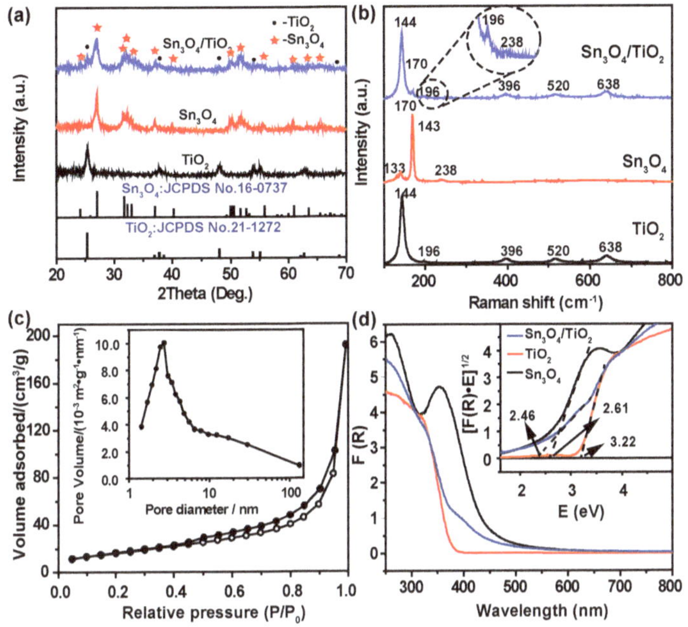

Figure 1. (**a**) XRD patterns, (**b**) Raman spectra of TiO_2, Sn_3O_4 and TiO_2/Sn_3O_4 nanocomposites, (**c**) N_2 adsorption-desorption isotherms of TiO_2/Sn_3O_4 nanocomposites and (**d**) UV-visible diffuse reflectance spectra and plots of $(F(R)h\nu)^{1/2}$ versus photo energy (right insert) of TiO_2, Sn_3O_4 and TiO_2/Sn_3O_4 nanocomposites.

The morphology and microstructure of the as-prepared TiO_2/Sn_3O_4 nanocomposites were carefully analyzed by microscopy. Generally, lots of interleaved Sn_3O_4 nanoplates are able to self-assemble into an ordinary flower-like nanostructure, as shown in Figure 3a. Similarly, by introducing highly monodispersed TiO_2 nanospheres of ~130 nm diameter (Figure 3b) into the growth environment, the homogeneous Sn_3O_4 nanoparticles started to grow on the surface of each individual TiO_2 core with intimate contact, thus forming an interface of two different semiconductors that would facilitate photo-excited electron transfer and photon-generated carrier separation (Figure 3c,d). Noticeably, the composites inherit a favorable dispersion in the solution, and fully contact with the absorbate, which is positive for the outstanding photocatalytic performance. However, it should be noted that the TiO_2/Sn_3O_4 nanocomposites are not completely covered by Sn_3O_4 nanocrystals. Furthermore, these advantageous heterojunctions with 10–20 nm sizes wrapping uniformly onto the surface of TiO_2 nanospheres were verified by TEM observation, as shown in Figure 1e,f. The well-resolved lattice fringes from the core and shell regions manifestly correspond to the (101) planes of anatase TiO_2 and the $(\bar{2}10)$ planes of Sn_3O_4, respectively, clearly revealing the phase distribution of the TiO_2/Sn_3O_4 nanocomposites again.

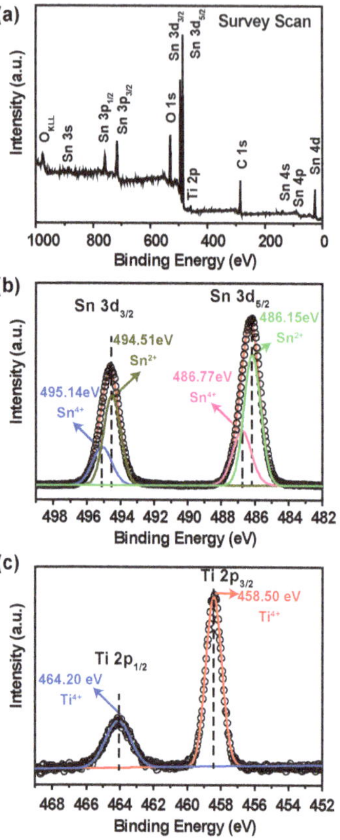

Figure 2. (**a**) Survey scan of XPS, (**b**) Sn 3d core level XPS spectra, and (**c**) Ti 2p core level XPS spectra of the TiO_2/Sn_3O_4 nanocomposites.

Figure 3. SEM images of the (**a**) Sn_3O_4, (**b**) TiO_2, and (**c**,**d**) TiO_2/Sn_3O_4 nanocomposites; (**e**) TEM and (**f**) HRTEM images of the TiO_2/Sn_3O_4 nanocomposites.

The photocatalytic activities of P25, TiO_2, Sn_3O_4 and TiO_2/Sn_3O_4 heterostructures were evaluated by the degradation of MO in water under UV- and visible-light irradiation (Figure 4a,b).The degradation of the MO solution under identical experimental conditions, but with no photocatalyst, is provided for comparison. The degradation efficiency of the as-synthesized TiO_2/Sn_3O_4 heterostructures was defined as C/C_0, where C_0 is the initial concentration of MO after equilibrium adsorption, and C is the concentration during the reaction. Both blank experiment results showed that MO could not be decomposed without photocatalyst under UV- or visible-light irradiation. In contrast, the photodegradation efficiency of TiO_2/Sn_3O_4 nanocomposites was 95% within 30 min under UV-light irradiation, which is superior to the as-prepared TiO_2 nanospheres (62%) and Sn_3O_4 nanoplates (70%). Furthermore, the MO decomposition efficiency found for the TiO_2/Sn_3O_4 photocatalyst was comparable to that determined under the same experimental conditions for the reference P25 catalyst; that is, 99% after 10 min. Additionally, in the visible-light irradiation experiment (Figure 4b), the TiO_2/Sn_3O_4 nanocomposites (81%) exhibited significantly higher photocatalytic activity than P25 (9%), TiO_2 (0.4%) and Sn_3O_4 (60%) at 30 min. Finally, MO was completely degraded within 80 min. The results show that the TiO_2/Sn_3O_4 heterostructures exhibited improved photocatalytic activity.

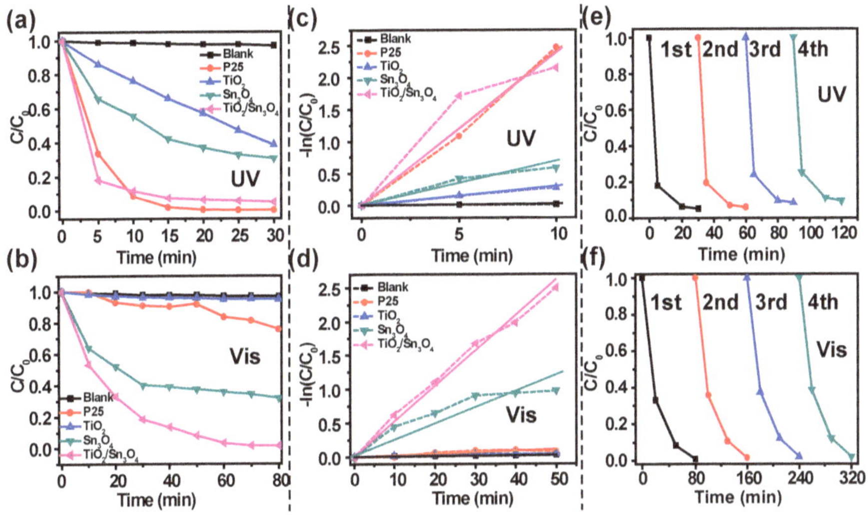

Figure 4. The photocatalytic activity (**a**,**b**), plots of -ln[C/C_0] versus irradiation time (**c**,**d**), and stability for MO photo-degradation (**e**,**f**) of TiO_2, P25, Sn_3O_4 and TiO_2/Sn_3O_4 nanocomposites under UV- and visible-light irradiation, respectively. The corresponding curves of MO without photocatalyst under UV and visible light irradiation are provided for comparison.

For a better understanding, the photocatalytic kinetics of the samples was analyzed using the Langmuir-Hinshelwood model, as shown in Figure 4c,d. All of the data follow a first-order reaction model, and the calculated apparent kinetic rate constants (κ) are summarized in Table 1. We found that, under UV irradiation, the TiO_2/Sn_3O_4 exhibited a much faster photo-decomposition activity ($\kappa = 0.24\ min^{-1}$) than the pure TiO_2 ($0.028\ min^{-1}$) and Sn_3O_4 ($0.064\ min^{-1}$), and was as fast as the P25 ($0.24\ min^{-1}$). Furthermore, under visible-light irradiation, the calculated value of κ for the TiO_2/Sn_3O_4 sample ($\kappa = 0.052\ min^{-1}$) was twice as high as that for the neat Sn_3O_4 ($0.024\ min^{-1}$), and more than a dozen times higher than that for the single TiO_2 ($0.0010\ min^{-1}$) and the P25 ($\kappa = 0.0023\ min^{-1}$). In addition, the TiO_2/Sn_3O_4 heterostructures could be recycled and reused at least four times without significant loss of efficiency (Figure 4e,f), which demonstrates its great potential as an efficient and stable photocatalytic material. These remarkably good performances can be attributed to the improved UV- and visible-light absorption efficiency, and the high photo-excited carrier-separation rate resulting from the novel TiO_2/Sn_3O_4 heterostructures.

Based on all of the results above, a possible mechanism for charge transfer and photocatalytic process can be proposed (Scheme 1). As illustrated in Figure 1d, the diffusion reflectance spectra (DRS) and plots of $(F(R)h\nu)^{1/2}$ versus photo energy ($h\nu$) indicate that the bandgap of Sn_3O_4 (2.61 eV) is smaller than that of TiO_2 (3.22 eV). Additionally, the potentials of the valence band (VB) and conduction band (CB) of Sn_3O_4 are higher than those of TiO_2, so the heterostructure of TiO_2/Sn_3O_4 belongs to typical type-II heterojunction [9]. When Sn_3O_4 contacts TiO_2 cores to form a heterojunction, the difference in chemical potential causes band bending at the interface of the junction [28], which drives photoexcited electrons to transfer from Sn_3O_4 to TiO_2, and photoexcited holes to migrate in the opposite direction, until the Fermi levels of TiO_2 and Sn_3O_4 reach equilibrium. The possible mechanisms for charge transfer and hydroxyl radical (•OH) generation under UV- and visible-light irradiation will be discussed separately. (1) Upon UV illumination, electrons in the VB could be excited to the CB of both oxides, simultaneously forming the same number of holes in the VB. This is due to the fact that the Sn_3O_4 nanoparticles were not fully coated as a shell onto the TiO_2 nanospheres (Figure 3c,d) and that the suitable bandgap of TiO_2 and Sn_3O_4 is lower than the energy of ultraviolet photons. Next, the photo-generated electrons were collected by the TiO_2 particles and the holes by the Sn_3O_4 particles; that is, electrons transferred from Sn_3O_4 to TiO_2, and holes migrated from TiO_2 to Sn_3O_4 (compare Scheme 1a with Figure 1d). The unique behavior that electrons and holes preferentially accumulate on different materials would result in a great separation of photo-generated carriers, and thus reduce the charge recombination rate, ultimately increasing carrier lifetime. As a consequence, the formation efficiency of hydroxyl radicals (•OH)—a strong oxidant for most pollutants [9,20,29]—by the reaction of holes with surface hydroxyl groups or physisorbed water molecules at the Sn_3O_4 surface and the production rates of •OH and superoxide radicals (O_2^-) radicals resulting from the reactions of electrons with dissolved oxygen molecules and water molecules will be massively enhanced; this will increase the volume of oxidant inside the system. (2) Under visible-light irradiation (Scheme 1b), electrons in the VB could be exclusively excited to the CB of Sn_3O_4, with a concomitant formation of the same number of holes in the VB. Due to the type II band alignment of the as-prepared sample, the photoexcited electrons in the Sn_3O_4 CB will be easily injected into the TiO_2 CB, where the electrons could reduce surface-absorbed O_2 over TiO_2 active sites to form superoxide radicals (O_2^-), and the new species can further yield •OH by reacting with water or oxidize MO. On the other hand, holes remaining in Sn_3O_4 could react with surface-absorbed H_2O to generate more •OH. Hydroxyl radicals (•OH) and superoxide radicals (O_2^-) stemming from the above procedure will degrade MO into colorless chemicals, and even CO_2 and H_2O, which is similar under UV illumination. All in all, the enhanced charge separation related to the TiO_2/Sn_3O_4 heterojunction favors the interfacial charge transfer to physisorbed species, forming •OH radicals and reducing possible back reactions, and therefore accounts for the higher activity of the TiO_2/Sn_3O_4 nanocomposites.

The photocatalytic oxidation of dyes occurs through the reactive species, which came into being after the light absorption and electron-hole formation by the photocatalyst [30]. Terephthalic acid photoluminescence probing technique (TAPL) was employed to examine the generation of active •OH radicals [31]. Figure 5a,b gives the •OH-trapping photoluminescent spectra of TiO_2/Sn_3O_4 nanocomposites in TA solution with UV- and visible-light irradiation, respectively. The increased photoluminescence intensity confirms that the •OH radicals are mainly responsible for the photodegradation process, and it also verifies the photocatalytic activity of the TiO_2/Sn_3O_4 nanocomposites.

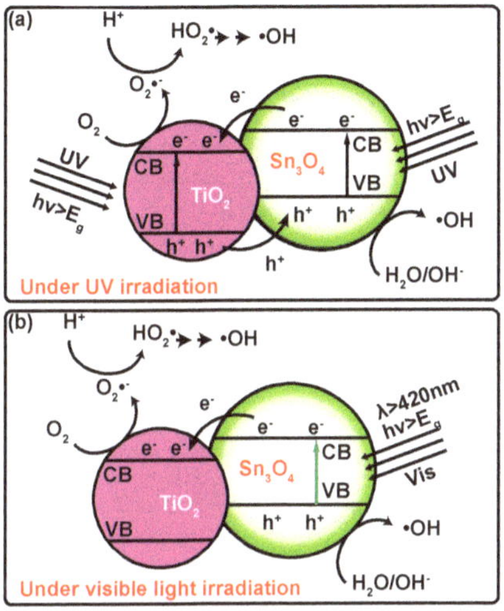

Scheme 1. Illustration of photo-induced charge transfer and separation at the interface of TiO_2/Sn_3O_4 hierarchical hybrid nanostructures under (**a**) UV- and (**b**) visible-light irradiation.

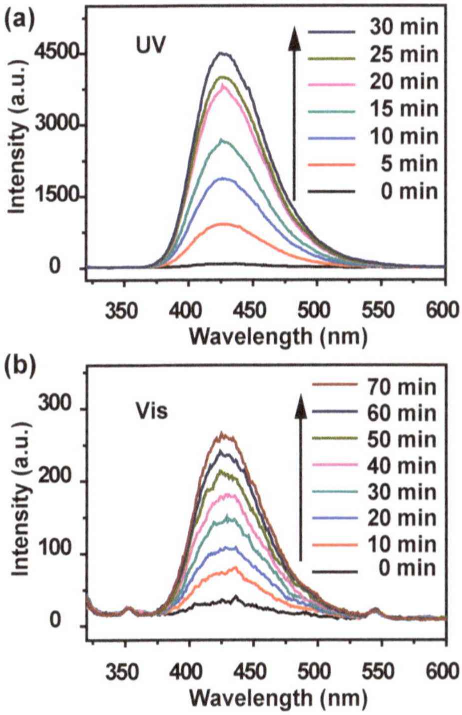

Figure 5. The •OH-trapping photoluminescence spectra of TiO_2/Sn_3O_4 nanocomposites under (**a**) UV- and (**b**) visible-light irradiation, respectively.

4. Conclusions

In summary, this study demonstrates a facile route to synthesizing TiO_2/Sn_3O_4 nanocomposites that not only display enhanced photocatalytic performance in UV irradiation, but also allow a significant level of visible light photocatalytic activity. The large surface area derived from the monodispersed mesopore TiO_2/Sn_3O_4 nanospheres and unique TiO_2/Sn_3O_4 heterojunctions are considered to be major contributions to supplying abundant active sites and separating photogenerated carriers, respectively. The strengthened photocatalytic performances will greatly promote the practical application of the TiO_2/Sn_3O_4 nanocomposites in eliminating organic pollutants from wastewater, and producing hydrogen by splitting.

Acknowledgments: We appreciate the financial support of the National Natural Science Foundation of China (Grant No. 61373072).

Author Contributions: F.W. and J.H. conceived and designed the experiments; J.H. and J.T. performed the experiments; J.H., J.T., X.L. and Z.W. analyzed the data; F.W., Q.L. and Y.L. contributed reagents/materials/analysis tools; J.H., J.T. and. X.L. wrote the paper with input from all authors.

Conflicts of Interest: The authors declare no conflict of interest.

References

1. Chen, X.; Liu, L.; Peter, Y.Y.; Mao, S.S. Increasing solar absorption for photocatalysis with black hydrogenated titanium dioxide nanocrystals. *Science* **2011**, *331*, 746–750. [CrossRef] [PubMed]
2. Sun, Z.; Liao, T.; Sheng, L.; Kou, L.; Kim, J.H.; Dou, S.X. Deliberate design of TiO_2 nanostructures towards superior photovoltaic cells. *Chem.-A Eur. J.* **2016**, *22*, 11357–11364. [CrossRef] [PubMed]
3. Sun, Z.; Liao, T.; Kou, L. Strategies for designing metal oxide nanostructures. *Sci. China Mater.* **2017**, *60*, 1–24. [CrossRef]
4. Asahi, R.; Morikawa, T.; Ohwaki, T.; Aoki, K.; Taga, Y. Visible-light photocatalysis in nitrogen-doped titanium oxides. *Science* **2001**, *293*, 269–271. [CrossRef] [PubMed]
5. Zhao, Z.; Tian, J.; Sang, Y.; Cabot, A.; Liu, H. Structure, synthesis, and applications of TiO_2 nanobelts. *Adv. Mater.* **2015**, *27*, 2557–2582. [CrossRef] [PubMed]
6. Manikandan, M.; Tanabe, T.; Li, P.; Ueda, S.; Ramesh, G.V.; Kodiyath, R.; Wang, J.; Hara, T.; Dakshanamoorthy, A.; Ishihara, S. Photocatalytic water splitting under visible light by mixed-valence Sn_3O_4. *ACS Appl. Mater. Interfaces* **2014**, *6*, 3790–3793. [CrossRef] [PubMed]
7. He, Y.; Li, D.; Chen, J.; Shao, Y.; Xian, J.; Zheng, X.; Wang, P. Sn_3O_4: A novel heterovalent-tin photocatalyst with hierarchical 3D nanostructures under visible light. *RSC Adv.* **2014**, *4*, 1266–1269. [CrossRef]
8. Berengue, O.; Simon, R.; Chiquito, A.; Dalmaschio, C.; Leite, E.; Guerreiro, H.; Guimarães, F.E.G. Semiconducting Sn_3O_4 nanobelts: Growth and electronic structure. *J. Appl. Phys.* **2010**, *107*, 033717. [CrossRef]
9. Chen, G.; Ji, S.; Sang, Y.; Chang, S.; Wang, Y.; Hao, P.; Claverie, J.; Liu, H.; Yu, G. Synthesis of scaly Sn_3O_4/TiO_2 nanobelt heterostructures for enhanced UV-visible light photocatalytic activity. *Nanoscale* **2015**, *7*, 3117–3125. [CrossRef] [PubMed]
10. Wang, Y.; Wang, Q.; Zhan, X.; Wang, F.; Safdar, M.; He, J. Visible light driven type II heterostructures and their enhanced photocatalysis properties: A review. *Nanoscale* **2013**, *5*, 8326–8339. [CrossRef] [PubMed]
11. Ye, W.; Shao, Y.; Hu, X.; Liu, C.; Sun, C. Highly Enhanced Photoreductive Degradation of Polybromodiphenyl Ethers with g-C_3N_4/TiO_2 under Visible Light Irradiation. *Nanomaterials* **2017**, *7*, 76. [CrossRef] [PubMed]
12. Low, J.; Yu, J.; Jaroniec, M.; Wageh, S.; Al-Ghamdi, A.A. Heterojunction photocatalysts. *Adv. Mater.* **2017**, *43*, 5234–5244. [CrossRef] [PubMed]
13. Linsebigler, A.L.; Lu, G.; Yates, J.T., Jr. Photocatalysis on TiO_2 surfaces: Principles, mechanisms, and selected results. *Chem. Rev.* **1995**, *95*, 735–758. [CrossRef]
14. Liu, Q.; Sun, Z.; Dou, Y.; Kim, J.H.; Dou, S.X. Two-step self-assembly of hierarchically-ordered nanostructures. *J. Mater. Chem. A* **2015**, *3*, 11688–11699. [CrossRef]
15. Sotelo-Vazquez, C.; Quesada-Cabrera, R.; Ling, M.; Scanlon, D.O.; Kafizas, A.; Thakur, P.K.; Lee, T.L.; Taylor, A.; Watson, G.W.; Palgrave, R.G. Evidence and Effect of Photogenerated Charge Transfer for

Enhanced Photocatalysis in WO_3/TiO_2 Heterojunction Films: A Computational and Experimental Study. *Adv. Funct. Mater.* **2017**, *27*. [CrossRef]
16. Nanakkal, A.; Alexander, L. Photocatalytic activity of graphene/ZnO nanocomposite fabricated by two-step electrochemical route. *J. Chem. Sci.* **2017**, *129*, 95–102. [CrossRef]
17. Yu, X.; Wang, L.; Zhang, J.; Guo, W.; Zhao, Z.; Qin, Y.; Mou, X.; Li, A.; Liu, H. Hierarchical hybrid nanostructures of Sn_3O_4 on N doped TiO_2 nanotubes with enhanced photocatalytic performance. *J. Mater. Chem. A* **2015**, *3*, 19129–19136. [CrossRef]
18. Nanakkal, A.; Alexander, L. Graphene/$BiVO_4$/TiO_2 nanocomposite: Tuning band gap energies for superior photocatalytic activity under visible light. *J. Mater. Sci.* **2017**, *52*, 7997–8006. [CrossRef]
19. Jiang, X.; Herricks, T.; Xia, Y. Monodispersed spherical colloids of titania: Synthesis, characterization, and crystallization. *Adv. Mater.* **2003**, *15*, 1205–1209. [CrossRef]
20. Martínez, D.S.; Martínez-De La Cruz, A.; Cuéllar, E.L. Photocatalytic properties of WO_3 nanoparticles obtained by precipitation in presence of urea as complexing agent. *Appl. Catal. A Gen.* **2011**, *398*, 179–186. [CrossRef]
21. Cullity, B. *Elements of X-Ray Diffractions*; Addison-Wesley: Reading, MA, USA, 1978; p. 102.
22. Sang, Y.; Yu, D.; Avdeev, M.; Qin, H.; Wang, J.; Liu, H.; Lv, Y. Yttrium aluminum garnet Nanoparticles with low antisite Defects studied with neutron and X-ray diffraction. *J. Solid State Chem.* **2012**, *192*, 366–370. [CrossRef]
23. Khan, M.M.; Ansari, S.A.; Pradhan, D.; Ansari, M.O.; Lee, J.; Cho, M.H. Band gap engineered TiO_2 nanoparticles for visible light induced photoelectrochemical and photocatalytic studies. *J. Mater. Chem. A* **2014**, *2*, 637–644. [CrossRef]
24. Wang, F.; Zhou, X.; Zhou, J.; Sham, T.-K.; Ding, Z. Observation of single tin dioxide nanoribbons by confocal Raman microspectroscopy. *J. Phys. Chem. C* **2007**, *111*, 18839–18843. [CrossRef]
25. Pawar, R.; Lee, C.S. *Heterogeneous Nanocomposite-Photocatalysis for Water Purification*; William Andrew: New York, NY, USA, 2015; pp. 68–76.
26. Wang, J.; Lu, C.; Liu, X.; Wang, Y.; Zhu, Z.; Meng, D. Synthesis of tin oxide (SnO & SnO_2) micro/nanostructures with novel distribution characteristic and superior photocatalytic performance. *Mater. Des.* **2017**, *115*, 103–111.
27. Wagner, C.D. *Handbook of X-Ray Photoelectron Spectroscopy*; A Reference Book of Standard Data for Use in X-ray Photoelectron Spectroscopy; Physical Electronics Division, Perkin-Elmer Corporation: Eden Prairie, MN, USA, 1979; pp. 68–69.
28. McDaniel, H.; Heil, P.E.; Tsai, C.-L.; Kim, K.; Shim, M. Integration of type II nanorod heterostructures into photovoltaics. *ACS Nano* **2011**, *5*, 7677–7683. [CrossRef] [PubMed]
29. Dong, W.; Pan, F.; Xu, L.; Zheng, M.; Sow, C.H.; Wu, K.; Xu, G. Q.; Chen, W. Facile synthesis of CdS@ TiO_2 core-shell nanorods with controllable shell thickness and enhanced photocatalytic activity under visible light irradiation. *Appl. Surf. Sci.* **2015**, *349*, 279–286. [CrossRef]
30. Malik, V.; Pokhriyal, M.; Uma, S. Single step hydrothermal synthesis of beyerite, $CaBi_2O_2(CO_3)_2$ for the fabrication of UV-visible light photocatalyst BiOI/$CaBi_2O_2(CO_3)_2$. *RSC Adv.* **2016**, *6*, 38252–38262. [CrossRef]
31. Cao, J.; Xu, B.; Luo, B.; Lin, H.; Chen, S. Novel BiOI/BiOBr heterojunction photocatalysts with enhanced visible light photocatalytic properties. *Catal. Commun.* **2011**, *13*, 63–68. [CrossRef]

nanomaterials

MDPI

Article

Study of the Photodynamic Activity of N-Doped TiO_2 Nanoparticles Conjugated with Aluminum Phthalocyanine

Xiaobo Pan [1,†], Xinyue Liang [1,†], Longfang Yao [1], Xinyi Wang [1], Yueyue Jing [1], Jiong Ma [1], Yiyan Fei [1], Li Chen [2] and Lan Mi [1,*]

1 Department of Optical Science and Engineering, Shanghai Engineering Research Center of Ultra-Precision Optical Manufacturing, Green Photoelectron Platform, Fudan University, 220 Handan Road, Shanghai 200433, China; 11110720002@fudan.edu.cn (X.P.); 14307130398@fudan.edu.cn (X.L.); 17110720023@fudan.edu.cn (L.Y.); 16210720013@fudan.edu.cn (X.W.); yyjing16@fudan.edu.cn (Y.J.); jiongma@fudan.edu.cn (J.M.); fyy@fudan.edu.cn (Y.F.)

2 School of Arts and Sciences, MCPHS University, 179 Longwood Ave, Boston, MA 02115, USA; lichenphy@gmail.com

* Correspondence: lanmi@fudan.edu.cn; Tel.: +86-21-6564-2092

† These authors contributed equally to this work.

Received: 29 September 2017; Accepted: 17 October 2017; Published: 20 October 2017

Abstract: TiO_2 nanoparticles modified with phthalocyanines (Pc) have been proven to be a potential photosensitizer in the application of photodynamic therapy (PDT). However, the generation of reactive oxygen species (ROS) by TiO_2 nanoparticles modified with Pc has not been demonstrated clearly. In this study, nitrogen-doped TiO_2 conjugated with Pc (N-TiO_2-Pc) were studied by means of monitoring the generation of ROS. The absorbance and photokilling effect on HeLa cells upon visible light of different regions were also studied and compared with non-doped TiO_2-Pc and Pc. Both N-TiO_2-Pc and TiO_2-Pc can be activated by visible light and exhibited much higher photokilling effect on HeLa cells than Pc. In addition, nitrogen-doping can greatly enhance the formation of $^1O_2$ and $\bullet O_2^-$, while it suppresses the generation of OH•. This resulted in significant photodynamic activity. Therefore, N-TiO_2-Pc can be an excellent candidate for a photosensitizer in PDT with wide-spectrum visible irradiation.

Keywords: titanium dioxide; phthalocyanine; reactive oxygen species; photodynamic therapy

1. Introduction

Titanium dioxide (TiO_2) nanoparticles have been widely studied in many fields such as solar cells, electrochromic devices, environment, and biomedicine [1,2]. Recently, researchers have focused on the application of photodynamic therapy (PDT) due to its low toxicity, high stability, excellent biocompatibility, and unique photocatalytic properties. When TiO_2 is photoexcited upon UV irradiation, hole-electron pairs are generated, which result in the formation of reactive oxygen species (ROS) via the redox reactions of oxygen or water molecules at the TiO_2 surface. The generated ROS can induce a remarkable photokilling effect against cancer cells [3–5]. Furthermore, when doped or modified with different methods, TiO_2 nanoparticles may become an attractive photosensitizer (PS) under visible light irradiation. In particular, TiO_2 nanoparticles modified with phthalocyanine have been proven to be promising as PSs with enhanced absorption in the visible region [6,7].

Phthalocyanine and its derivatives, as a second generation of PSs, are known to generate singlet oxygen ($^1O_2$) via energy transfer [8]. Also, there have been some studies showing that TiO_2 nanoparticles can generate specific ROS such as hydroxyl radicals (OH•) [9] and superoxide anion radicals ($\bullet O_2^-$) [10]. However, little work has been conducted to investigate the generation of ROS by TiO_2 nanoparticles modified with Pc.

In our previous work [11], nitrogen-doped TiO_2 nanoparticles (N-TiO_2) conjugated with aluminum phthalocyanine (Pc) were synthesized by a two-step surface modification method, and this novel material, N-TiO_2-Pc, exhibited significant photokilling efficiency on cancer cells. The photodynamic activity of N-TiO_2-Pc is the primary driving force underlying the PDT application, so it is important to demonstrate the photo-induced active species clearly. In this study, the photodynamic activity of N-TiO_2-Pc was studied by monitoring the generation of ROS and evaluating the photokilling effect upon light in different regions. These results are compared with Pc and non-doped TiO_2-Pc to reveal the roles of nitrogen-doping and Pc.

2. Results

2.1. Absorption Spectrum

The absorption spectra of N-TiO_2-Pc, TiO_2-Pc, and Pc in aqueous solutions are shown in Figure 1. The concentration of Pc in all three samples is the same, which is associated with the similar absorbance around 670 nm of all the samples. Meanwhile, the conjugates of N-TiO_2-Pc and TiO_2-Pc both demonstrate higher absorbance in the region of 400–500 nm compared with Pc. It is well known that pure anatase TiO_2 can only absorb UV light with a wavelength shorter than 387 nm [12]. When TiO_2 nanoparticles were modified with the amino silanization method, the absorbance in the visible region could be enhanced [13], especially in the blue and green regions [14]. N-TiO_2-Pc and TiO_2-Pc were both synthesized based on the amino silanization of TiO_2 nanoparticles, which leads to enhanced absorbance in the region of 400–500 nm. In addition, N-TiO_2-Pc shows higher visible absorbance than TiO_2-Pc due to nitrogen doping, which is in agreement with our previous report [15]. The higher absorption in the visible light region may induce a greater production of ROS and thus a higher photokilling effect on cancer cells.

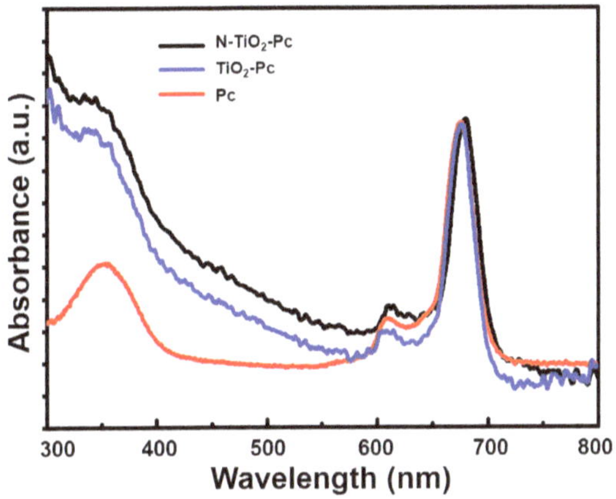

Figure 1. Absorption spectra of N-TiO_2-Pc (black), TiO_2-Pc (blue), and Pc (red) in aqueous solutions.

2.2. Production of ROS

The ROS generated by N-TiO_2-Pc, TiO_2-Pc, and Pc in aqueous suspensions under visible light irradiation were monitored by different ROS-sensitive fluorescence probes. The fluorescence intensities indicated the production of total ROS, $\bullet O_2^-/H_2O_2$, and $OH\bullet$, respectively. The production of ROS increased as a function of light exposure time ranging from 0 to 5 min (Figure 2). For comparison, the concentration of Pc is the same in all three samples.

Under 420–800 nm irradiation, the total ROS production by N-TiO_2-Pc was higher than those of TiO_2-Pc and Pc (Figure 2a). The total ROS production of N-TiO_2-Pc was about 1.8 times than of TiO_2-Pc and about 2.4 times that of Pc, which agrees well with the visible light absorbance result. N-TiO_2-Pc induced more $\bullet O_2^-/H_2O_2$, while TiO_2-Pc generated less $\bullet O_2^-/H_2O_2$ than Pc (Figure 2b). As shown in Figure 2c, TiO_2-Pc generated more OH•, was about twice of that of Pc, while N-TiO_2-Pc produced less OH• than TiO_2-Pc.

To further study the effect of 420–575 nm irradiation, a 575 nm-shortpass filter was added. It was determined that the power density of the lamp in the range of 420–800 nm was 17.8 mW·cm^{-2}, and that in the range of 420–575 nm was 8.4 mW·cm^{-2}, about half of 420–800 nm. Under 420–575 nm irradiation, Pc barely produced detectable ROS. This result indicates that Pc has no absorption in the range of 420–575 nm. Compared with TiO_2-Pc, the total ROS production of N-TiO_2-Pc was much higher, around 3.4 times that of TiO_2-Pc (Figure 2d). This indicates that the photoactivity of N-TiO_2-Pc is more efficient under this range of visible light. The $\bullet O_2^-/H_2O_2$ productions by N-TiO_2-Pc and TiO_2-Pc were similar, as shown in Figure 2e. Among the various reactive species, it seems the generation of OH• was not favored (Figure 2f). The reported ROS generated by TiO_2 included $\bullet O_2^-$, H_2O_2, OH•, and $^1O_2$ [10]. From Figure 2d–f, it can be seen that neither $\bullet O_2^-$, H_2O_2, or OH• produced by N-TiO_2-Pc represent the main contribution of the total ROS upon 420–575 nm irradiation. So, it can be assumed that $^1O_2$ may be the major composition of the various reactive species.

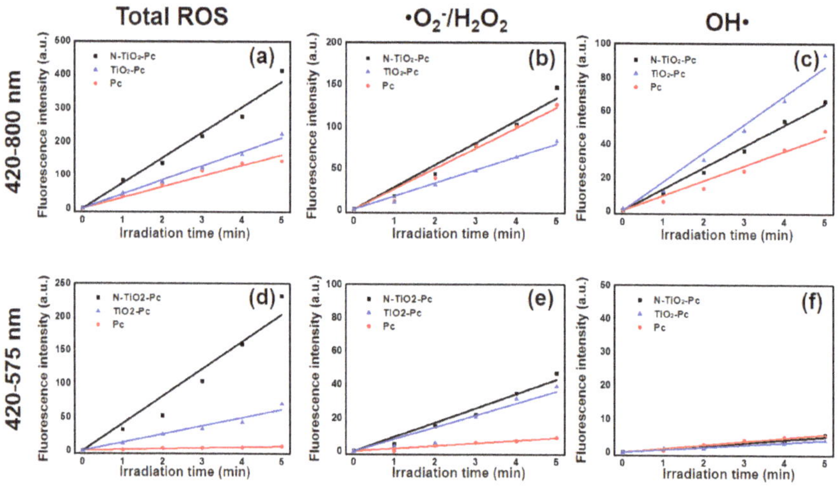

Figure 2. Comparison of photo-induced reactive oxygen species (ROS) by N-TiO_2-Pc, TiO_2-Pc, and Pc in aqueous solutions under light irradiation of (**a–c**) 420–800 nm (17.8 mW·cm^{-2}) and (**d–f**) 420–575 nm (8.4 mW·cm^{-2}), where the concentration of Pc is the same in all three samples. Fluorescence intensities indicate the production of (**a**,**d**) total ROS, (**b**,**e**) $\bullet O_2^-/H_2O_2$, and (**c**,**f**) OH• as a function of irradiation time.

To further study the contribution of different specific ROS generated by the samples, superoxide dismutase (SOD) and glycerol were used as $\bullet O_2^-$ and $^1O_2/\bullet O_2^-$ scavengers [16,17]. In the presence of specific ROS scavengers, the amount of eliminated $\bullet O_2^-$ and $^1O_2/\bullet O_2^-$ were monitored by the intensity decrease of the fluorescent probe. Then, the corresponding percentages were calculated using the intensity decrease compared with the fluorescence intensity measured without scavengers, and listed in Table 1. It can be seen that the nature of ROS is essentially $^1O_2$ rather than $\bullet O_2^-$, which is similar to the results of zinc oxide nanoparticles [18]. Since the highly reactive oxidative specie $^1O_2$ played a significant role in the generated ROS, the samples are supposed to have great photodynamic efficiency.

Table 1. Specific ROS percentage (%) of total ROS under different irradiation wavelengths.

	$\bullet O_2^-$		$^1O_2/\bullet O_2^-$	
Excitation Range	**420–800 nm**	**420–575 nm**	**420–800 nm**	**420–575 nm**
N-TiO_2-Pc	12.6 ± 0.3	20.9 ± 0.7	52.3 ± 1.8	66.6 ± 1.6
TiO_2-Pc	7.6 ± 0.2	10.0 ± 0.1	69.8 ± 1.7	63.7 ± 0.1
Pc	7.5 ± 0.1	-	65.6 ± 0.9	-

2.3. Photokilling Effects of Samples on HeLa Cells

The photokilling effects of samples on human cervical carcinoma cells (HeLa) were measured under different irradiation. The HeLa cells were first incubated with a medium containing 5.5–21.9 μg·mL^{-1} N-TiO_2-Pc/TiO_2-Pc (containing 0.48–1.9 μg·mL^{-1} Pc) for 1 h in the dark. For comparison, cells incubated with the same amount of 0.48–1.9 μg·mL^{-1} Pc were incubated as well. The irradiation time was the same for 420–800 nm and 420–575 nm, hence the contribution of 420–575 nm could be estimated with the same irradiation conditions except the wavelength range.

Under the irradiation of 420–800 nm (15.9 J·cm^{-2}), the surviving fractions of cells were decreased with the increased concentration of samples, as shown in Figure 3a. Pc showed weak photokilling effect with survival fractions of >83% for all the concentrations. TiO_2-Pc exhibited higher photokilling effect with the cell survival fractions in the range of 27–83%. N-TiO_2-Pc showed the highest photokilling effect. The cell survival fraction was below 46% when treated with 5.5 μg·mL^{-1} N-TiO_2-Pc, and the cell viability was as low as 14% when incubated with 21.9 μg·mL^{-1} N-TiO_2-Pc.

Under the irradiation of 420–575 nm (7.5 J·cm^{-2}), Pc did not show great photokilling effect. However, the cell viability dropped to 70% and 78% when treated with 21.9 μg·mL^{-1} N-TiO_2-Pc and TiO_2-Pc, respectively (Figure 3b). This indicated that both N-TiO_2-Pc and TiO_2-Pc can be activated by 420–575 nm irradiation, while nitrogen-doping can enhance the photodynamic activity of N-TiO_2-Pc.

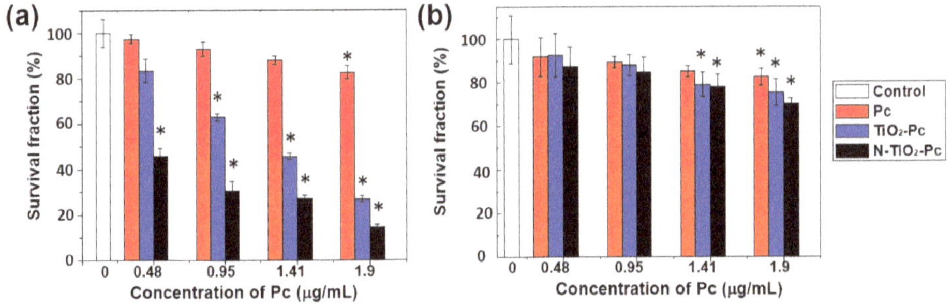

Figure 3. The photokilling effect on HeLa cells treated with 5.5–21.9 μg·mL^{-1} N-TiO_2-Pc or TiO_2-Pc with (**a**) 420–800 nm; (**b**) 420–575 nm light irradiation. For comparison, cells incubated with the same amount of 0.48–1.9 μg·mL^{-1} Pc were studied as well. * represents significant difference from the control group ($p < 0.05$).

3. Discussion

From Figure 2a,d, it can be seen that the total ROS production irradiated by 420–575 nm light was about half that irradiated by 420–800 nm light. This was because the power density of the lamp in the range of 420–575 nm was about half of that in the range of 420–800 nm. If the different irradiation light dose of 420–575 nm was same as 420–800 nm, the cell viability was expected to be 36% with

21.9 μg·mL^{-1} N-TiO_2-Pc. There was a gap of about 22% cell viability for N-TiO_2-Pc under 420–575 nm light compared with 420–800 nm. This could be explained by the notion that the photokilling effect was determined not only by the total ROS production, but also by the ROS type. Various species contribute differently depending on their lifetimes and diffusion lengths. The natures of ROS were different between 420–575 nm and 575–800 nm. On the other hand, specific ROS were studied in aqueous solutions, as shown in Figure 2 and Table 1, but the ROS would not be the same in the culture medium. In this case, the photokilling effect under 420–575 nm light was not significant (Figure 3b).

Figure 4 demonstrates a proposed mechanism of ROS production by N-TiO_2-Pc (or TiO_2-Pc) under light irradiation. The phthalocyanines in the solid state behave as *p*-type semiconductors with the energy of the band gap at about 1.9 eV [19], which can be excited by red light and mainly generate $^1O_2$ through energy transfer. Meanwhile, the bandgap of TiO_2 was narrowed and isolated states of N *2p* were located in the bandgap of N-TiO_2 due to the nitrogen-doping. As suggested in the theoretical study [20], doping with a 1–2% N concentration could result in a bandgap narrowing of 0.11–0.13 eV, and some N *2p* isolated states lying at 0.25–1.05 eV above the valence-band maximum of TiO_2. Therefore, the visible light of $\lambda \leq 575$ nm can excite the N-TiO_2 nanoparticles effectively, and it is more prone to transfer energy from N-TiO_2 to Pc compared with non-doped TiO_2. It has been shown through extensive studies that higher N doping amounts narrow the bandgap of TiO_2 and enhance the visible light absorption. The ROS generation is determined by both the light absorption ability and the quantum efficiency. Since the doped N atoms can serve as electron traps to inhibit the recombination of electrons and holes, the quantum efficiency of photoactivity could be promoted. The photogenerated electrons in the conduction band (CB) can react with oxygen molecules to generate $\bullet O_2^-$ and $^1O_2$. The production of $\bullet O_2^-$ and $^1O_2$ by N-TiO_2-Pc is significantly promoted by nitrogen-doping. On the other hand, the photogenerated positive holes in the valence band (VB) can oxidize water molecules to generate OH•. The results of Reeves proved OH• formation at nanoparticulate TiO_2 by electron spin resonance (ESR) studies [9]. This was also substantiated by experiments showing that TiO_2 generated more OH• than N-TiO_2 [10], which may be attributed to the low mobility of the photogenerated holes trapped in N *2p* levels of N-TiO_2 [21]. Since OH• contributes less to the photodynamic activity due to its shorter lifetime and lower diffusion length in comparison to $\bullet O_2^-$ [22], it can be understood that TiO_2-Pc exhibits less photodynamic activity than N-TiO_2-Pc. Therefore, the results suggest that N-TiO_2-Pc can be an excellent candidate for a photosensitizer in PDT with wide-spectrum visible irradiation.

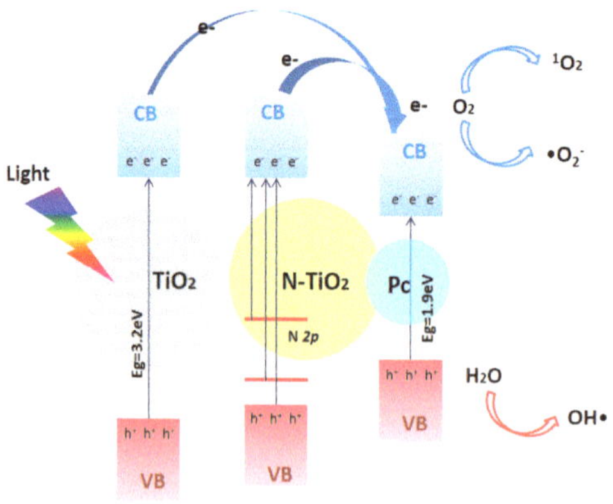

Figure 4. Schematic illustration of a proposed mechanism of ROS production by N-TiO_2-Pc (or TiO_2-Pc) under irradiation.

4. Materials and Methods

4.1. Preparation and Characterization of Samples

The chemical agents used in the preparation of N-TiO_2-Pc or TiO_2-Pc were anatase TiO_2 nanoparticles (<15 nm, Sigma-Aldrich Inc., St. Louis, MO, USA), gaseous ammonia (99.999%, Pujiang Inc., Jinhua, China), APTES (3-aminopropyl triethoxysilane, 99%; Aladdin Inc., Astoria, NY, USA), ammonia solution (25%-28%, Tongsheng Inc., Jiangsu, China), methanol (99.5%, Lingfeng Inc., Shanghai, China), and Pc (aluminum phthalocyanine chloride tetrasulfonate; Frontier Scientific Inc., Logan, UT, USA). Other chemical agents included 2-(9H-fluoren-9-ylmethoxycarbonylamino) oxyacetic acid (Fmoc-Aoa, Chem-Impex International, Inc., Bensenville, IL, USA), dimethylformamide (DMF, 98%, Sigma-Aldrich Inc., St. Louis, MO, USA), *N,N*-diisopropylethylamine (DIPEA, 99.5%, Sigma-Aldrich Inc., St. Louis, MO, USA), (benzotriazole-1-yloxy) tripyrrolidinophosphonium hexafluorophosphate (PyBOP, 98%, EMD Chemicals, Inc., Gibbstown, NJ, USA), and piperidine ($\geq$99.5%, Sigma-Aldrich Inc., St. Louis, MO, USA).

N-TiO_2-Pc nanoparticles were synthesized as described in our previous works [11]. Briefly, nitrogen-doped titanium dioxides (N-TiO_2) were obtained through the calcination of anatase TiO_2 in an ammonia atmosphere, and the N-dopant concentration was estimated to be 1.3%, as reported in our previous study [15,23]. Then, the NPs were modified with the amino silanization method [14] and coupled with Pc [11]. TiO_2-Pc nanoparticles were synthesized following the same procedure except for the calcination of anatase TiO_2 in an ammonia atmosphere. As previously reported [11], every 21.9 μg N-TiO_2-Pc or TiO_2-Pc contains 1.9 μg Pc, and the nanoparticles can be stably dispersed in aqueous solution.

The ultraviolet-visible (UV/Vis) absorption spectra of the N-TiO_2-Pc, TiO_2-Pc, and Pc samples were measured with a UV/Vis spectrometer (Shimadzu, UV3101pc, Tokyo, Japan).

4.2. Measurement of Reactive Oxygen Species (ROS)

The photo-induced generations of ROS in N-TiO_2-Pc, TiO_2-Pc, and Pc solutions were measured via 2′7′-dichlorofluorescein (DCFH). With light irradiation, the non-fluorescent DCFH reacts quickly with photo-induced ROS to form fluorescent DCF (2′7′-dichlorofluorescein). Thus, by measuring the fluorescence intensity of DCF, the relative yield of the produced ROS could be estimated. The DCFH solutions were prepared from the diacetate form DCFH (DCFH-DA) (Sigma-Aldrich Inc., St. Louis, MO, USA) by adding 0.5 mL of 1 mM DCFH-DA in methanol into 2 mL of 0.01 M NaOH. The mixture was kept in the dark for 30 min at room temperature before it was neutralized with 10 mL sodium phosphate buffer (pH = 7.2) [10,24]. Then, samples in phosphate buffered saline (PBS) solutions were individually mixed with DCFH (25 μM) before irradiation.

To evaluate the generations of specific reactive species, an OH•-sensitive fluorescence probe, 2-[6-(4-aminophenoxy)-3-oxo-3H-xanthen-9-yl]-benzoic acid (APF, Cayman Chemical, Ann Arbor, MI, USA) [25], and an $\bullet O_2^-/H_2O_2$ sensitive fluorescence probe, dihydrorhodamine 123 (DHR, Sigma-Aldrich Inc., St. Louis, MO, USA) [26], were used. Samples were mixed with APF (50 μM) or DHR (125 μM) before irradiation. When photo-induced OH• or $\bullet O_2^-/H_2O_2$ reacts with the non-fluorescent APF or DHR, the two probes can be converted to fluorescents quickly.

The sample solutions mixed with probes (or quenchers) were irradiated by a 150-W tungsten halogen lamp with different light filters for 5 min, respectively. The light of 420–800 nm was obtained with a 420 nm-longpass filter and an 800 nm-shortpass filter. The light of 420–575 nm was obtained with a 420 nm-longpass filter and a 575 nm-shortpass filter. The irradiation power densities were 17.8 $mW \cdot cm^{-2}$ (420–800 nm) and 8.4 $mW \cdot cm^{-2}$ (420–575 nm). At the same time, the fluorescence spectra were recorded by a fluorescence photometer (Hitachi, F-2500, Tokyo, Japan) with an interval of 1 min, and the fluorescent intensities were compared. The fluorescence intensities increased with the irradiation time, and the lines in Figure 2 were fitted linearly.

To further study the proportion of different specific ROS generated by the samples with light irradiation, several quenchers for specific ROS were used, including superoxide dismutase (SOD, 3IU, Beyotime, Jiangsu, China) for $\bullet O_2^-$ [17] and glycerol (99%, Sangon Biotech, Shanghai, China) for $^1O_2/\bullet O_2^-$ [16]. First, the fluorescence intensities of DCF with samples under irradiation were recorded as references, and the intensity versus time was a linear line with a slope noted as S_{REF}. Then, specific quenchers were respectively added into the DCFH and sample solutions before irradiation, where the concentrations of DCFH and samples were the same as that of the references. During irradiation, the fluorescence intensities of DCF in the presence of specific ROS quenchers were also recorded with an interval of 1 min. The lines of intensity with quenchers versus time were linear as well, and the slopes were noted as S_Q. Hence, the specific ROS percentages were obtained as $1 - \frac{S_Q}{S_{REF}}$, and are listed in Table 1.

4.3. Cell Culture and Cytotoxicity Assay

HeLa cells were seeded in 96-well plates containing Dulbecco's modified Eagle's medium (DMEM) (Gibco, Waltham, MA, USA) with 10% (*v*/*v*) fetal bovine serum (Sijiqing Inc., Hangzhou, China), and incubated in a fully humidified incubator at 37 °C with 5% CO_2 until reaching 80% confluence. Cells were incubated with a medium containing 5–20 μg·mL^{-1} N-TiO_2-Pc or TiO_2-Pc (containing 0.48–1.9 μg·mL^{-1} Pc) for 1 h in the dark. For comparison, cells incubated with 0.48–1.9 μg·mL^{-1} Pc were studied as well. Then, cells were incubated in fresh medium after washing three times and irradiated by the 150-W tungsten halogen lamp with different light filters, respectively. The irradiation time was same for 420–800 nm and 420–575 nm, therefore the visible-light illumination doses for cells were 15.9 J·cm^{-2} with 420–800 nm, and 7.5 J·cm^{-2} with 420–575 nm. The cells were incubated in the dark for 24 h before the cell viability study.

The cell viability assays were conducted by a modified 3-(4,5-dimethyl-2-thiazolyl)-2,5-diphenyl-2-H-tetrazolium bromide (MTT) method using WST-8 (2-(2-methoxy-4-nitrophenyl)-3-(4-nitrophenyl)-5-(2,4-disulfophenyl)-2H tetrazolium, monosodium salt) (Beyotime, Jiangsu, China). To each well, 100 μL culture medium with 10 μL of WST-8 solution was added. The cells were then incubated at 37 °C with 5% CO_2 for 2 h, and the absorbance of each well at 450 nm was recorded using a microplate reader (Bio-Tek Instruments Inc., Winooski, VT, USA). The absorbance at 450 nm before adding WST-8 was measured, and needed to be deducted to avoid any influence from nanoparticle samples. Cells incubated in DMEM medium without any treatment were used as control groups. Each experiment was conducted and measured independently at least three times.

Acknowledgments: The authors are grateful to the support of National Natural Science Foundation of China (11574056, 61505032, 61575046, 31500599), the Shanghai Rising-Star Program (16QA1400400), and CURE (Hui-Chun Chin and Tsung-Dao Lee Chinese Undergraduate Research Endowment) (16927).

Author Contributions: X.P. and X.L. performed the experiments; L.Y. and X.W. contributed by assisting in the experimental setup; X.L. and Y.J. analyzed the data; J.M., Y.F., and L.C. contributed instruction of data collection and interpretation; L.M. conceived and designed the experiments, contributed as the advisor to the research, and wrote the manuscript.

Conflicts of Interest: The authors declare no conflict of interest.

References

1. Dawson, A.; Kamat, P.V. Semiconductor-metal Nanocomposites. Photoinduced Fusion and Photocatalysis of Gold Capped TiO_2 (TiO_2/Au) Nanoparticles. *J. Phys. Chem. B* **2001**, *105*, 960–966. [CrossRef]
2. Yin, Z.F.; Wu, L.; Yang, H.G.; Su, Y.H. Recent progress in biomedical applications of titanium dioxide. *Phys. Chem. Chem. Phys.* **2013**, *15*, 4844–4858. [CrossRef] [PubMed]
3. Lagopati, N.; Kitsiou, P.V.; Kontos, A.I.; Venieratos, P.; Kotsopoulou, E.; Kontos, A.G.; Dionysiou, D.D.; Pispas, S.; Tsilibary, E.C.; Falaras, P. Photo-induced treatment of breast epithelial cancer cells using nanostructured titanium dioxide solution. *J. Photoch. Photobiol. A* **2010**, *214*, 215–223. [CrossRef]

4. Ghosh, S.; Das, A.P. Modified titanium oxide (TiO_2) nanocomposites and its array of applications: A review. *Toxicol. Environ. Chem.* **2015**, *97*, 1–43. [CrossRef]
5. Fan, W.; Huang, P.; Chen, X. Overcoming the Achilles' heel of photodynamic therapy. *Chem. Soc. Rev.* **2016**, *45*, 6488–6519. [CrossRef] [PubMed]
6. Lopez, T.; Ortiz, E.; Alvarez, M.; Navarrete, J.; Odriozola, J.A.; Martinez-Ortega, F.; Páez-Mozo, E.A.; Escobar, P.; Espinoza, K.A.; Rivero, I.A. Study of the stabilization of zinc phthalocyanine in sol-gel TiO_2 for photodynamic therapy applications. *Nanomed.* **2010**, *6*, 777–785. [CrossRef] [PubMed]
7. Jang, B.U.; Choi, J.H.; Lee, S.J.; Lee, S.G. Synthesis and characterization of Cu-phthalocyanine hybrid TiO_2 sol. *J. Porphyr. Phthalocyanines* **2009**, *13*, 779–786. [CrossRef]
8. Sun, Q.; Xu, Y. Sensitization of TiO_2 with Aluminum Phthalocyanine: Factors Influencing the Efficiency for Chlorophenol Degradation in Water under Visible Light. *J. Phys. Chem. C* **2009**, *113*, 12387–12394. [CrossRef]
9. Reeves, J.F.; Davies, S.J.; Dodd, N.J.F.; Jha, A.N. Hydroxyl radicals (•OH) are associated with titanium dioxide (TiO_2) nanoparticle-induced cytotoxicity and oxidative DNA damage in fish cells. *Mutat. Res. Fundam. Mol. Mech. Mutagen.* **2008**, *640*, 113–122. [CrossRef] [PubMed]
10. Li, Z.; Pan, X.; Wang, T.; Wang, P.-N.; Chen, J.-Y.; Mi, L. Comparison of the killing effects between nitrogen-doped and pure TiO_2 on HeLa cells with visible light irradiation. *Nanoscale Res. Lett.* **2013**, *8*, 96. [CrossRef] [PubMed]
11. Pan, X.; Xie, J.; Li, Z.; Chen, M.; Wang, M.; Wang, P.-N.; Chen, L.; Mi, L. Enhancement of the photokilling effect of aluminum phthalocyanine in photodynamic therapy by conjugating with nitrogen-doped TiO_2 nanoparticles. *Colloids Surf. B* **2015**, *130*, 292–298. [CrossRef] [PubMed]
12. Reddy, K.M.; Manorama, S.V.; Reddy, A.R. Bandgap studies on anatase titanium dioxide nanoparticles. *Mater. Chem. Phys.* **2003**, *78*, 239–245. [CrossRef]
13. Ukaji, E.; Furusawa, T.; Sato, M.; Suzuki, N. The effect of surface modification with silane coupling agent on suppressing the photo-catalytic activity of fine TiO_2 particles as inorganic UV filter. *Appl. Surf. Sci.* **2007**, *254*, 563–569. [CrossRef]
14. Xie, J.; Pan, X.; Wang, M.; Ma, J.; Fei, Y.; Wang, P.N.; Mi, L. The role of surface modification for TiO_2 nanoparticles in cancer cells. *Colloids Surf. B* **2016**, *143*, 148–155. [CrossRef] [PubMed]
15. Li, Z.; Mi, L.; Wang, P.-N.; Chen, J.-Y. Study on the visible-light-induced photokilling effect of nitrogen-doped TiO_2 nanoparticles on cancer cells. *Nanoscale Res. Lett.* **2011**, *6*, 356. [CrossRef] [PubMed]
16. Clejan, L.A.; Cederbaum, A.I. Role of Iron, Hydrgen Peroxide and Reactive Oxygen Species in Microsomal Oxidation of Glycerol to Formaldehyde. *Arch. Biochem. Biophys.* **1991**, *285*, 83–89. [CrossRef]
17. Tai, Y.; Inoue, H.; Sakurai, T.; Yamada, H.; Morito, M.; Ide, F.; Mishima, K.; Saito, I. Protective Effect of Lecithinized SOD on Reactive Oxygen Species-Induced Xerostomia. *Radiat. Res.* **2009**, *172*, 331–338. [CrossRef] [PubMed]
18. Sardar, S.; Chaudhuri, S.; Kar, P.; Sarkar, S.; Lemmens, P.; Pal, S.K. Direct observation of key photoinduced dynamics in a potential nano-delivery vehicle of cancer drugs. *Phys. Chem. Chem. Phys.* **2015**, *17*, 166–177. [CrossRef] [PubMed]
19. Wang, Q.; Wu, W.; Chen, J.; Chu, G.; Ma, K.; Zou, H. Novel synthesis of ZnPc/TiO_2 composite particles and carbon dioxide photo-catalytic reduction efficiency study under simulated solar radiation conditions. *Colloids Surf. A* **2012**, *409*, 118–125. [CrossRef]
20. Mi, L.; Zhang, Y.; Wang, P.N. First-principles study of the hydrogen doping influence on the geometric and electronic structures of N-doped TiO_2. *Chem. Phys. Lett.* **2008**, *458*, 341–345. [CrossRef]
21. Tafen, D.N.; Wang, J.; Wu, N.; Lewis, J.P. Visible light photocatalytic activity in nitrogen-doped TiO_2 nanobelts. *Appl. Phys. Lett.* **2009**, *94*, 093101–093103. [CrossRef]
22. Dimitrijevic, N.M.; Rozhkova, E.; Rajh, T. Dynamics of Localized Charges in Dopamine-Modified TiO_2 and their Effect on the Formation of Reactive Oxygen Species. *J. Am. Chem. Soc.* **2009**, *131*, 2893–2899. [CrossRef] [PubMed]
23. Mi, L.; Xu, P.; Wang, P.-N. Experimental study on the bandgap narrowings of TiO_2 films calcined under N_2 or NH_3 atmosphere. *Appl. Surf. Sci.* **2008**, *255*, 2574–2580. [CrossRef]
24. Cathcart, R.; Schwiers, E.; Ames, B.N. Detection of picomole levels of hydroperoxides using a fluorescent dichlorofluorescein assay. *Anal. Biochem.* **1983**, *134*, 111–116. [CrossRef]

25. Cossu, A.; Le, P.; Young, G.; Nitin, N. Assessment of sanitation efficacy against *Escherichia coli* O157:H7 by rapid measurement of intracellular oxidative stress, membrane damage or glucose active uptake. *Food Control* **2017**, *71*, 293–300. [CrossRef]
26. Sumitomo, K.; Shishido, N.; Aizawa, H.; Hasebe, N.; Kikuchi, K.; Nakamura, M. Effects of MCI-186 upon neutrophil-derived active oxygens. *Redox Rep.* **2007**, *12*, 189–194. [CrossRef] [PubMed]

nanomaterials

MDPI

Article

Photoelectrochemical Water Splitting Properties of Ti-Ni-Si-O Nanostructures on Ti-Ni-Si Alloy

Ting Li [1], Dongyan Ding [1,*], Zhenbiao Dong [1] and Congqin Ning [2]

1 Institute of Electronic Materials and Technology, School of Materials Science and Engineering, Shanghai Jiao Tong University, Shanghai 200240, China; litingstar@sjtu.edu.cn (T.L.); dzb0312@126.com (Z.D.)
2 State Key Laboratory of High Performance Ceramics and Superfine Microstructure, Shanghai Institute of Ceramics, Chinese Academy of Sciences, Shanghai 200050, China; cqning@mail.sic.ac.cn
* Correspondence: dyding@sjtu.edu.cn; Tel.: +86-21-3420-2741

Received: 31 August 2017; Accepted: 25 September 2017; Published: 31 October 2017

Abstract: Ti-Ni-Si-O nanostructures were successfully prepared on Ti-1Ni-5Si alloy foils via electrochemical anodization in ethylene glycol/glycerol solutions containing a small amount of water. The Ti-Ni-Si-O nanostructures were characterized by field-emission scanning electron microscopy (FE-SEM), energy dispersive spectroscopy (EDS), X-ray diffraction (XRD), and diffuse reflectance absorption spectra. Furthermore, the photoelectrochemical water splitting properties of the Ti-Ni-Si-O nanostructure films were investigated. It was found that, after anodization, three different kinds of Ti-Ni-Si-O nanostructures formed in the α-Ti phase region, Ti_2Ni phase region, and Ti_5Si_3 phase region of the alloy surface. Both the anatase and rutile phases of Ti-Ni-Si-O oxide appeared after annealing at 500 °C for 2 h. The photocurrent density obtained from the Ti-Ni-Si-O nanostructure photoanodes was 0.45 mA/cm^2 at 0 V (vs. Ag/AgCl) in 1 M KOH solution. The above findings make it feasible to further explore excellent photoelectrochemical properties of the nanostructure-modified surface of Ti-Ni-Si ternary alloys.

Keywords: anodization; TiO_2 nanostructure; doping; photoelectrochemical water splitting

1. Introduction

Titanium dioxide (TiO_2) has been intensively investigated as a favorable, eco-friendly photocatalyst owing to its relatively low cost, nontoxicity, and stable chemical properties [1,2]. In 1972, TiO_2 was used as a photochemical water splitting catalyst for the first time [3]. Recently, TiO_2 was demonstrated to be a promising photocatalyst for photocatalytic water splitting and solar energy conversion with high efficiency and photochemical stability [4–9]. However, the wide energy band gap (3.2 eV for anatase and 3.0 eV for rutile) and the fast recombination of photogenerated electrons and holes are the main drawbacks of TiO_2-based photoanodes [10]. Therefore, modification strategies including foreign element doping, surface decoration, and sensitization with dye have been adopted to overcome these drawbacks over the last 30 years [11–16]. One of the most studied methods is the doping of TiO_2 materials with metal ions or nonmetallic elements such as Ni, Ta, Nb, Fe, Zn, C, N, and so on [17–26].

Ti-alloy-based oxide nanotubes were fabricated through a direct anodization of TiNi binary alloy [17,18]. To date, few studies have been conducted on the anodic fabrication of Ti-Ni-Si-O nanostructures on Ti-Ni-Si alloy substrates. Si has a much lower density than Ti (2.33 g/cm^3 for Si vs. 4.54 g/cm^3 for Ti) as well as vast natural abundance, and it is environmentally friendly. Zhang et al. [27] found that the presence of Si could impair the recombination of photogenerated electrons and holes effectively. Also, the photocurrent density of Si-doped TiO_2 electrodes was 2–3 times higher than that of undoped TiO_2 electrodes. In this work, Ti-Ni-Si-O nanostructures were successfully grown on Ti-Ni-Si ternary alloy substrates via electrochemical anodization in ethylene glycol/glycerol solutions containing a small amount of water. The microstructures and photoelectrochemical properties, especially the photochemical water splitting of Ti-Ni-Si-O nanostructures, were investigated.

2. Results and Discussion

Figure 1 presents the typical microstructural features of as-cast Ti-1 wt % Ni-5 wt % Si alloy. Figure 1a shows the presence of multiphase, while Figure 1b shows a higher magnification image of different phases. EDS (energy dispersive spectroscopy) was used to test the compositions in the different phase regions. The EDS results are shown in Table 1. It was found that the gray region was α-Ti matrix, and the average composition of the black network-like region was 76.42 wt % Ti, 0.10 wt % Ni, and 23.48 wt % Si. Combined with the phase diagram calculated by Thermo-Calc software, it could be concluded that they were Ti_5Si_3 structures. In addition, the bright strip-like region was identified as the Ti_2Ni phase [28]. It is noticeable that the quantity of the Ti_5Si_3 phase was much more than that of the Ti_2Ni phase.

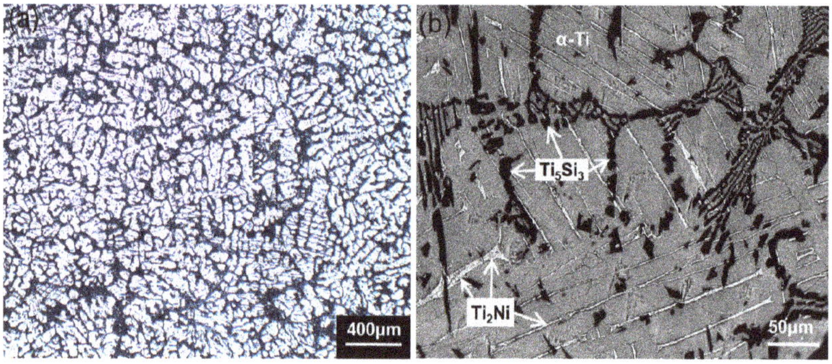

Figure 1. Typical microstructure of Ti-1Ni-5Si alloy: (**a**) Optical micrograph; (**b**) SEM image.

Table 1. Compositions of the α-Ti phase, Ti_2Ni phase, and Ti_5Si_3 phase of the alloy.

EDS Testing Areas	Elements (wt %)		
	Ti	Ni	Si
α-Ti phase	98.81	0.12	1.07
Ti_2Ni phase	88.02	11.89	0.09
Ti_5Si_3 phase	76.42	0.10	23.48

For the multi-phase Ti-1Ni-5Si alloy, the anodization process was not a uniform one due to the different anodization characteristics of different phases. Figure 2 shows SEM (scanning electron microscopy) images of different Ti-Ni-Si-O nanostructures grown in the α-Ti phase, Ti_2Ni phase, and Ti_5Si_3 phase regions. Obviously, three kinds of nanostructures formed on the surface of the alloy films. One was a self-organized nanotube array formed in the α-Ti phase region. The second was a nanotube array under the corrosion pits in the Ti_2Ni phase region. The third constituted irregular nanopores formed in the Ti_5Si_3 phase region. The Ti-Ni-Si-O nanotubes formed in the α-Ti phase region and the nanopores formed in the Ti_5Si_3 phase region had a pore diameter of about 64 nm. Table 2 shows the compositions tested by EDS for the α-Ti phase, Ti_2Ni phase, and Ti_5Si_3 phase regions after anodization. It is noticeable that the Si element was still rich in the Ti_5Si_3 phase regions while the Ni element was relatively rich in the Ti_2Ni phase regions.

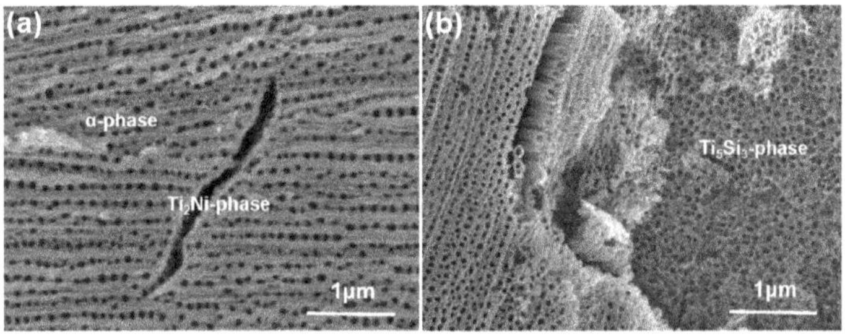

Figure 2. SEM images of scratched Ti-Ni-Si-O nanostructures showing: (**a**) nanotubes grown in the α-Ti phase region and Ti_2Ni phase region; (**b**) nanopores grown in the Ti_5Si_3 phase region.

Table 2. Compositions in the α-Ti phase, Ti_2Ni phase, and Ti_5Si_3 phase regions after anodization.

EDS Testing Areas	Elements (wt %)			
	Ti	Ni	Si	O
α-Ti phase region	56.48	–	1.06	42.46
Ti_2Ni phase region	66.77	1.94	0.82	30.47
Ti_5Si_3 phase region	60.35	–	9.74	29.91

The formation of TiO_2 nanotubes by anodization can be roughly divided into two steps. In the first step, an initial barrier layer is formed on the electrolyte-metal interface. Then, an oxide barrier layer is randomly distributed by the chemical etching action of fluoride ions, resulting in the growth of nanotubes under the top oxide layer [29,30]. During the final step, the pore growth morphology gradually changes to a homogeneous and self-organized morphology. Thus, a competition between the formation and the dissolution of the oxides always takes place during the anodization process [31]. For the anodization of the Ti-Ni-Si alloy here, the Ti_2Ni phase region and Ti_5Si_3 phase region should have a much quicker dissolution rate in the anodization electrolyte than the α-Ti phase region. For the Ti_2Ni phase, the dissolution rate of the oxides was so fast that there was no time to form any nanostructures. Thus, only etching pits were left in this region. In the Ti_5Si_3 phase region, the dissolution rate was faster than the formation rate of the oxides; thus, it was difficult to form nanotube structures. Instead, nanopores formed in this region. With a slower dissolution rate in the α-Ti phase region, the formation of stable Ti-Ni-Si-O nanotubes became easier than that in the other phase regions. Our previous literature [32] reported the similar phase-dependent anodization of the two-phase Ti_6Al_4V alloy. Ti-Al-V-O nanotube arrays formed in the α-phase region and irregular Ti-Al-V-O nanopores formed in the V-riched β-phase region of the Ti_6Al_4V alloy. The solubility of vanadium oxide in the F^--containing electrolyte played an important role in the competition between the formation and dissolution of the oxides. It could be concluded that for the present anodization system, the phase-dependent anodization was hard to control for a uniform formation of nanotube arrays on top of the multiphase substrate.

The as-anodized Ti-Ni-Si-O nanostructures were found to be amorphous, and they could crystallize after the annealing process. XRD was adopted to determine the crystal structure and possible phases during annealing. Figure 3 presents the XRD patterns of the as-anodized and the annealed Ti-Ni-Si-O nanostructures. In the diffraction pattern of the annealed sample, two sharp diffraction peaks centered at 2θ angles of 40.2° and 62.9° were assigned to Ti metal (JCPDS card No. of 65-9622, Jade 5.0) from the substrate. The diffraction peaks at 25.3° and 75.0° could be assigned to the anatase phase (JCPDS card No. 21-1272, Jade 5.0) of TiO_2. The peaks at 27.4° and 69.0° represented the rutile phase (JCPDS card No. 21-1276, Jade 5.0) of TiO_2. The diffraction peaks at 36.8°, 40.8°, 41.9°, 42.6°, 61.2°, and 66.4° were indexed to the characteristic peaks of Ti_5Si_3 (JCPDS No. 65-3597, Jade 5.0) from the substrate. No diffraction peaks related to the Ti_2Ni phase could be detected by XRD. In the diffraction pattern of the as-anodized sample, neither the anatase phase nor rutile phase could be observed. As shown in Figure 3, the amorphous structure of Ti-Ni-Si-O nanostructures had transformed into both anatase and rutile structures after annealing at 500 °C for 2 h. The anatase phase was found to be the major oxide phase.

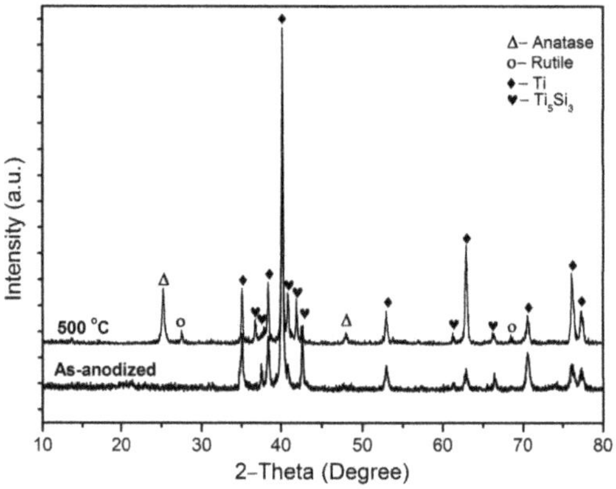

Figure 3. XRD patterns of the as-anodized and the annealed Ti-Ni-Si-O nanostructures.

Figure 4 shows the UV-Vis diffuse reflectance absorption spectra of the annealed Ti-Ni-Si-O photoanode. The band gap energy of the photoanode was estimated by using Tauc's method. It was observed that the Ti-Ni-Si-O photoanode showed an absorption edge at 402 nm. The band gap value was 3.08 eV, which was between the anatase band gap (3.2 eV) and the rutile band gap values (3.0 eV). The presence of both the anatase structure and rutile structure was attributed to the obtained band gap value [11].

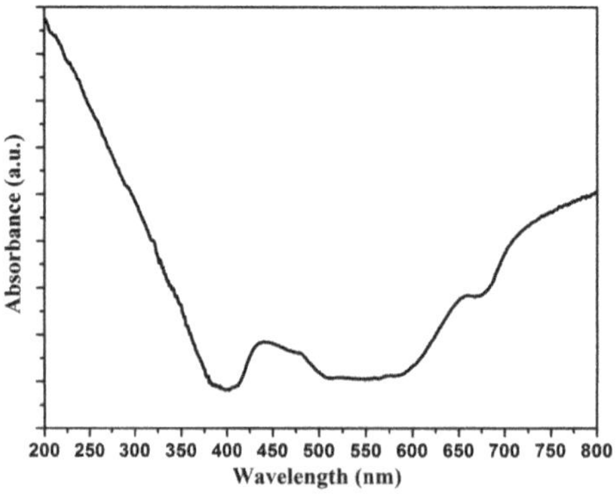

Figure 4. UV-Vis diffuse reflectance absorption spectra of the annealed Ti-Ni-Si-O photoanode.

The photoelectrochemical water splitting behavior of Ti-Ni-Si-O nanostructures is shown in Figure 5. The linear sweep was collected for the Ti-Ni-Si-O photoanodes with a scan rate of 50 mV/s. The photocurrent density was 0.45 mA/cm^2 at 0 V (vs. Ag/AgCl). The photocurrent under illumination was distinguishable from the dark current. Figure 4b presents photocurrent density vs. time scans for the Ti-Ni-Si-O photoanodes measured at 0 V (vs. Ag/AgCl). It could be seen that the photocurrent density was 0.45 mA/cm^2, which was in accordance to the results of the linear sweep experiment. The samples demonstrated stable and instantaneous changes as well as reproducible responses in the photocurrent after many illumination on/off cycles.

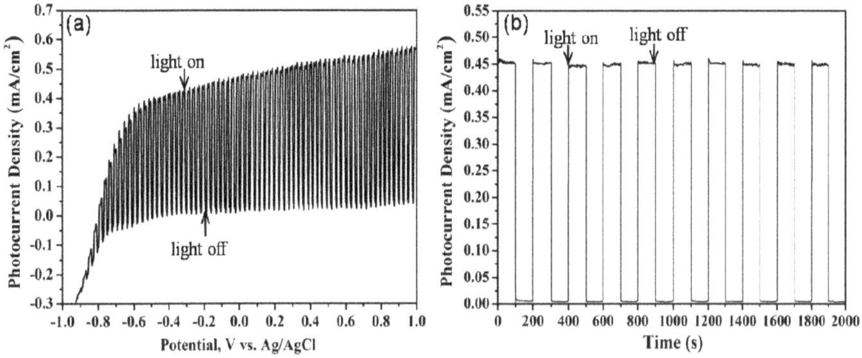

Figure 5. Photoelectrochemical water splitting behavior of Ti-Ni-Si-O nanostructures: (**a**) *I-V* curves in dark and under illumination; (**b**) transient photocurrent responses.

Electrochemical properties for Ti-Ni-Si-O nanostructure photoanodes annealed at 500 °C for 2 h were investigated, and the corresponding results are shown in Figure 6. Figure 6a shows the open-circuit potential (OCP) of Ti-Ni-Si-O photoanodes with time upon turning off the illumination. Without illumination, the OCP was about −0.50 V (vs. Ag/AgCl). As soon as the light was switched on, the OCP rapidly shifted negatively to a value of −0.80 V (vs. Ag/AgCl) due to the photogeneration of electron-hole pairs [33]. When turning off the illumination, the OCP gradually shifted positively to a steady state. These results indicated that the Ti-Ni-Si-O photoanodes had remarkable photoelectric conversion characteristics. The difference between the dark potential and light potential was about 0.3 V, which was the inherent characteristic of TiO_2 [34].

The carrier concentration (N_d) and flat-band potential (V_{FB}) can be calculated from the Mott-Schottky equation [35,36]:

$$\frac{1}{C^2} = \left(\frac{2}{e_0 \varepsilon \varepsilon_0 N_d}\right)\left[(V - V_{FB}) - \frac{kT}{e_0}\right]$$

where C is the capacitance of the space-charge region, e_0 is the electron charge (1.602×10^{-19} C), ε is the dielectric constant of TiO_2 (ε = 41.4 for anatase TiO_2 and 154.2 for rutile TiO_2 [37]), ε_0 is the permittivity of free space (8.854×10^{-12} F/m), N_d is the donor density of N-type semiconductor (carriers/cm^3), V is the applied potential bias at the electrode, k is the Boltzmann's constant (1.38×10^{-23} J/K), and T is the absolute temperature. It can be seen that there is a linear relationship between $1/C^2$ and V_{FB}. Furthermore, the flat-band potential V_{FB} can be calculated from the extrapolation of the line to $1/C^2 = 0$. Moreover, the carrier concentration can be obtained from the slope of the Mott-Schottky equation. Figure 6b presents Mott-Schottky plots of Ti-Ni-Si-O photoanodes with a frequency of 1000 Hz. It was calculated that the flat-band potential was −0.625 V (vs. Ag/AgCl). The carrier concentration was in the range of 2.13×10^{16}/cm^3 to 8.57×10^{16}/cm^3 for Ti-Ni-Si-O photoanodes, which was comparable with those of pure TiO_2 photoanodes [23]. Simelys et al. [38] reported that a higher carrier concentration could facilitate the charge separation at the semiconductor-electrolyte interface, and the carrier concentration reached up to 7.05×10^{19}/cm^3 for the TiO_2 nanotubes with a thickness of about 1.5 μm. The samples here showed a positive slope in the Mott-Schottky plots, as expected for an N-type semiconductor.

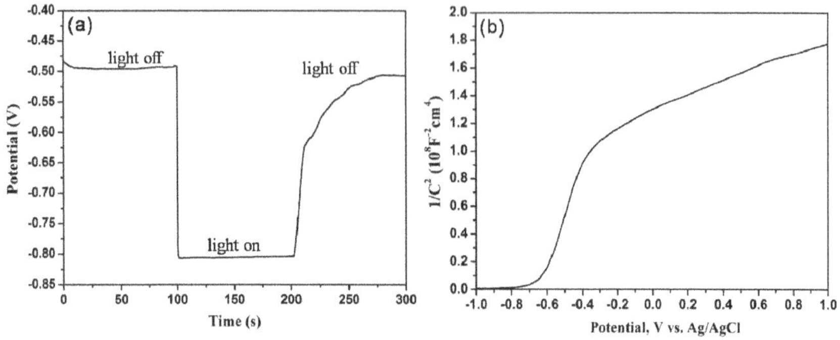

Figure 6. (**a**) Open-circuit potential of Ti-Ni-Si-O nanostructure photoanodes; (**b**) Mott-Schottky plots of Ti-Ni-Si-O nanostructure photoanodes with a frequency of 1000 Hz.

3. Materials and Methods

Ti-Ni-Si-O oxide films on the alloy substrate were synthesized through a direct anodic oxidation process. Prior to the anodization, the Ti-1Ni-5Si alloy foils with a size of 20 mm × 10 mm × 1 mm were mechanically polished and ultrasonically degreased in acetone and ethanol, rinsed with deionized water, and finally dried in air. The anodization was carried out in a conventional two-electrode electrochemical cell with the alloy foil as a working electrode and the platinum foil as a counter electrode at room temperature. All of the samples were anodized at a pulse voltage of 40 V with a constant frequency of 4000 Hz and a duty cycle of 50% for 90 min in an electrolyte of 5 vol % ethylene glycol/glycerol (Shanghai Lingfeng Chemical Reagent Co., Ltd., Shanghai, China) containing 0.30 M $(NH_4)_2SO_4$ and 0.4 M NH_4F (Sinopharm Chemical Reagent Co., Ltd., Shanghai, China) as well as 3 vol % deionized water. After anodization, the samples were immediately rinsed with deionized water and subsequently dried in air. All of the samples were annealed at 500 °C for 2 h in air to transform amorphous oxide into crystalline phases.

The structure and morphology of the oxide film were characterized through field emission scanning electron microscopy (SEM, FEI SIRION 200, Hillsboro, OR, USA). The chemical compositions were analyzed by energy dispersive spectroscopy (EDS, INCA X-ACT, Oxford, UK). The crystalline phase was characterized with an X-ray diffractometer (Rigaku Ultima IV, Tokyo, Japan) with Cu K_α radiation (λ = 0.15406 nm) at 40 kV and 30 mA with a scan speed of 5°/min over a 2θ range from 10° to 80°. Diffuse reflectance absorption spectra were collected by a UV-visible spectrometer (Perkin Elmer Inc., Lambda 750S, Waltham, MA, USA) with $BaSO_4$ as a reference. The photoelectrochemical measurement of different photoanodes was performed in 1 M KOH solution using a typical three-electrode system with oxide photoanode as a working electrode, Pt as a counter electrode, and Ag/AgCl as a reference electrode. A 150 W Xe lamp (Lanpu XQ350W, Shanghai, China) was used as a light source and the intensity of light illumination was controlled at 100 mW/cm^2. The illuminated area of the working electrode was 1 cm^2.

4. Conclusions

In summary, Ti-Ni-Si-O nanostructures were successfully fabricated through electrochemical anodization for photoelectrocatalytic water splitting. It was found that after anodization, three kinds of Ti-Ni-Si-O nanostructures grew in the α-Ti phase region, Ti_2Ni phase region, and the Ti_5Si_3 phase region of the alloy surface. Both anatase and rutile structures of Ti-Ni-Si-O oxide appeared after annealing at 500 °C for 2 h. The photocurrent density obtained from the Ti-Ni-Si-O nanostructure photoanodes was 0.45 mA/cm^2 at 0 V (vs. Ag/AgCl) in 1 M KOH solution. The above findings make it feasible to further explore the excellent photoelectrochemical properties of the nanostructure-modified surfaces of Ti-Ni-Si ternary alloys.

Acknowledgments: This work was supported by the National Natural Science Foundation of China (No. 51572170). We thank the contribution from the SEM lab at Instrumental Analysis Center of SJTU.

Author Contributions: The experimental design was planned by Dongyan Ding and Ting Li. The manuscript was written with contributions from all authors. All authors have given approval to the final version of the manuscript.

Conflicts of Interest: The authors declare no conflict of interest.

References

1. Lee, K.; Mazare, A.; Schmuki, P. One-dimensional titanium dioxide nanomaterials: Nanotubes. *Chem. Rev.* **2014**, *114*, 9385–9454. [CrossRef]
2. Kudo, A.; Miseki, Y. Heterogeneous photocatalyst materials for water splitting. *Chem. Soc. Rev.* **2009**, *38*, 253–278. [CrossRef] [PubMed]
3. Fujishima, A.; Honda, K. Electrochemical photolysis of water at a semiconductor electrode. *Nature* **1972**, *238*, 37–38. [CrossRef] [PubMed]
4. Gong, J.; Lai, Y.; Lin, C. Electrochemically multi-anodized TiO_2 nanotube arrays for enhancing hydrogen generation by photoelectrocatalytic water splitting. *Electrochim. Acta* **2010**, *55*, 4776–4782. [CrossRef]
5. Sreethawong, T.; Junbua, C.; Chavadej, S. Photocatalytic H_2 production from water splitting under visible light irradiation using Eosin Y-sensitized mesoporous-assembled Pt/TiO_2 nanocrystal photocatalyst. *J. Power Sources* **2009**, *190*, 513–524. [CrossRef]
6. Wang, J.; Lin, Z. Freestanding TiO_2 nanotube arrays with ultrahigh aspect ratio via electrochemical anodization. *Chem. Mater.* **2008**, *20*, 1257–1261. [CrossRef]
7. Wang, J.; Zhao, L.; Lin, V.S.Y.; Lin, Z. Formation of various TiO_2 nanostructures from electrochemically anodized titanium. *J. Mater. Chem.* **2009**, *19*, 3682–3687. [CrossRef]
8. Ye, M.; Xin, X.; Lin, C.; Lin, Z. High efficiency dye-sensitized solar cells based on hierarchically structured nanotubes. *Nano Lett.* **2011**, *11*, 3214–3220. [CrossRef] [PubMed]
9. Xin, X.; Wang, J.; Han, W.; Ye, M.; Lin, Z. Dye-sensitized solar cells based on a nanoparticle/nanotube bilayer structure and their equivalent circuit analysis. *Nanoscale* **2012**, *4*, 964–969. [CrossRef] [PubMed]
10. Liu, Q.; Ding, D.Y.; Ning, C.Q.; Wang, X.W. α-Fe_2O_3/Ti-Nb-Zr-O composite photoanode for enhanced photoelectrochemical water splitting. *Mater. Sci. Eng. B* **2015**, *196*, 15–22. [CrossRef]
11. Mohamed, A.E.R.; Rohani, S. Modified TiO_2 nanotube arrays (TNTAS): Progressive strategies towards visible light responsive photoanode, a review. *Energy Environ. Sci.* **2011**, *4*, 1065. [CrossRef]
12. Yu, K.P.; Yu, W.Y.; Ku, M.C.; Liou, Y.C.; Chien, S.H. Pt/titania-nanotube: A potential catalyst for CO_2 adsorption and hydrogenation. *Appl. Catal. B* **2008**, *84*, 112–118. [CrossRef]
13. Xing, L.; Jia, J.B.; Wang, Y.Z.; Zhang, B.L.; Dong, S.J. Pt modified TiO_2 nanotubes electrode: Preparation and electrocatalytic application for methanol oxidation. *Int. J. Hydrog. Energy* **2010**, *35*, 12169–12173. [CrossRef]
14. Gao, Z.D.; Qu, Y.F.; Zhou, X.M.; Wang, L.; Song, Y.Y.; Schmuki, P. Pt-decorated g-C_3N_4/ TiO_2 nanotube arrays with enhanced visible–light photocatalytic activity for H_2 evolution. *Chem. Open* **2016**, *5*, 197–200. [CrossRef]
15. Lee, W.; Kang, S.H.; Min, S.K.; Sung, Y.E.; Han, S.H. Co-sensitization of vertically aligned TiO_2 nanotubes with two different sizes of cdse quantum dots for broad spectrum. *Electrochem. Commun.* **2008**, *10*, 1579–1582. [CrossRef]
16. Baker, D.R.; Kamat, P.V. Photosensitization of TiO_2 nanostructures with CdS quantum dots: Particulate versus tubular support architectures. *Adv. Funct. Mater.* **2009**, *19*, 805–811. [CrossRef]
17. Ding, D.Y.; Ning, C.Q.; Huang, L.; Jin, F.C.; Hao, Y.Q.; Bai, S.; Li, Y.; Li, M.; Mao, D.L. Anodic fabrication and bioactivity of Nb-doped TiO_2 nanotubes. *Nanotechnology* **2009**, *20*, 305103. [CrossRef] [PubMed]
18. Qin, R.; Ding, D.Y.; Ning, C.Q.; Liu, H.G.; Zhu, B.S.; Li, M.; Mao, D.L. Ni-doped TiO_2 nanotube arrays on shape memory alloy. *Appl. Surf. Sci.* **2011**, *257*, 6308–6313. [CrossRef]
19. John, K.A.; Naduvath, J.; Mallick, S.; Pledger, J.W.; Remillard, S.K.; DeYoung, P.A.; Thankamoniamma, M.; Shripathi, T.; Philip, R.R. Electrochemical synthesis of novel Zn-doped TiO_2 nanotube/ZnO nanoflake heterostructure with enhanced DSSC efficiency. *Nano-Micro Lett.* **2016**, *8*, 381–387. [CrossRef]
20. Altomare, M.; Lee, K.; Killian, M.S.; Selli, E.; Schmuki, P. Ta–doped TiO_2 nanotubes for enhanced solar-light photoelectrochemical water splitting. *Chem.-A Eur. J.* **2013**, *19*, 5841–5844. [CrossRef] [PubMed]

21. Kim, H.I.; Monllor-Satoca, D.; Kim, W.; Choi, W. N-doped TiO_2 nanotubes coated with a thin TaO_xN_y layer for photoelectrochemical water splitting: Dual bulk and surface modification of photoanodes. *Energy Environ. Sci.* **2015**, *8*, 247–257. [CrossRef]
22. Ahmed, S.A. Ferromagnetism in Cr-, Fe- and Ni-doped TiO_2 samples. *J. Magn. Magn. Mater.* **2017**, *442*, 152–157. [CrossRef]
23. Liu, Q.; Ding, D.Y.; Ning, C.Q.; Wang, X.W. Reduced N/Ni-doped TiO_2 nanotubes photoanodes for photoelectrochemical water splitting. *RSC Adv.* **2015**, *5*, 95478–95487. [CrossRef]
24. Ansari, S.A.; Khan, M.M.; Ansari, M.O.; Cho, M.H. Nitrogen-doped titanium dioxide (N-doped TiO_2) for visible light photocatalysis. *New J. Chem.* **2016**, *40*, 3000–3009. [CrossRef]
25. Saharudin, K.A.; Sreekantan, S.; Lai, C.W. Fabrication and photocatalysis of nanotubular C-doped TiO_2 arrays: Impact of annealing atmosphere on the degradation efficiency of methyl orange. *Mater. Sci. Semicond. Process.* **2014**, *20*, 1–6. [CrossRef]
26. Kyeremateng, N.A.; Hornebecq, V.; Martinez, H.; Knauth, P.; Djenizian, T. Electrochemical fabrication and properties of highly ordered Fe-doped TiO_2 nanotubes. *ChemPhysChem* **2012**, *13*, 3707–3713. [CrossRef] [PubMed]
27. Zhang, Y.; Li, X.; Chen, D.; Ma, N.; Hua, X.; Wang, H. Si doping effects on the photocatalytic activity of TiO_2 nanotubes film prepared by an anodization process. *Scr. Mater.* **2009**, *60*, 543–546. [CrossRef]
28. Lu, B.C.; Wang, Y.L.; Xu, J. Revisiting the glass-forming ability of Ti-Ni-Si ternary alloys. *J. Alloys Compd.* **2009**, *475*, 157–164. [CrossRef]
29. Cai, Q.Y.; Yang, L.X.; Yu, Y. Investigations on the self–organized growth of TiO_2 nanotube arrays by anodic oxidization. *Thin Solid Film* **2006**, *515*, 1802–1806. [CrossRef]
30. Abdullah, M.; Kamarudin, S.K. Titanium dioxide nanotubes (TNT) in energy and environmental applications: An overview. *Renew. Sustain. Energy Rev.* **2017**, *76*, 212–225. [CrossRef]
31. Roy, P.; Berger, S.; Schmuki, P. TiO_2 nanotubes: Synthesis and applications. *Angew. Chem. Int. Ed.* **2011**, *50*, 2904–2939. [CrossRef] [PubMed]
32. Li, Y.; Ding, D.Y.; Ning, C.Q.; Bai, S.; Huang, L.; Li, M.; Mao, D.L. Thermal stability and in vitro bioactivity of Ti-Al-V-O nanostructures fabricated on Ti_6Al_4V alloy. *Nanotechnology* **2009**, *20*, 065708. [CrossRef] [PubMed]
33. Sun, Y.; Wang, G.X.; Yan, K.P. TiO_2 nanotubes for hydrogen generation by photocatalytic water splitting in a two-compartment photoelectrochemical cell. *Int. J. Hydrog. Energy* **2011**, *36*, 15502–15508. [CrossRef]
34. Liu, Z.; Pesic, B.; Raja, K.S.; Rangaraju, R.R.; Misra, M. Hydrogen generation under sunlight by self-ordered TiO_2 nanotube arrays. *Int. J. Hydrog. Energy* **2009**, *34*, 3250–3257. [CrossRef]
35. Kunadian, I.; Lipka, S.M.; Swartz, C.R.; Qian, D.; Andrews, R. Determination of carrier densities of boron–and nitrogen-doped multiwalled carbon nanotubes using Mott–Schottky plots. *J. Electrochem. Soc.* **2009**, *156*, K110–K115. [CrossRef]
36. Palmas, S.; Polcaro, A.M.; Ruiz, J.R.; Da Pozzo, A.; Mascia, M.; Vacca, A. TiO_2 photoanodes for electrically enhanced water splitting. *Int. J. Hydrog. Energy* **2010**, *35*, 6561–6570. [CrossRef]
37. Kim, J.Y.; Jung, H.S.; No, J.H.; Kim, J.R.; Hong, K.S. Influence of anatase–rutile phase transformation on dielectric properties of sol–gel derived TiO_2 thin films. *J. Electroceram.* **2006**, *16*, 447–451. [CrossRef]
38. Hernández, S.; Hidalgo, D.; Sacco, A.; Chiodoni, A.; Lamberti, A.; Cauda, V.; Tresso, E.; Saracco, G. Comparison of photocatalytic and transport properties of TiO_2 and ZnO nanostructures for solar-driven water splitting. *Phys. Chem. Chem. Phys.* **2005**, *17*, 7775–7786. [CrossRef] [PubMed]

nanomaterials

MDPI

Article

Microstructuring of Mesoporous Titania Films Loaded with Silver Salts to Enhance the Photocatalytic Degradation of Methyl Blue under Visible Light

Nicolas Crespo-Monteiro *, Anthony Cazier, Francis Vocanson, Yaya Lefkir, Stéphanie Reynaud, Jean-Yves Michalon, Thomas Kämpfe , Nathalie Destouches and Yves Jourlin

Univ Lyon, UJM-Saint-Etienne, CNRS, Institut d'Optique Graduate School, Laboratoire Hubert Curien UMR 5516, F-42023 Saint-Etienne, France; anthony.cazier@univ-st-etienne.fr (A.C.); francis.vocanson@univ-st-etienne.fr (F.V.); yaya.llefkir@univ-st-etienne.fr (Y.L.); stephanie.reynaud@univ-st-etienne.fr (S.R.); jean.yves.michalon@univ-st-etienne.fr (J.-Y.M.); thomas.kampfe@univ-st-etienne.fr (T.K.); nathali.destouche@univ-st-etienne.fr (N.D.); yves.jourlin@univ-st-etienne.fr (Y.J.)

* Correspondence: nicolas.crespo.monteiro@univ-st-etienne.fr

Received: 22 September 2017; Accepted: 5 October 2017; Published: 17 October 2017

Abstract: The microstructuring of the distribution of silver nanoparticles (NPs) in mesoporous titania films loaded with silver salts, using two-beam interference lithography leading to 1 Dimension (1D) grating, induces variations in the photocatalytic efficiency. The influence of the structuration was tested on the degradation of methyl blue (MB) under ultraviolet (UV) and visible illumination, giving rise to a significant improvement of the photocatalytic efficiency. The periodic distribution of the NPs was characterized by transmission electron microscopy (TEM), high-angle annular dark field scanning transmission electron microscopy (HAADF-STEM) and scanning electron microscopy (SEM).

Keywords: microstructuring; titania; visible; nanoparticles; sol-gel; photocatalysis

1. Introduction

Titanium dioxide (TiO_2) is one of the most investigated and widely used photocatalyzers for the photodegradation of pollutants in water and air [1]. Nevertheless, as a result of its large band gap, the photocatalytic activity of TiO_2 is only activated under ultraviolet (UV) light, which limits its use for indoor-environment applications [2]. Therefore, the development of photocatalysts showing a high activity under visible light irradiation is needed in order to be able to use the sunlight or rays from artificial sources more effectively in photocatalytic reactions. Many studies have attempted to develop visible light-sensitive photocatalysts using, for example, TiO_2 doped with metallic nanoparticles (NPs) such as Au, Ag or Cu [2–5]. This approach seems to be of interest for improving the photocatalytic effect in visible light due to the localized surface plasmon resonance band (LSPR) of the metallic NPs, which induces a high absorption in the visible range. At the interface between metallic NPs and TiO_2, there is a potential barrier (Schottky barrier) that is low enough to allow the excitation of electrons at the surface of the metallic NPs in the conduction band of TiO_2. The released charges lead to the same photocatalytic effect as in TiO_2 excited by UV. However, any change in the size and/or shape of the metallic NPs and their distribution and concentration, as well as the optical properties of their environment, will have significant effects on their LSPR and consequently on the overall efficiency of the device [5–7]. This can even lead to a counterproductive effect; the non-resonant metallic NPs can act as recombination sites for the photo-generated electrons [5]. Despite extensive investigations, most of the developed systems are not suitable for practical indoor applications because of both the small amount of light present in this environment and the low light-harvesting efficiency of the devices [2,8].

In this paper, we propose to increase the photocatalytic efficiency of TiO_2 films loaded with silver salts, using a direct film microstructuration and periodic NP distribution close to the films' surface using UV laser interference lithography. The structures are characterized by transmission electron microscopy (TEM), high-angle annular dark field scanning transmission electron microscopy (HAADF-STEM) and scanning electron microscopy (SEM). Their photocatalytic efficiency has been obtained from the degradation of methyl blue (MB) during UV and visible illumination.

2. Results

The films used in this work were mesoporous TiO_2 xerogel films deposited on glass substrates. After deposition, the films were doped with silver salts by soaking them for 1 h in an aqueous ammoniacal silver nitrate solution at 0.75 M. After synthesizing the films, different treatments were carried out on their surface. The first was a homogeneous UV illumination for 30 min with two lamps emitting at 254 nm and each delivering a power of 15 W. This illumination allowed the silver salts to be reduced and a homogenous distribution of silver NPs in the uppermost 80–100 nm layer of the film to be produced (Figures 1a,d and 2b). Silver NP growth inside the mesoporous TiO_2 matrix [9] leads to a certain diameter distribution of the silver NPs, which has been estimated by image processing [10] to be in the range of 1 to 12 nm in diameter with a maximum particle size distribution of around 5 nm in diameter (Figure 1g). Before illumination, the films were transparent and the spectroscopic measurement (PerkinElmer Lambda 900) showed no absorption bands in the UV or visible range, but after illumination, a large absorption band centered around 460 nm appeared (Figure 3a). This absorption band was due to the LSPR resulting from the presence of silver NPs in the TiO_2 matrix.

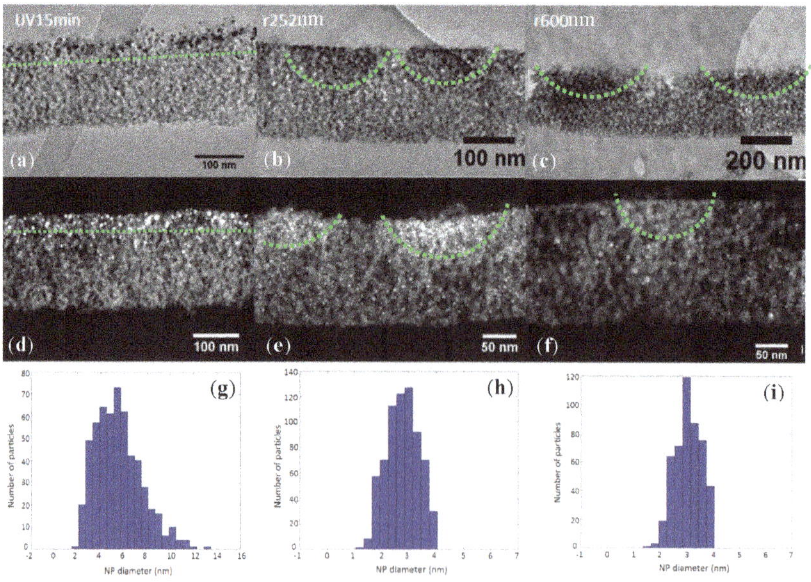

Figure 1. Transmission electron microscopy (TEM) (**a–c**) and high-angle annular dark field scanning transmission electron microscopy (HAADF-STEM) (**d–f**) images of cross-section of samples UV15min (**a**,**d**), r252nm (**b**,**c**) and r600nm (**e**,**f**). Nanoparticle (NP)-size histograms deduced from image processing carried out on the HAADF-STEM images of samples UV15min (**g**), r252nm (**h**) and r600nm (**i**).

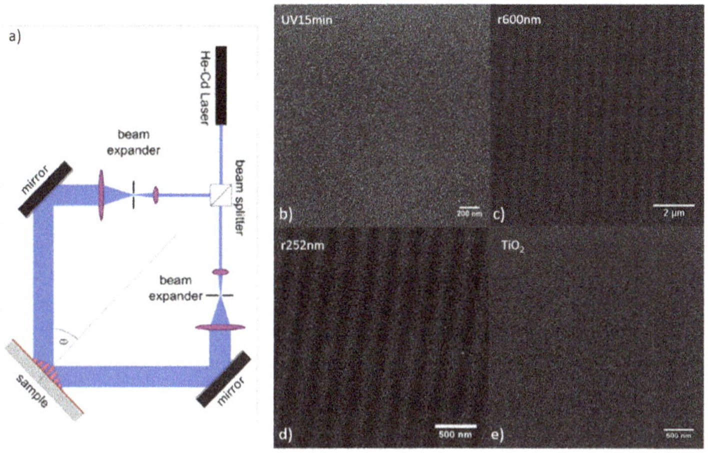

Figure 2. (**a**) Sketch of the holographic bench. Scanning electron microscopy (SEM) images of samples UV15min (**b**), r600nm (**c**), r252nm (**d**) and TiO_2 (**e**).

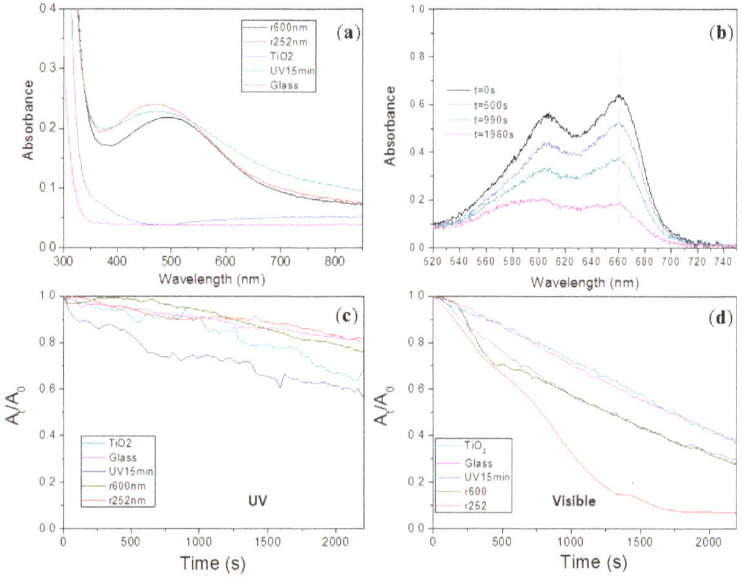

Figure 3. (**a**) Absorbance spectra of glass substrate (glass), TiO_2, UV15min, r252nm and r600nm samples. (**b**) Absorbance variation of methyl blue (MB) solution during the visible illumination of the sample r600nm. (**c**,**d**) Absorbance variation at 660 nm normalized by the absorbance at the initial state (t = 0 s) for the different samples under ultraviolet (UV) or visible illumination, respectively.

In order to test the influence of microstructuring on the film of silver NPs, two different gratings were realized on the film's surface, corresponding to a 1 Dimension (1D) periodic index variation close to the film's surface. The design of the grating was optimized to couple the desired incident wavelength into the film. The first grating coupled the light at a wavelength inside the LSPR (~430 nm), whilst the second coupled the light at a wavelength outside the LSPR (~1020 nm).

The gratings were fabricated by two-beam interference lithography (Figure 2a), using a laser of 325 nm wavelength delivering a continuous wave power of 100 mW. The polarized laser beam was split into two arms, which were recombined on the sample, where the overlap created an interferogram during the illumination in the form of a 1D periodic intensity modulation, whose fringe period (Λ) was fixed by the angle (θ) and the wavelength (λ) according to the equation $\Lambda = \lambda/(2 \cdot \sin(\theta))$. The mirrors were mounted on rotation stages and the substrate holder was placed on a rail to allow for automatic tuning of the period. The exposure time for each structure was very long at about 1.5 h, and vibrations had to be kept to a minimum by mounting the optical bench on an air cushion table to filter low-frequency vibrations. Additionally, one of the interferometer arms used a mirror mounted on a piezoelectric motor controlled in a feedback loop by a measurement of the interference fringe positions allowing for a stabilization of the fringe pattern against air convection effects and higher-frequency vibrations. Considering an average refractive index of 1.7 for mesoporous TiO_2 films (index obtained by ellipsometry measurement), the fringe period required to obtain a grating of NPs capable of coupling the light at the wavelengths of 430 and 1020 nm into the TiO_2 matrix were 430/1.7 = 252 nm (sample r252nm) and 1020/1.7 = 600 nm (sample r600nm), respectively.

Figure 1 shows images of the realized gratings, obtained by TEM (Figure 1b,c) and HAADF-STEM (Figure 1e,f), and Figure 2 shows the images of gratings by SEM (Figure 2c,d). For the two structures, most of the silver NPs were localized in the first 80–100 nm from the film's surface and had a diameter of around 2–4 nm with a maximum at 3 nm (Figure 1h,i). The period obtained for the sample r252nm was ~252 nm and this was ~600 nm for the sample r600nm. The NPs induced film coloration and an absorption band in the visible range. The absorbance spectra recorded in the visible range (Figure 3a) showed a similar level of absorbance for all samples with NPs (r252nm, r600nm and UV15min). For these different samples, the degradation of MB was studied over 40 min under UV light by monitoring the absorbance of the MB solution at a wavelength of 660 nm (Figure 3c). After examining the behavior for UV illumination, the photocatalytic activity of the fabricated films was also tested under visible light exposure by investigating the MB degradation using a solar simulator lamp with a cut-off filter at 400 nm and monitoring the absorbance of the MB solution (Figure 3b). Figure 3d shows the absorbance variation at a wavelength of 660 nm during the 40 min of illumination.

3. Discussion

The results obtained during UV illumination (Figure 3c) show that the samples with NP gratings had a degradation rate similar to a non-photocatalytic sample (glass substrate), which means that the samples were essentially not photocatalytic under UV light. The mesoporous TiO_2 film, which had a size of pores between 5 and 15 nm (Figure 2e), had a slightly higher degradation rate. The low degradation efficiency of the TiO_2 film could be explained by the fact that it was mainly amorphous [11] and therefore not very catalytic under UV light. A better result was obtained for the sample with a homogeneous distribution of silver (UV15min sample). It is well known that the quantity of silver NPs present inside the TiO_2 matrix can improve [12] (by limiting the recombination of electron–hole pairs photogenerated in the TiO_2) or inhibit [13] (NPs serving as a recombination site for photo-induced charges) the photocatalytic effect of TiO_2. If we assume that the presence of silver NPs improves the photocatalytic effect of the TiO_2 matrix, the samples with silver NPs (r252nm and r600nm films) should have been more photocatalytic than samples without them (TiO_2). However, the yield was lower than for the pure TiO_2 films. Consequently, silver NPs seem to inhibit the photocatalytic effect of TiO_2 rather than improve it. The improvement of the photocatalytic effect could also be due to the increase in crystallization of the TiO_2 matrix (TiO_2 crystallized in anatase phase is more photocatalytic than amorphous TiO_2). Initially, the used films are amorphous, but during UV or visible illumination, even if a low intensity is used, the growth or oxidation of silver NPs can induce a temperature rise that can initiate in their vicinity the crystallization of TiO_2 in nanocrystals of anatase, brookite or rutile phase [11]. The increase in the number of silver NPs can thus lead to a greater number of anatase

TiO_2 nanocrystals in the film and consequently improves its photocatalytic efficiency. Currently, the influence of each mechanism is unclear and remains to be studied.

During the visible illumination, the sample with a NP grating period of 252 nm showed a better photocatalytic activity (Figure 3d); it allowed for an increasing of the degradation rate of MB by a factor of 2.5 compared to the pure TiO_2 sample, and by a factor of 1.75 compared to the UV/TiO_2 sample. The sample with a NP grating period of 600 nm showed a MB degradation similar to that of the sample with the homogeneous distribution of silver NPs. The degradation rates obtained were not only due to the films but also to the light irradiation conditions. If a sample with no photocatalytic activity (glass substrate) was exposed to the same illumination conditions (Figure 3d), one could also observe a degradation of MB, but in this case the degradation time was shorter (the same overall absorbance variation of MB was obtained 7 min later). If a mesoporous TiO_2 film without silver salts (Figure 3d) was exposed to the same illumination condition, the degradation rate of MB was the same as for a glass substrate; that is to say that the mesoporous TiO_2 films had no photocatalytic activity under visible irradiation. This confirmed, as it has already been shown in several articles, that the presence of silver NPs in the TiO_2 matrix is essential to its photocatalytic activity in the visible range [5]. Furthermore, these results show that the distribution of the silver NPs in the TiO_2 matrix also has an impact on the degradation of MB under visible light. A suitable structuring of the distribution of the silver NPs in the form of a periodic grating allows for increasing (sample r252nm) the photocatalytic efficiency of the TiO_2/Ag films; however, the parameters need to be chosen carefully, as it has been shown that it is also possible to decrease (sample r600nm) the photocatalytic efficiency. For the sample r252nm, the periodic structuring allowed for an increase in the amount of light absorbed by the films at wavelengths comprised in the LSPR band, increasing the number of available charges. However, the structuring could also change the photocatalytic efficiency by influencing the electron–hole recombination, which is an important point that needs to be clarified in subsequent studies.

4. Materials and Methods

The process to synthetize mesoporous TiO_2 films is detailed in [14]. Their thicknesses were estimated by profilometry (Dektak XT, Bruker, Wissembourg, France) and ellipsometry (SEMILAB GES5-E, Semilab, Budapest, Hungary) to be 150 ± 50 nm, and the porosity volume fraction was estimated at 27%. The size of their pores varied between 5 and 15 nm. Initially the films were mostly amorphous and transparent [11].

The photocatalytic reaction system for UV illumination was composed of two UV lamps of 15 W emitting at 254 nm. For illumination in the visible spectrum, the system consisted of a solar simulator lamp (Newport 94011A Sol series Solar Simulator, Newport, Irvine, CA, USA) equipped with a cut-off filter at 400 nm. A drop of 500 μL of MB (Aldrich, Saint Louis, MO, USA) at 10^{-4} M was deposited on the films' surface, using substrates of size 2.5×2.5 cm^2. After 30 min in the dark, at room temperature in air, the films were exposed in the UV and in the visible spectrum. During the exposition, the absorbance of the MB solution was recorded every 2 min by a UV–visible spectrometer (Ocean Optics HR2000+, Ocean Optics, Winter Park, FL, USA). MB degradation was detected by measuring the absorption at a wavelength of 660 nm. For UV illumination, a white lamp with an intensity of 1 $\mu W \cdot cm^{-2}$ was added to allow for an absorbance measurement at 660 nm. All experiments were conducted at room temperature in air.

5. Conclusions

In summary, we have shown that microstructuring of a TiO_2/Ag film influences the photocatalytic efficiency of films on the degradation of MB under visible light. A film for which silver NPs are distributed in the form of a periodic 1D grating fabricated by two-beam interference lithography, whose period has been chosen to couple more light in the film at the wavelength of the localized surface plasmon resonance of the silver NPs allows for an increasing of the photocatalytic efficiency of the TiO_2 films in white light by a factor of 2.5 compared to a film without NPs and by a factor of 1.75 with respect to a film with a homogeneous distribution of the NPs. For the opposite, a film with an unsuitable structuring (wavelength coupled outside the LSPR of silver NPs) decreases the photocatalytic efficiency of the device relative to a film with a homogeneous distribution. We also showed that under UV illumination, the structuring did not have a measurable effect, and the photocatalytic behavior on the degradation of the MB was not very effective overall. The increase in the photocatalytic efficiency shown in this work can make it possible to consider microstructuring of TiO_2/Ag films for applications in the field of indoor air treatment, which is currently limited by the low efficiency of the homogeneous TiO_2/Ag films, in particular as a result of light sources present in this environment, whose intensities are comparably low with a weak amount of UV light. Instead of using a linear grating, the TiO_2/Ag film can be microstructured with a two-dimensional grating of NPs and/or a two-dimensional topographic grating, which will allow a further increase in the photocatalytic efficiency in the visible range by taking into account the (usually) non-polarized nature of natural light.

Acknowledgments: The authors thank the French Centre National de la Recherche Scientifique (CNRS) for its financial support in the framework of PEPS project MINATAIR.

Author Contributions: N.C.-M. conducted the study; F.V. synthesized and deposited the films; A.C. performed the laser exposures. The different characterizations were performed as follows: N.C.-M. and J.-Y.M. performed optical spectroscopy, S.R. performed SEM, and Y.L. performed TEM and STEM-HAADF. N.D. and Y.J. offered suggestions and commented on data analyses. T.K. realized E.M. simulations. The paper was written by N.C.-M. and T.K. with contributions from Y.J.

Conflicts of Interest: The authors declare no conflict of interest.

References

1. Carp, O.; Huisman, C.L.; Reller, A. Photoinduced reactivity of titanium dioxide. *Prog. Solid State Chem.* **2004**, *32*, 33–177. [CrossRef]
2. Qiu, X.; Miyauchi, M.; Sunada, K.; Minoshima, M.; Liu, M.; Lu, Y.; Li, D.; Shimodaira, Y.; Hosogi, Y.; Kuroda, Y.; et al. Hybrid CuxO/TiO_2 nanocomposites as risk-reduction materials in indoor environments. *ACS Nano* **2012**, *6*, 1609–1618. [CrossRef] [PubMed]
3. Klein, M.; Grabowska, E.; Zaleska, A. Noble metal modified TiO_2 for photocatalytic air purification. *Physicochem. Probl. Miner. Process.* **2015**, *51*, 49–57.
4. Ramírez, R.J.; Arellano, C.A.P.; Varia, J.; Martínez, S.S. Visible light-induced photocatalytic elimination of organic pollutants by TiO_2: A review. *Curr. Org. Chem.* **2015**, *19*, 540–555. [CrossRef]
5. Jaafar, N.F.; Jalil, A.A.; Triwahyono, S. Visible-light photoactivity of plasmonic silver supported on mesoporous TiO_2 nanoparticles (Ag-MTN) for enhanced degradation of 2-chlorophenol: Limitation of Ag-Ti interaction. *Appl. Surf. Sci.* **2017**, *392*, 1068–1077. [CrossRef]
6. Kowalska, E.; Rau, S.; Ohtani, B.; Kowalska, E.; Rau, S.; Ohtani, B. Plasmonic Titania Photocatalysts Active under UV and Visible-Light Irradiation: Influence of Gold Amount, Size, and Shape. *J. Nanotechnol.* **2012**, *2012*, 361853. [CrossRef]
7. Banerjee, A.N. The design, fabrication, and photocatalytic utility of nanostructured semiconductors: Focus on TiO_2-based nanostructures. *Nanotechnol. Sci. Appl.* **2011**, *4*, 35–65. [CrossRef] [PubMed]
8. Paz, Y. Application of TiO_2 photocatalysis for air treatment: Patents' overview. *Appl. Catal. B Environ.* **2010**, *99*, 448–460. [CrossRef]
9. Crespo-Monteiro, N.; Destouches, N.; Bois, L.; Chassagneux, F.; Reynaud, S.; Fournel, T. Reversible and Irreversible Laser Microinscription on Silver-Containing Mesoporous Titania Films. *Adv. Mater.* **2010**, *22*, 3166–3170. [CrossRef] [PubMed]
10. Liu, Z.; Epicier, T.; Lefkir, Y.; Vitrant, G.; Destouches, N. HAADF-STEM characterization and simulation of nanoparticle distributions in an inhomogeneous matrix. *J. Microsc.* **2017**, *266*, 60–68. [CrossRef] [PubMed]
11. Crespo-Monteiro, N.; Destouches, N.; Epicier, T.; Balan, L.; Vocanson, F.; Lefkir, Y.; Michalon, J.-Y. Changes in the Chemical and Structural Properties of Nanocomposite Ag: TiO_2 Films during Photochromic Transitions. *J. Phys. Chem. C* **2014**, *118*, 24055–24061. [CrossRef]

12. Zhang, L.; Yu, J.C.; Yip, H.Y.; Li, Q.; Kwong, K.W.; Xu, A.-W.; Wong, P.K. Ambient Light Reduction Strategy to Synthesize Silver Nanoparticles and Silver-Coated TiO_2 with Enhanced Photocatalytic and Bactericidal Activities. *Langmuir* **2003**, *19*, 10372–10380. [CrossRef]
13. Sung-Suh, H.M.; Choi, J.R.; Hah, H.J.; Koo, S.M.; Bae, Y.C. Comparison of Ag deposition effects on the photocatalytic activity of nanoparticulate TiO_2 under visible and UV light irradiation. *J. Photochem. Photobiol. Chem.* **2004**, *163*, 37–44. [CrossRef]
14. Nadar, L.; Sayah, R.; Vocanson, F.; Crespo-Monteiro, N.; Boukenter, A.; Joao, S.S.; Destouches, N. Influence of reduction processes on the colour and photochromism of amorphous mesoporous TiO_2 thin films loaded with a silver salt. *Photochem. Photobiol. Sci.* **2011**, *10*, 1810–1816. [CrossRef] [PubMed]

nanomaterials

MDPI

Article

Piezoelectric Potential in Single-Crystalline ZnO Nanohelices Based on Finite Element Analysis

Huimin Hao [1,2,*], Kory Jenkins [2], Xiaowen Huang [3], Yiqian Xu [1], Jiahai Huang [1] and Rusen Yang [2,4,*]

1. Key Lab of Advanced Transducers and Intelligent Control System, Ministry of Education and Shanxi Province, Taiyuan University of Technology, Taiyuan 030024, China; xu276816496@sina.com (Y.X.); huangjiahai@tyut.edu.cn (J.H.)
2. Department of Mechanical Engineering, University of Minnesota, Minneapolis, MN 55455, USA; jenk0131@umn.edu
3. Department of Applied Physics, The Hong Kong Polytechnic University, Hong Kong, China; huangxiaowen2013@gmail.com
4. School of Advanced Materials and Nanotechnology, Xidian University, Xi'an 710071, China

* Correspondence: haohuimin@tyut.edu.cn (H.H.); yangr@umn.edu (R.Y.); Tel.: +86-35-1601-4551 (H.H.); +1-612-626-4318 (R.Y.)

Received: 7 October 2017; Accepted: 1 December 2017; Published: 7 December 2017

Abstract: Electric potential produced in deformed piezoelectric nanostructures is of significance for both fundamental study and practical applications. To reveal the piezoelectric property of ZnO nanohelices, the piezoelectric potential in single-crystal nanohelices was simulated by finite element method calculations. For a nanohelix with a length of 1200 nm, a mean coil radius of 150 nm, five active coils, and a hexagonal coiled wire with a side length 100 nm, a compressing force of 100 nN results in a potential of 1.85 V. This potential is significantly higher than the potential produced in a straight nanowire with the same length and applied force. Maintaining the length and increasing the number of coils or mean coil radius leads to higher piezoelectric potential in the nanohelix. Appling a force along the axial direction produces higher piezoelectric potential than in other directions. Adding lateral forces to an existing axial force can change the piezoelectric potential distribution in the nanohelix, while the maximum piezoelectric potential remains largely unchanged in some cases. This research demonstrates the promising potential of ZnO nanohelices for applications in sensors, micro-electromechanical systems (MEMS) devices, nanorobotics, and energy sciences.

Keywords: piezotronic; numerical simulation; nanohelix; FEM

1. Introduction

Helical structures have been widely used in industry due to their low stiffness and superior capability to resist large axial strain, while helical structures are also the fundamental configuration for DNA and many other biomolecules. Three-dimensional helical nanostructures of zinc oxide (ZnO) have also been investigated, such as nanohelices [1–3], nanorings formed by self-coiling nanobelts [4,5], and nanosprings [6–8]. ZnO helical nanostructures showed superelastic behavior, and their spring constant increased continuously up to 300–800% when they were stretched [2]. As a piezoelectric material, ZnO nanostructures generate piezoelectric potential when exposed to physical stimulation, such as stretching, compression, and bending. Taking advantage of this phenomenon, ZnO has been used to fabricate nanogenerators for energy harvesting [9–12]. ZnO is also a semiconductor material. Piezoelectric potential can alter electronic transport in ZnO nanostructures, which has resulted in novel devices, e.g., piezoelectric field effect transistors [13], strain sensors [14,15], programmable electromechanical memories [16], and logic circuits [17].

Considering the importance of piezoelectric potential in ZnO nanostructures for their applications in electronics, sensors, actuators, and nanogenerators, the distribution and effects of piezoelectric potential in ZnO nanowires have been studied. Lippman theory [18–20] has been applied to predict the distribution of equilibrium potential in a deformed ZnO nanowire. Wang calculated the piezoelectric potential in a bent ZnO nanowire, and a numerical calculation of the piezoelectric potential distribution in a ZnO nanowire without doping was carried out. The potential difference was around 0.4 V in a ZnO nanowire grown along the *c*-axis with a length of 1200 nm and a hexagonal side length of 100 nm under a compressing force of 85 nN [9]. The calculation results were verified by later measurements of the asymmetric voltage distribution on the tensile and compressive side surfaces of a ZnO nanowire [9]. Experimental measurements and numerical modeling works revealed that the piezoelectric potential can cause a re-distribution of free charge carriers in ZnO nanowires and modify the electronic transport of ZnO nanowire-based devices. However, for the piezoelectric potential of ZnO, current studies are mainly limited to nanowires.

Owing to the enriched physical and chemical properties, the unique spiral geometry of helical nanostructures was studied to discover novel properties for new nanodevice design and fabrication. Chen studied the mechanics of carbon nanocoils. They found that the spring constant K stays constant with increasing elongation when the spring had lower helical angles, and that the carbon nanocoil returned completely to its relaxed geometry after loading without apparent plastic deformation [21]. For potential device applications, the mechanical properties [22,23], electrical properties [24–26], optical properties [27,28], and magnetic properties [29] of helical nanostructures of several materials were studied. Nevertheless, the study of piezoelectric potential in helical nanostructures is still lacking.

Ultrasmall, deformation-free, single-crystal nanohelices of piezoelectric ZnO have been reported [1]. It has been shown that the electrostatic interaction between polar surfaces plays an important role in forming the deformation-free nanohelices. In this work, we used a three-dimensional (3D) finite element method (FEM) to examine the piezoelectric potential in ZnO nanohelices. We discussed the effects of the number of coils, mean radius of the nanohelix, and applied forces on the piezoelectric potential. We focused on the equilibrium piezoelectric potential and showed that nanohelices could produce significantly higher potential than a nanowire with the same height under the same force. These results indicate that ZnO nanohelices can be excellent candidates for fabricating piezoelectric nanodevices, such as nanogenerators, actuators, and nanosensors.

2. Model and Method

2.1. Model Configuration

Figure 1 shows the structure of a ZnO nanowire and a ZnO nanohelix. The ZnO nanowire in Figure 1a is modeled as a cylinder with a hexagonal cross-section in the a-b crystallographic plane and its *c*-axis along the *z*-axis; Figure 1b shows a deformation-free, single-crystal nanohelix of piezoelectric ZnO with a hexagonal cross-section in the a-c crystallographic plane and its *c*-axis along the *z*-axis. Nanowires are normally grown along their *c*-axis, which is along the *z*-axis in Figure 1a. The ZnO nanohelix in Figure 1b is formed by a wire with a hexagonal cross-section. The center line of the ZnO nanohelix coincides with the *z*-axis. A sequential rotation in the growth direction results in a non-twisted single-crystal structure for the entire nanohelix [1]. The details of the growth and structure characterization of the nanohelix are reported in Reference [1]. The hexagonal side length D and total length along the major axial direction L are assumed to be the same for the nanowire and the nanohelix. The mean radius of the coil and the number of coils of the nanohelix are represented by R and T, respectively. Studies of right-handed nanohelices and left-handed nanohelices produced similar results, and the results of right-handed nanohelices are presented here.

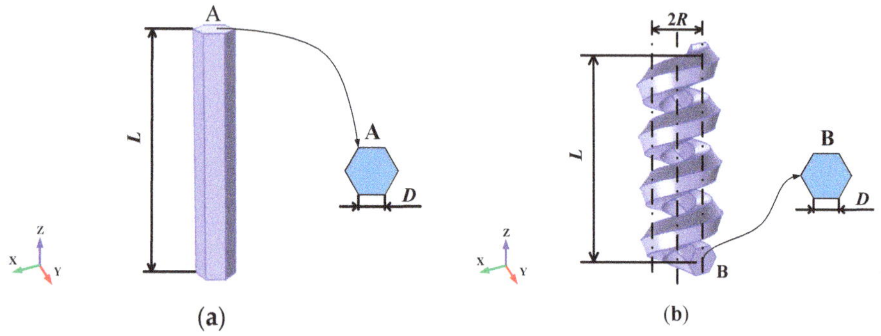

Figure 1. Schematic illustration of (**a**) ZnO nanowire and (**b**) ZnO nanohelix. Both nanostructures have a hexagonal cross-section with the same side length D and height L. The nanohelix has a number of coils T and a mean radius of coil R.

We assumed that there is no body force f_e^b and no free charge carriers ρ_e^b in order to simplify the computation and focus on the piezoelectric potential created in the nanowires and nanohelices. The effect of free charge carriers in nanowires has been investigated and reported [19,30]. A more comprehensive evaluation of piezoelectric nanohelices with free charge carriers considered is the subject of a future study. The modeling parameters are given in Table 1 [31].

Table 1. List of parameters for modeling.

Parameters	Value
Density (kg/m^3)	5680
Elastic constants	
c_{11} (GPa)	209.7
c_{12} (GPa)	121.1
c_{13} (GPa)	105.1
c_{33} (GPa)	211.3
c_{44} (GPa)	42.3
c_{55} (GPa)	43.6
Piezoelectric constants	
e_{31} (C/m^2)	−0.57
e_{33} (C/m^2)	1.32
e_{15} (C/m^2)	−0.48
Relative dielectric constants	
$\kappa_{\perp}^{r}$	8.54
$\kappa_{\parallel}^{r}$	10.20

2.2. FEM Modeling of Nanowires and Nanohelices

Since there is no body force in the nanostructures, the divergence of the stress tensor σ should be zero in the static piezoelectric problem [19,20]:

$$\nabla \cdot \sigma = \vec{f}_e^{\,b} = 0 \tag{1}$$

where σ is the stress tensor.

The constitutive relation between stress σ and electric displacement $\vec{D}$ is governed by the fundamental piezoelectric equations:

$$\begin{cases} \sigma_p = c_{pq}\varepsilon_q - e_{kp}E_k \\ D_i = e_{iq}\varepsilon_q + \kappa_{ik}E_k \end{cases} \tag{2}$$

where c_{pq} is the linear elastic constant, ε is the strain, e_{kp} is the linear piezoelectric coefficient, κ_{ik} is the dielectric constant, and E is the electric field.

By assuming no free charge ρ_e^b in the nanowire or the nanohelix, the Gauss equation must be satisfied:

$$\nabla \cdot \vec{D} = \rho_e^b = 0 \tag{3}$$

and the compatibility equation should be satisfied:

$$\nabla \times \nabla \times \varepsilon = 0 \tag{4}$$

We calculated the piezoelectric potential in nanowires and nanohelices by solving the above nonlinear partial differential Equations (1)–(4) with the program COMSOL Multiphysics®. The bottom face of the nanowire and the nanohelix was fixed and electrically grounded in our model. A force was applied only to the top end, and the piezoelectric potential was numerically calculated. A nanowire and a nanohelix with the same side length and height were simulated to compare their piezoelectric potentials when exposed to the same force. A series of nanohelices with different numbers of coils and mean radii of the coil were calculated to reveal the change of piezoelectric potential in a helical structure. The effect of force direction was also investigated.

3. Simulation Results and Discussion

3.1. Pizeoelectric Potential and Displacement of the ZnO Nanohelix and Nanowire

The piezoelectric potential has been extensively calculated for ZnO nanowires, and the fabrication of nanowire-based nanogenerators has seen great success. It is of great interest to know the piezoelectric potential in other nanostructures. In our initial study, we considered a nanowire and a nanohelix with the same side length D = 100 nm and height L = 1200 nm. The nanohelix had five coils and a mean radius of coil of R = 150 nm. A compressing force of F = 100 nN was applied to the nanowire at the top surface along the z-axis, and the same force of F was also applied on the upper cross-section of the nanohelix parallel to the z-axis, such that the nanowire and the nanohelix were compressed. Calculated piezoelectric potentials in the nanowire and the nanohelix are shown in Figure 2.

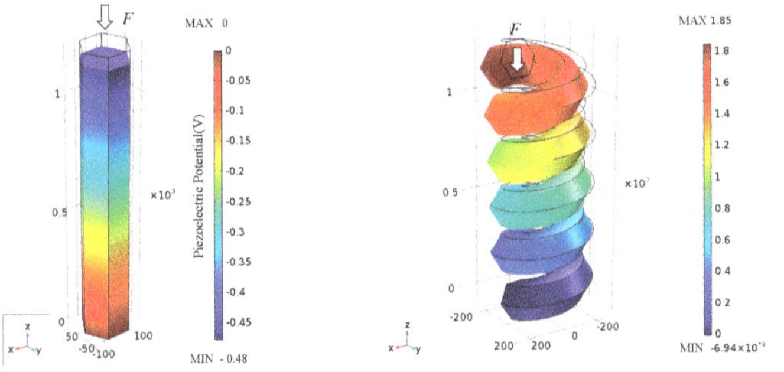

(a) **(b)**

Figure 2. The piezoelectric potential distribution in (**a**) ZnO nanowire and (**b**) nanohelix under a compressing force of 100 nN along the z-axis.

In Figure 2a, the red side is ground and the blue side is the negative potential side. The nanowire in Figure 2a shows a maximum piezoelectric potential of 0.48 V in the nanowire and the top surface shows a maximum displacement of 0.03 nm. A maximum piezoelectric potential around 0.4 V was obtained when we applied an 85 nN compressing force in the same way. This is consistent with a previous work, and lends confidence to our simulation results [9]. The nanohelix has a greatly reduced stiffness due to its helical structure. In Figure 2b, the blue side is ground and the red side is the positive potential side. The maximum displacement of the nanohelix in Figure 2b reached 10.2 nm under the same 100 nN compressing force. The numerical computation revealed a maximum piezoelectric potential of 1.85 V at the top of the nanohelix, which is significantly greater than the piezoelectric potential of 0.48 V found in the nanowire. Owing to the single-crystal structure, the same height ZnO nanohelix can be thought of as a longer ZnO nanowire. By the piezoelectric potential, the piezoelectric field is created through the constructive add-up of the dipole moments created by all units in the crystal. More units will create a higher piezoelectric potential. Therefore, the piezoelectric potential continuously drops from one end of the nanohelix to the other. Meanwhile, numerical calculation of the piezoelectric potential distribution in a ZnO nanohelix at a stretching force of 100 nN was calculated. Note that the stretching force generated the same continuous piezoelectric potential in the ZnO nanohelix with reversed polarity.

This work shows that the nanohelix can produce higher potential than the nanowire when they are exposed to the same force. In addition, the helical structure has a much lower resonant frequency along its central axial direction than that of the nanowire. Consequently, nanohelix-based nanogenerators may perform better than nanowire-based nanogenerators to harvest energy from the environment where low-frequency vibration is more frequently observed.

3.2. *Effect of the Number of Coils and the Mean Radius of the Coil on the Pizeoelectric Potential of a Nanohelix*

A nanohelix is normally formed from a nanowire that grows in a specific crystallographic direction and follows a helical path [1]. Different growth conditions can result in nanohelices of different dimensions. For the research of a greater number of coils, we used a second model, similar to the first but with an increased length of 1900 nm. We thus studied the deformation and the piezoelectric potential produced from different nanohelices with the same length of 1900 nm. We kept the mean coil radius at 150 nm and changed the number of coils from four to 10. The spring pitch as well as the stiffness decreased as the number of coils increased. We applied a compressive force of 100 nN along the z-axis at the top face of the nanohelix. The piezoelectric potential in the nanohelices was calculated using FEM simulation. The maximum potential was found at the top part of the nanohelix in this study, and the potential increased from 0.63 V to 4.01 V as the number of coils increased. The displacement increased from 12.1 nm to 28.2 nm with the increasing number of coils, as shown in Figure 3a. We then kept the number of coils at five and changed the mean coil radius of the nanohelix. Our simulation of nanohelices with different mean coil radius revealed a similar trend. As the mean coil radius increased from 150 nm to 200 nm under the same applied force of 100 nN, the piezoelectric potential in the nanohelix increased from 1.85 V to 2.90 V, while the displacement of the nanohelix increased from 10.02 nm to 19.40 nm, as shown in Figure 3b.

Figure 3a shows that the displacement and the piezoelectric potential increased linearly with the number of coils, as expected. Figure 3b shows that the piezoelectric potential increased linearly with the mean coil radius, while nonlinearity was found in the displacement. Due to the special helical structure, the total displacement of the nanohelix in Figure 3b consists of three components including the displacement in the x-direction, displacement in the y-direction, and displacement in the z-direction. Previous researchers showed that the spring constant K of a linear elastic spring was directly proportional to $1/R^3$ (where R is the coil radius), if the force was applied along the central axis and only the torsion generated from the extension of the spring in the low-strain regime was considered [32,33]. Consequently, when a force F along the z-axis is applied to a typical spring with $d \ll R$, a nonlinear behavior was observed in displacements ($\Delta = F/K \propto R^3$), which is consistent with our result shown in Figure 3b. However, the diameter of the coil wire was comparable to the mean coil radius of the helix in this work, and the location of the force applied was not on the central axis of the helix. The difference in the structure and the force applied in this work compared to an earlier study [32,33] resulted in a small deviation of the behavior shown in Figure 3b from the behavior of a typical spring.

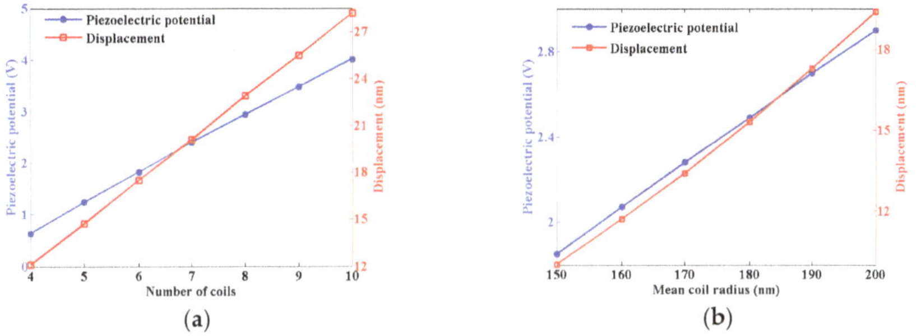

Figure 3. The change of the maximum piezoelectric potential and displacement with the number of coils in (**a**) and the mean coils radius in (**b**) of ZnO nanohelices with a constant length of 1900 nm.

3.3. Effect of Acting Forces on the Pizoelectric Potential and Displacement

Different forces may cause nanowires to be bent, stretched, or compressed. The piezoelectric potential in a bent nanowire was first used in nanogenerators and piezoelectric potential field effect transistors [9,10,13]. Later, the piezoelectric potential in a stretched or compressed nanowire was used for the fabrication of nanogenerators [11,34], tactile imaging [35], sensors [36], and other devices [37]. Nanohelices produced greater piezoelectric potential than nanowires as they were compressed along the axial direction. We further studied the piezoelectric potential in a nanohelix as different forces were applied.

To simplify the calculation, all of the compressing forces were uniformly applied at the upper face of the nanohelix. We calculated the piezoelectric potential of the nanohelix for separate cases where a force of 100 nN was applied along the x-axis, y-axis, or z-axis. We also calculated the piezoelectric potential of the nanohelix under different combinations of these forces. The maximum potential in each case is summarized in Table 2. The force along the z-axis clearly produced higher piezoelectric potential than the same amount force applied in other directions, and the displacement was also at its minimum when the force was only along the z-axis. When a lateral force was applied, the nanohelix was bent significantly. The force in the x-direction pushed the top end towards the central axis of the spring and the produced maximum piezoelectric potential was lower than that produced by a force in the y-axis. When two or more force components were involved, the maximum piezoelectric potential was normally higher with a force along the z-axis included.

Table 2. Maximum piezoelectric potential and displacement under forces in different directions.

Applied Force Components (nN)			Piezoelectric Potential (V)	Displacement (nm)
x-Axis	y-Axis	z-Axis		
0	0	100	1.85	10.2
0	100	0	0.35	48.3
100	0	0	0.29	47.3
100	100	0	0.48	67.8
100	0	100	1.85	37.6
0	100	100	1.60	49.4
100	100	100	1.60	61.6

When a lateral bending force was added to the existing compressing force along the z-axis, the piezoelectric potential distribution was significantly changed. However, the maximum piezoelectric potential in the nanohelix could change in any direction because the maximum and minimum potentials due to each individual force component occurred at different locations. For example, the maximum piezoelectric potential was 1.85 V when a force of 100 nN was applied along the z-axis. The potential remained at 1.85 V as an additional force of 100 nN along the x-axis was added. In comparison, the maximum piezoelectric potential decreased to 1.60 V when a force of 100 nN along the y-axis was added. This result indicates a coupling effect between the three applied force components on the generation of piezoelectric potential.

The displacement of the nanohelix is dependent on its mechanical property and the boundary condition. In our case, the fixed constraint is on the x-z plane. Owing to the stiffness of the nanohelix, the displacement along the x-axis and y-axis are greater than that along the z-axis. In particular, deformation along the y-axis is most prone to deform the spring. Similarly, the coupling effect between the three directions of applied force components on the total displacement cannot be ignored, even though the total displacement is the synthesis of the x-axis, y-axis, and z-axis components.

4. Conclusions

In conclusion, FEM simulation has been used to study the piezoelectric potential in a ZnO nanohelix, and it predicted much higher piezoelectric potential than that in a nanowire with the same length and applied force. Increasing the number of coils or mean coil radius of a nanohelix of a constant height results in a higher maximum piezoelectric potential when the force is kept constant. Both lateral bending force and vertical compressing force can create piezoelectric potentials. Applied forces in different directions have a coupling effect on the piezoelectric potential. A force along the z-axis produces a higher maximum piezoelectric potential that favors energy harvesters and other piezotronic devices. Adding a lateral force to the existing vertical force can change the distribution of the piezoelectric potential while the maximum potential may not be greatly changed. Meanwhile, the mechanical property of a ZnO nanohelix is also studied. The force component on the y-axis showed the biggest effect on displacement. This work not only demonstrated a new and excellent candidate for nanogenerators and other electronic devices, but it is also expected to lead to new exploration of other piezoelectric nanostructures.

Acknowledgments: The authors are grateful for financial support from the Department of Mechanical Engineering and the College of Science and Engineering of the University of Minnesota. Research is supported in part by NSF (ECCS-1150147) and by NSF IGERT grant DGE-1069104. This work was also supported by Shanxi Province Key Research and Development Program (International Cooperation) (Grant No. 201603D421009).

Author Contributions: Huimin Hao, Kory Jenkins and Rusen Yang conceived the project and decided the research plan in this study. Humin Hao and Yiqian Xu performed numerical simulations. Humin Hao and Rusen Yang carried out the data analysis. Huimin Hao, Kory Jenkins and Rusen Yang wrote the manuscript. All authors discussed the results and commented on the manuscript.

Conflicts of Interest: The authors declare no conflict of interest.

References

1. Yang, R.S.; Ding, Y.; Wang, Z.L. Deformation-free single-crystal nanohelixes of polar nanowires. *Nano Lett.* **2004**, *4*, 1309–1312. [CrossRef]
2. Gao, P.X.; Mai, W.; Wang, Z.L. Superelasticity and nanofracture mechanics of zno nanohelices. *Nano Lett.* **2006**, *6*, 2536–2543. [CrossRef] [PubMed]
3. Gao, H.; Zhang, X.T.; Zhou, M.Y.; Zhang, Z.G.; Wang, X.Z. Growth of novel zno nanohelices modified by sio 2-sheathed zno discs. *Nanotechnology* **2007**, *18*, 065601. [CrossRef]
4. Kong, X.Y.; Ding, Y.; Yang, R.; Wang, Z.L. Single-crystal nanorings formed by epitaxial self-coiling of polar nanobelts. *Science* **2004**, *303*, 1348–1351. [CrossRef] [PubMed]
5. Kong, X.Y.; Wang, Z.L. Spontaneous polarization-induced nanohelixes, nanosprings, and nanorings of piezoelectric nanobelts. *Nano Lett.* **2003**, *3*, 1625–1631. [CrossRef]

6. Gao, P.X.; Wang, Z.L. High-yield synthesis of single-crystal nanosprings of zno. *Small* **2005**, *1*, 945–949. [CrossRef] [PubMed]
7. Yang, R.; Wang, Z.L. Springs, rings, and spirals of rutile-structured tin oxide nanobelts. *J. Am. Chem. Soc.* **2006**, *128*, 1466–1467. [CrossRef] [PubMed]
8. Huang, T.; Liu, Z.; Huang, G.; Liu, R.; Mei, Y. Grating-structured metallic microsprings. *Nanoscale* **2014**, *6*, 9428–9435. [CrossRef] [PubMed]
9. Wang, Z.L.; Song, J.H. Piezoelectric nanogenerators based on zinc oxide nanowire arrays. *Science* **2006**, *312*, 242–246. [CrossRef] [PubMed]
10. Wang, X.D.; Song, J.H.; Liu, J.; Wang, Z.L. Direct-current nanogenerator driven by ultrasonic waves. *Science* **2007**, *316*, 102–105. [CrossRef] [PubMed]
11. Yang, R.; Qin, Y.; Dai, L.; Wang, Z.L. Power generation with laterally packaged piezoelectric fine wires. *Nat. Nanotechnol.* **2009**, *4*, 34–39. [CrossRef] [PubMed]
12. Zhu, R.; Zhang, W.G.; Yang, R.S. High output piezoelectric nanogenerator: Development and application. *Sci. Adv. Mater.* **2012**, *4*, 798–804. [CrossRef]
13. Wang, X.D.; Zhou, J.; Song, J.H.; Liu, J.; Xu, N.S.; Wang, Z.L. Piezoelectric field effect transistor and nanoforce sensor based on a single zno nanowire. *Nano Lett.* **2006**, *6*, 2768–2772. [CrossRef] [PubMed]
14. Kory, J.; Rusen, Y. Mechanical transfer of zno nanowires for a flexible and conformal piezotronic strain sensor. *Semicond. Sci. Technol.* **2017**, *32*, 074004.
15. Zhou, J.; Gu, Y.; Fei, P.; Mai, W.; Gao, Y.; Yang, R.; Bao, G.; Wang, Z.L. Flexible piezotronic strain sensor. *Nano Lett.* **2008**, *8*, 3035–3040. [CrossRef] [PubMed]
16. Wu, W.Z.; Wang, Z.L. Piezotronic nanowire-based resistive switches as programmable electromechanical memories. *Nano Lett.* **2011**, *11*, 2779–2785. [CrossRef] [PubMed]
17. Wu, W.Z.; Wei, Y.G.; Wang, Z.L. Strain-gated piezotronic logic nanodevices. *Adv. Mater.* **2010**, *22*, 4711–4715. [CrossRef] [PubMed]
18. Gao, Y.; Wang, Z.L. Electrostatic potential in a bent piezoelectric nanowire. The fundamental theory of nanogenerator and nanopiezotronics. *Nano Lett.* **2007**, *7*, 2499–2505. [CrossRef] [PubMed]
19. Gao, Z.; Zhou, J.; Gu, Y.; Fei, P.; Hao, Y.; Bao, G.; Wang, Z.L. Effects of piezoelectric potential on the transport characteristics of metal-zno nanowire-metal field effect transistor. *J. Appl. Phys.* **2009**, *105*, 113707. [CrossRef] [PubMed]
20. Sun, C.L.; Shi, J.A.; Wang, X.D. Fundamental study of mechanical energy harvesting using piezoelectric nanostructures. *J. Appl. Phys.* **2010**, *108*, 034309. [CrossRef]
21. Chen, X.Q.; Zhang, S.L.; Dikin, D.A.; Ding, W.Q.; Ruoff, R.S.; Pan, L.J.; Nakayama, Y. Mechanics of a carbon nanocoil. *Nano Lett.* **2003**, *3*, 1299–1304. [CrossRef]
22. Wang, J.-S.; Feng, X.-Q.; Wang, G.-F.; Yu, S.-W. Twisting of nanowires induced by anisotropic surface stresses. *Appl. Phys. Lett.* **2008**, *92*, 191901. [CrossRef]
23. Da Fonseca, A.F.; Galvão, D.S. Mechanical properties of nanosprings. *Phys. Rev. Lett.* **2004**, *92*, 175502. [CrossRef] [PubMed]
24. Smith, D.; Mailhiot, C. Theory of semiconductor superlattice electronic structure. *Rev. Modern Phys.* **1990**, *62*, 173. [CrossRef]
25. Clarke, D.R. Varistor ceramics. *J. Am. Ceram. Soc.* **1999**, *82*, 485–502. [CrossRef]
26. Hwang, G.; Hashimoto, H.; Bell, D.J.; Dong, L.; Nelson, B.J.; Schön, S. Piezoresistive ingaas/gaas nanosprings with metal connectors. *Nano Lett.* **2009**, *9*, 554–561. [CrossRef] [PubMed]
27. Zhang, Z.-Y.; Zhao, Y.-P. Optical properties of helical Ag nanostructures calculated by discrete dipole approximation method. *Appl. Phys. Lett.* **2007**, *90*, 221501. [CrossRef]
28. Zhang, Z.-Y.; Zhao, Y.-P. Optical properties of helical and multiring Ag nanostructures: The effect of pitch height. *J. Appl. Phys.* **2008**, *104*, 013517. [CrossRef]
29. Kamata, K.; Suzuki, S.; Ohtsuka, M.; Nakagawa, M.; Iyoda, T.; Yamada, A. Fabrication of left-handed metal microcoil from spiral vessel of vascular plant. *Adv. Mater.* **2011**, *23*, 5509–5513. [CrossRef] [PubMed]
30. Gao, Y.; Wang, Z.L. Equilibrium potential of free charge carriers in a bent piezoelectric semiconductive nanowire. *Nano Lett.* **2009**, *9*, 1103–1110. [CrossRef] [PubMed]
31. Wang, Z.L. Progress in piezotronics and piezo-phototronics. *Adv. Mater.* **2012**, *24*, 4632–4646. [CrossRef] [PubMed]
32. Granet, I. *Strength of Materials for Engineering Technology*, 2nd ed.; Reston Pub. Co.: Reston, VA, USA, 1980.

33. Wahl, A. *Mechanical Springs*; McGraw-Hill: New York, NY, USA, 1963; p. 119.
34. Nguyen, V.; Zhu, R.; Jenkins, K.; Yang, R. Self-assembly of diphenylalanine peptide with controlled polarization for power generation. *Nat. Commun.* **2016**, *7*, 13566. [CrossRef] [PubMed]
35. Wu, W.Z.; Wen, X.N.; Wang, Z.L. Taxel-addressable matrix of vertical-nanowire piezotronic transistors for active and adaptive tactile imaging. *Science* **2013**, *340*, 952–957. [CrossRef] [PubMed]
36. Jenkins, K.; Nguyen, V.; Zhu, R.; Yang, R. Piezotronic effect: An emerging mechanism for sensing applications. *Sensors* **2015**, *15*, 22914–22940. [CrossRef] [PubMed]
37. Wang, Z.L.; Wu, W.Z. Piezotronics and piezo-phototronics: Fundamentals and applications. *Natl. Sci. Rev.* **2014**, *1*, 62–90. [CrossRef]

nanomaterials

MDPI

Article

Preparation of Nano-TiO_2-Coated SiO_2 Microsphere Composite Material and Evaluation of Its Self-Cleaning Property

Sijia Sun, Tongrong Deng, Hao Ding *, Ying Chen and Wanting Chen

Beijing Key Laboratory of Materials Utilization of Nonmetallic Minerals and Solid Wastes, National Laboratory of Mineral Materials, School of Materials Science and Technology, China University of Geosciences, Xueyuan Road, Haidian District, Beijing 100083, China; ssjcugb@163.com (S.S.); 18010157480@163.com (T.D.); chenying@cugb.edu.cn (Y.C.); wantingchen123@163.com (W.C.)

* Correspondence: dinghao@cugb.edu.cn; Tel.: +86-010-82322982

Received: 28 September 2017; Accepted: 30 October 2017; Published: 3 November 2017

Abstract: In order to improve the dispersion of nano-TiO_2 particles and enhance its self-cleaning properties, including photocatalytic degradation of pollutants and surface hydrophilicity, we prepared nano-TiO_2-coated SiO_2 microsphere composite self-cleaning materials (SiO_2–TiO_2) by co-grinding SiO_2 microspheres and TiO_2 soliquid and calcining the ground product. The structure, morphology, and self-cleaning properties of the SiO_2–TiO_2 were characterized. The characterization results showed that the degradation efficiency of methyl orange by SiO_2–TiO_2 was 97%, which was significantly higher than that obtained by pure nano-TiO_2. The minimum water contact angle of SiO_2–TiO_2 was 8°, indicating strong hydrophilicity and the good self-cleaning effect. The as-prepared SiO_2–TiO_2 was characterized by the nano-TiO_2 particles uniformly coated on the SiO_2 microspheres and distributed in the gap among the microspheres. The nano-TiO_2 particles were in an anatase phase with the particle size of 15–20 nm. The nano-TiO_2 particles were combined with SiO_2 microspheres via the dehydroxylation of hydroxyl groups on their surfaces.

Keywords: SiO_2; nano-TiO_2; self-cleaning; hydrophilicity

---

1. Introduction

Nano-titanium dioxide (TiO_2) is a typical semiconductor material with excellent properties. Moreover, it is stable, cheap, and non-toxic [1,2]. Therefore, it has been widely applied in the environmental protection [3], energy [4], and other fields [5,6]. In addition to the photocatalytic activity of TiO_2 under ultraviolet (UV) irradiation, the self-cleaning effect due to photoinduced hydrophilic properties of TiO_2 has always been one of the hotspots [7,8]. Its self-cleaning mechanism is generally ascribed to two effects [9,10]. Firstly, under the irradiation of ultraviolet light or ultraviolet in sunlight, the active components induced by the photocatalytic action of TiO_2 on the TiO_2 self-cleaning film can react with the pollutants adhering to the surface, thus achieving the decomposition of pollutants. Secondly, due to the super-hydrophilicity of the self-cleaning film, the decomposed products can be washed away by rain, so as to maintain the clean material surface [11]. In China and other developing countries, the contents of dust and oily dirt are high in the urban atmosphere and dust and oily dirt tend to adhere to building walls and glass surface to make the surface dirty. Nano-TiO_2 self-cleaning materials may be used to coat such surfaces [12,13].

However, some factors restrict the application scope of nano-TiO_2 self-cleaning materials. For example, the agglomeration phenomenon and poor dispersivity of TiO_2 particles in the application system significantly reduces its self-cleaning effect [14,15]. Coating TiO_2 particles on the matrix surface can significantly improve the dispersibility of TiO_2 particles and enhance the photocatalytic efficiency and self-cleaning performance under the synergistic effect of the matrix [16,17]. In this way, the aforementioned problems may be solved. Many silicon materials are used as substrates to prepare nano-TiO_2 coated composite catalysts, such as quartz tube [18], glass fibers [19], and nano-silica [20]. These catalysts all exhibit the good photocatalytic activity with different functional characteristics. Meanwhile, the micro-nano-morphology of the carrier-based nanoparticles, which are constructed from the surface of the composite self-cleaning material, can also increase the roughness of the self-cleaning film and further improve the super-hydrophilicor super-hydrophobic properties [21,22] Prabhu [23] prepared the reduced graphene oxide (rGO)/TiO_2 composite self-cleaning material according to the solvothermal method and improved the visible light absorption efficiency of the composite self-cleaning materials, which exhibited the good photocatalytic efficiency and super-hydrophilic performance under light irradiation. Zhou [24] added the prepared SiO_2–TiO_2 composite colloidal particles into the fluorocarbon coating and realized more stable self-cleaning performance than that of adding single nano-TiO_2 particles under ultraviolet light irradiation, thus suggesting its possible industrial application in outdoor environments. Zhang [25] and Ciprian [26] prepared SiO_2–TiO_2 composite films by the sol-gel impregnation and freeze-drying deposition method and realized the excellent self-cleaning performance and high transmittance to visible light. In general, the abovementioned preparation methods of nano-TiO_2 composite self-cleaning material have some problems, such as the high cost, the complicated process and the difficulty in large-scale production and application. Therefore, it is necessary to select cheap matrix materials and simple composite process. Surolia [27] prepared the TiO_2-coated fly ash photocatalyst via the sol-gel method with the cheap fly ash as substrate, exhibiting well photocatalysis degradation performance. Therefore, it is an effective way to improve the efficiency of resource utilization by using natural mineral or industrial by-product as substrate to prepare nano-TiO_2-coated photocatalytic material.

In this study, with SiO_2 microspheres as the matrix, nano-TiO_2-coated SiO_2 microsphere composite self-cleaning materials (SiO_2–TiO_2) were prepared by the wet grinding of SiO_2 microspheres and nano-TiO_2 soliquid and the subsequent calcination of the ground product. Then, we determined the photocatalytic activity and photoinduced hydrophilicity of SiO_2–TiO_2, analyzed the structure and morphology, and discussed the mechanism of the interaction between TiO_2 and SiO_2 particles. The SiO_2 microspheres used in this study were recovered from the by-product, silica fume, which was produced during the industrial production of fused zirconia. The SiO_2 microspheres mainly exist in the amorphous phase and have regular morphology, high surface activity, and low cost [28,29]. However, during the past years, silica fume was usually applied in cement, concrete and refractory products as an additive and its use efficiency was low [30,31]. To the best of our knowledge, the preparation of functional materials including composite photocatalytic materials with SiO_2 microsphere as a matrix was seldom reported. In the study, the spherical shape of the SiO_2 microspheres can increase the fluidity of SiO_2–TiO_2 and promote the film formation process and the micrometer size of the SiO_2 can improve the recyclability of nano-TiO_2. It is expected that the SiO_2 microspheres can exert a synergistic effect on the performance of SiO_2–TiO_2 and reduce the cost of composite self-cleaning materials [25]. Meanwhile, the mechanical-chemical grinding method used in this study is a simple and non-pollution particle compound method. We prepared the SiO_2–TiO_2 composite materials with the good photocatalysis activity and self-cleaning effect via a simple composite process with cheap matrix materials. The preparation process exhibits significant economic and environmental values.

2. Methods

2.1. Raw Materials and Reagents

The SiO_2 microspheres used in this study were recovered from the by-product, silica fume, which was produced during the industrial production of fused zirconia and was provided by a zirconia production enterprise in Jiaozuo (Jiaozuo, China). The main chemical constituents (mass fraction, %)

of SiO_2 microspheres were 93.78% SiO_2 and 4.96% ZrO_2. SiO_2 is mainly composed of amorphous phase, exhibiting the microsphere morphology with the particle size of 1–3 μm. The SiO_2 particles are aggregated to form the aggregates with the larger particle size. After depolymerizing the aggregates, the SiO_2 microspheres exist in a dispersed state.

Tetrabutyl titanate ($C_{16}H_{36}O_4Ti$) from Beijing Chemical Industry Group Co., Ltd. (Beijing, China) was used as the titanium source. Acetylacetone ($C_5H_8O_2$) supplied by Xi Long Chemical Co., Ltd. (Guangzhou, China) was used as a hydrolysis control agent. Methyl orange ($C_{14}H_{14}N_3SO_3Na$) from Beijing Chemical Industry Group Co., Ltd. (Beijing, China) was used as a target pollution for photocatalytic degradation. Ethanol and deionized water are also used as solvents throughout the preparation process.

2.2. *Preparation Method*

2.2.1. Depolymerization of SiO_2 Microspheres

Considering the agglomeration effect of particles in the raw SiO_2 microspheres, SiO_2 microspheres need to be depolymerized and dispersed before compositing with nano-TiO_2. The depolymerization method was described as follows: The SiO_2 microsphere materials were added into the ethanol solution to form a suspension. After adding ceramic grinding balls (the ratio of ball to material, 3:1), the suspension was then ground in the mixing mill (CSDM-S3, Beijing Paleozoic Powder Technology Co., Ltd., Beijing, China) for 60 min. Finally, the dispersed SiO_2 microspheres were obtained after ball-material separation, filtration, and desiccation.

2.2.2. Preparation of Nano-TiO_2 Soliquid

Firstly, 8.5 mL of tetrabutyl titanate was dissolved into 10 mL of ethanol solution. The mixed solution was stirred evenly and marked as Solution A. Then, 1.3 mL of acetylacetone was dissolved into 10 mL of ethanol solution, and the obtained solution was marked as Solution B. Then, Solution B was slowly added into Solution A and 19.35 mL of the mixture of ethanol and water (water 0.85 mL) was also added into Solution A. Afterwards, the mixture was stirred vigorously at room temperature for 12 h and the stirred mixture was aged for 48 h to obtain the nano-TiO_2 soliquid. The viscosity of the nano-TiO_2 soliquid obtained after 48-h aging was measured to be 2×10^{-3} Pa·s by a digital display viscometer (NDJ-8S, Shanghai Precision Instrument and Meter Co., Ltd., Shanghai, China). For comparison, partial nano-TiO_2 soliquid was dried and calcined to prepare TiO_2 nanoparticles. According to the X-ray diffraction (XRD) data and the Scherrer Equation, the grain size of nano-TiO_2 was calculated to be 15–20 nm.

2.2.3. Preparation of SiO_2–TiO_2

Firstly, the dispersed SiO_2 microspheres were added into the ethanol solution, which was stirred to form a suspension. Secondly, the suspension was added into the aged nano-TiO_2 soliquid to form the SiO_2/TiO_2 mixture. Thirdly, the SiO_2/TiO_2 mixture were stirred by a CSDM-S3 mixing mill (Beijing Gosdel Powder&Technology Co., Ltd., Beijing, China) for 90 min after the addition of a certain amount of grinding balls to obtain the SiO_2/TiO_2 soliquid composites. Then, the SiO_2/TiO_2 soliquid composites were put in a SRJX-5-13 chamber electric furnace (Tianjin Taisite Instrument Co., LTD, Tianjin, China) and calcined at 500 °C for 2 h. Finally, the SiO_2–TiO_2 was prepared.

2.3. *Characterization*

2.3.1. Evaluation of Self-Cleaning Performance

Photocatalytic Activity

The photocatalytic degradation performance of SiO_2–TiO_2 was tested with the methyl orange as the target degradation pollutant. The system was irradiated by a mercury lamp (100 W, the main wavelength of 254 nm). Then, 40 mg of SiO_2–TiO_2 was added to 50 mL of prepared methyl orange dilution (concentration 10 mg/L). In order to reduce the measurement error caused by sample adsorption, the dark reaction was carried out for 0.5 h and then the concentration of methyl orange (C_0) in the solution was measured. After turning on the light source, the concentration of methyl orange (C) in solution was measured every 20 min. The photocatalytic degradation performance of the samples was characterized and evaluated based on the change of C/C_0.

The concentration of methyl orange was measured according to the following procedure. Firstly, the solution was centrifuged and the absorbance of the supernatant was measured with a Cary 5000 UV–VIS spectrophotometer (USA Varian, Palo Alto, CA, USA). The concentration of methyl orange in the solution was calculated according to the relationship between absorbance and concentration.

Hydrophilicity

The hydrophilicity of the SiO_2–TiO_2 particles was characterized based on the wetting degree of water on its surface. The wetting degree was reflected by the measured water contact angle on its surface. The SiO_2–TiO_2 composite powder was pressed into a sheet-like sample by a tableting machine and then the water contact angle was measured by a contact angle meter (JC2000D, Shanghai Zhongchen Digital Technic Apparatus Co. Ltd., Shanghai, China) three times. The measurement results were averaged.

2.3.2. Characterization of Structure and Morphology

We observed the morphology of SiO_2–TiO_2 by scanning electron microscope (SEM) (S-3500N, Hitachi, Ltd., Tokyo, Japan) and transmission electron microscope (TEM) (FEI Tecnai G2 F20, Portland, OR, USA). The surface functional groups were examined by an infrared spectroscope (Spectrum 100, PerkinElmer Instruments (Shanghai) Co., Ltd., Shanghai, China) with KBr as the medium, and the weights of each sample and KBr were, respectively, 1 and 200 mg. The phase analysis was carried out with an X-ray diffractometer (D/MAX2000, Rigaku Corporation, Tokyo, Japan).The specific surface areas of SiO_2 and SiO_2–TiO_2 were tested by the QuadraSorb SI specific surface area analyzer (Quantachrome Instrument Company, Boynton Beach, FL, USA). In addition, the surface roughness of SiO_2 microspheres and SiO_2–TiO_2 were evaluated using a Mutimode VIII atomic force microscope (Bruke, Fremont, CA, USA).

3. Results and Discussion

3.1. *Properties of* SiO_2–TiO_2

3.1.1. Photocatalytic Properties of SiO_2–TiO_2

Figure 1a represents the degradation behaviors of methyl orange dye during irradiation as a function of time (min) in the presence of SiO_2–TiO_2 with different TiO_2 ratios (the mass ratio of TiO_2 to SiO_2–TiO_2). As shown in Figure 1a, the SiO_2 microspheres exhibit no degradation effect on methyl orange, whereas pure TiO_2 has a certain degradation effect on methyl orange. All of the prepared SiO_2–TiO_2 materials exhibit the significantly higher photocatalytic degradation efficiency on methyl orange dye than that of pure nano-TiO_2. Among these SiO_2–TiO_2 samples, with SiO_2–TiO_2-40 (TiO_2 ratio is 40%) as the photocatalyst, after the solution was irradiated for 40 min, the C/Co was reduced to about 0.1 and the degradation efficiency reached 90%. After the 120 min irradiation, the degradation efficiency reached 97%. With the pure nano-TiO_2 as the photocatalyst, the degradation efficiencies after 40 and 120 min respectively reached 50% and 90%. The abovementioned results indicated that the photocatalytic activity of nano-TiO_2 had been greatly improved when TiO_2 coated the surface of SiO_2 microspheres. In addition, the TiO_2 ratio had a significant effect on the degradation efficiency of SiO_2–TiO_2. With the increase in the TiO_2 ratio from 20% to 40%, the photocatalytic degradation efficiency gradually increased and finally reached its maximum value. When the mass ratio of TiO_2 increased to 50%, the degradation efficiency decreased. However, the degradation efficiency of SiO_2–TiO_2 with different TiO_2 ratios was always higher than that of pure nano-TiO_2. The phenomenon might be interpreted in two aspects: Firstly, the coating of nano-TiO_2 on SiO_2 microsphere surface could improve the dispersibility of nano-TiO_2, thus resulting in an increase in the number of reactive groups under irradiation and increasing the quantum efficiency. Secondly, SiO_2 had a high reflection efficiency on ultraviolet radiation, and the light reflected by SiO_2 could be absorbed by TiO_2, thus improving the absorption of ultraviolet light by SiO_2–TiO_2. The specific surface area analysis results showed that the surface area of SiO_2 had been significantly incresed from its original value of 5.698 to 44.410 m^2/g after TiO_2 coating. This result also comfirmed that the SiO_2 microspheres had been coated by nano TiO_2 effectively.Figure 1b shows the influence of the ratio of grinding ball to materials (B-M) in the grinding process on the photocatalytic activity of SiO_2–TiO_2. The degradation efficiency of SiO_2–TiO_2 samples prepared with grinding balls was significantly higher than that of the SiO_2–TiO_2 prepared without grinding balls (B-M is 0). The degradation effect was the best when the B-M ratio was 5. After 120 min irradiation, the highest degradation efficiency was 95% (C/C_0 = 0.05) at the B-M ratio of 5% and 65% at the B-M of 0. The above results showed that the grinding process had an important effect on the performance of SiO_2–TiO_2. Therefore, the proper B-M ratio should be selected. As shown in Figure 1b, the degradation effect of SiO_2–TiO_2 is stronger than that of pure nano-TiO_2. The result is consistent with the results shown in Figure 1a.

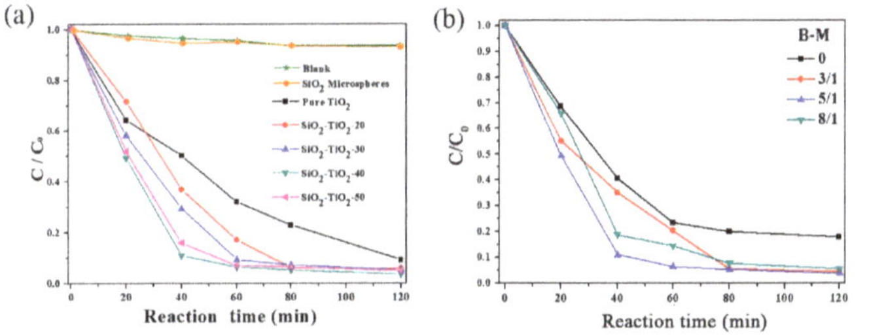

Figure 1. Influences of (**a**) TiO_2 ratio and (**b**) B-M ratio on the photocatalytic performance of SiO_2–TiO_2. (**a**) SiO_2–TiO_2-20, 30, 40, 50 represent the mass ratio of TiO_2 to SiO_2–TiO_2 is 20%, 30%, 40% and 50%; and (**b**) B-M represents the mass ratio of grinding balls to the materials.

The UV–VIS absorption spectra of bare SiO_2 microspheres, nano-TiO_2, and SiO_2–TiO_2-50 were obtained for comparison (Figure 2). The light absorption of SiO_2 in a wavelength range between 300 and 400 nm was insignificant, whereas TiO_2 absorbed light with the wavelength below 400 nm. The SiO_2–TiO_2 exhibited the higher light absorption in a wavelength range from 200 to 400 nm than that of pure nano-TiO_2, which was completely different from bare SiO_2 microspheres. The results indicated that the SiO_2–TiO_2 had the higher UV absorption due to the high reflection efficiency on ultraviolet radiation by SiO_2 microspheres, confirming that SiO_2 microspheres were coated by nano-TiO_2 particles with similar light absorption properties to TiO_2. Meanwhile, this results contribute to the good photocatalytic activity of SiO_2–TiO_2.

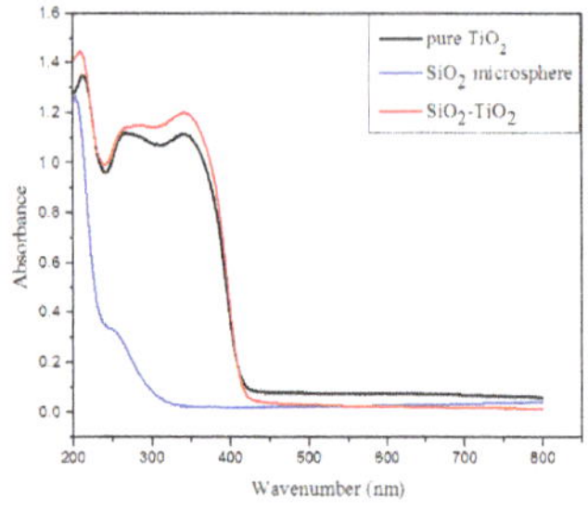

Figure 2. UV–VIS absorption spectra of pure TiO_2, SiO_2 microsphere and SiO_2–TiO_2.

3.1.2. Hydrophilic Properties of SiO_2–TiO_2

Figure 3 shows the change of water contact angle of SiO_2–TiO_2 particles with different TiO_2 ratios after irradiation by ultraviolet light for 2 h. For the SiO_2 microsphere materials, the contact angle was maintained to be 28° after UV irradiation, indicating that the UV light had no effect on its hydrophilicity. The water contact angle of pure TiO_2 is 26° before UV irradiation, which is higher than that of SiO_2–TiO_2, indicating that the coating of TiO_2 on SiO_2 surface can improve the hydrophilicity of TiO_2. The improvement effect may be interpreted as follows. The dispersion of nano-TiO_2 was improved and then more active hydroxyl groups on TiO_2 surface were exposed. Meanwhile, the water contact angle of pure TiO_2 decreased from 26° to 10° after UV irradiation, indicating the photoinduced hydrophilicity of TiO_2. The water contact angle of SiO_2–TiO_2 was 15–18° and decreased to 8–13° after UV irradiation, showing the strong hydrophilicity. The SiO_2–TiO_2-40 (TiO_2 ratio is 40) showed the strongest hydrophilicity and its water contact angles were 17° and 8° before and after UV irradiation respectively. The strong photo-induced hydrophilicity and photocatalytic activity of SiO_2–TiO_2 indicate its good self-cleaning performance.

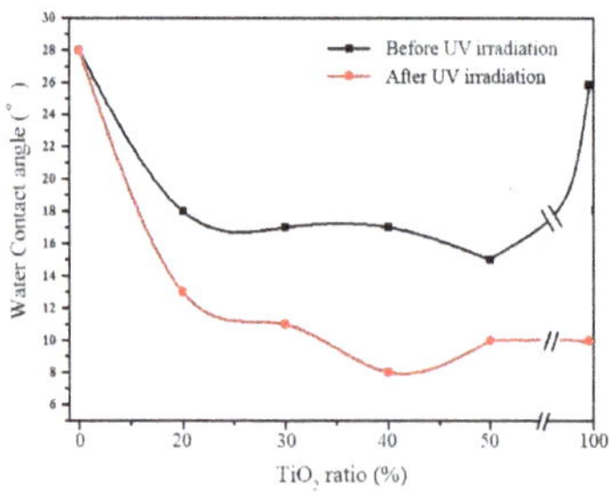

Figure 3. Relationship between the water contact angle and the content of TiO_2.

To investigate the mechanism of the photoinduced hydrophilicity of SiO_2–TiO_2, the infrared spectral analysis was carried out. Figure 4 shows the Fourier transform infrared spectroscopy (FT-IR) spectra of SiO_2–TiO_2-20 and SiO_2–TiO_2-30 before and after UV irradiation. The characteristic absorption peaks in the range of 2800–3800 cm^{-1} and 1620 cm^{-1} in all the samples were ascribed to the vibration of the hydroxyl groups on the SiO_2–TiO_2 surface. When the TiO_2 ratio was 30%, after the UV irradiation (b2 in Figure 4), the intensity of the absorption peak in the range of 2800–3800 cm^{-1} in the FTIR spectrum of SiO_2–TiO_2 was higher than that in the spectrum b1 (before the UV irradiation) and the peak was shifted to the higher wavenumber. Meanwhile, the absorption peak at 1620 cm^{-1} in b2 was sharper than that in b1. The abovementioned results indicated that the number of hydroxyl groups on the surface of SiO_2–TiO_2 increased after UV irradiation and that the SiO_2–TiO_2 exhibited the reaction activity with water. We believed that the production of hydroxyl groups was induced by the photoinduced action of TiO_2. The change was consistent with the remarkable enhancement of the surface hydrophilicity of SiO_2–TiO_2 after UV irradiation in Figure 3.

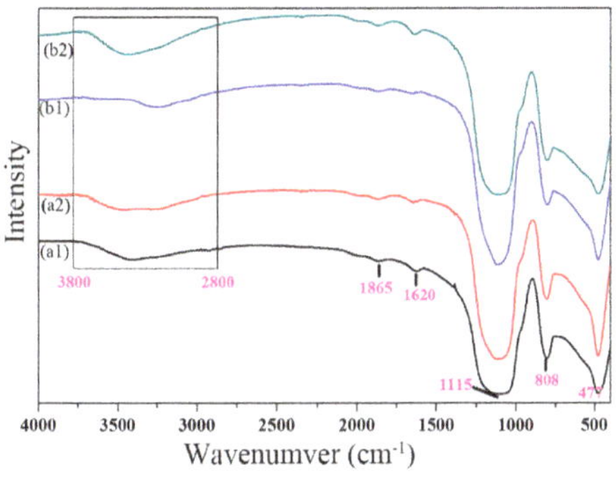

Figure 4. Fourier transform infrared spectroscopy (FT-IR) spectrum of the SiO_2–TiO_2 with different TiO_2 ratios. (**a1**) SiO_2–TiO_2-20, before UV irradiation; (**a2**) SiO_2–TiO_2-20, after UV irradiation; (**b1**) SiO_2–TiO_2-30, before UV irradiation; and (**b2**) SiO_2–TiO_2-30, after UV irradiation; The black rectangle region represents the absorption bands caused by the vibration of the hydroxyl radical

3.2. *Structure and Morphology of SiO_2–TiO_2*

3.2.1. XRD Analysis

Figure 5 shows the XRD patterns of SiO_2–TiO_2 with different TiO_2 ratios. In addition to the diffraction peak of amorphous SiO_2 microspheres, the diffraction peaks of the anatase phase also appeared in the XRD patterns of all SiO_2–TiO_2 samples, and the intensity of diffraction peaks of the anatase phase increased with the increase in the TiO_2 ratio. Especially, when the TiO_2 ratio was 50%, the complete anatase diffraction peak (JCPDS 21-1272) appeared in the XRD pattern of SiO_2–TiO_2-50 (Figure 5c) [32]. The abovementioned results indicated that nano-TiO_2 existed as an anatase phase. Among all the TiO_2 crystal phases, the anatase exhibited the highest photocatalytic activity, which was consistent with the results of photocatalytic activity and photoinduced hydrophilicity of SiO_2–TiO_2.

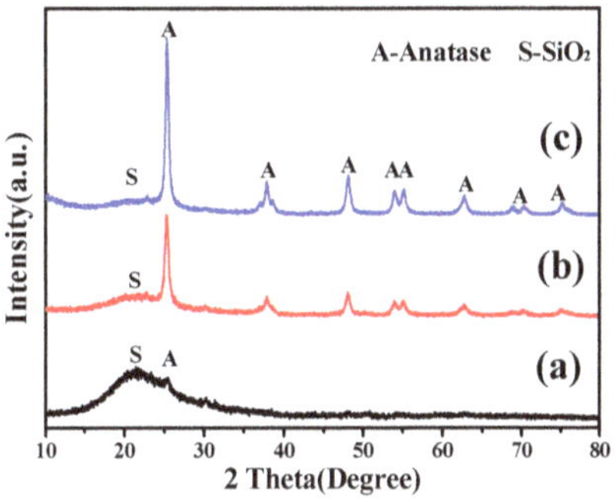

Figure 5. XRD patterns of SiO_2–TiO_2 with different TiO_2 ratios. (**a**) 30% TiO_2; (**b**) 40% TiO_2; and (**c**) 50% TiO_2.

3.2.2. Morphology and Element Analysis

Figure 6 shows the SEM images of SiO_2–TiO_2 with different TiO_2 ratios. In Figure 6a, the exposed surfaces of SiO_2 microspheres are smooth without covering. However, the micron-submicron hierarchical structure morphology can be observed in Figure 6b–d. The surface of the SiO_2 microspheres became rough and was covered with a certain amount of irregular particles. Meanwhile, with the increase in the TiO_2 ratio, the roughness and coverage area of the SiO_2 microsphere surface increased accordingly. According to the preparation process, it was presumed that the coating on the surface of the microspheres should be nano-TiO_2 particles. The surface roughness of SiO_2 microspheres and SiO_2–TiO_2-50 were evaluated using an atomic force microscope, and the corresponding atomic force microscope (AFM) images were shown in Figure 6a,d (see the built-in images). The tested surface roughness of SiO_2 microspheres and SiO_2–TiO_2 were 1.63 and 18.4 nm, respectively. These results show that the surface roughness of SiO_2 increased significantly after it was coated by nano-TiO_2, indicating that the surface structure of SiO_2 has changed. Additionally, in the magnification image of SiO_2–TiO_2 shown in Figure 6b, the nano-TiO_2 particles not only uniformly coated the surface of the SiO_2 microspheres, but also exist in the gap among SiO_2 microspheres. In this way, several microspheres were connected together as a whole.

Figure 6. Scanning electron microscope (SEM) and atomic force microscope (AFM) images of (**a**) SiO_2 microsphere and (**b–d**) SiO_2–TiO_2 with different ratios. (**b**) 30% TiO_2, and the inset image is a high magnification image; (**c**) 40% TiO_2; and (**d**) 50% TiO_2; the inset images in (**a**,**d**) are AFM images.

To confirm the composition of the coating on the surface of SiO_2 microsphere, a surface scanning analysis of the main elements in the selected part of the SiO_2–TiO_2 SEM was carried out (Figure 7). The Ti element was almost distributed throughout the scan area, like the distribution of Si element. The distribution density of Ti element is proportional to the TiO_2 ratio. This confirmed that the nano-TiO_2 particles had coated the surface and were distributed in the gap among SiO_2 microsphere. The results were consistent with SEM results (Figure 6).

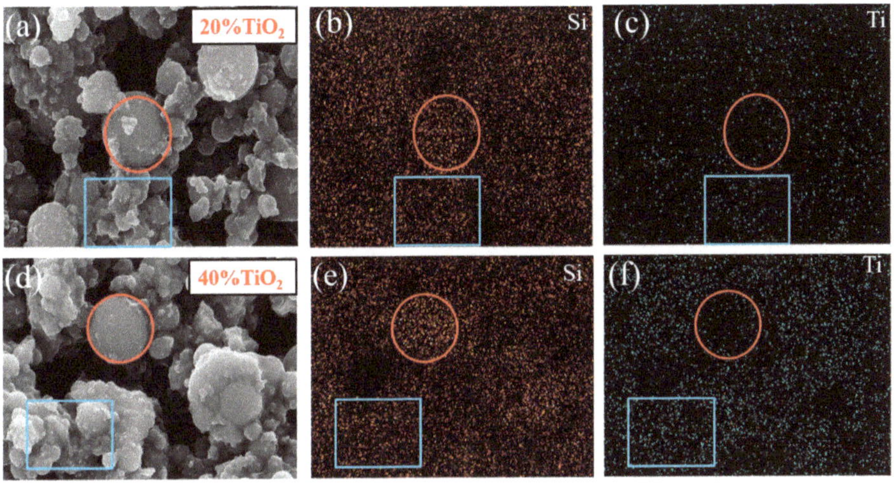

Figure 7. Scanning results of surface elements of SiO_2–TiO_2 with (**a**–**c**) 20% TiO_2 and (**d**–**f**) 40% TiO_2.

Figure 8 shows the TEM and high resolution transmission electron microscopy (HRTEM) images of the SiO_2–TiO_2 samples (TiO_2 ratio is 40%). Circular SiO_2 microspheres and irregular nano-TiO_2 particles surrounding the SiO_2 microspheres are observed in Figure 8a, confirming that the nano-TiO_2 particles has coated the surface of SiO_2 microspheres. In the HRTEM (Figure 8c), the interplanar spacing of the three major facets were measured to be d = 0.352 nm [33], which was consistent with the (101) crystal face of anatase (JCPDS 21-1272). The above results indicated that the nano-TiO_2 coating on the surface of SiO_2 microspheres was anatase and that the mainly exposed crystal face was (101).

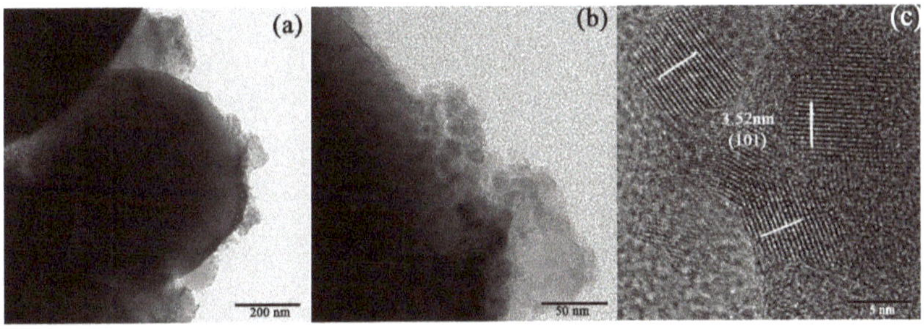

Figure 8. (**a**,**b**) Transmission electron microscope (TEM) and (**c**) high resolution transmission electron microscopy (HRTEM) images of SiO_2–TiO_2 at different scales.

3.3. Mechanism of the Interaction between SiO_2 and TiO_2 Particles

Figure 9 shows the FT-IR spectra of SiO_2 and SiO_2–TiO_2 with different TiO_2 ratios. The absorption bands at 1115, 808, and 477 cm^{-1} are typical absorption bands of Si–O bonds, indicating that the main component of the composite is SiO_2 [34].With the increase in the TiO_2 ratio, the intensity of absorption bands corresponding to SiO_2 decreased, indicating that the nano-TiO_2 coated the SiO_2 surface. In addition, the absorption bands (3200–3550 cm^{-1}) derived from Si–OH and Ti–OH showed the significant displacement and broadening phenomena when the SiO_2 was coated by the nano-TiO_2, indicating that the chemical environment had been changed and the association degree of hydroxyl groups on particles surface had increased. It was obviously caused by the formation of hydrogen bonds between Si–OH and Ti–OH or the further dehydroxylation reaction. It should be inferred that the chemical combination between SiO_2 microspheres and nano-TiO_2 particles was formed through the interaction of hydroxyl groups on their surfaces.

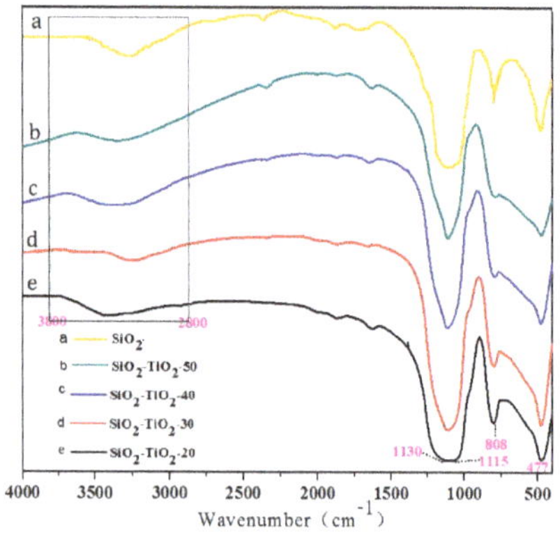

Figure 9. FT-IR of SiO_2–TiO_2 with different TiO_2 ratios. SiO_2–TiO_2-20, 30, 40, 50 represent the mass ratio of TiO_2 to SiO_2–TiO_2 is 20%, 30%, 40% and 50%; The black rectangle region represents the absorption peak caused by the vibration of the hydroxyl radical.

Figure 10 shows the schematic diagram of the bonding mechanism of SiO_2–TiO_2. Based on the above results, the bonding mechanism can be described as follows: firstly, the SiO_2 microspheres were ground in the ethanol medium with grinding balls. The strong grinding force made SiO_2 microspheres depolymerization and exposed more hydroxyl groups, thus displaying the higher reactivity. Secondly, the prepared nano-TiO_2 soliquid was ground with the activated SiO_2 violently, so that the collision probability between particles increased and lead to the contact and reactions between the hydroxyl groups on the SiO_2 and TiO_2 surfaces. Finally, water produced by the dehydroxylation of the particles was further removed by calcination. The SiO_2 and TiO_2 particles were bounded by –Si–O–Ti– bonds. The strength of the chemical bond was stronger than that of van der Waals forces and other physical forces, so the coating of nano-TiO_2 on SiO_2 surface was firm.

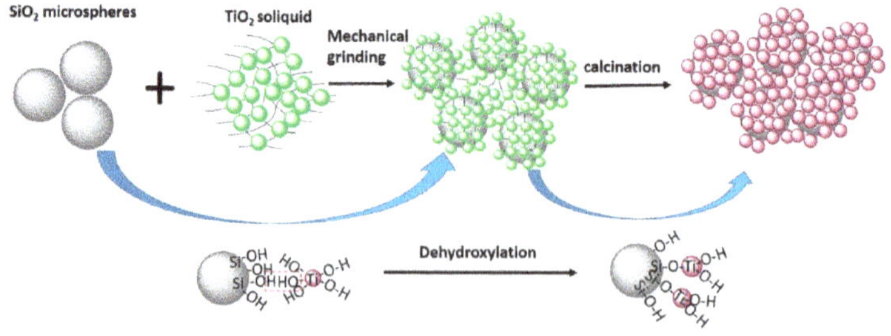

Figure 10. Schematic diagram of the bonding mechanism of SiO_2–TiO_2.

4. Conclusions

In the study, with the by-product SiO_2 microspheres produced during the industry production of fused-zirconia as the substrates, SiO_2–TiO_2 particles were prepared by the wet-grinding of SiO_2 microspheres and nano-TiO_2 and calcination of the ground product. The degradation efficiency of SiO_2–TiO_2 on methyl orange reached 97%, which was significantly higher than that of pure nano-TiO_2.

The water contact angle of SiO_2–TiO_2 was 8°, indicating the strong photoinduced hydrophilicity and the good self-cleaning effect.

The SiO_2–TiO_2 particles were characterized by the nano-TiO_2 uniformly coated on the SiO_2 microspheres and distributed in the microsphere gap. The nano-TiO_2 particles existed in an anatase phase with the particle size of 15–20 nm and are combined with SiO_2 microspheres by the dehydration of hydroxyl groups on particle surfaces.

Author Contributions: Hao Ding, Tongrong Deng and Sijia Sun conceived and designed the experiments; Sijia Sun and Tongrong Deng performed the experiments; Sijia Sun, Tongrong Deng and Wanting Chen analyzed the data; Ying Chen contributed reagents/materials/analysis tools; Sijia Sun and Hao Ding wrote the paper. Authorship must be limited to those who have contributed substantially to the work reported.

Conflicts of Interest: The authors declare no conflict of interest.

References

1. Liu, L.; Chen, X. Titanium dioxide nanomaterials: Self-structural modifications. *Chem. Rev.* **2014**, *114*, 9890–9918. [CrossRef] [PubMed]
2. Kormann, C.; Bahnemann, D.W.; Hoffmann, M.R. Preparation and characterization of quantum-size titanium dioxide. *J. Phys. Chem. C* **1988**, *92*, 5196–5201. [CrossRef]
3. Pu, S.; Zhu, R.; Ma, H.; Deng, D.; Pei, X.; Qi, F.; Chu, W. Facile in-situ design strategy to disperse TiO_2 nanoparticles on graphene for the enhanced photocatalytic degradation of rhodamine 6G. *Appl. Catal. B Environ.* **2017**, *218*, 208–219. [CrossRef]
4. Preethi, L.K.; Mathews, T.; Nand, M.; Jha, S.N.; Gopinath, C.S.; Dash, S. Band alignment and charge transfer pathway in three phase anatase-rutile-brookite TiO_2, nanotubes: An efficient photocatalyst for water splitting. *Appl. Catal. B Environ.* **2017**, *218*, 9–19. [CrossRef]
5. Ye, Z.; Tai, H.; Guo, R.; Yuan, Z.; Liu, C.; Su, Y.; Chen, Z.; Jiang, Y. Excellent ammonia sensing performance of gas sensor based on graphene/titanium dioxide hybrid with improved morphology. *Appl. Surf. Sci.* **2017**, *419*, 84–90. [CrossRef]
6. Fu, H.; Yang, X.; An, X.; Fan, W.; Jiang, X.; Yu, A. Experimental and theoretical studies of V_2O_5@TiO_2 core-shell hybrid composites with high gas sensing performance towards ammonia. *Sens. Actuators B Chem.* **2017**, *252*, 103–115. [CrossRef]
7. Jalvo, B.; Faraldos, M.; Bahamonde, A.; Rosal, R. Antimicrobial and antibiofilm efficacy of self-cleaning surfaces functionalized by TiO_2 photocatalytic nanoparticles against staphylococcus aureus and pseudomonas putida. *J. Hazard. Mater.* **2017**, *340*, 160–170. [CrossRef] [PubMed]
8. Tan, B.Y.L.; Tai, M.H.; Juay, J.; Liu, Z.; Sun, D. A study on the performance of self-cleaning oil–water separation membrane formed by various TiO_2, nanostructures. *Sep. Purif. Technol.* **2015**, *156*, 942–951. [CrossRef]
9. Ganesh, V.A.; Raut, H.K.; Nair, A.S.; Ramakrishna, S. A review on self-cleaning coatings. *J. Mater. Chem. A* **2011**, *21*, 16304–16322. [CrossRef]
10. Kim, S.M.; In, I.; Park, S.Y. Study of photo-induced hydrophilicity and self-cleaning property of glass surfaces immobilized with TiO_2 nanoparticles using catechol chemistry. *Surf. Coat. Technol.* **2016**, *294*, 75–82. [CrossRef]
11. Petica, A.; Gaidau, C.; Ignat, M.; Sendrea, C.; Anicai, L. Doped TiO_2 nanophotocatalysts for leather surface finishing with self-cleaning properties. *J. Coat. Technol. Res.* **2015**, *12*, 1153–1163. [CrossRef]
12. Diamanti, M.V.; Paolini, R.; Rossini, M.; Aslan, A.B.; Zinzi, M.; Poli, T.; Pedeferri, M.P. Long term self-cleaning and photocatalytic performance of anatase added mortars exposed to the urban environment. *Constr. Build. Mater.* **2015**, *96*, 270–278. [CrossRef]
13. Rao, X.; Liu, Y.; Fu, Y.; Liu, Y.; Yu, H. Formation and properties of polyelectrolytes/TiO_2 composite coating on wood surfaces through layer-by-layer assembly method. *Holzforschung* **2016**, *70*, 361–367. [CrossRef]
14. Lei, M.; Li, F.S.; Tanemura, S.; Fisher, C.A.J.; Li, L.Z.; Liang, Q.; Xu, G. Cost-effective nanoporous SiO_2–TiO_2, coatings on glass substrates with antireflective and self-cleaning properties. *Appl. Energy* **2013**, *112*, 1198–1205.

15. Diamanti, M.V.; Gadelrab, K.R.; Pedeferri, M.P.; Stefancich, M.; Pehkonen, S.O.; Chiesa, M. Nanoscale investigation of photoinduced hydrophilicity variations in anatase and rutile nanopowders. *Langmuir* **2013**, *29*, 14512–14518. [CrossRef] [PubMed]
16. Calia, A.; Lettieri, M.; Masieri, M. Durability assessment of nanostructured TiO_2, coatings applied on limestones to enhance building surface with self-cleaning ability. *Build. Environ.* **2016**, *110*, 1–10. [CrossRef]
17. Jo, W.K.; Tayade, R.J. Facile photocatalytic reactor development using nano-TiO_2 immobilized mosquito net and energy efficient UVLED for industrial dyes effluent treatment. *J. Environ. Chem. Eng.* **2016**, *4*, 319–327. [CrossRef]
18. Natarajan, K.; Natarajan, T.S.; Tayade, R.J. Photocatalytic reactor based on UV-LED/TiO_2 coated quartz tube for degradation of dyes. *Chem. Eng. J.* **2011**, *178*, 40–49. [CrossRef]
19. Yang, S.B.; Chun, H.H.; Tayade, R.J.; Jo, W.K. Iron-functionalized titanium dioxide on flexible glass fibers for photocatalysis of benzene, toluene, ethylbenzene, and *o*-xylene (BTEX) under visible- or ultraviolet-light irradiation. *J. Air Waste Manag.* **2015**, *65*, 365–373. [CrossRef] [PubMed]
20. Smitha, V.S.; Manjumol, K.A.; Baiju, K.V.; Ghosh, S.; Perumal, P.; Warrier, K.G.K. Sol–gel route to synthesize titania-silica nano precursors for photoactive particulates and coatings. *J. Sol-Gel Sci. Technol.* **2010**, *54*, 203–211. [CrossRef]
21. Shi, G.; Chen, J.; Wang, L.; Wang, D.; Yang, J.; Li, Y.; Zhang, L.; Ni, C.; Chi, L. Titanium oxide/silicon moth-eye structures with antireflection, p–n heterojunctions and superhydrophilicity. *Langmuir* **2016**, *32*, 27666724. [CrossRef] [PubMed]
22. Pakdel, E.; Daoud, W.A. Self-cleaning cotton functionalized with TiO_2/SiO_2 focus on the role of silica. *J. Colloid Interface Sci.* **2013**, *401*, 1–7. [CrossRef] [PubMed]
23. Prabhu, S.; Cindrella, L.; Kwon, O.J.; Mohanraju, K. Superhydrophilic and self-cleaning RGO–TiO_2, composite coatings for indoor and outdoor photovoltaic applications. *Sol. Energy Mater. Sol. Cells* **2017**, *169*, 304–312. [CrossRef]
24. Zhou, J.; Tan, Z.; Liu, Z.; Jing, M.; Liu, W.; Fu, W. Preparation of transparent fluorocarbon/TiO_2-SiO_2, composite coating with improved self-cleaning performance and anti-aging property. *Appl. Surf. Sci.* **2017**, *396*, 161–178. [CrossRef]
25. Zhang, H.; Fan, D.; Yu, T.; Wang, C. Characterization of anti-reflective and self-cleaning SiO_2–TiO_2, composite film. *J. Sol-Gel Sci. Technol.* **2013**, *66*, 274–279. [CrossRef]
26. Ciprian, M.; Alexandru, E.; Anca, D. SiO_2/TiO_2 multi-layered thin films with self-cleaning and enhanced optical properties. *Bull. Mater. Sci.* **2017**, *40*, 1–10. [CrossRef]
27. Surolia, P.K.; Tayade, R.J.; Jasra, R.V. TiO_2-coated cenospheres as catalysts for photocatalytic degradation of methylene blue, *p*-nitroaniline, *n*-decane, and *n*-tridecane under solar irradiation. *Ind. Eng. Chem. Res.* **2010**, *49*, 8908–8919. [CrossRef]
28. Mavukkandy, M.O.; Bilad, M.R.; Kujawa, J.; Al-Gharabli, S.; Arafat, H.A. On the effect of fumed silica particles on the structure, properties and application of PVDF membranes. *Sep. Purif. Technol.* **2017**, *187*, 365–373. [CrossRef]
29. Peng, Y.; Zhang, J.; Liu, J.; Ke, J.; Wang, F. Properties and microstructure of reactive powder concrete having a high content of phosphorous slag powder and silica fume. *Constr. Build. Mater.* **2015**, *101*, 482–487. [CrossRef]
30. Pedro, D.; Brito, J.D.; Evangelista, L. Mechanical characterization of high performance concrete prepared with recycled aggregates and silica fume from precast industry. *J. Clean. Prod.* **2017**, *164*, 939–949. [CrossRef]
31. Liu, J.; Wang, D. Influence of steel slag-silica fume composite mineral admixture on the properties of concrete. *Powder Technol.* **2017**, *320*, 230–238. [CrossRef]
32. Wang, S.; Zheng, W.T.; Lian, J.S.; Jiang, Q. Photocatalytic property of Fe doped anatase and rutile TiO_2 nanocrystal particles prepared by sol–gel technique. *Appl. Surf. Sci.* **2012**, *263*, 260–265. [CrossRef]
33. Zhou, X.; Wu, J.; Zhang, J.; He, P.; Ren, J.; Zhang, J.; Lu, J.; Liang, P.; Xu, K.; Shui, F. The effect of surface heterojunction between (001) and (101) facets on photocatalytic performance of anatase TiO_2. *Mater. Lett.* **2017**, *205*, 173–177. [CrossRef]
34. Chen, Y.; Ding, H.; Sun, S. Preparation and characterization of ZnO nanoparticles supported on amorphous SiO_2. *Nanomaterials* **2017**, *7*, 217. [CrossRef] [PubMed]

nanomaterials

MDPI

Article

Resistive Switching of Sub-10 nm TiO_2 Nanoparticle Self-Assembled Monolayers

Dirk Oliver Schmidt [1,2], Nicolas Raab [3,4], Michael Noyong [1,2], Venugopal Santhanam [5], Regina Dittmann [3,4] and Ulrich Simon [1,2,*]

1 JARA-FIT, 52056 Aachen, Germany; oliver.schmidt@ac.rwth-aachen.de (D.O.S.); michael.noyong@ac.rwth-aachen.de (M.N.)
2 Institute of Inorganic Chemistry, RWTH Aachen University, 52074 Aachen, Germany
3 JARA-FIT, 52425 Jülich, Germany; n.raab@gmx.net (N.R.); r.dittmann@fz-juelich.de (R.D.)
4 Peter Grünberg Institut 7, Forschungszentrum Jülich GmbH, 52428 Jülich, Germany
5 Department of Chemical Engineering, Indian Institute of Science, Bangalore 560012, India; venu@chemeng.iisc.ernet.in
* Correspondence: ulrich.simon@ac.rwth-aachen.de; Tel.: +49-241-80-94644

Received: 28 September 2017; Accepted: 31 October 2017; Published: 4 November 2017

Abstract: Resistively switching devices are promising candidates for the next generation of non-volatile data memories. Such devices are up to now fabricated mainly by means of top-down approaches that apply thin films sandwiched between electrodes. Recent works have demonstrated that resistive switching (RS) is also feasible on chemically synthesized nanoparticles (NPs) in the 50 nm range. Following this concept, we developed this approach further to the sub-10 nm range. In this work, we report RS of sub-10 nm TiO_2 NPs that were self-assembled into monolayers and transferred onto metallic substrates. We electrically characterized these monolayers in regard to their RS properties by means of a nanorobotics system in a scanning electron microscope, and found features typical of bipolar resistive switching.

Keywords: TiO_2 nanoparticles; self-assembly; resistive switching

1. Introduction

The astounding developments in information technology over the last few decades reliably obeyed Moore's law [1,2]. However, this predicted trend of miniaturization is coming to an end due to physical limitations [3]. At the same time, an increasing demand for digital data storage is anticipated, which will require new, non-volatile data storage technologies in the near future.

Resistive random access memories (RRAM) are promising candidates for data storage applications [4,5]. They rely on resistive switching (RS), which results from a resistance change of a functional layer sandwiched between metal electrodes. RRAM devices are typically composed of a metal–insulator–metal layer structure, mainly in the form of thin films that are structured by means of lithographic (top–down) techniques. As an alternative approach, nanoparticle (NP) thin films formed via chemical synthesis and assembly can be utilized as a functional layer in RS devices. Such a bottom–up approach in principle allows the fabrication of cell dimensions that exceed the size limits of top–down approaches [6]. From a technological point of view, NPs can be synthesized via inexpensive methods and under mild reaction conditions [7]. Subsequently, the NPs can be deposited on the electrodes using solution-based techniques that are suitable for organic or polymeric substrates, thus leading to flexible memory devices [8].

The resistive switching of NPs is often investigated in a configuration that is similar to conventional thin film cells, wherein NP assemblies are the functional layer of the device. Typically, most of these RS cells based on NP assemblies were fabricated as follows: firstly, NPs are deposited on a bottom electrode via spin-coating or dip-coating methods, which enable control of the NP assembly thickness. Secondly, top electrodes are deposited on the NP assemblies. One of the first reports on RS of a NP assembly as a functional layer utilizing Fe_3O_4 NPs was given by Kim et al. in 2009 [9]. In the following years, the RS of iron oxide-based NPs were further investigated, e.g., consisting of Fe_2O_3 NP assemblies [8,10,11], Pt-Fe_2O_3 core–shell NP assemblies [12], or of mixed Pt–Fe_2O_3 core–shell/Fe_2O_3 NP assemblies [13]. Besides iron oxide NPs, RS behavior was also reported for CdS NPs [7], CeO_2 nanocubes [14,15], $BaTiO_3$ NPs [16], ZnO NPs [17], NiO NPs [18], Ge–GeO_x nanowires (NWs) [19], and for In_2O_3 nanorods [20]. The RS of assemblies consisting of spherical 3 nm TiO_x NPs was demonstrated by Goren et al. in a Co–TiO_x NPs–Co structure [21]. The TiO_x NPs were synthesized by a sol–gel method, and the as-synthesized NPs were amorphous. NP films with a thickness of 55 nm were prepared by spin-coating, and the cells showed bipolar resistive switching (BRS). The NP devices were compared with a TiO_x thin film device, and while the film device exhibited only switching at one interface, the authors report switching at both interfaces for the NP device.

However, despite the very small size of the individual NPs (e.g., 3 nm), the thickness of the NP assemblies is often quite high, and thicker than typical thin film structures. As an exception, Uenuma et al. demonstrated BRS for a monolayer of 6 nm magnetite NPs in a metal–NPs–metal structure [22].

Recently, we reported the RS of individual TiO_2 NPs with sizes of approximately 350 nm as well as 50 nm [23]. In order to continue down-scaling the RS devices composed of TiO_2 NPs, we chemically synthesized sub-10 nm TiO_2 NPs by a solvothermal method. We chose TiO_2 as a model material because its RS properties are investigated in single crystals [24] as well as thin films [25,26]. Furthermore, the complementary metal-oxide-semiconductor (CMOS) compatibility of TiO_2 [27], as well as the abundance of Ti in the earth's crust [28], ensures its economic viability. The immobilization of the TiO_2 NPs acting as switching units is necessary in order to integrate NPs into resistive switching devices. Therefore, the self-assembly of NPs into a well-ordered, hexagonally packed monolayer is desirable. In the literature, different methods are reported to obtain self-assembled NP monolayer. One is the drop-drying of a colloidal solution [29], which is assisted by an electric field [30] or by molecule interactions [31]. Another is by spreading a colloidal solution of hydrophobic NPs onto a water surface [32–34]. A review summarizing the various methods can be found in reference [35]. However, to the best of our knowledge, up until now, the self-assembly of sub-10 nm TiO_2 NPs into monolayer films has not been reported.

In this paper, we present the synthesis of sub-10 nm TiO_2 NPs and their characterization by means of powder X-ray diffraction (XRD) as well as transmission electron microscopy (TEM). Self-assembly experiments were performed in order to obtain hexagonally close-packed TiO_2 NP films on a water surface, and we obtained TiO_2 NP monolayer films with lateral dimensions of 1 μm^2. We transferred the self-assembled films to planar Pt–Ir surfaces, which function as bottom electrodes, via two different approaches. The transferred films were characterized using a scanning electron microscope (SEM), atomic force microscope (AFM), and transmission electron microscope (TEM). We performed localized electrical measurements by means of a nanorobotics setup in a SEM, as well as by means of local conductive atomic force microscopy (LC-AFM). Finally, we investigated the RS properties of the films in a SEM.

2. Results

2.1. *Solvothermal Synthesis of TiO_2 Nanoparticles*

In order to synthesize TiO_2 NPs with a diameter of sub-10 nm, we adapted the solvothermal synthesis methods of Dinh et al. [36]. Titanium butoxide was used as titanium precursor, oleylamine and oleic acid were used as capping agents. The ratio of the two capping ligands allowed the authors to control the particle morphology. Rhombic NP shapes were obtained with a titanium butoxide/oleic acid/oleylamine ratio of 1:4:6; truncated rhombic NP shapes were obtained with a ratio of 1:5:5, and spherical NP shapes were obtained with a ratio of 1:6:4. With regard to the envisaged self-assembly of the TiO_2 NPs as monolayers, as densely as possible, the spherical morphology is desirable, as it allows for a hexagonal close packing, thereby covering approximately 91% of the available surface [37].

In a series of syntheses, we obtained the highest amount of spherical shaped NPs with a titanium butoxide/oleic acid/oleylamine molar ratio of 1:3:2, and avoiding ethanol solvent. The synthesized TiO_2 NPs are shown in Figure 1a,b. The ratio of spherical to non-spherical shaped NPs was ca. 27:1, meaning that 96% of the yielded NPs presented a spherical morphology. A mean particle longitudinal of (5.7 ± 1.1) nm and a mean particle transversal of (4.6 ± 0.8) nm were determined (Figure 1c; for the corresponding histograms, see Figure S1), which implied a nearly spherical morphology. Recorded powder XRD reflection patterns of the obtained NPs showed broadened reflexes, according to the minute particles' size, and matched the simulated anatase reflection patterns (Figure 1d) [38].

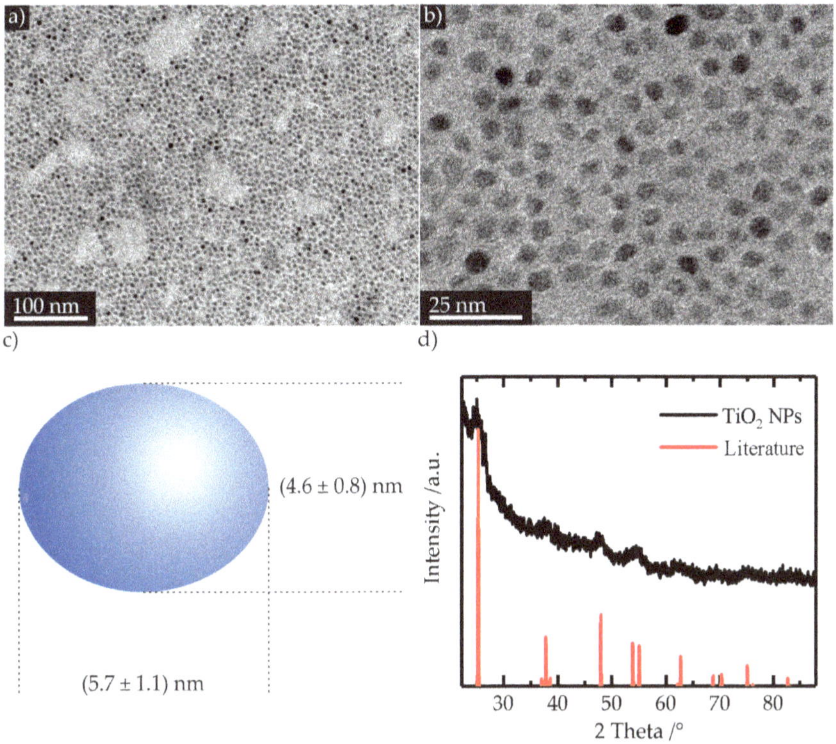

Figure 1. Representative TEM images of the synthesized TiO_2 nanoparticles (NPs) (**a**,**b**). Schematic illustration of the NP with the corresponding mean longitudinal and transversal (**c**). Powder XRD reflection patterns of the NPs (black), and simulated literature anatase reflection patterns (red) (**d**).

2.2. Formation of Self-Assembled TiO_2 Nanoparticle Monolayers

In order to obtain TiO_2 NP monolayer films, we followed the method of Santhanam et al., in which the authors applied for the monolayer formation of hydrophobic gold NPs [33]. Briefly, the synthesized sub-10 nm TiO_2 NPs were dispersed in an organic solvent and were dropped on a water surface with a controlled surface curvature. Due to the evaporation of the organic solvent, a self-assembled monolayer was formed on the water surface. In order to perform TEM investigations of the self-assembled film immobilized at the water/air surface, the surface was touched with a carbon coated TEM grid. A schematic drawing of the method is shown in Figure 2a,b.

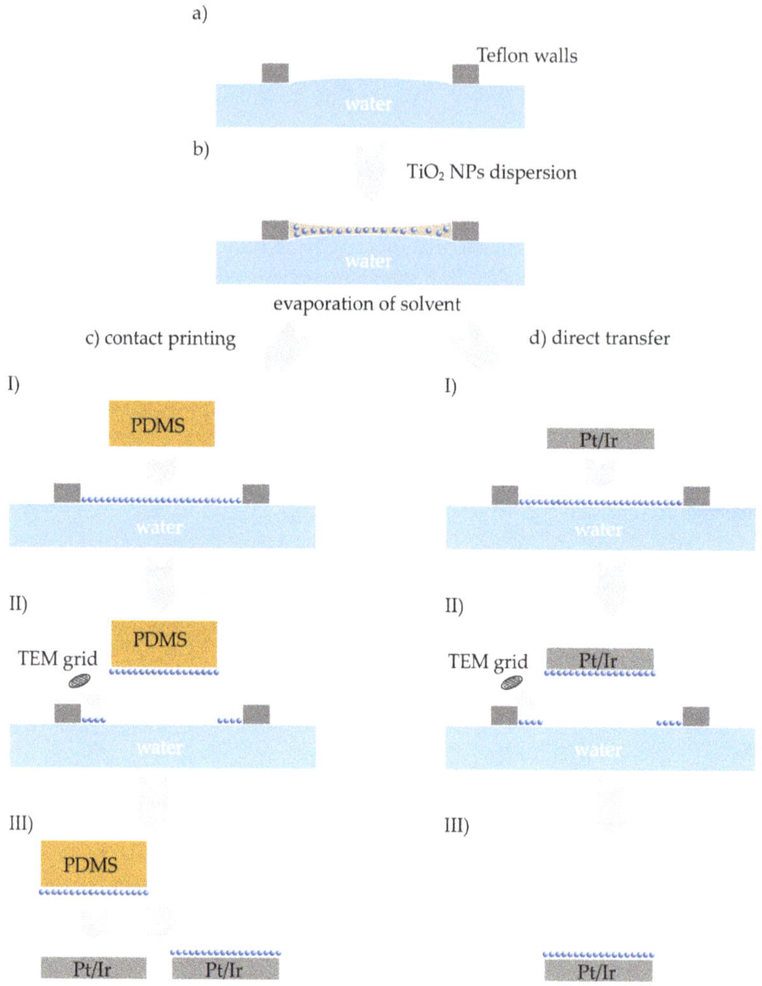

Figure 2. Schematic drawing for the self-assembly of a TiO_2 NP monolayer (not drawn to scale) (**a**,**b**), for the two-step, microcontact printing method (**c** I to III), and for the one-step method for the direct transfer of self-assembled TiO_2 films (**d** I to III).

The formation of well-ordered monolayer films is challenging, because it depends on the following experimental parameters. First, the NPs have to be spherical and monodisperse; otherwise, a hexagonally close packing of the NPs is not possible. Second, the TiO_2 NPs need to be functionalized with hydrophobic ligands for the formation of a stable colloidal dispersion in organic solvents. Furthermore, the concentration of TiO_2 NPs in the organic solvent influences the self-assembly. In this context, too low concentrations lead to small and isolated regions of closed packed NPs, whereas too high concentrations lead to multilayers of NPs. Furthermore, the organic solvent itself has to fulfill certain requirements. Most importantly, the solvent must allow for a stable dispersion of TiO_2 NPs, and the density of the solvent must be lower than that of water. Finally, the evaporation rate of the chosen solvent determined by its volatility is crucial, since too fast as well as too slow evaporation rates induce the formation of multilayers instead of monolayers. Mixtures of different solvents can be used to precisely adjust the properties and meet these requirements. Additionally, the rate of evaporation of the organic solvents depends upon the air velocity and the temperature in the laboratory hood, which thus also influence the NP monolayer formation. In the scope of this work, the colloid concentration, the solvent composition, and the evaporation rate were investigated. We performed the experiments under air/ambient conditions.

In a series of experiments, we obtained the largest continuous self-assembled TiO_2 NP monolayer film with NPs presenting a mean particle longitudinal of (5.7 $\pm$ 1.1) nm and a mean particle transversal of (4.6 $\pm$ 0.8) nm, as well as a spherical NP to irregular-shaped NP ratio of 27:1 (see Figure 1). Since the as-synthesized TiO_2 ligands were functionalized with oleic acid and oleylamine as hydrophobic ligands, no additional ligand exchange reactions were necessary. We prepared the self-assembled film with a solvent mixture of pentane/dichloromethane 3:1, 0.56 g/mL TiO_2 NPs concentration, and the addition of 0.0076 mol/L oleylamine solution in hexane to the dispersion. The corresponding TEM images (Figure 3a,b) revealed that an area of approximately 1 μm^2 was covered with mostly a monolayer of TiO_2 NPs. The dense, close packing is clearly visible in the images. Fast Fourier transformation was performed in these regions with help of the Software ImageJ (Version 1.43u) showing the reconstructed hexagonal patterns (see Figure 3b, inset). The center-to-center spacing of the NPs amounted to ca. (9 $\pm$ 1) nm, which corresponds to the dimensions of the NPs, plus the approximately 2 nm length of the oleic acid and oleylamine ligands, assuming that a monolayer of the ligands has formed on the TiO_2 NP surface. A similar spacing of 2 nm was reported by Sun et al. for monodispersed, hexagonally packed FePt NPs capped with oleic acid and oleylamine [29]. Additionally, several patches of bi- and multilayers were formed, as indicated by the areas showing a lower brightness in the TEM images compared with those of the TiO_2 NP monolayers (Figure 3a,b). The formation of well-ordered monolayers only took place in confined regions, preferably at the edges of multilayers. The formation of the multilayer domains can be mainly attributed to fluctuations along the retracting contact line during the evaporation of the solvent, as well as due to tearing of the film during the transfer process. Experiments revealed that the adjustable parameters of colloid concentration and solvent could be controlled well. However, the evaporation rate could not be fully managed due to the random temperature and air velocity of the fume hood. Nevertheless, in this work, we produced a well-ordered, self-assembled TiO_2 NP monolayer, and the observed dimensions of the TiO_2 NP monolayers are sufficient enough for RS experiments performed by means of the nanorobotics setup in SEM.

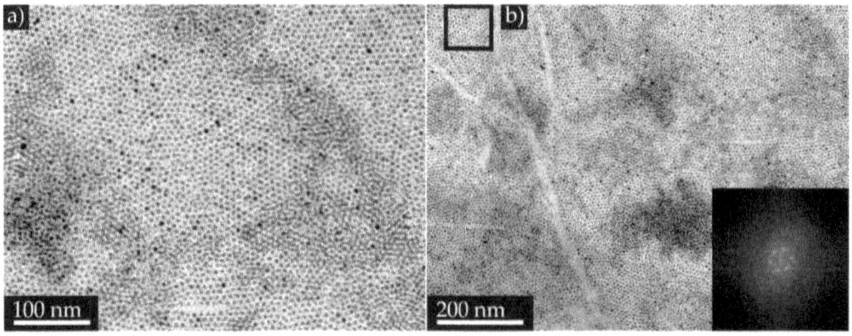

Figure 3. Representative TEM images of the self-assembled TiO_2 NP film with decreasing magnification (**a**,**b**). The inset in (**b**) shows the fast Fourier transformation of the black highlighted area of the monolayer.

2.3. *Preparation of Resistive Switching Devices*

The self-assembled TiO_2 NP films were formed on a water surface, and had to be transferred onto a metallic surface as the bottom electrode in order to allow the investigation of their RS behavior. A 1 cm^2 silicon wafer with a native oxide layer, which was coated with a homogenous 180 nm thick Pt/Ir alloy (80% Pt, 20% Ir) metal film, was used as the support. In order to transfer the assembled TiO_2

NP films to a metal surface, the film was carefully brought into contact with a polydimethylsiloxane (PDMS) stamp, or directly with a Pt/Ir surface. A schematic drawing of the two methods is illustrated in Figure 2.

The two-step method was performed following a published microcontact printing procedure [39]. In contrast to the direct, one-step method, the TiO_2 NP film was first transferred to a stamp, and afterwards the TiO_2 NP film was printed to any desired surface. The application of stamps with nanoscaled features would allow for the preparation of nanoscaled TiO_2 NP patterned surfaces as resistive switching devices [39]. The TiO_2 NP film shown in the TEM images (Figure 4) was prepared with NPs that had a mean longitudinal of (7.9 ± 2.2) nm, a mean transversal of (4.8 ± 0.8) nm, and a spherical NP to non-spherical shaped NP ratio of 5:1 (for the characterization of these NPs, see Figure S2). We lifted the TiO_2 NP film from the water surface with a planar polydimethylsiloxane (PDMS) stamp (Figure 2c I and II). After the evaporation of the residual water droplet, the dried films on the planar side of the PDMS stamp were transferred to a Pt/Ir surface by pressing the stamp onto the surface (Figure 2c III). We characterized the formed film via TEM (Figure 4a,b), and the transferred film via AFM (Figure 4c,d). The TEM images as well as AFM images show similar TiO_2 NP mono-, bi-, and multilayers, as well as voids between the layers. Since we observed no macroscopic wrinkles or cracks in the TEM images, we conclude that the self-assembled TiO_2 NP film was transferred from the water surface to the carbon film of the TEM substrate without any changes of the film, at least in the dimensions shown in the TEM images. The recorded height profile revealed a variation of height of approximately 5 nm (Figure 4e), which corresponds well to the dimensions of the NPs determined by TEM analysis. Based on the resemblance of AFM and TEM images, we concluded that the self-assembled TiO_2 NP film was successfully transferred onto the Pt/Ir surface. We obtained similar results for the one-step method (see Figure 2d I to III). For detailed results, see Figure S3 in the Supporting Information. Hence, we successfully prepared RS devices composed of TiO_2 NP films by means of two different methods for the subsequent electrical characterization.

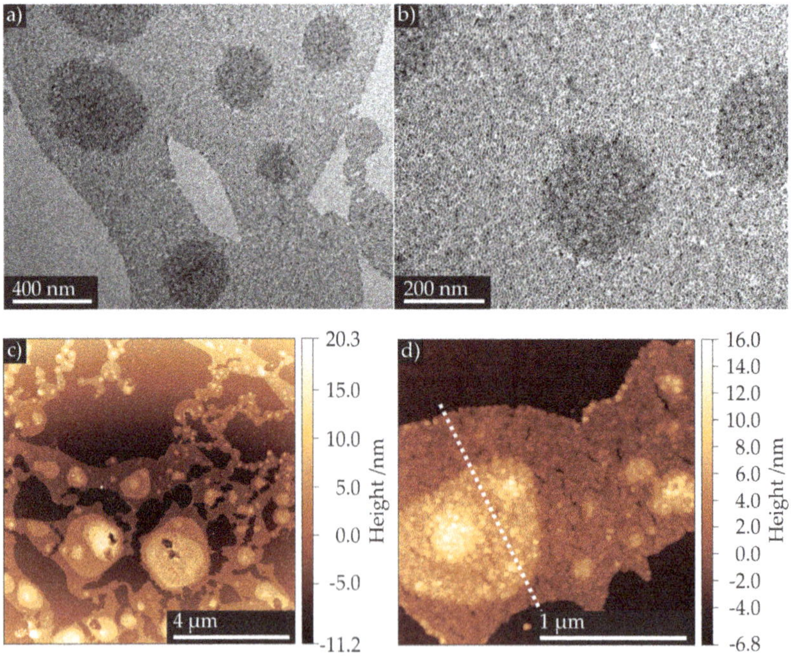

Figure 4. *Cont.*

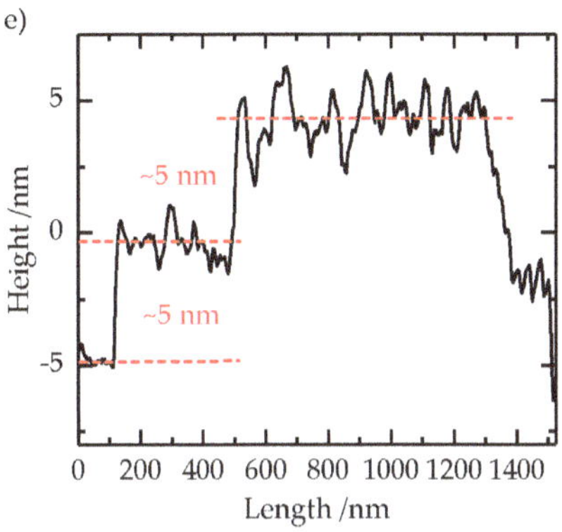

Figure 4. Exemplary TEM images of the self-assembled TiO_2 NP film (**a**,**b**). Tapping mode atomic force microscope (AFM) images of the TiO_2 NP film transferred by the microcontact printing method onto a Pt/Ir surface (**c**,**d**). Corresponding height profile (**e**) taken along the white line in (**d**) showing height differences of ca. 5 nm.

2.4. Electrical Characterization

We performed RS experiments by means of a nanorobotics system for local in situ electrical measurements in SEM [40]. Prior to the in situ electrical characterization in SEM, we performed an oxygen plasma cleaning step with the (Pt/Ir)/TiO_2 NP film substrates to widely remove the oleylamine, as well as the oleic acid ligands. As top electrodes, we utilized Pt/Ir coated AFM probes with a radius of curvature of approximately 100 nm, and a special elongated tip in the front part of the cantilever; thus, they are visible from the top in the SEM. This setup enables the flexible addressing of certain locations on the thin films. While the voltage was applied to the tip, the planar Pt/Ir bottom electrode was set to ground. We monitored the movement of the tip on the NP films, as well as the structural changes of the tip during the resistive switching experiments. After one measurement, the tip electrode was lifted off and moved to the next point of interest, which allowed successive characterization under identical experimental conditions [40]. Schematic illustrations of the experimental setup and an exemplary SEM image are displayed in Figure 5a,b, respectively.

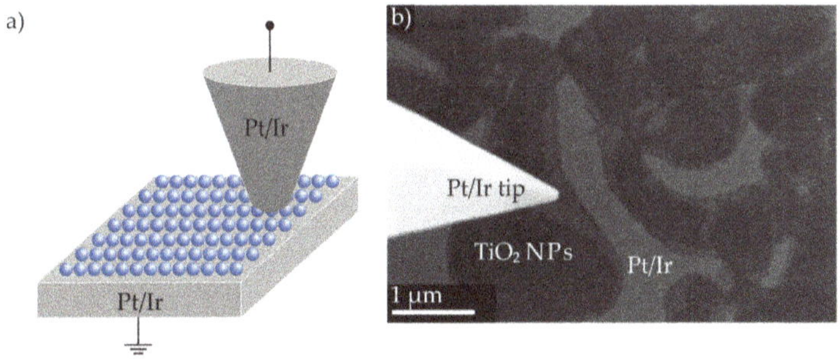

Figure 5. Schematic illustrations of the (Pt/Ir)/TiO_2 NP film/(Pt/Ir tip) device (**a**). Exemplary SEM image of a TiO_2 NP film on the Pt/Ir surface transferred by the one-step method; on the left hand side, the Pt/Ir coated tip electrode is visible (**b**) (contrast of the SEM image was increased after the measurement; for the original SEM image, see Figure S4).

Directly before the experiment, we cleaned the SEM chamber, the measurement tips, and the TiO_2 NP films with Ar plasma to further eliminate contaminations. Due to NP diameters below 10 nm and weak material contrast, individual TiO_2 NPs immobilized on the Pt/Ir surface could not be resolved in SEM during the electrical characterization, which requires a large working distance due to the presence of the tip. However, comparing the AFM, TEM, and SEM images, we assume that the bright regions are the Pt/Ir surface (Figure 5b). Furthermore, we assume that the areas exhibiting a slightly lower brightness correspond to TiO_2 NP monolayers, while the darkest areas correspond to TiO_2 NP multilayers. At the left hand side of the SEM image, the measurement probe that was brought into mechanical contact with a TiO_2 NP monolayer is visible.

In order to test different areas, brighter and darker regions discernible in the SEM images (see Figure 5b) were contacted with the Pt/Ir tip. The recorded *I–V* curves showed a linear behavior and resistances of ca. 300 Ω, which matched the resistances determined by addressing a pristine metallic surface. Hence, the electrical characterization confirmed that the bright regions do correspond to the Pt/Ir bottom electrode. By positioning the tip on regions exhibiting a lower brightness, strictly non-linear *I–V* curves revealing a high resistance were recorded. These regions are identified as corresponding as expected to the TiO_2 NP layer. Therefore, during the in situ electrical characterization in SEM, SEM images and electrical responses allowed for facile differentiation between the Pt/Ir bottom electrode and the TiO_2 NP layer.

In the SEM images (Figure 5b), stepwise brightness differences are visible within the TiO_2 NP layer. Based on the TEM and AFM analysis results, the TiO_2 NP layer with a higher brightness is assumed to correspond to a monolayer, while the TiO_2 NP layer with a lower brightness is assumed to correspond to a multilayer. Different spots of the TiO_2 NP monolayer or multilayer were brought into contact with the tip, and non-linear *I–V* curves without any hysteretic behavior were recorded. The tip diameter of the Pt/Ir coated measurement probes was approximately 100 nm. Hence, multiple sub-10 nm TiO_2 NPs were simultaneously addressed. However, we did not find a clear dependence between layer thickness and resistance. In multiple layers, the absolute number of particles that contributed to the conducting path varied. Moreover, different numbers of resistances by each particle in series and in parallel lead to varying overall resistance.

Additionally, we performed LC-AFM measurements. The LC-AFM allows the simultaneous measurement of topography and current through the sample, and is operated in the contact mode to record the current distribution of the scanned region. We utilized conductive diamond tips AppNano Doped Diamond with a radius of curvature of 100–300 nm for the experiments. In order to perform the electrical measurements without inducing a resistance change of the NPs, a voltage of 20 mV was applied to the tip, and the topography and the current were recorded simultaneously. The contact mode topography image (Figure 6a) revealed TiO_2 mono-, bi-, and multilayers similar to those observed in tapping mode. In the current mapping image (Figure 6b), the TiO_2 NP layer can be clearly distinguished from the Pt/Ir surface, as the latter exhibited currents ranging from approximately 100 nA to 340 nA, while the areas with TiO_2 NP layers exhibited no current flow.

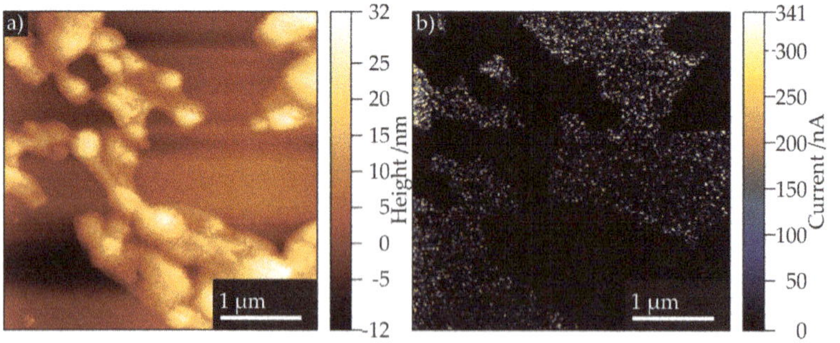

Figure 6. Contact mode AFM images of a TiO_2 NP film on a Pt/Ir surface (**a**) and corresponding distribution of the current (**b**), scanned simultaneously with the topography.

These findings by means of SEM and LC-AFM are in agreement with the metallic, highly conducting character of the Pt/Ir surface and the insulating character of the anatase TiO_2 NPs [41].

In order to study the RS behavior of our devices in the SEM, we applied write voltage sweeps from 0 V $\rightarrow$ X V $\rightarrow$ 0 V $\rightarrow$ X V $\rightarrow$ 0 V, or from 0 V $\rightarrow$ $-$X V $\rightarrow$ 0 V $\rightarrow$ X V $\rightarrow$ 0 V. We set a current compliance of 1 μA up to 10 μA to protect the TiO_2 NP layer, as well as the metal coating of the measurement tips. We identified a current compliance of 10 μA to be suitable for the resistive switching experiments. The *I*–*V* curve shown in Figure 7a was recorded on a TiO_2 NP monolayer area contacted by the measurement tip. The *I*–*V* curve showed a typical BRS behavior, with a SET process of the device from the high resistance state (HRS) into the low resistance state (LRS) at a voltage of ca. -2.5 V. The RESET process, the switching of the device from the LRS to the HRS, took place over a voltage range from 1.0 V to approximately 2.8 V, and switched the device back into the HRS. Hence, for the *I*–*V* curves shown in Figure 7a, we observed the counter eightwise switching polarity. The hysteresis and the current are larger at a negative voltage polarity compared with the positive voltage polarity. The recorded *I*–*V* curve shown in Figure 7b, which was also recorded on a TiO_2 NP monolayer, demonstrated BRS behavior exhibiting the SET process at a positive voltage, and the RESET process at a negative polarity; hence, eightwise switching polarity is observed. In general, the switching polarity of a BRS device is determined by a microstructurally asymmetric cell design, or a voltage/current-controlled electroforming process. The underlying switching mechanism for valence change memories is generally explained by a formation and rupture of a conductive filament inside the insulating TiO_2 matrix due to the redistribution of oxygen vacancies under an applied electric field, and the effect of Joule heating [42]. This gives rise to a resistance hysteresis exhibiting the counter eightwise polarity [43]. The resistance hysteresis showing an "eightwise" switching polarity was recently investigated in $SrTiO_3$ thins films by means of detailed in situ TEM analysis. Electrochemical oxygen evolution and oxygen reduction reactions were found to be responsible for the resistance change [44]. In order to decipher the underlying switching mechanism for our TiO_2 NP devices, comparable elaborate analysis would be required, which goes far beyond the scope of this paper. For LC-AFM measurements, the switching polarity of a Fe:$SrTiO_3$ film could be adjusted via the switching voltage [43]. The measurement tip was in contact with the TiO_2 NP layer as briefly as possible to keep the thermal drift and creep effects in the piezoelectric control elements of the nanorobotics setup, as well as the specimen stage, as low as possible during the electrical characterization. Nonetheless, the position and the contact force of the tip changed during the application of voltage sweeps. Hence, the switching polarity could not be controlled in our experiments.

The device exhibiting the *I*–*V* curve, as shown in Figure 7b, was switched between 1 V and -1 V. The current reached the current compliance of 10 μA at both voltage polarities. Upon repeating the voltage sweep, the size of the hysteresis and the current values changed. After three cycles, the BRS (see Figure S5a) behavior was no longer observed. Instead, the current showed linear dependence on the applied voltage (see Figure S5b).

Additionally, in some cases, we observed a structural change of the contacted TiO_2 NP layer via SEM (see Figure S6). It is possible that the high voltage and current, accompanied by Joule heating, induced the structural change, leading to a direct contact between the measurement tip and the Pt/Ir surface, and thereby resulting in the observed linear behavior. During the prolonged application of voltage sweep, the position and the contact force of the tip changed due to the thermal drift and creep effects in the piezoelectric control elements of the nanorobotics setup and specimen stage, which also resulted in a possible direct contact. However, for most of the measurements, we observed a short circuit, although in the SEM images no clear damage of the NP layer was visible. Alternatively, it may be possible that individual sub-10 nm TiO_2 NPs could be reduced to a better conducting state, e.g., Ti_4O_7 Magnéli phases [45]. These phases show metallic conductivity, and thus may be responsible for the observed linear behavior of the *I*–*V* curves [42]. However, a detailed analysis to identify Magnéli phases would go far beyond the scope of this work. We found during the measurements that voltages above ± 3 V tend to cause a short circuit of the RS devices. In total, we electrically characterized

14 spots of a TiO_2 NP monolayer, as well as 12 spots of a TiO_2 NP multilayer. Overall, six spots showed a BRS behavior.

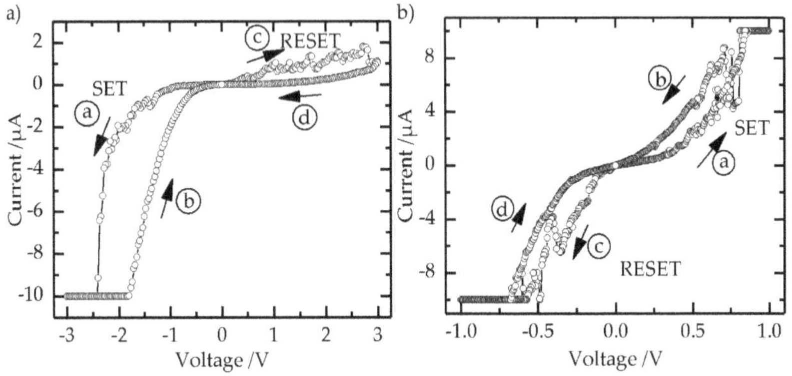

Figure 7. Two *I–V* curves recorded on different TiO_2 NP monolayers exhibiting bipolar resistive switching (BRS) behavior (**a**,**b**) (arrows and small letters depict voltage sweep sequence).

2.5. Summary

In order to obtain sub-10 nm resistive switching units, we synthesized TiO_2 NPs with a size below 10 nm by a solvothermal method. Self-assembly of a TiO_2 NP monolayer film on a water surface was prepared following the method of Santhanam et al. We achieved dense packed monolayers in an area of 1 μm^2, which up until now had not been accomplished elsewhere. Since the TiO_2 NP films were prepared on a water surface, they had to be transferred to metallic surfaces in order to subsequently electrically characterize the NPs. We successfully executed a one-step method, as well as a two-step microcontact printing method to transfer the self-assembled film to Pt/Ir surfaces that acted as bottom electrodes during resistive switching experiments. The microcontact printing method especially paves the way for the preparation of structured devices. The electrical characterization of the self-assembled TiO_2 NP films on Pt/Ir bottom electrodes was performed by means of the nanorobotics setup SEM, as well as by LC-AFM. With both methods, we could unambiguously distinguish the Pt/Ir surface from the TiO_2 NP layer by their different electrical response. BRS-like behavior of the TiO_2 NP monolayer films was observed in SEM.

3. Materials and Methods

Solvothermal synthesis. TiO_2 NPs were synthesized modifying a solvothermal approach known from the literature, which allows the control of the NP morphology by adjusting the molar ratio of the $Ti(OBu)_4$/oleic acid/oleylamine (TB/OA/OAM) [36]. Oleic acid was purchased from Sigma Aldrich (Taufkirchen, Germany), oleylamine (C-18 content 80%–90%) from Acros Organics (Schwerte, Germany, and titanium butoxide ($Ti(OBu)_4$) (97% purity) from Sigma Aldrich (Taufkirchen, Germany). Typically, OAM and OA, as well as ethanol (EtOH) (absolute, 99% purity, Fisher Chemicals, Schwerte, Germany), were added in the Teflon inset, and stirred with a magnetic stirrer. After the addition of TB, stirring was continued for 10 min. The vessel was sealed with a Teflon lid, and set into a stainless steel autoclave. The autoclave was heated in a furnace to the reaction temperature for a determined time. Afterwards, the autoclave was cooled down to room temperature, and the product was transferred into 50 mL polystyrene tubes. Particle solutions were centrifuged and purified by washing with EtOH by suspension and centrifugation cycles. The obtained powder was dried at room temperature and characterized by powder XRD and TEM measurements. For the transmission electron microscopy analysis, the samples were dispersed in *n*-hexane (Riedel de Haen, 99% purity, Seelze, Germany) in an ultrasonic bath. The suspension was deposited on a carbon film copper mesh and dried. Best results were obtained with 1.44 mL TB (4.25 mmol), 4.04 mL OA (12.73 mmol), 2.79 mL OAM (8.48 mmol), and without EtOH. The stirring time was 8 min, and the autoclave was heated 18 h at 180 °C. Measurements were performed on the ZEISS LIBRA 200FE microscope (ZEISS, Oberkochen, Germany) in transmission mode operated at 200 kV. For the determination of particle size, at least 200 particles were counted, and the statistical analysis of the images was performed with the software ImageJ (Version 1.43u).

Formation of self-assembled TiO_2 NP monolayers. Self-assembled monolayer arrays of TiO_2 NPs were formed following an approach within the literature [33]. TiO_2 NPs prepared by solvothermal synthesis were dispersed in different non-polar solvents, or mixtures of the solvents, by ultrasonication. If the solution was turbid, despite continued ultrasonication, solutions of oleylamine or oleic acid in hexane were added until clear NP dispersions were obtained. A Teflon disk with a thickness of ca. 2 mm, an outer diameter of ca. 5 cm, and an inner circular hole with a diameter of ca. 2 cm was utilized for the self-assembly. The inner circular hole had to exhibit a sharp edge. The Teflon disk was placed on two 1 cm-high Al cubes standing in a Petri dish, and the whole setup was carefully leveled. Subsequently, tap water was filled into the Petri dish until the water surface contacted the underside of the disk. At this point, further tap water was slowly added with a Pasteur pipette until a concave upward curvature of the water surface was visible inside the inner circular hole of the Teflon disk. Drops of water were added until the water curvature changed to a slight convex upward curvature. The Petri dish was protected by a glass cylinder with a height of ca. 5 cm to minimize the influence of air currents on the surface. Approximately 0.4 mL of the NPs solution was gently dropped on the water surface, and the organic solvent was allowed to evaporate in the closed laboratory hood for 10 min. The air flow of the laboratory hood was measured with an anemometer. The colloid concentration, as well as the solvent or solvent mixtures, were investigated. The temperature and air velocity of the laboratory hood could not be controlled during the experiments. Best results were obtained with 0.56 mg/mL TiO_2 NPs in 2.07 mL pentane, 0.85 mL DCM (100% purity, VWR Chemicals, Langenfeld, Germany), and 50 μL of 7.6 mmol oleylamine in hexane solution.

For TEM characterization of the self-assembled TiO_2 NPs, the film on the water surface was lightly touched with the carbon-coated side of a carbon-coated cooper grid (S160, Plano, Wetzlar, Germany). Residual water was carefully removed with a tissue.

RS devices were prepared by two different approaches. For the one-step method, the TiO_2 NP films floating on the water surface were gently touched with the Pt/Ir surface, and afterwards allowed to dry. For the other approach, a microcontact printing two-step method was applied from the literature. For the preparation of the PDMS stamp, glass microscope slides were placed in a plastic weighing dish and covered with Canada Balsam (Sigma Life Science, Taufkirchen, Germany) and nail polish 3 in 1 XXXL shine (Essence Multi Dimensions, Sulzbach, Germany). Subsequently, a silicon wafer with a native SiO_2 oxide layer was glued to the glass substrate with the non-polished side, and dried for 30 min at 70 °C. Silicon oligomer and the catalyst of the SYLGARD® 184 Silicone Elastomer Kit (Dow Corning, Wiesbaden, Germany) were mixed in a 10:1 weight ratio and filled into the weighing dish, covering the polished SiO_2 surface, under stirring for ca. 10–15 min. After aging for 20 min, the polymerization process was continued at 70 °C for 3 h. Before usage, the PDMS stamps were peeled of the SiO_2 wafer, and cut into the corresponding shape for the Pt/Ir electrodes. Directly before usage, the stamps were immersed into hexane, and subsequently in EtOH, for 5 min each, and dried with a stream of N_2. A PDMS stamp was used to transfer the NP films from the water surface onto the Pt/Ir electrodes. The stamp was pressed lightly onto the films on the water surface, residual water was wiped off with a tissue, and finally, the stamp was pressed onto the Pt/Ir surface to transfer the TiO_2 NP films. The transferred films on the Pt/Ir surface were characterized by AFM. Prior to the electrical characterization in SEM, the (Pt/Ir)/TiO_2 NP film substrates were treated with O_2 plasma to remove residual oleylamine or oleic acid ligands. The electrical characterization was performed by means of the nanorobotics setup in SEM, as well as by means of LC-AFM. The SEM chamber, and thus the measurement tips, as well as the TiO_2 NP film, were treated with Ar plasma prior to the experiments.

Preparation of Pt/Ir bottom electrodes. Silicon wafers were cleaned in an ultrasonic bath with ultrapure water first, followed by EtOH, and then dried with N_2. A Ti adhesion layer with a thickness of 10 nm, and a Pt/Ir (80:20) alloy layer with a thickness of approximately 160 nm, were deposited on the wafers by direct current (DC) sputtering (0.01 mbar Ar/100 W).

Preparation of Pt/Ir coated tips. AFM tips with a spring constant of approximately 40 N m^{-1} and with a special geometry were purchased from ATEC-NC, Nanosensors, Wetzlar, Germany. The front part of the cantilever is visible from the top, and thus can be monitored in the SEM. The tips were isotropically coated with Pt/Ir by radio frequency (RF) sputtering (0.017 mbar Ar/40 W). Metal-coated tips were freshly prepared before the measurements, and measured in the SEM to exclude contamination or damage of the tips. The obtained coated probes had a radius of curvature of approximately 100 nm.

Electrical characterization with nanorobotics setup in SEM. The electrical characterization was performed in situ in a field-emission scanning electron microscope ZEISS Supra 35-VP (ZEISS, Oberkochen, Germany) using a nanorobotics setup (Klocke Nanotechnik GmbH, Aachen, Germany) and a semiconductor analyzer (Agilent 4156C, bsw TestSystem & Consulting AG, Ismaning, Germany). Detailed information about the setup is given elsewhere [40]. Prior to the measurement, the electric conductivity of the tips was determined by contacting two tips with each other and measuring voltage sweeps from -10 mV $\rightarrow$ 10 mV $\rightarrow$ -10 mV. Experiments were only continued if a linear *I–V* behavior was observed, and a resistance below 1000 Ω was measured. Typically, the probe/probe resistance was ca. 400–600 Ω. Additionally, before addressing a TiO_2 NP film spot, the probe was brought into contact with the Pt/Ir bottom electrode. Again, measurements were only continued if a linear *I–V* behavior was observed, and a resistance below 1000 Ω was measured. This control was repeated during the measurements. The voltage was applied to the Pt/Ir tip electrode, while the Pt/Ir film was grounded, and voltage sweeps from 0 V $\rightarrow$ X V $\rightarrow$ 0 V $\rightarrow$ $-$X V $\rightarrow$ 0 V, 0 V $\rightarrow$ $-$X V $\rightarrow$ 0 V $\rightarrow$ X V $\rightarrow$ 0 V were applied under high vacuum conditions (10^{-6} mbar). *I–V* curves were recorded with a current compliance (CC) to protect the metal coating of the tip electrode. Voltages and CC were varied during the experiments.

Electrical characterization with local conductive atomic force microscopy. LC-AFM measurements were performed at ambient pressure with a Cypher AFM from Asylum Research, Wiesbaden, Germany. Conductive diamond tips AppNano Doped Diamond with a radius of curvature of 100–300 nm were utilized.

Supplementary Materials: The following are available online at www.mdpi.com/2079-4991/7/11/370/s1, Figure S1: Histograms of the synthesized TiO_2 NPs' longitudinal and transversal (a,b), Figure S2: Exemplary TEM images of the synthesized TiO_2 NPs (a,b) and corresponding histograms of NPs' longitudinal (7.9 ± 2.2) nm and transversal (4.8 ± 0.8) nm (c,d), resulting in a mean particle diameter of (6.3 ± 1.9) nm. Powder XRD patterns of the NPs (black) and literature anatase data (red) (e) [38], Figure S3: Exemplary TEM images of the self-assembled TiO_2 NP film (a,b). Tapping mode AFM images of the TiO_2 NP film transferred to a Pt/Ir surface by the one-step method. (c,d). Corresponding height profile (e) taken along the white line in (d) showing the height differences of approximately 8 nm, Figure S4: SEM image without enhanced contrast of a TiO_2 NP film on the Pt/Ir surface, and on the left-hand side, the Pt/Ir coated tip electrode is visible, Figure S5: Three consecutive switching cycles recorded on a TiO_2 NP monolayer (a), and subsequent permanent LRS (b), Figure S6: SEM image of a TiO_2 NP layer after a SET process. The tip was lifted off the TiO_2 NP layer, revealing a morphologicaly change of the layer.

Acknowledgments: This work was financially supported by the DFG within SFB 917 "Nanoswitches". Venugopal Santhanam acknowledges Alexander von Humboldt foundation for a research fellowship. TEM measurements were performed by Frank Schiefer and Felix Schrader (Institute of Inorganic Chemistry, RWTH Aachen University 52074 Aachen, Germany) at the GFE—Gemeinschaftslabor für Elektronenmikroskopie der RWTH Aachen University, Ahornstrasse 55, 52074 Aachen.

Author Contributions: U.S., R.D., M.N., V.S., D.O.S. and N.R. conceived and designed the experiments; D.O.S., N.R. performed the experiments; D.O.S. and N.R. analyzed the data; D.O.S., N.R., M.N., V.S., R.D. and U.S. wrote the paper.

Conflicts of Interest: The authors declare no conflict of interest. The founding sponsors had no role in the design of the study; in the collection, analyses, or interpretation of data; in the writing of the manuscript, and in the decision to publish the results.

References

1. Moore, G.E. Cramming more components onto Integrated circuits. *Electronics* **1965**, *38*, 114–117. [CrossRef]
2. Moore, G.E. Progress in Digital Integrated Electronics. *IEDM Tech. Dig.* **1975**, *21*, 11–13.
3. Waldrop, M.M. The chips are down for Moore's law. *Nature* **2016**, *530*, 144–147. [CrossRef] [PubMed]
4. Waser, R.; Dittmann, R.; Staikov, G.; Szot, K. Redox-Based Resistive Switching Memories—Nanoionic Mechanisms, Prospects, and Challenges. *Adv. Mater.* **2009**, *21*, 2632–2663. [CrossRef]
5. Ielmini, D.; Waser, R. *Resistive Switching: From Fundamentals of Nanoionic Redox Processes to Memristive Device Applications*, 1st ed.; Wiley-VCH: Weinheim, Germany, 2016.
6. Ielmini, D.; Cagli, C.; Nardi, F.; Zhang, Y. Nanowire-based resistive switching memories: Devices, operation and scaling. *J. Phys. D Appl. Phys.* **2013**, *46*, 074006. [CrossRef]
7. Ju, Y.C.; Kim, S.; Seong, T.-G.; Nahm, S.; Chung, H.; Hong, K.; Kim, W. Resistance Random Access Memory Based on a Thin Film of CdS Nanocrystals Prepared via Colloidal Synthesis. *Small* **2012**, *8*, 2849–2855. [CrossRef] [PubMed]
8. Kim, J.-D.; Baek, Y.-J.; Jin Choi, Y.; Jung Kang, C.; Ho Lee, H.; Kim, H.-M.; Kim, K.-B.; Yoon, T.-S. Investigation of analog memristive switching of iron oxide nanoparticle assembly between Pt electrodes. *J. Appl. Phys.* **2013**, *114*, 224505. [CrossRef]
9. Kim, T.H.; Jang, E.Y.; Lee, N.J.; Choi, D.J.; Lee, K.-J.; Jang, J.; Choil, J.; Moon, S.H.; Cheon, J. Nanoparticle Assemblies as Memristors. *Nano Lett.* **2009**, *9*, 2229–2233. [CrossRef] [PubMed]
10. Hu, Q.; Jung, S.M.; Lee, H.H.; Kim, Y.-S.; Choi, Y.J.; Kang, D.-H.; Kim, K.-B.; Yoon, T.-S. Resistive switching characteristics of maghemite nanoparticle assembly. *J. Phys. D Appl. Phys.* **2011**, *44*, 085403. [CrossRef]
11. Yoo, J.W.; Hu, Q.; Baek, Y.-J.; Choi, Y.J.; Kang, C.J.; Lee, H.H.; Lee, D.-J.; Kim, H.-M.; Kim, K.-B.; Yoon, T.-S. Resistive switching characteristics of maghemite nanoparticle assembly on Al and Pt electrodes on a flexible substrate. *J. Phys. D Appl. Phys.* **2012**, *45*, 225304. [CrossRef]
12. Baek, Y.-J.; Hu, Q.; Yoo, J.W.; Choi, Y.J.; Kang, C.J.; Lee, H.H.; Min, S.-H.; Kim, H.-M.; Kim, K.-B.; Yoon, T.-S. Tunable threshold resistive switching characteristics of Pt-Fe_2O_3 core-shell nanoparticle assembly by space charge effect. *Nanoscale* **2013**, *5*, 772–779. [CrossRef] [PubMed]
13. Lee, J.-Y.; Baek, Y.-J.; Hu, Q.; Choi, Y.J.; Kang, C.J.; Lee, H.H.; Kim, H.-M.; Kim, K.-B.; Yoon, T.-S. Multimode threshold and bipolar resistive switching in bi-layered Pt-Fe_2O_3 core-shell and Fe_2O_3 nanoparticle assembly. *Appl. Phys. Lett.* **2013**, *102*, 122111. [CrossRef]
14. Younis, A.; Chu, D.; Mihail, I.; Li, S. Interface-Engineered Resistive Switching: CeO_2 Nanocubes as High-Performance Memory Cells. *ACS Appl. Mater. Interfaces* **2013**, *5*, 9429–9434. [CrossRef] [PubMed]
15. Younis, A.; Chu, D.; Li, C.M.; Das, T.; Sehar, S.; Manefield, M.; Li, S. Interface Thermodynamic State-Induced High-Performance Memristors. *Langmuir* **2014**, *30*, 1183–1189. [CrossRef] [PubMed]
16. Chu, D.; Lin, X.; Younis, A.; Li, C.M.; Dang, F.; Li, S. Growth and self-assembly of $BaTiO_3$ nanocubes for resistive switching memory cells. *J. Solid State Chem.* **2014**, *214*, 38–41. [CrossRef]
17. Li, C.; Beirne, G.J.; Kamita, G.; Lakhwani, G.; Wang, J.; Greenham, N.C. Probing the switching mechanism in ZnO nanoparticle memristors. *J. Appl. Phys.* **2014**, *116*, 114501. [CrossRef]
18. Kim, H.J.; Baek, Y.-J.; Choi, Y.J.; Kang, C.J.; Lee, H.H.; Kim, H.-M.; Kim, K.-B.; Yoon, T.-S. Digital versus analog resistive switching depending on the thickness of nickel oxide nanoparticle assembly. *RSC Adv.* **2013**, *3*, 20978–20983. [CrossRef]
19. Prakash, A.; Maikap, S.; Rahaman, S.Z.; Majumdar, S.; Manna, S.; Ray, S.K. Resistive switching memory characteristics of Ge/GeO_x nanowires and evidence of oxygen ion migration. *Nanoscale Res. Lett.* **2013**, *8*, 220. [CrossRef] [PubMed]
20. Younis, A.; Chu, D.; Li, S. Tuneable resistive switching characteristics of In_2O_3 nanorods array via Co doping. *RSC Adv.* **2013**, *3*, 13422–13428. [CrossRef]
21. Goren, E.; Ungureanu, M.; Zazpe, R.; Rozenberg, M.; Hueso, L.E.; Stoliar, P.; Tsur, Y.; Casanova, F. Resistive switching phenomena in TiO_x nanoparticle layers for memory applications. *Appl. Phys. Lett.* **2014**, *105*. [CrossRef]
22. Uenuma, M.; Ban, T.; Okamoto, N.; Zheng, B.; Kakihara, Y.; Horita, M.; Ishikawa, Y.; Yamashita, I.; Uraoka, Y. Memristive nanoparticles formed using a biotemplate. *RSC Adv.* **2013**, *3*, 18044–18048. [CrossRef]
23. Schmidt, D.O.; Hoffmann-Eifert, S.; Zhang, H.; La Torre, C.; Besmehn, A.; Noyong, M.; Waser, R.; Simon, U. Resistive Switching of Individual, Chemically Synthesized TiO_2 Nanoparticles. *Small* **2015**, *11*, 6444–6456. [CrossRef] [PubMed]

24. Szot, K.; Rogala, M.; Speier, W.; Klusek, Z.; Besmehn, A.; Waser, R. TiO_2—A prototypical memristive material. *Nanotechnology* **2011**, *22*, 254001–254022. [CrossRef] [PubMed]
25. Yang, J.J.; Inoue, I.H.; Mikolajick, T.; Hwang, C.S. Metal oxide memories based on thermochemical and valence change mechanisms. *MRS Bull.* **2012**, *37*, 131–137. [CrossRef]
26. Strachan, J.P.; Yang, J.J.; Montoro, L.A.; Ospina, C.A.; Ramirez, A.J.; Kilcoyne, A.L.D.; Medeiros-Ribeiro, G.; Williams, R.S. Characterization of electroforming-free titanium dioxide memristors. *Beilstein J. Nanotechnol.* **2013**, *4*, 467–473. [CrossRef] [PubMed]
27. Hermes, C.; Bruchhaus, R.; Waser, R. Forming-Free TiO_2-Based Resistive Switching Devices on CMOS-Compatible W-Plugs. *IEEE Electron Device Lett.* **2011**, *32*, 1588–1590. [CrossRef]
28. Warneck, P.; Williams, J. *The Atmospheric Chemist's Companion, Numerical Data for Use in the Atmospheric Sciences*; Springer: New York, NY, USA, 2012.
29. Sun, S.; Murray, C.B.; Weller, D.; Folks, L.; Moser, A. Monodisperse FePt Nanoparticles and Ferromagnetic FePt Nanocrystal Superlattices. *Science* **2000**, *287*, 1989–1992. [CrossRef] [PubMed]
30. Giersig, M.; Mulvaney, P. Preparation of ordered colloid monolayers by electrophoretic deposition. *Langmuir* **1993**, *9*, 3408–3413. [CrossRef]
31. Huie, J.C. Guided molecular self-assembly: A review of recent efforts. *Smart Mater. Struct.* **2003**, *12*, 264. [CrossRef]
32. Wen, T.; Majetich, S.A. Ultra-Large-Area Self-Assembled Monolayers of Nanoparticles. *ACS Nano* **2011**, *5*, 8868–8876. [CrossRef] [PubMed]
33. Santhanam, V.; Liu, J.; Agarwal, R.; Andres, R.P. Self-Assembly of Uniform Monolayer Arrays of Nanoparticles. *Langmuir* **2003**, *19*, 7881–7887. [CrossRef]
34. Davi, M.; Keßler, D.; Slabon, A. Electrochemical oxidation of methanol and ethanol on two-dimensional self-assembled palladium nanocrystal arrays. *Thin Solid Films* **2016**, *615*, 221–225. [CrossRef]
35. Grzelczak, M.; Vermant, J.; Furst, E.M.; Liz-Marzán, L.M. Directed Self-Assembly of Nanoparticles. *ACS Nano* **2010**, *4*, 3591–3605. [CrossRef] [PubMed]
36. Dinh, C.-T.; Nguyen, T.-D.; Kleitz, F.; Do, T.-O. Shape-Controlled Synthesis of Highly Crystalline Titania Nanocrystals. *ACS Nano* **2009**, *3*, 3737–3743. [CrossRef] [PubMed]
37. Shircliff, R.A.; Stradins, P.; Moutinho, H.; Fennell, J.; Ghirardi, M.L.; Cowley, S.W.; Branz, H.M.; Martin, I.T. Angle-Resolved XPS Analysis and Characterization of Monolayer and Multilayer Silane Films for DNA Coupling to Silica. *Langmuir* **2013**, *29*, 4057–4067. [CrossRef] [PubMed]
38. *Inorganic Crystal Structure Database*; ICSD #9852; FIZ Karlsruhe.
39. Santhanam, V.; Andres, R.P. Microcontact Printing of Uniform Nanoparticle Arrays. *Nano Lett.* **2004**, *4*, 41–44. [CrossRef]
40. Noyong, M.; Blech, K.; Rosenberger, A.; Klocke, V.; Simon, U. In-situ nanomanipulation system for electrical measurements in SEM. *Meas. Sci. Technol.* **2007**, *18*, N84–N89. [CrossRef]
41. Xiao, P.F.; Lai, M.O.; Lu, L. Electrochemical properties of nanocrystalline TiO_2 synthesized via mechanochemical reaction. *Electrochim. Acta* **2012**, *76*, 185–191. [CrossRef]
42. Kwon, D.-H.; Kim, K.M.; Jang, J.H.; Jeon, J.M.; Lee, M.H.; Kim, G.H.; Li, X.-S.; Park, G.-S.; Lee, B.; Han, S.; et al. Atomic structure of conducting nanofilaments in TiO_2 resistive switching memory. *Nat. Nanotechnol.* **2010**, *5*, 148–153. [CrossRef] [PubMed]
43. Dittmann, R.; Muenstermann, R.; Krug, I.; Park, D.; Menke, T.; Mayer, J.; Besmehn, A.; Kronast, F.; Schneider, C.M.; Waser, R. Scaling Potential of Local Redox Processes in Memristive $SrTiO_3$ Thin-Film Devices. *Proc. IEEE* **2012**, *100*, 1979–1990. [CrossRef]
44. Cooper, D.; Baeumer, C.; Bernier, N.; Marchewka, A.; La Torre, C.; Dunin-Borkowski, R.E.; Menzel, S.; Waser, R.; Dittmann, R. Anomalous Resistance Hysteresis in Oxide ReRAM: Oxygen Evolution and Reincorporation Revealed by In Situ TEM. *Adv. Mater.* **2017**, *29*, 1700212. [CrossRef] [PubMed]
45. Andersson, S.; Magnéli, A. Diskrete Titanoxydphasen im Zusammensetzungsbereich $TiO_{1,75}$-$TiO_{1,90}$. *Naturwissenschaften* **1956**, *43*, 495–496. [CrossRef]

nanomaterials

MDPI

Article

Thermally Stimulated Currents in Nanocrystalline Titania

Mara Bruzzi [1,*], Riccardo Mori [2], Andrea Baldi [3], Ennio Antonio Carnevale [3], Alessandro Cavallaro [4] and Monica Scaringella [5]

1 Dipartimento di Fisica e Astronomia, Università di Firenze, Via G. Sansone 1, Sesto Fiorentino, 50019 Firenze, Italy
2 Albert-Ludwigs-Universität Freiburg, Experimentelle Teilchenphysik, Physikalisches Institut, Hermann-Herder Straße 3, 79104 Freiburg im Breisgau, Germany; riccardo.mori@physik.uni-freiburg.de
3 Dipartimento di Ingegneria Industriale, Università di Firenze, Via S. Marta 3, 50139 Firenze, Italy; andrea.baldi@unifi.it (A.B.); ennio.carnevale@unifi.it (E.A.C.)
4 LBT Observatory, University of Arizona, 933 N. Cherry Ave, Tucson, AZ 85721, USA; cavallaro@lbto.org
5 Dipartimento di Ingegneria dell'Informazione, Università di Firenze, Via S. Marta 3, 50139 Firenze, Italy; monica.scaringella@unifi.it
* Correspondence: mara.bruzzi@unifi.it; Tel.: +39-055-4572291

Received: 31 October 2017; Accepted: 12 December 2017; Published: 5 January 2018

Abstract: A thorough study on the distribution of defect-related active energy levels has been performed on nanocrystalline TiO_2. Films have been deposited on thick-alumina printed circuit boards equipped with electrical contacts, heater and temperature sensors, to carry out a detailed thermally stimulated currents analysis on a wide temperature range (5–630 K), in view to evidence contributions from shallow to deep energy levels within the gap. Data have been processed by numerically modelling electrical transport. The model considers both free and hopping contribution to conduction, a density of states characterized by an exponential tail of localized states below the conduction band and the convolution of standard Thermally Stimulated Currents (TSC) emissions with gaussian distributions to take into account the variability in energy due to local perturbations in the highly disordered network. Results show that in the low temperature range, up to 200 K, hopping within the exponential band tail represents the main contribution to electrical conduction. Above room temperature, electrical conduction is dominated by free carriers contribution and by emissions from deep energy levels, with a defect density ranging within 10^{14}–10^{18} cm^{-3}, associated with physio- and chemi-sorbed water vapour, OH groups and to oxygen vacancies.

Keywords: thermally stimulated currents; photocurrent; titanium dioxide; hopping; nanoporous film; desorption current; chemisorbed current

1. Introduction

The study of the electronic transport in nanocrystalline Titanium dioxide (nc-TiO_2) is motivated by its wide range of application, from catalysis to green energy systems such as Dye Sensitized Solar Cells (DSSCs) [1] and toxic gas sensing devices [2]. In fact, performances of devices based on nc-TiO_2 strongly depend on electron transport mechanisms, which can be very different from those dominating in the bulk single-crystal semiconductor, due to the complex morphology structure and to the huge active surface of this porous material with respect to its volume. Investigation of the electrical transport properties of nc-TiO_2 and its relationship with surface and bulk defects is thus strategic in the perspective of increasing their performance. A model of the electrical conductivity in nc-TiO_2 that takes into account all the complexities of this material (disorder, fractional dimensionality of the nanoporous material and potential barriers between the constituent nanoparticles) is still lacking; its implementation is a cumbersome task. In order to progress in this direction, a systematic characterisation of defects states, in the widest range of energy and an evaluation of their concentration in the material is mandatory.

The thermally stimulated current technique is one of the most effective tool for characterizing electrical defects in semiconductors. In this paper, we apply this method to get an overall picture of defect distribution in nanocrystalline Titania used in state-of-art DSSC devices. For ordinary semiconductors the well established models of the Thermally Stimulated Currents (TSC) give a quantitative description of the experimental results and then allow the extraction of the parameter values; for disordered semiconductors in general, and more specifically for nanocrystalline mesoporous materials, only a few studies are available and, as a consequence of their complexity, of difficult implementation for a numerical fit of the experimental data [3]. These latter give evidence that TiO_2 nanoparticles have a great number of bulk and surface defects inducing a structural and energetic disorder, and, as a consequence, a continuum of levels into the band gap [4]; but the experimental frame is even more complex because when a film is deposited, the nanoparticles give rise to a porous network, with a porosity typically ranging between 0.5% and 0.7% [5], where a large amount of them are separated by potential barriers [6]. A theory of the electrical conduction in disordered and nanocrystalline materials is the basis for a model of the conductivity in the mesoporous TiO_2 [7] and several studies have been devoted to establish if the hopping or multi-trapping mechanism prevails. The conduction in the hopping mechanism is given by the carrier tunneling between different sites, and in the multi-trapping mechanism by a series of successive events where the carriers are temporarily promoted from localized levels to the conduction band; unfortunately the experimental signatures of the two mechanisms are indistinguishable and then what is the prevailing mechanism is still debated [8–11]. To take into account the porosity of the medium or the interparticle barrier get more complex the theoretical description of the conductivity and we must refer ourselves to models of a disordered medium of fractional dimensionality, i.e., intermediate between 2D and 3D, or to models of a network of perfect nanoparticles; specifically as far as TiO_2 is concerned, to our knowledge, these aspects have been considered only at a qualitative level [5]. In consideration of the above sketched complexities some strong simplification is necessary in order to extract quantitative information concerning the carrier transport, and specifically the TSC.

The model of Thermally Stimulated Currents (TSC) used to interpret our experimental results, takes account of the heavily disordered microscopic nature of nc-TiO_2. In this work, we adopt the model usually considered for electrical transport in nc-TiO_2 which treats this material as a strongly disordered 3D medium and neglects the potential barriers at the grain interface [8,12]; this, despite the strong simplification that it implies, is able to give reason of the main features of the electrical transport and is simple enough to allow the implementation of a numerical algorithm useful for fitting the experimental data. In our model we avoid any a priori assumption about the conduction mechanism, and this is essential because, as we will show, in general both the hopping and free carrier, i.e., the multitrapping, contributions must be taken into account for the electrical conductivity [13]: the possible prevalence of the one or the other depends on the temperature and the Fermi level position. Developed from reference models given in [13–15] and first applied in [16] to get an insight about the prevalence of hopping against multitrapping conductivity in nanocrystalline titania below 200 K, in this paper the model has been further extended to determine main parameters of the band tail, as energy decay constant, defect and recombination centres concentrations, hopping frequency factors. Moreover, our results, extended up to 630 K, show that hopping conductivity can be neglected over 200 K, where conductivity appears to be dominated by discrete energy levels in the forbidden gap, probably related to water, OH groups and oxygen vacancies, which role in transport properties is nowadays still a matter of intense debate [17,18].

2. Mixed Conduction Model in nc-TiO_2

In heavily disordered semiconductors, significant carrier hopping can occur between localized sites, whose energies fall within the band gap, giving a non-negligible contribution to the electrical conductivity. As a result, mixed conduction mechanisms with hopping and free carriers contributions should be taken into account when discussing the electrical transport in such a material, especially at low-to-moderate temperatures and high carrier densities.

As it is usually done, we consider the nc-TiO_2 as a strongly disordered intrinsic semiconductor whose charge carriers are electrons: then the Density of States (DoS) is modelled in the simplest way as the sum of the extended state contribution (i.e., the conduction band), with the typical square root dependence on energy, and the localized state contribution characterized by an exponential tail, below the conduction band, as for amorphous materials [19]. In fact, this model neglects the DoS features in the intermediate region connecting the exponential Urbach tail and the conduction band, whose description has been object of several studies [20]. Such expressions are rather complex and their use, even at the cost of the computation time, do not provide substantial improvements to the fitting procedure. Moreover, in our measurements the Fermi level never attaints this region and the hopping, as well the multi-trapping, take place over an extended part of the DoS, then coming rather insensitive to the DoS details. Then, the density of states in the overall energy range, putting at $\varepsilon = 0$ the border between extended and localized states, is given by the function:

$$g(\varepsilon) = \frac{N_L}{E_L}\theta(-\varepsilon)e^{\frac{\varepsilon}{E_L}} + \theta(\varepsilon)N_{C0}\sqrt{\varepsilon + E_B}, \tag{1}$$

with $\theta(\varepsilon)$ Heaviside function, E_L tailing factor, $N_L = \int_{-\infty}^{0} g(\varepsilon)d\varepsilon$ total density of localized states within the band gap. N_{C0} is related to the effective density of states at the conduction band minimum. The value $N_{C0} = 4 \times 10^{19}$ cm^{-3} · meV$^{-3/2}$, used for all subsequent calculations, has been chosen in order to get at room temperature $N_C = 5.2 \times 10^{21}$ cm^{-3}, typical for TiO_2. The energy parameter:

$$E_B = \left(\frac{N_L}{N_{C0}E_L}\right)^2, \tag{2}$$

is introduced so as to have the continuity of $g(\varepsilon)$ at $\varepsilon = 0$. The DoS obtained with this procedure is shown in Figure 1.

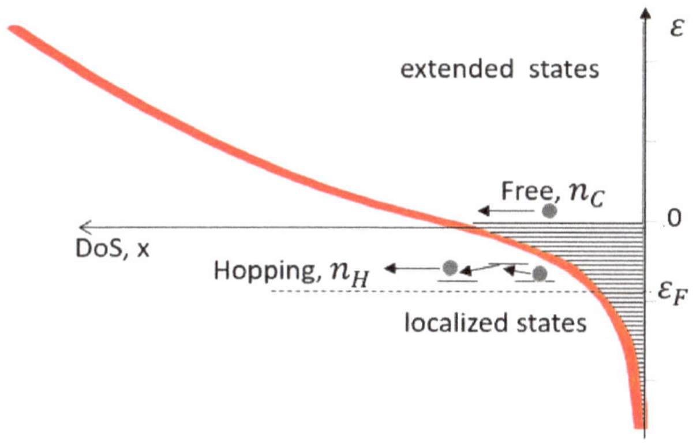

Figure 1. Density of states (DoS) as obtained with an exponential decay tail in the forbidden gap and a square root dependence in the conduction band region. The boundary is kept at $\varepsilon_C = 0$, minimum of the conduction band. Conduction is due to both free electrons in the conduction band and to hopping at localized states.

Mixed electrical conductivity, in case of a n-type nc-TiO_2 layer, is the result of transport processes of both hopping electrons within the band tail and free electrons in the conduction band, as depicted in Figure 1, then for the total conductivity we have:

$$\sigma = \sigma_H + \sigma_C = q\mu_H n_H + q\mu_C n_C, \tag{3}$$

where σ_H and σ_C, μ_H and μ_C, n_H and n_C, are conductivity, mobility and electron concentration of hopping carriers and free carriers respectively; q is the electron charge. In (3) μ_H is an effective mobility that must be calculated taking into account that the hopping probability depends on the site. Assuming the validity of the Einstein relation, the hopping mobility is related to the carrier energy through the diffusivity $D(\varepsilon)$:

$$\mu(\varepsilon) = \frac{q}{K_B T} D(\varepsilon) = \frac{q}{K_B T} r(\varepsilon)^2 \nu, \tag{4}$$

with ν hopping rate and $r(\varepsilon)$ average distance between hopping sites available for hopping for carrier with energy ε. The rate for hopping from site i to site j, respectively characterized by energies ε_i and ε_j, is given by the Miller-Abrahams [21] model:

$$\begin{aligned} \nu &= \nu_0 e^{-2\frac{r_{ij}}{\alpha} - \frac{\varepsilon_j - \varepsilon_i}{K_B T}}, \quad \varepsilon_i < \varepsilon_j \\ \nu &= \nu_0 e^{-2\frac{r_{ij}}{\alpha}}, \quad \varepsilon_i > \varepsilon_j \end{aligned} \tag{5}$$

where r_{ij} is the distance between sites at ε_i and at ε_j, α is the localization radius of the electron and ν_0 is the hopping frequency coefficient.

The localization radius can be assumed as a constant, of the order of a few angstrom, for states deeply localized, but when approaching the boundary $\varepsilon = 0$ with the extended states this length diverges. Then, as the hopping probability depends exponentially on this length, the description of the conduction for carriers localized in states near the mobility edge must take into account this behavior. This divergence, according to several models, behaves as $\varepsilon^{-\gamma}$ with γ ranging between 0.5 and 1.5 [19]. Here, we assume for the localization length α:

$$\alpha(\varepsilon) = \alpha_0 \left(1 - \frac{z}{\varepsilon}\right)^{\gamma}, \tag{6}$$

with α_0 carrier localization length of the deep states and z the energy to which α starts to diverge. The evaluation of the hopping conductivity requires some kind of approximation in order to get an expression useful for fitting experimental data. A considerable simplification is usually obtained referring to the "transport energy" level [8,14,22], that plays for hopping the same role of the mobility edge for the free carriers, but for a quantitave description of the TSC is necessary to go beyond this approach [13]. Then, to this purpose, we will follow the approach proposed by Nagy [15] where the hopping conductivity can be calculated as:

$$\sigma_H(T) = q \int \mu \, dn = \nu_0 \frac{q^2}{K_B T} \int_{-\infty}^{0} g(\varepsilon) f(e, E_F, T) r(\varepsilon)^2 e^{-2\frac{r(\varepsilon)}{\alpha(\varepsilon)}} d\varepsilon, \tag{7}$$

where $r(\varepsilon)$ is an average hopping distance for a site of energy ε, defined in term of the concentration of the available (unoccupied) final states:

$$\begin{aligned} \frac{1}{r(\varepsilon)^3} &= \frac{4\pi}{3BR_L^3} \left(\int_{-\infty}^{\varepsilon} \frac{g(\varepsilon')}{N_L} [1 - f(\varepsilon', E_F, T)] d\varepsilon' + \int_{\varepsilon}^{0} \frac{g(\varepsilon')}{N_L} [1 - f(\varepsilon', E_F, T)] e^{\frac{\varepsilon - \varepsilon'}{KT}} d\varepsilon' \right), \\ \frac{1}{R_L{}^3} &= \frac{4\pi}{3} \int_{-\infty}^{0} g(\varepsilon') d\varepsilon' = \frac{4\pi}{3} N_L \end{aligned} \tag{8}$$

$$f(\varepsilon, E_F, T) = 1 - e^{\frac{\varepsilon - \varepsilon_F}{KT}} + \frac{1}{2} e^{\frac{3(\varepsilon - \varepsilon_F)}{2KT}} \quad \varepsilon \le \varepsilon_F$$

$$f(\varepsilon, E_{\mathrm{F}}, T) = e^{-\frac{\varepsilon-\varepsilon_{\mathrm{F}}}{KT}} - \frac{1}{2}e^{-\frac{3(\varepsilon-\varepsilon_{\mathrm{F}})}{2KT}} \quad \varepsilon \geq \varepsilon_{\mathrm{F}}$$

with $f(\varepsilon, E_{\mathrm{F}}, T)$, Fermi-Dirac distribution function, α given by Equation (6) and B the percolative limit factor, namely the average number of site links, a parameter dependent on the system dimensionality (B = 2.7 for 3D hopping [14]). The integral in Equation (8) is splitted in two terms the first one taking into account, for the level of energy ε , the number of sites reachable with a down conversion process, having probability 1, the second the number of the sites reachable with a thermally activated up conversion process whose probability is given by the Boltzmann factor.

The Fermi energy E_F is obtained by evaluating the total electron concentration n as the sum of the electron density in the band tail ($\varepsilon < 0$), n_{H}, and in the conduction band ($\varepsilon > 0$), n_{C}; then $n = n_{\mathrm{C}} + n_{\mathrm{H}}$, where:

$$n_{\mathrm{H}} = \int_{-\infty}^{0} g(\varepsilon) f(\varepsilon, E_{\mathrm{F}}, T) d\varepsilon, \quad n_{\mathrm{C}} = \int_{0}^{\infty} g(\varepsilon) f(\varepsilon, E_{\mathrm{F}}, T) d\varepsilon, \tag{9}$$

Finally, the contribution of free electrons to the electrical conductivity is given by the typical expression of the conductivity for electrons in the conduction band:

$$\sigma_{\mathrm{C}}(T) = q\mu_{\mathrm{C}} n_{\mathrm{C}} = q\frac{\mu_{\mathrm{C0}}}{(K_{\mathrm{B}}T)^{3/2}} n_{\mathrm{C}}, \tag{10}$$

with n_{C} as calculated from Equation (9). A typical value for the TiO_2 mobility constant is μ_{C0} = 1 cm^2 meV$^{3/2}$/Vs, which gives a room temperature mobility $\mu_{\mathrm{C}} = 7.5 \times 10^{-3}$ cm^2/Vs [23].

3. Rate Equations for Priming and Thermally Stimulated Process

The rate equation for hopping and free carriers is:

$$\frac{dn}{dt} = -\left(\frac{dn_{\mathrm{H}}}{dt}\right) - \frac{n_{\mathrm{C}}}{\tau_{\mathrm{C}}} + S(t), \tag{11}$$

where $S(t)$ is the generation rate during priming (e.g., light exposure, null during the TSC scan). Here conduction carriers decay is considered via annihilation on recombination centers or trapping from deep levels, characterized by an active energy level within the gap. The free electron lifetime, τ_{C}, is typically dependent on the defect capture cross section σ_{t} and concentration N_{t}:

$$\frac{1}{\tau_{\mathrm{C}}} = N_{\mathrm{t}}\sigma_{\mathrm{t}}\langle v_{\mathrm{th}}\rangle = N_{\mathrm{t}}\sigma_{\mathrm{t}}\sqrt{\frac{3K_{\mathrm{B}}T}{m}}, \tag{12}$$

where m is the effective mass of the free carriers (in TiO_2 it is about 7 times the electron mass m_0 [24]) and $\langle v_{\mathrm{th}}\rangle$ their average thermal velocity. In general, in a disordered semiconductor, defects may have a spread in energy, so its concentration is calculated through a gaussian distribution:

$$N_{\mathrm{t}} = \frac{N_{\mathrm{t0}}}{\sqrt{2\pi}\sigma_{E_{\mathrm{t}}}} \int e^{-\frac{(\varepsilon-E_{\mathrm{t}})^2}{2\sigma_{E_{\mathrm{t}}}{}^2}} d\varepsilon \tag{13}$$

The decay of the hopping carriers, in turn, is given as [14]:

$$\left(\frac{dn_{\mathrm{H}}}{dt}\right)_{dec} = -\int_{-\infty}^{0} \frac{g(\varepsilon)f(\varepsilon)}{\tau(T,\varepsilon)} d\varepsilon, \tag{14}$$

$$\frac{1}{\tau(T,\varepsilon)} = N_{\mathrm{t}}\frac{r(\varepsilon)^2 D(\varepsilon)}{\alpha} = \nu_0 N_{\mathrm{t}}\frac{r(\varepsilon)^4}{\alpha} e^{-2\frac{r(\varepsilon)}{\alpha}}, \tag{15}$$

In the following, we will describe results of decayed/fractionated Thermally Stimulated Currents (TSC) experiments [25] analyzed with the mixed conduction model described above. In this method, after priming the sample only once at a low temperature, successive cycles of heating/cooling are applied to fractionally deplete levels of lowering energy. The sample is heated up to a first maximum temperature T_{stop} and cooled down to a first minimum temperature T_{start} then it is heated and cooled again to higher T values and so on, eventually up to the final temperature.

At first, we have used the decayed TSC method to get useful information about the DoS shape. For each scan of the decayed TSC, we extracted the couple (E_{act}, Q) with E_{act} = activation energy as determined from the initial rise of the *TSC* and Q = emitted charge calculated by integrating the TSC of the corresponding scan [25]. In this way, we obtained the DoS shape as a function of the energy by plotting the charge released at each step as a function of the activation energy E_{act}.

Then, delayed TSC measurements have been best fitted using the mixed conduction model, considering a constant heating/cooling rate: $\beta = \frac{dT}{dt}$, starting from the initial condition due to priming, stated that in between two scans there is a time delay in which the electrical state of the sample evolve very slowly. So, assuming the initial state got from priming, it is possible to fit the entire sequence of the delayed TSC at once. The rate equation is solved starting from an evaluation of the Fermi level, then the distance between hopping sites, the hopping carrier density and the average hopping rate are obtained and finally iterating for each temperature the calculation is performed for the entire experimental data set.

In the high temperature range, above room temperature, when conductivity is mainly due to free carriers and hopping can be neglected, in our model TSC is considered as dominated by deep centers with discrete energy levels E_t in the forbidden gap, characterized by a capture cross section σ_n (for electrons) and a trap N_t concentration. So, we consider the standard TSC expression as [26]:

$$I_{TSC}(T) = q\mu_C(T) \sum N_t F e_n(T) e^{-\frac{1}{\beta}\int_{T_i}^{T} e_n(T)dT}, \tag{16}$$

with $e_n(T) = N_C \sigma_n v_{th} e^{-\frac{E_C - E_t}{k_B T}}$ emission constant, Σ surface normal to electric field F, T_i initial temperature of the scan. Due to the disorder in the nanocrystalline material, the TSC peak usually results in a peak broader in temperature than the standard one. This is due to the fact that the energy E_t of a defect varies within a certain range due to local morphological changes. We have taken into account that by convoluting the TSC peak with a gaussian distribution, as given in Equation (13).

4. Experimental Set-Up and Procedure

To manufacture our samples, we used a colloidal system produced by Solaronix (Aubonne, Switzerland), containing about 11 wt %. nanocrystalline titanium dioxide mixed with optically dispersing anatase particles (13/400 nm, Ti-Nanoxide D/SP). This commercial product, specifically developed for prototypal electrodes in DSSCs is an anatase titania particle paste for the deposition of active opaque layers. A mixing of large, 400 nm average size, and small, 13 nm average size, nano-particles ensures both very high surface area and efficient light diffusion. We deposited the nc-TiO_2 paste on alumina substrates having two parallel gold contacts, 7 mm long and spaced 0.8 mm; thickness of the film is about 1 µm. A picture of the sample is shown in Figure 2. After deposition, the films have been syntherized in two steps, 30 min each, first at 280 °C and then at 450 °C. The current-voltage characteristics of the sample showed an ohmic behaviour in the overall investigated range (0–100 V) with room temperature resistivity of the order of 2×10^8 Ωm [27]. In a typical DSSCs, with a 2 µm-thick nc-TiO_2 film, a voltage of about 0.5 V is applied and an average electric field of about 2.5×10^5 V/m is settled. In our TSC measurements we therefore chose to apply a bias of 100 V across the sample, in view to get an electric field of the same order of magnitude, considering the increased distance between our planar electrodes.

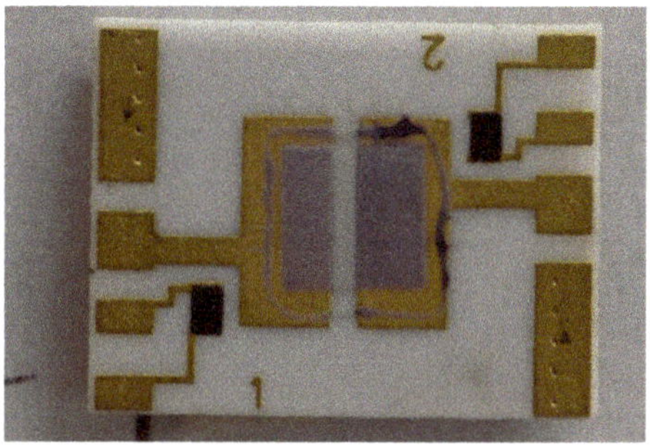

Figure 2. Nc-TiO_2 film deposited on an alumina chip for Thermally Stimulated Currents (TSC) measurements.

To perform TSC measurements in the temperature range 5–300 K, the alumina substrates coated with the nc-TiO_2 have been placed in a sample holder equipped with a 4 Ω wounded-wire heating resistor and a silicon temperature sensor (Leybold GmbH, Köln, Germany). The sample-holder has been inserted into a dewar containing liquid He (LHe) and positioned over the LHe vapours to ensure stable temperatures down to 4.2 K, minimize thermal inertia and reduce possible mismatch between the sample and the thermometer. Details of the experimental setup are given in [28]. Polarization of the sample and current reading was performed by a Keithley 6517 electrometer (Tektronix Ltd., Berkshire, UK), the heater was biased by a TTi QL564P power supply (TTI, Inc., Maisach-Gernlinden, Germany) and temperature was read by a DRC91C temperature controller (Lake Shore Cryotronics, Inc., Westerville, OH, USA). Priming was performed by a Light Emitting Diode (LED) source placed in front of the sample inside the sample holder. We used two priming sources: a 400 nm Ultra Violet (UV) and a 355 nm UV LED (Roithner-Lasertechnik, Vienna, Austria), having 12 mW (typical) and 8.4 mW (maximum) output power, respectively. LEDs were driven by a Systron Donner 110D pulse generator (Systron Dr, Concord, CA, USA). The light spot on the sample during illumination has a diameter of about 2 mm.

Delayed TSC measurements have been performed as follows. We primed the sample at a low temperature T_0 with the LED source, biasing the sample at 100 V. Then, we waited a time interval to make fast transient effects relaxing and to get a constant temperature on the whole sample. Then, fractionated TSC analysis has been carried out performing different heating/cooling cycles up to 300 K. TSC has been also studied in the temperature range from 300 to 630 K using a different chamber where heating/cooling is performed by a system controlling temperature, pressure and gas composition. During each TSC measurement, both in the low and high temperature ranges, the scan rate was fixed at 0.1 K/s.

5. Experimental Results and Discussion

Figure 3 shows a typical TSC spectrum observed in the overall range 5–630 K obtained with the procedure described in the previous section. A fractionated TSC is performed up to 300 K, after priming at 5 K. Then, a second priming is carried out at 300 K inside the high temperature TSC setup and a second TSC analysis is performed up to 630 K. The low temperature analysis is divided into 4 TSC fractions in the ranges: 5–20, 20–80, 80–180, 180–300 K, then, a unique TSC curve is measured after priming at 300 K up to 630 K. At last, a final cooling step from 300 to 250 K is measured to close the whole cycle.

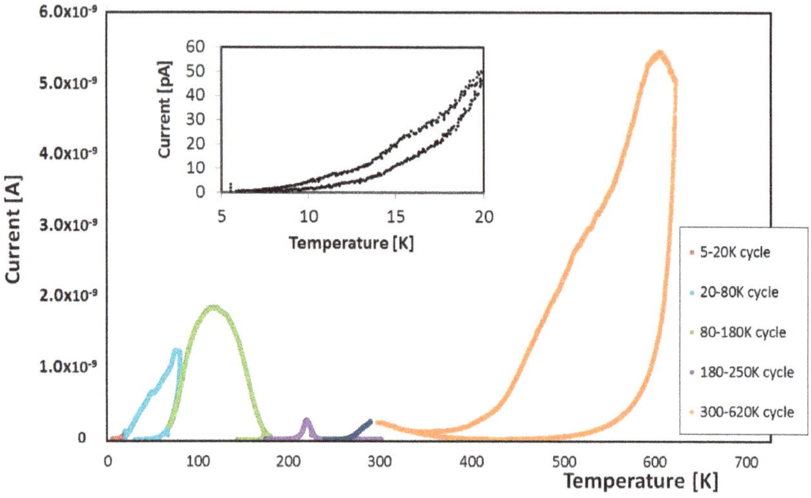

Figure 3. Fractionated TSC analysis performed in different cycles of measurements after ultraviolet (UV) priming at 5 K. Inset on the left shows an enlarged view of the 5–20 K temperature range.

Main conductivity processes in the two temperature ranges 5–300 K and 300–630 K are different. In fact, in the low T range, hopping conduction is non- negligible against free conduction. Conversely, in the high T range conductivity is mainly due to free carriers. Moreover, in this latter case, the influence of the surrounding gas atmosphere to the charge state of deep discrete levels in the forbidden gap cannot be neglected. Thus, in the following, we will discuss separately results measured below and above room temperature.

5.1. TSC Analysis below Room Temperature

Figure 4 compares the TSC spectrum measured in the low T range together with TSC data reported in past for single crystal TiO_2 [29]. The experimental peaks for sc-TiO_2 have been multiplied respectively by 1700 (120 K) and 500 (230 K) to compare with those measured with the nc-TiO_2 ones. The lower TSC signal observed in single-crystal TiO_2 indicates that defect concentrations here is far below those encountered in the nanocrystalline morphology.

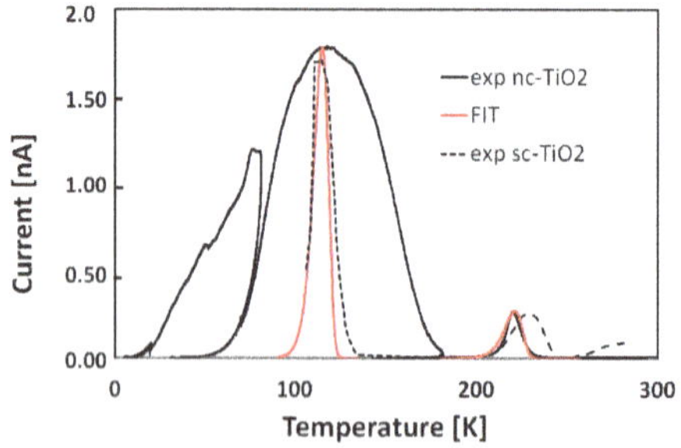

Figure 4. Fractional TSC measured with the nc-TiO_2 film in the range 5–300 K compared with experimental TSC peaks reported in [29] for single crystal TiO_2 and with a best fit obtained considering a standard TSC emission from discrete energy levels. The experimental peaks for sc-TiO_2 have been multiplied respectively by 1700 (120 K) and 500 (230 K) to compare with those measured with the nc-TiO_2 ones.

In single crystal TiO_2, two peaks related to two discrete energy levels at 120 and 230 K are present, which can be described in terms of standard TSC emissions [26] as given in Equation (16). In nanostructured TiO_2, we observe a much broader peak at 120 K, which cannot be described in terms of standard single-level *TSC* emission, and a peak at 220 K, rather similar to the one measured in single crystal TiO_2. In the figure, a best fit of this latter peak is shown, obtained with a standard TSC analysis (Equation (16)), a very high capture cross section, $\sigma_n \approx 10^{-10}$ m^2 and $E_t \approx 0.8$ eV, $N_t \approx 10^{14}$ cm^{-3}.

To investigate the origin of the broad band peaked at 120 K, we then performed a measurement in the same temperature range with more delayed heating steps (up to 10). Results are shown in the inset of Figure 5. We note that, in a standard thermally activated emission where carriers are emitted from discrete energy levels towards the corresponding extended band, the current measured at the foot of the TSC curve in each heating/cooling step has a dependence on temperature given by: $I(T) \propto T^2 e^{-\frac{E_{act}}{K_B T}}$, while if hopping conduction dominates, the dependence should be (Mott's expression): $I(T) \propto e^{-\left(\frac{T_0}{T}\right)^4}$ [30]. Inset of Figure 5 shows current measured in the range 5–150 K in a Mott plot. Indeed, the foot of the logarithmic plots at each heating/cooling stage is linear with $T^{-1/4}$, in agreement with the fact that hopping conduction is prevailing in this temperature range and that, as suggested by [13], it should be related to a band tail deforming the DoS shape close to the mobility edge.

As described in the previous section, to investigate the DoS shape we evaluated, for each scan, the activation energy in the rising foot range of each peak and the corresponding total emitted charge. The result is shown in Figure 5, where the total charge is plotted as a function of the activation energy. Results evidence a mono-exponential DoS: $f(\varepsilon) = Q_L e^{\frac{\varepsilon}{E_L}}$ (E), in the energy range 0.1–0.6 eV with E_L tailing factor.

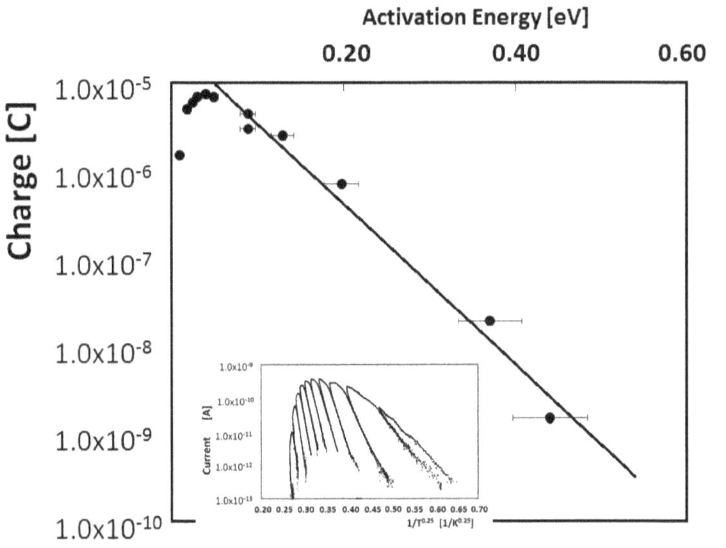

Figure 5. Emitted charge plotted as a function of the activation energy measured in the rising range of each peak of the fractionated TSC experiment shown in the inset. The exponential fit reflects the trend of the band tail in the deepest range within the forbidden gap. Inset: Delayed TSC in the range 5–150 K plotted as a function of $1/T^{1/4}$ to evidence the contribution of hopping conduction in the electrical transport.

Best fit gives: E_L = 47.5 meV, $Q_L = 2.8 \times 10^{-5}$ C, values fairly in agreement with literature [31]. We note that in the low energy range of the plot, the charge is lower than the expected value indicated by the exponential trend. This can be explained considering that priming could not fill all the states in the highest part of the band tail, close to the boundary point at $\varepsilon = 0$. As a strong evidence of this, using a deeper UV LED (355 nm) the drop appears for shallower energies with respect to the shallower UV LED (400 nm). Considering a 2 mm diameter light spot and a sample thickness of about 1 μm, the effective volume involved in this process is: $Vol \approx \pi \times 10^{-12}$ m^3 and a rough estimate of the density of states is: $N_L = \frac{Q_L}{qVol} \approx 5 \times 10^{19}$ cm^{-3}.

To best-fit our TSC measurement, we considered as fixed a group of parameters, related to the crystalline quality of our sample, mainly N_{C0}, μ_{C0}, α and γ, already given in [13]. Then, we determined first the best-fit of a TSC peak from a unique scan through a χ-square procedure by opportunely changing variable parameters starting from values of E_L and N_L obtained from Figure 5. Significant parameters as frequency factor ν_0, concentration of recombination centres, N_{rec}, initial Fermi level position E_{F0} (this latter due to the filling procedure and calculated from the conduction band miminum) have been opportunely changed to improve our simulated peak. Figure 6 shows, as an example, how these variable parameters can affect the process. In the figure, the red curve is our best fit, obtained with E_L = 60 meV, $N_L = 10^{20}$ m^{-3}, $N_{rec} = 7.5 \times 10^{21}$ m^{-3}, frequency factor $\nu_0 = 4 \times 10^{12}$ s^{-1}, E_{F0} = 66.5 meV. Other curves in the plot are obtained with same parameters apart from one that has been intentionally changed to evidence its influence on the simulation. Green line, uses a higher recombination centre concentration, $N_{rec} = 6.0 \times 10^{22}$ m^{-3}, this fasten the decay of the TSC peak in the high temperature region, due to a reduced charge lifetime. Violet curve comes from a lower energy decay factor, E_L = 50 meV, enhancing emission of charges at low temperature. Orange line is characterised by a lower frequency factor, $\nu_0 = 2 \times 10^{12}$ s^{-1}, bringing to a smoothed peak. Finally, the blue curve has been calculated using a deeper initial Fermi level, E_{F0} = 72.2 meV, corresponding to less initial charge within the hopping band tail after the priming process. In this case, a lower emission especially in the low temperature range is observed, as expected. So, each parameter is influencing a particular range of temperature and a peculiarity of the complex shape of the emission and, even if many parameters are inter-playing in the formation of the whole peak, each can be optimized almost individually to improve the overall simulation.

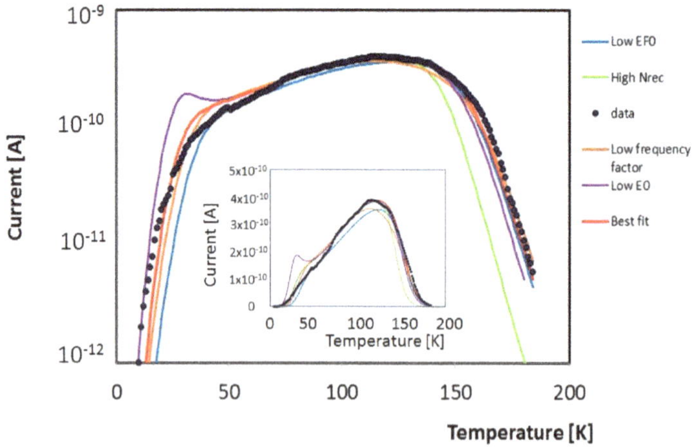

Figure 6. Fit procedure of data measured during a unique *TSC* emission in the range 5–200 K (convoluted from data of Figure 3). Red: best fit obtained with $N_{rec} = 75 \times 10^{20}$ m^{-3}, frequency factor $\nu_0 = 4 \times 10^{12}$ s^{-1}, E_{F0} = 66.5 meV, E_L = 60 meV, $N_L = 10^{20}$ m^{-3}. Green: same with higher concentration of recombination centres $N_{rec} = 6 \times 10^{22}$ m^{-3}, violet: same with lower E_L = 50 meV, orange: same with lower frequency factor $\nu_0 = 2 \times 10^{12}$ s^{-1}, blue same with lower initial Fermi level E_{F0} = 72.2 meV.

Best-fit of TSC experimental data shown in Figure 4 in the low temperature range, obtained using our mixed conductivity model taking care of the band tail in the range down to 0.6 eV plus a discrete level at 0.8 eV, is shown in Figure 7. Of note the agreement between numerical and experimental data in the overall range, up to almost 4 orders of magnitude of the current. A disagreement is observed at high temperature, where the experimental current stabilizes itself on the pAs range, while in the numerical model it decreases to lower values, as a consequence of the decrease of the Fermi level towards midgap. The pAs contribution to the current could be due to the residual presence of water vapor physisorbed on the film surface, as discussed in the next section.

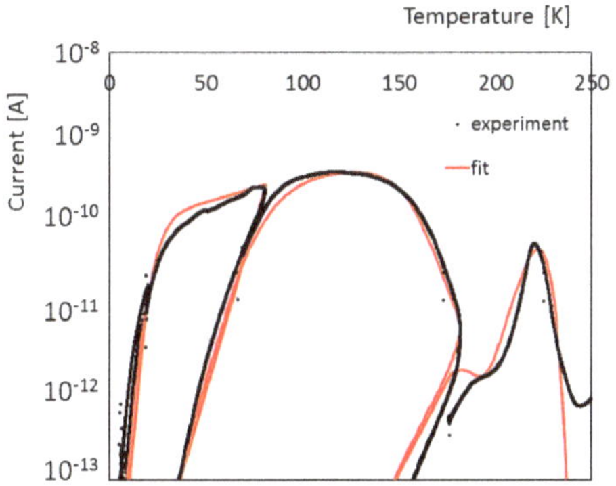

Figure 7. Experimental TSC measured in the 5–250 K range and best fit obtained considering the mixed conductivity model taking care of both hopping and free carriers conduction plus emission from a discrete energy level at 0.8 eV.

5.2. TSC Analysis above Room Temperature

In this temperature range, the effect of hopping conduction should become more and more negligible against free carrier one. Moreover, physisorption and chemisorption mechanisms at surface should also participate to conduction. In particular, dangling bonds at the nc-TiO_2 surface are capturing and releasing oxygen depending on pressure and relative humidity, these effects should be possibly investigated separately. To this purpose, we performed different sets of measurements as follows.

5.2.1. Current vs. Temperature with No Priming

Measurements are performed without previous priming. Using a low heating/cooling rate (0.1 K/s) as a first approximation we can assume a quasi-stationary equilibrium. The sample is kept in dark with air at a pressure of 1100 mb, slightly higher than ambient pressure. A typical measurement is presented in Figure 8.

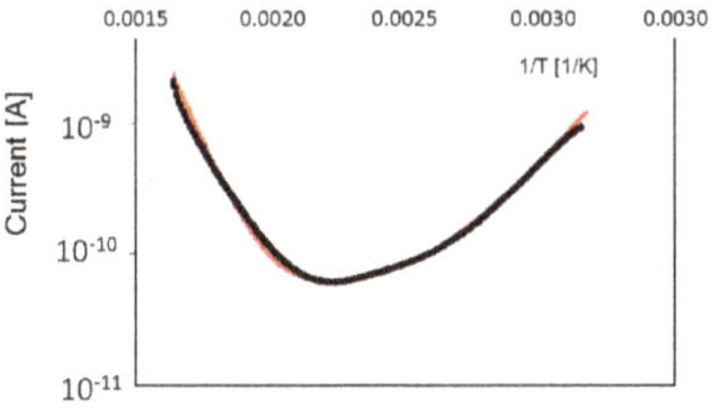

Figure 8. Current measured as a function of reciprocal temperature during a quasi-stationary heating process, in the range 300–600 K with no priming. Dark: data, red: best-fit.

The Arrhenius plot shows two distinct ranges: up to about 400 K the current decreases increasing the temperature T, then it increases with T. To explain this behavior we can consider the model proposed by [32], taking account of two dominant defects, one acting as a trap, the other as a recombination center (probably associated to dangling bonds at surface, releasing holes via a thermally activating process). Neglecting the small contribution of the hopping, the rate equation for the charged carriers (free electrons and holes concentrations are denoted by n, p) is then given by:

$$\frac{dn}{dt} = N_{t1}c_1 e^{-E_{t1}/K_B T} - Bnp - (n - n_0)\gamma, \tag{17}$$

The first term of the right side is due to emission of electrons from the trap of energy E_{t1}, concentration N_{t1} and frequency factor c_1. The second term describes the recombination of electrons with holes at the recombination center. We here assume that also hole capture is a thermally activated process: $p = N_{t2}c_2 e^{-E_{t2}/K_B T}$, with N_{t2}, E_{t2}, c_2 respectively concentration, energy and frequency factor of the recombination center and B a probability coefficient. The third term in Equation (17) takes account of other possible free electron removal mechanisms, with a coefficient γ, as trapping from deeper levels, acting on the excess concentration, n_0 being the equilibrium electron concentration.

During the current temperature measurements in dry fluxed air the system is actually in a quasi-stationary regime, so we can reasonably consider $\frac{dn}{dt} = 0$, then:

$$n = \frac{n_0 + \frac{N_{t1}c_1}{\gamma} e^{-\frac{E_{t1}}{K_B T}}}{1 + \frac{BN_{t2}c_2}{\gamma} e^{-\frac{E_{t2}}{K_B T}}} \tag{18}$$

which gives a current dependence with temperature as:

$$I_{\text{fit}} = F\left(\frac{a + be^{-\frac{E_{t1}}{K_B T}}}{1 + de^{-\frac{E_{t2}}{K_B T}}}\right) \tag{19}$$

with $a = q\Sigma\mu_C n_0$; $b = q\Sigma\mu_C \frac{N_{t1}c_1}{\gamma}$; $d = \frac{BN_{t2}c_2}{\gamma}$ and where again Σ is the surface normal to electric field F. Best-fit of our data with Equation (19), shown in Figure 8, is obtained with energy values: $E_{t1} = 1.30 \pm 0.05$ eV and $E_{t2} = 0.40 \pm 0.05$ eV. To briefly comment on these values, we observe that e.g., in [33] shallower defect levels at $\sim$0.24−0.4 eV in anatase TiO_2 were attributed to Ti interstitials, while deeper ones at $\sim$0.9−1.1 eV to oxygen-vacancies.

5.2.2. *TSC* after Storage in Dark and Humid Environment

To analyze the effect of water vapor on TSC data, we first primed the sample by keeping it in a controlled humid environment (rh = 20%) in dark at room temperature (T = 300 K) for selected time intervals, up to 14 days. Then, a TSC heating/cooling cycle has been performed by fluxing dry air with a pressure slightly higher than atmosphere, in view to measure only emissions originated during charging in the storage period. As an example, Figure 9a,b show TSC experimental data obtained after 2 and 4 days storage respectively. Curves labeled "data" report experimental measurements, while those labeled TSC show the thermally stimulated currents obtained by subtracting the current measured during the cooling scan from that measured during the heating scan. TSC emissions are observed within two distinct ranges of temperatures: one from ambient temperature, up to 400 K, the other one above 400 K.

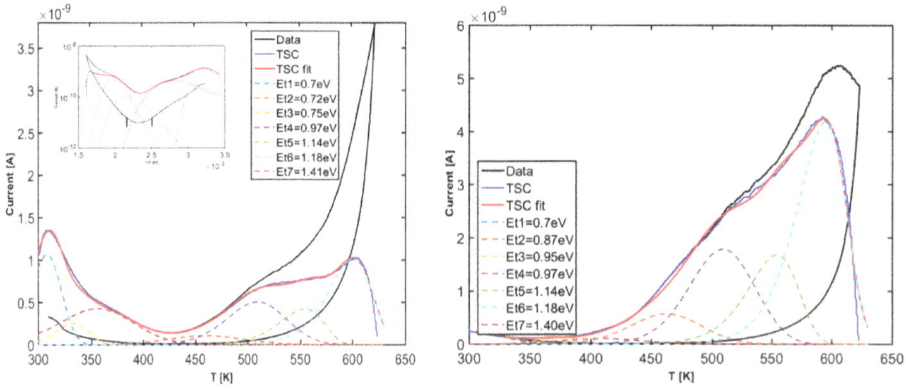

(a) (b)

Figure 9. TSC after storage in ambient air (T = 300 K) in humid environment (rh = 20%) for (**a**) two days (inset: logarithmic plot) experimental data (black); TSC with background subtracted (blue); TSC best fit (red curve) obtained with a set of 7 peaks with energy E_{t0} given in the legend; (**b**) TSC measurements after four days storage and best fit with same E_{t0} parameters as in (**a**).

An evaluation of the main TSC components involved in these measurements has been carried out in order to identify the origin of the emissions. Measurements show statistically broadened emissions They can be fitted using TSC peaks convoluted with a gaussian as given in Equation (13). Best fits have been obtained by opportunely changing trap concentration N_t for a same set of (E_t, σ_{E_t}, σ_n), within errors, best fitting the two measurements. Up to seven energy levels are required to fit our data. Parameters are shown in Table 1, energy levels are peaked at $E_{to} = 0.7$–1.14 eV and are characterized by an energy spread σ_{E_t} up to 70 meV. As a general trend, increasing the storage time, peaks at low temperatures decrease their concentration N_t, while those at high temperatures increase N_t. A source of uncertainty in the determination of concentration for the peak at ambient temperature is due to the increasing background current observed during the cooling stage, observed especially in Figure 9a. This effect can indicate reversible charging/discharging of the involved energy states, maybe due to adsorbing/desorbing from the porous alumina substrate.

Table 1. Best fit trap parameters of the *TSC* measured after storage in humid air environment at room temperature.

Peak #	E_t [eV]	σ_t [eV]	σ_n [cm^2]	N_t [10^{16} cm^{-3}] 2 Days	N_t [10^{16} cm^{-3}] 4 Days
1–3	0.70–75	0.070	10^{-20}–10^{-18}	0.34	0.06
4	0.97	0.070	4×10^{-20}	0.42	0.07
5	1.14	0.020	2×10^{-19}	0.3	1.43
6	1.18	0.007	5×10^{-20}	0.82	1.15
7	1.41	0.007	7×10^{-19}	0.25	1.14

To comment on the origin of these peaks, we observe that our measurements are in agreement e.g., with temperature programmed desorption (TPD) analyses measured in past with TiO_2 after exposure to water. Four peaks at 155, 190, 295, and 490–540 K were observed by [34–37], the first three assigned to molecular desorption from multilayer, second layer, and first layer states, while the higher temperature feature was assigned to recombinative desorption. Effusion peaks from water were also observed in [34] by thermal desorption in the 150–350 K range from porous nanostructured TiO_2 and attributed to physiosorbed H_2O, while in [35], two H_2O effusion peaks were detected around 440 and 650 K. Moreover, physically adsorbed and dissociated H_2O molecules in nanostructured anatase TiO_2 have been studied by Fourier Transform InfraRed (FTIR) emission spectroscopy at different temperatures in the range 100–300 °C [38]. A 3665 cm^{-1} band assigned to OH hydrogen bonded (adjacent) OH groups was observed to considerably decrease when the sample was heated from 373 to 573 K, while one at 3705 cm^{-1}, attributed to isolated OH groups, more difficult to remove from the surface than adjacent OH groups, only slightly changed. A 3250 cm^{-1} component attributed to the stretching vibration of water molecules that are hydrogen bonded was considerably weakened when heated up to 423 K, while the one at 3400 cm^{-1}, related to hydrogen-bound surface OH groups (Ti OH), became visible at this temperature. This latter and the 1625 cm^{-1} band (identified as the water bond-bending vibration mode) finally disappeared at 573 K.

Although a direct correlation between TPD, FTIR and our TSC measurements should be performed when same kind of samples and same experimental conditions are considered, we can qualitatively conclude that TSC peaks in the range up to 400 K should be related to adsorbed molecular water, while dissociated species, as hydrogen-bonded OH groups, should be involved in emissions in the range 400–600 K. In our measurements, increasing the storage time, peaks at low temperature are decreasing, while those at high temperature are increasing, and eventually saturating. This can be explained considering that molecular adsorbed water, in time, is slowly evolving into the formation of hydrogen bonded OH species, more stable at room temperature.

5.2.3. TSC after Illumination in He Atmosphere with Different Pressures

To evidence the effect on TSC of the oxygen-exchange at surface, we performed a set of measurements were the sample was primed in a dry He atmosphere at different pressure, from 10^{-6} to 1 bar, at room temperature (T = 300 K), during illumination with a Xe lamp for a selected time interval. Then, TSC heating/cooling cycles were carried out by fluxing dry air with a pressure slightly higher than atmosphere, to measure only emissions originated during charging in the storage period. In fact, it is known that oxygen vacancies can be created by annealing TiO_2 at elevated temperatures in an oxygen-poor environment, such as a pure He gas atmosphere or vacuum condition [39].

Results of photocurrent measurements during priming are shown in Figure 10a. TSC curves after priming in these conditions are shown in Figure 10b.

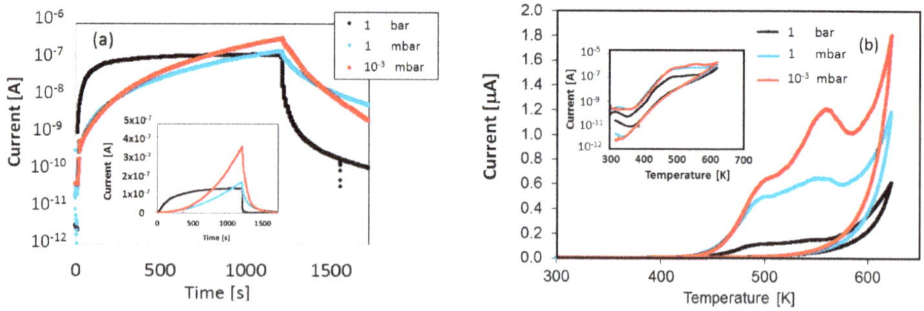

Figure 10. (**a**) Photocurrent measured during priming by illuminating with a Xe lamp the nc-TiO_2 sample in an He atmosphere of different pressures at T = 300 K; (**b**) TSC measured after priming with 1 bar,1 mbar, 10^{-3} mbar. Heating/cooling cycles are performed in He atmosphere with 1 bar pressure.

While at 1 bar the photocurrent is almost saturating during priming, at low pressure it increases superlinearly, a fact that can be explained considering the creation of extra oxygen-vacancies, which are releasing an ever increasing free carriers concentration, so favoring the passivation of deep traps during priming. To comment on TSC curves reported in Figure 10b (inset: logarithmic plot) we observe that, in the case of vacuum priming, the tail in the cooling stage of increasing current below 400 K observed in case of humid environment (Figures 8 and 9) is almost absent. Then, higher TSC emissions are observed in the high temperature range in vacuum, showing an increasing number of passivated deep traps in the priming stage. Best fits of the TSC measurements in He atmosphere have been performed starting with the same set of energy levels used in the previous section. They are shown in Figure 11a–c. Measured emissions have been calculated considering TSC peaks statistically broadened as given in Equation (13). Results are shown in Figure 11a–c respectively for the cases of 1, 10^{-3}, 10^{-6} bar. Best-fit procedure turns out in a six-fold emission, with trap parameters listed in Table 2. The fit has been performed considering the same set of values (E_t, σ_{E_t}, σ_n), within errors, for the three measurements, and best-fitting the TSC scans obtained by subtracting the response during cooling to the one measured during heating and by opportunely changing the trap concentration values N_t. Trap parameters given in Table 2 are in agreement with the model in [40] indicating that localized donor states originating from oxygen vacancies are located at 0.75–1.18 eV below the conduction band of Titania.

Table 2. Best fit trap parameters of the TSC measured after storage in different pressures of He atmosphere at room temperature during illumination with a Xe lamp.

Peak #	E_t [eV]	σ_t [eV]	σ_n [cm^2]	N_t [10^{18} cm^{-3}] 1 Bar	N_t [10^{18} cm^{-3}] 10^{-3} Bar	N_t [10^{18} cm^{-3}] 10^{-6} Bar
1	0.750	0.040	1×10^{-18}	1.8×10^{-4}	0.7×10^{-3}	0.5×10^{-3}
2	0.870	0.001	3×10^{-18}	0.4×10^{-4}	0.2×10^{-3}	0.3×10^{-3}
3	1.031	0.015	3×10^{-20}	0.31	1.35	1.35
4	1.166	0.030	1×10^{-18}	0.37	1.53	1.08
5	1.177	0.007	5×10^{-20}	0.38	2.87	6.21
6	1.411	0.009	7×10^{-19}	0.94	2.88	4.77

The six peaks used to best-fit our TSC curves are characterized by the same (E_t, σ_{Et}, σ_n), within errors, used to fit TSC measurements in the humid environment (apart of the shallowest two levels that here are present in only one component). Here, at every pressure analyzed, shallowest levels have negligible concentrations with respect to deepest levels. Looking to plots in Figure 11, we observe a good agreement between fit and data on a four-orders of magnitude scale. Logarithmic plots are shown as a function of 1/T: the observed linear trend is in favour with our previous observation, that hopping conduction in this high-temperature range is negligible.

To comment on these measurements, we observe that in low pressure/high temperature conditions the probability to form oxygen-vacancies on the surface of the nanostructured titania increases, an evidence widely discussed, e.g., in [41]. It is thus reasonable to hypothesize that TSC peaks shown in Figures 9–11 are related to such a phenomenon, as oxygen-atoms released from the surface leave behind electrons in the conduction band which are collected at electrodes and participate to the emission process. The fact that defects related to these TSC peaks are the same measured after exposure to humid environment, (apart from the shallowest levels which we attributed to adsorbed molecular water) evidences that also OH groups are more likely to be associated to oxygen vacancies, as e.g., suggested in [41].

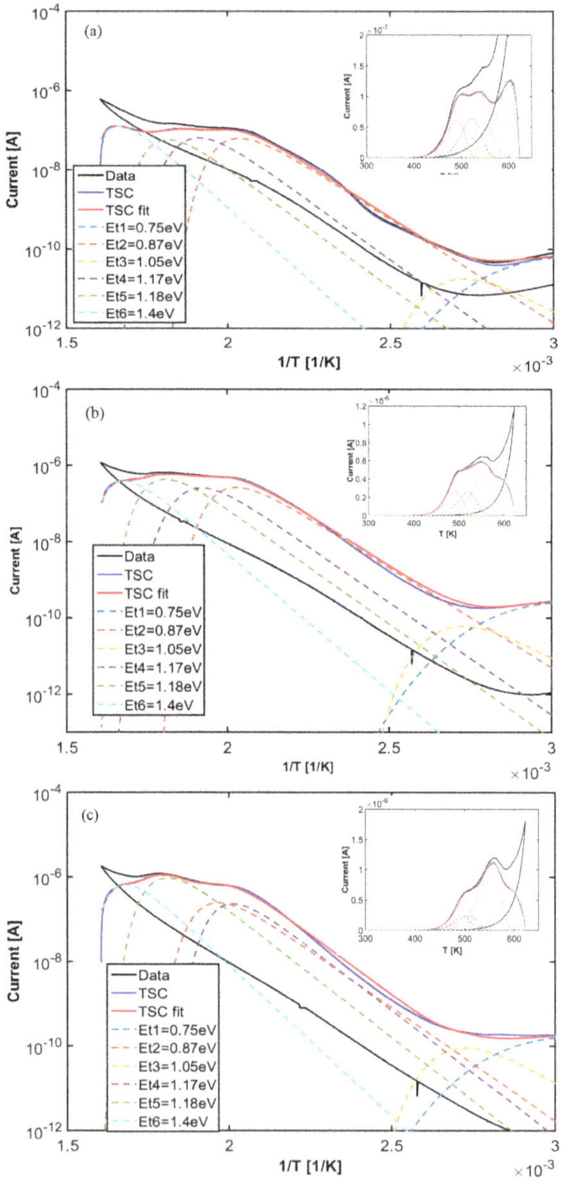

Figure 11. TSC response after priming in He atmosphere with a pressure in chamber (**a**) 1 bar; (**b**) 10^{-3} bar (**c**) 10^{-6} bar. TSC experimental data (black); Blue: TSC heating scan subtracted from background (cooling stage); dotted curves: TSC peaks convoluted with a gaussian, parameters are given in Table 2.

6. Conclusions

Nanocrystalline Titanium dioxide is widely applied as a high gap semiconducting material in many optoelectronic devices, from solar cells to gas sensors, where its peculiar electronic properties and high chemical reactivity play a crucial role. To attain good performances in terms of efficiency and photoactivity, Titania is mostly used in form of porous nanocrystalline thin films, a material characterized by a high degree of microstructural disorder, which detrimental effect in transport properties should be taken in great care and possibly minimized.

The thermally stimulated current technique is one of the most effective tool for characterizing electrical defects in semiconductors. In this paper, we have used this method to get an overall picture of the defect distribution in nanocrystalline Titania used in state-of-art DSSC devices. The model of TSC used to interpret our experimental results, briefly described in this work, takes account of the heavily disordered microscopic nature of nc-TiO_2. Mixed conductivity with non-negligible contributions from hopping between localized defects grouped in a band-tail below the conduction band is considered. Moreover, a broadening of the energy levels associated to discrete defects in the forbidden gap, has been accounted for, by convoluting the TSC standard emission with a gaussian distribution. Shallow-to deep energy levels ranging from 0.1 to 1.4 eV have been studied via a thermal spectroscopy spanning from 5 to 630 K and interpreted with this model: main results of our analysis can be summarized as follows.

An exponential DoS tail within the forbidden gap has been observed in the range 0.1–0.6 eV from the bottom of the conduction band, with energy tailing factor of 50–60 meV, characterized by density of states of the order of 1×10^{20} cm^{-3}. This tail is responsible for a large TSC emission visible after priming with a UV source at 5 K for temperatures up to approximately 150 K. At higher temperatures, up to room temperature and above, the hopping contribution to conduction becomes more and more negligible against free carrier one and contributions from discrete energy levels emitting in the conduction band become visible. Similar to single crystal TiO_2, a sharp TSC peak at 220 K is observed, with energy 0.8 eV, probably related to water adsorption. Above room temperature, dark current measured as a function of the temperature without any priming reveals to be non-negligible, as it should be due in pure intrinsic TiO_2 material. So, we studied it separately before any TSC analysis. Measurements as a function of temperature shows a double exponential trend, with a minimum at about 400 K. To explain this behavior a model taking account of two dominant defects, one acting as a recombination center, the other as a trap has been considered. The activation energy measured in this experiment in dark, for the recombination center, is about 0.4 eV a value compatible with dangling bonds at surface [22]. The trap energy, evaluated as about 1.3 eV, is in the highest energy range found for trap states observed in our further analyses.

Then, TSC response above background current has been studied as a function of the temperature, as a spontaneous emission observed after a prolonged storage of the sample in moderately ($rh = 20\%$) humid ambient air. They are best-fitted using gaussian-broadened TSC emissions, with average energy 0.7–1.4 eV, characterised by uncertainties up to 70 meV. Two groups of peaks are measured, respectively below and above temperature. Peaks up to 350 K are probably related to molecular desorption from multilayer, second layer, and first layer states, while the higher temperature features should be assigned to desorption of OH groups [42]. Our measurements show that increasing the storage time molecolar adsorbed water slowly evolves into recombinative species. Finally, to evidence the relationship between TSC emissions and vacancy-oxygen defects, measurements have been carried out after priming in inert atmosphere (He) at different pressures. Main TSC emissions, observed in the high temperature range, 400–630 K, are best-fitted considering energy levels for localized energy states in the range 0.75–1.18 eV, in agreement with a model from [41] for donor states related to oxygen-vacancies below the conduction band.

Acknowledgments: Authors wish to heartily acknowledge Franco Bogani, from University of Florence (now retired) for helpful discussions and precious guidance. This study has been performed within the SEAR laboratory of CERTUS, University of Florence, and within ICAD Consortium, c/o Dipartment of Industrial Engineering, University of Florence, Florence, Italy.

Author Contributions: Andrea Baldi, experimental set-up and samples preparation; Alessandro Cavallaro and Monica Scaringella, electronic read-out and data acquisition, Mara Bruzzi and Riccardo Mori modelling and data analysis; Mara Bruzzi, paper writing, team coordination, Ennio Antonio Carnevale funding and team director.

Conflicts of Interest: The authors declare no conflict of interest.

References

1. Gratzel, M. Dye-sensitized solar cells. *J. Photochem. Photobiol. C* **2003**, *4*, 145–153. [CrossRef]
2. Lin, H.-M.; Keng, C.-H.; Tung, C.-Y. Gas-sensing properties of nanocrystalline TiO_2. *Nanostruct. Mater.* **1997**, *9*, 747–750. [CrossRef]
3. Anta, J.A.; Morales-Florez, V. Combined Effect of Energetic and Spatial Disorder on the Trap-Limited Electron Diffusion Coefficient of Metal-Oxide Nanostructures. *J. Phys. Chem. C* **2008**, *112*, 10287–10293. [CrossRef]
4. Movilla, J.L.; Garcia-Belmonte, G.; Bisquert, J.; Planelles, J. Calculation of electronic density of states induced by impurities in TiO_2 quantum dots. *Phys. Rev. B* **2005**, *72*, 153313. [CrossRef]
5. Ofir, A.; Dor, S.; Grinis, L.; Zaban, A.; Dittrich, T.; Bisquert, J. Porosity dependence of electron percolation in nanoporous TiO_2 layers. *J. Phys. Chem. B* **2008**, *128*, 064703. [CrossRef] [PubMed]
6. Martin, N.; Besnard, A.; Sthal, F.; Vaz, F.; Nouveau, C. The contribution of grain boundary barriers to the electrical conductivity of titanium oxide thin films. *Appl. Phys. Lett.* **2008**, *93*, 064102. [CrossRef]
7. Nelson, J. Continuous-time random-walk model of electron transport in nanocrystalline TiO_2 electrodes. *Phys. Rev. B* **1999**, *59*, 15374. [CrossRef]
8. Bisquert, J. Hopping Transport of Electrons in Dye-Sensitized Solar Cells. *J. Phys. Chem. C* **2007**, *111*, 17163–17168. [CrossRef]
9. Kopidakis, N.; Benkstein, K.D.; van de Lagemaat, J.; Frank, A.J.; Yuan, Q.; Schiff, E.A. Temperature dependence of the electron diffusion coefficient in electrolyte-filled TiO_2 nanoparticle films: Evidence against multiple trapping in exponential conduction-band tails. *Phys. Rev. B* **2006**, *73*, 045326. [CrossRef]
10. Hartenstein, B.; Bassler, H.; Jakobs, A.; Kehr, K.W. Comparison between multiple trapping and multiple hopping transport in a random medium. *Phys. Rev. B* **1996**, *54*, 8574. [CrossRef]
11. Bisquert, J. Interpretation of electron diffusion coefficient in organic and inorganic semiconductors with broad distributions of states. *Phys. Chem. Chem. Phys.* **2008**, *10*, 3175. [CrossRef] [PubMed]
12. Nelson, J.; Haque, S.A.; Klug, D.R.; Durrant, J.R. Trap-limited recombination in dye-sensitized nanocrystalline metal oxide electrodes. *Phys. Rev. B* **2001**, *63*, 205321. [CrossRef]
13. Gu, B.; Xu, Z.; Dong, B. A theoretical interpretation of thermostimulated conductivity in amorphous semiconductors. *J. Non-Cryst. Solids* **1987**, *97–98*, 479–482. [CrossRef]
14. Baranovski, S.D.; Zhu, M.; Faber, T.; Hensel, F.; Thomas, P.; von der Linden, M.B.; van der Weg, W.F. Thermally stimulated conductivity in disordered semiconductors at low temperatures. *Phys. Rev. B* **1997**, *55*, 16227–16232. [CrossRef]
15. Nagy, A.; Hundhausen, M.; Ley, L.; Brunst, G.; Holzenkämpfer, E. Steady-state hopping conduction in the conduction-band tail of a-Si:H studied in thin-film transistors. *Phys. Rev. B* **1995**, *52*, 11289–11295. [CrossRef]
16. Bruzzi, M.; Mori, R.; Carnevale, E.; Scaringella, M.; Bogani, F. Low temperature Thermally Stimulated Currents (TSC) characterization of nanoporous TiO_2 films. *Phys. Status Solidi A* **2014**, *211*, 1691–1697. [CrossRef]
17. Diebold, U. Perspective: A controversial benchmark system for water-oxide interfaces: $H_2O/TiO_2(110)$. *J. Chem. Phys.* **2017**, *147*, 040901. [CrossRef] [PubMed]
18. Setvin, M.; Hulva, J.; Parkinson, G.S.; Schmid, M.; Diebold, U. Electron transfer between anatase TiO_2 and an O_2 molecule directly observed by atomic force microscopy. *Proc. Natl. Acad. Sci. USA* **2017**, *114*, E2556–E2562. [CrossRef] [PubMed]
19. Tiedje, T.; Cebulka, J.M.; Morel, D.L.; Abeles, B. Evidence for exponential band tails in amorphous silicon hybride. *Phys. Rev. Lett.* **1981**, *46*, 21–25. [CrossRef]
20. O'Leary, S.K.; Lim, P.K. Influence of the kinetic energy of localization on the distribution of electronic states in amorphous semiconductors. *Appl. Phys. A* **1998**, *66*, 53–58. [CrossRef]
21. Miller, A.; Abrahams, E. Impurity Conduction at Low Concentrations. *Phys. Rev.* **1960**, *120*, 745–755. [CrossRef]
22. Shklovskii, B.I.; Efros, A.L. *Electronic Properties of Doped Semiconductors*, 1st ed.; Springer: Heidelberg, Germany, 1984; ISBN 978-3-662-02403-4.
23. Bak, T.; Nowotny, J.; Rekas, M.; Sorrell, C.C. Defect chemistry and semiconducting properties of titanium dioxide: III. Mobility of electronic charge carriers. *J. Phys. Chem. Solids* **2003**, *64*, 1069–1087. [CrossRef]
24. Yagi, E.; Hasiguti, R.R.; Aono, M. Electronic conduction above 4 K of slightly reduced oxygen-deficient rutile TiO_2-x. *Phys. Rev. B* **1996**, *54*, 7945–7956. [CrossRef]

25. Steiger, J.; Schmechel, R.; von Seggern, H. Energetic trap distributions in organic semiconductors. *Synth. Mater.* **2002**, *129*, 1–7. [CrossRef]
26. Blood, P.; Orton, J.W. *The Electrical Characterization of Semiconductors: Majority Carriers and Electron States*, 2nd ed.; Academic Press: London, UK, 1992; pp. 393–397, 469–473. ISBN 9780125286275.
27. Mori, R.; Scaringella, M.; Cavallaro, A.; Bogani, F.; Bruzzi, M. Low temperature thermally stimulated current analysis of nanocrystalline Titanium dioxide. In Proceedings of the 10th International Conference on Large Scale Applications and Radiation Hardness of Semiconductor Detectors-Rd11, Florence, Italy, 6–8 July 2011; PoS: Florence, Italy, 2011.
28. Baldini, A.; Bruzzi, M. Thermally stimulated current spectroscopy: Experimental techniques for the investigation of silicon detectors. *Rev. Sci. Instrum.* **1993**, *64*, 932–936. [CrossRef]
29. Wakim, F.G. Some effects of trapping on the Photoelectronic properties of TiO_2 single crystals. *Phys. Stat. Sol. A* **1970**, *1*, 479–485. [CrossRef]
30. Baranovski, S.; Rubel, O. Description of Charge transport in amorphous semiconductors. In *Charge Transport in Disordered Solids with Applications in Electronics*, 1st ed.; Baranovski, S., Ed.; John Wiley & Sons: Chichester, UK, 2006; p. 61. [CrossRef]
31. Van de Lagemaat, J.; Frank, A.J. Nonthermalized Electron Transport in Dye-Sensitized Nanocrystalline TiO_2 Films: Transient Photocurrent and Random-Walk Modeling Studies. *J. Phys. Chem. B* **2001**, *105*, 11194–11205. [CrossRef]
32. Nelson, J.; Eppler, A.M.; Ballard, I.M. Photoconductivity and charge trapping in porous nanocrystalline titanium dioxide. *J. Photochem. Photobiol. A* **2002**, *148*, 25–31. [CrossRef]
33. Weiler, B.; Gagliardi, A.; Lugli, P. Kinetic Monte Carlo Simulations of Defects in Anatase Titanium Dioxide. *J. Phys. Chem. C* **2016**, *120*, 10062–10077. [CrossRef]
34. Henderson, M.A. Structural Sensitivity in the Dissociation of Water on TiO_2 Single-Crystal Surfaces. *Langmuir* **1996**, *12*, 5093–5098. [CrossRef]
35. Henderson, M.A.; Otero-Tapia, S.; Castro, M.E. The chemistry of methanol on the TiO_2(110) surface: The influence of vacancies and coadsorbed species. *Faraday Discuss.* **1999**, *114*, 313–329. [CrossRef]
36. Diebold, U. The surface science of titanium dioxide. *Surf. Sci. Rep.* **2003**, *48*, 59–229. [CrossRef]
37. Weidmann, J.; Dittrich, T.; Kostantinovac, E.; Lauermann, I.; Uhlendorfa, I.; Koch, F. Influence of oxygen and water related surface defects on the dye sensitized TiO_2 solar cell. *Sol. Energy Mater. Sol. Cells* **1999**, *56*, 153–165. [CrossRef]
38. Zheng, Z.; Teo, J.; Chen, X.; Liu, H.; Yuan, Y.; Waclawik, E.R.; Zhong, Z.; Zhu, H. Correlation of the Catalytic Activity for Oxidation Taking Place on Various TiO_2 Surfaces with Surface OH Groups and Surface Oxygen Vacancies. *Chem. Eur. J.* **2010**, *16*, 1202–1211. [CrossRef] [PubMed]
39. Pan, X.; Yang, M.; Fu, X.; Zhanga, N.; Xu, Y. Defective TiO_2 with oxygen vacancies: Synthesis, properties and photocatalytic applications. *Nanoscale* **2013**, *5*, 3601–3614. [CrossRef] [PubMed]
40. Nakamura, I.; Negishi, N.; Kutsuna, S.; Ihara, T.; Sugihara, S.; Takeuchi, K. Role of oxygen vacancy in the plasma-treated TiO_2 photocatalyst with visible light activity for NO removal. *J. Mol. Catal. A Chem.* **2000**, *161*, 205–212. [CrossRef]
41. Bowker, M. The surface structure of titania and the effect of reduction. *Curr. Opin. Solid State Mater. Sci.* **2006**, *10*, 153–162. [CrossRef]
42. Brookes, I.M.; Muryn, C.A.; Thornton, G. Imaging Water Dissociation on TiO_2(110). *Phys. Rev. Lett.* **2001**, *87*, 266103. [CrossRef] [PubMed]

nanomaterials

MDPI

Article

In Vitro Sonodynamic Therapeutic Effect of Polyion Complex Micelles Incorporating Titanium Dioxide Nanoparticles

Satoshi Yamamoto, Masafumi Ono, Eiji Yubaand Atsushi Harada *

Department of Applied Chemistry, Graduate School of Engineering, Osaka Prefecture University, 1-1 Gakuen-cho, Naka-ku, Sakai, Osaka 599-8531, Japan; st108078@edu.osakafu-u.ac.jp (S.Y.); Masafumi_Ono@hisamitsu.co.jp (M.O.); yuba@chem.osakafu-u.ac.jp (E.Y.)

* Correspondence: harada@chem.osakafu-u.ac.jp; Tel.: +81-72-254-9328

Received: 1 August 2017; Accepted: 6 September 2017; Published: 11 September 2017

Abstract: Titanium dioxide nanoparticles (TiO_2 NPs) can act as sonosensitizers, generating reactive oxygen species under ultrasound irradiation, for use in sonodynamic therapy. For TiO_2 NPs delivery, we prepared polyion complex micelles incorporating TiO_2 NPs (TiO_2 NPs-PIC micelles) by mixing TiO_2 NPs with polyallylamine bearing poly(ethylene glycol) grafts. In this study, the effects of polymer composition and ultrasound irradiation conditions on the sonodynamic therapeutic effect toward HeLa cells were evaluated experimentally using cell viability evaluation, intracellular distribution observation, and a cell staining assay. TiO_2 NPs-PIC micelles with widely distributed features induced a significant decrease in cell viability under ultrasound irradiation. Furthermore, prolonging the irradiation time killed cells more effectively than did increasing the ultrasound power. The combination of TiO_2 NP-PIC micelles and ultrasound irradiation was confirmed to induce apoptotic cell death.

Keywords: Titanium dioxide nanoparticles; sonodynamic therapy; polyion complex micelles

1. Introduction

Titanium dioxide (TiO_2) can act as a photosensitizer and is known to generate reactive oxygen species (ROS), including OH and HO_2 radicals, superoxide anions (O^{2-}), hydrogen peroxide (H_2O_2), and $^1O_2$, under ultraviolet (UV) irradiation (less than 390 nm) [1–4]. UV-irradiated TiO_2 nanoparticles (NPs) have shown a cell-killing effect toward HeLa cells [5]. However, the clinical use of TiO_2 NPs is hampered because UV light cannot deeply penetrate human tissue, and TiO_2 NPs have poor dispersion stability at physiological pH [6,7]. Shimizu et al. found that TiO_2 generates ROS under ultrasound irradiation (39 kHz) [8], although the ultrasound frequency (39 kHz) was too low for clinical applications. Additionally, sonicating TiO_2 NPs at a clinically appropriate frequency that allows deep body invasion (1 MHz) also showed an effective decrease in cell viability and inhibited tumor growth in vivo when TiO_2 NP suspension was directly injected into tumor [9]. This indicated the availability of TiO_2 NPs in sonodynamic therapy (SDT) and showed that developing a carrier system that could deliver TiO_2 NPs into cells by improving their dispersion stability under physiological conditions was required for effective SDT.

We have focused on the charge properties of surface OH groups on TiO_2 NPs. The isoelectric point of TiO_2 NPs with an anatase crystal structure is 6.2, meaning that TiO_2 NPs are negatively charged at neutral pH [10]. Polyion complex (PIC) micelles incorporating TiO_2 NPs (TiO_2 NP-PIC micelles) were successfully prepared using polyallylamine bearing poly(ethylene glycol) grafts (PAA-g-PEG) [11], in which the micelles were formed through electrostatic interaction as a driving force, and van der Waals force and hydrophobic interaction were also stabilized as a result of polyion complex formation.

Although bare TiO_2 NPs have poor solubility against water at physiological pH, the incorporation of TiO_2 NPs into the micelles provided a remarkable improvement in dispersion stability. It was confirmed that ultrasound irradiation to HeLa cells treated by TiO_2 NP-PIC micelles induced a decrease in cell viability through $^1O_2$ generation. The cell viability decreased as irradiation time increased. When other irradiation conditions were kept constant, the decrease in cell viability was dependent on irradiation time. Furthermore, this decrease in cell viability was completely inhibited by the presence of glutathione, which is a radical scavenger, demonstrating that the cell-killing effect was due to ROS generated by ultrasound irradiation of the TiO_2 NP-PIC micelles. In this study, we evaluated the effects of polymer composition and sonication time on the cell-killing effect of the TiO_2 NP-PIC micelles. Furthermore, it was confirmed that the cell-killing effect of the TiO_2 NP-PIC micelles was induced by apoptotic cell death.

2. Results and Discussion

TiO_2 NP-PIC micelles were prepared using four types of PAA-g-PEG bearing PEGs of different molecular weights (Mn = 2000 and 5000) and grafting densities (13 and 26 mol % for PEG2000, and 12 and 21 mol % for PEG5000), named 2k13, 2k26, 5k12, and 5k21, respectively. The mean diameter, polydispersity index, zeta potential, and composition (polymer/TiO_2 *w*/*w*) were determined, as summarized in Table 1. For the mean diameter and zeta-potential, it was difficult to compare with bare TiO_2 NPs due to their poor dispersion stability at physiological pH. The compositions were controlled by the molecular weight (Mn) of the PEG grafts, with PEG grafts of Mn 2000 and 5000 giving polymer/TiO_2 *w*/*w* ratios of 2 and 4, respectively. The mean diameter tended to grow large, such that the PEG graft Mn was also large and the PEG graft content was high. Importantly, all prepared TiO_2 NP-PIC micelles had almost neutral zeta potentials, suggesting that electrically neutral PEG grafts surrounded the micellar surface.

Table 1. Characterization of TiO_2 NP-PIC micelles prepared using various kinds of poly(ethylene glycol) grafts (PAA-g-PEG).

PAA-g-PEG	Mean Diameter (nm) [1]	Zeta Potential (mV) [2]	Composition (Polymer/TiO_2 *w*/*w*) [3]
2k13	61	−0.1	2.1
2k26	86	1.1	1.8
5k12	89	2.9	4.0
5k21	132	1.6	4.6

[1] Values determined using dynamic light scattering (DLS). [2] Values determined using laser-Doppler electrophoresis. [3] Values calculated using thermogravimetric/differential thermal analysis (TG/DTA).

The cell-killing effect of the TiO_2 NP-PIC micelles toward HeLa cells was evaluated by MTT (3-(4,5-di-methylthiazol-2-yl)-2,5-diphenyltetrazolium bromide) assay at various ultrasound irradiation times (Figure 1a). Micelles without ultrasound irradiation (ultrasound irradiation time = 0) maintained high cell viability and demonstrated negligible cytotoxicity. Prolonging ultrasound irradiation induced a decrease in cell viability, showing an obvious difference in the cell-killing effect among the micelles. Additionally, ultrasound irradiation to the cells without the treatment of the TiO_2 NP-PIC micelles did not induce the decrease in cell viability as shown in Figure 1a, although ultrasound irradiation to a solvent without TiO_2 NPs induce the generation of solvent radicals, i.e., H and OH radicals that can combine to give hydrogen and hydrogen peroxide in the case of water [12,13]. This suggests that the main cytotoxic species in ROS generated by ultrasound irradiation might be singlet oxygen. The half maximal inhibitory time of ultrasound irradiation (IT50) values, as an indication of the cell-killing effect under ultrasound irradiation, were determined from Figure 1a for each micelle, as shown in Figure 1b. There was a 5.7-fold difference between the IT50 values of the most effective 5k12 micelles and the least effective 2k13 micelles.

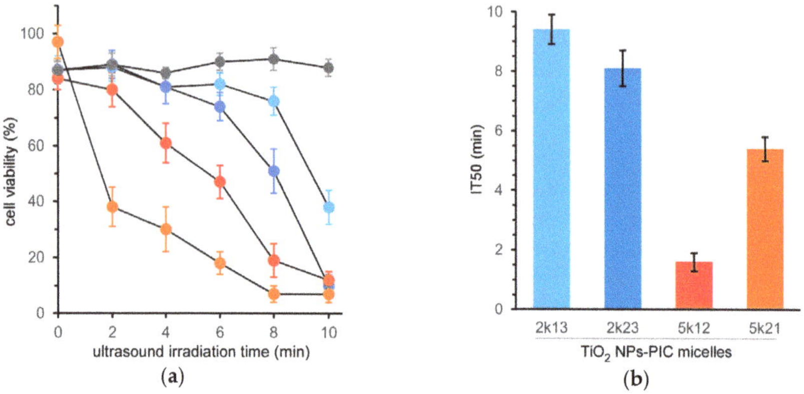

Figure 1. Cell-killing effect of TiO_2 NP-PIC micelles under ultrasound irradiation. (**a**) Effect of ultrasound irradiation time on viability of HeLa cells treated with TiO_2 NP-PIC micelles. (**b**) Half maximal inhibitory time of ultrasound irradiation (IT50) of TiO_2 NP-PIC micelles. 2k13, 2k26, 5k12, and 5k21 micelles and without the micelles are represented by light blue, blue, orange, red, and gray symbols, respectively. Ultrasound irradiation was performed for varying times (frequency: 1.0 MHz; power: 0.5 W/cm^2; duty cycle: 10%).

The difference in the cell-killing effect among the micelles might be due to the difference in ROS generation, cellular uptake, and intracellular distribution. Singlet oxygen sensor green (SOSG) has been used as a probe to confirm that the ultrasound irradiation of TiO_2 NP-PIC micelles increased the amount of $^1O_2$ generation in proportion to the irradiation time and the ultrasound power, and that there was no difference among the micelles [11]. The amounts of $^1O_2$ generation for 5k12 micelles and 2k13 micelles were compared using SOSG by flow cytometry (Figure 2). For both micelles, the fluorescence intensity of HeLa cells treated by the mixture of micelles and SOSG were slightly increased even without ultrasound irradiation, suggesting that the comparable amount of SOSG were taken up into the cells. The ultrasound irradiation to HeLa cells treated with the mixture of micelles and SOSG provided significant increase in fluorescence intensity, indicating that TiO_2 NP-PIC micelles could generate $^1O_2$ in the cells, and there was no difference in the fluorescence intensity between 5k12 micelles and 2k13 micelles. Furthermore, the cellular uptake of micelles was already evaluated by flow cytometry [11]. The TiO_2 NPs uptake increased in an incubation time-dependent manner, suggesting that cellular uptake occurred via an endocytosis pathway, with no meaningful difference in TiO_2 NPs uptake among the micelles. Consequently, TiO_2 NP-PIC micelles could generate $^1O_2$ in the cells even after 24 hours of incubation with HeLa cells, and the generated amount of $^1O_2$, which is the main cytotoxic species among ROS, might be comparable among micelles. Therefore, to explain the difference in the cell-killing effect among the micelles, the intracellular distribution of the micelles, especially their distribution to mitochondria, was compared between the most effective 5k12 micelles and least effective 2k13 micelles using fluorescein 5-isothiocyanate (FITC)-labeled TiO_2 NPs. ROS damage to mitochondria is known to effectively induce cellular apoptosis [14]. The intracellular distributions of the micelles were compared by laser scanning microscopy (Figure 3), in which the mitochondria were stained and observed. Mitochondria, identified by red fluorescence, were widely distributed in the cytoplasm. For both micelles, green fluorescence dots were observed in the cytoplasm, with most green fluorescence overlapping with red fluorescence. However, it should be noted that 5k12 micelles were more widely distributed in the cytoplasm than the 2k13 micelles despite the comparable amount of TiO_2 NP-PIC micelle uptake into the cells. This meant that 5k12 micelles were likely to cause ROS damage to more mitochondria. The difference in the cell-killing effect among the micelles shown in Figure 1 might be due to the difference in intracellular distribution of the TiO_2 NP-PIC micelles.

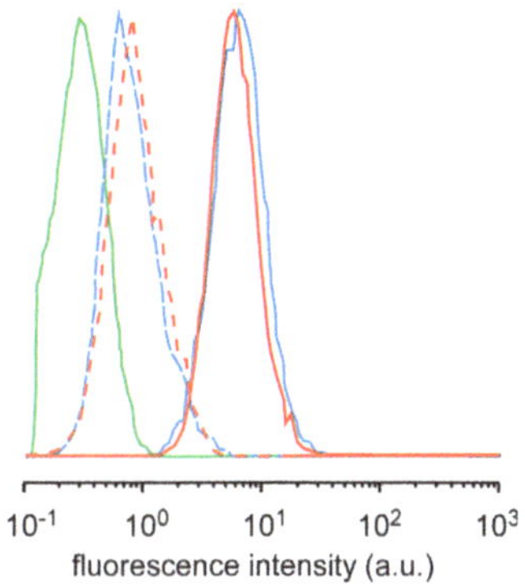

Figure 2. Flow cytometry analysis of $^1O_2$ generation of TiO_2 NP-PIC micelles (red line, 5k12 micelles; light blue line, 2k13 micelles) with (solid line) and without (dashed line) ultrasound irradiation. (frequency: 1.0 MHz; power: 1.0 W/cm^2; irradiation time: 2 min; duty cycle: 10%).

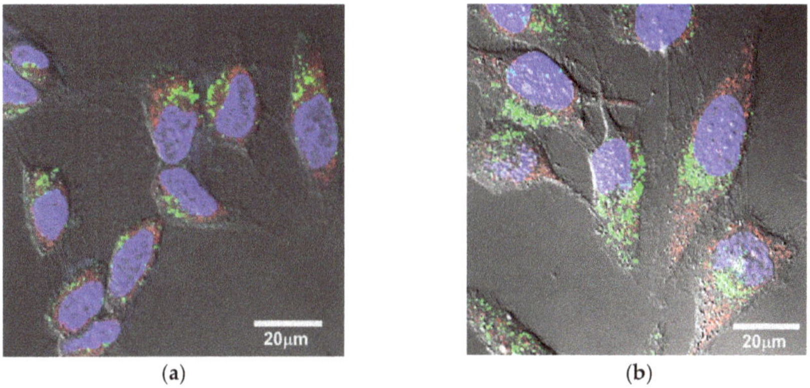

Figure 3. Confocal laser scanning microscopy images overlaid with differential interference contrast images of HeLa cells treated with (**a**) 2k13 and (**b**) 5k12 micelles for 24 h of incubation. Micelles were prepared using FITC-labeled TiO_2 NPs. Nuclei and mitochondria were stained with Hoechst and MitoTracker Red, respectively.

The effect of ultrasound power on the cell-killing effect of 5k21 micelles was evaluated by MTT assay (Figure 4). The cell viability at a power of 0.5 W/cm^2 after 2 min of irradiation was the same as that shown in Figure 1a. Increasing the ultrasound power resulted in a decrease in cell viability. However, the effect of ultrasound power was weak compared with that of irradiation time, with half the cells remaining alive at an ultrasound power of 5.0 W/cm^2. As described above, the amount of $^1O_2$ generated by ultrasound irradiation to TiO_2 NP-PIC micelles increased in proportion with both the irradiation time and the ultrasound power [11]. By increasing the ultrasound power from 0.5 to 5.0 W/cm^2, the generated amount of $^1O_2$ increased 10-fold, but the cell viability decreased by approx. 50%. Furthermore, the cell viability in Figure 4 stopped falling at approx. 50%, with little decrease in cell viability observed when further increasing the ultrasound power. In contrast, prolonging the irradiation time from 2 to 10 min increased the amount of $^1O_2$ generated five-fold and effectively decreased the cell viability to approx. 10%. These results indicated that prolonging the ultrasound irradiation time was more effective than increasing the ultrasound power to increase the sonodynamic therapeutic effect of TiO_2 NP-PIC micelles. The diffusion of the micelles in the cytoplasm might participate in this difference between the effects of irradiation time and power. Due to high reactivity, the lifetime and diffusion distance of ROS in the cytoplasm are 10–40 ns and 10–20 nm, respectively [15], and these values were determined for ROS generated by photo-irradiation to photosensitizer. The reactivity of ROS is the same as those generated by ultrasound irradiation to TiO_2 NPs, and ROS generated by ultrasound irradiation to TiO_2 NP-PIC micelles inside the cells have comparable lifetime and diffusion distance of ROS generated by photo-irradiation to photosensitizer. Therefore, TiO_2 NP-PIC micelles can only damage the limited mitochondria existing near them during ultrasound irradiation. Accordingly, prolonging the irradiation time, which increases the diffusion area, might result in an increased cell-killing effect, while the increase in the ultrasound power might not be effective. This agreed with the results shown in Figures 1 and 3, in which micelles distributed widely in the cytoplasm exhibited effective cell-killing.

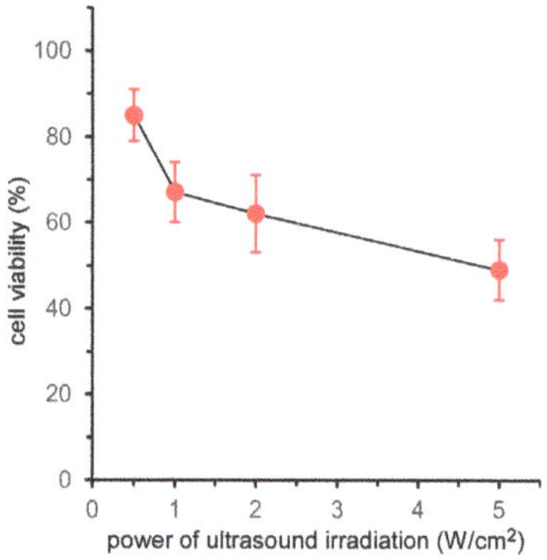

Figure 4. Effect of ultrasound irradiation power on the viability of HeLa cells treated by 5k21 micelles. Ultrasound irradiation was performed using varying ultrasound power (frequency: 1.0 MHz; irradiation time: 2 min; duty cycle: 10%).

Finally, the mechanism of cell death induced by ultrasound irradiation of the TiO_2 NP-PIC micelles was evaluated by an annexin V and propidium iodide (PI) double staining assay using flow cytometry. In the case of apoptotic cells, their phospholipid membrane asymmetry is lost, leading to the exposure of phosphatidylserine (PS) at the cellular surface, a process that can be monitored using annexin V. Annexin V can identify apoptotic cells with the exposed PS, since annexin V is a Ca^{2+}-dependent phospholipid-binding protein with a high affinity for PS. The stage of apoptosis can be distinguished using both FITC-labeled annexin V and PI. At the late stage of apoptosis, the permeability of the plasma membrane increases, and PI can bind to cellular DNA by moving across the cell membrane. Therefore, late-stage cells are stained with both PI and annexin V, whereas early-stage cells are stained with only annexin V. Figure 5a shows the flow cytometry of HeLa cells under various treatments. The cell count in the lower right region, in which the cells were stained with only annexin V, increased when treated with the micelles and further increased under ultrasound irradiation. This increase in cell count in the lower right region under ultrasound irradiation indicated that ultrasound irradiation induced apoptosis in HeLa cells treated with TiO_2 NP-PIC micelles. Cell death induced by a combination of TiO_2 NPs and ultrasound irradiation has been reported to be due to apoptosis. Yamaguchi et al. reported that water-dispersed TiO_2-PEG induced apoptotic cell death in human glioblastoma cell line U251 under ultrasound irradiation [16]. Furthermore, Ninomiya et al. reported that TiO_2 NPs modified with pre-S1/S2 proteins, which are part of the L protein from the hepatitis B virus with a high affinity toward hepatocyte, induced apoptotic cell death in human hepatoma HepG2 cells under ultrasound irradiation [17]. Therefore, it is fitting that cell death induced by TiO_2 NP-PIC micelles under ultrasound irradiation was due to apoptosis, not necrosis.

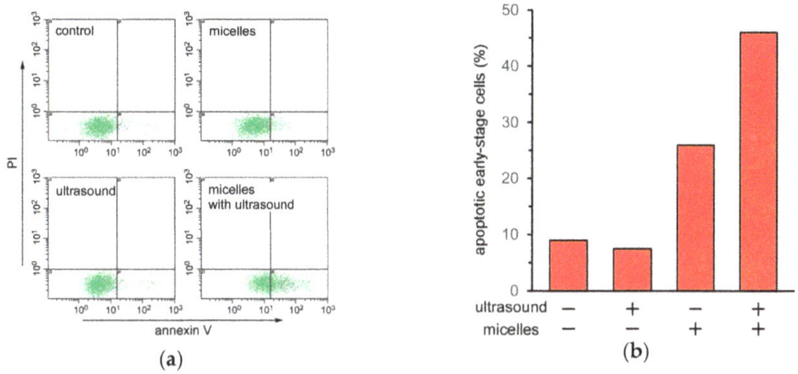

Figure 5. Evaluation of cell death mechanism for HeLa cells treated with 5k21 micelles with and without ultrasound irradiation. (**a**) Flow cytometry of HeLa cells doubly stained with annexin V-FITC and PI. (**b**) Percentages of apoptotic early-stage cells under various treatments. HeLa cells were incubated with the micelles for 24 h, after which ultrasound irradiation (frequency: 1.0 MHz; irradiation time: 6 min; power: 0.5 W/cm^2; duty cycle: 10%) was performed.

3. Materials and Methods

3.1. Materials

Four kinds of PAA-g-PEG with the same PAA main chain (DP = 160), bearing PEGs of different molecular weights (Mn = 2000 and 5000) and grafting densities (13 and 26 mol % for PEG2000, and 12 and 21 mol % for PEG5000), were synthesized according to our previous report [18]. These graft copolymers were abbreviated as 2k13, 2k26, 5k12, and 5k21, respectively. TiO_2 NP dispersion with an anatase crystal structure (10 nm, pH < 3) was purchased from Ishihara Sangyo Kaisha, Ltd. (Osaka, Japan). Fluorescein 5-isothiocyanate (FITC)-labeled TIO_2 NPs were prepared according to our previous report [11]. MitoTracker Red CMXRos and Hoechst 33342 were purchased from Thermo Fisher Scientific Inc. (Waltham, MA, USA). An Annexin-V-FLUOS staining kit was purchased from Roche Diagnostics GmbH (Mannheim, Germany). Fetal calf serum (FCS) was purchased from Biowest (Riverside, MO, USA). Dulbecco's modified Eagle's medium (DMEM) was purchased from Nissui Pharmaceutical (Tokyo, Japan). Singlet oxygen sensor green (SOSG) was purchased from Invitrogen (Eugene, OR, USA). All reagents were used without further purification.

3.2. Preparation of TiO_2 NP–PIC Micelles

TiO_2 NP dispersion and PAA-g-PEG aqueous solutions were mixed with the weight ratio of polymer to TiO_2 (polymer/TiO_2) fixed at 10. The mixing solutions (pH < 3) were neutralized using aqueous NaOH. Free polymer was removed by ultrafiltration using a USY-20 ultrafiltration unit (molecular weight cut off: 200,000; Toyo Roshi, Ltd. (Tokyo, Japan), and the solvent was also exchanged with phosphate-buffered saline (PBS). The final micelle composition, i.e., the weight ratio of polymer and TiO_2, was determined using thermogravimetric/differential thermal analysis.

3.3. Physicochemical Characterization of TiO_2 NP-PIC Micelles

DLS and laser-Doppler electrophoresis measurements were carried out at 25 °C using an ELS-8000 (Otsuka Electronics Co., Ltd., Osaka, Japan) instrument equipped with an He/Ne ion laser (λ = 633 nm). In DLS measurements, and the mean diameter was calculated using the Stokes-Einstein equation [19]. Laser-Doppler electrophoresis (Hercules, CA, USA) was employed as a technique to measure particle velocity. The electrophoretic mobility was determined from frequency shifts, which is the difference between scattered light and original beam, caused by the Doppler effect. The zeta potential was calculated using the Smoluchowski equation [19]. TG/DTA measurements were carried out using a TG8120 instrument (Rigaku, Tokyo, Japan). The samples were measured under an N_2 atmosphere from room temperature to 550 °C at a heating rate of 10 °C/min and calibrated using Al_2O_3 as a standard sample.

3.4. Experiments Using Cultured Cells

HeLa cells were seeded in 100 µL of DMEM supplemented with 10% FCS in each well of a 96-well plate at 1×10^4 cells for 1 day. Micelle solutions were gently added to the cells and incubated at 37 °C for 24 h. In the case of the confirmation of $^1O_2$ generation, the mixture of micelle solutions including SOSG were gently added to the cells. The cells were washed with PBS and 100 µL of DMEM supplemented with 10% FCS. Ultrasound irradiation was performed using an ultrasound probe (Φ 6 mm) with a size similar than that of a well in the 96-well plate. The probe was immersed into culture media, and the distance between probe and the bottom of 96-well plate was fixed to 7 mm. After sonication, 6 µL of MTT (3-(4,5-dimethylthiazol-2-yl)-2,5-diphenyl tetrazolium bromide) solution was added to each well, and the plates were incubated 37 °C for 3 h, followed by the addition of 100 µL of 2-isopropanol containing 0.1 M HCl. The number of viable cells was determined by absorbance at 570 nm. For the annexin V and propidium iodide (PI) double staining assay, the cells were incubated for 6 h after ultrasound irradiation and then stained using an Annexin-V-FLUOS staining kit (Mannheim, Germany). After staining, the cells were detached from the surface of the dish using trypsin, and the cellular fluorescence was evaluated by flow cytometry (EPICS XL, Beckman Coulter, Inc. Brea, CA, USA).

4. Conclusions

TiO_2 NP-PIC micelles exhibited a cell-killing effect toward HeLa cells through $^1O_2$ generation under ultrasound irradiation. The wide intracellular distribution of TiO_2 NP-PIC micelles and prolonged ultrasound irradiation time provided effective cell-killing corresponding to the wide distribution of mitochondria in the cytoplasm, suggesting that TiO_2 NP-PIC micelles might induce apoptosis through singlet oxygen generation by ultrasound irradiation. It is expected that TiO_2 NP-PIC micelles might become a clinically available sonodynamic therapy system via intravenous injection through the combination with high intensity focused ultrasound irradiation.

Acknowledgments: The authors thank to Kenji Kono, who passed away last year, for valuable discussion and warm support. This research was partly supported by the Terumo Foundation for Life Sciences and Arts (Atsushi Harada) and a Grant-in-Aid for JSPS Research Fellow from Japan Society for the Promotion of Science (Satoshi Yamamoto). We thank Simon Partridge, from Edanz Group (www.edanzediting.com/ac) for editing a draft of this manuscript.

Author Contributions: Atsushi Harada conceived and designed the experiments; Satoshi Yamamoto and Masafumi Ono performed the experiments; Eiji Yuba discussed and commented on the experimental data; Satoshi Yamamoto wrote the paper.

Conflicts of Interest: The authors declare no conflicts of interest.

References

1. Fujishima, A.; Honda, K. Electrochemical Photolysis of Water at a Semiconductor Electrode. *Nature* **1972**, *238*, 37–38. [CrossRef] [PubMed]
2. Hoffmann, M.R.; Martin, S.T.; Choi, W.; Bahnemann, D.W. Environmental Applications of Semiconductor Photocatalysis. *Chem. Rev.* **1995**, *95*, 69–96. [CrossRef]
3. Mills, A.; Le Hunte, S. An Overview of Semiconductor Photocatalysis. *J. Photochem. Photobiol. A Chem.* **1997**, *108*, 1–35. [CrossRef]
4. Schwartz, P.F.; Turro, N.J.; Bossmann, S.H.; Braum, A.M.; Wahab, A.M.A.A.; Durr, H. A New Method to Determine the Generation of Hydroxyl Radicals in Illuminated TiO_2 Suspensions. *J. Phys. Chem. B* **1997**, *101*, 7127–7134. [CrossRef]

5. Cai, R.; Kubota, Y.; Shuin, T.; Sakai, H.; Hashimoto, K.; Fujishima, A. Induction of Cytotoxicity by Photoexcited TiO_2 Particles. *Cancer Res.* **1992**, *52*, 2346–2348. [PubMed]
6. Yaremko, Z.M.; Tkachenko, N.H.; Bellmann, C.; Pich, A. Redispergation of TiO_2 Particles in Aqueous Solutions. *J. Colloid Interface Sci.* **2006**, *296*, 565–571. [CrossRef] [PubMed]
7. French, R.A.; Jacobson, A.R.; Kim, B.; Isley, S.L.; Penn, R.L.; Baveye, P.C. Influence of ionic strength, pH, and cation valence on aggregation kinetics of titanium dioxide nanoparticles. *Environ. Sci. Technol.* **2009**, *43*, 1354–1359. [CrossRef] [PubMed]
8. Shimizu, N.; Ogino, C.; Dadjour, M.F.; Murata, T. Sonocatalytic Degradation of Methylene Blue with TiO_2 Pellets in Water. *Ultrason. Sonochem.* **2007**, *14*, 184–190. [CrossRef] [PubMed]
9. Harada, Y.; Ogawa, K.; Irie, Y.; Endo, H.; Feruil, L.B., Jr.; Uemura, T.; Tachibana, K. Ultrasound Activation of TiO_2 in Melanoma Tumors. *J. Control. Release* **2011**, *149*, 190–195. [CrossRef] [PubMed]
10. Keller, A.A.; Wang, H.; Zhou, D.; Lenihan, H.S.; Cherr, G.; Cardinale, B.J.; Miller, R.; Ji, Z. Stability and Aggregation of Metal Oxide Nanoparticles in Natural Aqueous Matrices. *Environ. Sci. Technol.* **2010**, *44*, 1962–1967. [CrossRef] [PubMed]
11. Harada, A.; Ono, M.; Yuba, E.; Kono, K. Titanium dioxide nanoparticles-entrapped polyion complex micelles generating singlet oxygen in the cells by ultrasound irradiation for sonodynamic therapy. *Biomater. Sci.* **2013**, *1*, 65–73. [CrossRef]
12. Cravotto, G.; Cintas, P. Power ultrasound in organic synthesis: Moving cavitational chemistry from academia to innovative and large-scale applications. *Chem. Soc. Rev.* **2006**, *35*, 180–196. [CrossRef] [PubMed]
13. Silva, R.; Ferreira, H.; Cavaco-Paulo, A. Sonoproduction of Liposomes and Protein Particles as Templates for Delivery Purposes. *Biomacromolecules* **2011**, *12*, 3353–3368. [CrossRef] [PubMed]
14. Simon, H.U.; Haj-Yehia, A.; Levi-Schaffer, F. Role of rective oxygen species (ROS) in apoptosis induction. *Apoptosis* **2000**, *5*, 415–418. [CrossRef] [PubMed]
15. Redmond, R.W.; Kochevar, I.E. Spatially Resolved Cellular Response to Singlet Oxygen. *Photochem. Photobiol.* **2006**, *82*, 1178–1186. [CrossRef] [PubMed]
16. Yamaguchi, S.; Kobayashi, H.; Narita, T.; Kanehira, K.; Sonezaki, S.; Kudo, N.; Kubota, Y.; Terasaka, S.; Houkin, K. Sonodynamic therapy using water-dispersed TiO_2-polyethylene glycol compound on glioma cells: Comparison of cytotoxic mechanism with photodynamic therapy. *Ultrason. Sonochem.* **2011**, *18*, 1197–1204. [CrossRef] [PubMed]
17. Ninomiya, K.; Ogino, C.; Oshima, S.; Sonoke, S.; Kuroda, S.; Shimizu, N. Targeted sonodynamic therapy using protein-modified TiO_2 nanoparticles. *Ultrason. Sonochem.* **2012**, *19*, 607–614. [CrossRef] [PubMed]
18. Kawamura, A.; Kojima, C.; Iijima, M.; Harada, A.; Kono, K. Polyion Complex Micelles Formed from Glucose Oxidase and Comb-Type Polyelectrolyte with Poly(ethylene glycol) Grafts. *J. Polym. Sci. Pol. Chem.* **2008**, *46*, 3842–3852. [CrossRef]
19. Harada, A.; Kataoka, K. Formation of Polyion Complex Micelles in an Aqueous Milieu from a Pair of Oppositely-Charged Block Copolymers with Poly(ethy1ene glycol) Segments. *Macromolecules* **1995**, *28*, 5294–5299. [CrossRef]

nanomaterials

MDPI

Article

Lipid-Coated Zinc Oxide Nanoparticles as Innovative ROS-Generators for Photodynamic Therapy in Cancer Cells

Andrea Ancona [1], Bianca Dumontel [1], Nadia Garino [1,2], Benjamin Demarco [3], Dimitra Chatzitheodoridou [3], Walter Fazzini [1], Hanna Engelke [3,*] and Valentina Cauda [1,2,*]

1 Department of Applied Science and Technology, Politecnico di Torino, Corso Duca degli Abruzzi 24, 10129 Turin, Italy; andrea.ancona@polito.it (A.A.); bianca.dumontel@polito.it (B.D.); nadia.garino@polito.it (N.G.); s223910@studenti.polito.it (W.F.)

2 Center for Sustainable Future Technologies—CSFT@POLITO, Istituto Italiano di Tecnologia, Corso Trento 21, 10129 Turin, Italy

3 Department of Chemistry, Ludwig-Maximilians-University of Munich, Butenandtstrasse 11E, 81377 Munich, Germany; benjamindemarcov@gmail.com (B.D.); dimitra.ch93@googlemail.com (D.C.)

* Correspondence: hanna.engelke@cup.uni-muenchen.de (H.E.); valentina.cauda@polito.it (V.C.); Tel.: +49-89-2180-77626 (H.E.); +39-011-090-7389 (V.C.)

Received: 9 February 2018; Accepted: 28 February 2018; Published: 2 March 2018

Abstract: In the present paper, we use zinc oxide nanoparticles under the excitation of ultraviolet (UV) light for the generation of Reactive Oxygen Species (ROS), with the aim of further using these species for fighting cancer cells in vitro. Owing to the difficulties in obtaining highly dispersed nanoparticles (NPs) in biological media, we propose their coating with a double-lipidic layer and we evaluate their colloidal stability in comparison to the pristine zinc oxide NPs. Then, using Electron Paramagnetic Resonance (EPR) coupled with the spin-trapping technique, we demonstrate and characterize the ability of bare and lipid-coated ZnO NPs to generate ROS in water only when remotely actuated via UV light irradiation. Interestingly, our results reveal that the surface chemistry of the NPs greatly influences the type of photo-generated ROS. Finally, we show that lipid-coated ZnO NPs are effectively internalized inside human epithelial carcinoma cells (HeLa) via a lysosomal pathway and that they can generate ROS inside cancer cells, leading to enhanced cell death. The results are promising for the development of ZnO-based therapeutic systems.

Keywords: zinc oxide nanoparticle; supported lipidic bilayer; reactive oxygen species; electron paramagnetic spectroscopy; photodynamic therapy; colloidal stability; 5,5-dimethyl-L-pyrroline-*N*-oxide (DMPO)

1. Introduction

In recent years, major efforts have been devoted to the development of novel suitable nanotechnological platforms to improve medical diagnosis and therapies. Thanks to their higher area to volume ratio compared to larger particles, several nanomaterials have been exploited to deliver drugs specifically towards target regions or to elicit cytotoxic effects only when externally stimulated [1]. Among these, nanostructured zinc oxide (ZnO) has been widely studied, due to its interesting properties, such as piezoelectric and pyroelectric behaviors and biocompatible features, being classified as a "GRAS" (generally recognized as safe) substance by the Food and Drug Administration (FDA) [2,3]. As a semiconductor, it shows a band gap of 3.3 eV, thus it can adsorb ultraviolet (UV) light and emit it in the visible range, showing potential capabilities as light emitting diode (LED) or as a reporter in bio-imaging applications. The optical properties of ZnO can be efficiently exploited for the generation of Reactive Oxygen Species (ROS). Actually, in the presence of a radiation having a wavelength of less than 400 nm, the electrons (e^-) are excited from the valence (VB) to conduction band (CB), generating positive holes

(h^+). In aqueous environment, the photo-generated e^- can reduce oxygen molecules, forming superoxide radical anion (O_2^-) [4,5], while the h^+ can oxidize water molecules and hydroxide ions, generating hydroxyl radicals and hydrogen peroxide (H_2O_2) molecules [6]. Furthermore, the recombination of the electron–hole pair can generate emission of a photon (radiative recombination), which in turn can excite ground state oxygen generating singlet oxygen. All these ROS, when produced intracellularly, can exert highly cytotoxic effects [7] and one can take advantage of their generation for killing tumor cells. Actually, an overproduction of intracellular ROS produces oxidative stress [8], altering the cell cycle [9] and promoting cell death through apoptosis [5] or autophagy [10]. Moreover, ROS can induce lipid peroxidation, associated with impairment of cell membrane structure [11], protein denaturation and different types of DNA damage [12].

Therefore, several studies have proposed the photoexcitation of nanostructured ZnO to produce intracellular ROS as an effective therapeutic strategy, called Photodynamic Therapy (PDT), causing severe toxicity in different cancer cell lines [13,14]. This therapy promises better selectivity and fewer side-effects compared to most traditional chemo- and radio-therapies. ZnO nanoparticles (NPs) can indeed accumulate specifically within the tumor region thanks to the Enhanced Permeation and Retention effect [15]. In this way, when the light is directly focused in the region of interest, the therapeutic effect is highly localized. However, the PDT shows limited tissue penetration owing to the small penetration depth of UV light (less than 1 mm) used to excite the NPs. Therefore, PDT can be efficiently exploited for superficial tumors or with optical wave-guide irradiation in the case of deeper but accessible cancer tissues.

Using the Electron Paramagnetic Resonance (EPR) technique, the ROS species generated by ZnO NPs in aqueous solutions can be efficiently recognized [16]. It was reported that the ROS generated and thus the PDT effects are size dependent in the range from 20 to 100 nm [17], where smaller NPs present higher cytotoxicity. Further efforts are thus needed to unravel the effective production mechanism of ROS from nanostructured ZnO, in particular when the ZnO particle surface is modified by other molecules, thus functionalized.

According to the above considerations, the present paper reports on the preparation of ZnO NPs of about 20 nm in diameter and their surface functionalization to improve their biological stability and biocompatibility. The final aim is to study the effect of the surface molecular groups on the final ROS production and species. A self-assembly of phospholipids was achieved forming a supported lipid bilayer on the surface of the bare ZnO NPs. The results show that highly stabilized NPs without aggregation are obtained in Phosphate Buffered Saline (PBS) when the lipidic bilayer is formed. A deep investigation through the EPR technique is here proposed on the ROS produced under UV irradiation by bare and lipid-coated NPs in different aqueous, physiological and biological fluids. The results reveal that there is an important variability of the ROS production types depending on the NPs surface chemistry and dispersion media. The internalization into HeLa cancer cells and the intracellular ROS production are studied. Finally, the cytotoxic activity of both bare and lipid-coated NPs at different concentrations and under UV-stimulus activation is evaluated, leading to efficient cancer cell death. Therefore, the reported nanosystem, conjugating high stability in biological media and remote-activated cytotoxicity, can be a novel powerful tool to nanomedicine therapy against cancer.

2. Results and Discussion

2.1. ZnO Nanoparticles Synthesis and Characterization

ZnO nanoparticles were synthesized by a simple wet chemical method, as reported in detail in the Materials and Methods Section. The morphology and particle size of the as-prepared ZnO nanoparticles were characterized by Field Emission Scanning Electron Microscopy (FESEM), as shown in Figure 1a. The FESEM image shows that these nanoparticles have a spherical shape with an average diameter of 14 ± 2 nm (as measured from FESEM images using Fiji software, Open source, $n = 50$). Analysis of the X-ray diffraction pattern confirmed the crystalline structure of the ZnO NPs.

The diffraction peaks matched well with the characteristic peaks of the single-phase wurtzite crystalline structure, as shown in Figure 1c (black curve). Applying the Debye–Scherrer equation to the broad diffraction peaks, an average size of 15 nm of the nanocrystallites was obtained, in fair agreement with the electron imaging results.

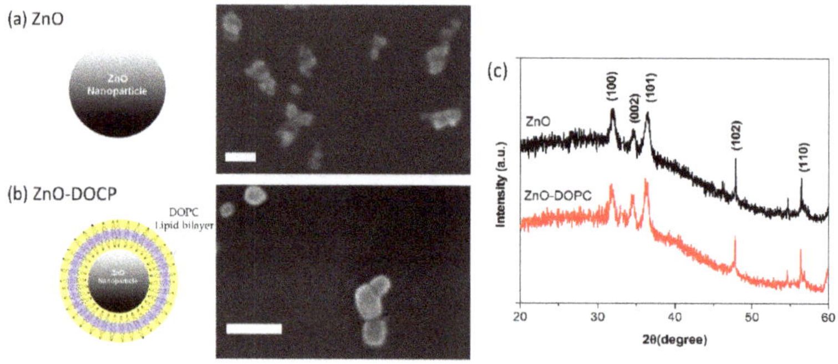

Figure 1. Morphology and crystalline structure of bare and lipid-coated zinc oxide NPs. Scheme and FESEM images of: bare ZnO NPs (**a**); and lipid-coated ZnO NPs (**b**). For FESEM images, all the NPs were coated by a thin layer of Pt. Scale bare is 30 nm in both images. (**c**) Representative X-ray diffractograms of the ZnO NPs: pristine (black curve) and lipid-coated (red curve) ones. Non-indexed peaks derive from the silicon wafer used as sample substrate.

The lipid-coated nanoparticles were prepared by a solvent-exchange method using the commercial phospholipid DOPC (1,2-dioleoyl-sn-glycero-3-phosphocholine). As shown in Figure 1b,c, Field Emission Scanning Electron Microscope (FESEM) and X-ray Powder diffraction (XRD) analyses confirmed that the morphology and the crystalline structure of the NPs was not modified after the functionalization, except in the size of lipid-coated ZnO NPs (average diameter of 21 ± 5 nm, measured from FESEM images using Fijisoftware, n = 20).

Dynamic Light Scattering (DLS) experiments were performed for the two samples to evaluate their hydrodynamic diameters and stability in water (Figure 2a). The absence of micrometer-scale aggregates in all measurements suggests good dispersion and low aggregation behavior for both samples. Interestingly, the lipidic functionalization of the ZnO nanoparticle surface contributed to a larger mean hydrodynamic diameter (110 nm for ZnO-DOPC NPs) than the one obtained for the pristine ZnO NPs (55 nm).

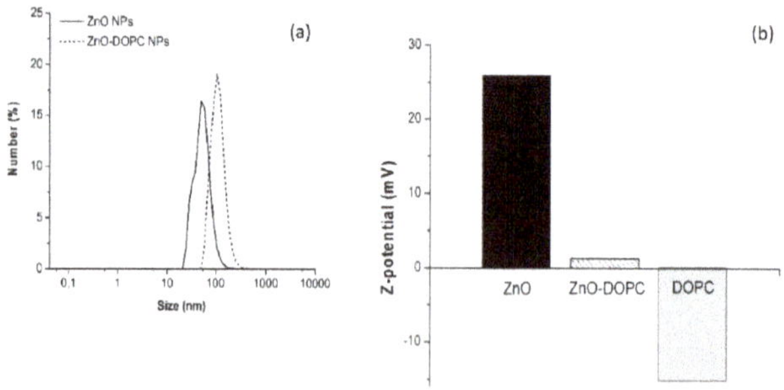

Figure 2. Hydrodynamic diameters and Z potentials of bare and lipid-coated ZnO NPs. Dynamic Light Scattering (**a**); and Z-potential (**b**) measurements of the two samples compared to DOPC lipids micelles.

Moreover, the Z-potential of lipids micelles and of bare and lipid-coated ZnO NPs was evaluated in water maintaining neutral pH by titration with NaOH and HCl 1 M. As shown in Figure 2b, the DOPC micelles present a negative Z potential, equal to −15 mV. The obtained value is in fair agreement with literature studies, that attribute it to a characteristic orientation of lipids polar head in solutions with low ionic strength [18].

Concerning the nanoparticles samples, different values were obtained depending on the surface properties of the ZnO NPs. For pristine ZnO NPs, the measured positive Z-potential (26 mV) is in good agreement with the literature values [19] and it is due to the protonation of the hydroxyl groups at the nanoparticle surface. A strong decrease of the Z-potential value was obtained for lipid-coated ZnO nanoparticles: the DOPC phospholipid shell neutralized the positive charges of the ZnO surface and lowered the Z-potential down to 1.3 mV. Showing different behaviors for pristine and functionalized ZnO nanoparticles, the DLS and Z-potential measurements together clearly suggest that the lipid functionalization worked successfully, almost completely shielding the ZnO NP in a protective lipid shell.

To further confirm the formation of the lipid bilayer on the ZnO NPs surface, fluorescence microscopy co-localization experiments in wide-field configuration were performed (Figure 3). The DOPC shell was marked with 1% Bodipy FL DHPE lipid, characterized by a fluorescent excitation maximum at 488 nm, while ZnO NPs, functionalized with amino-propyl groups, were marked with Atto550-NHS ester dye, having a fluorescence excitation at 550 nm. Merging the two distinct images, green (lipid shell) and red (ZnO NPs) spots overlap forming a yellow spot corresponding to the co-localized lipid shell on the nanoparticle surface, thus further confirming the successful coverage by phospholipids of the ZnO nanoparticles.

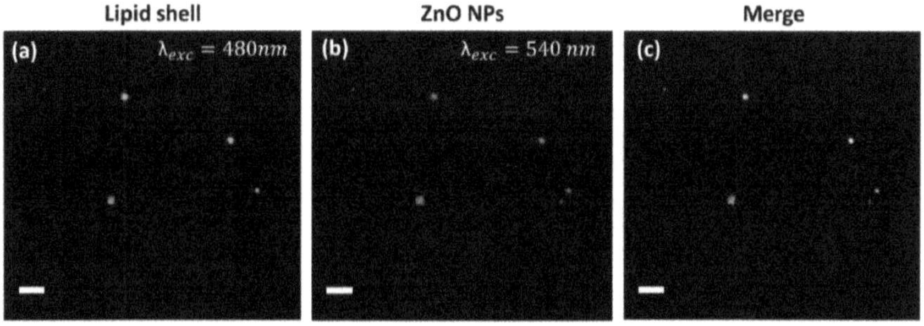

Figure 3. Co-localization of the lipid-shell with ZnO NPs. Wide-Field Fluorescence images of: (**a**) lipid-shells labeled by 1% Bodipy-DHPE; (**b**) amino-propyl functionalized (ZnO-NH_2) nanoparticles marked with Atto550-NHS ester; and (**c**) the merged images showing the co-localization of the lipid-shell with the ZnO NPs. Scale bar: 10 μM.

It is worth mentioning that the scale bar in the optical fluorescence images of Figure 3 is 10 μM. Thus, the resulting size of the nanoparticles can seem in disagreement with the sizes measured by DLS and FESEM. However, one should note that the magnification used for fluorescence imaging (in this case, 60× objective) is not enough to resolve sizes of 20 nm and will always be constrained by Abbé's limit, in this case about 200 nm. This resolution limit indicates that it is not possible to distinguish whether there is one single or more particles within each bright spot of 200 nm. Owing to the high particles brightness and their strong dilution, we assume that mainly single particles were imaged.

2.2. Biostability of Lipid-Coated ZnO Nanoparticles in Physiological Media

The efficient delivery of nanoparticles to the pathological site of interest is a crucial step to achieve an effective photodynamic therapeutic effect. In the case of injection into living systems, this could be hindered by aggregation of the ZnO NPs in contact with plasma fluids: several studies have indeed shown that both circulation time in the blood stream and cellular uptake are strongly influenced by the nanoparticles' colloidal stability in these media [20,21].

As a first step towards the comprehension of the behavior of ZnO NPs in biological media, further Dynamic Light Scattering (DLS) experiments were performed in Phosphate Buffered Saline (PBS). Moreover, the effect of the lipid-coating on the colloidal stability of ZnO NPs has been evaluated. As shown in Figure 4a, a striking difference between the behavior of the two samples was noted when suspended in PBS. Bare ZnO NPs showed a strong aggregation behavior, forming micrometer-scale aggregates, while lipid-coated ZnO NPs did not show any aggregation, confirming the hydrodynamic size obtained in water suspension.

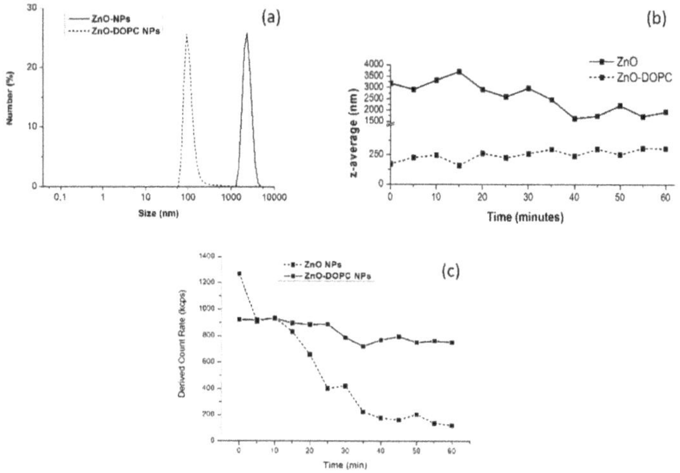

Figure 4. The lipid-shell increases colloidal stability of ZnO NPs in PBS: (**a**) Dynamic Light Scattering (DLS) measurements in number percent of the bare (ZnO) and lipid-coated (ZnO-DOPC) nanoparticles in Phosphate Buffered Saline (PBS); (**b**) mean hydrodynamic size (Z-average) of bare ZnO NPs (solid curve) and lipid-coated ZnO NPs (dotted curve) in PBS over time; and (**c**) derived count rate of the two samples over time.

To further study this behavior, the mean hydrodynamic radius (z-average) of ZnO and ZnO-DOPC NPs was recorded in real-time for 1 h (Figure 4b). As soon as bare ZnO NPs were exposed to the PBS solution (time 0), the measured z-average was relatively high (3200 nm), confirming that this sample promptly formed huge aggregates. Over time, the z-average moderately decreases (down to 1890 nm), suggesting that the NPs maintain their aggregated form and partially precipitate. The marked decrease of the derived count rate over time (Figure 4c) strongly supports this hypothesis. On the contrary, lipid-coated ZnO NPs did not form micrometer-scale aggregates over time when suspended in PBS, maintaining a Z-average size between 100 and 250 nm, thus confirming their higher colloidal stability compared to the bare ZnO NPs.

The same improvement of colloidal and chemical stability of lipid-coated ZnO nanoparticles in biological media was studied in detail in a previous work [22]. As already observed for biostability assays performed in cell culture media (EMEM) and simulated human plasma (Simulated Body Fluid, SBF), this improved colloidal behavior can be attributed to the lipid shell stabilization, shielding the ZnO NPs. In particular, for PBS, the phosphate ions contained in large quantity in the buffer solution are strongly reactive towards ZnO leading to the formation of poorly soluble zinc-phosphate precipitates [23]. Thus, these data suggest that the lipid coating can limit the contact of solution's components with the ZnO surface, preventing NPs' aggregation.

2.3. *Reactive Oxygen Species Generation*

2.3.1. ROS Generation in the Absence of External Actuation

In photodynamic therapy, it is crucial to avoid the activation of photosensitizer in the absence of light irradiation that would lead to uncontrollable and dangerous side effects. To investigate this possibility, the ROS generation by pristine ZnO nanoparticles was first studied without external actuation (ambient light) in different water and physiological media. The formation of hydroxyl and superoxide anion radicals by ZnO NPs was characterized using the Electron Paramagnetic Resonance (EPR) technique coupled with the DMPO (5,5-dimethyl-pyrroline *N*-oxide) spin trap. This compound can trap both hydroxyl and superoxide anion radicals, thus suitable for studying ROS generation by NPs. A suspension of ZnO NPs (500 μg/mL) in the tested medium was kept at 37 °C for 1 h under continuous stirring, inserted into the EPR cavity, as described in detail the Materials and Methods Section, and the EPR spectrum was recorded.

In Figure 5, the EPR spectra corresponding to the tested suspensions in the absence (black curve) and in the presence (red curve) of pristine ZnO NPs are shown. Using water as dispersing medium, no spin adduct was detected in the obtained spectra as shown in Figure 5a. The same type of noisy signals was obtained using Phosphate Buffered Saline (PBS) as medium (Figure 5b). On the contrary, signals composed by several spin adducts were detected in Simulated Body Fluid (SBF) suspensions, as shown in Figure 5c. The presence of the ZnO nanoparticles in the medium did not modify the recorded EPR spectrum, thus indicating that the detected spin adducts were not due to ROS generation by ZnO NPs themselves, but to the reaction of the DMPO spin trap with medium components. Indeed, the high concentration of metal ions in SBF can react with DMPO forming EPR-detectable compounds [24]. Finally, when EPR spectra were recorded using the cell culture medium (EMEM), a triplet signal was detected in both control and ZnO NPs experiments. Similar to the SBF experiments, this signal is due to the reaction of medium components with the DMPO spin trap. Cell culture medium, indeed, includes several autoxidizable compounds, such as the amino-acids L-tyrosine and L-triptophan, that can oxidize the DMPO spin trap giving rise to spin adducts detectable by EPR [25].

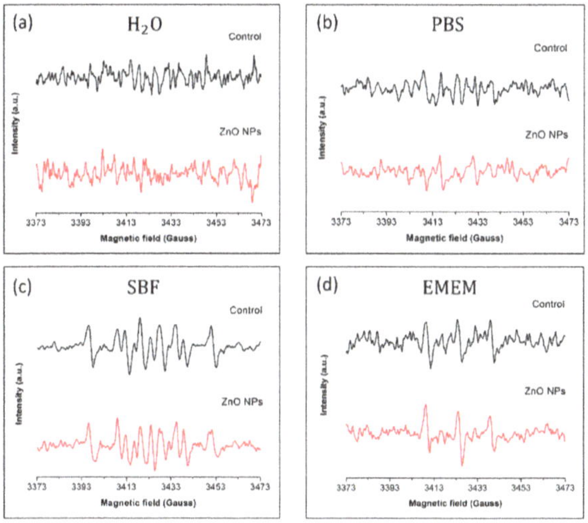

Figure 5. EPR spectra bare ZnO NP suspensions (500 μg/mL) in different media using DMPO (50 mM) as a spin trap: (**a**) Water; (**b**) Phosphate Buffered Saline (PBS); (**c**) Simulated Body Fluid (SBF); and (**d**) Eagle's Minimum Essential Medium (EMEM). Black spectra: medium without ZnO NPs, Red spectra: medium with 500 μg/mL ZnO NPs.

Taken together, these measurements demonstrate that, in the absence of any external actuation such as UV irradiation, our pristine ZnO NPs cannot generate hydroxyl and superoxide anion radicals when suspended in different physiological media. This is the first step towards the synthesis of an effective photosensitizer able to work as a remotely-activated ROS generator.

2.3.2. ROS Generation under UV Illumination by BARE ZnO Nanoparticles

To evaluate the capability of ROS generation by our pristine ZnO nanoparticles under UV irradiation, water suspensions of bare ZnO NPs were irradiated with UV light for 5 min and, using DMPO as a spin trap, the EPR spectrum was recorded as described in the Materials and Methods Section.

As shown in Figure 6a, a characteristic DMPO-OH spin adduct giving rise to four resolved peaks was obtained when ZnO NPs were present in the water solution. The control experiment, performed in the absence of particles, gave rise to far smaller DMPO-OH signals due to water photolysis. These signals suggest that our pristine ZnO nanoparticles greatly enhanced the generation of hydroxyl radicals under UV illumination in water.

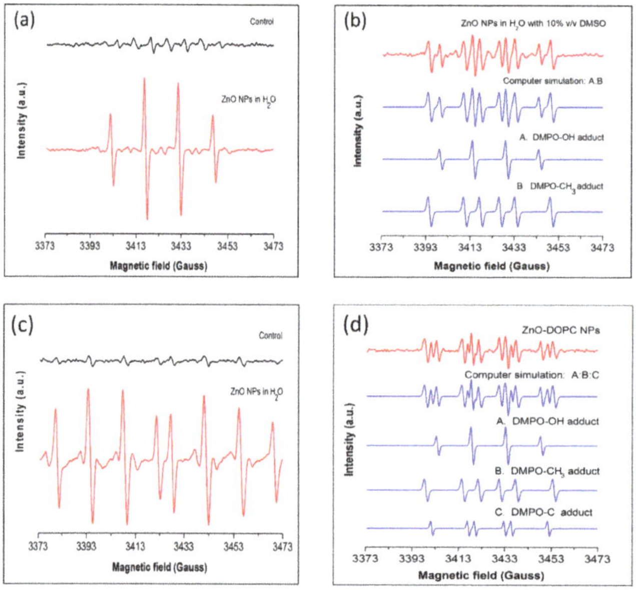

Figure 6. ROS formation in aqueous suspensions of bare and lipid-coated ZnO nanoparticles (500 µg/mL) irradiated with UV light. (**a**) EPR spectrum obtained after the irradiation with UV light of an aqueous solution with (red) and without (black) ZnO NPs using DMPO as spin trap. The characteristic four peaks of the DMPO-OH adduct are detected. (**b**) Effect of the addition of 10% *v*/*v* Dimethyl Sulfoxide (DMSO) to the solution. New peaks corresponding to the DMPO-CH_3 spin adduct are detected (red curve), as confirmed by the computer simulation of the experimental spectrum (blue curves). (**c**) Verification of the absence of superoxide anion radical generation using DEPMPO as spin trap. Only the characteristic peaks of DEPMPO-OH spin adduct are detected with (red) and without (black) ZnO NPs in suspension. (**d**) Effect of surface functionalization on ROS generation by ZnO NPs under UV illumination detected using DMPO as spin trap. Using lipid-shell functionalized ZnO NPs, a complex EPR spectrum is obtained (red curve). Computer simulation reveals that, together with the detection of the DMPO-OH and DMPO-CH_3 spin adducts, a short-chain carbon-centered radical is detected (blue curves).

Since DMPO can also trap superoxide anion radicals to produce the spin adduct DMPO-OOH, which is not stable and readily decomposes to the DMPO-OH adduct [26,27], $O_2\cdot$ generation could not be excluded. To determine whether the spectrum presented in Figure 6a was due to the generation of superoxide and/or hydroxyl radical, DMSO was added to the ZnO water suspension to scavenge the photo-generated hydroxyl radical. The resulting EPR spectrum is shown in Figure 6b. The addition of DMSO significantly reduced the DMPO-OH quartet signal, thus suggesting that OH· radicals are actually generated by ZnO nanoparticles. Computer simulations were performed to identify the new peaks present in the detected signal (blue curves). The simulated spectrum matched well with the experimental one: together with the DMPO-OH adduct, the DMPO-CH_3 adduct was detected. This is formed by the reaction of the photo-generated OH radical with the DMSO molecule [28], thus further confirming the hydroxyl radical generation.

To verify the absence of the superoxide anion (O_2), the dedicated spin trap DEPMPO was used. This can trap hydroxyl and superoxide anion radicals forming radical-specific stable spin adducts, thus enabling the direct detection of the superoxide anion [29]. In Figure 7, the EPR spectra obtained after UV illumination of water suspensions are shown. In the presence of ZnO NPs in solution (red curve), eight resolved peaks were detected, corresponding to the characteristic peaks of the DEPMPO-OH spin adduct. These were also detected in the control experiment, but with much lower intensities, and are thus attributed to water photolysis. The absence of the DEPMPO-$O_2\cdot$spin adduct strongly suggests that superoxide anion radicals are not photo-generated by ZnO NPs.

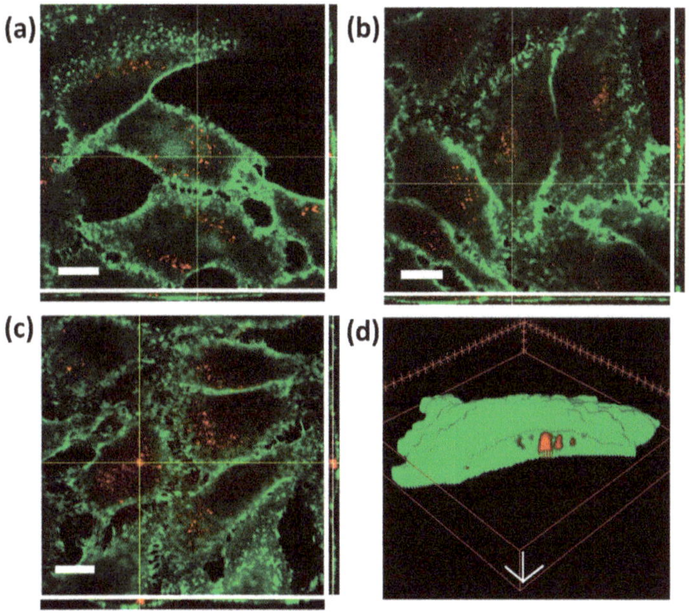

Figure 7. Lipid-shell functionalized ZnO NPs are successfully internalized by HeLa cells after 24 h incubation. (**a**–**c**) Representative fluorescent image of HeLa cells (membranes in green) incubated for 24 h with fluorescent ZnO-DOPC NPs (in red). Rectangles on the right side and bottom of each image show orthogonal views of the z-stack (along the yellow lines) of images proving that the particles are inside the cells. Scale bar: 10 μM. (**d**) 3D representation of ZnO-DOPC nanoparticles taken up into the HeLa cell. ZnO-DOPC NPs (in red) are clearly visible inside the cellular membrane (in green). Z-stacks images were processed with Particle_in_Cell-3D Macro [30] using Fiji software. Scale bar: 5 μM.

2.3.3. Effect of Surface Functionalization on ROS Generation

Since the photo-generation of an electron–hole pair and the subsequent formation of hydroxyl radicals happens at the surface of the ZnO nanoparticle, it can be expected that modifying the chemistry of the surface, i.e., by phospholipid bilayer coating, would influence its ROS generation ability. To investigate this hypothesis, EPR spin-trapping experiments were performed using lipid-coated ZnO nanoparticles and DMPO as spin trap. Water suspension of ZnO-DOPC NPs were irradiated with UV for 5 min and the EPR spectrum recorded.

As shown in Figure 6d, a far more complex EPR spectrum was recorded compared to the spectrum obtained for bare ZnO NPs (Figure 6a, red curve). Indeed, the new spectrum is a composition of several DMPO spin adducts. To distinguish the different spin adducts, computer simulation was performed to fit the experimental spectrum with characteristic DMPO adducts spectra. The simulated spectra (blue curves) matched well with the experimental one (red curve), and the single contributions could be distinguished. Indeed, together with the DMPO-OH spin adduct, DMPO-CH_3 and a short-chain carbon-centered radical adduct were detected. Importantly, the presence of the DMPO-OH spin adduct in the experimental spectrum suggests that phospholipid bilayer coating did not prevent the hydroxyl radical photo-generation. The other two spin adducts derive from the reaction of DMPO with short-chain carbon-centered radicals probably formed by fragmentation of the phospholipid chain undergoing lipid peroxidation. Indeed, this could be initiated by the oxidative attack of the photo-generated hydroxyl radicals that, being highly reactive species, can oxidize unsaturated fatty acids in lipid membranes starting the lipid peroxidation process [31].

Together, these results confirm the possibility to use the lipid-coated ZnO NPs as nano ROS-generator under UV irradiation and reveal that the surface functionalization can clearly influence the type of photo-generated ROS.

2.4. Cellular Uptake and Intracellular ROS Generation

2.4.1. Uptake and Internalization Pathway in Cancer Cells

In photodynamic therapy, the internalization of the photosensitizer by the target malignant cells is a crucial step toward an effective therapeutic effect. ROS have short lifetimes and diffuse poorly in biological environment: therefore, they can be effective in exerting a cytotoxic effect only if photo-generated intracellularly [32].

To determine if cancer cells could effectively internalize lipid-coated ZnO NPs, human epithelial carcinoma cells (HeLa) were used as target cells. Cancer cells were treated with 18 μg/mL of ZnO-DOPC NPs and after 24 h of incubation the cellular uptake was qualitatively assessed by fluorescence images using Spinning Disk Microscopy, as shown in Figure 7. Z-stacks of live-cell images allowed the direct visualization of the lipid-coated nanoparticles localized inside the cellular membrane, thus confirming the successful uptake.

Several studies demonstrated that, in addition to an efficient cellular uptake, the intracellular localization of the photosensitizer can greatly affect the therapeutic outcome of photodynamic therapy. ROS have a short life-time and react close to their site of generation, therefore the photodamage can depend on the precise subcellular localization of the NPs [32]. Thus, a careful investigation of the intracellular localization of our lipid-coated ZnO NPs was performed by fluorescent microscopy co-localization experiments using a fluorescent lysosomal marker, as described in detail in the Materials and Methods Section and shown in Figure 8. Co-localization of the lysosomal fluorescent spots (red) with the ZnO-DOPC NPs ones (green) suggests that the cancer cells internalized the lipid-coated ZnO NPs through an endosomal-lysosomal pathway, which eventually led to the internalization of the nanoparticles inside the cancer cell lysosomes (Figure 8c).

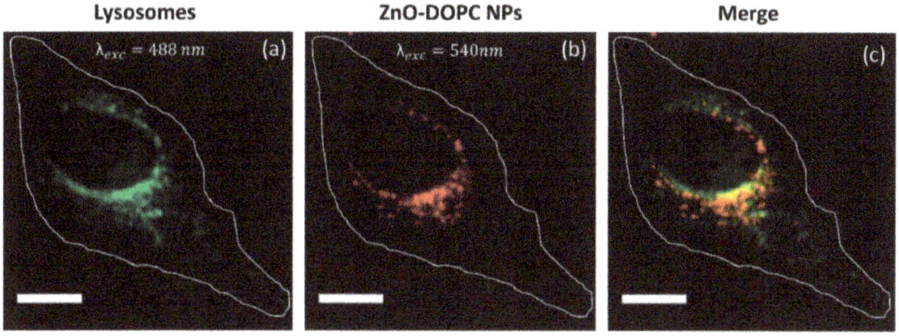

Figure 8. Cancer cells internalized lipid-coated ZnO NPs through an endosomal-lysosomal pathway. Wide-Field Fluorescence images of: (**a**) Lysosomes marked with CellLight Lysosomes-GFP; (**b**) Lipid-coated ZnO NPs marked with Atto550-NHS ester; and (**c**) The merged images showing the co-localization of the ZnO-DOPC NPs with the intracellular lysosomes. Scale bar: 5 μM.

2.4.2. Intracellular ROS Generation

Finally, to evaluate if the lipid-coated ZnO NPs preserved their ability to photo-generate ROS after their internalization inside cancer cells, fluorescence microscopy experiments based on the oxidation of 2′-7′-dichlorofluorescein (DCF) were performed. HeLa cells were incubated with DCF and lipid-coated ZnO NPs, and after 24 h irradiated with UV light. As shown in Figure 9, the control experiments did not show any green fluorescence due to DCF oxidation, suggesting that the exposure of cancer cells to the UV irradiation did not generate any ROS in the absence of ZnO-DOPC NPs. On the contrary, when HeLa cells were incubated with ZnO-DOPC NPs (red spots) and exposed to UV light for 30 s, green fluorescence due to the oxidation of DCF appeared close to the internalized NPs. This indicates that lipid-coated ZnO nanoparticles were able to generate ROS inside cancer cells when irradiated with UV light, thus suggesting ZnO-DOPC NP as innovative ROS-generator agent for photodynamic therapy.

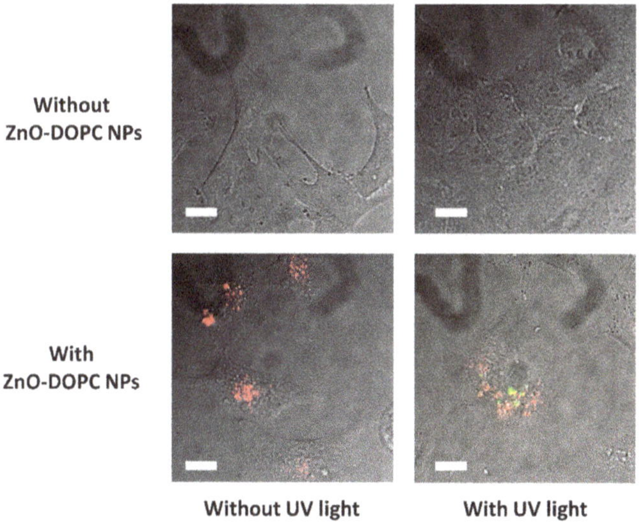

Figure 9. Intracellular ROS generation by lipid-coated ZnO nanoparticles. Scale bar: 5 μM.

2.5. *Preliminary Nanoparticles Cytotoxicity and Photodynamic Effect Study*

Cytotoxicity and photodynamic effect of both bare ZnO and lipid-coated ZnO nanoparticles were investigated on HeLa cells. It is worth mentioning that, in coherence with both the internalization and the intra-cellular ROS generation ability experiments, amine-functionalized ZnO nanoparticles were used in the experiments (EPR and Z-potential measurements of amine-functionalized ZnO NPs are shown in the Supplementary Materials).

The effects of different concentrations of ZnO and ZnO–DOPC nanoparticles on HeLa cell culture for 24 h are shown in Figure 10. From a quantitative point of view, these data confirm the cytotoxic behavior (in absence of UV light activation) of both ZnO and ZnO-DOPC NPs only at high NPs concentration. The two kinds of ZnO NPs had no visible cytotoxic effect up to a concentration of about 18 μg/mL, while both showed a significant cytotoxic effect for higher concentrations.

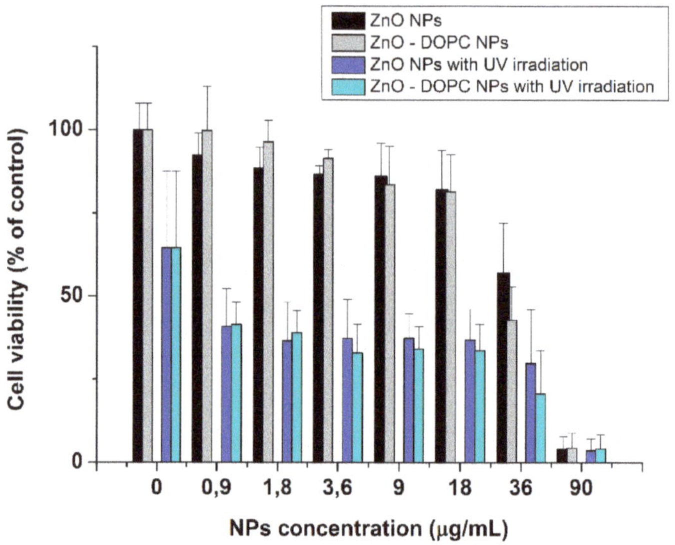

Figure 10. Cytotoxicity and photodynamic effect of different concentrations of ZnO nanoparticles and lipid-coated ZnO nanoparticles with and without UV irradiation (30 s).

Moreover, HeLa cell cultures were irradiated with UV light (at a wavelength of 255 nm) for 30 s to preliminary evaluate the ability of NPs-induced photo-generated ROS to induce cytotoxic effects in vitro. Cancer cells treated with both kind of ZnO NPs showed a marked decrease of cell viability after 24 h from the UV-exposure, compared to the control without NPs-treatment (first columns of the graph in Figure 10 reporting the concentration of 0 μg/mL of nanoparticles). These preliminary data suggest that ZnO, used at non-toxic concentrations, can induce cytotoxic effects when irradiated with UV light. Moreover, the lipid-coating, while increasing NPs stability in biological media, does not decrease its photodynamic therapeutic effect. Future studies will focus on the optimization of the light exposure parameters to enhance the cancer cell killing photodynamic efficacy of ZnO-DOPC NPs.

3. Materials and Methods

3.1. *Synthesis and Functionalization of ZnO Nanoparticles*

Zinc oxide nanoparticles were prepared by applying a similar wet chemical method as that described in the previous report [33]. Zinc acetate dihydrate ($Zn(CH_3COO)_2 \cdot H_2O$ (3.73 mmol) was dissolved in methanol (42 mL) and heated under continuous stirring. As the temperature reached

60 °C, 318 µL of bi-distilled water and a solution containing 7.22 mmol of NaOH in 23 mL of methanol were added drop by drop to the zinc acetate solution. This was maintained at 60 °C for 2.15 h and then washed two times with fresh ethanol centrifuging at 3046 rcf for 5 min.

ZnO NPs were functionalized according to previously reported method [34]. In brief, the amino-propyl functionalized zinc oxide nanoparticles (ZnO-NH_2 NPs) used for fluorescence microscopy experiments were obtained exploiting the reaction between bare ZnO nanoparticles and 3-aminopropyltrimethoxysilane (APTMS), while the lipid-coated zinc oxide nanoparticles (ZnO-DOPC NPs) were prepared by a solvent exchange method using the commercial phospholipid DOPC (1,2-dioleoyl-sn-glycero-3-phosphocholine), as described in previous reports [35].

3.2. *Characterization of Zinc Oxide Nanoconstructs*

The morphology of ZnO and ZnO-DOPC nanoparticles was studied by Field Emission Scanning Electron Microscopy (FESEM, Auriga and Merlin, Karl Zeiss, Oberkochen, Germany). The diluted samples were spotted on a silica wafer and coated by a thin layer of Pt for further imaging. The particle size and Zeta potential of the three samples were determined using the Dynamic Light Scattering (DLS) technique (Zetasizer Nano ZS90, Malvern, Worcestershire, UK), while the crystalline structure was analyzed by X-ray diffraction with a X'Pert diffractometer in configuration θ–2θ Bragg-Brentano using a Cu-Kα radiation (λ = 1.54 Å, 40 kV and 30 mA).

To confirm the formation of the supported lipid bilayer on the surface of ZnO-DOPC nanoparticles, fluorescence co-localization experiments were performed. The DOPC shell was labeled with 1% Bodipy-DHPE lipid by incubating this dye (0.2 µg per mg of lipids) with the dispersed DOPC lipids prior to assembly. ZnO nanoparticles, after amine functionalization using APTMS, were labeled with Atto550-NHS ester dye (2 µg per mg of NPs) overnight under stirring at RT and then washed twice with fresh ethanol. A fully-motorized wide-field inverted microscope Nikon Eclipse TiE (Nikon, Tokyo, Japan), in combination with a high resolution sCMOS camera (Zyla 4.2 Plus from Andor) and an immersion 60× oil objective was used.

3.3. *Bio-Stability Assay*

For the bio-stability assay, bare and lipid-coated ZnO nanoparticles were tested in Phosphate Buffered Saline (PBS, Sigma Aldrich, St. Louis, MO, USA). Dynamic Light Scattering (DLS) measurements were performed suspending 500 µg of nanoparticles in 1 mL of Phosphate Buffered Saline (PBS).

3.4. *Spin Trapping Measurements Coupled with EPR Spectroscopy*

The EPR-spin trapping technique coupled with the spin traps 5,5-dimethyl-L-pyrroline-*N*-oxide (DMPO, 50 mM) and 5-(diethoxyphosphoryl)-5-methyl-L-pyrroline-*N*-oxide (DEPMPO, 40 mM) (Sigma, St. Louis, MO, USA) was used to detect hydroxyl radicals and superoxide radicals.

To verify the ROS generation by bare ZnO NPs (500 µg/mL) in different biological media, 10 µL of 1 M DMPO were mixed with 190 µL of the various tested media (water, PBS, Simulated Body Fluid (SBF) or Cell Culture Media (EMEM)) containing ZnO NPs. The resulting solution was maintained under continuous stirring (150 rpm) at constant temperature of 37 °C for 1 h and transferred into a quartz microcapillary tube for the EPR measurement. For the detection of UV-induced ROS generation, 100 µL of the selected spin trap were mixed with 100 µL of the appropriate ZnO NPs suspension. The solution was mixed with a vortex, irradiated with UV light (wavelength 350–450 nm, Intensity: 150 mW/cm^2) for 5 min and then promptly transferred into a quartz microcapillary tube. In all EPR measurements, 25 µM of Diethylenetriaminepentaacetic acid (DTPA) was added to the solution as metal ion chelator. For experiments labeled as "Absence of external actuation", there was an ambient light illumination.

The microcapillary tubes were then inserted in the EPR cavity and the spectra were recorded on a Bruker EMXnano X-Band spectrometer (Bruker, Billerica, MA, USA). The EPR measurement conditions were as follows: Frequency, 9.74 GHz; scan width, 100 G; receiver gain, 60 dB; time constant, 1.28 ms; sweep time, 160 s; scans, 10.

After acquisition, the spectrum was processed using the Bruker Xenon software (Bruker, Billerica, MA, USA) for baseline correction. Simulation of the recorded spectra was performed using the Bruker SpinFit software.

3.5. Internalization Experiments in HeLa Cells

HeLa cells were cultured in Dulbecco's Modified Eagle Medium (DMEM) with 10% FBS and 1% PenStrep. They were seeded into ibiTreat μ-slides (ibidi) at a concentration of 5000 cells per well in 300 μL of DMEM. The day after seeding, cells were incubated with particles for 24 h and for the lysosome experiments also with 8 μL of CellLight Lysosomes-GFP (Thermo fisher scientific, Waltham, MA, USA). The particles were labeled with Atto633-NHS overnight and then prepared as for the characterization described above. For internalization experiments cells were stained with WGA488 (Thermo fisher scientific, Waltham, MA, USA), and washed with DMEM prior to imaging. For imaging we used a spinning disk microscope (Zeiss Cell Observer SD with a Yokogawa spinning disk unit CSU-X1). Lysosome-GFP and WGA were excited with a 488 nm laser and the particles with a 639 nm laser. Band-pass filters 525/50 and 690/60 (both Semrock) were used in the detection path for Lysosome-GFP/WGA and the particles, respectively.

3.6. Detection of Intracellular ROS Generation

HeLa cells were cultured, seeded and incubated with particles as described above for the internalization experiments. 2′,7′-dichlorofluorescein diacetate (DCFDA) was dissolved in DMSO shortly before use and added to the cells yielding a final concentration of 0.13 μM. After an incubation time of 30 min cells were imaged to obtain the control before UV illumination. After 30 s of UV illumination and another 30 min incubation, cells were imaged for the UV-illumination. Microscopy was performed as for the internalization experiments and a 488 nm laser and a BP 525/50 filter were used to image the DCFDA.

3.7. Cytotoxicity and Photodynamic Experiments

HeLa cells were seeded at a concentration of 5000 cells per well into 96 well plates (Corning) containing a final volume of 100 mL of medium. One day after seeding they were incubated with NPs at the desired concentration. Twenty-four hours after incubation with NPs, MTT assay were performed according to the standard protocol.

For photodynamic experiments, after 5 h of incubation with the desired NP concentrations, cells were exposed to 30 s of UV illumination and MTT were carried out after 24 h to UV exposure.

4. Conclusions

In the present study, we propose the synthesis and characterization of lipid-coated ZnO nanoparticles as new photosensitizer for PDT against cancer. First, we show that the lipid-coating increases the colloidal stability of the ZnO NPs in Phosphate Buffered Saline (PBS). Then, we demonstrate that bare and lipid-coated ZnO nanoparticles generate hydroxyl radicals only when irradiated with UV light. Moreover, the phospholipid bilayer coating induces the photo-generation of short-chain carbon centered free radicals, thus suggesting that the nanoparticle surface chemistry plays a crucial role in determining the type of photo-generated free radicals. Finally, we show that lipid-coated NPs are effectively internalized by HeLa cells through an endosomal-lysosomal pathway and that they can generate ROS even once internalized and kill cancer cell at non-toxic concentration thanks to the UV-stimuli activation.

The herein reported results pave the way for a more conscious design of nanoparticles for PDT treatment, with the surface chemistry being an important factor to be considered for an efficient ROS production, and imply the potential of lipid-coated NPs as innovative ROS-generators for therapeutic activity against cancer.

Supplementary Materials: The following are available online at www.mdpi.com/2079-4991/8/3/143/s1, Figure S1: ROS formation in aqueous suspensions of amine-functionalized ZnO nanoparticles (500 μg/mL) irradiated with UV light). Computer simulation reveals that DMPO-OH and DMPO-CH_3 spin adducts are detected (blue curves). Figure S2: Z-potential measurement of pristine ZnO NPs and amine-functionalized NPs.

Acknowledgments: This work has received funding from the European Research Council (ERC) under the European Union's Horizon 2020 research and innovation program (grant agreement No. 678151—Project Acronym "TROJANANOHORSE"—ERC starting Grant) and by a Regional program entitled "Attrarre Docenti di Qualità con Starting Grant" from the Compagnia di Sanpaolo, D.R. 349-2016.

Author Contributions: The manuscript was written through contributions of all authors. All authors have given approval to the final version of the manuscript.

Conflicts of Interest: The authors declare no conflict of interest.

References

1. Marchesan, S.; Prato, M. Nanomaterials for (Nano)medicine. *ACS Med. Chem. Lett.* **2013**, *4*, 147–149. [CrossRef] [PubMed]
2. Cauda, V.; Gazia, R.; Porro, S.; Stassi, S.; Canavese, G.; Roppolo, I.; Chiolerio, A. Nanostructured ZnO Materials: Synthesis, Properties and Applications. In *Handbook of Nanomaterials Properties*; Springer: Berlin/Heidelberg, Germany, 2014; pp. 137–177. ISBN 978-3-642-31106-2.
3. Xu, S.; Wang, Z.L. One-dimensional ZnO nanostructures: Solution growth and functional properties. *Nano Res.* **2011**, *4*, 1013–1098. [CrossRef]
4. Sirelkhatim, A.; Mahmud, S.; Seeni, A.; Kaus, N.H.M.; Ann, L.C.; Bakhori, S.K.M.; Hasan, H.; Mohamad, D. Review on Zinc Oxide Nanoparticles: Antibacterial Activity and Toxicity Mechanism. *Nano-Micro Lett.* **2015**, *7*, 219–242. [CrossRef]
5. Bisht, G.; Rayamajhi, S. ZnO Nanoparticles: A Promising Anticancer Agent. *Nanobiomedicine* **2016**, *3*, 9. [CrossRef]
6. Bogdan, J.; Pławińska-Czarnak, J.; Zarzyńska, J. Nanoparticles of Titanium and Zinc Oxides as Novel Agents in Tumor Treatment: A Review. *Nanoscale Res. Lett.* **2017**, *12*, 225. [CrossRef] [PubMed]
7. Fu, P.P.; Xia, Q.; Hwang, H.-M.; Ray, P.C.; Yu, H. Mechanisms of nanotoxicity: Generation of reactive oxygen species. *J. Food Drug Anal.* **2014**, *22*, 64–75. [CrossRef] [PubMed]
8. Namvar, F.; Rahman, H.S.; Mohamad, R.; Azizi, S.; Tahir, P.M.; Chartrand, M.S.; Yeap, S.K. Cytotoxic Effects of Biosynthesized Zinc Oxide Nanoparticles on Murine Cell Lines. Available online: https://www.hindawi.com/journals/ecam/2015/593014/ (accessed on 19 September 2017).
9. Liu, J.; Feng, X.; Wei, L.; Chen, L.; Song, B.; Shao, L. The toxicology of ion-shedding zinc oxide nanoparticles. *Crit. Rev. Toxicol.* **2016**, *46*, 348–384. [CrossRef] [PubMed]
10. Yu, K.-N.; Yoon, T.-J.; Minai-Tehrani, A.; Kim, J.-E.; Park, S.J.; Jeong, M.S.; Ha, S.-W.; Lee, J.-K.; Kim, J.S.; Cho, M.-H. Zinc oxide nanoparticle induced autophagic cell death and mitochondrial damage via reactive oxygen species generation. *Toxicol. In Vitro* **2013**, *27*, 1187–1195. [CrossRef] [PubMed]
11. Pelicano, H.; Carney, D.; Huang, P. ROS stress in cancer cells and therapeutic implications. *Drug Resist. Updates* **2004**, *7*, 97–110. [CrossRef] [PubMed]
12. Saliani, M.; Jalal, R.; Kafshdare Goharshadi, E. Mechanism of oxidative stress involved in the toxicity of ZnO nanoparticles against eukaryotic cells. *Nanomed. J.* **2016**, *3*, 1–14.
13. Hackenberg, S.; Scherzed, A.; Kessler, M.; Froelich, K.; Ginzkey, C.; Koehler, C.; Burghartz, M.; Hagen, R.; Kleinsasser, N. Zinc oxide nanoparticles induce photocatalytic cell death in human head and neck squamous cell carcinoma cell lines in vitro. *Int. J. Oncol.* **2010**, *37*, 1583–1590. [PubMed]
14. Zhang, H.; Shan, Y.; Dong, L. A comparison of TiO_2 and ZnO nanoparticles as photosensitizers in photodynamic therapy for cancer. *J. Biomed. Nanotechnol.* **2014**, *10*, 1450–1457. [CrossRef] [PubMed]
15. Maeda, H. Toward a full understanding of the EPR effect in primary and metastatic tumors as well as issues related to its heterogeneity. *Adv. Drug Deliv. Rev.* **2015**, *91*, 3–6. [CrossRef] [PubMed]
16. Lipovsky, A.; Tzitrinovich, Z.; Friedmann, H.; Applerot, G.; Gedanken, A.; Lubart, R. EPR Study of Visible Light-Induced ROS Generation by Nanoparticles of ZnO. *J. Phys. Chem. C* **2009**, *113*, 15997–16001. [CrossRef]
17. Li, J.; Guo, D.; Wang, X.; Wang, H.; Jiang, H.; Chen, B. The Photodynamic Effect of Different Size ZnO Nanoparticles on Cancer Cell Proliferation in Vitro. *Nanoscale Res. Lett.* **2010**, *5*, 1063–1071. [CrossRef] [PubMed]
18. Chibowski, E.; Szcześ, A. Zeta potential and surface charge of DPPC and DOPC liposomes in the presence of PLC enzyme. *Adsorption* **2016**, *22*, 755–765. [CrossRef]

19. Degen, A.; Kosec, M. Effect of pH and impurities on the surface charge of zinc oxide in aqueous solution. *J. Eur. Ceram. Soc.* **2000**, *20*, 667–673. [CrossRef]
20. Albanese, A.; Tang, P.S.; Chan, W.C.W. The effect of nanoparticle size, shape, and surface chemistry on biological systems. *Annu. Rev. Biomed. Eng.* **2012**, *14*, 1–16. [CrossRef] [PubMed]
21. Illes, B.; Hirschle, P.; Barnert, S.; Cauda, V.; Wuttke, S.; Engelke, H. Exosome-Coated Metal–Organic Framework Nanoparticles: An Efficient Drug Delivery Platform. *Chem. Mater.* **2017**, *29*, 8042–8046. [CrossRef]
22. Dumontel, B.; Canta, M.; Engelke, H.; Chiodoni, A.; Racca, L.; Ancona, A.; Limongi, T.; Canavese, G.; Cauda, V. Enhanced biostability and cellular uptake of zinc oxide nanocrystals shielded with a phospholipid bilayer. *J. Mater. Chem. B* **2017**, *5*, 8799–8813. [CrossRef] [PubMed]
23. Reed, R.B.; Ladner, D.A.; Higgins, C.P.; Westerhoff, P.; Ranville, J.F. Solubility of nano-zinc oxide in environmentally and biologically important matrices. *Environ. Toxicol. Chem.* **2012**, *31*, 93–99. [CrossRef] [PubMed]
24. Makino, K.; Hagi, A.; Ide, H.; Murakami, A.; Nishi, M. Mechanistic studies on the formation of aminoxyl radicals from 5,5-dimethyl-L-pyrroline-*N*-oxide in Fenton systems. Characterization of key precursors giving rise to background ESR signals. *Can. J. Chem.* **1992**, *70*, 2818–2827. [CrossRef]
25. Grzelak, A.; Rychlik, B.; Bartosz, G. Light-dependent generation of reactive oxygen species in cell culture media. *Free Radic. Biol. Med.* **2001**, *30*, 1418–1425. [CrossRef]
26. Finkelstein, E.; Rosen, G.M.; Rauckman, E.J. Spin trapping of superoxide and hydroxyl radical: Practical aspects. *Arch. Biochem. Biophys.* **1980**, *200*, 1–16. [CrossRef]
27. Finkelstein, E.; Rosen, G.M.; Rauckman, E.J. Production of hydroxyl radical by decomposition of superoxide spin-trapped adducts. *Mol. Pharmacol.* **1982**, *21*, 262–265. [PubMed]
28. Baptista, L.; da Silva, E.C.; Arbilla, G. Oxidation mechanism of dimethyl sulfoxide (DMSO) by OH radical in liquid phase. *Phys. Chem. Chem. Phys.* **2008**, *10*, 6867–6879. [CrossRef] [PubMed]
29. Mojović, M.; Vuletić, M.; Bacić, G.G. Detection of oxygen-centered radicals using EPR spin-trap DEPMPO: The effect of oxygen. *Ann. N. Y. Acad. Sci.* **2005**, *1048*, 471–475. [CrossRef] [PubMed]
30. Torrano, A.A.; Blechinger, J.; Osseforth, C.; Argyo, C.; Reller, A.; Bein, T.; Michaelis, J.; Bräuchle, C. A fast analysis method to quantify nanoparticle uptake on a single cell level. *Nanomedicine* **2013**, *8*, 1815–1828. [CrossRef] [PubMed]
31. Qian, S.Y.; Wang, H.P.; Schafer, F.Q.; Buettner, G.R. EPR detection of lipid-derived free radicals from PUFA, LDL, and cell oxidations. *Free Radic. Biol. Med.* **2000**, *29*, 568–579. [CrossRef]
32. Castano, A.P.; Demidova, T.N.; Hamblin, M.R. Mechanisms in photodynamic therapy: Part one—Photosensitizers, photochemistry and cellular localization. *Photodiagn. Photodyn. Ther.* **2004**, *1*, 279–293. [CrossRef]
33. Pacholski, C.; Kornowski, A.; Weller, H. Self-Assembly of ZnO: From Nanodots to Nanorods. *Angew. Chem. Int. Ed.* **2002**, *41*, 1188–1191. [CrossRef]
34. Cauda, V.; Engelke, H.; Sauer, A.; Arcizet, D.; Rädler, J.; Bein, T. Colchicine-Loaded Lipid Bilayer-Coated 50 nm Mesoporous Nanoparticles Efficiently Induce Microtubule Depolymerization upon Cell Uptake. *Nano Lett.* **2010**, *10*, 2484–2492. [CrossRef] [PubMed]
35. Datz, S.; Engelke, H.; Schirnding, C.V.; Nguyen, L.; Bein, T. Lipid bilayer-coated curcumin-based mesoporous organosilica nanoparticles for cellular delivery. *Microporous Mesoporous Mater.* **2016**, *225*, 371–377. [CrossRef]

nanomaterials

MDPI

Article

Ti-Based Biomedical Material Modified with TiO_x/TiN_x Duplex Bioactivity Film via Micro-Arc Oxidation and Nitrogen Ion Implantation

Peng Zhang [1], Xiaojian Wang [1], Zhidan Lin [1], Huaijun Lin [1], Zhiguo Zhang [1], Wei Li [1], Xianfeng Yang [2,*] and Jie Cui [2,*]

1 Institute of Advanced Wear & Corrosion Resistant and Functional Materials, Jinan University, Guangzhou 510632, China; tzhangpeng@jnu.edu.cn (P.Z.); xiaojian.wang@jnu.edu.cn (X.W.); linzd@jnu.edu.cn (Z.L.); hjlin@jnu.edu.cn (H.L.); zhigzhang@jnu.edu.cn (Z.Z.); liweijn@aliyun.com (W.L.)

2 Analytical and Testing Center, South China University of Technology, Guangzhou 510640, China

* Correspondence: czxfyang@scut.edu.cn (X.Y.); czcuijie@scut.edu.cn (J.C.); Tel./Fax: +86-20-8711-1074 (X.Y. & J.C.)

Received: 29 September 2017; Accepted: 18 October 2017; Published: 23 October 2017

Abstract: Titanium (Ti) and Ti-based alloy are widely used in the biomedical field owing to their excellent mechanical compatibility and biocompatibility. However, the bioinert bioactivity and biotribological properties of titanium limit its clinical application in implants. In order to improve the biocompatibility of titanium, we modified its surface with TiO_x/TiN_x duplex composite films using a new method via micro-arc oxidation (MAO) and nitrogen ion implantation (NII) treatment. The structural characterization results revealed that the modified film was constructed by nanoarrays composed of TiO_x/TiN_x composite nanostitches with a size of 20~40 nm. Meanwhile, comparing this with pure Ti, the friction property, wear resistance, and bioactivity were significantly improved based on biotribological results and in vitro bioactivity tests.

Keywords: titanium; titania; micro-arc oxidation; ion implantation; biotribological properties

1. Introduction

Titanium alloys have been widely used in the skeletal system of the human body as constituents of reconstructive devices (e.g., hip or knee join replacement implant) or fracture fixation products (e.g., bone plates, screws, and nails) [1–5]. The stability of the implant-bone interface is of great importance for a successful bone restoration or bone replacement [6–9]. The oxide films covering titanium implant surfaces have attracted extensive research interest as they are demonstrated to be crucial for fast osseointegration [10–12]. In particular, titanium dioxide (TiO_2) thin films on Ti alloys could lead to many desirable properties such as excellent biocompatibility, blood compatibility, corrosion resistance, excellent bonding strength with the substrate, and negative surface charge in physiological solution [13].

To obtain a ceramic-like TiO_2 film on the implant surface, micro-arc oxidation (MAO) is one of the most economic choices; thus it is easy to adopt for mass production. Another advantage of MAO is that it can be applied to an implant with complex structures and leads to a uniform oxide layer. By applying a positive voltage to Ti alloy implants, a TiO_2 layer could be obtained in an electrolyte [14,15]. This layer is beneficial to cell attachment and bone growth and also shows a better apatite forming ability than nature oxide film on a pure titanium surface [16]. The characteristics of the titanium surface can affect cell proliferation and differentiation; therefore, the choice of the surface modification is crucial to ensure the quality of the process of the formation of new tissues [17,18]. The surfaces of titanium biomaterials with different porosity or structures would offer a superior performance in supporting cell growth than a common tissue culture plate [19].

In current practice, many failures of titanium implants have been found to be related to excessive wear of the implant material [20]. Clinical experience showed that pure Ti and its alloy were known to be more susceptible to wear than stainless steels, which resulted in greater amounts of metallic particles for a loose functional implant [21]. The worn metallic particles might cause local irritations or systemic effects and even the removal of implants [22–24]. To improve the wear resistance of the pure titanium surface, titanium nitride (TiN) has been proposed for orthopedic and dental implants due to its high hardness and remarkable resistance to wear and corrosion [25,26]. Moreover, TiN surface has been found to beneficial for the spontaneous nucleation of calcium phosphate [27].

To enhance titanium implants' mechanical, tribological, and biological properties and to improve their friction and wear properties in the human body and their long-term performance, it is proposed to coat a film with multifunctional properties by combining different surface modification techniques. TiO_2/TiN duplex films combine the advantages of TiO_2 and TiN. On one side, a titanium oxide layer on the outer surface could get high blood compatibility. On the other side, the TiN films between the TiO_2 and the titanium alloy substrate might improve the wear resistance and the adherence strength between the deposited films on the titanium implants.

In this paper, a Ti-based bioactive material with TiO_x/TiN_x duplex bioactivity films was synthesized by a new modification method via micro-arc oxidation and nitrogen ion implantation (MAO-NII). Firstly, porous ceramic-like TiO_2 films were formed on the titanium substrate by MAO treatment. The TiO_2 films were then treated by nitrogen ion implantation with different nitrogen ion doses. The structural characteristics of the TiO_x/TiN_x composite modified film (including morphology, phase component, and element composition) were studied. The in vitro bioactivities of the coated specimens were investigated by immersing them in simulated body fluid (SBF) and by examining the apatite formation on their surfaces. Additionally, cell culturing was carried out to study the cyto-compatibility of the duplex films.

2. Experimental Procedure

2.1. Preparation (MAO and NII Treatments)

Commercially available pure titanium alloys (TA2, purchased from the Northwest Non-ferrous Institute of Technology in Xi'an, China), were used as substrates in the current study. The titanium samples were cut into plates with a size of 15 mm × 10 mm × 2 mm. The surfaces of the plates were abraded with silica papers of 200, 400, 600, 800, and 1200 grit in turn and washed in an ultrasonic bath for 20 min with acetone, ethanol, and de-ionized water, respectively. The titanium plates were then dried in an oven at 40 °C. Micro-arc oxidation (MAO) was carried out using an alternating current-type high power supply (PN-III). The Ti plates served as the anode electrodes, and a stainless steel plate was used as the counter electrode. 0.2 M of calcium acetate monohydrate ($(CH_3COO)_2Ca{\cdot}H_2O$, CA) and 0.02 M of β-glycerophosphoric acid disodium salt pentahydrate ($C_3H_7Na_2O_6P{\cdot}5H_2O$, β-GP) were used as electrolytes. After being treated at 350 V for 5 min, a porous TiO_2 film formed on the surface of the Ti substrate.

Nitrogen ion implantation (NII) treatment was then performed by using ion implantation equipment with a Kaufman gas ion source (Southwestern Institute of Physics, Chengdu, China). The initial gas pressure in the implantation chamber was under 3×10^{-3} Pa. Nitrogen was implanted into the above TiO_2 films to produce MAO-NII modified samples with an acceleration energy of 80 keV and implantation doses of 0.1, 0.5, 1.0, 5.0, 10, and 20 × 10^{17} ions/cm^2. The corresponding samples were denoted as Ti-MAO-N0.1 Ti-MAO-N0.5, Ti-MAO-N1.0, Ti-MAO-N5.0, Ti-MAO-N10, and Ti-MAO-N20, respectively.

2.2. Structure Characterization

The morphologies of the specimen before and after soaking in SBF were examined by scanning electron microscopy (SEM, ZEISS SUPRA 40, Oberkochen, Germany) and transmission electron microscopy (TEM, JEM-2100F, Tokyo, Japan). For the as prepared sample with kind conductivity, the specimens, after being rinsed and dried, were directly sent for SEM observation. A small amount of powder scraped from the specimen by a diamond knife was dispersed on a micro grid and sent for TEM observation under 200 kV. After SBF and cell tests with poor conductivity, the samples must be rinsed, dried, and sprayed with gold. The phase compositions were characterized by X-ray diffractions (XRD, D/Max 2400 V, Rigaku, Tokyo, Japan) using Cu K_α radiation in the regular range of $2\theta = 20^\circ \sim 80^\circ$, with an accelerating voltage of 36 kV and a current of 100 mA.

2.3. Friction and Wear Test Bioactivity Evaluation

A wet friction and wear test was carried on a ball-on-disc high speed reciprocation friction and wear tester (MFT-R4000, Lan zhou, China, Lanzhou Institute of Chemical Physics, Chinese Academy of Sciences). An Al_2O_3 ceramic ball with a diameter of 4.0 mm and a radius of 0.032 μm was used as the friction match pair. 50% calf serum (the ratio of calf serum with demonized water is 1:1) was chosen as the lubricant. During the friction tests, the applied normal load was 200 g, with a reciprocation frequency of 2 Hz. The reciprocation distance was fixed as 5 mm, and the friction time was 1 h. Each material is tested for five parallel samples, and the average value is taken by removing the highest and lowest values.

The SBF solution was prepared on the basis of Kokubo's recipe [22] for the bioactivity evaluation. The SBF tests were referenced according to Hiroaki's method [28]. The volume of SBF that is used for testing was determined by Equation (1).

$$Vs = Sa/10 \tag{1}$$

where Vs is the volume of SBF (mL) and Sa is the apparent surface area of the specimen (mm^2) [28]. Put the calculated volume of SBF into a plastic bottle or beaker. After heating the SBF to 36.5 °C, the specimens were submerged pensile in the SBF to avoid over-saturation. After soaking for 24 days, the specimens were washed gently with distilled water and dried at room temperature.

The MC3T3 E1 cell line from embryonic osteoblasts of mouse embryos was used for cytotoxicity tests. The culture medium consisted of alpha-minimum essential medium (α-MEM) supplemented with 10% fetal bovine serum (FBS), 100 $U \cdot mL^{-1}$ of penicillin, and 100 $\mu g \cdot mL^{-1}$ of streptomycin sulfate. The experiments were conducted in an incubator at 37 °C, with a humidified atmosphere of 95% air and 5% CO_2 for two days. The specimens were sterilized by heating at 180 °C for 1 h. The cells were fixed with 5 mL of 10% formalin for 30 min, stained with 8 mL of 0.15% methylene blue for an additional 30 min, washed thoroughly with different concentrations of alcohol, and dried [29,30].

3. Results and Discussion

The XRD patterns of all the specimens treated with the MAO-NII procedure are shown in Figure 1. In addition to the anatase and rutile TiO_2 obtained in the MAO process, unsaturated titanium oxide and titanium nitride were also detected. The XRD peaks marked with rhombi in all patterns, as shown in Figure 1, could be indexed to an unsaturated titania of Ti_5O_9. The XRD peaks marked with stars corresponded to $Ti_3N_{1.29}$. It can be concluded that titanium nitrides and various titanium oxides coexist in the modified layer after nitrogen ion implantation. Namely, a TiO_x/TiN_x composite film on the surface of Ti substrate has been obtained.

Then the morphology of the composite films was investigated by SEM. The top-view surface morphology of the samples implanted with different nitrogen doses is shown in Figure 2. The typical characteristics of MAO-NII modified films with micron-sized pores are shown in Figure 2a. The wall surfaces of the holes are very slick when the N implantation dose is very low (0.1×10^{17} ions/cm^2). By increasing the N implantation dose, the surface morphology was changed in varying degrees. Some small pores appeared on the surface of the MAO modified layer, as shown in Figure 2b,c. When the N implantation dose was 5.0×10^{17} ions/cm^2, most of the surface was not slick anymore, as shown in Figure 2d. This rough morphological feature of the surface is made more obvious in Ti-MAO-N10 and Ti-MAO-N20 by continuously increasing the N implantation dose to 10 and 20×10^{17} ions/cm^2 (Figure 2e,f). Thus we found that the surface would become rougher with increasing N implantation doses.

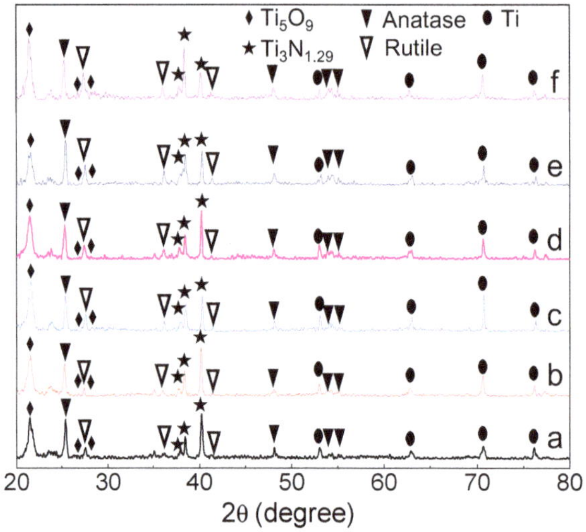

Figure 1. (**a–f**) X-ray diffraction (XRD) patterns of the micro-arc oxidation (MAO) specimens with different N implantation doses: Ti-MAO-N0.1, Ti-MAO-N0.5, Ti-MAO-N1.0, Ti-MAO-N5.0, Ti-MAO-N10, and Ti-MAO-N20.

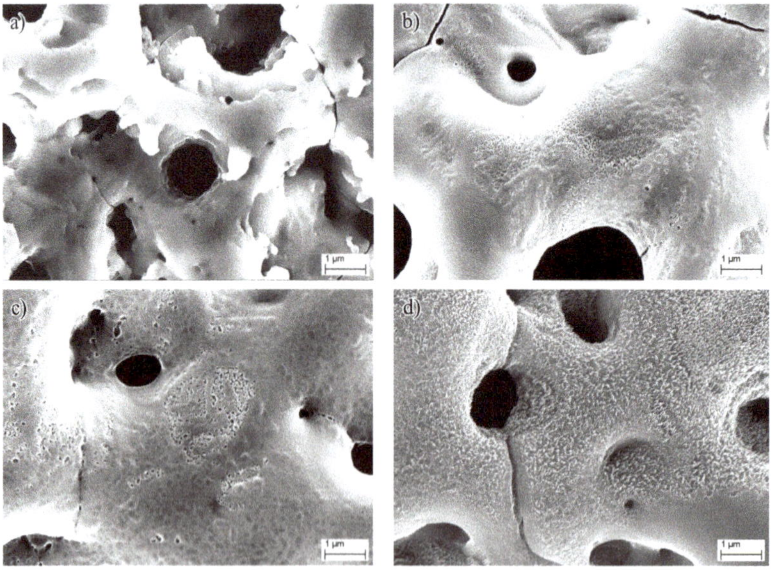

Figure 2. *Cont.*

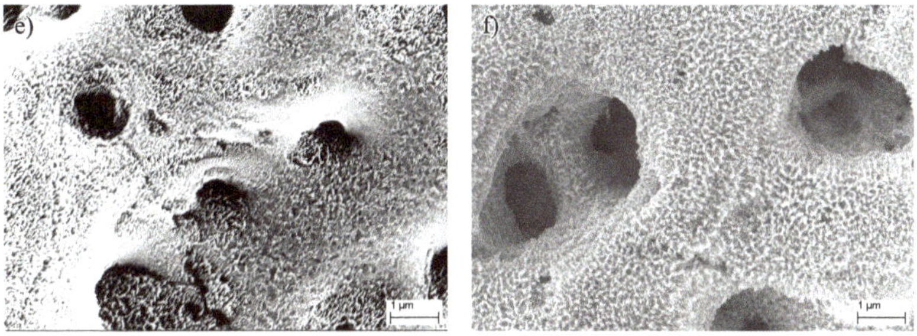

Figure 2. Surface morphologies of the micro-arc oxidation and nitrogen ion implantation (MAO-NII) modified specimens with different N ion implantation doses: (**a**) Ti-MAO-N0.1; (**b**) Ti-MAO-N0.5; (**c**) Ti-MAO-N1.0; (**d**) Ti-MAO-N5.0; (**e**) Ti-MAO-N10; and (**f**) Ti-MAO-N20.

Interestingly, unique nanoarrays composed of vertically aligned nanostitches with a size of 20 to 40 nm were found in Ti-MAO-N10 by a high magnification morphological observation under SEM, as shown in Figure 3a. Moreover, the sharp tips of the nanostitches were only about 5~10 nm. However, the nanostitch array structure disappeared when the N implantation dose was increased to 20×10^{17} ions/cm^2, as shown in Figure 3b, and it was replaced by many small pores with a size of 30~60 nm. Presumably, the nanostitch array structure collapsed due to the high ion implant energy from the increase of the implantation dose.

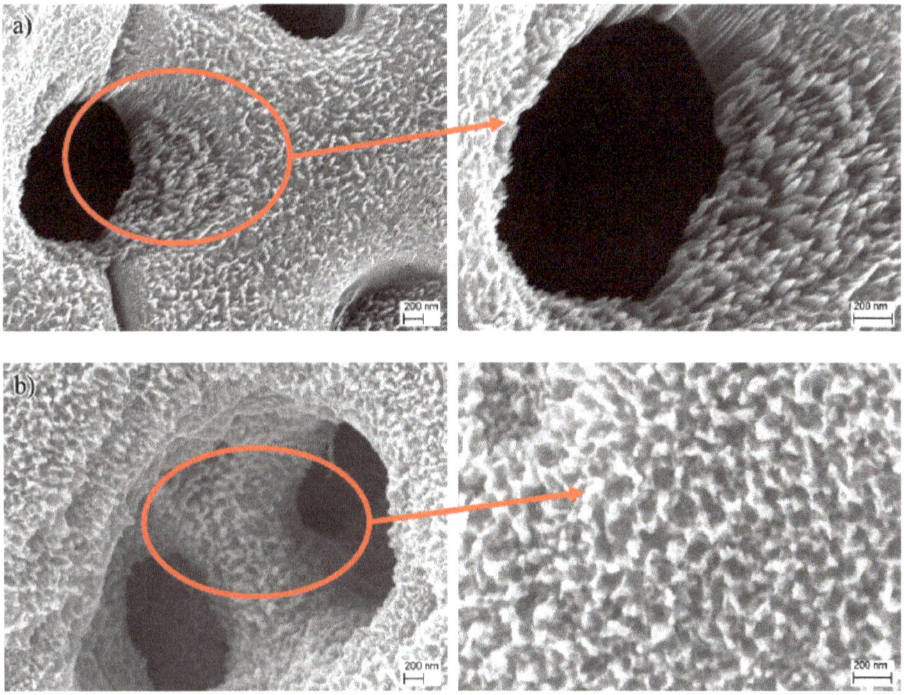

Figure 3. Surface images of nitrogen ion implantation MAO specimen under high magnification morphology: (**a**) Ti-MAO-N10 and (**b**) Ti-MAO-N20.

To further study the crystallographic structure of the TiO_x/TiN_x composite film, a small amount of powder was scraped from the surface of Ti-MAO-N1.0 for TEM observation. The result is shown in Figure 4. It is proved again the nanostitch structure has a tip diameter of 5 nm and a bottom diameter of about 30 nm. As shown in Figure 4a, the main diffraction spots in the selected area of the electron diffraction (SAED) pattern from a circle area of the film correspond to anatase TiO_2 and $Ti_3N_{1.29}$ phases. High resolution TEM (HRTEM) suggests the presence of $Ti_3N_{1.29}$ and Ti_5O_9 respectively within the top and bottom of the nanostitch. Hence, the formation mechanism of the composite modified film with the nanostitch array structure can be speculated based on the above results. As shown in Figure 5, lots of defects such as nanopores appeared with the bombardment of implanted ionic fluxes towards the surfaces at first. Then nanostricks formed and grew to form nanoarrays, which should be mainly composed of unsaturated titania due to the O atoms in TiO_2 being be partly removed by implanted N ions during the NII process. Meanwhile, some of the O atoms of titanium oxide on the surface nanostitches, especially on the top surface, were replaced by N, resulting in the nucleation of nano-sized titanium nitride on the tips. However, if the NII dose is too high, the nanostitch array structure will collapse due to the high implantation energy and superheat.

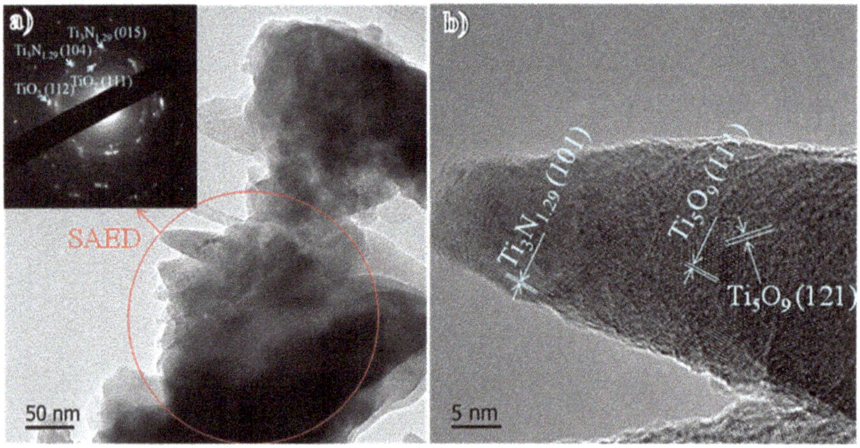

Figure 4. (**a**) Bright field transmission electron microscopy (TEM) image with selected area of the electron diffraction (SAED) pattern inset and (**b**) high resolution TEM (HRTEM) image of TiO_x/TiN_x film of Ti-MAO-N1.0.

The friction factors for specimens after MAO and NII treatment are obtained under 50% mavericks serum lubrication conditions, as shown in Figure 6. It can be found that the friction factors for the treated specimens were all lower than that of the pure titanium (0.404). In addition, the results showed that Ti-MAO-N1.0 (10^{17} ions/cm^2) exerted the lowest average friction factor, which may lead to the best biological friction performance. As discussed above, the composite modified layer is composed of antatase and rutile TiO_2, Ti_5O_9 and $Ti_3N_{1.29}$. Ceramic TiO_2 and titanium nitride are hard phases, which show much higher hardness than pure Ti. As well, the surface hardness increased with the amount of hard nitrides and oxides, which resulted in the reduction of the friction factor. However, the nitrogen ion implantation exists as a saturation injection dose. The increasing amount of implantation will not enhance the surface hardness unlimitedly. In this study, 10^{17} ions/cm^2 was the best implantation dose, when considering the biological friction performance.

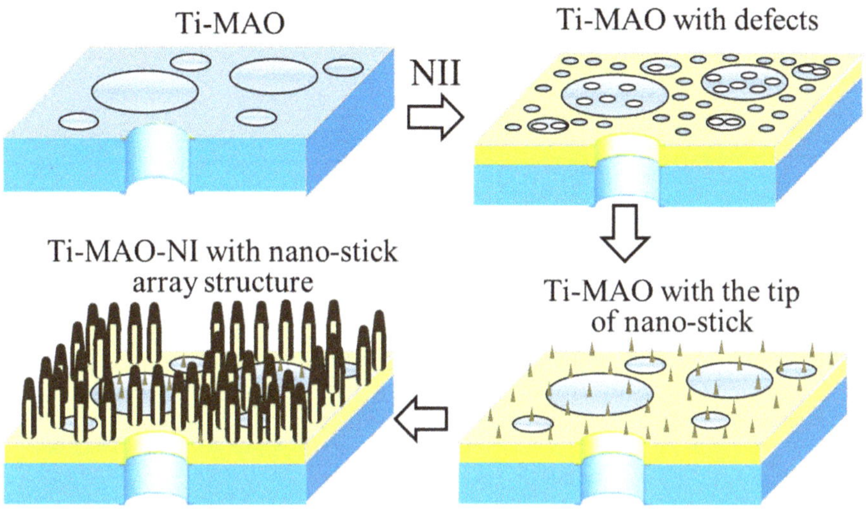

Figure 5. Schematic diagram of the growth mechanism of the TiO_x/TiN_x composite modification layer with a nanostitch array structure during NII.

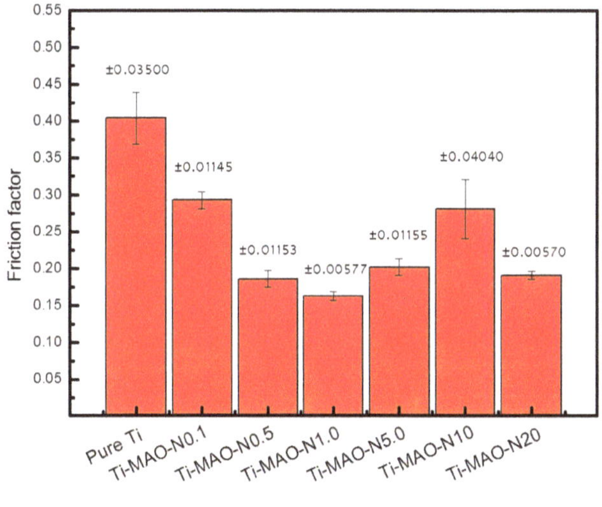

Figure 6. Friction coefficient of MAO specimens with different N doses under 50% mavericks serum lubrication conditions.

In addition to biological friction, biological activity is also concerned in this study. The specimens treated by MAO and NII were immersed in biomimetic mineralization solution for 24 days. The samples after soaking were examined by SEM, and the results are shown in Figure 7. SEM images showed that all the sample surfaces were almost fully covered with ball–shaped particles. There are some nano-flakes on the surfaces of the globular objects, which is similar to the urchin shown with the red arrow. Compared with the SBF soaking result of pure Ti, shown in Figure S1a in the supporting information, the bioactivity of the Ti-MAO-NI (Figure S1b,c) samples is improved. Additionally, combed with the energy dispersive spectrometer (EDS) results shown in Table 1, it can be predicted that Ca and P are fully deposited on the surfaces of the samples after MAO and NII. It can be predicted that apatite was deposited, which is one of the important human bone inorganic compositions [30,31]. Hence, this illustrates a good ability to inducer phosphate deposition, which also leads to good in vitro bioactivity and also indicates that all the samples after MAO and NII modification show good bioactivity. Furthermore, it can also found that the change of the content of Ca and P is consistent with the content of N implantation. Hence, it can be suspected that titanium nitride is beneficial to apatite deposition and facilitates bioactivity.

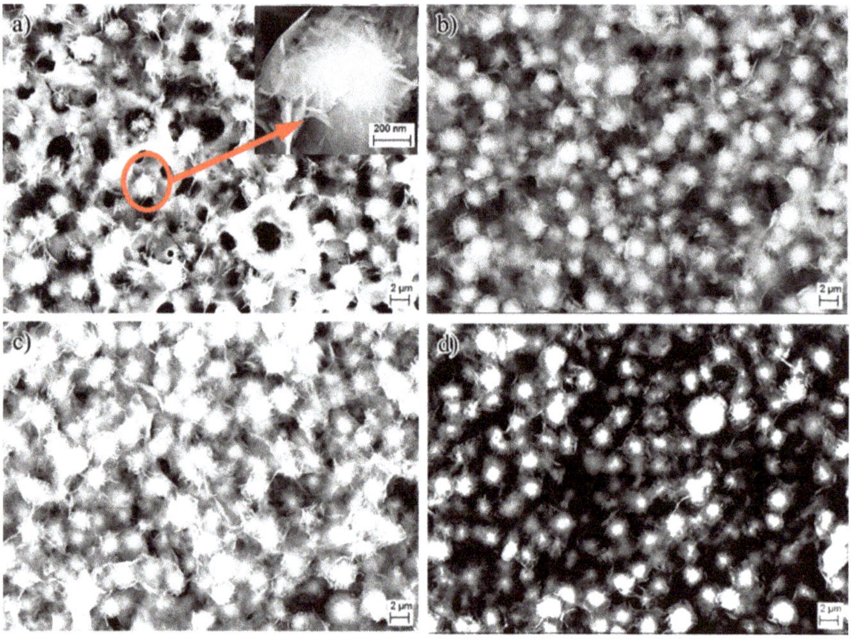

Figure 7. Surface morphology of the samples after soaking in biomimetic mineralization solution for 24 days: (**a**) Ti-MAO-N0.1; (**b**) Ti-MAO-N1.0; (**c**) Ti-MAO-N5.0; and (**d**) Ti-MAO-N20.

Table 1. Energy dispersive spectrum (EDS) analysis results of samples with different doses of N ion implantation after soaking in biomimetic mineralization solution for 24 days.

Sample Name	Element Content (wt %)				
	Ti	O	Ca	P	N
Ti-MAO-N0.1	41.12	43.39	10.30	3.47	1.72
Ti-MAO-N1.0	42.30	41.19	11.01	3.48	2.02
Ti-MAO-N5.0	36.79	40.52	14.43	4.94	3.32
Ti-MAO-N20	51.89	36.05	5.87	4.14	2.05

The cyto-compatibility of the samples was also evaluated by cell cultures on the different surfaces. The surface morphology of the Ti-MAO-NI specimens after cell culturing for two days is shown in Figure 8 and Figure S2 in the supporting information. The alkaline phosphatase activity of osteoblasts on pure Ti and Ti-MAO-N1.0 surface was shown in Figure S3. The cells spread well and showed a good attachment, with a plumpness and a polygon shape on the specimen surface. The cells also developed numerous filopodia, sensing the different surface specimens. Compared with the Ti-MAO-N0.1, Ti-MAO-N0.5, Ti-MAO-N1.0, Ti-MAO-N10, and Ti-MAO-N20, the amount of cells in Ti-MAO-N5.0 is the most. Thus, it can be concluded that cyto-compatibility can be promoted by NII treatment.

The porosity and nanostitch array structures of the samples can affect cell proliferation and differentiation, which is crucial to ensuring the quality of the process of the formation of new tissues. The morphology of these surfaces could have a mechanical influence on the cells used. It should be further investigated [32].

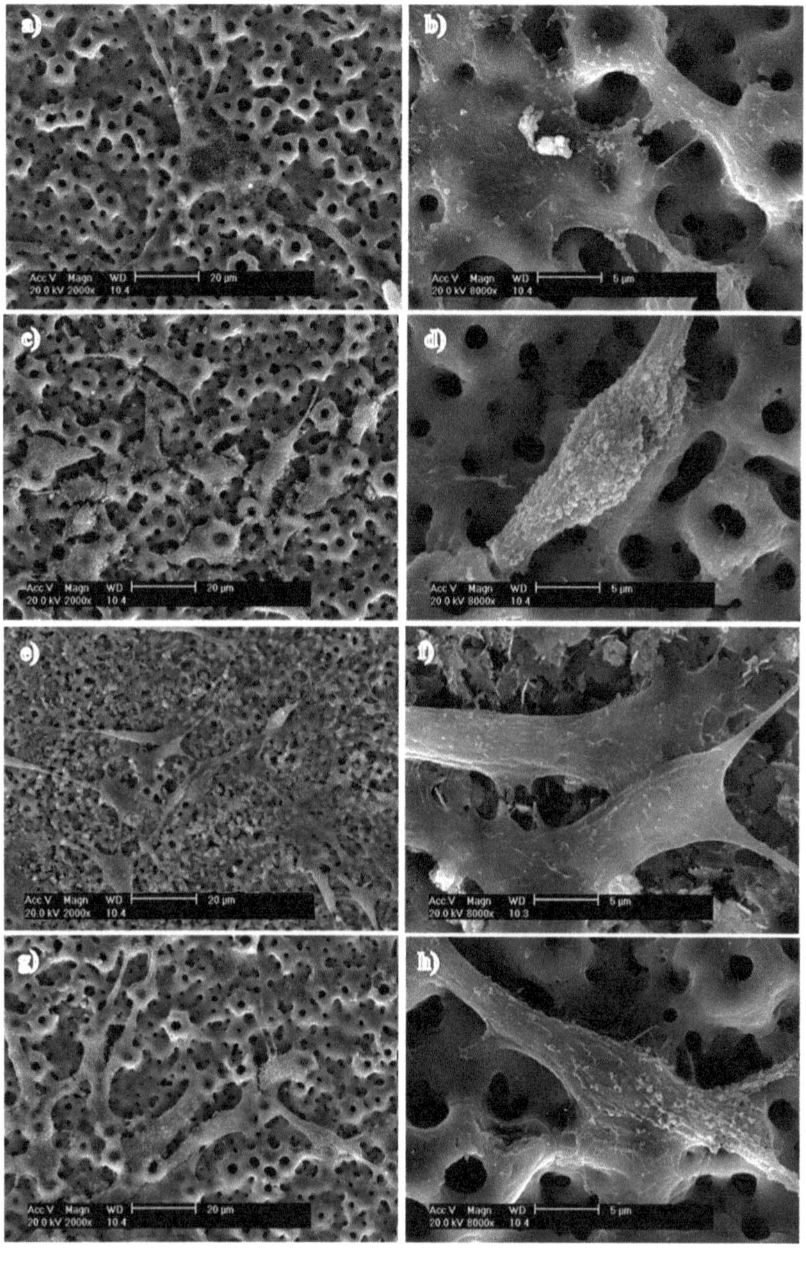

Figure 8. The surface morphology of MAO specimens with different N implantation doses after cell culturing for two days: (**a**,**b**) Ti-MAO-N0.1; (**c**,**d**) Ti-MAO-N1.0; (**e**,**f**) Ti-MAO-N5.0; (**g**,**h**) Ti-MAO-N20.

4. Conclusions

A composite modified film has been successfully synthesized on the surface of Ti alloy by MAO-NII treatment. The as-prepared film is composed of titania (anatase and rutile TiO_2), unsaturated titania (Ti_5O_9), and titanium nitride ($Ti_3N_{1.29}$). The surface exhibits a unique nanostitch array structural feature, the growth mechanism of which has been studied by detailed TEM characterization. The friction factor of the composite modified film was much lower than that of pure titanium (0.404). Bioactivity and cellular compatibility have improved greatly with the MAO and NII treatment compared with pure Ti. Additionally, it was also found that the nanostitch array structure collapsed due to the high implantation energy and superheat when the implantation dose rose to 20×10^{17} ions/cm^2.

Supplementary Materials: The following are available online at http://www.mdpi.com/2079-4991/7/10/343/s1, Figure S1: Surface morphology of pure Ti (a), Ti-MAO-N0.5(b) and Ti-MAO-N10 (c) after soaking in biomimetic mineralization solution for 24 days, Figure S2: The surface morphology of MAO specimens with different N implantation dose after cell cultured for 2 days: Ti-MAO-N0.5 (a,b), Ti-MAO-N10 (c,d), Figure S3: The alkaline phosphatase activity of osteoblasts on pure Ti and Ti-MAO-N1.0 surface.

Acknowledgments: This research was supported by the Guangdong Province Science and Technology Project (No. 2015A030310488) and the Scientific Cultivation and Innovation Fund Project of Jinan University (No. 21617427). Xianfeng Yang thanks the New Faculty Start-up for support in the form of funding and the Fundamental Research Funds for the Central Universities (No. 2015ZZ046) from South China University of Technology. Jie Cui thanks the Fundamental Research Funds for the Central Universities (No. 2017BQ046) from South China University of Technology.

Author Contributions: Peng Zhang conceived and designed the experiments, performed the experiments, collected the data, analyzed the data and wrote the manuscript; Xiaojian Wang designed the study and reviewed the manuscript. Zhidan Lin, Huaijun Lin, Zhiguo Zhang and Wei Li designed the study, contributed the reagents, the materials and the analysis tools; Xianfeng Yang analysed the data, interpreted data, reviewed and edited the manuscript. Jie Cui conceived and designed the experiments, analyzed the data, wrote and reviewed the manuscript.

Conflicts of Interest: The authors declare no conflicts of interest.

References

1. Rack, H.J.; Qazi, J.I. Titanium alloys for biomedical applications. *Mater. Sci. Eng. C Mater. Biol. Appl.* **2006**, *26*, 1269–1277. [CrossRef]
2. Kokubo, T.; Kim, H.M.; Kawashita, M. Novel bioactive materials with different mechanical properties. *Biomaterials* **2003**, *24*, 2161–2175. [CrossRef]
3. Hotchkiss, K.M.; Reddy, G.B.; Hyzy, S.L.; Schwartz, Z.; Boyan, B.D.; Olivares-Navarrete, R. Titanium surface characteristics, including topography and wettability, alter macrophage activation. *Acta Biomater.* **2016**, *31*, 425–434. [CrossRef] [PubMed]
4. Feng, B.; Chen, J.Y.; Qi, S.K.; He, L.; Zhao, J.Z.; Zhang, X.D. Characterization of surface oxide films on titanium and bioactivity. *J. Mater. Sci. Mater. Med.* **2002**, *13*, 457–464. [CrossRef] [PubMed]
5. Jung, H.D.; Jang, T.S.; Wang, L.; Kim, H.E.; Koh, Y.H.; Song, J. Progress in Research on the Surface/Interface of Materials for Hard issue Implant. *Biomaterials* **2015**, *37*, 49–61. [CrossRef] [PubMed]
6. Singh, R.; Lee, P.D.; Dashwood, R.J.; Lindley, T.C. Titanium foams for biomedical applications: A review. *Mater. Technol.* **2010**, *25*, 127–136. [CrossRef]
7. Niinomi, M.; Akahori, T. Improvement of the fatigue life of titanium alloys for biomedical devices through microstructural control. *Expert Rev. Med. Devices* **2010**, *7*, 481–488. [CrossRef] [PubMed]
8. Niinomi, M. Metallic biomaterials. *J. Artif. Organs* **2008**, *11*, 105–110. [CrossRef] [PubMed]
9. Oshida, Y.; Tuna, E.B.; Aktören, O.; Gençay, K. Dental Implant Systems. *Int. J. Mol. Sci.* **2010**, *11*, 1580–1678. [CrossRef] [PubMed]
10. Subramani, K.; Jung, R.E.; Molenberg, A. Biofilm on Dental Implants—A Review of the Literature. *Int. J. Oral Maxillofac. Implants* **2009**, *24*, 616–626. [PubMed]

11. Graham, W.G.; Stalder, K.R. Plasmas in liquids and some of their applications in nanoscience. *J. Phys. D Appl. Phys.* **2011**, *44*, 17. [CrossRef]
12. Dzhurinskiy, D.; Gao, Y.; Yeung, W.K.; Strumban, E.; Leshchinsky, V.; Chu, P.J.; Matthews, A.; Yerokhin, A.; Maev, R.G. Characterization and corrosion evaluation of TiO_2:n-HA coatings on titanium alloy formed by plasma electrolytic oxidation. *Surf. Coat. Technol.* **2015**, *269*, 258–265. [CrossRef]
13. Wang, Q.; Cheng, M.Q.; He, G.; Zhang, X.L. Surface mdification of porous titanium with microarc oxidation and its effects on osteogenesis activity in vitro. *J. Nanomater.* **2015**, *16*, 408634.
14. Walsh, F.C.; Low, C.T.J.; Wood, R.J.K.; Stevens, K.T.; Archer, J.; Poeton, A.R.; Ryder, A. Plasma electrolytic oxidation (PEO) for production of anodised coatings on lightweight metal (Al, Mg, Ti) alloys. *Trans. Inst. Met. Finish.* **2009**, *87*, 122–135. [CrossRef]
15. Beline, T.; Marques, I.D.V.; Matos, A.O.; Ogawa, E.S.; Ricomini, A.P.; Rangel, E.C.; da Cruz, N.C.; Sukotjo, C.; Mathew, M.T.; Landers, R. Production of a biofunctional titanium surface using plasma electrolytic oxidation and glow-discharge plasma for biomedical applications. *Biointerphases* **2016**, *11*, 11–13. [CrossRef] [PubMed]
16. Marin, E.; Diamanti, M.V.; Boffelli, M.; Sendoh, M.; Pedeferri, M.P.; Mazinani, A.; Moscatelli, M.; Del Curto, B.; Zhu, W.; Pezzotti, G. Effect of etching on the composition and structure of anodic spark deposition films on titanium. *Mater. Des.* **2016**, *108*, 77–85. [CrossRef]
17. Paduanoa, F.; Marrelli, M.; Alom, N.; Amer, M.; White, J.L.; Shakesheff, M.K.; Tatulloa, M. Decellularized bone extracellular matrix and human dental pulp stem cells as a construct for bone regeneration. *J. Biomater. Sci. Polym. Ed.* **2017**, *28*, 730–748. [CrossRef] [PubMed]
18. Marrelli, M.; Tatullo, M. Influence of PRF in the healing of bone and gingival tissues. Clinical and histological evaluations. *Eur. Rev. Med. Pharmacol. Sci.* **2013**, *17*, 1958–1962. [PubMed]
19. Marrelli, M.; Falisi, G.; Apicella, A.; Apicella, D.; Amantea, M.; Cielo, A.; Bonanome, L.; Palmieri, F.; Santacroce, L.; Giannini, S.; et al. Behaviour of dental pulp stem cells on different types of innovative mesoporous and nanoporous silicon scaffolds with different functionalizations of the surfaces. *J. Biol. Regul. Homeost. Agents* **2015**, *29*, 217–223.
20. Lai, Y.K.; Huang, J.Y.; Cui, Z.Q.; Ge, M.Z.; Zhang, K.Q.; Chen, Z.; Chi, L.F. Recent advances in TiO_2-based nanostructured surfaces with controllable wettability and adhesion. *Small* **2016**, *12*, 2203–2224. [CrossRef] [PubMed]
21. Krasimir, V.; Melanie, M.R. Plasma Nanoengineering and Nanofabrication. *Nanomaterials* **2016**, *6*, 122.
22. Nabavi, H.F.; Aliofkhazraei, M.; Rouhaghdam, A.S. Morphology and corrosion resistance of hybrid plasma electrolytic oxidation on CP-Ti. *Surf. Coat. Technol.* **2017**, *322*, 59–69. [CrossRef]
23. Aleksandra, R.; Adrian, T.; Tomasz, J.; Wiesław, K.; Beata, S.; Marzena, W.S.; Magdalena, S.; Ewa, T.; Lars, P.N.; Piotr, P. The Bioactivity and Photocatalytic Properties of Titania Nanotube Coatings Produced with the Use of the Low-Potential Anodization of Ti6Al4V Alloy Surface. *Nanomaterials* **2017**, *7*, 197.
24. Geetha, M.; Singh, A.K.; Asokamani, R.; Gogia, A.K. Ti based biomaterials, the ultimate choice for orthopaedic implants—A review. *Prog. Mater. Sci.* **2009**, *54*, 397–425. [CrossRef]
25. Liu, J.D.; Zhao, S.P.; Wang, H.L.; Cui, Y.X.; Jiang, W.W.; Liu, S.M.; Wang, N.; Liu, C.Q.; Chai, W.P.; Ding, W.Y. Study on the chemical bond structure and chemical stability of N doped into TiO_2 film by N ion beam implantation. *Micro Nano Lett.* **2016**, *11*, 758–761. [CrossRef]
26. Piscanec, S.; Colombi Ciacchi, L.; Vesselli, E.; Comelli, G.; Sbaizero, O.; Meriani, S.; De Vita, A. Bioactivity of TiN-coated titanium implants. *Acta Mater.* **2004**, *52*, 1237–1245. [CrossRef]
27. Hamidia, M.F.F.A.; Harunb, W.S.W.; Samykanoc, M.; Ghanib, S.A.C.; Ghazallib, Z.; Ahmadd, F.; Sulonge, A.B. A review of biocompatible metal injection moulding process parameters for biomedical applications. *Mater. Sci. Eng. C* **2017**, *78*, 1263–1276. [CrossRef] [PubMed]
28. Tadashi, K.; Hiroaki, T. How useful is SBF in predicting in vivo bone bioactivity? *Biomaterials* **2006**, *27*, 2907–2915.
29. Shirai, T.; Shimizu, T.; Ohtani, K.; Zen, Y.; Takaya, M.; Tsuchiya, H. Antibacterial iodine-supported titanium implants. *Acta Biomater.* **2011**, *7*, 1928–1933. [CrossRef] [PubMed]
30. Zhang, P.; Zhang, Z.G.; Li, W.; Zhu, M. Effect of Ti-OH groups on microstructure and bioactivity of TiO_2 coating prepared by micro-arc oxidation. *Appl. Surf. Sci.* **2013**, *268*, 381–386. [CrossRef]

31. Bose, S.; Tarafder, S. Calcium phosphate ceramic systems in growth factor and drug delivery for bone tissue engineering: A review. *Acta Biomater.* **2012**, *8*, 1401–1421. [CrossRef] [PubMed]
32. Tatullo, M.; Marrelli, M.; Falisi, G.; Rastelli, C.; Palmieri, F.; Gargari, M.; Zavan, B.; Paduano, F.; Benagiano, V. Mechanical influence of tissue culture plates and extracellular matrix on mesenchymal stem cell behavior: A topical review. *Int. J. Immunopathol. Pharmacol.* **2016**, *29*, 3–8. [CrossRef] [PubMed]

nanomaterials

MDPI

Article

ZnO Nanoparticles Protect RNA from Degradation Better than DNA

Jayden McCall [1], Joshua J. Smith [2], Kelsey N. Marquardt [2], Katelin R. Knight [2], Hunter Bane [2], Alice Barber [2] and Robert K. DeLong [1,*]

[1] Nanotechnology Innovation Center Kansas State (NICKS), Department of Anatomy and Physiology, College of Veterinary Medicine, Manhattan, KS 66506, USA; pcf17@ksu.edu

[2] Department of Biomedical Sciences, College of Health and Human Services, Missouri State University, Springfield, MO 65897, USA; JoshuaJSmith@MissouriState.edu (J.J.S.); Marquardt410@live.missouristate.edu (K.N.M.); Katelin126@live.missouristate.edu (K.R.K.); Bane3@live.missouristate.edu (H.B.); AliceBarber@UTA.edu (A.B.)

* Correspondence: robertdelong@ksu.edu; Tel.: +1-785-532-6313

Received: 22 September 2017; Accepted: 31 October 2017; Published: 8 November 2017

Abstract: Gene therapy and RNA delivery require a nanoparticle (NP) to stabilize these nucleic acids when administered in vivo. The presence of degradative hydrolytic enzymes within these environments limits the nucleic acids' pharmacologic activity. This study compared the effects of nanoscale ZnO and MgO in the protection afforded to DNA and RNA from degradation by DNase, serum or tumor homogenate. For double-stranded plasmid DNA degradation by DNase, our results suggest that the presence of MgO NP can protect DNA from DNase digestion at an elevated temperature (65 °C), a biochemical activity not present in ZnO NP-containing samples at any temperature. In this case, intact DNA was remarkably present for MgO NP after ethidium bromide staining and agarose gel electrophoresis where these same stained DNA bands were notably absent for ZnO NP. Anticancer RNA, polyinosinic-polycytidylic acid (poly I:C) is now considered an anti-metastatic RNA targeting agent and as such there is great interest in its delivery by NP. For it to function, the NP must protect it from degradation in serum and the tumor environment. Surprisingly, ZnO NP protected the RNA from degradation in either serum-containing media or melanoma tumor homogenate after gel electrophoretic analysis, whereas the band was much more diminished in the presence of MgO. For both MgO and ZnO NP, buffer-dependent rescue from degradation occurred. These data suggest a fundamental difference in the ability of MgO and ZnO NP to stabilize nucleic acids with implications for DNA and RNA delivery and therapy.

Keywords: DNase; RNase; DNase activity; RNase activity; metal oxide nanoparticle (MONP); gel electrophoresis; MgO; ZnO; DNA stability; RNA stability

1. Introduction

In nanomedicine, the specific anticancer activity of zinc oxide (ZnO) nanoparticle (NP) is believed to be due to its reactive oxygen species (ROS) generation and tumor pH-dependent ion disassociation as well as the inhibition of various kinases important in cancer cell signaling [1–4]. Poly inosinic:poly cytidilic acid (poly I:C) is among the most well-characterized anticancer RNA, possessing RNA targeting and anti-metastatic activity [5–7]. Our group has studied the interaction and delivery of poly I:C by ZnO NP [8,9]. Magnesium is best known to stabilize RNA structure-function; however, ironically, in the nanoscale (<100 nm), the effect of the corresponding antibacterial metal oxide, magnesium oxide (MgO) NP [10,11], on nucleic acids is unknown. RNA is particularly susceptible to hydrolysis catalyzed rapidly by nucleases (RNases) present in biological fluids (e.g., serum) and tissues (e.g., tumor). Similarly, DNA is susceptible to degradation by DNase enzymes. Previously, we have shown that complexation of plasmid DNA vaccine or siRNA (small interfering RNA) to certain types of organic NP or organic-inorganic hybrid NP is protective [12,13]; however, the influence of ZnO or MgO NP chemistries on DNA and RNA degradation is an important unanswered question and has never before been compared. Further, it was recently discovered that ZnO NP acts as a biochemical inhibitor of the bacterial enzyme, beta-galactosidase (β-Gal) which has been linked to its antibacterial activity [14]. The presence of such hydrolytic and degradative enzymes in serum or tumor environment may indicate cancer cell escape from the tumor compartment or otherwise contribute to metastasis or immuno-suppression. Thus, the impact that NP have on enzyme activity in these environments is important, and further their ability to inhibit biomolecular degradation will likely influence the extent and time-course of the DNA or RNA biological activity. Here, we show for the first time that when DNase is heat treated, the presence of MgO but not ZnO NP protects against DNA degradation. However, for RNA (poly I:C), the integrity of the full-length RNA as measured by gel electrophoresis is best protected in the presence of ZnO NP but not MgO, after exposure to serum-containing media and mouse melanoma tumor homogenate. These data have dramatic ramifications for the delivery of DNA and RNA in vivo and to the development of inorganic metamaterials or composite NP for anticancer applications.

2. Results and Discussion

2.1. DNA Stability in the Presence of MgO NP

The effect of the MgO NP on DNase digestion of plasmid DNA was examined. In this experiment, DNase was incubated at 4, 85, 65, 45, or 21 °C (rt: room temperature), then combined with MgO, stained with ethidium bromide (ETBr) and analyzed by agarose gel electrophoresis. The results are shown in Figure 1.

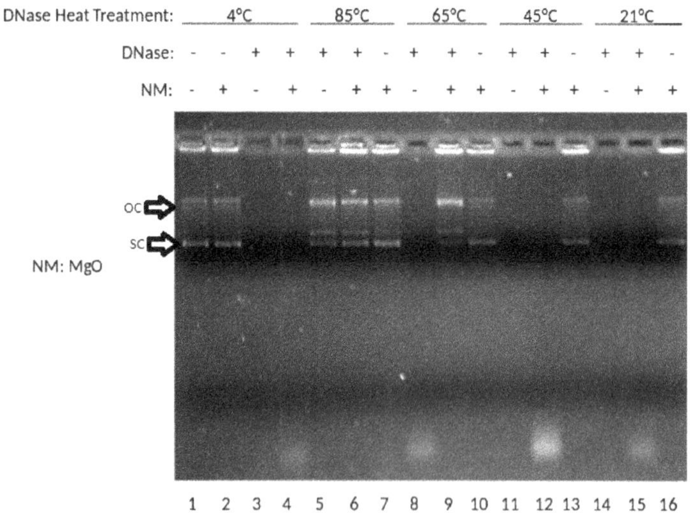

Figure 1. Effects of MgO on DNase activity at various temperatures shown by gel electrophoresis.

As shown in Figure 1, lane 1 shows the results for DNA without DNase or nanoparticle treatment where the super-coiled (SC) DNA migrates furthest in the gel and a stable open circle (OC) form runs midway down the gel followed by higher order species retained in the well. Lanes 2, 7, 10, 13, and 16 show DNA treated with MgO. Lanes 3, 5, 8, 11, and 14 show results for DNA treated with DNase. Lanes 4, 6, 9, 12, and 15 show results for DNA treated with DNase and MgO. At 85 °C, the OC band becomes more prevalent and the linear band (not labeled) appears running between OC and SC. At 65 °C, the presence of MgO clearly protects against degradation with SC, OC and Lin bands present in the lane with nanomaterial (NM) present but no stained bands in the absence of NP. At 45 or 21 °C, intact DNA is only present when incubated with MgO NP. These data indicated a temperature-dependence of DNA stability to DNase degradation provided by MgO NP. The DNA stabilizing effect was buffer-dependent, especially at high non-physiological concentrations (100 or 400 mM) of either $MgCl_2$ or $ZnCl_2$ (data not shown).

2.2. DNA Stability in the Presence of ZnO NP

The effect of the ZnO NP on DNase digestion of plasmid DNA was examined next. In this experiment, DNase was incubated at 4, 85, 65, 45, or 21 °C (rt), then combined with ZnO, stained with ETBr and analyzed by agarose gel electrophoresis similarly. The results are shown in Figure 2.

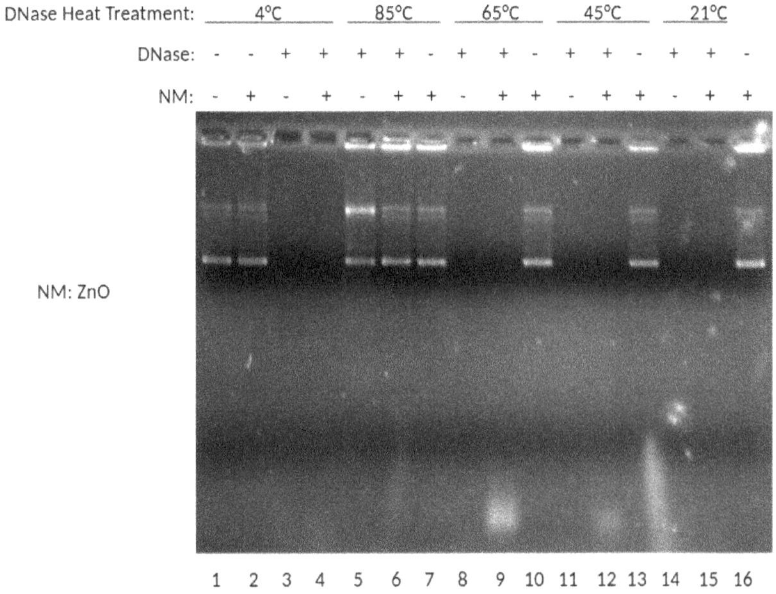

Figure 2. Effects of ZnO on DNase activity at various temperatures shown by gel electrophoresis.

As shown in Figure 2, this gel shows the effects of ZnO on degradation of DNA when treated with DNase previously incubated at varying temperatures. In this experiment, DNase was incubated at 4, 85, 65, 45, and 21 °C (room temperature), then combined with ZnO. Lane 1 shows the results for DNA without DNase or nanoparticle treatment at 4 °C. Lanes 2, 7, 10, 13, and 16 show DNA treated with ZnO. Lanes 3, 5, 8, 11, and 14 show results for DNA treated with DNase. Lanes 4, 6, 9, 12, and 15 show results for DNA treated with DNase and ZnO. By marked contrast, the presence of ZnO NP was not protective of DNA in this experiment, unlike MgO shown above. Thus, at neither temperature were intact DNA bands shown for DNase incubated with ZnO NP in the presence of DNA. DNA and RNA are known to interact with ZnO NP [9,10] and it is interesting to note that in comparison to MgO NP shown in Figure 1, the ZnO NP ratio of SC to OC appears to be slightly greater. Interestingly, ZnO NP was much more buffer-dependent than MgO and could be induced to provide DNase protection at higher buffer concentrations (data not shown). Taken together, these data would suggest that ZnO NP is able to interact with both DNA and protein consistent with previous observations [9–12], but that the MgO effect may be more driven by nucleic acid interaction.

2.3. *RNA Stability in the Presence of Serum Provided by ZnO but Not MgO NP*

Poly inosinic:poly cytidilic acid (poly I:C) is a well-characterized anticancer RNA [5–7]. Our group has recently shown that intratumoral co-administration of poly I:C with ZnO NP has potent antitumor activity in a mouse model of experimental melanoma, inducing an antimetastatic biochemical and immunological profile [8]. For pharmacological activity, maintaining RNA stability in biological fluids and the tumor micro-environment is critical. Thus, the effect of MgO and ZnO NP on poly I:C digestion after exposure to fetal bovine serum (FBS) [12,13] was examined next. The results are shown in Figure 3.

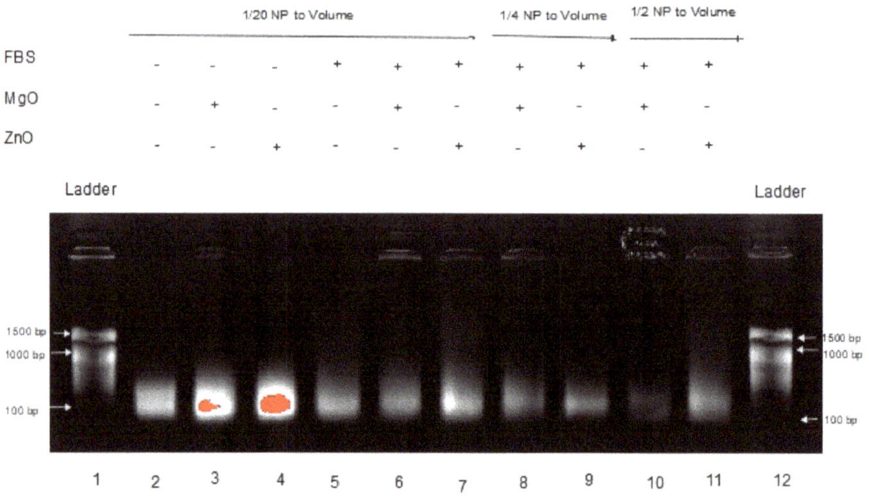

Figure 3. Effects of MgO and ZnO at various concentrations in Fetal Bovine Serum (FBS).

This gel shows the effects of MgO and ZnO in FBS, an environment known to have RNase activity [6,7]. Lanes 1 and 12 are DNA ladders for comparison. Lanes 2 through 4 are Poly I:C alone, with MgO, and ZnO, respectively, in water. Lanes 5 through 7 are Poly I:C with MgO, and ZnO, respectively, in FBS. Lanes 8 and 9 are Poly I:C with MgO and ZnO, respectively, at a higher concentration (1/4 of volume sample) in FBS. Lanes 10 and 11 are Poly I:C with MgO and ZnO, respectively, at a higher ratio of FBS (1/2 of volume sample). At each ratio, it is clear that the intensity of the poly I:C RNA band is greatest in the presence of ZnO than MgO NP. Using a separate sample of RNA from Torula yeast, MgO NP accelerated the rate of fluorescence band intensity loss relative to ZnO NP similarly (data not shown). These data suggested that ZnO, but not MgO NP, protects RNA from hydrolysis.

2.4. *RNA Stability in the Presence of Tumor Homogenate Provided by ZnO but Not MgO NP*

Previously, we have shown that complexation to nanoparticle can protect RNA from degradation in tissue homogenate [13]. For an unmodified RNA such as long non-coding RNA, mRNA, RNA vaccines, etc., to exert robust function within the tumor environment, RNA stability is critical. Thus, poly I:C RNA stability in melanoma tumor homogenate for ZnO was compared to MgO NP. The results are shown in Figure 4.

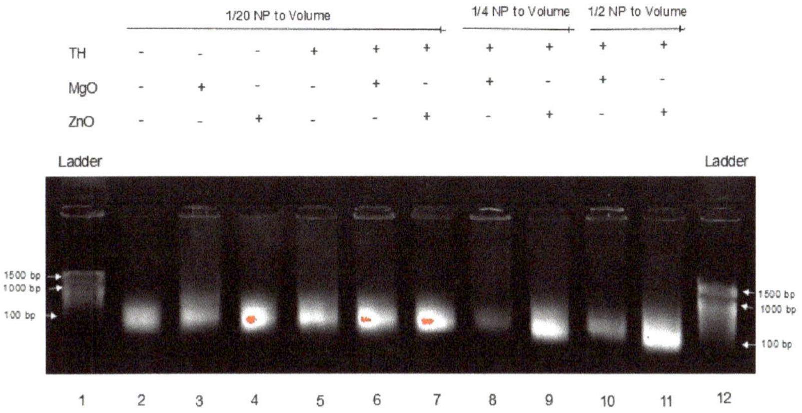

Figure 4. Effects of MgO and ZnO at various concentrations in Tumor Homogenate.

This gel shows the effects of MgO and ZnO in tumor homogenate, an environment known to have RNase activity. Lanes 1 and 12 are DNA ladders for comparison. Lanes 2 through 4 are Poly I:C alone, with MgO, and ZnO, respectively, in water. Lanes 5 through 7 are Poly I:C alone, with MgO, and ZnO, respectively, in tumor homogenate. Lanes 8 and 9 are Poly I:C with MgO and ZnO, respectively, at a higher concentration (1/4 of volume sample) in tumor homogenate. Lanes 10 and 11 are Poly I:C with MgO and ZnO, respectively, at a higher ratio (1/2 of volume sample) of tumor homogenate. The poly I:C band intensity was again greatest in the presence of ZnO in comparison to MgO NP. These data indicate that ZnO, but not MgO NP, protects poly I:C RNA from degradation in the tumor homogenate.

2.5. RNA Compatibility of NP

Metal oxide NP generate reactive oxygen species (ROS) such as hydroxide radical and superoxides and ZnO NP are known to form cationic hydrate species in aqueous buffer [12]. Prolonged exposure to these may therefore, in addition to hydrolysis, increase the rate of RNA degradation and certain silica or nitride NP are also known to split water or to effect RNA stability [13,14]. In this case, a pure macromolecular RNA from Torula yeast obtainable in bulk was used and exposed to NP in water and the RNA compatibility was assessed similarly by removing aliquots and analyzing them by RNA gel electrophoresis where the relative fluorescence intensity of the intact RNA band was plotted over time, as shown in Figure 5.

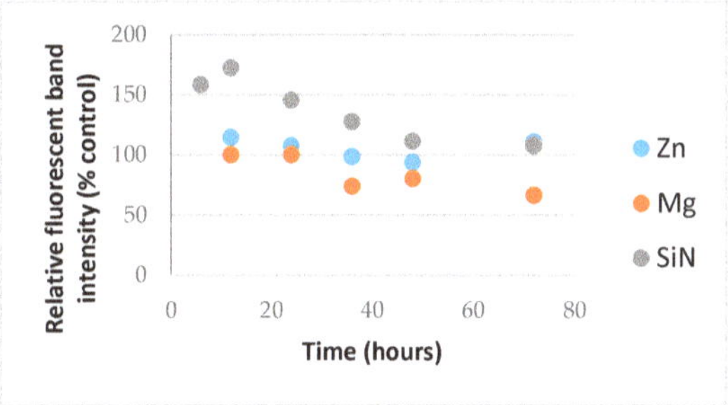

Figure 5. RNA compatibility of ZnO (**blue**) and MgO (**orange**) relative to silica nitride (SiN_4) control (**grey**).

As shown in Figure 5, the control NP led to an initial increase in RNA stain intensity likely reflecting some denaturation of the RNA as the NP binds, allowing access to more dye molecules and then a gradual decline in the RNA band intensity as the RNA degrades. In the presence of MgO, there was no initial increase but instead a loss of RNA from the supernatant likely caused by the aggregation of RNA and its loss from the supernatant consistent with bands in the well-being higher-order species observed in the previous gels. This is in contrast to chemical degradation inferred from the control. Importantly, in the presence of ZnO NP, however, no physicochemical alteration in the RNA, at least by relative fluorescence intensity of the full-length RNA difference from control RNA receiving no NP, was detected during 72 h of exposure.

3. Materials and Methods

3.1. Materials

Nanoscale ZnO (<100 nm) and MgO (<50 nm), poly inosinic:poly cytidilic acid (poly I:C), torula yeast RNA and DNase were obtained from Sigma-Aldrich (St. Louis, MO, USA). The nanoparticle size and zeta potential were confirmed on a Malvern Zetananosizer (Westborough, MA, USA) as per our previous reports [4,8,9]. The NP were washed with double-distilled water (ddH_2O) and precipitated from alcohol and air-dried to a powder within a biological safety cabinet prior to use. Stock samples were generally prepared at 1 mg/mL in sterile ddH_2O except for tumor homogenate which was prepared from sterile PBS (phosphate-buffered saline) and FBS was obtained from Hyclone (South Logan, UT, USA), thawed and used directly.

3.2. DNA Degradation with Heat-Killed DNase

Gel electrophoresis was conducted to test the DNA degradation with heat-killed DNase. In this experiment, 20-μL samples were created consisting of 1 μL of 1.0 μg/μL DNA (puc118 plasmid; Clontech Cat.# 3318, Takara Bio USA Inc., Mountain View, CA, USA) and varying concentrations of DNase to make up the volume. The DNase was used at a concentration of 1:40 and the NP were at a concentration of 1 mg/mL. One sample of each concentration of DNase was heat denatured by incubation for 30 min at 75 °C. Then, all samples were incubated at 25 °C for 30 min. A volume of 2 μL loading dye was added to each sample (after NanoDrop of both). Samples were then assayed in 1% agarose gel (100 mA for 45 min), followed by staining using ethidium bromide (EtBr).

3.3. RNA Degradation

In both gels, Poly I:C was used at a stock concentration of 355 μg/mL. Each sample had 1 μL of this solution for a 1:19 Poly I:C to total volume ratio. Both the MgO and ZnO came from stock solutions of 1 mg/mL. Depending on the sample, each sample either had 1 μL (1:19), 5 μL (1:3), or 10 μL (1:1) of the MgO or ZnO NP stock suspension, sonicated and vortexed prior to its addition to DNase, or to RNA or DNA in solution. The FBS and tumor homogenate were 3 μL in all the samples in which they were used (3:17). Poly I:C, MONP (metal oxide nanoparticle), and FBS or tumor homogenate were combined with ultrapure water (enough to create 18 μL of solution so that when the 40% w/v sucrose and Safestain were added, there would be 20 μL per sample/well) in microcentrifuge tubes and incubated at 37 °C for 1 h. After 1 h, they were combined with 1 μL of sucrose and 1 μL of Bullseye DNA Safestain (C138) (Midwest Scientific, Valley Park, MO, USA), and then electrophoresed on an 1% agarose/TAE (Tris base, acetic acid, and EDTA) gel at 100 V for 20 min. They were then imaged using a Bio-Rad Molecular Imager GelDoc™ XR+ Imaging System (Hercules, CA, USA).

3.4. RNA Compatibility

Torula yeast (Sigma-Aldrich, St. Louis, MO, USA) was dissolved in RNase-free double-distilled deionized water at 1 mg/mL. To an equal volume of RNA was added a 1 mg/mL suspension also in RNase-free double-distilled water. The two were vortex mixed briefly for 10 to 15 s and exposed to physiological temperature and at 6, 12, 24, 36, 48 and 72 h, samples were removed and electrophoresed on 1% agarose/TAE, fluorescently stained with ethidium bromide and the band intensities determined on a Kodak Gel Logic 200 Imaging System (Rochester, NY, USA).

4. Conclusions

The results suggest that RNA is best protected by ZnO NP, and DNA by MgO NP. Recently, our group elucidated the molecular mechanism by which RNA interacts to ZnO NP which involved interaction both to the phosphodiester and base [9]. In the absence of direct protein interaction which we know is possible for ZnO NP [15], this would suggest that its RNA interaction either restricts RNase access or sterically blocks the enzyme from being able to cleave the phosphodiester bond. The thermal protection provided by MgO to DNA from DNase digestion, however, suggests a different mechanism is operative, whereby its protein interaction is able to preserve protein structure and hence function under conditions which would normally denature it. Results from Figure 1 suggest that MgO can decrease DNase degradation of DNA under certain parameters. From Figure 1, it can be determined that MgO decreases DNase activity when the DNase is incubated at 65 °C—a temperature that is known to be denaturing to the DNase enzyme, but at which it can still function. The results from Figure 2 regarding ZnO treatment did not show a significant decrease in DNase activity under any temperature tested. This can be seen because in Lanes 5, 6, 7, and 9 of Figure 1, three distinct bands can be seen. These bands represent, from the top, supercoiled DNA that is still intact, open circle DNA where one strand has been damaged, and linear DNA where both strands have been damaged. In Figure 2, only Lane 5 shows the same three bands, indicating that the ZnO added in the other lanes did not protect the DNA from degradation.

In Figure 3, a dimmer signature can be seen in the MgO samples in FBS versus the ZnO samples in FBS, indicating that the RNA in those samples either experienced more degradation or were tied up in higher-order complexes or species, resulting in less free RNA able to migrate through the gel. This is seen even more severely in Figure 4 when the samples are in tumor homogenate rather than FBS. Even at higher concentrations of MgO, it is unable to protect the RNA from degradation by its RNase-active environment. With fewer and less concentrated proteins expected from tumor homogenate than serum-containing media, this suggests that this affect is more nuclease-protective than an aggregation phenomenon triggered by the protein corona to MgO or lack thereof from ZnO NP, although more research would be required to conclusively demonstrate this.

Acknowledgments: This work was supported by NIH/NCI R15 grant "Anticancer RNA Nanoconjugates" (7R15CA139390-03) to R.K.D.

Author Contributions: Jayden McCall: Performed the RNA degradation gels in Figures 3 and 4 and wrote much of the manuscript. Joshua J. Smith: Mentored Marquardt, Knight, and Barber and edited the manuscript. Kelsey N. Marquardt: Performed the DNA degradation gels in Figures 1 and 2 with Knight. Katelin R. Knight: Performed the DNA degradation gels in Figures 1 and 2 with Marquardt. Hunter Bane: Performed the RNA stability with ZnO, MgO, and SiN in Figure 5. Alice Barber: Did preliminary research to get the DNA degradation experiments started. Robert K. DeLong: Mentored McCall, Bane, and Barber and assisted McCall in compiling information for the manuscript and co-edited it.

Conflicts of Interest: The authors declare no conflict of interest.

References

1. Zhang, H.; Ji, Z.; Xia, T.; Meng, H.; Low-Kam, C.; Liu, R.; Pokhrel, S.; Lin, S.; Wang, X.; Liao, Y.P.; et al. Use of metal oxide nanoparticle band gap to develop a predictive paradigm for oxidative stress and acute pulmonary inflammation. *ACS Nano* **2012**, *6*, 4349–4368. [CrossRef] [PubMed]
2. Moon, S.H.; Choi, W.J.; Choi, S.W.; Kim, E.H.; Kim, J.; Lee, J.O.; Kim, S.H. Anti-cancer activity of ZnO chips by sustained zinc ion release. *Toxicol. Rep.* **2016**, *3*, 430–438. [CrossRef] [PubMed]
3. Roy, R.; Singh, S.K.; Chauhan, L.K.; Das, M.; Tripathi, A.; Dwivedi, P.D. Zinc oxide nanoparticles induce apoptosis by enhancement of autophagy via PI3K/Akt/mTOR inhibition. *Toxicol. Lett.* **2014**, *227*, 29–40. [CrossRef] [PubMed]

4. DeLong, R.K.; Mitchell, J.A.; Morris, R.T.; Comer, J.; Hurst, M.N.; Ghosh, K.; Wanekaya, A.; Mudge, M.; Schaeffer, A.; Washington, L.L.; et al. Enzyme and cancer cell selectivity of nanoparticles: Inhibition of 3-D metastatic phenotype and experimental melanoma by zinc oxide. *J. Biomed. Nanotechnol.* **2017**, *13*, 221–231. [CrossRef]
5. Cheng, Y.S.; Xu, F. Anticancer function of polyinosinic-polycytidylic acid. *Cancer Biol. Ther.* **2010**, *10*, 1219–1223. [CrossRef] [PubMed]
6. Cobaleda-Siles, M.; Henriksen-Lacey, M.; Ruiz de Angulo, A.; Bernecker, A.; Gómez Vallejo, V.; Szczupak, B.; Llop, J.; Pastor, G.; Plaza-Garcia, S.; Jauregui-Osoro, M.; et al. An iron oxide nanocarrier for dsRNA to target lymph nodes and strongly activate cells of the immune system. *Small* **2014**, *10*, 5054–5067. [CrossRef] [PubMed]
7. Forte, G.; Rega, A.; Morello, S.; Luciano, A.; Arra, C.; Pinto, A.; Sorrentino, R. Polyinosinic-polycytidylic acid limits tumor outgrowth in a mouse model of metastatic lung cancer. *J. Immunol.* **2012**, *188*, 5357–5364. [CrossRef] [PubMed]
8. Ramani, M.; Mudge, M.C.; Morris, R.T.; Zhang, Y.; Warcholek, S.A.; Hurst, M.N.; Riviere, J.E.; DeLong, R.K. Zinc Oxide Nanoparticle-Poly I:C RNA Complexes: Implication as Therapeutics against Experimental Melanoma. *Mol. Pharm.* **2017**, *14*, 614–625. [CrossRef] [PubMed]
9. Ramani, M.; Nguyen, T.D.T.; Aryal, S.; Ghosh, K.C.; DeLong, R.K. Elucidating the RNA Nano–Bio Interface: Mechanisms of Anticancer Poly I:C RNA and Zinc Oxide Nanoparticle Interaction. *J. Phys. Chem. C* **2017**, *121*, 15702–15710. [CrossRef]
10. He, Y.; Ingudam, S.; Reed, S.; Gehring, A.; Strobaugh, T.P., Jr.; Irwin, P. Study on the mechanism of antibacterial action of magnesium oxide nanoparticles against foodborne pathogens. *J. Nanobiotechnol.* **2016**, *14*, 54. [CrossRef] [PubMed]
11. Leung, Y.H.; Ng, A.M.; Xu, X.; Shen, Z.; Gethings, L.A.; Wong, M.T.; Chan, C.M.; Guo, M.Y.; Ng, Y.H.; Djurišić, A.B.; et al. Mechanisms of antibacterial activity of MgO: Non-ROS mediated toxicity of MgO nanoparticles towards *Escherichia coli*. *Small* **2014**, *10*, 1171–1183. [CrossRef] [PubMed]
12. Kang, H.; DeLong, R.; Fisher, M.H.; Juliano, R.L. Tat-conjugated PAMAM dendrimers as delivery agents for antisense and siRNA oligonucleotides. *Pharm. Res.* **2005**, *22*, 2099–2106. [CrossRef] [PubMed]
13. Reyes-Reveles, J.; Sedaghat-Herati, R.; Gilley, D.R.; Schaeffer, A.M.; Ghosh, K.C.; Greene, T.D.; Gann, H.E.; Dowler, W.A.; Kramer, S.; Dean, J.M.; et al. mPEG-PAMAM-G4 nucleic acid nanocomplexes: Enhanced stability, RNase protection, and activity of splice switching oligomer and poly I: C RNA. *Biomacromolecules* **2013**, *14*, 4108–4115. [CrossRef] [PubMed]
14. Cha, S.H.; Hong, J.; McGuffie, M.; Yeom, B.; Van Epps, J.S.; Kotov, N.A. Shape-Dependent Biomimetic Inhibition of Enzyme by Nanoparticles and Their Antibacterial Activity. *ACS Nano* **2015**, *9*, 9097–9105. [CrossRef] [PubMed]
15. Zheng, D.W.; Li, B.; Li, C.X.; Fan, J.X.; Lei, Q.; Li, C.; Xu, Z.; Zhang, X.Z. Carbon-Dot-Decorated Carbon Nitride Nanoparticles for Enhanced Photodynamic Therapy against Hypoxic Tumor via Water Splitting. *ACS Nano* **2016**, *10*, 8715–8722. [CrossRef] [PubMed]

nanomaterials

MDPI

Article

ZnO Interactions with Biomatrices: Effect of Particle Size on ZnO-Protein Corona

Jin Yu, Hyeon-Jin Kim, Mi-Ran Go, Song-Hwa Bae and Soo-Jin Choi *

Department of Applied Food System, Major of Food Science & Technology, Seoul Women's University, Seoul 01797, Korea; ky5031@swu.ac.kr (J.Y.); kimhj043@naver.com (H.-J.K.); miran8190@naver.com (M.-R.G.); songhwa29@naver.com (S.-H.B.)

* Correspondence: sjchoi@swu.ac.kr; Tel.: +82-2-970-5634; Fax: +82-2-970-5977

Received: 27 September 2017; Accepted: 2 November 2017; Published: 6 November 2017

Abstract: Zinc oxide (ZnO) nanoparticles (NPs) have been widely used for food fortification, because zinc is essential for many enzyme and hormone activities and cellular functions, but public concern about their potential toxicity is increasing. Interactions between ZnO and biomatrices might affect the oral absorption, distribution, and toxicity of ZnO, which may be influenced by particle size. In this study, ZnO interactions with biomatrices were investigated by examining the physicochemical properties, solubility, protein fluorescence quenching, particle–protein corona, and intestinal transport with respect to the particle size (bulk vs. nano) in simulated gastrointestinal (GI) and plasma fluids and in rat-extracted fluids. The results demonstrate that the hydrodynamic radii and zeta potentials of bulk ZnO and nano ZnO in biofluids changed in different ways, and that nano ZnO induced higher protein fluorescence quenching than bulk ZnO. However, ZnO solubility and its intestinal transport mechanism were unaffected by particle size. Proteomic analysis revealed that albumin, fibrinogen, and fibronectin play roles in particle–plasma protein corona, regardless of particle size. Furthermore, nano ZnO was found to interact more strongly with plasma proteins. These observations show that bulk ZnO and nano ZnO interact with biomatrices in different ways and highlight the need for further study of their long-term toxicity.

Keywords: zinc oxide; interaction; gastrointestinal fluid; plasma; particle size

---

1. Introduction

Zinc oxide (ZnO) is widely utilized in industry because of its ultraviolet (UV) protective, nutritional, and anti-microbial properties [1–3]. ZnO nanoparticles (NPs) are currently used in cosmetics, sunscreen products, the agricultural industry, food additives, and packaging [4,5]. In particular, ZnO NPs have been used as food fortifications and agricultural fertilizers, because zinc plays an important role in the metabolism and protein synthesis, and in the regulation of gene expression and enzyme and hormone activities [6–8]. However, NPs have a large surface area to volume ratio, which results in high reactivity and behaviors unlike those of micro-sized materials in biological systems. Hence, NPs might induce unexpected biological responses and biokinetic behaviors, and this raises public concerns about their potential toxicity.

Many studies have been performed on the toxicity of ZnO NPs in cell lines and animal models [9–12], and some conflicting results have been reported. The use of NPs with different physicochemical properties or different experimental conditions are likely to produce different results. Interaction between particles and biomatrices is another important factor for toxicological consideration. Particles administered orally encounter diverse biological matrices, such as gastrointestinal (GI) fluids and blood, and these interactions lead to the formation of particle–biomatrix corona, which can alter their physicochemical property, biological interaction, and biological fate [13,14]. In particular, NPs–protein corona formation has been well reported to affect cellular responses. For example, it was reported that the ZnO NPs–serum protein interaction influences cytotoxicity, showing lower or higher cytotoxicity when NPs were dispersed in serum proteins [11,15,16]. The aggregation of di-block copolymer NPs was found to be induced by fibrinogen, while the adsorption of albumin and complement component 3 (C3) protein on the surface of NPs triggered the activation of the immune complement cascade [17]. In addition, it was reported that the NPs–plasma protein interaction can be implicated in immunological recognition, molecular targeting, biodistribution, and intracellular uptake [18]. The majority of studies on this topic have focused on the determination of NPs interactions with plasma proteins, though these interactions are surely affected by particle size. Indeed, ZnO NPs were reported to exhibit a higher cytotoxicity and inflammation response than micro ZnO in human monocytes, and their size-dependent cytotoxicity toward human lung epithelial cells was also demonstrated [19–21]. Moreover, the question as to whether NPs interactions with biomatrices lead to positive or negative effects on biological systems remains to be answered.

In the present study, we explored the interactions between ZnO particles and biological fluids (gastric fluids, intestinal fluids, and plasma) with respect to particle size (bulk vs. nano), and examined the physicochemical characteristics (hydrodynamic radius, zeta potential, and dissolution property) of bulk and nano ZnOs in vitro simulated biological fluids and ex vivo rat-extracted fluids. Particle–protein interactions were evaluated by measuring the protein fluorescence quenching ratio in the presence of particles, and proteomic analysis was further conducted to identify the plasma proteins adsorbed on the surface of bulk and nano ZnOs, respectively. Finally, we investigated the effect of particle size on the intestinal transport mechanism.

2. Results

2.1. Characterization of Bulk and Nano ZnOs

The particle size, morphology, and size distribution of ZnOs were determined by scanning electron microscopy (SEM). The images showed that nano ZnO particles were spherical and had a narrow size distribution, whereas bulk ZnO particles were more irregular with a rectangle- or square-like shape and wider size distribution (Figure 1). The average primary particle sizes of bulk and nano ZnOs, as determined from the SEM images, were 289.6 ± 68.1 and 28.2 ± 8.2 nm, respectively. However, dynamic light scattering (DLS) data revealed that both particles agglomerated or aggregated when suspended in distilled water (DW), showing 3453.3 ± 278.0 and 1976.0 ± 198.7 nm for bulk and nano ZnOs, respectively (Table 1, Figure S1). On the other hand, the zeta potential values of the bulk and nano ZnOs were similar, showing 17.5 ± 1.6 and 16.0 ± 1.0 mV for the former and the latter, respectively, without statistical difference ($p > 0.05$, Table 2).

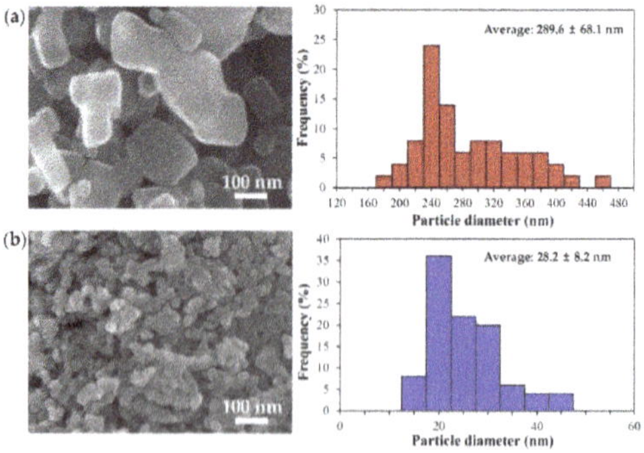

Figure 1. Scanning electron microscopic (SEM) images and size distribution of (**a**) bulk ZnO and (**b**) nano ZnO. Particle size distribution was determined by randomly selecting 200 particles from the SEM images.

Table 1. Hydrodynamic radii of bulk and nano ZnOs in different simulated biological fluids.

	Size	Time	DW	Gastric Fluid	Intestinal Fluid	Plasma
Hydrodynamic radius (nm)	Bulk ZnO	1 min	$3453.3 \pm 278.0^{A,a}$	$2161.0 \pm 257.0^{B,b}$	$3224.5 \pm 180.0^{A,a}$	$3327.3 \pm 268.8^{A,a}$
		1 h		$2463.0 \pm 235.3^{B,b,c}$	$2828.5 \pm 158.7^{A,b}$	$2184.3 \pm 203.8^{B,c}$
		6 h		$3176.0 \pm 272.4^{A,a}$	$2585.0 \pm 84.9^{B,b}$	$1735.8 \pm 114.0^{C,c}$
		24 h		$3237.3 \pm 81.8^{A,a}$	$2227.5 \pm 139.0^{C,b}$	$1473.8 \pm 51.4^{C,c}$
	Nano ZnO	1 min	$1976.0 \pm 198.7^{A,b}$	$2060.8 \pm 41.4^{A,B,a}$	$3009.8 \pm 200.4^{B,c}$	$2485.3 \pm 226.9^{B,b,*}$
		1 h		$2365.0 \pm 188.9^{B,a,b}$	$3079.8 \pm 206.5^{B,c}$	$2439.0 \pm 144.8^{B,b}$
		6 h		$2993.8 \pm 203.2^{C,c}$	$2435.0 \pm 162.3^{C,b}$	$1761.8 \pm 250.0^{A,C,a}$
		24 h		$3180.8 \pm 81.3^{C,c}$	$2300.5 \pm 131.7^{A,C,b}$	$1502.3 \pm 187.8^{C,a}$

Different capital (A,B,C) and lower case (a,b,c) letters indicate significant differences between incubation times and between simulated biofluid types, respectively ($p < 0.05$). * indicates significant differences compared to bulk ZnO ($p < 0.05$).

Table 2. Zeta potentials of bulk and nano ZnOs in different simulated biological fluids.

	Size	Time	DW	Gastric Fluid	Intestinal Fluid	Plasma
Zeta potential (mV)	Bulk ZnO	1 min	$17.5 \pm 1.6^{A,a}$	$-17.4 \pm 0.5^{B,b}$	$-23.4 \pm 0.6^{B,c}$	$-27.5 \pm 0.9^{B,d}$
		1 h		$-16.9 \pm 0.6^{B,b}$	$-25.9 \pm 1.8^{C,c}$	$-27.9 \pm 0.9^{B,c}$
		6 h		$-16.9 \pm 0.9^{B,b}$	$-27.1 \pm 0.9^{C,D,c}$	$-28.5 \pm 0.4^{B,c}$
		24 h		$-16.6 \pm 0.4^{B,b}$	$-28.6 \pm 0.7^{D,c}$	$-27.9 \pm 0.7^{B,c}$
	Nano ZnO	1 min	$16.0 \pm 1.0^{A,a}$	$-14.8 \pm 0.7^{B,b,*}$	$-19.0 \pm 0.6^{B,c,*}$	$-23.1 \pm 1.0^{B,d,*}$
		1 h		$-13.5 \pm 0.8^{B,b,*}$	$-20.6 \pm 0.9^{C,c,*}$	$-22.8 \pm 0.9^{B,d,*}$
		6 h		$-13.8 \pm 0.7^{B,b,*}$	$-20.7 \pm 0.6^{C,c,*}$	$-24.6 \pm 0.8^{B,C,d,*}$
		24 h		$-13.6 \pm 1.0^{B,b,*}$	$-17.2 \pm 2.1^{B,c,*}$	$-26.2 \pm 0.7^{C,d,*}$

Different capital (A,B,C,D) and lower case (a,b,c,d) letters indicate significant differences between incubation times and between simulated biofluid types, respectively ($p < 0.05$). * indicates significant differences compared to bulk ZnO ($p < 0.05$).

2.2. Changes in the Physicochemical Properties of ZnOs in Simulated Biofluids

The zeta potentials and hydrodynamic radii of ZnO particles in simulated biological fluids were measured in order to investigate changes in their physicochemical properties after interactions. Table 1 shows that the hydrodynamic radii of bulk ZnO gradually decreased in simulated intestinal and plasma fluids, while a significant decrease was only found in simulated gastric fluid during 1 h. In particular, a remarkable decrease was observed under the simulated plasma condition upon incubation. On the other hand, the overall increase in the hydrodynamic radii of nano ZnO was found under simulated gastric and intestinal conditions, whereas the hydrodynamic size decreased with incubation in plasma fluid versus that in DW.

The negative zeta potential values for both bulk and nano ZnOs were measured in all biological fluids (Table 2), though negative surface charges were significantly greater for bulk ZnO than for nano ZnO under all conditions tested.

2.3. Dissolution Properties of ZnOs in Simulated Biofluids

The solubility of ZnO particles was evaluated in simulated gastric, intestinal, and plasma fluids in order to elucidate their biological fate when administered orally. It was found that 24.5% and 24.2% of bulk and nano ZnOs, respectively, dissolved into zinc ions in simulated gastric fluid (Figure 2a). Meanwhile, the respective solubilities of ZnOs were ~0.2% and 2.8% under simulated intestinal and plasma conditions (Figure 2b,c), respectively. In all cases, no significant differences between particle sizes were found ($p > 0.05$).

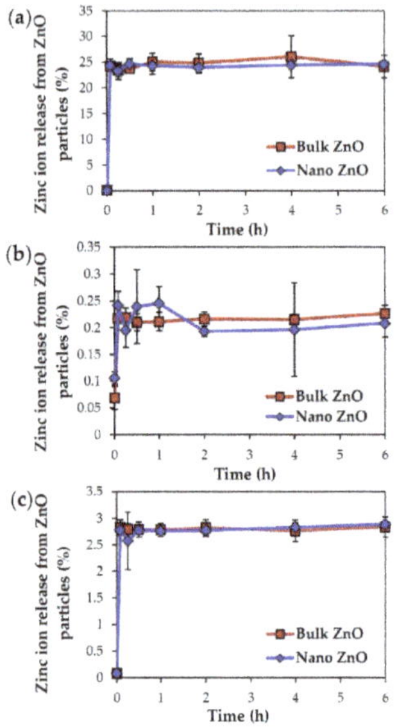

Figure 2. In vitro dissolution properties of ZnOs in simulated (**a**) gastric, (**b**) intestinal, and (**c**) plasma fluids. No significant differences between particle sizes were found ($p > 0.05$).

On the other hand, the ex vivo solubility of ZnO particles was also investigated in rat-extracted biological fluids; ~12%, ~9%, and 2% solubilities were found in gastric fluid, intestinal fluid, and plasma, respectively (Figure 3). Particle size was not found to influence ex vivo solubility ($p > 0.05$).

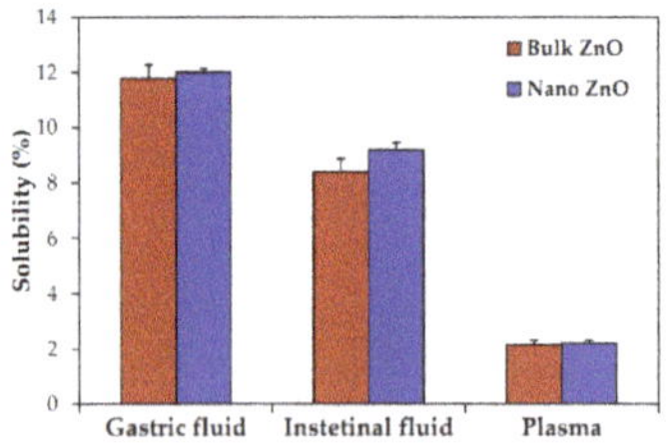

Figure 3. Ex vivo dissolution properties of ZnOs in rat-extracted gastrointestinal fluids and plasma after incubation for 30 min. No significant differences between particle sizes were found ($p > 0.05$).

2.4. ZnO Interactions with Proteins in Simulated Biofluids

When particle interactions with proteins were evaluated in simulated biofluids by measuring the protein fluorescence quenching ratio, a gradual and dramatic increase in fluorescence quenching by bulk and nano ZnOs was found in the simulated gastric fluid upon incubation time, but no significant difference was observed between particle sizes (Figure 4a) ($p > 0.05$). The fluorescence ratios of both bulk and nano ZnO reached ~64% after incubation for 24 h. Relatively high fluorescence quenching (~66%) was induced just after adding both particle types to simulated intestinal fluid and this was maintained for 24 h (Figure 4b). On the other hand, particle interactions with plasma were simulated using phosphate buffered saline (PBS) containing bovine serum albumin (BSA) or fibrinogen, the most abundant plasma proteins. A gradual increase in the fluorescence quenching ratio was observed under simulated plasma condition containing BSA (Figure 4c).

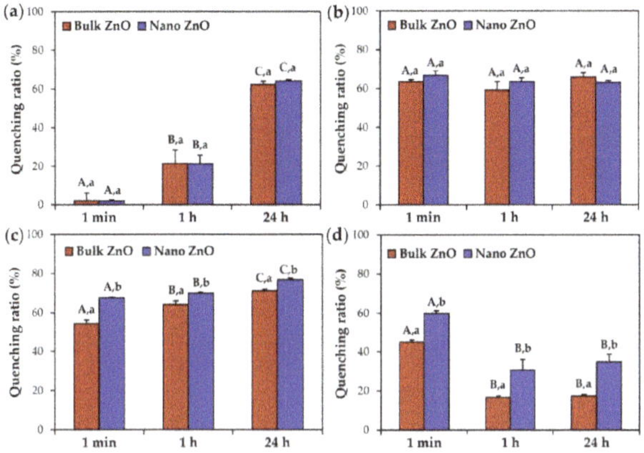

Figure 4. Protein fluorescence quenching ratios of bulk and nano ZnOs in simulated (**a**) gastric and (**b**) intestinal fluids and in (**c**) plasma containing bovine serum albumin (BSA) or (**d**) fibrinogen. Different letters capital (A,B,C) and lower case (a,b) letters indicate significant differences between incubation times and between bulk and nano ZnOs, respectively ($p < 0.05$).

In particular, a statistically high fluorescence quenching ratio was found in the presence of nano ZnO versus bulk ZnO. Nano ZnO also interacted more strongly with simulated plasma fluid containing fibrinogen than bulk ZnO, but the highest fluorescence quenching was observed immediately after adding particles and subsequently decreased (Figure 4d). Meanwhile, slight red shifts by both bulk and nano ZnOs were observed in simulated gastric and intestinal fluids after incubation for 24 h (Figure S2).

2.5. ZnO Plasma–Protein Corona

Rat plasma proteins bound to ZnO particles were quantitatively analyzed by Bradford assay before gel electrophoresis. Higher amount of proteins was found to be adsorbed on nano ZnO than bulk ZnO, demonstrating a total of 1544 and 2152 μg adsorbed proteins for bulk and nano ZnOs, respectively. Plasma proteins adsorbed on the surface of ZnO particles were determined by one-dimensional (1D) gel electrophoresis (Figure 5a). The results show that the patterns of proteins bound to bulk and nano ZnOs differ. In particular, a much larger amount of proteins was adsorbed on nano ZnO. Further protein analysis by two-dimensional (2D) gel electrophoresis showed different binding profiles between two particles (Figure 5b). As expected, a more intense interaction between nano ZnO and plasma proteins were found.

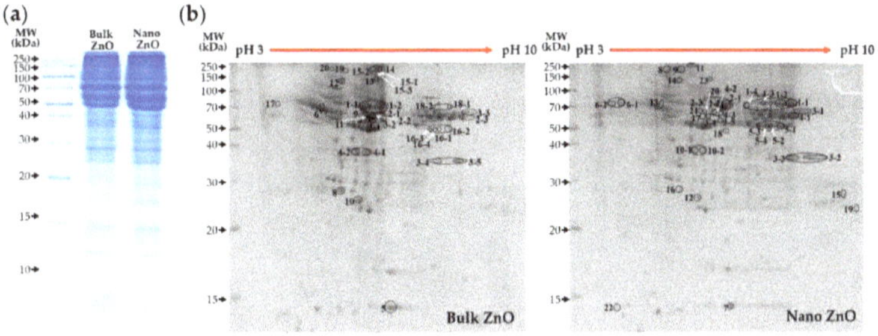

Figure 5. Plasma protein-binding profiles of bulk and nano ZnOs separated by (**a**) one-dimensional and (**b**) two-dimensional gel electrophoresis.

The protein corona that formed around the particles were identified by mass spectrophotometry (MS) according to protein molecular weight (MW) and isoelectric point (pI). The most abundant plasma proteins identified in the particle–protein coronas are listed in Table 3. A total of 20 and 23 proteins were determined to be adsorbed on bulk and nano ZnOs, respectively. Among them, 19 proteins were commonly found in coronas, regardless of particle size. Serum albumin and fibrinogen were the most abundant two proteins in the corona formed on both bulk and nano ZnOs. Fibronectin was also commonly and frequently found in particle–protein corona, regardless of particle size. However, the plasma protein-binding profiles between bulk and nano ZnOs differed. In particular, fibronectin 1 isoform CRA-b was only detected in the bulk ZnO-protein corona, while complement C1q subcomponent subunit B precursor, complement C3 precursor, SWItch/sucrose non-fermentable (SWI/SNF)-related matrix-associated actin-dependent regulator of chromatin subfamily D member 3, and keratin K6 were only identified in nano ZnO-protein corona.

Table 3. List of the most abundant plasma proteins adsorbed on bulk and nano ZnOs as determined by liquid chromatography-mass spectrometry/mass spectrometry (LC-MS/MS).

MW (kDa)	pI	Bulk ZnO	No.	Nano ZnO	pI	MW (kDa)
68.7	6.09	Serum albumin	1	Serum albumin precursor	6.09	68.8
54.3	7.89	Fibrinogen B beta chain	2	Serum albumin	6.09	68.7
60.5	7.56	Fibrinogen alpha subunit	3	Fibrinogen alpha subunit	7.56	60.5
167.2	6.46	Alpha-1-macroglobulin	4	Fibrinogen B beta chain	7.89	54.3
15.7	5.77	Prealbumin	5	Fibrinogen gamma chain precursor	5.85	49.7
45.7	5.37	Serine protease inhibitor A3K	6	Vitronectin	5.68	54.7
60.7	6.56	Fibrinogen alpha chain precursor	7	Prealbumin	5.77	15.7
26.2	5.50	Serum amyloid P-component precursor	8	Alpha-1-inhibitor 3	5.70	163.8
87.0	5.57	Fibrinogen alpha chain isoform X2	9	Fibronectin isoform X3	5.61	262.8
103.6	6.08	Inter-alpha-inhibitor H4 heavy chain	10	Alpha-1-macroglobulin	6.46	167.2
54.2	7.90	Fibrinogen beta chain precursor	11	Fibronectin isoform X2	5.54	262.8
254.4	5.27	Fibronectin isoform X6	12	Inter-alpha-inhibitor H4 heavy chain	6.08	103.6
272.6	5.50	Fibronectin	13	Serine protease inhibitor A3K	5.37	45.7
262.8	5.54	Fibronectin isoform X2	14	Fibronectin isoform X6	5.27	254.4
253.2	5.47	Fibronectin 1 isoform CRA-b	15	Complement C1q subcomponent subunit B precursor	9.13	26.6
49.7	5.85	Fibrinogen gamma chain precursor	16	Serum amyloid P-component precursor	5.50	26.2
54.7	5.68	Vitronectin	17	Fibrinogen alpha chain precursor	6.56	60.7
68.8	6.09	Serum albumin precursor	18	Complement C3 precursor	6.06	186.4
262.8	5.61	Fibronectin isoform X3	19	SWI/SNF-related matrix-associated actin-dependent regulator of chromatin subfamily D member 3	9.36	53.6
163.8	5.70	Alpha-1-inhibitor 3	20	Fibrinogen beta chain precursor	7.90	54.2
			21	Fibrinogen alpha chain isoform X2	5.57	87.0
			22	Keratin K6	3.10	5.6
			23	Fibronectin	5.50	272.6

Proteins marked in bold font were only found in bulk or nano ZnO corona. The list of proteins above is presented in intensity order as determined by 2D gel electrophoresis. Abbreviations: MW, molecular weight; kDa, kilo dalton; pI, isoelectric point; No, number.

2.6. Intestinal Transport Mechanism

The intestinal transport mechanism of bulk and nano ZnOs was determined using an in vitro human follicle-associated epithelium (FAE) model and Caco-2 monolayers, which represent microfold (M) cells in Peyer's patches and intestinal tight junction barriers, respectively. The result demonstrated that both different-sized ZnOs were primarily transported by M cells, and a slight increase in both particle transports through Caco-2 monolayers was also found (Figure 6). However, transported amounts were not found to be dependent on particle size ($p > 0.05$).

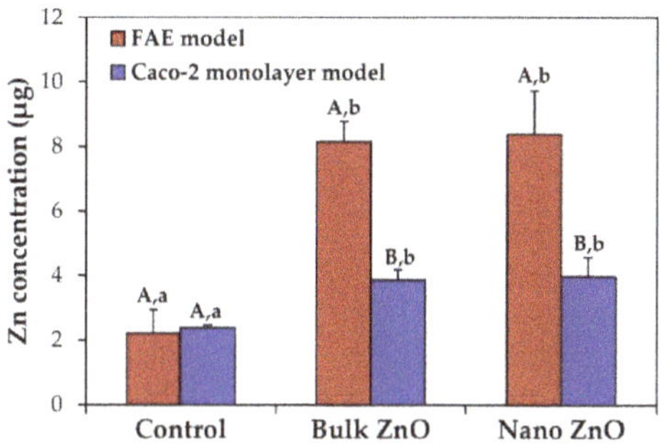

Figure 6. Intestinal transport of bulk and nano ZnOs using an in vitro human follicle-associated epithelium (FAE) model and a Caco-2 monolayer model. Different capital (A,B) and lower case (a,b) letters indicate significant differences between the FAE and Caco-2 monolayer models and among untreated control, bulk ZnO, and nano ZnO, respectively ($p < 0.05$).

3. Discussion

In the present study, we evaluated ZnO interactions with in vitro and ex vivo biological matrices, such as GI fluids and plasma, with respect to particle size (bulk vs. nano) in order to elucidate the effects of particle size on biological interactions. The particle size (289.6 ± 68.1 nm) and hydrodynamic size (3453.3 ± 278.0 nm) of bulk ZnO were quite different from those of nano ZnO (28.2 ± 8.2 and 1976.0 ± 198.7 nm), and both particles formed agglomerates or aggregates in aqueous solution (Table 1, Figure S1).

The hydrodynamic radii and zeta potential values of the bulk and nano ZnOs changed in all biological fluids, indicating particle interaction with biomatrices (Tables 1 and 2). In particular, a significant decrease in the hydrodynamic radii of both particles under simulated plasma conditions observed after incubation for 24 h suggests that plasma proteins facilitate particle dispersion (Table 1) [22–24]. It is worth noting that bulk and nano ZnOs formed agglomerates or aggregates in DW and biofluids. ZnO NPs were reported to agglomerate or aggregate in aqueous solution, but their hydrodynamic size remarkably decreased in the presence of BSA or serum protein [15,25–27]. Hence, it seems that ZnO particles form agglomerates, not aggregates, and their dispersion can be enhanced by particle–protein interaction. Meanwhile, the positive zeta potential values of bulk and nano ZnOs in DW became negative in all simulated biofluids (Table 2), indicating the interaction effect on surface characteristics of particles. In addition, significantly more negative zeta potentials were found for bulk ZnO than for nano ZnO in all biofluids, implying different interaction between particle sizes.

However, particle size did not affect solubility, as demonstrated by dissolutions of ~24%, 0.2%, and 2.8% in simulated gastric fluid, intestinal fluid, and plasma, respectively, for both particle sizes (Figure 2). The high dissolution properties of ZnO NPs under acidic and gastric conditions [28] and low solubility in neutral fluids have been well reported [29,30], which is in good agreement with our results. Further investigation on ex vivo solubility showed ~12%, ~9%, and 2% dissolution of both bulk and nano ZnOs in rat-extracted gastric, intestinal, and plasma fluids, respectively, without statistical significances between particle sizes (Figure 3). The in vitro and ex vivo solubility of both ZnOs differed, except in plasma. This lower ex vivo solubility of ZnOs in rat-extracted gastric fluid than in vitro may be due to the comparatively high pH of rat gastric fluid (pH~3.2 in rats and pH~1.5 in man) [31]. On the other hand, higher solubility was found in rat intestinal fluid than in simulated fluid, which suggests that the total amount of acidic gastric fluid in rat intestinal fluid was greater than in simulated intestinal fluid. Taken together, these results suggest that ZnO particles are primarily present in particulate forms after oral administration, regardless of particle size, although some portion can be dissolved into zinc ions. Interestingly, it would appear that particle size does not critically affect the solubility and biological fate of ZnOs, which is in line with the previously published finding that ZnO particle size (bulk vs. nano) had no effect on absorption following oral administration to rats [32].

Since all biofluids contain proteins, ZnO interaction with proteins was evaluated by measuring the protein fluorescence quenching ratio in the presence of particles. Fluorescence quenching was observed in all biological fluids (Figure 4), suggesting clear particle–protein interactions, although the quenching ratios depended on biofluid type. Similar particle interactions with GI fluids, regardless of particle size, imply a low effect of particle–protein interaction on oral absorption. On the other hand, the fact that red shifts were observed for both different-sized ZnOs in GI fluids (Figure S1) suggests conformational or structural changes of digestion enzymes, which could potentially affect the digestion and utilization of nutrients. It is worth noting that the quenching ratios of nano ZnO were significantly higher than those of bulk ZnO under plasma conditions containing BSA or fibrinogen, suggesting particle-size-dependent interactions in plasma. It seems that particle size can more critically affect particle–protein corona formation in plasma, than in GI fluid. Moreover, both ZnO size types interacted more strongly with BSA than with fibrinogen, which is in line with our proteomic results (Table 3).

When the particle–plasma protein corona was further investigated using a proteomic approach, an obvious particle-size-dependent difference was observed by 1D and 2D gel electrophoresis (Figure 5). Furthermore, a much larger amount and more protein types were found to be adsorbed on nano ZnO (23 kinds) than on bulk ZnO (20 kinds). MS results in the identification of particle–protein corona revealed that serum albumin was the most important protein forming the corona, followed by fibrinogen, regardless of particle size (Table 3). Fibronectin was also frequently found in the coronas of both bulk and nano ZnOs. Indeed, albumin, fibrinogen, and fibronectin have been reported to be abundantly adsorbed on NPs [33–35]. Albumin is the most abundant plasma protein and is responsible for colloid osmotic pressure, transportation, and detoxification [36,37]. Hence, the formation of ZnO–albumin corona could affect particle toxicity, distribution, and circulation time. Fibrinogen and fibronectin are both glycoproteins; fibrinogen is involved in the blood coagulation system [38,39] and fibronectin plays a role in cell adhesion, growth, and wound healing [35,40]. Thus, particle interactions with fibrinogen or fibronectin could affect innate immune response. The complementary system is an essential part of the immune system [41,42], which was only found in nano ZnO–plasma protein corona. These findings show that ZnO–plasma protein interactions are dependent on particle size and suggest that nano ZnO is more likely to affect immune response than bulk ZnO.

On the other hand, the intestinal transport mechanism was not influenced by particle size (Figure 6), indicating that bulk and nano ZnOs were transported through M cells and Caco-2 monolayers. ZnO particles were primarily transported by M cells, regardless of particle size. M cells found in Peyer's patches in the small intestine are implicated in the transport of various molecules, including macromolecules and particles [43,44]. Indeed, the intestinal transportation of diverse NPs by M cells has been recently reported [45–47], which concurs with our results. It should be noted that ZnO particles can be also transported through intestinal tight junction barriers, as demonstrated by our Caco-2 monolayer model. This may be associated with their partial ionized fate, because non-ionized NPs under physiological conditions tend not to be transported through tight junctions [48,49]. Moreover, no significant differences in the intestinal transportation amount between particle sizes were found, which could explain the similar oral absorption between bulk ZnO and nano ZnO [32]. It appears that particle–protein interactions did not remarkably affect the intestinal transport mechanism and oral bioavailability, regardless of particle size, but particle size plays a role in interactions with plasma proteins in blood. Further study is required to elucidate the impact of ZnO NP–protein interactions on potential long-term toxicity.

4. Materials and Methods

4.1. Materials

Nano ZnO (20 nm) and bulk ZnO (5 μm) were purchased from Sumitomo (Tokyo, Japan) and Sigma-Aldrich (St Louis, MO, USA), respectively. Each particle was dispersed in DW (5 mg/mL) for 30 min, just prior to experiments.

4.2. Characterization

Primary particle size and morphology were examined using SEM (FEIQUANTA 250 FEG, Hillsboro, OR, USA). The hydrodynamic radius and zeta potential were determined with a Zetasizer Nano Series (Malvern, Westborough, MA, USA). The measurements were performed at 25 °C by dispersing particles in DW or biofluids.

4.3. Preparation of Simulated Biofuids

Simulated gastric, intestinal, and plasma fluids were prepared for in vitro studies as previously described [50]. The simulated gastric fluid was prepared by dissolving 2 g sodium chloride (Samchun Chemical Co., Ltd., Pyeongtaek, Korea) and 3.2 g pepsin (Sigma-Aldrich, St Louis, MO, USA) in DW, and the pH was adjusted to 1.5 with 1 N hydrochloric acid (Duksan Pure Chemicals Co., Ltd., Ansan, Gyeonggi-do, Korea), and then made up to 1000 mL. For the simulated intestinal fluid, bile salt (87.5 mg) (Sigma-Aldrich, St Louis, MO, USA) and pancreatin (25 mg) (Sigma-Aldrich, St Louis, MO, USA) were added to the simulated gastric fluid, and then the pH was adjusted to 6.8 with saturated sodium bicarbonate solution (Sigma-Aldrich, St Louis, MO, USA). The simulated plasma fluid was prepared by dissolving 50 g of BSA (Sigma-Aldrich, St Louis, MO, USA) in PBS solution (NaCl 137 mM, KCl 2.7 mM, Na_2HPO_4 10 mM, KH_2PO_4 1.8 mM; Dongin Biotech. Co., Ltd., Seoul, Korea), and then made up to 1000 mL.

4.4. Animals and Preparation of Rat-Extracted Biofluids

Five-week-old female Sprague Dawley (SD) rats weighing around 150 g were obtained from Nara Biotech Co., Ltd. (Seoul, Korea) and acclimated to environments for seven days before experiment. Animals were housed in plastic animal cages in a ventilated room maintained at 20 ± 2 °C and 60 ± 10% relative humidity under a 12 h light/dark cycle. Water and commercial laboratory complete feed for rats were provided *ad libitum*. All animal experiments were performed in accordance with the approved animal protocol and guideline established by the Animal and Ethics Review Committee of Seoul Women's University (IACUC-2016A-3).

Rat biofluids, such as GI fluids and plasma, were obtained as previously reported [50]. Briefly, stomachs and small intestines were collected and rinsed with saline, and then gastric and intestinal fluids were obtained by centrifugation at 16,000× *g* for 15 min at 4 °C. To obtain plasma, blood sample was collected via tail vein using a catheter, and centrifuged at 16,000× *g* for 3 min at 4 °C.

4.5. In Vitro and Ex Vivo Dissolution Properties of ZnO in Biofluids

Each particle (bulk and nano ZnOs) suspension was added to simulated and rat-extracted biofluids (5 mg/mL) and incubated with gentle shaking (180 rpm) at 37 °C. After designated incubation times, supernatants were collected by centrifugation (16,000× *g*) for 15 min. Pre-digestion of the collected supernatants was performed with 10 mL of 60% ultrapure nitric acid and 0.5 mL of H_2O_2 at 180 °C until samples were completely digested. Solutions were diluted with 2.5 mL of distilled and deionized water (DDW) and quantitative analysis of the dissolved Zn from ZnO was carried out using inductively coupled plasma–atomic emission spectroscopy (ICP-AES, JY2000 Ultrace, HORIBA Jobin Yvon, Longjumeau, France).

4.6. Fluorescence Quenching Measurement

To determine particle–protein interaction, a fluorescence quenching measurement was performed in two different simulated plasma fluids. BSA (Sigma-Aldrich, St Louis, MO, USA) or fibrinogen (Sigma-Aldrich, St Louis, MO, USA) was added to PBS at concentration of 1 mg/mL. ZnO particle suspensions (100 μL of 5 mg/mL) were added in 1 mL of simulated solutions and incubated for 1 min, 1 h, and 24 h at 37 °C with gentle shaking (180 rpm). A protein fluorescence quenching assay was carried out using a spectrophotometer (Molecular Devices, LLC., Sunnyvale, CA, USA). Fluorescence spectra were measured at 300–420 nm using an excitation wavelength of 280 nm. Protein fluorescence quenching ratios were calculated as (I0-I)/I0, where I0 and I stand for fluorescence intensities in the absence and presence of ZnO particle suspensions, respectively.

4.7. 1D and 2D Gel Electrophoresis

Particle suspensions of bulk and nano ZnOs (100 μL of 50 mg/mL) were incubated with 1 mL of rat plasma at 37 °C with gentle shaking (180 rpm) for 1 h, and then centrifuged at 23,000× *g* for 1 h at 4 °C in order to separate unbound proteins. The precipitates were washed three times with DDW. The 1D gel electrophoresis was carried out by suspending samples in rehydration buffer (7 M urea, 2 M thio-urea, 4% 3-[(3-cholamido-propyl)dimethylammonio]-1-propanesulfonate (CHAPS), 2.5% dithiothreitol) containing protease inhibitor cocktail (Roche Molecular Biochemicals, Indianapolis, IN, USA), and by vortexting overnight. After centrifugation at 15,000× *g* for 20 min, the lysates were re-suspended in sodium dodecyl sulfate (SDS) sample buffer (2% SDS, 0.1% bromophenol blue, 10% glycerol, 0.5% β-mercaptoethanol in 50 mM Tris-HCl, pH 6.8), and then heated at 95 °C for 5 min. After cooling to room temperature, samples (30 μg of protein) were loaded into 15% SDS-polyacrylamide (PAGE) gels. Protein concentrations were determined by Bradford method (Bio-Rad, Hercules, CA, USA).

For 2D gel electrophoresis, pH 3–10 immobilized pH gradient (IPG) strips (GE Healthcare Life Sciences, Pittsburgh, PA, USA) were rehydrated in swelling buffer (7 M urea, 2 M thiourea, 0.4% (*w*/*v*) dithiothreitol, and 4% (*w*/*v*) CHAPS). The protein lysates (700 μg) were loaded into the rehydrated IPG strips using an IPGphor III (GE Healthcare Life Sciences). 2D separation was performed on 14% (*v*/*v*) SDS-PAGE gels. After gel fixation for 1 h in 40% (*v*/*v*) methanol containing 5% (*v*/*v*) phosphoric acid, the gels were stained with Coomassie blue G-250 solution (ProteomeTech, Seoul, Korea), and destained in 1% (*v*/*v*) acetic acid. Images were acquired with an Image Scanner III (Bio-Rad, Hercules, CA, USA).

4.8. Identification of Proteins by Liquid Chromatography-Mass Spectrometry/Mass Spectrometry

Image analysis was carried out using an Image Master^{TM} 2D Platinum software (GE Healthcare Life Science, Pittsburgh, PA, USA). To compare the densities of protein spots induced by bulk or nano ZnOs, more than 25 spots were landmarked and normalized. In-gel digestion of protein spots from Coomassie Blue stained gels was performed as previously described [51]. Prior to mass spectrometric analysis, the peptide solutions were desalted using a reversed-phase column [52]. The eluted peptides were analyzed by liquid chromatography-mass spectrometry/mass spectrometry (LC-MS/MS) on a nano ACQUITY UPLC (Waters, Milford, MA, USA) directly coupled to a Finnigan LCQ DECA iontrap mass spectrometer (Thermo Scientific, Waltham, MA, USA). Spectra were acquired and processed using the MASCOT software (Matrix Science, London, UK). The individual spectra from MS/MS were processed using a SEQUEST software (Thermo Quest, San Jose, CA, USA). Only significant hits as defined by the MASCOT software (Matrix Science, London, UK) probability analysis were taken.

4.9. Three Dimensional (3D) Cell Culture for FAE Model

ZnO particle transport by M cells was investigated using an in vitro human intestinal FAE model, as previously described [53]. Human intestinal epithelial Caco-2 cells and non-adherent human Burkitt's lymphoma Raji B cells (Korean Cell Line Bank, Seoul, Korea) were grown in minimum essential medium (MEM; Welgene Inc., Gyeongsangbuk-do, Korea) and Roswell Park Memorial Institute (RPMI) 1640 medium (Welgene Inc.), respectively, supplemented with 10% fetal bovine serum, 1% non-essential amino acids, 1% L-glutamine, 100 units/mL penicillin, and 100 μg/mL streptomycin at 37 °C under 5% CO_2 atmosphere. After coating Transwell® polycarbonate inserts (SPL Lifescience, Gyeonggi-do, Korea) with Matrigel™ matrix (Becton Dickinson, Bedford, MA, USA) for 2 h, supernatants were removed, and then inserts were washed with DMEM. Caco-2 cells (1×10^6 cells/well) were seeded in the apical side and grown for 14 days. Lymphoma Raji B cells (1×10^6 cells/well) were added to the basolateral side, and maintained for five days. Apical medium was then replaced with ZnO suspensions (16.25 μg/mL), and incubation continued for 6 h. The transported amounts of ZnO particles were determined by measuring total Zn levels in the basolateral side using ICP-AES (JY2000 Ultrace, HORIBA Jobin Yvon). Pre-digestion for the ICE-AES analysis was performed in the same manner as described in "In vitro and ex vivo dissolution properties".

4.10. 3D Cell Culture for Intestinal Epithelial Monolayers

The transport of ZnO particles by the intestinal epithelium monolayer was evaluated using an in vitro Caco-2 monoculture system. Caco-2 cells (4.5×10^5 cells/well) were seeded on the upper insert side in the same manner as described in the FAE model, and then cultured for 21 days. After replacing the apical medium of cell monolayers with ZnO suspensions (16.25 μg/mL), incubation continued for 6 h. The transported concentrations of ZnO particles were determined in the same manner as described in the FAE model.

4.11. Statistical Analysis

Results were presented as means ± standard deviation. Experimental values were compared with each other. One-way analysis of variance (ANOVA) with Tukey's Test in SAS Ver.9.4 (SAS Institute Inc., Cary, NC, USA) was used to determine the significances between experimental groups. Statistical significance was accepted for p values < 0.05.

5. Conclusions

In this study, ZnO interactions with biological fluids were clearly demonstrated in terms of changes in the physicochemical properties, solubility, fluorescence quenching, and particle–plasma protein corona formation. Bulk and nano ZnO were found to interact differently with biomatrices, in particular, nano ZnO exhibited lower negative surface charges and had higher fluorescence quenching ratios under simulated plasma condition. More abundant plasma proteins were determined to be adsorbed on nano ZnO than on bulk ZnO. In particular, complement C was only identified in nano ZnO-plasma protein corona, while serum albumin, fibrinogen, and fibronectin seemed to play roles in corona formation, regardless of particle size. However, particle solubility and the intestinal transport mechanism did not appear to be influenced by particle size. Further study is required to elucidate the effect of NP interaction with biomatrices on potential toxicity and nutrient absorption after long-term exposure.

Supplementary Materials: The following are available online at www.mdpi.com/2079-4991/7/11/377/s1, Figure S1: Representative size distribution of (a,c,e) bulk ZnO and (b,d,f) nano ZnO, measured by dynamic light scattering (DLS), in simulated (a,b) gastric, (c,d) intestinal, and (e,f) plasma fluids; Figure S2: Fluorescence spectra of simulated (a,b) gastric fluid, (c,d) intestinal fluid, (e,f) plasma containing bovine serum albumin (BSA, plasma-BSA), and (g,h) plasma containing fibrinogen (plasma-fibrinogen) in the absence or presence of bulk ZnO or nano ZnOs.

Acknowledgments: This research was supported by the Basic Science Research Program through the National Research Foundation of Korea (NRF) funded by the Ministry of Education (2015R1D1A1A01057150) and by a research grant from Seoul Women's University (2017).

Author Contributions: Soo-Jin Choi conceived and designed the experiments, and wrote the paper; Jin Yu performed the solubility and protein experiments and analyzed the data; Hyeon-Jin Kim performed the physicochemical experiments using a Zetasizer; Mi-Ran Go and Song-Hwa Bae performed the cell experiments.

Conflicts of Interest: The authors declare no conflict of interest. The founding sponsors had no role in the design of the study; in the collection, analyses, or interpretation of data; in the writing of the manuscript, and in the decision to publish the results.

References

1. Burman, U.; Saini, M.; Praveen-Kumar. Effect of zinc oxide nanoparticles on growth and antioxidant system of chickpea seedlings. *Toxicol. Environ. Chem.* **2013**, *95*, 605–612. [CrossRef]
2. Pasquet, J.; Chevalier, Y.; Couval, E.; Bouvier, D.; Noizet, G.; Morliere, C.; Bolzinger, M.-A. Antimicrobial activity of zinc oxide particles on five micro-organisms of the challenge tests related to their physicochemical properties. *Int. J. Pharm.* **2014**, *460*, 92–100. [CrossRef] [PubMed]
3. Suresh, D.; Nethravathi, P.C.; Udayabhanu; Rajanaika, H.; Nagabhushana, H.; Sharma, S.C. Green synthesis of multifunctional zinc oxide (ZnO) nanoparticles using *Cassia fistula* plant extract and their photodegradative, antioxidant and antibacterial activities. *Mater. Sci. Semicond. Proc.* **2015**, *31*, 446–454. [CrossRef]
4. Espitia, P.J.P.; Soares, N.F.F.; dos Reis Coimbra, J.S.; de Andrade, N.J.; Cruz, R.S.; Medeiros, E.A.A. Zinc oxide nanoparticles: Synthesis, antimicrobial activity and food packaging applications. *Food Bioprocess Technol.* **2012**, *5*, 1447–1464. [CrossRef]
5. Lohani, A.; Verma, A.; Joshi, H.; Yadav, N.; Karki, N. Nanotechnology-based cosmeceuticals. *ISRN Dermatol.* **2014**, *2014*, 843687. [CrossRef] [PubMed]
6. Zhao, C.Y.; Tan, S.X.; Xiao, X.Y.; Qiu, X.S.; Pan, J.Q.; Tang, Z.X. Effects of dietary zinc oxide nanoparticles on growth performance and antioxidative status in broilers. *Biol. Trace Elem. Res.* **2014**, *160*, 361–367. [CrossRef] [PubMed]
7. Faiz, H.; Zuberi, A.; Nazir, S.; Rauf, M.; Younus, N. Zinc oxide, zinc sulfate and zinc oxide nanoparticles as source of dietary zinc: Comparative effects on growth and hematological indices of juvenile grass carp (*Ctenopharyngodon idella*). *Int. J. Agric. Biol.* **2015**, *17*, 568–574. [CrossRef]
8. Raya, S.D.H.A.; Hassan, M.I.; Farroh, K.Y.; Hashim, S.A.; Salaheldin, T.A. Zinc oxide nanoparticles fortified biscuits as a nutritional supplement for zinc deficient rats. *J. Nanomed. Res.* **2016**, *4*, 2.
9. Hong, T.-K.; Tripathy, N.; Son, H.-J.; Ha, K.-T.; Jeong, H.-S.; Hahn, Y.-B. A comprehensive in vitro and in vivo study of ZnO nanoparticles toxicity. *J. Mater. Chem. B* **2013**, *1*, 2985–2992. [CrossRef]
10. Paek, H.-J.; Lee, Y.-J.; Chung, H.-E.; Yoo, N.-H.; Lee, J.-A.; Kim, M.-K.; Lee, J.-K.; Jeong, J.; Choi, S.-J. Modulation of the pharmacokinetics of zinc oxide nanoparticles and their fates in vivo. *Nanoscale* **2013**, *5*, 11416–11427. [CrossRef] [PubMed]
11. Jo, M.-R.; Chung, H.-E.; Kim, H.-J.; Bae, S.-H.; Go, M.-R.; Yu, J.; Choi, S.-J. Effects of zinc oxide nanoparticle dispersants on cytotoxicity and cellular uptake. *Mol. Cell. Toxicol.* **2016**, *12*, 281–288. [CrossRef]
12. Li, M.; Zhu, L.; Lin, D. Toxicity of ZnO nanoparticles to *Escherichia coli*: Mechanism and the influence of medium components. *Environ. Sci. Technol.* **2011**, *45*, 1977–1983. [CrossRef] [PubMed]
13. Treuel, L.; Nienhaus, G.U. Toward a molecular understanding of nanoparticle-protein interactions. *Biophys. Rev.* **2012**, *4*, 137–147. [CrossRef] [PubMed]
14. Choi, S.-J.; Choy, J.-H. Biokinetics of zinc oxide nanoparticles: Toxicokinetics, biological fates, and protein interaction. *Int. J. Nanomed.* **2014**, *9*, 261–269.
15. Anders, C.B.; Chess, J.J.; Wingett, D.G.; Punnoose, A. Serum proteins enhance dispersion stability and influence the cytotoxicity and dosimetry of ZnO nanoparticles in suspension and adherent cancer cell models. *Nanoscale Res. Lett.* **2015**, *10*, 448. [CrossRef] [PubMed]
16. Zukiene, R.; Snitka, V. Zinc oxide nanoparticle and bovine serum albumin interaction and nanoparticles influence on cytotoxicity in vitro. *Colloids Surf. B* **2015**, *135*, 316–323. [CrossRef] [PubMed]
17. Vauthier, C.; Persson, B.; Lindner, P.; Cabane, B. Protein adsorption and complement activation for di-block copolymer nanoparticles. *Biomaterials* **2011**, *32*, 1646–1656. [CrossRef] [PubMed]
18. Wolfram, J.; Yang, Y.; Shen, J.; Moten, A.; Chen, C.; Shen, H.; Ferrari, M.; Zhao, Y. The nano-plasma interface: Implications of the protein corona. *Colloids Surf. B* **2014**, *124*, 17–24. [CrossRef] [PubMed]
19. Hsiao, I.-L.; Huang, Y.-J. Effects of serum on cytotoxicity of nano- and micro-sized ZnO particles. *J. Nanopart. Res.* **2013**, *15*, 1829. [CrossRef] [PubMed]

20. Lin, W.; Xu, Y.; Huang, C.-C.; Ma, Y.; Shannon, K.B.; Chen, D.-R.; Huang, Y.-W. Toxicity of nano- and micro-sized ZnO particles in human lung epithelial cells. *J. Nanopart. Res.* **2009**, *11*, 25–39. [CrossRef]
21. Sahu, D.; Kannan, G.M.; Vijayaraghavan, R. Size-dependent effect of zinc oxide on toxicity and inflammatory potential of human monocytes. *J. Toxicol. Environ. Health A* **2014**, *77*, 177–191. [CrossRef] [PubMed]
22. Bihari, P.; Vippola, M.; Schultes, S.; Praetner, M.; Khandoga, A.G.; Reichel, C.A.; Coester, C.; Tuomi, T.; Rehberg, M.; Krombach, F. Optimized dispersion of nanoparticles for biological in vitro and in vivo studies. *Part. Fibre Toxicol.* **2008**, *5*, 14. [CrossRef] [PubMed]
23. Ji, Z.; Jin, X.; George, S.; Xia, T.; Meng, H.; Wang, X.; Suarez, E.; Zhang, H.; Hoek, E.M.V.; Godwin, H.; et al. Dispersion and stability optimization of TiO_2 nanoparticles in cell culture media. *Environ. Sci. Technol.* **2010**, *44*, 7309–7314. [CrossRef] [PubMed]
24. Vranic, S.; Gosens, I.; Jacobsen, N.R.; Jensen, K.A.; Bokkers, B.; Kermanizadeh, A.; Stone, V.; Baeza-Squiban, A.; Cassee, F.R.; Tran, L.; et al. Impact of serum as a dispersion agent for in vitro and in vivo toxicological assessments of TiO_2 nanoparticles. *Arch. Toxicol.* **2017**, *91*, 353–363. [CrossRef] [PubMed]
25. Churchman, A.H.; Wallace, R.; Milne, S.J.; Brown, A.P.; Brydson, R.; Beales, P.A. Serum albumin enhances the membrane activity of ZnO nanoparticles. *Chem. Commun.* **2013**, *49*, 4172–4174. [CrossRef] [PubMed]
26. Sasidharan, N.P.; Chandran, P.; Khan, S.S. Interaction of colloidal zinc oxide nanoparticles with bovine serum albumin and its adsorption isotherms and kinetics. *Colloids Surf. B* **2013**, *102*, 195–201. [CrossRef] [PubMed]
27. Zhang, Y.; Chen, Y.; Westerhoff, P.; Hristovski, K.; Crittenden, J.C. Stability of commercial metal oxide nanoparticles in water. *Water Res.* **2008**, *42*, 2204–2212. [CrossRef] [PubMed]
28. Avramescu, M.-L.; Rasmussen, P.E.; Chenier, M.; Gardner, H.D. Influence of pH, particle size and crystal form on dissolution behaviour of engineered nanomaterials. *Environ. Sci. Pollut. Res.* **2017**, *24*, 1553–1564. [CrossRef] [PubMed]
29. Liu, Y.; Gao, W. Growth process, crystal size and alignment of ZnO nanorods synthesized under neutral and acid conditions. *J. Alloys Compd.* **2015**, *629*, 84–91. [CrossRef]
30. Gwak, G.-H.; Lee, W.-J.; Paek, S.-M.; Oh, J.-M. Physico-chemical changes of ZnO nanoparticles with different size and surface chemistry under physiological pH conditions. *Colloids Surf. B* **2015**, *127*, 137–142. [CrossRef] [PubMed]
31. McConnell, E.L.; Basit, A.W.; Murdan, S. Measurements of rat and mouse gastrointestinal pH, fluid and lymphoid tissue, and implications for in-vivo experiments. *J. Pharm. Pharmacol.* **2008**, *60*, 63–70. [CrossRef] [PubMed]
32. Kim, M.-K.; Lee, J.-A.; Jo, M.-R.; Choi, S.-J. Bioavailability of silica, titanium dioxide, and zinc oxide nanoparticles in rats. *J. Nanosci. Nanotechnol.* **2016**, *16*, 6580–6586. [CrossRef] [PubMed]
33. Lundqvist, M.; Stigler, J.; Elia, G.; Lynch, I.; Cedervall, T.; Dawson, K.A. Nanoparticle size and surface properties determine the protein corona with possible implications for biological impacts. *Proc. Natl. Acad. Sci. USA* **2008**, *105*, 14265–14270. [CrossRef] [PubMed]
34. Martel, J.; Young, D.; Young, A.; Wu, C.-Y.; Chen, C.-D.; Yu, J.-S.; Young, J.D. Comprehensive proteomic analysis of mineral nanoparticles derived from human body fluids and analyzed by liquid chromatography-tandem mass spectrometry. *Anal. Biochem.* **2011**, *418*, 111–125. [CrossRef] [PubMed]
35. Dobrovolskaia, M.A.; Neun, B.W.; Man, S.; Ye, X.; Hansen, M.; Patri, A.K.; Crist, R.M.; McNeil, S.E. Protein corona composition does not accurately predict hematocompatibility of colloidal gold nanoparticles. *Nanomedcine* **2014**, *10*, 1453–1463. [CrossRef] [PubMed]
36. Kratz, F. Albumin as a drug carrier: Design of prodrugs, drug conjugates and nanoparticles. *J. Control. Release* **2008**, *132*, 171–183. [CrossRef] [PubMed]
37. Peng, Q.; Zhang, S.; Yang, Q.; Zhang, T.; Wei, X.-Q.; Jiang, L.; Zhang, C.-L.; Chen, Q.-M.; Zhang, Z.-R.; Lin, Y.-F. Preformed albumin corona, a protective coating for nanoparticles based drug delivery system. *Biomaterials* **2013**, *34*, 8521–8530. [CrossRef] [PubMed]
38. Drew, A.F.; Liu, H.; Davidson, J.M.; Daugherty, C.C.; Degen, J.L. Wound-healing defects in mice lacking fibrinogen. *Blood* **2001**, *97*, 3691–3698. [CrossRef] [PubMed]
39. Smiley, S.T.; King, J.A.; Hancock, W.W. Fibrinogen stimulates macrophage chemokine secretion through toll-like receptor 4. *J. Immunol.* **2001**, *167*, 2887–2894. [CrossRef] [PubMed]
40. Grinnell, F.; Billingham, R.E.; Burgess, L. Distribution of fibronectin during wound healing in vivo. *J. Investig. Dermatol.* **1981**, *76*, 181–189. [CrossRef] [PubMed]

41. Moghimi, S.M.; Andersen, A.J.; Ahmadvand, D.; Wibroe, P.P.; Andresen, T.L.; Hunter, A.C. Material properties in complement activation. *Adv. Drug Deliv. Rev.* **2011**, *63*, 1000–1007. [CrossRef] [PubMed]
42. Moyano, D.F.; Liu, Y.; Peer, D.; Rotello, V.M. Modulation of immune response using engineered nanoparticle surfaces. *Small* **2016**, *12*, 76–82. [CrossRef] [PubMed]
43. Chen, M.-C.; Mi, F.-L.; Liao, Z.-X.; Hsiao, C.-W.; Sonaje, K.; Chung, M.-F.; Hsu, L.-W.; Sung, H.-W. Recent advances in chitosan-based nanoparticles for oral delivery of macromolecules. *Adv. Drug Deliv. Rev.* **2013**, *65*, 865–879. [CrossRef] [PubMed]
44. Gamboa, J.M.; Leong, K.W. In vitro and in vivo models for the study of oral delivery of nanoparticles. *Adv. Drug Deliv. Rev.* **2013**, *65*, 800–810. [CrossRef] [PubMed]
45. Fievez, V.; Plapied, L.; Plaideau, C.; Legendre, D.; des Rieux, A.; Pourcelle, V.; Freichels, H.; Jerome, C.; Marchand, J.; Preat, V.; et al. In vitro identification of targeting ligands of human M cells by phage display. *Int. J. Pharm.* **2010**, *394*, 35–42. [CrossRef] [PubMed]
46. Bae, S.-H.; Yu, J.; Go, M.-R.; Kim, H.-J.; Hwang, Y.-G.; Choi, S.-J. Oral toxicity and intestinal transport mechanism of colloidal gold nanoparticle-treated red ginseng. *Nanomaterials* **2016**, *6*, 208. [CrossRef] [PubMed]
47. Kim, H.-J.; Bae, S.-H.; Kim, H.-J.; Kim, K.-M.; Song, J.H.; Go, M.-R.; Yu, J.; Oh, J.-M.; Choi, S.-J. Cytotoxicity, intestinal transport, and bioavailability of dispersible iron and zinc supplements. *Front. Microbiol.* **2017**, *8*, 749. [CrossRef] [PubMed]
48. Lee, J.-A.; Kim, M.-K.; Song, J.H.; Jo, M.-R.; Yu, J.; Kim, K.-M.; Kim, Y.-R.; Oh, J.-M.; Choi, S.-J. Biokinetics of food additive silica nanoparticles and their interactions with food components. *Colloids Surf. B* **2017**, *150*, 384–392. [CrossRef] [PubMed]
49. Yang, Y.-X.; Song, Z.-M.; Cheng, B.; Xiang, K.; Chen, X.-X.; Liu, J.-H.; Cao, A.; Wang, Y.; Liu, Y.; Wang, H. Evaluation of the toxicity of food additive silica nanoparticles on gastrointestinal cells. *J. Appl. Toxicol.* **2014**, *34*, 424–435. [CrossRef] [PubMed]
50. Lee, J.A.; Kim, M.K.; Kim, H.M.; Lee, J.K.; Jeong, J.; Kim, Y.R.; Oh, J.M.; Choi, S.J. The fate of calcium carbonate nanoparticles administered by oral route: Absorption and their interaction with biological matrices. *Int. J. Nanomed.* **2015**, *10*, 2273–2293.
51. Bahk, Y.Y.; Kim, S.A.; Kim, J.S.; Euh, H.J.; Bai, G.H.; Cho, S.N.; Kim, Y.S. Antigens secreted from *Mycobacterium tuberculosis*: Identification by proteomics approach and test for diagnostic marker. *Proteomics* **2004**, *4*, 3299–3307. [CrossRef] [PubMed]
52. Gobom, J.; Nordhoff, E.; Mirgorodskaya, E.; Ekman, R.; Roepstorff, P. Sample purification and preparation technique based on nano-scale reversed-phase columns for the sensitive analysis of complex peptide mixtures by matrix-assisted laser desorption/ionization mass spectrometry. *J. Mass Spectrom.* **1999**, *34*, 105–116. [CrossRef]
53. Des Rieux, A.; Fievez, V.; Theate, I.; Mast, J.; Preat, V.; Schneider, Y.-J. An improved in vitro model of human intestinal follicle-associated epithelium to study nanoparticle transport by M cells. *Eur. J. Pharm. Sci.* **2007**, *30*, 380–391. [CrossRef] [PubMed]

nanomaterials

MDPI

Article

Crystallization of TiO_2 Nanotubes by *In Situ* Heating TEM

Alberto Casu [1,*], Andrea Lamberti [2], Stefano Stassi [2] and Andrea Falqui [1,*]

1 King Abdullah University of Science and Technology (KAUST), Biological and Environmental Sciences and Engineering (BESE) Division, NABLA Lab, 23955-6900 Thuwal, Saudi Arabia
2 Department of Applied Science and Technology, Politecnico di Torino, Corso Duca degli Abruzzi 24, 10129 Torino, Italy; andrea.lamberti@polito.it (A.L.); stefano.stassi@polito.it (S.S.)
* Correspondence: alberto.casu@kaust.edu.sa (A.C.); andrea.falqui@kaust.edu.sa (A.F.); Tel.: +966-54-6685104 (A.C.); +966-54-4700060 (A.F.)

Received: 27 September 2017; Accepted: 11 January 2018; Published: 14 January 2018

Abstract: The thermally-induced crystallization of anodically grown TiO_2 amorphous nanotubes has been studied so far under ambient pressure conditions by techniques such as differential scanning calorimetry and *in situ* X-ray diffraction, then looking at the overall response of several thousands of nanotubes in a carpet arrangement. Here we report a study of this phenomenon based on an *in situ* transmission electron microscopy approach that uses a twofold strategy. First, a group of some tens of TiO_2 amorphous nanotubes was heated looking at their electron diffraction pattern change versus temperature, in order to determine both the initial temperature of crystallization and the corresponding crystalline phases. Second, the experiment was repeated on groups of few nanotubes, imaging their structural evolution in the direct space by spherical aberration-corrected high resolution transmission electron microscopy. These studies showed that, differently from what happens under ambient pressure conditions, under the microscope's high vacuum ($p < 10^{-5}$ Pa) the crystallization of TiO_2 amorphous nanotubes starts from local small seeds of rutile and brookite, which then grow up with the increasing temperature. Besides, the crystallization started at different temperatures, namely 450 and 380 °C, when the *in situ* heating was performed irradiating the sample with electron beam energy of 120 or 300 keV, respectively. This difference is due to atomic knock-on effects induced by the electron beam with diverse energy.

Keywords: TiO_2 amorphous nanotubes; high resolution transmission electron microscopy; *in situ* transmission electron microscopy; amorphous-crystalline phase transition; electron beam effects; anodic oxidation

1. Introduction

Titanium dioxide (TiO_2), even in form of nanotubes, is a well-known material due to its physical and chemical properties and related applications in many diverse fields, such as non-linear optics, photocatalysis, energy storage, optoelectronics, transport-related phenomena, dye-synthesized solar cells, mesoporous structures and film formation [1–14]. TiO_2 can be found as an amorphous and in several, different crystalline polymorphs, being the most known anatase (tetragonal), rutile (tetragonal) and brookite (orthorhombic). Besides, TiO_2 amorphous nanotubes (TANs) prepared by anodic oxidation of a titanium surface also deserved large interest due to their further and diverse applications, again such as energy harvesting and storage, optoelectronics, photocatalysis, water splitting, and as an effective biomaterial [15–18]. Indeed, as-grown TANs can be easily crystallized in the desired phase by simple thermal treatment.

So far, few detailed studies of both amorphous TiO_2 films of different thickness and TAN crystallization have been published [19–23]. Most of them are based on techniques, such as X-ray diffraction (XRD) and differential scanning calorimetry (DSC), that show the overall behavior of a multitude of objects subject to a thermal ramp performed under standard pressure conditions (10^5 Pa). More in detail, the studies performed on amorphous TiO_2 films deposited by magnetron sputtering onto monocrystalline silicon substrate showed that both the initial crystallization temperature and the time needed to crystallize the whole film depend strongly on its features: the higher its thickness, the lower both the initial crystallization temperature and the time needed for the film's complete crystallization [19,20]. Among these works, the minimum transition temperature found for the crystallization of a thick TiO_2 film (800 nm-thick) was reported as being about 180 °C [10]. Moreover, once the amorphous films started to crystallize onto the silicon substrate, the zones close to it evolved to the rutile phase, and those far from it to anatase. Besides, after their formation the crystalline domains kept a small size (i.e., less than 10 nm) up to a quite high temperature; also, a substantial increase in temperature was needed to promote the transition from anatase to rutile for the portion of the films not in proximity of the substrate [21]. A quite similar XRD- and DSC-based approach was followed to investigate the crystallization of highly ordered TANs grown on a titanium substrate [22]. In the latter case, it was shown that at 350 °C TANs started to crystallize as anatase, with a mean crystalline domain size around 20 nm, while rutile appeared only for temperatures higher than 600 °C. Finally, an important work was published showing the crystallization of TANs studied by both thermal gravimetric analysis/differential thermal analysis, XRD and *ex situ* transmission electron microscopy (TEM) [23]. In this work, which concerns template-directed low temperature atomic layer deposition on an alumina template, physically confined TANs were prepared and thermally crystallized. In this case, it was clearly observed that TANs with a wall thickness of 5 nm, when confined, become crystalline at 400 °C taking the anatase structure, which is retained even if the nanotubes are heated up to 1000 °C. Besides, if the TANs are annealed with no further physical confinement thanks to the low-temperature programmed dissolution of the polymer template, they still keep their tubular shape (this phenomenon, due to their intrinsic curvature, being called "self-confinement") and again take the anatase crystal phase at the same temperature of the confined ones, i.e., 400 °C, with some minority traces of the rutile phase being found at 1000 °C.

Thus, even though the thermally-induced crystallization of TANs has been understood in terms of transition temperatures and final crystalline phases obtained, as in the case of the amorphous TiO_2 films, these investigations did not unveil the effect of a very low external pressure on this phenomenon, nor the mechanism that leads to the crystal formation at the local scale under the abovementioned high vacuum conditions.

Furthermore, even though some works were published concerning *in situ* studies of materials with some similarity to TANs for shape or composition [24–27], to our best knowledge the TANs thermally-induced crystallization was never investigated by an *in situ* TEM-based approach. Herein, the study of TAN crystallization has been then performed using this approach and following a twofold experimental strategy that allowed us to observe how this phenomenon occurs and evolves at a very local scale. The first part of this strategy consisted in heating *in situ* a group of several tens of TANs, while looking at their electron diffraction pattern and at its variation over time. This allowed the determination of both the starting temperature of crystallization and the evolution of the appearing crystalline phases. Second, the experiment was repeated on several groups of few nanotubes, imaging their structural change in the direct space by spherical aberration (C_s)-corrected high resolution transmission electron microscopy (HRTEM). These studies permitted the determination that, differently from what happens under standard pressure, under the microscope's vacuum ($p < 10^{-5}$ Pa) the crystallization of TiO_2 amorphous nanotubes starts from local small seeds of rutile and brookite, which then grow up with the increasing temperature. Besides, different starting temperatures of crystallization, namely 450 °C and 380 °C, were observed when the *in situ* heating was performed with electron beam energy of 100 and 300 keV, respectively. This difference is due to atomic knock-on effects induced by the electron beam with diverse energy.

2. Results

After deposition on the micro electro-mechanical system (MEMS) constituting the *in situ* holder grid, it was possible to determine that TANs displayed a mean diameter of 80 nm and mean length of 250 nm, as shown in Figure 1a,b, with a minority group of longer TANs (see Figure 1b). As a first experiment, the TANs' structural evolution was studied by recording the two-dimensional electron diffraction (ED) patterns of a several micron-sized aggregate of TiO_2 nanotubes, as displayed in Figure 1b, while increasing the temperature from room temperature (RT = 20 °C) to 800 °C. No structural evolution was observed in the amorphous TiO_2 sample up to 450 °C, as evidenced by the lack of diffraction spots in the selected area electron diffraction (SAED) pattern. Between 450 and 520 °C a full structural evolution was observed and is shown in Figure 1c: the first low-intensity diffraction spots appeared at 450 °C, with a subsequent evolution of the patterns towards a full SAED pattern showing diffraction rings and spots compatible with the rutile phase by 470 °C. A subsequent increase in temperature to 500 °C resulted in the appearance of additional diffraction spots and rings, which were attributed to the presence of both the rutile and brookite phases, until a stable diffraction pattern was reached at 520 °C. Additional heating treatments up to 800 °C did not result in any further modification of the SAED patterns, which were also maintained after decreasing the temperature to RT.

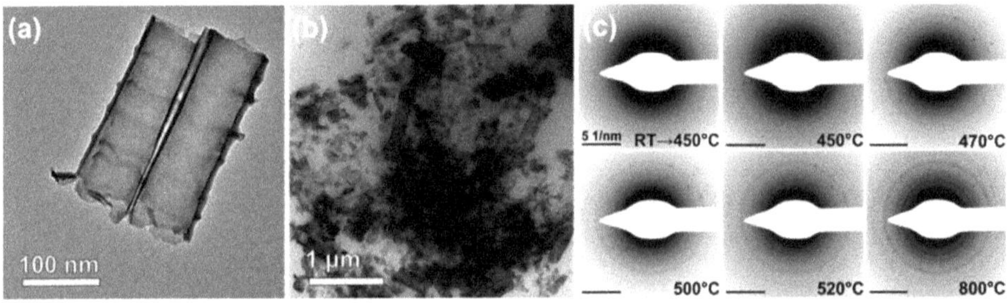

Figure 1. *In situ* selected area electron diffraction (SAED) heating experiment: (**a**) Representative (TiO2 amorphous nanotubes) TANs at room temperature RT; (**b**) The several micron-sized aggregate of TiO_2 TANs chosen for the *in situ* heating; (**c**) SAED patterns of the aggregate reported in panel (**b**), recorded at different temperatures during the *in situ* heating experiment.

In situ HRTEM analysis was then performed on a 300 kV Cs-corrected microscope, taking into account the structural information obtained by the previous experiment and studying the structural evolution of different aggregates against temperature at a very local scale. In fact, given the amorphous nature of TANs at room temperature, it was not possible to know *a priori* which ones would have been properly oriented for HRTEM imaging. Then, several aggregates constituted by a few TANs each were imaged and their positions were recorded before starting any heating ramp. Some of these aggregates are reported in Figure 2. Besides, the thermal treatment during the *in situ* HRTEM experiment was performed according to a slightly different procedure than that followed for the first ED-based experiment reported above. In fact, here the sample heating was performed up to a given temperature, then lowered to 100 °C during the acquisition of the corresponding HRTEM images and heated up again to a higher fixed temperature. This "sawtooth" thermal approach allowed the analysis of the structural evolution of the different zones under the same thermal conditions: the whole sample was subject to the temperature ramp, but the time needed to analyze the appropriately chosen zones was spent at low temperature, where no temperature-driven structural variation is expected, thus minimizing the occurrence of any further structural evolution in-between different zones.

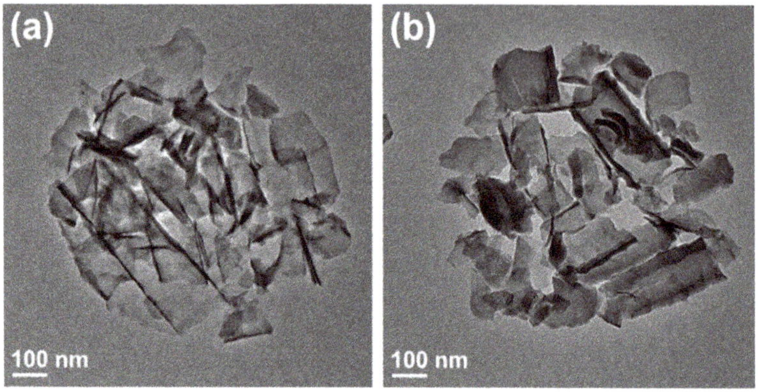

Figure 2. Low magnification transmission electron microscopy (TEM) images of micron-sized TANs: (**a**,**b**) Small aggregates of TANs recorded at RT and used for the *in situ* high resolution transmission electron microscopy (HRTEM) heating experiment.

The first evolution of the sample was already observed at 380 °C, with seeds around 10 nm in size appearing at the nanotubes, and indicated in the low magnification TEM images (shown in Figure 3a,b) by red arrows. It should be noticed that these seeds were not observed when the same aggregates were previously imaged by low magnification TEM, as reported in Figure 2. Moreover, two-dimensional fast Fourier transform (2D-FFT) analysis of the seeds' HRTEM images reported in Figure 3d,e confirmed that these seeds are generally polycrystalline in nature and sport lattice distances and angular relationships compatible with brookite and rutile (as shown in the lower part of panels Figure 3d,e). Moreover, a patched crystallization of the TANs was detected outside the seeds and can be observed in Figure 3c. Even if the seldom occurrence of seeds would not have been likely detected during the SAED experiment, the occurrence of zones with patched crystallization at a nominally lower temperature indicates a possible additional contribution induced by the more energetic electron beam with respect to the previous experiment.

Figure 3. *Cont.*

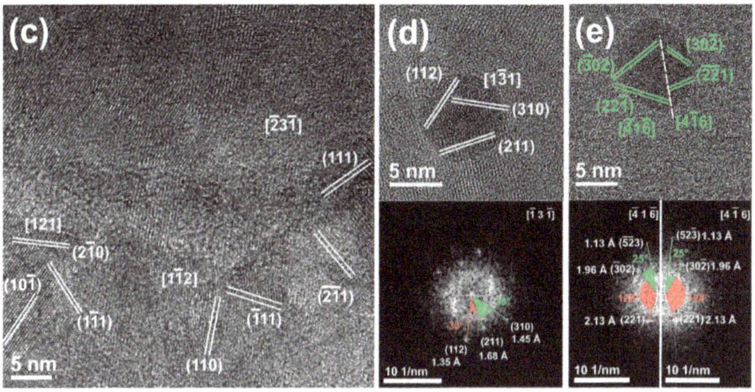

Figure 3. *In situ* HRTEM heating experiment of micron-sized TANs: (**a**,**b**) Evolution of the small aggregates of TANs reported in Figure 2, now recorded at 380 °C. The red arrows reported in panel **a** and **b** indicate the localization of several small crystalline seeds not observed in the panel (**a**,**b**) of Figure 2. The white dotted rectangles indicate the regions of interest analyzed by HRTEM and reported in panels **c**, **d** and **e**; (**c**) HRTEM image of a region with patched crystallization of rutile (in white) in the first aggregate (see the rectangle in panel 3**a**); (**d**) HRTEM image of a rutile seed (in white) in the second aggregate (see the rectangle on the left in panel and corresponding two-dimensional fast Fourier transform (2D-FFT) showing the expected interplanar distances and angular relationships, 3**b**); (**e**) HRTEM image of a polycrystalline brookite seed (in green) in the second aggregate (see the rectangle on the right in panel 3**b**). The two main crystalline domains are separated by a white dashed line, and the corresponding 2D-FFTs are presented below. Both show the interplanar distances and angular relationships expected from brookite.

Figure 4 shows the TANs imaged after the temperature was further increased to 500 °C, in order to reach conditions corresponding to the final crystallization of the sample. The seeds were still present and showing a slight increase in size and roundness, while the nanotubes had evolved to polycrystals. More in detail, the seeds were still polycrystalline, albeit with bigger-sized domains of rutile and brookite phases, while the nanotubes featured a disordered crystallization, featuring extended crystal domains along with smaller, disordered, multi-domain zones; the most extended crystal domains could be identified by 2D-FFT analysis as rutile, with smaller zones and islands also showing structural features of brookite and anatase.

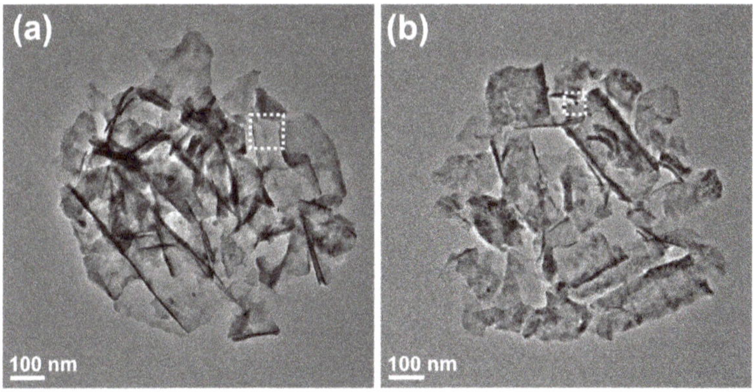

Figure 4. *Cont.*

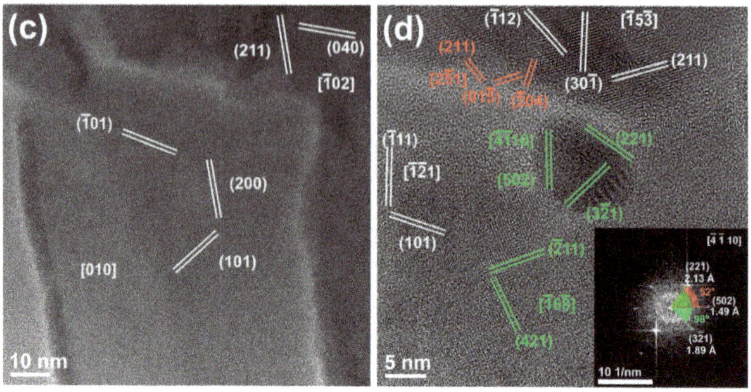

Figure 4. *In situ* HRTEM heating experiment of micron-sized TANs: (**a**,**b**) Evolution of the small aggregates of TANs reported in Figures 2 and 3, now recorded at 500 °C. The white dotted rectangles indicate the regions of interest analyzed by HRTEM and reported in panels **c** and **d**; (**c**) HRTEM image of a an extended crystal domain of rutile (in white) in the first aggregate (see the rectangle in panel 4**a**); (**d**) HRTEM image of region of the second aggregate with patched crystallization (see the rectangle in panel 4**b**). Distinct crystal domains of rutile (in white), brookite (in green) and anatase (in red) can be identified. The 2D-FFT of the brookite seed is presented in the inset, showing its interplanar distances and corresponding angular relationships for the observed zone axis.

Moreover, a group of bigger TANs, with a mean diameter of more than 200 nm and a length above one micron, showed a different structural evolution with respect to their smaller counterpart, displayed in Figure 5. These nanotubes did not present any structural evolution at 380 °C apart from showing very few 5-nm-sized seeds (Figure 5b,d), while at 500 °C they presented wider and more ordinate crystal domains of brookite and anatase (Figure 5c,e). The dissimilarity between their structural evolution and that of the nanotubes on the one hand, and of the general evolution of the aggregates on the other hand, suggest that the bigger TiO_2 tubes represent a minor part of the sample, while their crystallization in anatase and brookite phases over rutile and brookite could be attributed to a temperature driven, size-related effect. Finally, no trace of materials coming from reduced TiO_2 was found in any region of the examined aggregates. The thermal profiles used in both the SAED-based and HRTEM-based experiments are shown in the Scheme 1, where both the crystal phases appearance and their thermal evolution, as described above, are also summarized.

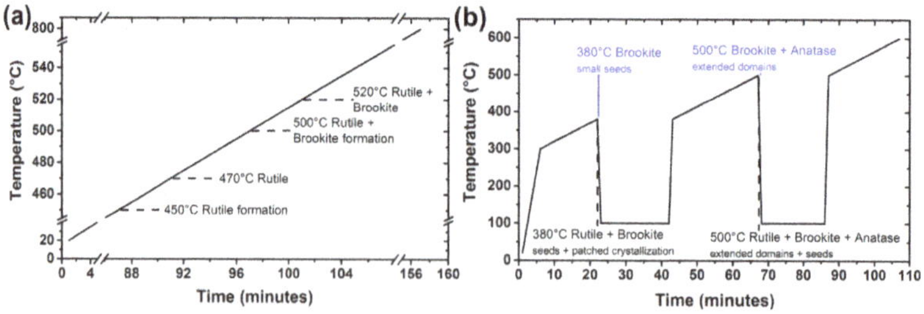

Scheme 1. Time vs. temperature schemes depicting the thermal ramps used during the *in situ* heating studies: (**a**) thermal ramp used during the *in situ* SAED heating experiment; (**b**) thermal ramp used during the *in situ* HRTEM heating experiment. Milestones in black at the bottom of panel (**b**) indicate the structural evolution of the main population of TANs (250 nm in length, see Figures 2–4); milestones in blue at the top of panel (**b**) indicate the structural evolution of the secondary population (around one micron in length, see Figure 5). The pressure value in both the experiments was less than 10^{-5} Pa.

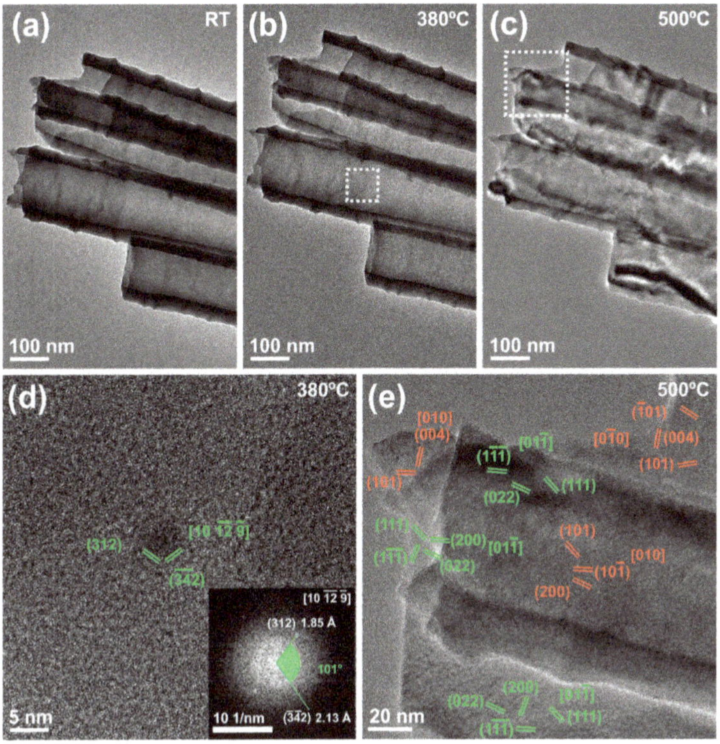

Figure 5. *In situ* HRTEM heating experiment of bigger TANs: (**a**–**c**) Evolution of the aggregates of TANs recorded at RT (**c**), 380 °C (**b**) and 500 °C (**c**), The white dotted rectangles indicate the regions of interest analyzed by HRTEM and reported in panels **d** and **e**; (**d**) HRTEM image of a brookite seed (in green) recorded at 380 °C (see the rectangle in panel 3**b**). The inset presents the corresponding 2D-FFT, showing the interplanar distances and angular relationships expected for the brookite; (**d**) HRTEM image of extended crystal domains of brookite (in green) and anatase (in red) recorded at 500 °C (see the rectangle in panel 4**c**).

3. Discussion

Recently, we carried out an in-depth study of the effect of TAN crystallization on their mechanical properties [28]. In that case the TAN heating was performed *ex situ*, i.e., under ambient pressure conditions outside the electron microscope, and the TANs were subsequently imaged by HRTEM at RT after they reached a temperature of 150 , 300 and 450 °C, respectively. The structural evolution measured at these temperatures was then compared with that studied by means of both HRTEM and Raman spectroscopy. We found that the TAN crystallization had just started at 150 °C, with the appearance of very small (2–3 nm) crystalline domains of anatase. At 300 °C, their size grew to about 10 nm, while some further and quite smaller domains of brookite also appeared. Finally, at 450 °C crystallization was observed to have further proceeded, with the TANs being constituted by large zones of anatase, even if brookite was still present. As the HRTEM provided a very local structural characterization obtained by 2D-FFT analysis, the Raman spectroscopy, performed over a much larger scale, allowed the determination of the anatase as the majority phase. These results basically confirm what was already reported both in [22,23], where TANs of similar size to ours were thermally crystallized under ambient pressure conditions, as reported in the introduction. It is noteworthy that in the two cases the observed evolution is very similar in terms of starting temperature of the amorphous-to-crystalline transition, as well as the kind of crystalline phase formed and the growth observed in the size of crystalline domains. Conversely, when performed *in situ*, i.e., inside the electron microscope, the same heating experiment leads to different outcomes due to the variation in vacuum conditions ($p < 10^{-5}$ Pa). First, under such a low pressure the TANs underwent a transition to rutile and brookite, with the crystallization in general starting at higher temperatures than those determined in all the previous heating experiments conducted under standard pressure. Then, even if the intrinsically very local nature of HRTEM imaging does not allow the provision of a precise quantitative determination of the majority crystal phase appearing as a consequence of the thermal treatment, some rough consideration may be attempted about the relative prominence of the crystal phases we observed. In fact, looking at the HRTEM results reported in Figures 2–4, it appears quite clearly that most of the smaller TANs crystallize in polycrystalline rutile nanotubes, with the SAED-based results reported in Figure 1 confirming what was locally observed by HRTEM. Conversely, the thermal evolution of the rarer and bigger TANs reported in Figure 5 lead to the crystallization in both the anatase and brookite phases, in a similar fashion to what observed in the case of physically confined and self-confined TANs under ambient pressure conditions [23]. To our best knowledge, this is the first case in which the TANs amorphous-to-crystalline phase transition has been investigated under such a low-pressure conditions, although similar phase variation processes have already been observed at room temperature under lower vacuum conditions ($p \approx 10^{-3}$ Pa) due to visible light irradiation [29,30].

Furthermore, changing the electron beam energy of our TEM experiment (from 120 keV for the first, ED-based observation to 300 keV for the C_s-corrected HRTEM imaging) gives rise to a pronounced decrease of the TANs' initial crystallization temperature, indicating an electron beam contribution to the phenomenon. The several effects caused by the electron beam irradiation of a sample have been extensively studied, see for instance [31,32], and among them specimen heating is the one that could be most immediately considered in our case. However, if this was the only effect to take into account, the sample should undergo a local thermal increase at least equal to the difference in temperature observed when the electron beam energy (E_0) changed from 120 to 300 keV (i.e., $\Delta T = 450\ ^\circ\mathrm{C} - 380\ ^\circ\mathrm{C} = 70\ ^\circ\mathrm{C}$ for $\Delta E = 180$ keV) in order to give rise to the lower crystallization temperature observed at 300 keV. Conversely, electron beam-induced sample heating is a well-known phenomenon caused by inelastic electron–electron scattering, capable of raising the sample temperature by only a few degrees [31]. Then, it cannot provide the additional thermal contribution needed to decrease the temperature of crystallization that we observed. Moreover, even if ΔT is usually considered directly proportional to ΔE, the total amount of energy lost by the electron beam decreases with the increasing energy, due to the reduced electron–atom scattering cross-section. For the same reason, the radiolysis effects, which consist in bond breaking, decrease when E_0 increases, and cannot be considered as playing a role in the lower crystallization temperature of the TANs. Conversely, such a decrease has to be ascribed to atomic displacement, also known as the knock-on effect, suffered by the sample atoms when the high-energy beam electrons hit them. In this case, the maximum amount of energy E_m transferred between fast electrons and atoms is related to E_0 by the following equation [33]:

$$E_m = 2E_0(E_0 + 2m_0c^2)/Mc^2, \quad (1)$$

where c is the light speed, m_0 the electron rest mass and M the target atom's mass. Then, the knock-on effect takes place whenever the displacement energy is lower than the transferred energy, with the mass of the target atom being the only sample-related parameter and atoms with lower mass being more likely to suffer from atomic displacement. The dependence of the starting temperature of crystallization on the electron beam energy E_0 was also previously observed for different materials, namely $Si_2Sb_2Te_5$ [34] and $Ge_2Sb_2Te_5$ [35]. In both cases, the materials amorphous-to-crystalline phase transition was promoted by an increase of E_0, as a consequence of atomic displacement induced by the electron beam irradiation. It is noteworthy to highlight that the change in E_0 affected just the starting temperature of crystallization, not the crystalline phases formed (brookite and rutile) nor the final polycrystalline nature of the crystallized nanotubes.

Also, our results are in good accordance with those obtained by non-thermal, optically-assisted crystallization of amorphized TiO_2 nanocrystals to the rutile phase under oxygen-poor conditions [30]. There, an adequately energetic irradiation gave rise to oxygen desorption processes that improved the superficial chemical reactivity of TiO_2 and triggered the crystallization of neighboring amorphous nanoparticles into bigger rutile nanocrystalline seeds. Here, the variation in the electron beam energy from 120 to 300 keV increases the displacement of oxygen atoms, thus affecting the chemical reactivity at the surface of TiO_2 nanotubes and leading to the decrease in crystallization temperature that we observed, while also providing an indication with regards to the rare formation of the nanoseeds observed in Figures 3–5. Local variations in the quantity of displaced oxygen atoms could render the amorphous TiO_2 locally unstable and trigger the formation of small nanocrystalline precursors (<2 nm) that act as a starting point for both nanoseeds and patched crystallization zones. Then, the occurrence of nanoseeds could derive from the fast sintering of two or more nearby precursors [36,37], that coalesce to reach an energetically favorable condition and are not further modified by the heating. The appearance of rutile nanoseeds might be caused by the direct nucleation of at least one rutile precursor, which will command the crystallization during the fast sintering [38], while brookite nanoseeds should form in presence of at least one anatase precursor [36]. Then the formation of brookite seeds, albeit unexpected according to the TiO_2 phase diagram under standard pressure conditions, suggests a local effect of the amorphous-to-crystalline transition under high vacuum conditions. These results require further experimental and theoretical studies to better comprehend how such low pressure conditions influence the phase diagram of TiO_2 at the nanosize.

Obviously, understanding why different crystal phases appear and stabilize in the shorter or longer TANs crystallized under high vacuum, or depending on the pressure conditions adopted during heating, is not trivial. In fact, it is well known that crystallization occurs via a collective atom displacement, following the two steps of nucleation and subsequent growth. Since the phenomenon is basically driven by the crystallization enthalpy and the free volume reduction, an in-depth understanding of their dependence on the external pressure would be needed to fully comprehend the differences observed when the TANs amorphous-to-crystalline phase transition is performed under high vacuum (i.e., lower than 10^{-5} Pa). With this aim, further theoretical studies will need to be performed to elucidate more in-depth the origin of the pressure-dependent differences we observed and these will be presented in a separate paper.

4. Materials and Methods

A Ti foil was used as a working electrode for the anodic growth of TiO_2 nanotube arrays in a two-electrode configuration. Anodic oxidation was carried out in an ethylene glycol electrolyte containing 0.5 wt % NH_4F and 2.5 vol % deionized water under a constant voltage of 60 V for 0.5 h using a Direct current (DC) power supply. The samples were then rinsed in Deionized Water (DI-water), dried and detached by ultra-sonication in ethanol for five minutes.

TiO_2 amorphous nanotubes were studied by *in situ* TEM heating through two different sets of experiments. For both the experiments, the amorphous nanotubes were dispersed in distilled water and then drop-casted on silicon nitride based-MEMS, which also acted as the TEM heating substrate. First, an electron diffraction imaging was conducted on nanotubes aggregates using a Tecnai TEM by FEI (Hillsboro, OR, USA), equipped with a lanthanum exaboride thermionic electron source, a twin objective lens and a Orius CCD camera by Gatan (Pleasanton, CA, USA). This microscope worked at an acceleration voltage of 120 kV. In this case the *in situ* heating (5 $^{\circ}C \cdot min^{-1}$) was performed in the RT-to-800 °C thermal range in order to assess the crystallization temperature. Second, the HRTEM imaging was performed using a spherical aberration (C_S) corrected Titan microscope by FEI (Hillsboro, OR, USA), equipped with a Schottky field emission gun (FEG), a CEOS spherical aberration corrector of the objective lens, a 2K × 2K CCD camera by Gatan (Pleasanton, CA, USA), and with the microscope working at an acceleration voltage of 300 kV. In this case the *in situ* heating (5 °C·min^{-1}) was performed in the 300-to-600 °C thermal range, after having fast reached (t < 300 s) the starting temperature of 300 °C. The temperature was fast lowered to 100 °C for the acquisition of the HRTEM images corresponding to each heating temperature and consequently restored before continuing the *in situ* heating. Both experiments were performed using a Wildfire MEMS-based *in situ* heating TEM sample holder by DENSsolutions (Delft, Netherlands). The pressure value in both the electron microscopes was less than 10^{-5} Pa. A structural characterization was performed to calculate the interplanar distances and identify crystal phases: in the case of the SAED patterns it was carried out by measuring the point-to-point distance of diffraction rings and diffraction spots and calculating the correspondent interplanar distances; in the case of the HRTEM images, interplanar distances and angular relationships were more precisely obtained by two-dimensional fast Fourier transform (2D-FFT) analysis of regions of interests.

5. Conclusions

In summary, our work showed how TAN crystallization occurs under high vacuum conditions by *in situ* heating them in a TEM. When irradiated with electrons accelerated up to the energy of 120 keV, the starting temperature of crystallization was equal to 450 °C, while if the same *in situ* heating was performed under an electron beam of 300 keV, this temperature decreased to 380 °C, due to the atomic displacement suffered by the sample atoms as a consequence of the higher electron energy, which did not affect the formation of different crystalline phases. Besides, it was also shown with unprecedented detail how TAN crystallization occurs, starting from small brookite and rutile crystalline seeds that then grow in size during the following further heating. A comparison with the same phenomenon studied under standard pressure conditions highlighted two main differences: first, the transition starting temperature was higher, and second, the crystalline phases formed were brookite and rutile rather than anatase. Furthermore, this study opens an unprecedented opportunity to study the TiO_2 crystallization under high-vacuum conditions. This could also allow preferential access to crystalline phases that are normally forbidden when the thermally-driven crystallization takes place under ambient pressure conditions.

Acknowledgments: The authors acknowledge financial support from the KAUST baseline funding of Andrea Falqui.

Author Contributions: Andrea Lamberti and Stefano Stassi prepared the samples. Andrea Falqui and Alberto Casu conceived and performed the *in situ* TEM experiments, analyzed the data and wrote the paper. All the authors discussed in-depth the paper.

Conflicts of Interest: The authors declare no conflict of interest. The founding sponsors had no role in the design of the study; in the collection, analyses, or interpretation of data; in the writing of the manuscript, and in the decision to publish the results.

References

1. Long, H.; Chen, A.; Yang, G.; Li, Y.; Lu, P. Third-order optical nonlinearities in anatase and rutile TiO_2 thin films. *Thin Solid Films* **2009**, *517*, 5601–5604. [CrossRef]
2. Dürr, M.; Schmid, A.; Obermaier, M.; Rosselli, S.; Yasuda, A.; Nelles, G. Low-temperature fabrication of dye-sensitized solar cells by transfer of composite porous layers. *Nat. Mater.* **2005**, *4*, 607–611. [CrossRef] [PubMed]
3. Yang, M.-Q.; Xu, Y.-J. Selective photoredox using graphene-based composite photocatalysts. *Phys. Chem. Chem. Phys.* **2013**, *15*, 19102–19118. [CrossRef] [PubMed]
4. Page, K.; Palgrave, R.G.; Parkin, I.P.; Wilson, M.; Savin, S.L.P.; Chadwick, A.V. Titania and silver–titania composite films on glass—Potent antimicrobial coatings. *J. Mater. Chem.* **2007**, *17*, 95–104. [CrossRef]
5. Deng, D.; Kim, M.G.; Lee, L.Y.; Cho, J. Green energy storage materials: Nanostructured TiO_2 and Sn-based anodes for lithium-ion batteries. *Energy Environ. Sci.* **2009**, *2*, 818–837. [CrossRef]

6. Crossland, E.J.W.; Noel, N.; Sivaram, V.; Leijtens, T.; Alexander-Webber, J.A.; Snaith, H.J. Mesoporous TiO_2 single crystals delivering enhanced mobility and optoelectronic device performance. *Nature* **2013**, *495*, 215–219. [CrossRef] [PubMed]
7. Guldin, S.; Hüttner, S.; Tiwana, P.; Orilall, M.C.; Ülgüt, B.; Stefik, M.; Docampo, P.; Kolle, M.; Divitini, G.; Ducati, C.; et al. Improved conductivity in dye-sensitised solar cells through block-copolymer confined TiO_2 crystallization. *Energy Environ. Sci.* **2011**, *4*, 225–233. [CrossRef]
8. Tétreault, N.; Horváth, E.; Moehl, T.; Brillet, J.; Smajda, R.; Bungener, S.; Cai, N.; Wang, P.; Zakeeruddin, S.M.; Forró, L.; et al. High-efficiency solid-state dye-sensitized solar cells: Fast charge extraction through self-assembled 3D fibrous network of crystalline TiO_2 nanowires. *ACS Nano* **2010**, *4*, 7644–7650. [CrossRef] [PubMed]
9. Adachi, M.; Murata, Y.; Takao, J.; Jiu, J.; Sakamoto, M.; Wang, F. Highly efficient dye-sensitized solar cells with a titania thin-film electrode composed of a network structure of single-crystal-like TiO_2 nanowires made by the "oriented attachment" mechanism. *J. Am. Chem. Soc.* **2004**, *126*, 14943–14949. [CrossRef] [PubMed]
10. Jennings, J.R.; Ghicov, A.; Peter, L.M.; Schmuki, P.; Walker, A.B. Dye-sensitized solar cells based on oriented TiO_2 nanotube arrays: Transport, trapping, and transfer of electrons. *J. Am. Chem. Soc.* **2008**, *130*, 13364–13372. [CrossRef] [PubMed]
11. Fabregat-Santiago, F.; Barea, E.M.; Bisquert, J.; Mor, G.K.; Shankar, K.; Grimes, C.A. High carrier density and capacitance in TiO_2 nanotube arrays induced by electrochemical doping. *J. Am. Chem. Soc.* **2008**, *130*, 11312–11316. [CrossRef] [PubMed]
12. Docampo, P.; Guldin, S.; Steiner, U.; Snaith, H.J. Charge transport limitations in self-assembled TiO_2 photoanodes for dye-sensitized solar cells. *J. Phys. Chem. Lett.* **2013**, *4*, 698–703. [CrossRef] [PubMed]
13. Choi, S.Y.; Mamak, M.; Speakman, S.; Chopra, N.; Ozin, G.A. Evolution of nanocrystallinity in periodic mesoporous anatase thin films. *Small* **2004**, *1*, 226–232. [CrossRef] [PubMed]
14. Kondo, J.N.; Domen, K. Crystallization of mesoporous metal oxides. *Chem. Mater.* **2008**, *20*, 835–847. [CrossRef]
15. Lu, X.; Wang, G.; Zhai, T.; Yu, M.; Gan, J.; Tong, Y.; Li, Y. Hydrogenated TiO_2 nanotube arrays for supercapacitors. *Nano Lett.* **2012**, *12*, 1690–1696. [CrossRef] [PubMed]
16. Liu, Z.; Zhang, X.; Nishimoto, S.; Jin, M.; Tryk, D.A.; Murakami, T.; Fujishima, A. Highly ordered TiO_2 nanotube arrays with controllable length for photoelectrocatalytic degradation of phenol. *J. Phys. Chem. C* **2008**, *112*, 253–259. [CrossRef]
17. Gui, Q.; Xu, Z.; Zhang, H.; Cheng, C.; Zhu, X.; Yin, M.; Song, Y.; Lu, L.; Chen, X.; Li, D. Enhanced photoelectrochemical water splitting performance of anodic TiO_2 nanotube arrays by surface passivation. *ACS Appl. Mater. Interfaces* **2014**, *6*, 17053–17058. [CrossRef] [PubMed]
18. Park, J.; Bauer, S.; Von Der Mark, K.; Schmuki, P. Nanosize and vitality: TiO_2 nanotube diameter directs cell fate. *Nano Lett.* **2007**, *7*, 1686–1691. [CrossRef] [PubMed]
19. Nicthová, L.; Kužel, R.; Matêj, Z.; Šícha, J.; Musil, J. Time and thickness dependence of crystallization of amorphous magnetron deposited TiO_2 thin films. *Z. Kristallogr. Suppl.* **2009**, *30*, 235–240. [CrossRef]
20. Kužel, R.; Nicthová, L.; Matêj, Z.; Herman, D.; Šícha, J.; Musil, J. Study of crystallization of magnetron sputtered TiO_2 thin films by X-ray scattering. *Z. Kristallogr. Suppl.* **2007**, *26*, 247–252. [CrossRef]
21. Kužel, R.; Nicthová, L.; Matêj, Z.; Herman, D.; Šícha, J.; Musil, J. Growth of magnetron sputtered TiO_2 thin films studied by X-ray scattering. *Z. Kristallogr. Suppl.* **2007**, *26*, 241–246. [CrossRef]
22. Oh, H.-J.; Lee, S.; Lee, B.; Jeong, Y.; Chi, C.-S. Surface characteristics and phase transformation of highly ordered TiO_2 nanotubes. *Met. Mater. Int.* **2011**, *17*, 613–616. [CrossRef]
23. Kim, M.; Bae, C.; Kim, H.; Yoo, H.; Montero Moreno, J.M.; Jung, H.J.; Bachmann, J.; Nielsch, K.; Shin, H. Confined crystallization of anatase TiO_2 nanotubes and their implications on transport properties. *J. Mater. Chem. A* **2013**, *1*, 14080–14088. [CrossRef]
24. Xue, C.; Narushima, T.; Ishida, Y.; Tokunaga, T.; Yonezawa, T. Double-wall TiO_2 nanotube arrays: Enhanced photocatalytic activity and *in situ* TEM observations at high temperature. *ACS Appl. Mater. Interfaces* **2013**, *6*, 19924–19932. [CrossRef] [PubMed]
25. Yuan, W.; Wang, Y.; Li, H.; Wu, H.; Zhang, Z.; Selloni, A.; Sun, C. Real-time observation of reconstruction dynamics on TiO_2 (001) surface under oxygen via an environmental transmission electron microscope. *Nano Lett.* **2015**, *16*, 132–137. [CrossRef] [PubMed]
26. Park, H.; Jie, H.S.; Kim, K.H.; Ahn, J.P.; Park, J.K. *In-Situ* TEM observation on phase formation of TiO_2 nanoparticle synthesized by flame method. *Materials Sci. Forum* **2007**, *534*, 81–84. [CrossRef]

27. Ghassemi, H.; Harlow, W.; Mashtalir, O.; Beidaghi, M.; Lukatskaya, M.R.; Gogotsi, Y.; Taheri, M.L. *In situ* environmental transmission electron microscopy study of oxidation of two-dimensional Ti_3C_2 and formation of carbon-supported TiO_2. *J. Mater. Chem. A* **2014**, *2*, 14339–14343. [CrossRef]
28. Stassi, S.; Lamberti, A.; Roppolo, I.; Casu, A.; Bianco, S.; Scaiola, D.; Falqui, A.; Pirri, C.F.; Ricciardi, C. Evolution of nanomechanical properties and crystallinity of individual titanium dioxide nanotube resonators. *Nanotechnology* **2017**. [CrossRef]
29. Ricci, P.C.; Casu, A.; Salis, M.; Corpino, R.; Anedda, A. Optically controlled phase variation of TiO_2 nanoparticles. *J. Phys. Chem. C* **2010**, *114*, 14441. [CrossRef]
30. Ricci, P.C.; Carbonaro, C.M.; Stagi, L.; Salis, M.; Casu, A.; Enzo, S. Delogu, anatase-to-rutile phase transition in TiO_2 nanoparticles irradiated by visible light. *J. Phys. Chem. C* **2013**, *117*, 7850. [CrossRef]
31. Egerton, R.F.; Li, P.; Malac, M. Radiation damage in the TEM and SEM. *Micron* **2004**, *35*, 399–409. [CrossRef] [PubMed]
32. Egerton, R.F. Control of radiation damage in the TEM. *Ultramicroscopy* **2013**, *127*, 100–108. [CrossRef] [PubMed]
33. Hobbs, L.W. Radiation effects in analysis of inorganic specimen by TEM. In *Introduction to Analytical Electron Microscropy*; Hren, J.J., Goldstein, J.I., Joy, D.C., Eds.; Plenum: New York, NY, USA, 1979; pp. 437–480, ISBN 978-1-4757-5583-1.
34. Zhang, T.; Song, Z.; Sun, M.; Liu, B.; Feng, S.; Chen, B. Investigation of electron beam induced phase change in $Si_2Sb_2Te_5$ material. *Appl. Phys. A* **2008**, *90*, 451–455. [CrossRef]
35. Kooi, B.J.; Groot, W.M.G.; De Hosson, J.T.M. *In situ* transmission electron microscopy study of the crystallization of $Ge_2Sb_2Te_5$. *J. Appl. Phys.* **2004**, *95*, 451–455. [CrossRef]
36. Mao, Q.; Ren, Y.; Luo, K.H.; Li, S. Sintering-induced phase transformation of nanoparticles: A molecular dynamics study. *J. Phys. Chem. C* **2015**, *119*, 28631–28639. [CrossRef]
37. Zhou, Y.; Fichthorn, K.A. Microscopic view of nucleation in the anatase-to-rutile transformation. *J. Phys. Chem. C* **2012**, *116*, 8314–8321. [CrossRef]
38. Koparde, V.N.; Cummings, P.T. Phase Transformations during sintering of titania nanoparticles. *ACS Nano* **2008**, *2*, 1620–1624. [CrossRef] [PubMed]

MDPI
St. Alban-Anlage 66
4052 Basel
Switzerland
Tel. +41 61 683 77 34
Fax +41 61 302 89 18
www.mdpi.com

Nanomaterials Editorial Office
E-mail: nanomaterials@mdpi.com
www.mdpi.com/journal/nanomaterials

www.ingramcontent.com/pod-product-compliance
Lightning Source LLC
Chambersburg PA
CBHW051709210326
41597CB00032B/5418